FIFTH EDITION

INTRODUCTION TO FLUID MECHANICS

FIFTH EDITION

INTRODUCTION TO FLUID MECHANICS

WILLIAM S. JANNA

CRC Press is an imprint of the
Taylor & Francis Group, an **informa** business

CRC Press
Taylor & Francis Group
6000 Broken Sound Parkway NW, Suite 300
Boca Raton, FL 33487-2742

Printed in Canada on acid-free paper
Version Date: 20150714

International Standard Book Number-13: 978-1-4822-1161-0 (Hardback)

Visit the Taylor & Francis Web site at
http://www.taylorandfrancis.com

and the CRC Press Web site at
http://www.crcpress.com

To Him who ordered the Universe by a word,
whose love and glory are seen in the
beauty of natural science and in
the accomplishments of simple men,
and to my lovely wife, Marla, who daily reflects
this love and glory to me.

Contents

Preface

Introduction to Fluid Mechanics, Fifth Edition is intended for use at the undergraduate level in a mechanical or civil engineering or applied sciences curriculum. It is assumed that the students have knowledge of calculus and physics so that learning to use mathematics to model physical principles in fluid mechanics proceeds without much difficulty.

The book is arranged into 13 chapters and is written using SI units as well as British gravitational units. To exclude either of these unit systems from any fundamental area of study would be premature at this time, although efforts have been made for many years at a national level to convert entirely to SI units. A brief description of the engineering system, complete with a discussion of g_c, is also included (for illustrative purposes only).

A BRIEF DESCRIPTION OF THE CHAPTERS

Chapter 1, which introduces the text, presents definitions appropriate to the study of fluid mechanics. Chapter 2 deals with fluid statics, including pressure measurement, forces exerted by fluids at rest, buoyancy, and stability. In Chapter 3, the basic equations of fluid mechanics are derived from a general conservation equation. The control volume concept is explained, and the continuity, momentum, energy, and Bernoulli equations are presented. Chapter 4 deals mainly with dimensional analysis and modeling and introduces the Rayleigh method, the Buckingham pi method, and the inspection method. This chapter also discusses dimensional homogeneity, illustrates by example the use of dimensional analysis to correlate data, and shows mathematical techniques for modeling prototypes.

Chapter 5 provides an important application of the basic concepts. This application—incompressible flow in conduits—is of great significance to mechanical and civil engineers. Topics include laminar and turbulent flow, nominal pipe sizes, standard tubing sizes, friction factor and pipe roughness, minor losses, noncircular cross sections, and pump sizing for piping systems. Friction factor equations for the Moody diagram are given, as are correlations for coiled tubes and internally finned tubes. Chapter 6 continues with applications to fluid flows past objects and discusses lift and drag forces. It also presents analyses for flows past tractor–trailer trucks, automobiles, and bicycle–rider combinations. Chapter 7, on open-channel flow, is primarily of importance to civil engineers, and the first few sections follow a format similar to that found in classic hydraulics texts. Chapter 8 is an introduction to compressible flow, covering the basic concepts that lead to the solution of practical problems in the field.

In a one-semester course, completion of Chapter 8 could coincide with the end of the semester. The rest of the text is intended for use at an intermediate or second level of study in fluid mechanics. Chapter 9 provides a study of turbomachinery, including design criteria, a description of commercially available machines and installations, and the method used to select pumps and hydraulic turbines for various situations, as well as a section on windmill propellers. Chapter 10 surveys measurement techniques commonly employed in fluid mechanics, such as fluid properties, closed-conduit flows, and open-channel flows.

Some of the more mathematically oriented and classical topics of fluid mechanics then follow. Chapter 11 is an introduction to the equations of motion for isothermal systems (the Navier–Stokes equations) and includes applications to a number of laminar flow problems. Chapter 12 presents simple solutions to the equations of inviscid flow and illustrates the method of superposition to obtain equations for more complex flows. Chapter 13 discusses boundary-layer flows; boundary-layer equations are derived and applied to the problem of flow over a flat plate. The momentum integral equation is derived and applied to both laminar and turbulent flows over a flat plate.

Each chapter concludes with a Problems section. The problems are arranged so that the easier ones are presented first, which helps in building the students' confidence and skill in learning the principles involved. The more difficult problems then allow the students to analyze the topic in more detail. The problems are designed to systematically improve the students' ability to understand and apply the equations of fluid mechanics to various practical problems such as a flow from a draining coffee pot or drag force exerted by a bicycle–rider combination.

In addition, the end-of-chapter problems have been grouped together by topic. This feature makes it easier for the instructor to select and assign problems that pertain to the specific area under study. It will also be easier for the students to review specific portions of the text by solving pertinent problems.

In adding new material, I have tried to achieve a thorough and comprehensible presentation of fluid mechanics from a practical viewpoint, without producing an encyclopedic and therefore inaccessible book.

Learning is enhanced and strengthened when we use equations to mathematically model phenomena that we see and interact with every day. In all chapters, therefore, a strong emphasis is placed on solving practical problems. This approach makes learning a visual experience and provides the students with an introduction to the types of problems they are likely to encounter in practice.

The text has been used successfully in three different courses: a first course in fluid mechanics covering Chapters 1 through 6; an intermediate course in fluid mechanics covering Chapters 7, 8, 10, and 11; and a course in turbomachinery covering Chapter 9, supplemented with information from other sources.

Regardless of how many times a manuscript is checked, the occasional mistake does seem to slip past. The author invites readers to report any errors to the publisher so that misconceptions are not taught as truths. The author also invites readers' comments, which will be gratefully accepted as advice on how to improve the text.

Acknowledgments

I thank the many faculty members and students who wrote letters of encouragement, made suggestions on how to improve the text, and pinpointed portions of the text that required clarification. I am also greatly indebted and extend my gratitude to my manuscript reviewers.

I extend my thanks also to CRC Press staff members, Jonathan Plant, engineering editor, for his guidance, patience, sense of humor, and ability to suggest things politely, and Jill Jurgensen, senior project coordinator, and to the University of Memphis for various forms of help. Finally, I acknowledge the encouragement and support of my lovely wife, Marla, who made many sacrifices during the preparation of the various editions of this text.

Author

William S. Janna received his BSME, MSME, and PhD from the University of Toledo, Toledo, Ohio. He joined the mechanical engineering faculty of the University of New Orleans in 1976, where he became department chair and served in that position for four years. Subsequently, he joined the University of Memphis in 1987 as chair of the Department of Mechanical Engineering. He served as associate dean for graduate studies and research in the Herff College of Engineering. His research interests include boundary-layer methods of solution for various engineering problems, modeling the melting of ice objects of various shapes, and the study of sublimation from various geometries. He is the author of three textbooks and teaches short courses for the American Society of Mechanical Engineers (ASME). He teaches courses in heat transfer, fluid mechanics, and design of fluid/thermal systems. He has designed and constructed a number of experiments in fluid mechanics and heat transfer laboratories.

Introduction

Other Books by William S. Janna

Engineering Heat Transfer, 2nd edition, CRC Press, Boca Raton, FL, 2000

Most of the texts on heat transfer have focused on the mathematics of the subject, typically at an advanced level. Engineers need a reference that provides a strong, practical foundation in heat transfer that emphasizes real-world problems and helps develop problem-solving skills.

The second edition of *Engineering Heat Transfer* fulfills that need. This book emphasizes effective, accurate modeling of heat transfer problems. It contains several real-world examples to amplify theory and to show how to use derived equations to model physical problems. Confidence-building exercises begin with easy problems and progress to more difficult ones. Problem-solving skills are developed methodically and thoroughly.

The text is concise and user friendly. It covers the topics of conduction, convection, and radiation heat transfer in a manner that does not overwhelm the reader. It contains a multitude of drawings, graphs, and figures to clearly convey information that is critical to envisioning the modeling of problems in an abstract study like heat transfer.

The text is uniquely suited to the actual practice of engineering.

Design of Fluid Thermal Systems, 2nd edition, Cengage Learning, Boston, MA, 1998

This book is intended for a capstone course in energy systems or thermal sciences that corresponds to the machine design course in mechanical systems. The text is divided into two major sections. The first is on piping systems, blended with the economics of pipe size selection and the sizing of pumps for piping systems. The second is on heat exchangers, or, more generally, devices available for the exchange of heat between two process streams.

"Show and Tell" exercises are provided in this text; these require students to provide brief presentations on various topics (e.g., various types of valves that are commonly used, venturi meters, and pump impellers).

The text also contains numerous design project descriptions. A student group can select one of these design projects and devote an entire semester to finishing it.

Project management methods are described and students are taught how to complete task planning sheets to keep track of the progress made on the designs. Project report writing is also discussed, and a suggested format is provided.

When a student completes the course material, he or she will have mastered some practical design skills. For example, a student will acquire the ability to size a pipeline to meet least annual cost criteria, to select a pump and ensure that cavitation is avoided, and to select and size a heat exchanger to provide a required fluid outlet temperature or a heat transfer rate. The student will also gain the experience of working in a group and in observing the effective planning and management of project design activities.

The text is comprehensible and gives much practical information on design in the fluid thermal systems area. It relates industrial practice to fundamental engineering concepts in a capstone design experience.

1 Fundamental Concepts

Fluid mechanics is the branch of engineering that deals with the study of fluids—both liquids and gases. Such a study is important because of the prevalence of fluids and our dependence on them. The air we breathe, the liquids we drink, the water transported through pipes, and the blood in our veins are examples of common fluids. Further, fluids in motion are potential sources of energy that can be converted into useful work—for example, by a waterwheel or a windmill. Clearly, fluids are important, and a study of them is essential to the engineer. The objectives of this chapter are to describe the unit systems used in the chapter, to define a fluid, to discuss common properties of fluids, to establish features that distinguish liquids from gases, and to present the concept of a continuum.

1.1 DIMENSIONS AND UNITS

Before we begin the exciting study of fluid mechanics, it is prudent to discuss dimensions and units. In this text, we use two unit systems: the British gravitational system and the international system (SI). Whatever the unit system, dimensions can be considered as either fundamental or derived. In the British system, the fundamental dimensions are length, time, and force. The units for each dimension are given in the following table:

British Gravitational System		
Dimension	**Abbreviation**	**Unit**
Length	L	foot (ft)
Time	T	second (s)
Force	F	pound-force (lbf)

Mass is a derived dimension with units of slug and defined in terms of the primary dimensions as

$$1 \text{ slug} = 1 \frac{\text{lbf} \cdot \text{s}^2}{\text{ft}} \tag{1.1}$$

Converting from the unit of mass to the unit of force is readily accomplished because the slug is defined in terms of the lbf (pound-force).

Example 1.1

An individual weighs 165 lbf.

a. What is the person's mass at a location where the acceleration due to gravity is 32.2 ft/s^2?
b. On the moon, the acceleration due to gravity is one-sixth of that on earth. What is the weight of this person on the moon?

Solution

a. Applying Newton's law, we write

$$F = ma$$

Substituting gives

$$165 \text{ lbf} = m \, (32.2 \text{ ft/s}^2)$$

Solving for mass, we obtain

$$m = \frac{165 \text{ lbf}}{32.2 \text{ ft/s}^2} = 5.12 \text{ lbf} \cdot \text{s}^2/\text{ft}$$

or

$$m = 5.12 \text{ slug}$$

b. The mass is the same on the moon as on the earth. Again we apply Newton's law,

$$F = ma$$

where $m = 5.12$ slug and the acceleration due to gravity is

$$a = \frac{1}{6} 32.2 \ \text{ft/s}^2 = 5.37 \ \text{ft/s}^2$$

Substituting, the weight on the moon becomes

$$F = 6.12 \text{ slug} \, (5.37 \text{ ft/s}^2)$$

or

$$F = 27.2 \text{ lbf}$$

The second unit system we use is the SI system. In this system, there are three fundamental dimensions, as shown in the following table:

SI Unit System		
Dimension	**Abbreviation**	**Unit**
Mass	M	kilogram (kg)
Length	L	meter (m)
Time	T	second (s)

In this system, force is a derived dimension given in newtons (abbreviated N). The newton is defined in terms of the other units as

$$1 \text{ N} = 1 \frac{\text{kg} \cdot \text{m}}{\text{s}^2} \tag{1.2}$$

As with the British gravitational system, force and mass units are defined in terms of one another. Conversion is thus a relatively simple task.

Example 1.2

a. What is the weight of 1 m^3 of water on Earth's surface if the water has a mass of 1 000 kg?
b. What is its weight on Mars, where the acceleration due to gravity is about two-fifths that on Earth?

Solution

a. We use Newton's law,

$$F = ma$$

where $m = 1\ 000$ kg

$$a = 9.81\ \text{m/s}^2$$

We then have

$$F = 1\ 000\ \text{kg}\frac{9.81\ \text{m}}{\text{s}^2} = 9\ 810\ \ \text{kg}\cdot\text{m/s}^2$$

or

$$F = 9\ 810\ \text{N}$$

b. On Mars, $a = \frac{2}{5}(9.81) = 3.92\ \text{m/s}^2$ and $m = 1\ 000\ \text{kg}$. Hence, we obtain

$$F = 1\ 000\ \text{kg}\ (3.92)\ \text{m/s}^2 = 3\ 920\ \text{kg}\cdot\text{m/s}^2$$

or

$$F = 3\ 920\ \text{N}$$

Certain conventions are followed in using SI. For instance, in the preceding example, four-digit numbers were written with a space where one might normally place a comma. Other conventions and definitions will be pointed out as we encounter them.

It is permissible in SI to use prefixes with the units for convenience. Table A.1 gives a complete listing of the names of multiples and submultiples of SI units. As an example of the usage of prefixes, consider the answer to Example 1.2(b):

$$F = 3\ 920\ \text{N} = 3.920\ \text{kN}$$

where, by definition, 10^3 is a factor by which the unit is multiplied and is represented by the lowercase letter k.

Mention has been made of derived dimensions. An example is area A, which has dimensions of L^2—units of square feet in the English system and square meters in SI. In general, all derived dimensions are made up of fundamental dimensions.

For purposes of illustration, let us briefly examine the English engineering system of units. In this system, mass, length, time, and force are fundamental dimensions with units of pound-mass, foot, second, and pound-force, respectively (see the following table):

English Engineering System		
Dimension	**Abbreviation**	**Unit**
Mass	M	pound-mass (lbm)
Length	L	foot (ft)
Time	T	second (s)
Force	F	pound-force (lbf)

These units are related by Newton's law, written as a proportionality, that force is proportional to the product of mass and acceleration:

$$F \propto ma$$

By introducing a constant of proportionality K, we obtain

$$F = Kma$$

Next let us define 1 lbf as the force required to accelerate a mass of 1 lbm at a rate of 32.2 ft/s². By substitution into the previous equation, we get

$$1 \text{ lbf} = K\,(1 \text{ lbm})(32.2 \text{ ft/s}^2)$$

Solving for the reciprocal of the constant yields

$$g_c = \frac{1}{K} = 32.2 \text{ lbm}\cdot\text{ft}/(\text{lbf}\cdot\text{s}^2) \tag{1.3}$$

where g_c is a constant that arises in equations in the English system to make them dimensionally correct. Thus, Newton's law in the English system is written as

$$F = \frac{ma}{g_c} \tag{1.4}$$

Remember that $1/g_c = K$ and K was introduced as a proportionality constant; g_c is not the acceleration due to gravity. When one is using the British gravitational system or the SI system with equations of fluid mechanics, the conversion factor g_c is not necessary, nor is it used. The advantage of working in the English engineering system is that the pound-mass and the pound-force are equal numerically at locations where the acceleration due to gravity g is 32.2 ft/s². In other words, an object having a mass of 10 lbm will weigh 10 lbf where the acceleration due to gravity is 32.2 ft/s². The main disadvantage of using the English engineering system is that the equations of fluid mechanics are correctly written with g_c appearing where appropriate. This usage can be a source of confusion to the student. In the British gravitational and SI unit systems, g_c is not necessary because its effect is already accounted for by the way mass and force are related. For reference purposes, it is useful to note that

$$32.174 \text{ lbm} = 1 \text{ slug} \tag{1.5}$$

Other unit systems that have been developed are the British absolute and the CGS absolute. The dimensions and units of each are given in Table 1.1 along with those of the British gravitational, SI,

TABLE 1.1
Conventional Systems of Units

Dimension	British Gravitational	SI	English Engineering	British Absolute	CGS Absolute
Mass (fundamental)		kg	lbm	lbm	g
Force (fundamental)	lbf		lbf		
Mass (derived)	slug				
Force (derived)		N		poundal	dyne
Length	ft	m	ft	ft	cm
Time	s	s	s	s	s
g_c (conversion factor)	—	—	$32.2\,\frac{\text{lbm}\cdot\text{ft}}{\text{lbf}\cdot\text{s}^2}$	—	—

and English engineering systems. Note that only in the English engineering system are both force and mass defined as fundamental dimensions. Thus, we must use the conversion factor g_c for converting mass units to force units and vice versa when working with the English engineering system. The other systems directly define these two variables in terms of each other.

Solving problems in fluid mechanics often requires us to convert from one set of units to another. For easing this burden, a set of conversion tables appears as Table A.2. The listing is alphabetical by physical quantity; factors are given as multipliers to change other units into SI units.

There is a controversy regarding the terms *weight* and *weighing*. When a substance is weighed, the result can be expressed as a force or as a mass. It is common to hear of something weighing 10 N. There are proponents of this type of reporting as well as proponents of the alternative—something weighing 10 kg. In this section, we shall take no formal stand; but to avoid confusion, we will arbitrarily interpret the act of weighing to produce a force and give the result with an accompanying value for gravity. When gravity is not noted, we assume 32.2 ft/s^2 (9.81 m/s^2).

The units for angular measurement are the radian (rad) or the degree (°). There are 2π radians per 360°. It is important to remember that the radian is a dimensionless unit but the degree is not. Rotational speed is correctly specified in terms of radians per second (rad/s) or, alternatively, in units of revolutions per unit time. Here again, radians are dimensionless but revolutions are not. There are 2π radians per revolution.

The unit for temperature (t) measurement in the British gravitational system is the degree Rankine (°R). The Rankine scale is an absolute temperature scale. The degree Fahrenheit (°F) is commonly used and is related to the degree Rankine by

$$t\,(^\circ\text{R}) = t\,(^\circ\text{F}) + 460 \tag{1.6}$$

In SI units, the unit for absolute temperature measurement is the degree Kelvin (K, correctly written without the ° symbol). Also used is degree Celsius (°C). These two temperature scales are related by

$$t\,(\text{K}) = t\,(^\circ\text{C}) + 273 \tag{1.7}$$

1.2 DEFINITION OF A FLUID

A fluid is a substance that deforms continuously under the action of an applied shear stress. This definition can be easily illustrated if a fluid is compared to a solid. Recall from strength of materials how a solid material deforms when a shear stress is applied. Figure 1.1 shows a planar element Δx by Δy that is acted upon by shear stress τ. The element is fixed at its base and will deflect a finite amount until an equilibrium position is reached. The final position depends on the magnitude of the shear stress.

Consider next a fluid-filled space formed by two horizontal parallel plates that are a distance Δy apart (Figure 1.2). The upper plate has an area A in contact with the fluid. The upper plate is moved

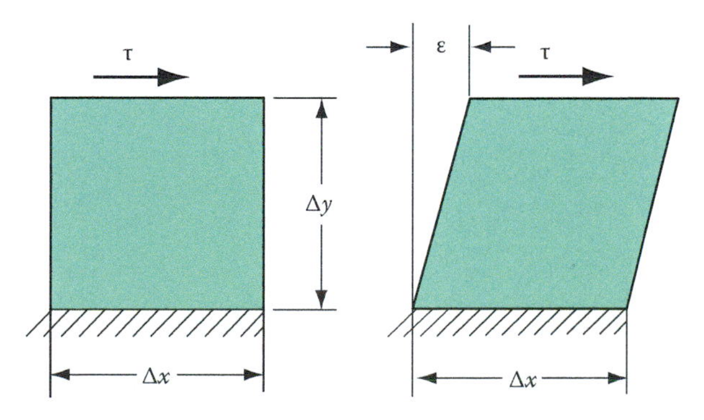

FIGURE 1.1 Deformation of a planar element.

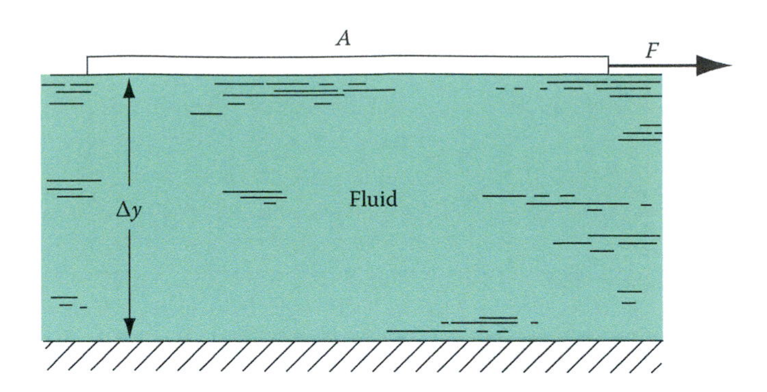

FIGURE 1.2 A fluid acted upon by an applied shear stress.

to the right when pulled with a force F; the lower plate is stationary. The applied shear stress then is $\tau = F/A$. As soon as the plate is pulled, it continues to move but, unlike the solid, never reaches a final equilibrium position. The fluid deforms continuously.

1.3 PROPERTIES OF FLUIDS

The equations of fluid mechanics allow us to predict the behavior of fluids in various flow situations. To use the equations, however, there must be information regarding properties. The properties we discuss in this chapter include density, viscosity, pressure, kinematic viscosity, surface tension, specific heat, internal energy, enthalpy, and compressibility.

1.3.1 Density

The **density** of a fluid is its mass per unit volume, represented by the letter ρ. If the mass of 1 ft^3 of water is 1.94 slug, its density is ρ = 1.94 slug/ft^3. If the mass of 1 m^3 of liquid is 820 kg, its density is ρ = 820 kg/m^3. Density has dimensions of M/L^3. The density of various substances is given in Tables A.3 and A.6.

One quantity of importance related to density is specific weight. Whereas density is mass per unit volume, **specific weight** is weight per unit volume. Specific weight is related to density by

$$SW = \rho g \tag{1.8}$$

with dimension F/L^3 (lbf/ft^3 or N/m^3).

Another useful quantity is specific gravity, which is also related to density of a substance. The **specific gravity** of a substance is the ratio of its density to the density of water at 4°C:

$$s = \frac{\rho}{\rho_w} \quad (\rho_w = \text{density of water}) \tag{1.9}$$

Values of specific gravity of various substances appear in the property tables of the appendix: Table A.4 for water at various temperatures, Table A.5 for common liquids, Table A.7 for various solids, and Table A.8 for various metals and alloys. For our purposes here, we will take ρ_w to be 1.94 slug/ft^3 or 1 000 kg/m^3.

Example 1.3 Laboratory Experiment: Density of Ocean Breeze Shampoo

An experimental method of finding the density of a liquid involves measuring the buoyant force exerted on an object of known volume while it is submerged, as illustrated in Figure 1.3.

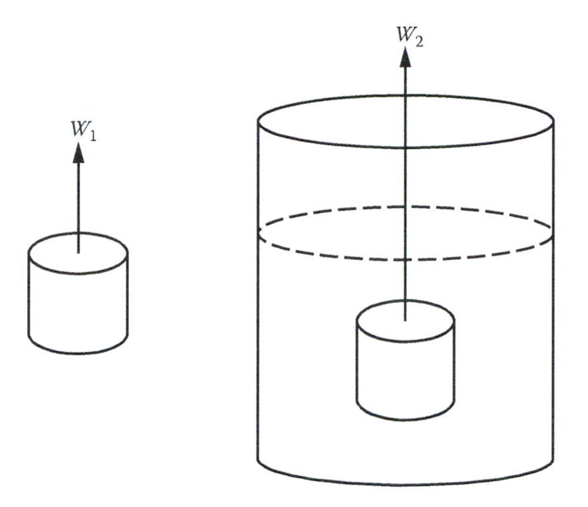

FIGURE 1.3 Measuring the buoyant force on an object with a hanging weight.

The difference between the weight of the object in air W_1 and the weight of the object while submerged W_2 is known as the buoyant force. Referring to the figure, the buoyant force B is

$$B = W_1 - W_2$$

Dividing this difference by the volume displaced by the solid gives the weight per unit volume from which density can be calculated.

Method

A stainless steel cylinder of radius 2.5 cm and 4 cm long is submerged in a quantity of Ocean Breeze shampoo. The specific gravity of the steel is 8. The weight of the cylinder while submerged is 5.3 N. Determine the density of the shampoo.

Solution

The volume of the cylinder is calculated to be

$$\forall_{\text{cyl}} = \pi R^2 L = \pi(0.025\,\text{m})^2(0.04\,\text{m}) = 7.85\times10^{-5}\text{m}^3$$

By definition, the specific gravity of the cylinder is

$$s = \frac{\rho}{\rho_w}$$

With the density of water as 1 000 kg/m^3, the density of the steel is

$$\rho_{\text{cyl}} = s\,\rho_w = 8(1\,000\text{ kg/m}^3) = 8\,000\text{ kg/m}^3$$

The weight of the cylinder in air is calculated to be

$$W_1 = (\rho\forall)_{\text{cyl}}\,g = (8\,000\text{ kg/m}^3)(7.85\times10^{-5}\text{m}^3)(9.81\text{m/s}^2)$$

$$W_1 = 6.16\text{ kg}\cdot\text{m/s}^2 = 6.16\text{ N}$$

The weight of the cylinder while submerged was measured as $W_2 = 5.3$ N. The buoyant force is the difference between these two weights:

$$B = W_1 - W_2 = 6.16\text{ N} - 5.3\text{ N} = 0.86\text{ N}$$

The specific weight of the liquid is

$$SW = \rho g = \frac{B}{V\!\!\!\!-} = \frac{0.86 \text{ N}}{7.85 \times 10^{-5} \text{m}^3} = 10\,998 \text{ N/m}^3$$

The density is finally determined as

$$\rho = \frac{10\,997 \text{ N/m}^3}{9.81 \text{ m/s}^2} = 1121 \text{ N} \cdot \text{s}^2/\text{m}^4 = \left(1121 \text{kg} \cdot \text{m/s}^2\right)\left(\text{s}^2/\text{m}^4\right)$$

$$\rho_{\text{lig}} = 1121 \text{ kg/m}^3$$

Without a detailed discussion of accuracy, precision, and significant digits, we accept this answer as an engineering approximation of the true value of the density of the shampoo. The equations used to arrive at this solution may be combined into a single expression:

$$\rho_{\text{liq}} = S_{\text{object}} \rho_w - \frac{W_2}{V\!\!\!\!-_{\text{object}} g}$$

1.3.2 Viscosity

One important property of a fluid is its **viscosity**, which is a measure of the resistance the fluid has to an externally applied shear stress. This property arises from the definition of a fluid, so we will examine it in that regard. Recall from Section 1.2 that a fluid is defined as a substance that deforms continuously under the action of an applied shear stress. Consider again a fluid-filled space formed by two horizontal parallel plates (Figure 1.4). The upper plate has an area A in contact with the fluid and is pulled to the right with a force F_1 at a velocity V_1. If the velocity at each point within the fluid could be measured, a velocity distribution like that illustrated in Figure 1.4 would result. The fluid velocity at the moving plate is V_1 because the fluid adheres to that surface. This phenomenon is called the nonslip condition. At the bottom, the velocity is zero with respect to the boundary, owing again to the nonslip condition. The slope of the velocity distribution is dV_1/dy.

If this experiment is repeated with F_2 as the force, a different slope or strain rate results: dV_2/dy. In general, to each applied force there corresponds only one shear stress and only one strain rate. If data from a series of these experiments were plotted as τ versus dV/dy, Figure 1.5 would result for a fluid such as water. The points lie on a straight line that passes through the origin. The slope of the resulting line in Figure 1.5 is the viscosity of the fluid because it is a measure of the fluid's resistance to shear. In other words, viscosity indicates how a fluid will react (dV/dy) under the action of an external shear stress (τ).

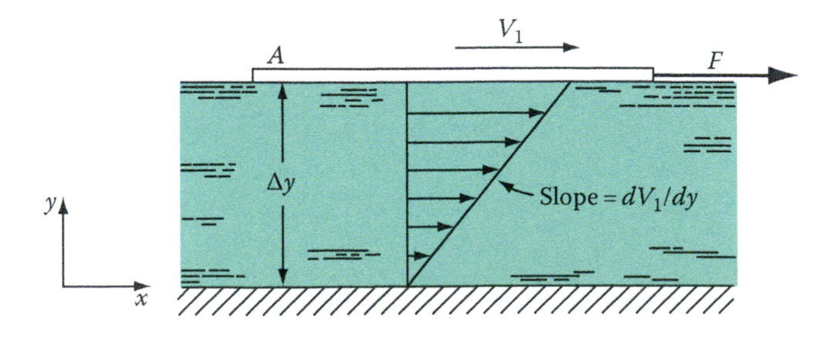

FIGURE 1.4 Shear stress applied to a fluid.

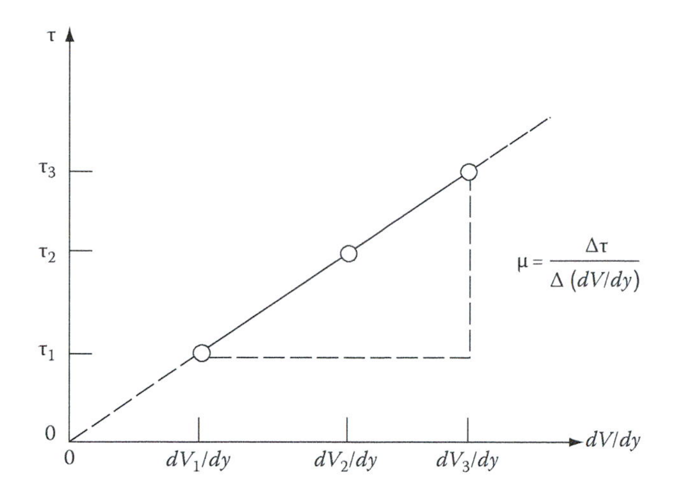

FIGURE 1.5 A plot of τ versus dV/dy (a rheological diagram) for Newtonian fluids.

The plot of Figure 1.5 is a straight line that passes through the origin. This result is characteristic of a *Newtonian fluid*, but there are other types of fluids called *non-Newtonian fluids*. A graph of τ versus dV/dy, called a **rheological diagram**, is shown in Figure 1.6 for several types of fluids. Newtonian fluids follow Newton's law of viscosity and are represented by the equation

$$\tau = \mu \frac{dV}{dy} \tag{1.11}$$

where

τ is the applied shear stress in dimensions of F/L^2 (lbf/ft^2 or N/m^2)
μ is the absolute or dynamic viscosity of the fluid in dimensions of $F \cdot T/L^2$ (lbf · s/ft^2 or N · s/m^2)
dV/dy = the strain rate in dimensions of $1/T$ (rad/s)

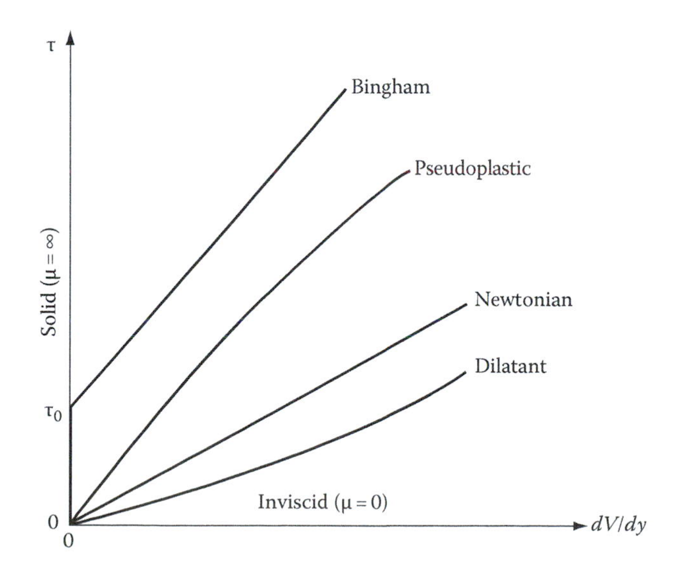

FIGURE 1.6 A rheological diagram for Newtonian and non-Newtonian time-independent fluids.

Examples of Newtonian fluids are water, oil, and air. If a fluid cannot be described by Equation 1.11, it is called a non-Newtonian fluid. On the basis of their behavior, these fluids are divided into three categories: time-independent, time-dependent, and viscoelastic.

1.3.2.1 Time-Independent Fluids

Wet beach sand and other water solutions containing a high concentration of powder are examples of **dilatant** fluids (Figure 1.6). These fluids exhibit an increase in viscosity with increasing shear stress. A power law equation (called the Ostwald–deWaele equation) usually gives an adequate description:

$$\tau = K\left(\frac{dV}{dy}\right)^n \qquad (n > 1) \tag{1.12}$$

where K is called a consistency index with dimensions of $(F \cdot T^n)/L^2$ (lbf · s^n/ft^2 or N · s^n/m^2) n is a flow behavior index.

Greases, mayonnaise, and starch suspensions are examples of **pseudoplastic** fluids (see Figure 1.6). These fluids exhibit a decrease in viscosity with increasing shear stress. Again a power law equation applies:

$$\tau = K\left(\frac{dV}{dy}\right)^n \qquad (n < 1) \tag{1.13}$$

Chocolate mixtures, drilling muds, greases, paint, paper pulp, soap, toothpaste, and sewage sludge are examples of **Bingham plastic** fluids (see Figure 1.6). These fluids behave as solids until an initial yield stress τ_0 is exceeded. Beyond τ_0, Bingham plastics behave like Newtonian fluids. The descriptive equation is

$$\tau = \tau_0 + \mu_0 \frac{dV}{dy} \tag{1.14}$$

1.3.2.2 Time-Dependent Fluids

A gypsum suspension is an example of a **rheopectic** fluid. A shear stress that increases with time gives the rheopectic fluid a constant strain rate. In Figure 1.4, τ_1 would have to increase with time to maintain a constant strain rate dV_1/dy.

Fast-drying paints, some liquid foods, and shortening are common examples of **thixotropic** fluids. These fluids behave in a manner opposite to rheopectic fluids. A shear stress that decreases with time gives a thixotropic fluid a constant strain rate. In Figure 1.4, τ_1 would have to decrease with time to maintain dV_1/dy.

1.3.2.3 Viscoelastic Fluids

Flour dough is an example of a **viscoelastic** fluid. Such fluids show both elastic and viscous properties. They partly recover elastically from deformations caused during the flow.

Most common fluids follow Newton's law of viscosity, however, and we will deal primarily with them in this text. The preceding paragraphs simply illustrate that many types of fluids exist. The absolute or dynamic viscosity of various Newtonian fluids appears in the property tables in the appendix—Table A.3 for air, Table A.4 for water, Table A.5 for common liquids, and Table A.6 for common gases.

Example 1.4

A fluid is placed in the area between two parallel plates. The upper plate is movable and connected to a weight by a cable as shown in Figure 1.7. Calculate the velocity of the plate for two cases.

a. Assume the fluid to be castor oil (Newtonian)
b. Assume the fluid to be grease having $\tau_0 = 4$ N/m^2 and $\mu_0 = 0.004$ N·s/m^2

In all cases, take $m = 0.05$ kg, $\Delta y = 5$ mm, $g = 9.81$ m/s^2, and the area of contact $A = 0.5$ m^2. Assume that steady state is achieved.

Solution

a. From Table A.5, μ for castor oil is 650×10^{-3} N·s/m^2. The force applied is $mg = 0.05$ kg (9.81 m/s^2) = 0.491 N, and the shear stress is $\tau = 0.491$ N/0.5 m^2 or $\tau = 0.981$ N/m^2. By definition,

$$\tau = \mu \frac{dV}{dy}$$

or

$$\tau = \mu \frac{\Delta V}{\Delta y}$$

Now at $y = 0$ we have $V = 0$, and at $y = \Delta y = 0.005$ m we have the velocity of interest. By substitution,

$$0.981 \text{ N/m}^2 = 0.650 \text{ N} \cdot \text{s/m}^2 \frac{V}{0.005 \text{ m}}$$

and

$$V = 0.0075 \text{ m/s} = 7.5 \text{ mm/s}$$

b. The applied shear stress is 0.981 N/m^2, and the initial shear stress of the grease is 4 N/m^2. The velocity of the plate is, therefore, $V = 0$ m/s. The plate will not move unless the applied shear stress exceeds 4 N/m^2.

Example 1.5 Laboratory Experiment: Viscosity of Peanut Butter

The first two columns of Table 1.2 shows actual viscosity data obtained on Jif® creamy peanut butter. The data were obtained with TA Rheometer, which consists of a flat stationary surface and a rotating disk. A tablespoon of the liquid of interest is placed on the flat surface, which is roughly 3 in. in diameter. The flat surface is then raised until the fluid

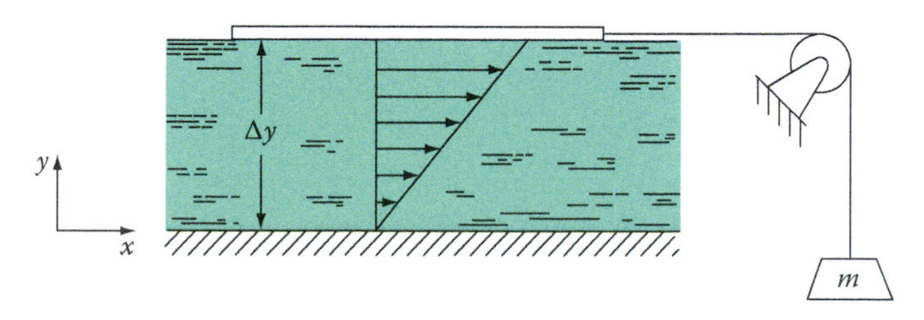

FIGURE 1.7 Sketch for Example 1.4.

TABLE 1.2
Viscosity Data for Jif® Creamy Peanut Butter

τ (Pa)	dV/dt (1/s)	μ (Pa·s)
268.4	0.06065	4426
310.5	0.5699	544.8
352.6	1.535	229.8
394.7	2.609	151.3
436.8	3.775	115.7
478.9	5.555	86.21
521.0	12.28	42.43
563.2	52.08	10.81
605.3	136.9	4.422

touches the disk. The apparatus begins rotating the disk and simultaneously measuring the rotational speed and torque required. The torque is related directly to the applied shear stress (column 1 of the table), and the rotational speed is directly related to the strain rate (column 2). Assuming a power law relationship, determine the equation that best describes the data and the type of fluid tested.

Solution

The data can be analyzed by hand with a calculator using equations for a least-squares analysis. Alternatively, the data can be entered onto a spreadsheet and a trend line equation is provided. The latter of these two methods yields the following power law equation:

$$\tau = K\left(\frac{dV}{dy}\right)^n = 395.9\left(\frac{dV}{dy}\right)^{0.1163}$$

A graph of the data is shown in Figure 1.8a. Comparing this graph with that of Figure 1.6, we conclude that the peanut butter is a **pseudoplastic fluid**.

The third column of Table 1.2 shows the apparent viscosity of the peanut butter and is merely the numbers in column 1 divided by those in column 2. A graph of the apparent viscosity versus strain rate for the last six data points is given in Figure 1.8b. We see that as the strain rate increases, the apparent viscosity decreases, and so the peanut butter is referred to as a *shear thinning* fluid. That is, as shear stress increases, the strain rate increases, and the apparent viscosity decreases.

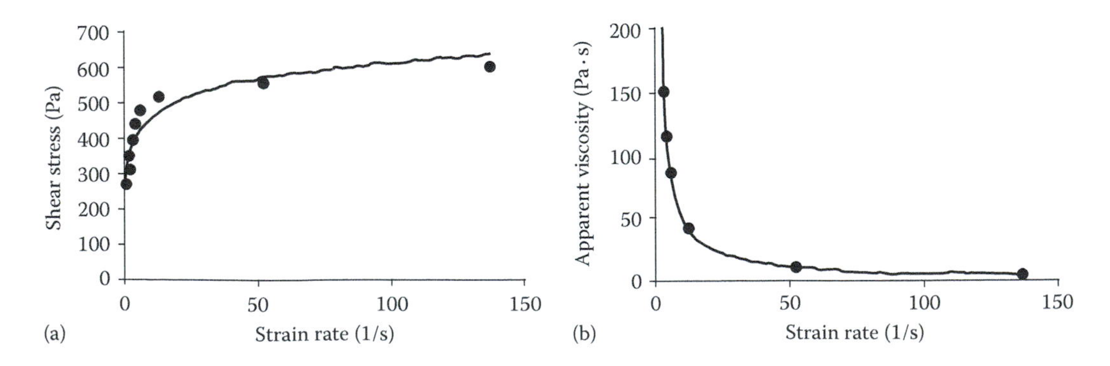

FIGURE 1.8 Graph of the viscosity data for Jif® creamy peanut butter.

Example 1.6 Laboratory Experiment: Viscosity of Ocean Breeze Shampoo

An inexpensive method of determining the viscosity of a transparent liquid is to use a falling sphere viscometer. The technique involves measuring the time it takes for a sphere to fall through a certain distance in the liquid. The distance traveled divided by the time gives what is known as the terminal velocity of the sphere. The apparatus consists of a cylinder made of clear plastic, and mounted vertically on a board. The cylinder is 0.6 m tall and has an inside diameter of 10 cm. Behind the cylinder is a scale that is divided into tenths of a foot. The cylinder is filled with Ocean Breeze shampoo. At the bottom of each cylinder is a 1/2 nominal pipe with two ball valves attached for retrieving the spheres at the conclusion of the experiment. See Figure 1.9.

Procedure

A sphere (a ball bearing) is dropped into the cylinder liquid and the time it takes for the sphere to fall a certain measured distance d is recorded. The distance divided by the measured time gives the terminal velocity of the sphere. The procedure can be repeated with different spheres and the results averaged. With the terminal velocity of a sphere measured and known, the absolute and kinematic viscosity of the liquid can be calculated.

Theory

An object falling through a fluid medium eventually reaches a constant final speed or terminal velocity. If this terminal velocity is sufficiently low, then the various forces acting on the object can be described with exact expressions. The forces acting on a sphere, for example, that is falling at terminal velocity through a liquid, are

$$\text{Weight} - \text{Buoyancy} - \text{Drag} = 0$$

The volume of a sphere is $\forall = \pi D^3/6$. The weight of the sphere is $m_s g = \rho_s \forall g = \rho_s g(\pi D^3/6)$. The buoyant force is the weight of the fluid displaced by the sphere, $mg = \rho \forall g = \rho g(\pi D^3/6)$ The drag force is due to a frictional effect that is related to the viscosity μ of the liquid.

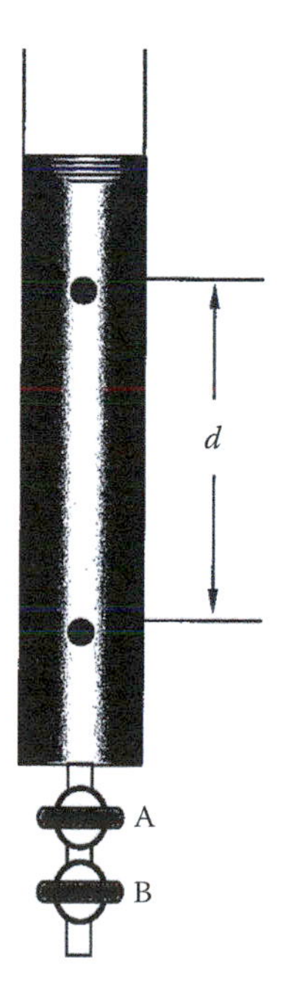

FIGURE 1.9 Terminal velocity measurement.

Under certain conditions, the drag force exerted on a sphere moving through a fluid may be written as $D_f = 3\pi\mu VD$. Combining these terms, we get

$$\rho_s g\frac{\pi D^3}{6} - \rho g\frac{\pi D^3}{6} - 3\pi\mu VD = 0 \tag{1.15}$$

All terms in this expression are known, except for the terminal velocity V, which is measured experimentally. Rearranging and simplifying,

$$\mu = \frac{\rho g D^2}{18V}\left(\frac{\rho_s}{\rho} - 1\right) \tag{1.16}$$

The expression for drag is valid only if the following equation is satisfied:

$$\frac{\rho VD}{\mu} < 1$$

Once the viscosity of the liquid is found, the aforementioned ratio *should* be calculated to be certain that the mathematical model gives an accurate description of a sphere falling through the liquid.

A ball bearing (s = 8) falls through a liquid—Ocean Breeze shampoo. It takes the sphere 11.03 s to fall 18.3 cm. The sphere diameter is 6.35 mm. Calculate the viscosity of the shampoo.

Solution

The viscosity is found with the equation derived previously:

$$\mu = \frac{\rho g D^2}{18V}\left(\frac{\rho_s}{\rho} - 1\right)$$

where

$\rho = 1\ 121\ \text{kg/m}^3$
$\rho_s = 8(1\ 000)\text{kg/m}^3$
$D = 0.006\ 35\ \text{m}$
$g = 9.81\ \text{m/s}^2$
$V = 18.3\ \text{cm}/11.03\ \text{s} = 1.66\ \text{cm/s} = 0.016\ 6\ \text{m/s}$

Substituting,

$$\mu = \frac{(1121\ \text{kg/m}^3)(9.81\ \text{m/s}^2)(0.006\ 35\ \text{m})^2}{18(0.016\ 6\ \text{m/s})}\left(\frac{8\ 000}{1121} - 1\right)$$

$$\mu = 9.11\frac{\text{kg}}{\text{m}^3}\frac{\text{m}}{\text{s}^2}\text{m}^2\frac{\text{s}}{\text{m}} = 9.11\frac{\text{kg}}{\text{s}\cdot\text{m}} = 9.11\frac{\text{kg}}{\text{s}\cdot\text{m}}\frac{\text{m}}{\text{m}}\frac{\text{s}}{\text{s}}$$

$$\mu = 9.11\frac{\text{kg}\cdot\text{m}}{\text{s}^2}\frac{\text{s}}{\text{m}^2}$$

$$\mu = 9.11\ \text{N}\cdot\text{s/m}^2 = 9.11\ \text{Pa}\cdot\text{s}$$

As a check on the validity of using the equation for the drag force, we calculate

$$\frac{\rho VD}{\mu} = \frac{1\ 121(0.016\ 6)(0.006\ 35)}{9.11} = 0.012$$

which is less than 1.

Manipulating the units in this fashion for every problem is unnecessary. If all parameters in an equation are expressed in terms of the fundamental dimensions/units (mass/kg, force/N, time/s, length/m), then the calculated result will also be in terms of the fundamental units.

1.3.3 Kinematic Viscosity

Absolute or dynamic viscosity μ has already been discussed. In many equations, however, the ratio of absolute viscosity to density often appears. This ratio is given the name of **kinematic viscosity** ν:

$$\nu = \frac{\mu}{\rho} \tag{1.17}$$

From Table A.5, for example, the density of ethylene glycol is 1 100 kg/m^3, whereas the dynamic viscosity is 16.2×10^{-3} N·s/m^2. Hence, the kinematic viscosity is

$$\nu = \frac{16.2\times 10^{-3}\mathrm{N\cdot s/m^2}}{1\ 100\ \mathrm{kg/m^3}} = 1.48\times 10^{-5}\ \mathrm{m^2/s}$$

The dimensions of kinematic viscosity are L^2/T (ft^2/s or m^2/s).

The terms centipoises and centistoke are sometimes used when referring to viscosity. A centipoises is abbreviated as cp and refers to the absolute viscosity of a fluid: 1 cp = 1×10^{-3} N·s/m^2. The centistoke refers to the kinematic viscosity 1 cs = 1×10^{-6} m^2/s.

1.3.4 Pressure

Pressure is defined as a normal force per unit area existing in the fluid. Pressure (p) has dimensions F/L^2 (lbf/ft^2 or N/m^2) and is treated in detail in Chapter 2. Fluids, whether at rest or moving, exhibit some type of pressure variation—either with height or with horizontal distance. In closed-conduit flow, such as flow in pipes, differences in pressure from beginning to end of the conduit maintain the flow. Pressure forces are significant in this application.

It is important to note that the viscosity and density of fluids both change with temperature. This can easily be seen in Table A.3 for air and in Table A.4 for water. Table A.3, for example, shows that over the range −23.15°C to 146.85°C, density changes from 1.413 to 0.840 kg/m^3. Viscosity varies from 15.99 to 23.66 N·s/m^2 over this same temperature range.

1.3.5 Surface Tension

Surface tension is a measure of the energy required to reach below the surface of a liquid and bring molecules to the surface to form a new area. Thus, surface tension arises from molecular considerations and has meaning only for liquid–gas, liquid–liquid, or liquid–vapor interfaces. In the liquid bulk sketched in Figure 1.10, compare molecules located at point A well below the surface to those at point B near the surface. A fluid particle at A is drawn uniformly in all directions to its neighboring particles. A particle at B is drawn more strongly to liquid particles in its vicinity than to those in the vapor above the surface. Consequently, a surface tension exists; in water, this surface tension is strong enough to support the weight of a needle. Figure 1.11 further illustrates this phenomenon as a platinum–iridium ring is being moved through a liquid–gas interface.

The previous discussion concerned the surface tension at a liquid–gas interface, where the forces of attraction were cohesive forces. If the liquid is in contact with a solid, however, adhesive forces must be considered as well. Some fluids, such as water, contain molecules that are drawn more closely to glass than to each other. This effect is illustrated in Figure 1.12a, which depicts water in a glass tube. Adhesive forces (water to glass) are greater than cohesive forces (water to water). Mercury exhibits the opposite effect, as shown in Figure 1.12b.

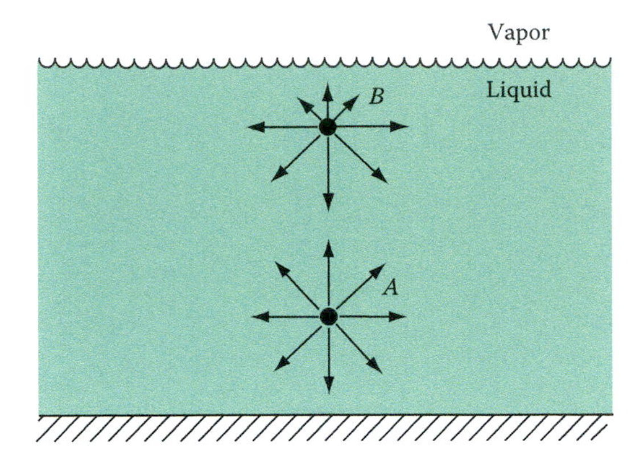

FIGURE 1.10 An illustration of differences in molecular attraction between two particles at separate locations.

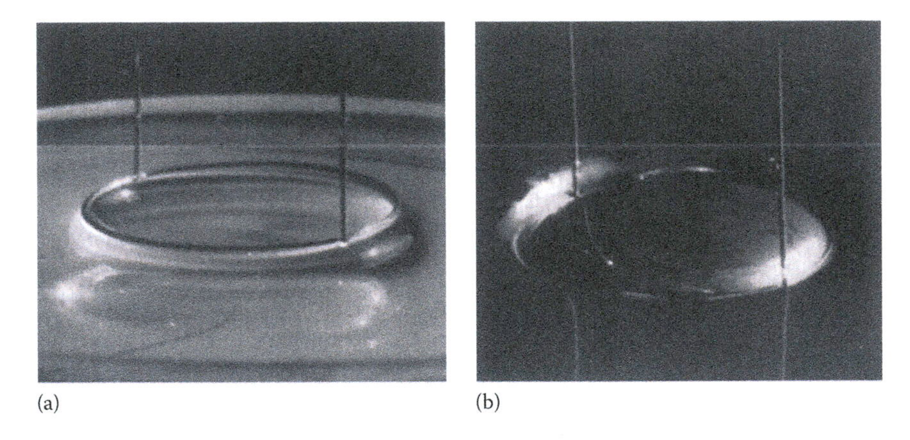

FIGURE 1.11 An illustration of surface tension. A platinum–iridium ring (ring circumference is 6 cm, wire diameter is 0.018 cm) (a) being pulled upward through the surface of a tap water–liquid soap mixture and (b) being pushed downward through the surface of liquid mercury of 90% purity.

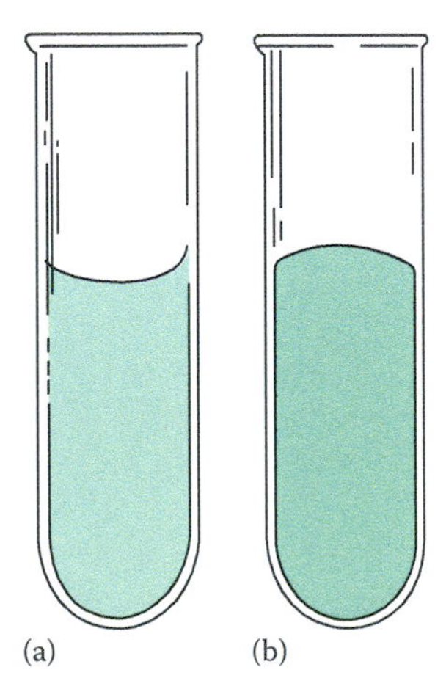

FIGURE 1.12 Capillary action in a glass tube showing a water meniscus and a mercury meniscus. (a) Water and (b) mercury.

Mercury-to-mercury forces are greater than those between mercury and glass. The resulting curved surface, called a **meniscus**, illustrates this phenomenon. The surface tension of various liquids is given in Tables A.5 and A.9.

As stated earlier, surface tension is the energy required to bring molecules to the surface to form a new area; thus, the dimension of surface tension is energy/area by definition: $F \cdot L/L^2 = F/L$ (lbf/ft or N/m). Surface tension, represented by the letter σ, is important in studies of droplets and jet flows. In the majority of problems we will investigate, surface tension forces are negligible in comparison to those due to pressure differences, gravity, and viscous forces.

Example 1.7

a. Develop an expression to calculate the pressure inside a droplet of liquid.
b. Determine the pressure inside of a 0.02 cm diameter water droplet exposed to atmospheric pressure (101 300 N/m²).

Solution

a. A sketch of a droplet in equilibrium is shown in Figure 1.13. The pressure difference from inside to outside tends to expand the droplet. The pressure force is constrained by the surface tension. The bursting force is pressure difference times area, and this force is balanced by the surface tension force. A force balance on the droplet of Figure 1.13 gives

$$\begin{pmatrix}\text{Pressure}\\ \text{difference}\end{pmatrix} \times \begin{pmatrix}\text{Cross-sectional}\\ \text{area}\end{pmatrix} = \begin{pmatrix}\text{Surface}\\ \text{tension}\end{pmatrix} \times \begin{pmatrix}\text{Surface}\\ \text{length}\end{pmatrix}$$

or

$$(p_i - p_o)\pi R^2 = \sigma(2\pi R)$$

Solving, we get

$$p_i - p_o = \frac{2\sigma}{R}$$

b. From Table A.5 for water, we read

$$\sigma = 71.97 \times 10^{-3} \text{ N/m}$$

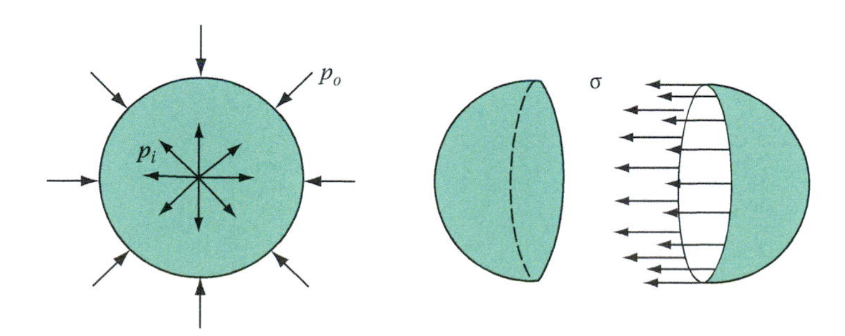

FIGURE 1.13 Sketch for Example 1.5.

The pressure outside the droplet is p_o = 101 300 N/m², and the radius is R = 0.02 = 0.01 cm = 0.0001 m. The pressure inside the droplet is thus found as

$$p_i = p_o + \frac{2\sigma}{R} = 101\,300\ \text{N/m}^2 + \frac{2\left(71.97\times10^{-3}\right)\text{N/m}}{0.0001\ \text{m}}$$

or

$$p_i = 102\,739\ \text{N/m}^2$$

Example 1.8

A capillary tube is one that has a very small inside diameter. When a capillary tube is immersed slightly in a liquid, the liquid will rise within the tube to a height that is proportional to its surface tension. This phenomenon is referred to as **capillary action**. Figure 1.14a depicts a glass capillary tube slightly immersed in water. As shown, the water rises by an amount h within the tube, and the angle between the meniscus and glass tube is θ. Perform a force balance on the system and develop a relationship between the capillary rise h and surface tension σ.

Solution

Figure 1.14b shows a cross section of the liquid column of height h and diameter $2R$. Also shown are the forces that act—the weight of W of the column of liquid and the surface tension force T. The weight is found with

$$W = mg = \frac{m}{\forall}\forall g$$

where $m/\forall$ is the density and volume is given by $\forall = \pi R^2 h$. Thus,

$$W = \rho(\pi R^2 h)g$$

The force due to surface tension acts at the circumference where the liquid touches the glass wall. The surface tension force is, therefore, the product of surface tension and circumference:

$$T = \sigma(2\pi R)$$

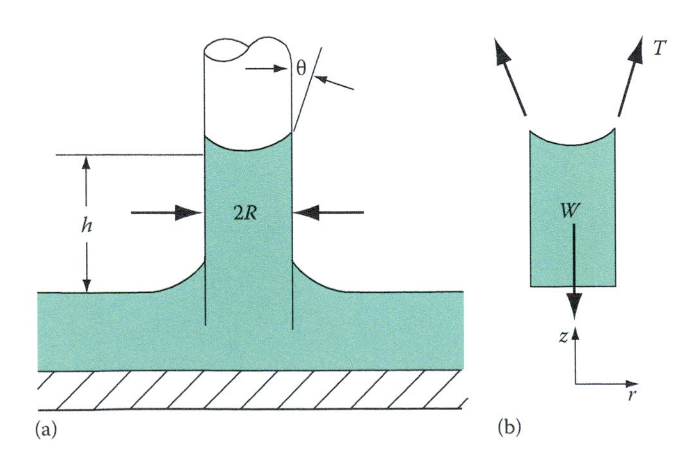

FIGURE 1.14 Capillary tube submerged in liquid and the resultant forces.

Summing forces in the vertical direction shows that the vertical component of the surface tension force must equal the weight of the liquid column:

$$W = T\cos\theta$$

or

$$\rho\pi R^2 hg = 2\pi R\sigma(\cos\theta)$$

Solving for capillary height gives

$$h = \frac{2\sigma}{\rho Rg}\cos\theta$$

1.3.6 Specific Heat

The **specific heat** of a substance is the heat required to raise a unit mass of the substance by 1°. The dimension of specific heat is energy/(mass · temperature): $F \cdot L/(M \cdot t)$.* The process by which the heat is added also makes a difference, particularly for a gas. The specific heat for a gas that undergoes a process occurring at constant pressure involves a different specific heat than that for a constant volume process. For example, the specific heat at constant pressure c_p for carbon dioxide is 0.205 Btu/(lbm · °R), or 876 J/(kg · K), and the specific heat at constant volume c_v is 0.158 Btu/(lbm · °R), or 674 J/(kg · K). (Table A.6 gives the specific heat of various gases.) Also of importance when dealing with these properties is the **ratio of specific heats**, defined as

$$\gamma = \frac{c_p}{c_v}$$

For carbon dioxide, the ratio of specific heats γ is 1.30. For air $\gamma = 1.4$.

The Btu (British thermal unit) is the unit of energy measurement in the English engineering system. One Btu is defined as the energy required to raise the temperature of 1 lbm of water by 1°F. However, because we are using the British gravitational system, the units we will encounter for specific heat are Btu/(slug · °R). Although both specific heats vary with temperature for real substances, they are in many cases assumed to be constant to simplify calculations.

Example 1.9

Suppose that 5 kg of air is placed in a cylinder-piston arrangement, as in Figure 1.15. The piston is weightless, is free to move vertically without friction, and provides a perfect seal.

a. How much heat must be added to increase the temperature of the air by 10 K?
b. How much heat must be added for the same temperature increase if the piston is constrained?

Solution

a. As heat is added, the air expands and forces the piston to move upward. The expansion therefore takes place at constant pressure if we assume a frictionless piston. From the definition of specific heat at constant pressure,

$$\tilde{Q} = mc_p(T_2 - T_1)$$

where
$\tilde{Q}$ = heat added
the specific heat is assumed to be constant
$T_2 - T_1$ is given as 10 K

* Note that when the denominator of a unit is written in shilling style (with a slash, /), to avoid confusion parentheses are used if a product is involved.

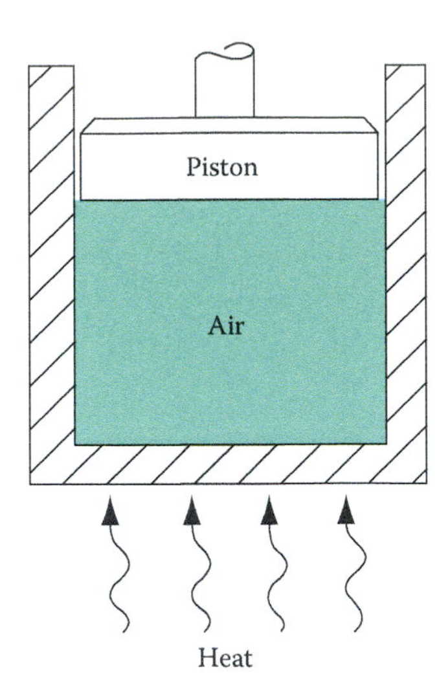

FIGURE 1.15 Sketch for Example 1.9.

The specific heat of air is 1 005 J/(kg·K) from Table A.6. By substitution,

$$\tilde{Q} = 5 \text{ kg}[1\,005 \text{ J/(kg}\cdot\text{K)}](10 \text{ K})$$

Solving, we get

$$\tilde{Q} = 50\,250 \text{ J} = 50.3 \text{ kJ}$$

b. With the piston held in place, the heat addition occurs at constant volume. Assuming constant specific heat, we have by definition

$$\tilde{Q} = mc_v(T_2 - T_1)$$

The specific heat at constant volume c_v is not found specifically in Table A.6 but can be determined from the data available. Table A.6 gives for air,

$$c_p = 1\,005 \text{ J/(kg}\cdot\text{K)}$$

and

$$\gamma = \frac{c_p}{c_v} = 1.4$$

Thus, the specific heat we are seeking is

$$c_v = \frac{c_p}{1.4} = \frac{1\,005}{1.4} = 718 \text{ J/(kg}\cdot\text{K)}$$

After substitution, we get

$$\tilde{Q} = 5 \text{ kg}[718 \text{ J/(kg}\cdot\text{K)}](10 \text{ K})$$

or

$$\tilde{Q} = 35\ 900\ \text{J} = 35.9\ \text{kJ}$$

More heat is required for the constant-pressure process because some of the added heat goes into work done in raising the piston. In the constant-volume process, by contrast, added heat goes directly into raising the temperature of the gas.

1.3.7 Internal Energy

Internal energy is the energy associated with the motion of the molecules of a substance. To illustrate, consider a quantity of gas. The gas can have three types of energy: energy of position (potential energy), energy of translation (kinetic energy), and energy of molecular motion (internal energy). In the preceding example, the heat added at constant volume went into internal energy of the oxygen. The added heat did not increase the potential or kinetic energies but instead affected the motion of the molecules. This effect is manifested as an increase in temperature. For a perfect gas with constant specific heats, it can be shown that

$$\frac{du}{dT} = c_v \tag{1.18}$$

Internal energy per unit mass has dimensions of energy/mass ($F \cdot L/M$).

1.3.8 Enthalpy

A quantity that appears often in equations is $u + p/\rho$; this quantity is given the special name **enthalpy**, h. The ratio of p/ρ is often referred to as flow work. In Example 1.9, the air was heated at constant pressure. The added heat went into increasing the internal energy of the gas and work done by raising the piston. Again for the case of a perfect gas with constant specific heats, it can be shown that

$$\frac{dh}{dT} = c_p \tag{1.19}$$

Enthalpy, like internal energy, has the dimension of energy/mass ($F \cdot L/M$).

Example 1.10

Two slugs of air at 14.7 lbf/in.^{2} and 70°F are placed in a constant-volume container. Heat is added until the final temperature is 100°F. Calculate the change in internal energy for the air.

Solution

Under the pressure and temperature conditions stated, the air can be assumed to have constant specific heats. From Table A.6 for air c_p = 7.72 Btu/(slug · °R) and c_p/c_v = 1.4. Therefore, c_v = 7.72/1.4 = 5.51 Btu/(slug · °R). The internal energy change is found with Equation 1.18 written as

$$\Delta u = c_v \Delta T = 5.51(100 - 70)$$

or

$$\Delta u = 165\ \text{Btu/slug}$$

Note that the units for specific heat contain °R in the denominator, but the air temperatures are given in °F. The temperature difference of the air, however, is still 40°, whether the temperatures are first converted to °R. A 1°F temperature difference equals a 1°R temperature difference. Likewise, a 1 K temperature difference equals a 1°C temperature difference.

1.3.9 Compressibility Factor/Bulk Modulus

Water at atmospheric pressure, 101.3 kN/m^2, and at room temperature has a density of about 1 000 kg/m^3. Suppose that the pressure exerted on the water is increased by a factor of 100 to 10 130 kN/m^2. The change in density, if measured, will be found to be less than 1% of the initial value, although the pressure exerted has been greatly increased. The property that describes this behavior is called the **compressibility factor**:

$$\beta = -\frac{1}{\forall}\left(\frac{\partial \forall}{\partial p}\right)_T \quad (1.20)$$

where
$\forall$ = volume
$\partial \forall / \partial p$ describes the change in volume with respect to pressure
the subscript notation indicates that the process is to occur at constant-temperature conditions

The reciprocal of the compressibility factor is the **isothermal bulk modulus**:

$$k = -\forall\left(\frac{\partial p}{\partial \forall}\right)_T \quad (1.21)$$

which can be written in difference form as

$$k = -\forall\left(\frac{\Delta p}{\Delta \forall}\right) = -\frac{\Delta p}{(\Delta \forall / \forall)} \quad (1.22)$$

The dimensions of the change in volume $\Delta \forall$ and the volume $\forall$ are the same, so the denominator of the aforementioned equation is dimensionless. The isothermal bulk modulus k therefore has the same units as pressure.

The isothermal bulk modulus can be measured, but it is usually more convenient to calculate it using data on the velocity of sound in the medium. The **sonic velocity** in a liquid (or a solid) is related to the isothermal bulk modulus and the density by

$$a = \sqrt{\frac{k}{\rho}} \quad (1.23)$$

where a is the sonic velocity in the medium. Sonic velocity of various liquids is provided in Table A.5.

Example 1.11

Determine the change in pressure required to decrease the volume of liquid acetone by 1% from its value at room temperature and pressure.

Solution

From Table A.5, we read

$$\rho = 0.787(1\ 000)\ \text{kg/m}^3$$

$$a = 1\ 174\ \text{m/s}$$

The isothermal bulk modulus is calculated as

$$k = \rho a^2 = [0.787(1\ 000)\ \text{kg/m}^3]\ (1\ 174\ \text{m/s})^2$$

or

$$k = 1.085 \times 10^9\ \text{N/m}^2$$

The 1% decrease in volume means that $-(\Delta \forall / \forall) = 0.01$. Equation 1.22 is rearranged to solve for the change in pressure:

$$\Delta p = -k\left(\frac{\Delta \forall}{\forall}\right) = -1.085 \times 10^9(-0.01)$$

Solving,

$$\Delta p = 1.085 \times 10^7\ \text{N/m}^2$$

Thus acetone initially at atmospheric pressure (101.3 kN/m^2) would have to experience a pressure increase of 10 850 kN/m^2 to change its volume by 1%. This represents a 100-fold pressure increase to yield a negligible change in volume. All liquids typically behave in this way under the action of external pressure changes. Consequently, we can consider liquids, in general, to be incompressible. Although these calculations were made for volume changes, the conclusions apply to density changes as well ($\rho = m/\forall$).

1.3.10 Ideal Gas Law

In the discussion of isothermal bulk modulus, a relationship was stated between changes in pressure and changes in volume for liquids. For gases, we use an **equation of state** to obtain such a relationship. Many gases under suitable conditions can be described by the **ideal gas law:**

$$p\forall = mRT$$

Using this equation, the density can be expressed as

$$\rho = \frac{p}{RT} = \frac{m}{\forall} \tag{1.24}$$

where
 T is the temperature in absolute units
 R is the gas constant (values of which are given in Table A.6 for various gases)

The gas constant for any gas can be obtained by dividing the universal gas constant by the molecular weight (called molecular mass in SI) of the gas. The universal gas constant is

$$\begin{aligned}\overline{R} &= 49{,}709\ \text{ft}\cdot\text{lbf/(slugmol}\cdot\text{R)} = 1545\ \text{ft}\cdot\text{lbf/(lbmol}\cdot\text{R)}\\ &= 8\ 312\ \text{N}\cdot\text{m}/(\text{mol K})\end{aligned} \tag{1.25}$$

(The molecular weight has units of slug/slugmol in the British gravitational system. In SI, the molecular mass has units of kg/mol. The mass unit does not appear in the denominator.) The ideal gas law closely approximates the behavior of most real gases at moderate pressures.

By way of comparison, let us calculate the percent change in density of a gas under a change of pressure. Consider a volume of air under a pressure of 100 kN/m^2 and again at a pressure of 500 kN/m^2. Thus,

$$\frac{p_1}{p_2} = \frac{100}{500} = \frac{1}{5}$$

From the ideal gas law applied to both states, and assuming constant temperature,

$$\frac{\rho_2}{\rho_1} = \frac{p_2}{p_1} = 5$$

and the percent change in density is

$$\left(\frac{\rho_2 - \rho_1}{\rho_1}\right)100 = 400\%$$

This result is in contrast to liquids, the density of which changes less than 1% for $p_2 = 100p_1$.

1.4 LIQUIDS AND GASES

Although liquids and gases are both fluids, there are significant differences between them. Liquids are incompressible, whereas gases are not. Another difference that is more obvious is that a liquid retains its own volume but takes the shape of its container. A gas takes the volume and the shape of its container. This difference is illustrated in Figure 1.16.

1.5 CONTINUUM

There are two possible mathematical approaches to the treatment of problems in fluid mechanics. In the **microscopic approach**, fluid behavior is described by specifying motion of the individual molecules. This is done traditionally using concepts from probability theory. In the **macroscopic approach**, fluid behavior is described by specifying motion of small fluid volumes containing many molecules. The macroscopic approach uses the average effect of many molecules and means that we are treating the fluid as a continuous medium or a **continuum**. In the study of fluid mechanics, we use the continuum approach throughout.

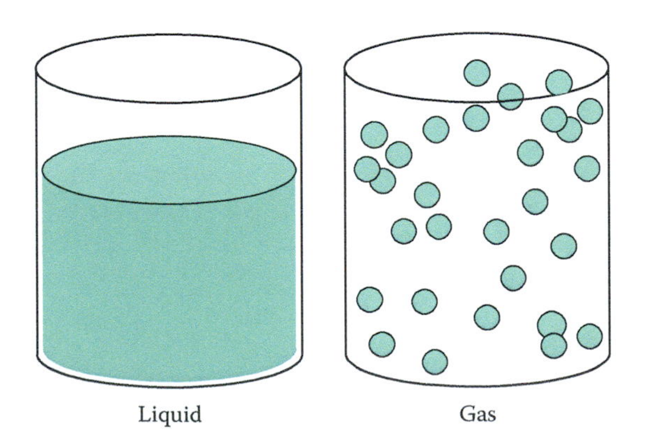

FIGURE 1.16 An illustration of one difference between a liquid and a gas.

Now let us redefine pressure with regard to the concept of a continuum. Consider a fluid particle made up of many molecules. We are interested in the average effects of these many molecules. Pressure at a point in a continuum is defined as a time-averaged normal force exerted by molecules on a unit surface. The area of the unit surface must be small, but large enough (compared to intermolecular distances) to yield representative results. If the unit area were too small, it is possible that no molecules would strike it. Thus a zero normal force would be obtained, and pressure at a point in this case would have no significance. Mathematically, pressure at a point in a continuum is defined as

$$p = \lim_{A \to A^*} \frac{F}{A}$$

Here A^* is a very small area not equal to zero but experiencing enough molecular collisions to be representative of the fluid bulk; F is the time-averaged normal force exerted by the collisions.

In a similar fashion, we can redefine density with regard to the concept of a continuum. Density is a mass per unit volume; mathematically,

$$\rho = \lim_{\mathcal{V} \to \mathcal{V}^*} \frac{M}{\mathcal{V}}$$

where

$\mathcal{V}$ is a volume that contains a mass M of fluid

$\mathcal{V}^*$ is a small volume not equal to zero but containing a large enough number of molecules that it is representative of the fluid

If $\mathcal{V}^*$ were allowed to shrink to zero, it would contain no molecules and density would have no significance.

To get an idea of how small $\mathcal{V}^*$ can be, consider a cube of air at room temperature and pressure that is 0.01 mm on each side. The 0.01 mm dimension is smaller than most measuring instruments and probes used in engineering, yet the cube contains 3×10^{10} molecules. So the 0.01 mm dimension is consistent with the continuum approach.

PROBLEMS

Dimensions, Units, Density

1.1 Use the Appendix tables to find the conversion factor between gallons per minute (gpm) to cubic feet per second (cfs or ft^3/s).

1.2 What is the conversion factor between BTU/hr and horsepower? Use the conversion factors from Table A.2 to determine the answer.

1.3 An object has a mass of 46 kg and weighs 450 N on a spring scale. Determine the acceleration due to gravity at this location.

1.4 A plastic milk container is labeled as containing 3.78 liters. If the specific gravity of milk is 1.03, determine the mass of the milk in the container and express it in kilograms.

1.5 The density of Ocean Breeze shampoo was determined by weighing an object of known volume in air and again by weighing it while submerged in the liquid. If the object was a 4 cm diameter sphere made of aluminum (specific gravity = 2.7), what is the expected weight of the object while submerged in the shampoo?

1.6 The density of Golden Apple Shampoo is to be determined by weighing an object of known volume in air, and again by weighing it while submerged in the liquid. A stainless steel cylinder of diameter 1 in. and length 2 in. is submerged in the shampoo. The measured weight while submerged is 0.39 lbf. What is the density of the shampoo?

1.7 The density of Strawberry Breeze Shampoo is to be determined by weighing an object of known volume in air and again by weighing it while submerged in the liquid. The object is a 4 cm diameter sphere made of brass (density = 8.4 g/cm^3), and the weight while submerged is 2.36 N. What is the density of the shampoo?

1.8 Water has a density of 1 000 kg/m^3. What is its density in lbm/ft^3, slug/ft^3, and g/cm^3?

1.9 The earth may be considered as a sphere whose diameter is 8000 miles and whose density is roughly approximated to be 6 560 kg/m^3. What is the mass of the earth in lbm and in kg?

1.10 It is commonly known that there are 16 ounces in one pound. However, Table A.2 lists the ounce as a unit of volume for liquids. A 1/2 lbf glass weighs 1 lbf when filled with 8 ounces of liquid. Determine the liquid density and the specific weight in SI units.

1.11 What is the weight in N of 1 ft^3 of kerosene?

1.12 What is the mass in kg of 5 ft^3 of acetone?

1.13 What is the density of ethylene glycol lbm/ft^3? What is the mass of 1 ft^3 of ethylene glycol in slugs?

1.14 Graph the density of air as a function of temperature.

1.15 In the petroleum industry, the specific gravity of a substance (usually an oil) is expressed in terms of degrees API; or °API (American Petroleum Institute). The specific gravity and the °API are related by

$$°\text{API} = \frac{141.5}{\text{specific gravity}} - 131.5$$

Graph °API (vertical axis) versus specific gravity ranging from 0.8 to 0.9 (typical for many oils).

1.16 Prepare a plot of specific gravity of water as a function of temperature. Let temperature vary in 10° increments.

Fluid Properties: Viscosity

1.17 In Example 1.4b, what is the minimum mass required to move the upper plate if the fluid has an initial yield stress of 4 N/m^2?

1.18 If the mass in example 1.4b is 0.025 kg, and the fluid is castor oil, determine the plate velocity.

1.19 Referring to Figure 1.7, assume that the fluid in the space is castor oil. What weight is required to move the plate at 5 cm/s?

1.20 A weightless plate is moving upward in a space as shown in Figure P1.20. The plate has a constant velocity of 2.5 mm/s, and kerosene is placed on both sides. The contact area for either side is 2.5 m^2. The plate is equidistant from the outer boundaries with Δy = 1.2 cm. Find the force F.

1.21 What is the kinematic viscosity of the Ocean Breeze shampoo in Example 1.6? Express the results in m^2/s and in centistokes (cs).

1.22 A falling sphere viscometer is used to measure the viscosity of shampoo whose density is 1 008 kg/m^3. A sphere of diameter 0.792 cm is dropped into the shampoo, and measurements indicate that the sphere travels 0.183 min 7.35 s. The density of the sphere is 7 940 kg/m^3. Calculate the absolute and kinematic viscosity of the shampoo.

1.23 Referring to Figure P1.20, the plate is being pulled upward in a space filled with chloroform. The plate velocity is 12 in./s and Δy = 0.05 in. The force is 2 lbf. Determine the area of contact of each side of the plate.

1.24 In Figure P1.24, the total space between stationary boundaries is 1 cm. Ethylene glycol is placed on the left side and propylene glycol on the right. When the infinite plate that separates the liquids is pulled upward, it finds an equilibrium position. Determine the lateral location of the plate if it has a thickness of 1 mm.

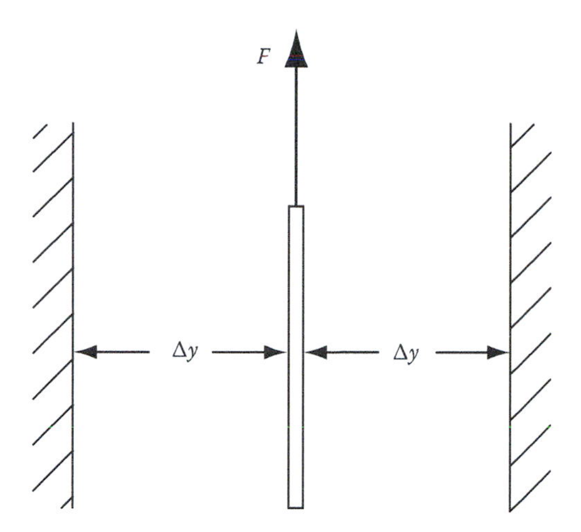

FIGURE P1.20

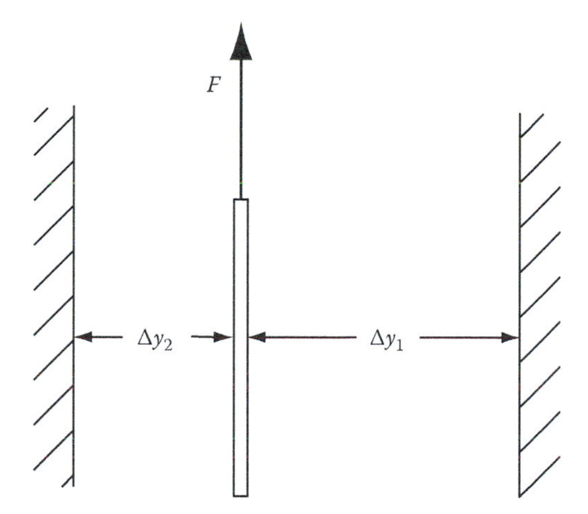

FIGURE P1.24

1.25 A Saybolt viscometer is used in the petroleum industry to measure viscosity of lubricating oils. The test oil is placed in a cup surrounded by a constant temperature bath. At time zero (a stopwatch is started), test oil is allowed to flow out of the bottom of the cup, through an orifice. The oil leaves in the form of a stream and is collected in a calibrated beaker. When 60 mL of oil flows through the orifice, the elapsed time is recorded. The time required for 60 mL of oil to flow through the orifice is thus experimentally determined. The viscosity of the oil is expressed in terms of the elapsed time; for example, one would say that the oil has "a viscosity of 100 Saybolt Universal Seconds." This is abbreviated as "100 SUS." The equation to convert SUS to units of m^2/s is given by

$$\nu\left(\mathrm{m^2/s}\right) = 0.224\times10^{-6}(\mathrm{SUS}) - \frac{185\times10^{-6}}{\mathrm{SUS}}$$

Graph this equation as ν on the vertical axis vs SUS. Allow SUS to vary from 30 to 120 SUS in increments of .10 SUS. (The equation is valid over only this range.)

1.26 Figure P1.24 illustrates an infinite plate being pulled upward in a space filled with ethyl alcohol on the right and an unknown fluid on the left. The plate is not equidistant from the boundaries; in fact, $\Delta y_1 = 2\Delta y_2$. Determine the viscosity of the unknown fluid.

1.27 What type of fluid is described by the following shear stress–strain rate data?

τ (N/m²)	0.4	0.82	2.50	5.44	8.80
dV/dy (rad/s)	0	10	50	120	200

1.28 What type of fluid is described by the following shear stress–strain rate data?

τ (N/m²)	0	0.005 8	0.008 9	0.010 7	0.011
dV/dy (rad/s)	0	25	50	75	100

1.29 What type of fluid is has the following shear stress–strain rate relationship?

τ (lbf/ft²)	0	4	9	14	19
dV/dy (rad/s)	0	12.5	25.0	37.5	50.0

1.30 Actual tests on Vaseline yielded the following data:

τ (N/m²)	0	200	600	1 000
dV/dy (1/s)	0	500	1 000	1 200

Graph the data and determine the fluid type.

1.31 Consider the act of spreading soft butter on bread with a knife. In essence, we have a stationary surface (the bread), a moving surface (the knife), and a newtonian fluid occupying the space in between. The part of the knife in contact with the butter has dimension of 9 cm × 1.6 cm. The knife is moved across the bread at a rate of 5 cm/s. The average thickness of the butter during the process is about 2 mm (a rough approximation). (a) Calculate the shear stress exerted on the butter if the force required to move the knife is 0.07 N. (b) Calculate the strain rate. (c) Calculate the absolute viscosity of the butter.

1.32 Figure P1.32 shows a shaft 4 in. in diameter moving through a well oiled sleeve that is 12 in. long. The force required to move the shaft is 25 lbf, and the shaft velocity is 5 in./s. The oil filled clearance between the shaft and sleeve is 0.005 in. Calculate the viscosity of the lubricating oil (a Newtonian fluid).

1.33 Mayonnaise is tested in the laboratory to obtain its rheological diagram. Two data points are

1. $\tau = 4.63 \times 10^{-2}$ lbf/ft² $\quad dV/dy = 25$ rad/s
2. $\tau = 6.52 \times 10^{-2}$ lbf/ft² $\quad dV/dy = 50$ rad/s

Determine the consistency index and the flow behavior index. Calculate the strain rate if the shear stress is increased to 7×10^{-2} lbf/ft².

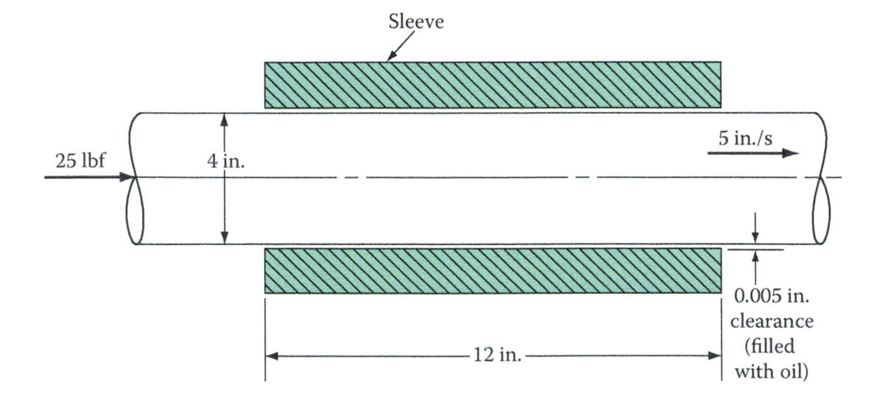

FIGURE P1.32

1.34 Two data points on a rheological diagram of a certain grease are

1. $dV/dy = 20$ rad/s $\tau = 8.72 \times 10^{-3}$ N/m^2
2. $dV/dy = 40$ rad/s $\tau = 2.10 \times 10^{-3}$ N/m^2

Determine the consistency index and the flow behavior index. Calculate the strain rate if the shear stress is increased to 3×10^{-2} N/m^2.

1.35 A highly viscous slow-drying paint has a viscosity μ_0 of 0.04 lbf·s/ft^2. At a shear stress of 2.7 lbf/ft^2, the strain rate is 70 rad/s. Calculate its initial yield stress.

1.36 A fluid with a viscosity of 8 centipoise has a density of 59 lbm/ft^3. What is its kinematic viscosity in the CGS system?

1.37 Graph the absolute viscosity of water as a function of temperature.

1.38 Graph the kinematic viscosity of water as a function of temperature.

Fluid Properties: Surface Tension

1.39 Calculate the pressure inside a 2 mm diameter drop of acetone exposed to atmospheric pressure (101.3 kN/m^2).

1.40 Determine the pressure inside a water droplet of diameter 500 μm in a partially evacuated chamber where $p = 70$ kN/m^2.

1.41 Calculate the pressure inside a 1/16 in. diameter drop of chloroform in contact with air at a pressure of 14.7 lbf/in^2.

1.42 A drop of benzene is 1 mm in diameter and is in contact with air at a pressure of 100 kN/m^2. (a) Calculate its internal pressure. (b) If the pressure difference (inside minus outside) for the benzene droplet is the same as that for a mercury droplet, what is the diameter of the mercury droplet?

1.43 If a small-diameter tube is immersed slightly in a liquid, the rise of a column of liquid inside is due to surface tension. This phenomenon is referred to as capillary action, examples of which are illustrated in Figure P1.43. The weight of the liquid column in the tube equals the product of force due to the pressure difference across the gas–liquid interface and the tube area. These are also equal to the peripheral force around the tube circumference due to surface tension. Thus we write

$$\rho g h(\pi R^2) = \sigma 2\pi R \cos\theta$$

The capillary rise can be solved for in terms of surface tension as

$$h = \frac{2\sigma}{\rho R g}\cos\theta$$

As shown in Figure P1.43, three cases can exist, depending on the value of the angle θ. When θ equals π/2, there is no rise in the tube. The angle θ is usually taken as 0° for water and 140°

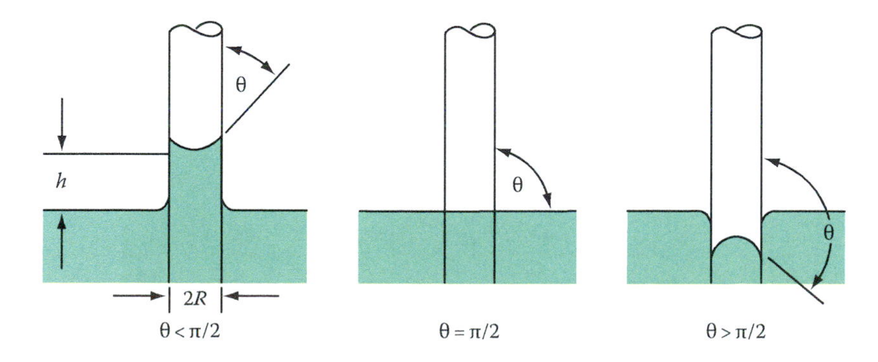

FIGURE P1.43

for mercury, if the tube is made of glass. (a) Determine the height h that water at room temperature would rise in a 4 mm diameter tube. (b) Determine the height h that mercury at room temperature would rise in a 4 mm diameter tube.

1.44 Following are surface tension data for water given as a function of temperature. If a 6 mm diameter tube is inserted into a sample at each temperature, the capillary rise as explained in Problem 1.40 would vary according to

$$h = \frac{2\sigma}{\rho R g}\cos\theta$$

Determine h for each temperature and plot h versus T.

T (°C)	σ (N/m)
0	75.6×10^{-3}
10	74.2×10^{-3}
20	73.1×10^{-3}
30	71.2×10^{-3}
40	69.6×10^{-3}
50	67.9×10^{-3}
60	66.2×10^{-3}
70	64.4×10^{-3}
80	62.9×10^{-3}
100	58.9×10^{-3}

1.45 An interesting variation of the capillary tube method of measuring surface tension is the hyperbola method. With the hyperbola method, two glass plates that have a small angle between them are positioned vertically as shown in Figure P1.45. The glass plates (5 in. × 5 in.) are separated by a small wedge at one end and held together by two binder clips at the other end. A transparency of graph paper is attached to one of the plates. If the plates are partially submerged, liquid rises between them and, when viewed from the side, the liquid surface forms a hyperbola.

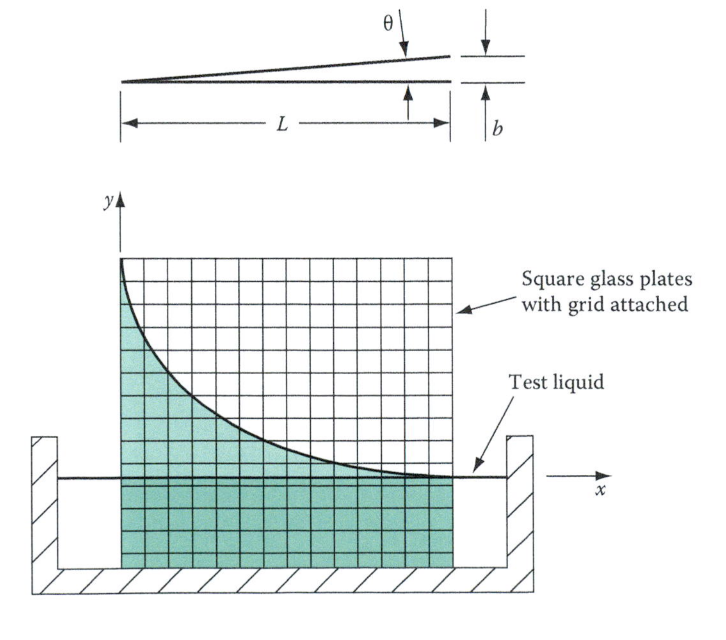

FIGURE P1.45

An x–y coordinate system is imposed on the graph paper. The x axis is the free surface of the liquid in the reservoir and the y axis is at the touching edges of the plates. Readings of x–y pairs of various points on the hyperbola are taken and can be used to calculate surface tension with

$$\sigma = \frac{xy\theta}{2}\rho g = (\text{geometry factors}) \cdot \rho g$$

Suppose that the fluid used in the hyperbola method is water and that the angle θ is 1°. Determine the equation of the hyperbola and sketch its shape. Is there any similarity between the aforementioned equation and that derived for the capillary tube?

1.46 A capillary tube that is 0.2 in. in diameter has its end submerged in mercury. The capillary depression (see Problem 1.43) is 0.052 in. Calculate the surface tension of the mercury.

1.47 Determine the height h that ethyl alcohol at room temperature would rise in a 5 mm diameter tube. The contact angle is 0°. (See Problem 1.43.)

1.48 The surface tension of benzene is measured with a capillary tube whose inside diameter is 4 mm. The contact angle is 0°. What is the expected height h that the benzene will rise in the tube? (See Problem 1.43.)

1.49 Determine the height h that carbon tetrachloride at room temperature would rise in a 3 mm diameter tube given that the contact angle is 0°. (See Problem 1.43.)

1.50 The surface tension of glycerin at room temperature is measured with a capillary tube. If the inside diameter of the tube is 2.5 mm and the contact angle is 0°, what is the expected rise of glycerin in the tube? (See Problem 1.43.)

1.51 Determine the height h that octane at room temperature would rise in a 2 mm diameter tube given that the contact angle is 0°. (See Problem 1.43.)

1.52 When glass tubes are used with mercury, instrumentation guides recommend tubes with a minimum bore of 10 mm to avoid capillary error. Estimate the height mercury rises in a tube of this diameter. (See Problem 1.43.)

Fluid Properties: Specific Heat, Internal Energy, Enthalpy

1.53 Figure P1.53 depicts 0.1 slugs of air in a piston-cylinder arrangement. Heat is removed from the air so that the air temperature is reduced by 25°F. Assuming a frictionless, movable piston, determine the amount of heat removed.

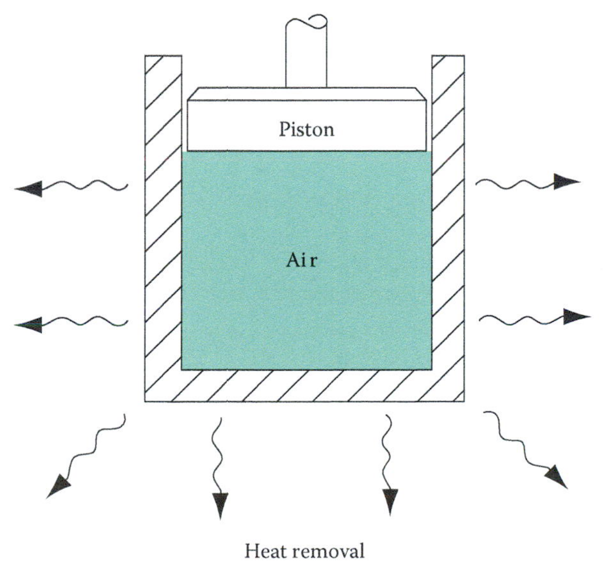

FIGURE P1.53

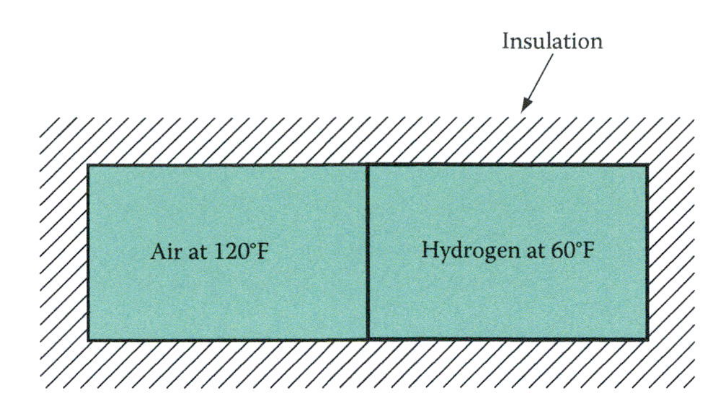

FIGURE P1.58

1.54 The vessel in Figure P1.53 contains 1 kg of carbon dioxide. The gas is cooled so that its temperature is decreased by 25°C. Determine the amount of heat removed per unit mass of carbon dioxide. Assume a frictionless, movable piston.

1.55 A rigid vessel contains 8 kg of argon heated by 50 kJ of energy. Determine the temperature change of the gas.

1.56 Carbon dioxide gas is inside of a constant volume container. Initially, the carbon dioxide is at 101.3 kN/m^2 and 25°C. The container is heated until the gas reaches 50°C. What is the change in the internal energy of the gas.

1.57 One kg of gas is in a piston cylinder arrangement. It is desired to raise the temperature of the gas by 25°C. When heat is added, the piston is free to move due to a frictionless seal. If the gas is helium, will more heat be required than if it were hydrogen? Calculate the heat required in both cases and also calculate the change in enthalpy for each gas.

1.58 Figure P1.58 shows two constant volume containers that are equal in size and they are in contact with one another. The air and its container are at 120°F while the hydrogen and its container are at 60°F. The containers are well insulated except where they touch. Heat is transferred from the air to the hydrogen and both fluids and container eventually reach the same final temperature. The mass of air is twice that of the hydrogen. Calculate the final temperature.

Fluid Properties: Isothermal Bulk Modulus

1.59 What change in pressure is required to decrease the volume of benzene by 1%?

1.60 The volume of glycerin changes by 2% under the action of a change in pressure. What is the change in pressure required to do this?

1.61 The pressure exerted on a liquid increases from 500 to 1 000 kPa. The volume decreases by 1%. Determine the bulk modulus of the liquid.

1.62 Determine the coefficient of compressibility for the liquid of Problem 1.61.

1.63 Water at 45°F has a bulk modulus of about 300,000 lbf/in^2. Determine the pressure rise required to decrease its volume by 1%.

1.64 Water at 20°C has a bulk modulus of 21.8×10^8 N/m^2. Determine the pressure change required to decrease its volume by 1%.

1.65 What is the bulk modulus of a liquid whose volume decreases by 0.5% for a pressure increase of 1000 lbf/in^2? For a density of 1.8 slug/ft^3, what is the sonic velocity in the liquid?

1.66 What is the bulk modulus of a liquid whose volume decreases by 0.5% for a pressure increase of 60 kN/m^2?

Fluid Properties: Ideal Gas Law

1.67 Carbon dioxide gas inside of a constant volume container is initially at 101.3 kN/m^2 and 25°C. The container is heated until the gas reaches 50°C. Calculate the final pressure of the gas.

1.68 What volume of air would have the same weight of 1 ft^3 of carbon dioxide if both are at room temperature and atmospheric pressure?

1.69 What is the density of air at 30°C and 300 kN/m^2?

1.70 With the ideal gas law, derive an expression that is useful for relating pressures and volumes at the beginning and end of a process that occurs at constant temperature and constant mass.

1.71 Use the ideal gas law to derive an expression that is useful for relating temperatures and volumes at the beginning and end of a process that occurs at constant pressure and constant mass.

1.72 Beginning with the universal gas constant as $\overline{R} = 49,709$ ft·lbf/(slugmol·°R). Use the conversion factor table to determine its value in SI units.

1.73 A piston-cylinder arrangement contains 0.07 slugs of air. Heat is removed from the air until the air temperature has been reduced by 15°F. For a frictionless piston, determine the change in volume experienced by the air if the pressure remains constant at 2500 lbf/ft^2.

1.74 Carbon dioxide gas exists in a chamber whose volume is 5 ft^3. The temperature of the gas is uniform at 40°C and the pressure is 3 atm. What is the mass of carbon dioxide contained?

1.75 A rigid vessel containing 8 kg of argon is heated until its temperature increases by 20°C. Determine the final pressure of the argon if its initial temperature is 28°C and its pressure is 150 kN/m^2.

1.76 A certain gas has a molecular mass of 40. It is under a pressure of 2.5 atm and a temperature of 25°C. Determine its density and pressure in g/cm^3.

1.77 Table A.3 gives properties of air at atmospheric pressure and for various temperatures. Atmospheric pressure is 101.3 kN/m^2. With this information, verify values of density in the table corresponding to any 4 temperatures of your choosing.

1.78 A gas mixture has a density of 0.01 kg/m^3 when a pressure of 4 atm is exerted on it. Determine its molecular mass if its temperature is 40°C.

2 Fluid Statics

In this chapter, we study the forces present in fluids at rest. Knowledge of force variations—or, more appropriately, pressure variations—in a static fluid is important to the engineer. Specific examples include water retained by a dam or bounded by a levee, gasoline in a tank truck, and accelerating fluid containers. In addition, fluid statics deals with the stability of floating bodies and submerged bodies and has applications in ship hull design and in determining load distributions for flat-bottomed barges. Thus, fluid statics concerns the forces that are present in fluids at rest, with applications to various practical problems.

The objectives of this chapter are to discuss pressure and pressure measurement, to develop equations for calculating forces on submerged surfaces, and to examine problems involving the stability of partially or wholly submerged bodies.

2.1 PRESSURE AND PRESSURE MEASUREMENT

Because our interest is in fluids at rest, let us determine the pressure at a point in a fluid at rest. Consider a wedge-shaped particle exposed on all sides to a fluid as illustrated in Figure 2.1a. Figure 2.1b is a free-body diagram of the particle cross section. The dimensions Δx, Δy, and Δz are small and tend to zero as the particle shrinks to a point. The only forces considered to be acting on the particle are due to pressure and gravity. On either of the three surfaces, the pressure force is $F = pA$. By applying Newton's second law in the x- and z-directions, we get, respectively,

$$\sum F_x = p_x \Delta z \Delta y - p_s \Delta s \Delta y \sin\theta = \left(\frac{\rho}{2}\Delta x \Delta y \Delta z\right) a_x = 0$$

$$\sum F_z = p_z \Delta x \Delta y - p_s \Delta s \Delta y \cos\theta - \rho g \frac{\Delta x \Delta y \Delta z}{2}$$

$$= \left(\frac{\rho}{2}\Delta x \Delta y \Delta z\right) a_z = 0$$

where

p_x, p_z, and p_s are average pressures acting on the three corresponding faces
a_x and a_z are the accelerations
ρ is the particle density

The net force equals zero in a static fluid. After simplification, with $a_x = \alpha_z = 0$, these two equations become

$$p_x \Delta z - p_s \Delta s \sin\theta = 0$$

and

$$p_z \Delta x - p_s \Delta s \cos\theta - \frac{\rho g}{2}\Delta x \Delta z = 0$$

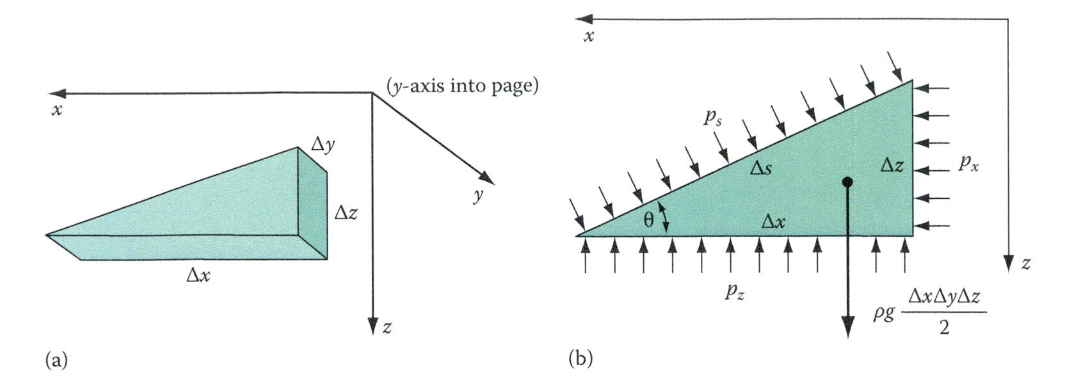

FIGURE 2.1 A wedge-shaped particle.

The third term on the left-hand side of the second equation can be neglected because it is a higher-order term containing $\Delta x \Delta z$, which is very small in comparison to the other terms. From the geometry of the wedge, we find that

$$\Delta z = \Delta s \sin \theta$$

and

$$\Delta x = \Delta s \sin \theta$$

with θ being arbitrarily chosen. Substituting into the pressure equations yields

$$\left.\begin{aligned} p_x &= p_s \\ p_z &= p_s \end{aligned}\right\} \quad p_x = p_z = p_s \tag{2.1}$$

which illustrates that pressure at a point is the same in all directions. This concept was shown for a two-dimensional model, but the proof is easily extended to three dimensions.

From the preceding paragraphs, we have seen that the forces acting on a fluid at rest are due to pressure and gravity. It is therefore important to learn how these forces vary in a static fluid. Consider an element of a fluid at rest, as illustrated in Figure 2.2a. The element chosen has a volume $dx\,dy\,dz$ and is sketched in a coordinate system where the positive z-direction is downward, coincident with the direction of the gravity force. Figure 2.2b is a view of the element looking in the positive y-direction; the force acting on the right face is $p\,dy\,dz$ and that on the left face is $[p + (\partial p/\partial x)dx]dy\,dz$, both normal to their respective surfaces. Summing forces in the x-direction, we have the following for a static fluid:

$$\sum F_x = 0 = p\,dy\,dz - \left(p + \frac{\partial p}{\partial x}dx\right)dy\,dz$$

Simplifying, we get

$$\frac{\partial p}{\partial x} = 0 \tag{2.2}$$

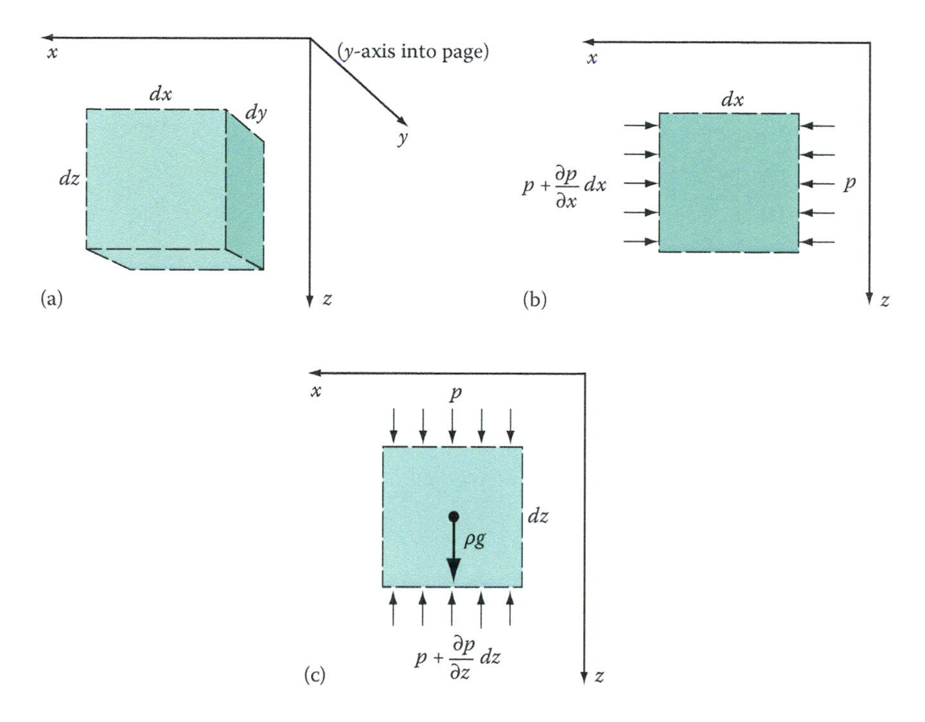

FIGURE 2.2 An element of fluid at rest.

which means that pressure does not vary with respect to x. A similar argument can be made for the forces in the y-direction, which would yield

$$\frac{\partial p}{\partial y} = 0 \tag{2.3}$$

Thus, Equations 2.2 and 2.3 show that there is no variation of pressure in any lateral direction.

Figure 2.2c gives a free-body diagram for the z-direction. Summing forces, we obtain

$$\sum F_z = 0 = p\,dx\,dy + \rho g\,dx\,dy\,dz - \left(p + \frac{\partial p}{\partial z}dz\right)dx\,dy$$

or

$$dp = \rho g\,dz \tag{2.4}$$

Therefore, pressure does vary in a static fluid in the z-direction—it increases with depth, as shown by Equation 2.4. Integrating both sides yields

$$\int_{p_1}^{p_2} dp = \int_{z_1}^{z_2} \rho g\,dz \tag{2.5}$$

where point 1 is a reference point such as the free surface of a liquid and point 2 is a point of interest. For incompressible fluids, the density is a constant, and Equation 2.5 can be easily evaluated to give

$$p_2 - p_1 = \rho g(z_2 - z_1) = \rho g \, \Delta z \tag{2.6}$$

where Δz is the depth below the liquid surface

This relationship is the basic equation of hydrostatics and is often written as

$$\Delta p = \rho g z \tag{2.7}$$

Example 2.1

A cylindrical open-topped tank that serves as a reservoir for kerosene before it is piped to another location is 140 ft in diameter. Determine the pressure difference between the top and the bottom of the walls due to the octane when the tank is filled to a depth of 25 ft.

Solution

We use the hydrostatic equation

$$p_2 - p_1 = \rho g(z_2 - z_1)$$

Section 1 refers to the free surface of the kerosene and section 2 is at the bottom. For kerosene, $\rho = 0.823(1.94 \text{ slug/ft}^3)$ (Table A.5). By substitution,

$$p_2 - p_1 = \Delta p = 0.823(1.94 \text{ slug/ft}^3)(32.2 \text{ ft/s}^2)(25 - 0) \text{ ft}$$

or

$$\Delta p = 1285 \text{ lbf/ft}^2$$

The result is independent of diameter.

Example 2.2

A cup of coffee is 7.5 cm in diameter and filled to a depth of 8.3 cm with coffee (assume properties are the same as for water). Calculate the pressure difference between the surface of the coffee and the bottom of the cup.

Solution

The hydrostatic equation applies with $\rho = 1\,000 \text{ kg/m}^3$ for water:

$$\Delta p = \rho g z = (1\,000 \text{ kg/m}^3)(9.81 \text{ m/s}^2)(0.083 \text{ m})$$

$$\Delta p = 814 \text{ N/m}^2$$

The result is independent of the cup diameter.

Equation 2.5 was integrated for the case of incompressible fluids (constant density), which is reasonable for liquids. The hydrostatic equation resulted. Gases, on the other hand, are compressible fluids with properties that are related by the ideal gas law, under certain simplifying conditions:

$$\rho = \frac{p}{RT}$$

By substitution into Equation 2.4, we get

$$dp = \frac{p}{RT} g\, dz$$

or

$$\frac{dp}{p} = \frac{g}{RT} dz$$

For constant temperature, this equation can be integrated from point 1 to point 2, yielding

$$\int_{p_1}^{p_2} \frac{dp}{p} = \frac{g}{RT} \int_{z_1}^{z_2} dz$$

$$\ln \frac{p_2}{p_1} = \frac{g}{RT}(z_2 - z_1)$$

Rearranging, we get

$$z_2 - z_1 = \frac{RT}{g} \ln \frac{p_2}{p_1} \tag{2.8}$$

In the most common example of a compressible fluid, our atmosphere, temperature is not a constant throughout but varies with height in the troposphere according to

$$T = T_0 - \alpha h \tag{2.9}$$

where

T is the temperature at any point from sea level (where $T = T_0$) to an altitude h of approximately ~36,000 ft, or 11 km

α is called a lapse rate (3.6°F/1000 ft or 6.5°C/km)

The stratosphere, the layer above the troposphere, can be described by the same ideal gas equation. A lapse rate equation is not necessary for the stratosphere because it is approximately isothermal.

Whether the fluid is compressible or incompressible, it is important to note that pressure variations in static fluids are the result of gravity. The pressure increases with depth in either case.

Example 2.3

Graph the relationship between pressure and elevation in the stratosphere assuming it to be isothermal at −57°C. The stratosphere begins at an altitude of approximately 11 000 m, where the pressure is 22.5 kPa. Extend the graph to an elevation of 20 000 m.

Solution

In this case, Equation 2.8 applies, but it must first be modified. Equation 2.8 was derived from Equation 2.4, but Equation 2.4 is based on the assumption that the positive z-direction is coincidental with the direction of the gravity force. In this example, we are dealing with the atmosphere, and we are quoting measurements of altitude using the Earth's surface as a reference.

So the positive z-direction is upward, which is opposite from the direction of the gravity force. Therefore, if we use Equation 2.8, we must account for this discrepancy. We can do so by using a negative g. Thus,

$$\Delta z = z_2 - z_1 = \frac{RT}{-g} \ln \frac{p_2}{p_1}$$

We select as our reference $z_1 = 11\ 000$ m, where $p_1 = 22.5$ kPa. From Table A.6 we read $R = 286.8$ J/(kg · K) for air, and we were given $T = 57°C = 216$ K. In applying the preceding equation, let $z_1 = z =$ any value ranging from 11 to 20 km and $p_2 = p =$ the corresponding pressure.

Substituting gives

$$z - 11\ 000 = -\frac{286.8(216)}{9.81} \ln \frac{p}{22\ 500}$$

or

$$z - 11\ 000 = -6\ 315 \ln\left(\frac{p}{22\ 500}\right)$$

Rearranging yields

$$\frac{11\ 000 - z}{6\ 315} = \ln \frac{p}{22\ 500}$$

Solving for pressure, we get

$$p = 22\ 500 \exp\left(\frac{11\ 000 - z}{6\ 315}\right)$$

A graph of this equation is provided in Figure 2.3.

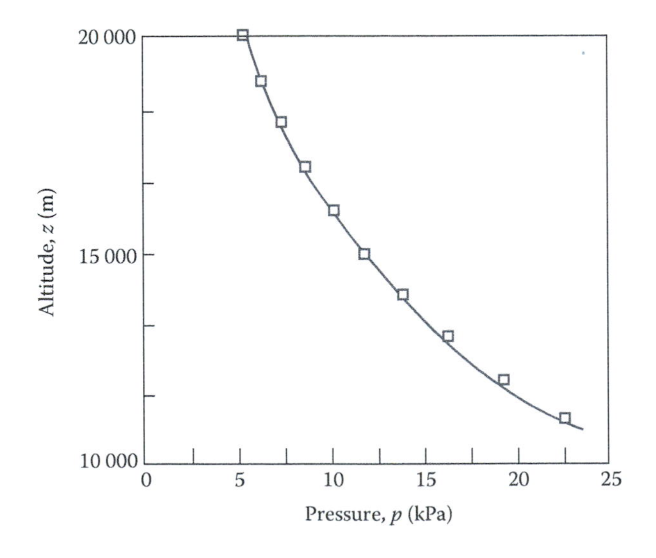

FIGURE 2.3 Pressure variation with elevation in the stratosphere.

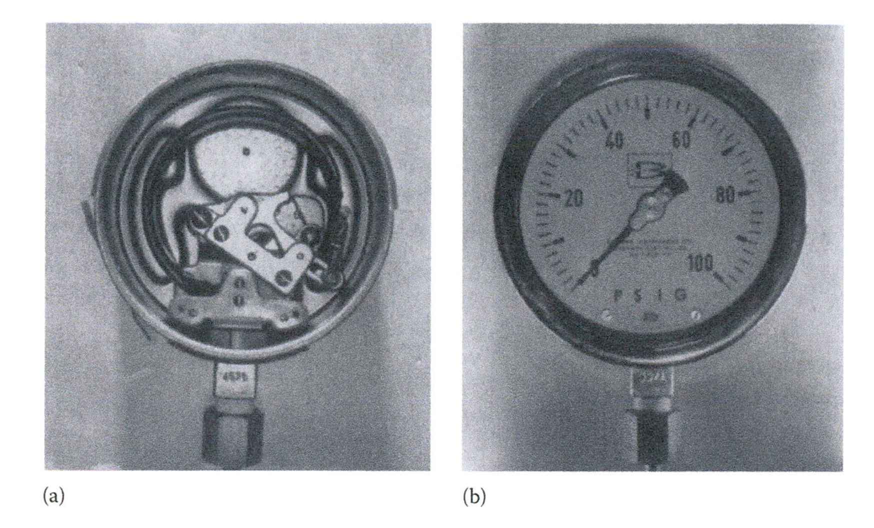

(a) (b)

FIGURE 2.4 A bourdon-tube pressure gauge. (a) Inside view and (b) outside view.

It is instructive at this point to examine techniques for measuring pressure. One common method in industry is the use of a **bourdon-tube pressure gauge** (Figure 2.4). The gauge consists of a housing that contains a fitting at the bottom for attachment to a pressure vessel. The fitting is connected to a curved tube that is elliptical in cross section. The other end of the bourdon tube is in turn connected to a rack-and-pinion assembly. The pinion shaft extends through the face of the gauge. A needle is pressed onto the pinion shaft outside the face. The face of the gauge is marked appropriately with numbers, and the gauge is calibrated. When a high-pressure fluid enters the fitting, the pressure is contained internally by the bourdon tube, which tends to straighten out. In the process, the rack is pulled and the pinion and needle rotate. When an equilibrium position is reached, the pressure is read directly on the face of the gauge. These gauges register the difference in pressure across the bourdon tube. When disconnected and exposed to atmospheric pressure, the dial is calibrated to read zero; readings from these instruments are therefore called **gauge pressures**.

Alternatively, one can measure **absolute pressure** as opposed to gauge pressure. Absolute pressure is zero only in a complete vacuum. Thus by definition,

$$p_g + p_{atm} = p \tag{2.10}$$

where

p_g is gauge pressure
p_{atm} is the atmospheric pressure
p is the absolute pressure

Commonly used British units of pressure are psig (lbf/in.^{2} gauge pressure) and psia (lbf/in.^{2} absolute). In the SI system, the unit of pressure is the pascal (Pa), where 1 Pa = 1 N/m^2.

It is important to be able to measure **atmospheric pressure** because it relates gauge pressure to absolute pressure. One technique is to use a barometer. This device consists of a tube that is inverted while submerged and full of liquid. For a sufficiently long tube, this operation results in the configuration represented schematically in Figure 2.5a. The space above the liquid in the tube is almost a complete vacuum. Because the reservoir is open to the atmosphere, the pressure at section 2 is atmospheric pressure. The difference in pressure between points 2 and 1 is

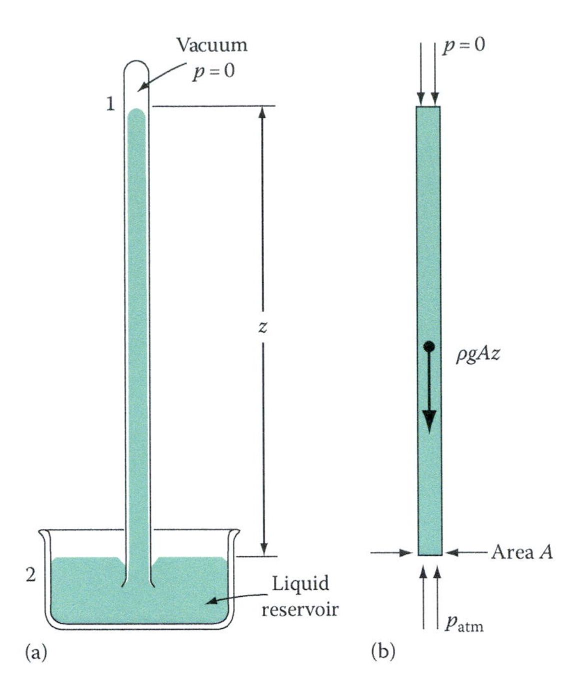

FIGURE 2.5 Schematic of a barometer.

sufficient to support the weight of the liquid column. A free-body diagram of the liquid is shown in Figure 2.5b. Summing forces gives

$$\Sigma F = 0 = p_{atm}A - \rho gAz - 0$$

or, for a barometer,

$$p_{atm} = \rho gz \tag{2.11}$$

Equation 2.11 is recognized as the hydrostatic equation. By experiment, it is known that when the barometer liquid is mercury (s = 13.6 from Table A.5), z = 760 mm. Atmospheric pressure is then calculated as

$$\begin{aligned} p_{atm} &= 13.6(1\,000\ \text{kg/m}^3)(9.81\ \text{m/s}^2)(0.760\ \text{m}) \\ &= 101.3\ \text{kPa} \end{aligned}$$

The preceding remarks concerning gauge pressure, absolute pressure, and atmospheric pressure are illustrated graphically in Figure 2.6.

Example 2.4

Atlanta, GA, has an elevation of roughly 1000 ft. A barometer located there would read ~30 in. (76.2 cm) of mercury. Determine the local atmospheric pressure at that location and express it in

a. psia
b. psig
c. kPa (absolute is implied)
d. Meters of water at 758°F

Assume standard atmospheric pressure as given in Figure 2.6.

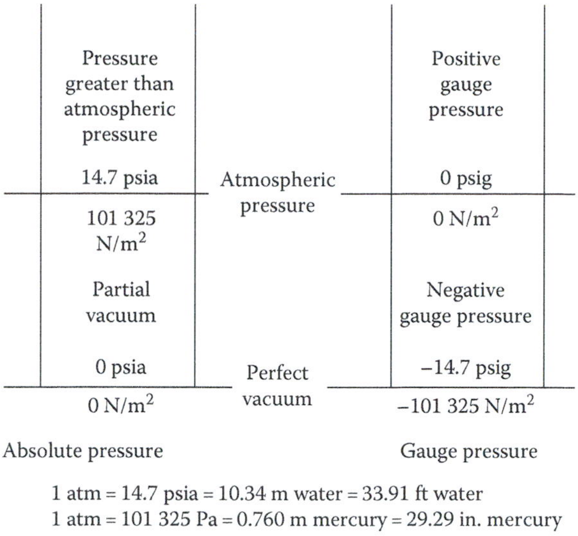

Pressure greater than atmospheric pressure

14.7 psia

101 325 N/m^2

Partial vacuum

0 psia

0 N/m^2

Atmospheric pressure

Perfect vacuum

Positive gauge pressure

0 psig

0 N/m^2

Negative gauge pressure

−14.7 psig

−101 325 N/m^2

Absolute pressure

Gauge pressure

1 atm = 14.7 psia = 10.34 m water = 33.91 ft water
1 atm = 101 325 Pa = 0.760 m mercury = 29.29 in. mercury

FIGURE 2.6 Pressure diagrams comparing absolute and gauge pressures.

Solution

a. The hydrostatic equation applies with the specific gravity of mercury, $s = 13.6$ (Table A.5):

$$p = \rho g z = 13.6(1.94\,\text{slug/ft}^3)(32.2\,\text{ft/s}^2)\left(\frac{30\,\text{in.}}{12\,\text{in./ft}}\right)$$

$$= 2\,124\,\text{lbf/ft}^2 \text{ absolute}$$

Dividing by 144 yields

$p = 14.75$ psia

b. If we use Equation 2.10, we obtain

$$p_g = 14.75 - 14.7$$

or

$p_g = 0.05$ psig

However, a bourdon-tube gauge would indicate 0 psig because it reads relative to local atmospheric pressure.

c. Applying the hydrostatic equation, we get

$$p = \rho g z = 13.6(1\,000\text{ kg/m}^3)(9.81\text{ m/s}^2)(0.76\text{ m})$$

Solving, we obtain

$p = 101.4$ kPa

d. Let us derive an equation to find the equivalent height in meters of water. Writing the hydrostatic equation for each liquid gives

$$p = \rho g z\big|_{Hg} = \rho g z\big|_{H_2O}$$

After simplification and rearrangement, we get

$$\frac{\rho_{Hg}}{\rho_{H_2O}} = s = \frac{z_{H_2O}}{z_{Hg}}$$

where the ratio of densities as written is the specific gravity of mercury. Thus,

$$z_{H_2O} = 13.6 z_{Hg} = 13.6(0.76\,\text{m})$$

$$z_{H_2O} = 10.34\,\text{m}$$

The analysis of the barometer shows that vertical columns of liquid can be used to measure pressure. One device used to effect this measurement is the **manometer**. A sketch of a U-tube manometer is given in Figure 2.7. One leg of the manometer is attached to a tank whose pressure is to be measured. The other leg is open to the atmosphere and is long enough to prevent the manometer liquid from overflowing. We can now use the hydrostatic equation to relate p to p_{atm}:

$$p_A + \rho_1 g z_1 = p_B$$

which states that the pressure at A plus the weight per unit area of a column of liquid equals the pressure at B. In other words, the difference in pressure between points A and B is the weight per unit area of the column of liquid with height z_1 and density ρ_1. Next we can write

$$p_B = p_C$$

because both are at the same elevation. Finally,

$$p_C = p_D + \rho_2 g z_2$$

Combining these three equations, we get

$$p_A + \rho_1 g z_1 = p_D + \rho_2 g z_2$$

Now $p_A = p$ = pressure of interest and p_D = atmospheric pressure. By substitution and after rearrangement, we have the following for the manometer of Figure 2.7:

$$p - p_{atm} = (\rho_2 z_2 - \rho_1 z_1) g$$

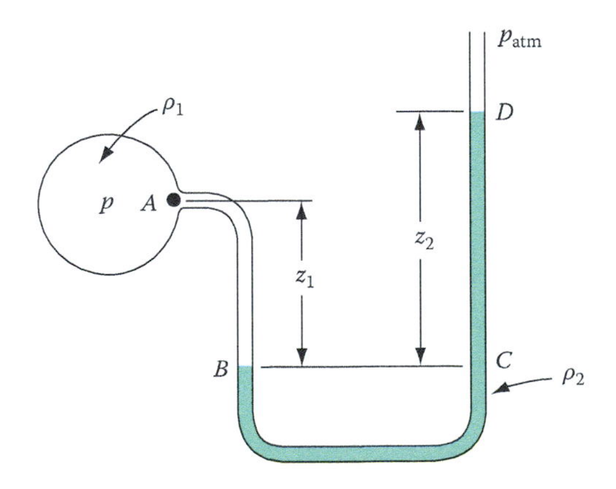

FIGURE 2.7 A simple U-tube manometer.

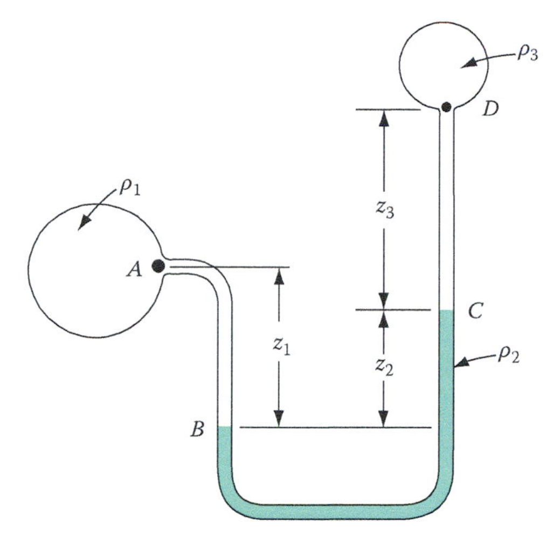

FIGURE 2.8 Use of a manometer to measure differential pressure.

Another application of manometers is in the measurement of pressure differential between two fluid reservoirs, as illustrated in Figure 2.8. Applying the hydrostatic equation to each leg yields

$$p_A + \rho_1 g z_1 = p_D + \rho_3 g z_3 + \rho_2 g z_2$$

and

$$p_A - p_D = (\rho_3 z_3 + \rho_2 z_2 - \rho_1 z_1) g$$

In many cases, an inverted U-tube is used as a manometer, as shown in Figure 2.9, where again the hydrostatic equation applies:

$$p_B + \rho_1 g z_1 = p_A$$

$$p_B + \rho_2 g z_2 = p_C$$

$$p_C + \rho_3 g z_3 = p_D$$

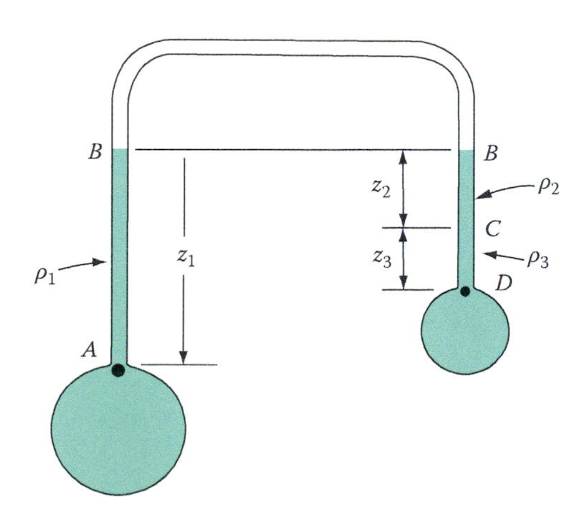

FIGURE 2.9 An inverted U-tube differential manometer.

By substitution, we obtain

$$p_D - \rho_3 g z_3 - \rho_2 g z_2 + \rho_1 g z_1 = p_A$$

After rearranging, we get

$$p_A - p_D = (\rho_1 z_1 - \rho_2 z_2 - \rho_3 z_3) g$$

Example 2.5

Figure 2.10 shows a portion of a pipeline that conveys benzene. A gauge attached to the line reads 200 kPa. It is desired to check the gauge reading with a benzene-over-mercury U-tube manometer. Determine the expected reading Δh on the manometer.

Solution

From Table A.5, we read

$$\rho = 876 \text{ kg/m}^3 \text{ (benzene)}$$

$$\rho = 13\ 000 \text{ kg/m}^3 \text{ (mercury)}$$

Applying the hydrostatic equation to the manometer gives

$$p_A + \rho_g(0.03) = p_D + \rho_{\text{Hg}} g \Delta h$$

Rearranging and solving for Δh gives

$$\Delta h = \frac{p_A - p_D + \rho g(0.03)}{\rho_{\text{Hg}} g}$$

The pressures are given as

$$p_A = 200\ 000 \text{ N/m}^2 \quad \text{and} \quad p_D = 0 \text{ N/m}^2$$

The 200 kPa is a pressure reading from a gauge. Substituting gives

$$\Delta h = \frac{200\ 000 \text{ N/m}^2 - 0 \text{ N/m}^2 + 876 \text{ kg/m}^3 (9.81 \text{ m/s}^2)(0.03 \text{ m})}{13\ 600 \text{ kg/m}^3 (9.81 \text{ m/s}^2)}$$

Solving,

$$\Delta h = 1.5 \text{ m}$$

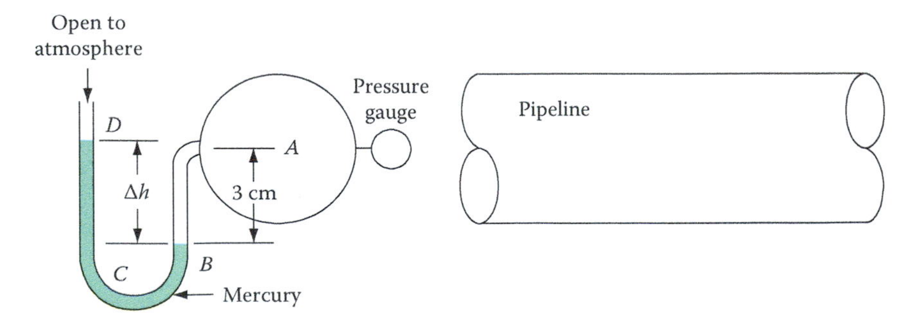

FIGURE 2.10 Sketch for Example 2.5.

Example 2.6

Figure 2.11 is a sketch of a device called a venturi meter, which is used in a pipeline to relate the volume flow rate in the pipeline to the pressure drop Δh. An air over liquid inverted U-tube manometer is attached to the meter and the reading Δh on the manometer is 4 in. What is the pressure drop if the liquid is water?

Solution

In pipeline flows, all measurements of vertical distance are made from the centerline of the pipe. We define a distance x from the centerline to the lowest liquid level in the manometer. Applying the hydrostatic equation from section 1 to 2, we get

$$p_1 - \rho_w g x - \rho_w g \Delta h = p_2 - \rho_w g x - \rho_{air} g \Delta h$$

where ρ_w is the density of water. The terms containing x cancel, so we conclude that the pressure drop $p_1 - p_2$ is not a function of x. Whether the manometer is 2 or 5 ft tall, the head change Δh is the same. Continuing, the density of air is negligible compare to that of water. Rearranging, the previous equation becomes

$$p_1 - p_2 = \rho_w g \Delta h$$

Substituting,

$$p_1 - p_2 = (1.94 \text{ slug/ft}^3)(32.2 \text{ ft/s}^2)(4/12 \text{ ft}) = 20.8 \text{ slug/(ft·s}^2) = 20.8 \text{ slug·ft/(ft}^2\text{·s}^2)$$

$$p_1 - p_2 = 20.8 \text{ lbf/ft}^2$$

Example 2.7

Two pipelines containing air are connected with two manometers as indicated in Figure 2.12. One of the manometers contains glycerine, and the head difference Δh_1 is 10 cm. The other manometer contains a liquid whose density is unknown, and it shows a difference Δh_2 of 13 cm. What is the density of the unknown liquid?

Solution

For glycerine, ρ_1 = 1.263(1 000) kg/m³ from Table A.5. Applying the hydrostatic equation to both manometers, we get

$$\Delta p = \rho_1 g \Delta h_1$$

$$\Delta p = \rho_2 g \Delta h_2$$

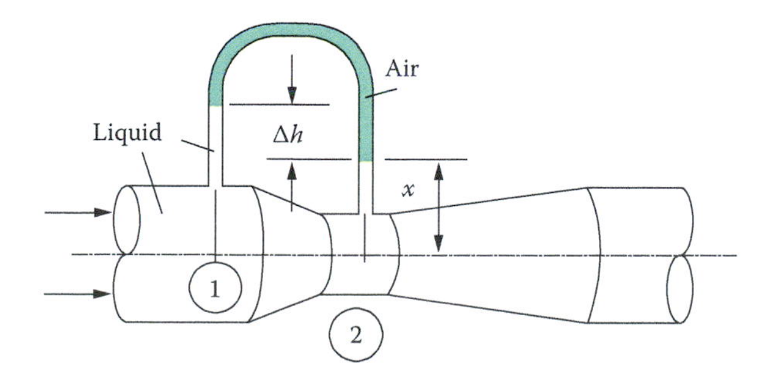

FIGURE 2.11 Inverted U-tube manometer attached to a venturi meter.

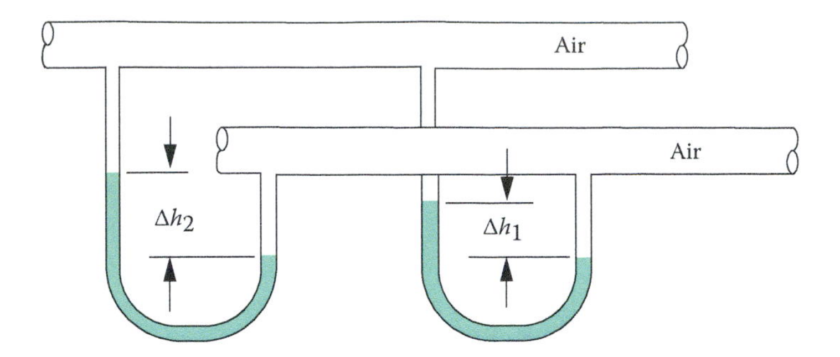

FIGURE 2.12 Manometers attached to air lines.

The pressure difference measured by both manometers is the same. Equating, and canceling the gravity term,

$$p_1 \Delta h_1 = \rho_2 \Delta h_2$$

Substituting, and solving,

$$1.263(1\,000)(0.10) = \rho_2(0.13)$$

$$\rho_2 = 972 \text{ kg/m}^3$$

2.2 HYDROSTATIC FORCES ON SUBMERGED PLANE SURFACES

Consider a vertical surface that is in contact with liquid on one side (Figure 2.13). If we apply the hydrostatic equation to each point of the wall and sketch the pressure variation, a distribution triangular in cross section results: a pressure prism. The pressure distribution is linear, and pressure increases with depth. Mathematically, it is more convenient for us to replace the distribution shown with a single force R_f acting at a distance z_r below the surface. Both R_f and z_r are yet to be determined. We will develop expressions for them from a general approach.

Figure 2.14a gives a profile view of a submerged, inclined plane surface having an irregular cross section (shown in Figure 2.14b). Also shown is the pressure distribution acting on the surface. An element of area dA has a force dR_f acting on it given by

$$dR_f = p\, dA$$

From the hydrostatic equation, p is found in terms of the space variable z as

$$p = \rho g z \sin \theta$$

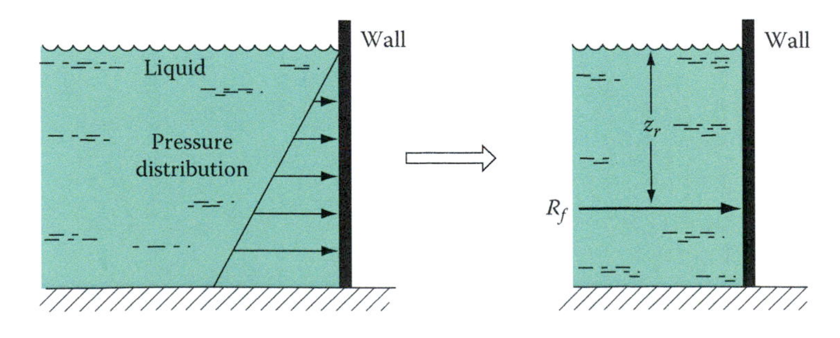

FIGURE 2.13 Pressure distribution and equivalent force on a submerged surface.

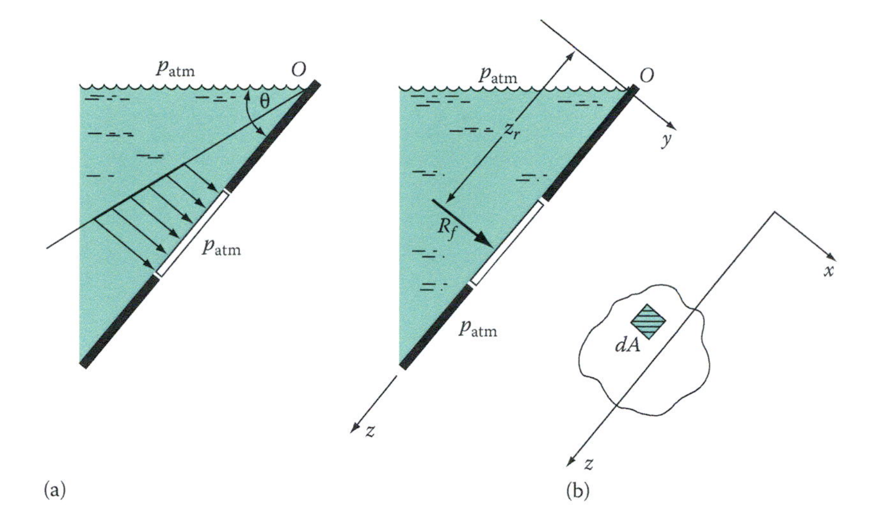

FIGURE 2.14 A submerged, inclined plane surface. (a) Profile views and (b) frontal view.

where $z \sin \theta$ is the vertical depth to dA. Note that atmospheric pressure acts on both sides and does not affect the resultant force or its location. The element of force acting on dA becomes

$$dR_f = (\rho g z \sin \theta) dA$$

Integration yields the total force exerted on the submerged plane as

$$R_f = \rho g \sin \theta \iint z\, dA \tag{2.12}$$

Recall from statics that the centroid of a plane area can be calculated with

$$z_c = \frac{\iint z\, dA}{A}$$

With this definition, Equation 2.12 now becomes

$$R_f = \rho g z_c A \sin \theta \tag{2.13}$$

where z_c is the distance to the centroid of the portion of the plane in contact with the liquid as measured from the liquid surface in the z-direction.

To find the location of R_f, let us write an expression for $z_r R_f$, the moment exerted about point O:

$$z_r R_f = \iint zp\, dA = \iint zp(\rho g z \sin \theta) dA$$

$$= \rho g z \sin \theta \iint z^2\, dA \tag{2.14}$$

The integral on the right-hand side is recognized as the second moment of area of the submerged plane about the x-axis (Figure 2.14):

$$I_{xx} = \iint z^2 dA$$

Rather than using I_{xx}, it is more convenient to use a moment of area that passes through the centroid of the submerged plane. With the parallel axis theorem, we get

$$I_{xx} = I_{xxc} + Az_c^2$$

Substituting into Equation 2.14 yields

$$z_r R_f = \rho g \sin\theta \left(I_{xxc} + Az_c^2 \right)$$

Combining with Equation 2.13, we get

$$z_r = \frac{(\rho g)(\sin\theta)\left(I_{xxc} + Az_c^2\right)}{(\rho g)(\sin\theta)\left(z_c A\right)}$$

or

$$z_r = z_c + \frac{I_{xxc}}{z_c A} \tag{2.15}$$

which is independent of liquid properties.

To determine the lateral location in the x-direction (Figure 2.14), we follow the same procedure. The moment about the z-axis is

$$\begin{aligned} x_r R_f &= \iint xp \, dA \\ &= \rho g \sin\theta \iint xz \, dA \end{aligned} \tag{2.16}$$

By definition, the product of inertia is

$$I_{xz} = \iint xz \, dA$$

Using the parallel axis theorem, we get

$$I_{xz} = I_{xzc} + Ax_c z_c$$

where I_{xzc} is a second moment about the centroid of the submerged plane and x_c is the distance from the z-axis to the centroid. Thus, Equation 2.16 becomes

$$x_r R_f = \rho g \sin\theta \, (I_{xzc} + Ax_c z_c)$$

Combining this result with Equation 2.13, we obtain

$$x_r = x_c + \frac{I_{xzc}}{z_c A} \tag{2.17}$$

Equation 2.13 provides a means for determining the magnitude of the resultant force acting on a submerged plane surface. Equations 2.15 and 2.17 give the location of that force vertically from the free surface and laterally from an assigned coordinate axis. For areas that are symmetric, x_r is coincident with the centroid because when one of the centroidal axes is an axis of symmetry, $I_{xzc} = 0$. The moments of inertia of various plane areas are given in Table B.1.

Example 2.8

Figure 2.15 shows a rectangular 3 ft × 4 ft gate hinged at the top, and filled to a depth of $h = 3.2$ ft of water. The gate is made of concrete whose density is 4.7 slug/ft³, and the gate width into the page is 2 ft. What is the magnitude of the clockwise moment about the hinge at O?

Figure 2.15b shows the pressure prism acting on the gate, and Figure 2.15c is a free body diagram showing relevant forces acting on the gate. The force due to the liquid is

$$R_f = \rho g z_c A$$

where

$z_c = h/2 = 3.2/2 = 1.6$ ft measured from the surface of the liquid
$A = 3.2(2) = 6.4$ ft²
$\rho = 1.94$ slug/ft³

Substituting,

$$R_f = 1.94(32.2)(1.6)(6.4) = 639.7 \text{ lbf}$$

The line of action of this force acts at

$$z_r = z_c + \frac{I_{xxc}}{z_c A}.$$

The second moment of the portion of the gate in contact with the liquid is

$$I_{xxc} = \frac{wh^3}{12} = \frac{2(3.2)^3}{12} = 5.45\,\text{ft}^4$$

The location of the force then is

$$z_r = 1.6 + \frac{5.46}{(1.6)(6.4)} = 2.13\,\text{ft}$$

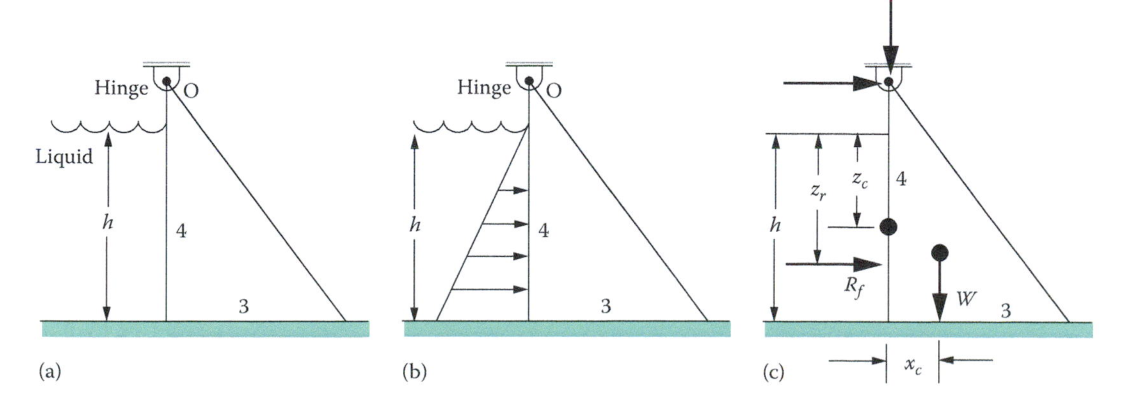

FIGURE 2.15 Rectangular gate holding water.

This distance is measured from the free surface of the liquid. From point O, the line of action of the force R_f is $(4 - 0.8 - z_r) = 3.2 - 2.13 = 1.07$ ft. The volume of concrete is $3(4)(2) = 24$ ft^3. So the weight W is $\rho \forall g = (4.7\,\text{slug/ft}^3)(24\,\text{ft}^3)(32.2\,\text{ft/s}^2)$ or $W = 3632.2$ lbf. This weight acts at a distance of $x_c = 1$ ft from the hinge. The moment then is

$$\Sigma M = Wx_c - R_f z_c = 3632(1) - 639.7(1.07)$$

$$\Sigma M = 2947.5\ \text{ft}\cdot\text{lbf CW}$$

Example 2.9

A container manufactured to store bags of ice is sketched in Figure 2.16a. As shown, the slanted surface contains an access door; how well it seals when closed must be tested. The method involves filling the container with water to a predetermined depth. Calculate the force exerted on the door and the location of the force for $H = 75$ cm and $\theta = 45°$.

Solution

Figure 2.16b shows the pressure prism that acts on the door. Figure 2.16c shows the equivalent force R_f and its location z_r. Note that the origin is found by extending the plane of the door to the surface level of the water. The force R_f is found with

$$R_f = \rho g z_c A \sin\theta$$

For water, Table A.5 gives $\rho = 1\,000$ kg/m^3. The distance to the centroid z_c from the origin is measured along the slant:

$$z_c = \frac{0.75}{\sin 45°} = 1.06\,\text{m}$$

The area of the access door is calculated to be

$$A = (0.3\ \text{m})(0.4\ \text{m}) = 0.12\ \text{m}^2$$

Substituting, we get

$$R_f = (1\,000\ \text{kg/m}^3)(9.81\ \text{m/s}^2)(1.06\ \text{m})(0.12\ \text{m}^2)\sin 30°$$

or

$$R_f = 883\ \text{N}$$

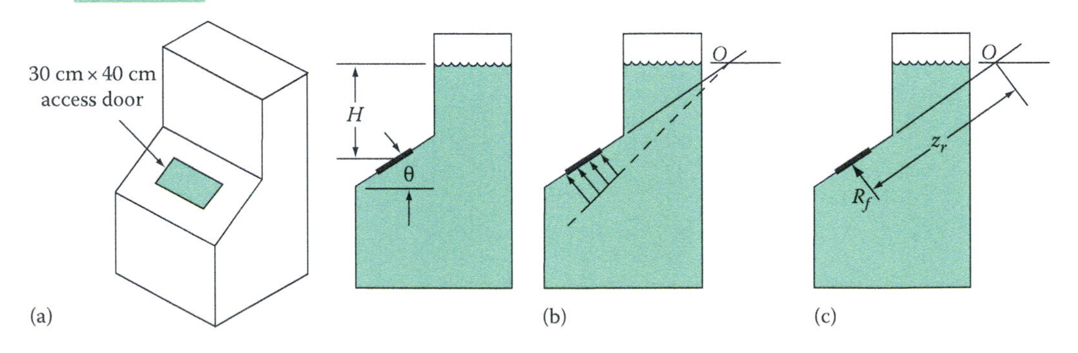

FIGURE 2.16 (a) Ice container filled with liquid for Example 2.9. (b and c) Profile views of container showing pressure prism, resultant force, and location.

The force acts at a distance z_r from the origin, which is given by

$$z_r = z_c + \frac{I_{xxc}}{z_c A}$$

For a rectangular door, Table B.1 gives

$$I_{xxc} = \frac{bz^3}{12} = \frac{0.4(0.3)^3}{12} = 9.0 \times 10^{-4} \text{ m}^4$$

Substituting,

$$z_r = 1.4 + \frac{9.0 \times 10^{-4}}{1.4(0.12)}$$

or

$$z_r = 1.41 \text{ m}$$

Example 2.10

The gas tank of an automobile is sketched in a profile view in Figure 2.17. The lower edge of a semicircular plug is located 1 cm from the tank floor. The tank is filled to a height of 20 cm with gasoline and pressurized to 150 kPa. Calculate the force exerted on the semicircular plug. Assume that gasoline properties are the same as those for octane.

Solution

A free-body diagram of the plug is shown in Figure 2.17. We identify three pressure distributions: one due to overpressure p_t of 150 kPa, one due to gasoline in contact with the plug, and one due to atmospheric pressure p_{atm}. Furthermore, the three pressure distributions can be replaced with three resultant forces.

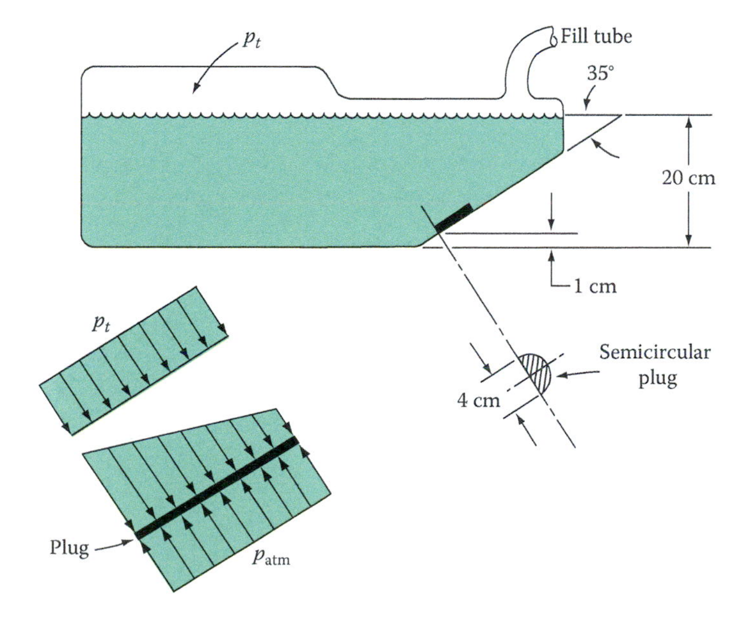

FIGURE 2.17 Sketch of a gas tank and plug for Example 2.10.

The first resultant force we seek is due to the overpressure and is given by

$$R_{f1} = p_t A$$

where

$$p_t = 150\,000 \text{ N/m}^2 \quad \text{and} \quad A = \pi D^2/8 = \pi(4)^2/8 = 6.28 \text{ cm}^2$$

Thus

$$R_{f1} = (150\,000 \text{ N/m}^2)(6.28 \times 10^{-4} \text{ m}^2)$$

or

$$R_{f1} = 94.2 \text{ N}$$

This result is independent of fluid density. Although the information is not asked for in the problem statement, this force acts at the centroid of the plug.

The second force is that associated with the gasoline in contact with the plug and is given by

$$R_{f2} = \rho g z_c A \sin\theta$$

Using the data of Table B.1, we write

$$z_c = \frac{20\,\text{cm}}{\sin\theta} - \frac{1 \text{ cm}}{\sin\theta} - \left[\frac{4(2\,\text{cm})}{3\pi}\right]$$

or

$$z_c = 32.3 \text{ cm}$$

From Table A.5 for octane, $\rho = 701$ kg/m³. Substituting, we get

$$R_{f2} = (701 \text{ kg/m}^3)(9.81 \text{ m/s}^2)(0.323 \text{ m})(6.28 \times 10^{-4} \text{ m}^2) \sin 35°$$

Solving,

$$R_{f2} = 0.8 \text{ N}$$

This force acts at a distance z_r from the free surface, as given by Equation 2.15.

The third force is due to atmospheric pressure and is given by

$$R_{f3} = p_{\text{atm}} A = (101\,300 \text{ N/m}^2)(6.28 \times 10^{-4} \text{ m}^2)$$

or

$$R_{f3} = 63.6 \text{ N}$$

Like R_{f1}, the force R_{f3} acts at the centroid of the plug.

The total resultant force acting on the plug is calculated with

$$R_f = R_{f1} + R_{f2} + R_{f3} = 81.6 + 0.8 - 63.6$$

or

$$R_f = 18.8 \text{ N}$$

Example 2.11 Laboratory Experiment: Force on a Submerged Plane Surface

Figure 2.18 is a sketch of an apparatus used to in this experiment. The apparatus consists of one-fourth of a torus of rectangular cross section and made of clear plastic. The torus is attached to a lever arm, which is free to rotate (within limits) about a pivot point. The torus has inside and outside radii, R_i and R_o, respectively, and it is constructed such that the center of these radii is at the pivot point of the lever arm. The torus is submerged in a liquid, and there will exist an unbalanced force that is exerted on the plane of dimensions $h \times w$. In order to bring the torus and lever arm back to their balanced position, a weight W must be added to the weight hanger. The force and its line of action can be found by summing moments about the pivot.

Following are data obtained on this apparatus, which shows among other things the actual weight required to balance the lever arm while the torus is submerged. Substitute appropriately, and determine if the preceding equations accurately predict the weight W.

$L = 33.6$ cm $h = 10.2$ cm $w = 7.6$ cm tank bottom to bottom of torus = 8.9 cm = d_2

Pivot to bottom of tank 31.8 cm = d_1 $W = 2.45$ N $z = 12.7$ cm

Solution

Figure 2.18b shows the pressure prism that acts on the submerged plane. Figure 2.18c shows the force R_f that acts at a distance z_r from the free surface.

The submerged plane area is $A = hw = 0.007\,742$ m^2. The distance from the free surface to the centroid of the plane is

$$z_c = d_1 - z - d_2 - (h/2) = 31.8 - 12.7 - 8.9 - (10.2/2) = 5.05 \text{ cm}$$

Applying the equations of this chapter, we write

$$R_f = \rho g z_c A = 1\,000(9.81)(0.050\,5)(0.007\,742) = 3.83 \text{ N}$$

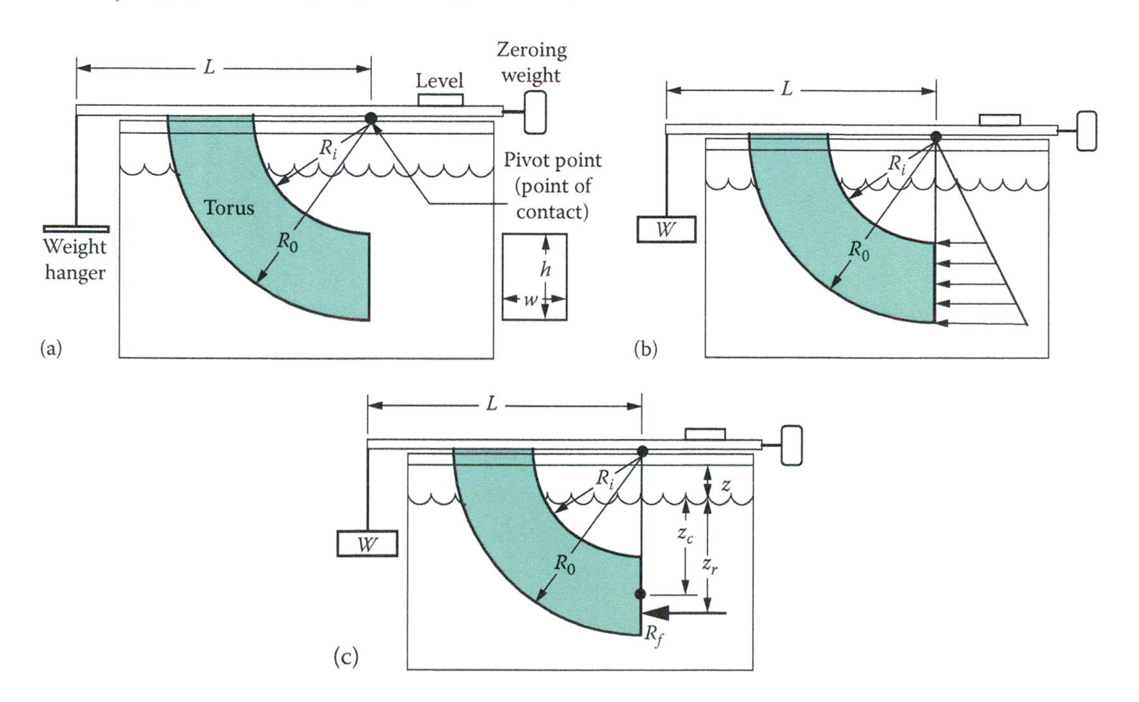

FIGURE 2.18 (a) A schematic of the center of pressure apparatus, (b) the pressure prism, and (c) the free body diagram.

The second moment is

$$I_{xxc} = \frac{wh^3}{12} = \frac{0.076(0.102)^3}{12} = 6.66 \times 10^{-6}\,\text{m}^4$$

The location of the force then is (from the free surface):

$$z_r = z_c + \frac{I_{xxc}}{z_c A} = 0.050\,5 + \frac{6.66 \times 10^{-6}}{0.050\,5(0.007\,742)} = 0.067\,5\,\text{m}$$

Summing moments about the hinge gives

$$W = \frac{R_f(z + z_r)}{L} = \frac{3.83(0.127 + 0.067\,5)}{0.336}$$

$$W = 2.22\ \text{N}$$

Compare this result to the measured weight of 2.45 N; the error is (2.45 – 2.22)/2.45 = 0.095 = 9.5%.

2.3 HYDROSTATIC FORCES ON SUBMERGED CURVED SURFACES

As we saw in previous examples, there are many cases of plane surfaces in contact with fluids. Another important area of interest involves determining forces on submerged curved surfaces. The hull of a floating ship is a curved surface in contact with liquid, as is the wall or sides of a drinking glass or funnel or culvert. To develop equations for these cases, consider the configuration illustrated in Figure 2.19a. A curved surface is shown in profile and projected frontal views. Let us examine the element of area d_A. The force acting is $p\,dA$ (Figure 2.19b). It is convenient to resolve this force into vertical and horizontal components, dR_v and dR_h, respectively. We write the horizontal component of this force directly as

$$dR_h = p\,dA\cos\theta$$

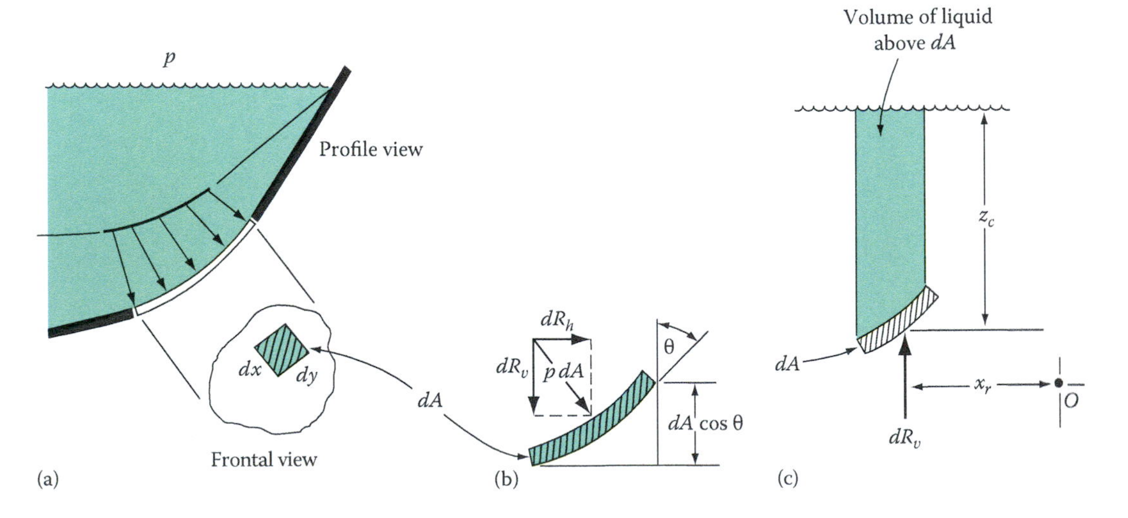

FIGURE 2.19 A submerged, curved surface.

where $dA \cos \theta$ is the vertical projection of dA. Integrating this expression gives a result similar to that for a submerged plane:

$$R_h = \rho g z_c A_v \tag{2.18}$$

where
$\rho g z_c$ is the pressure at the centroid of the surface
A_v is its area projected onto a vertical plane

The line of action, or location, of the horizontal force R_h is found by summing moments as was done for the plane surfaces in Section 2.2. The result is the same equation as was derived there except that we are now working with the vertical projection of the area—namely, A_v:

$$z_r = z_c + \frac{I_{xxc}}{z_c A_v}$$

Here z_c is the distance from the free surface to the centroid of the area A_v. The second moment of inertia I_{xxc} also applies to the vertical projected area A_v.

Next consider the vertical component of force, which is given by

$$dR_v = p\, dA \sin \theta$$

where $dA \sin \theta$ is the horizontal projection of dA. Combining this result with the hydrostatic equation, we obtain

$$dR_v = \rho g z_c\, dA \sin \theta \tag{2.19}$$

where again z_c, as shown in Figure 2.19c, is the vertical distance from the liquid surface to the centroid of dA. The quantity $z_c\, dA \sin \theta$ is the volume of liquid above dA. Equation 2.19 thus becomes

$$dR_v = \rho g\, d\mathcal{V}$$

and, after integration, yields

$$R_v = \rho g \mathcal{V} \tag{2.20}$$

Therefore, the vertical component of force acting on a submerged curved surface equals the weight of the liquid above it.

The location of R_v is found by first specifying an origin, point O in Figure 2.19c, and taking moments about it:

$$dR_v x_r = \rho g x\, d\mathcal{V}$$

Integrating over the entire submerged curved surface gives

$$R_v x_r = \rho g \iiint x\, d\mathcal{V}$$

Combining this equation with Equation 2.20, we get

$$x_r = \frac{(\rho g)\iiint x\,d\forall}{(\rho g)d\forall}$$
$$x_r = \frac{\iiint x\,d\forall}{\forall} \tag{2.21}$$

Thus, x_r is located at the centroid of the volume above the surface.

Example 2.12

A concrete culvert that contains water is 2.0 m in diameter. Determine the forces exerted on the portion labeled A–B in Figure 2.20 if the culvert is filled halfway. Determine also the location of the forces. Culvert length (into the paper) from joint to joint is 2.5 m.

Solution

The horizontal force is given by Equation 2.18:

$$R_h = \rho g z_c A_v$$

where z_c is the distance from the liquid surface to the centroid of the vertical projection of the surface in contact with the liquid

$$z_c = \frac{D}{4} = \frac{2.0}{4} \quad \text{or} \quad z_c = 0.5\,\text{m}$$

The vertical projection is a rectangle with area

$$A_v = \left(\frac{2.0}{2}\right)(2.5) = 2.5\,\text{m}^2$$

By substitution, we get

$$R_h = (1\,000\ \text{kg/m}^3)(9.81\ \text{m/s}^2)(0.5\ \text{m})(2.5\ \text{m}^2)$$

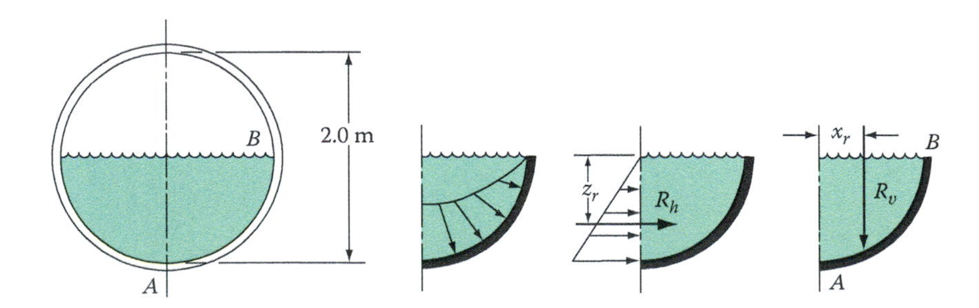

FIGURE 2.20 Cross section of a half-filled culvert.

Solving, we obtain

$R_h = 12.3$ kN

The vertical projected area is a rectangle whose second moment of inertia is given by

$$I_{xxc} = \frac{bz^3}{12} = \frac{2.5(2.0/2)^3}{12} = 0.208\,3\,\text{m}^4$$

The line of action of the horizontal force R_h then is

$$z_r = z_c + \frac{I_{xxc}}{z_c A_v} = 0.5 + \frac{0.208\,3}{0.5(2.5)}$$

or

$z_r = 0.667$ m below the free surface.

The vertical force is given by Equation 2.20:

$$R_v = \rho g \forall$$

and

$$\forall = \frac{1}{4}\left(\frac{\pi D^2}{4}\right) b = \frac{1}{4}\left[\frac{\pi(2.0)^2}{4}\right](2.5)\ \text{m}^3$$

so

$$\forall = 1.96\,\text{m}^3$$

Therefore,

$$R_v = (1\,000\ \text{kg/m}^3)(9.81\ \text{m/s}^2)(1.96\ \text{m}^3)$$

or

$R_v = 19.2$ kN

The line of action of the vertical force passes through the centroid of the quarter circle at a distance into the page of 1.25 m. Referring to Table B.1, the centroid of the quarter circle is given by

$$x_r = \frac{4R}{3\pi} = \frac{4(2.0/2)}{3\pi}$$

or

$x_r = 0.42$ m from the centerline

Thus, R_h acts at a depth z_r below the centroid of the plane area formed by the projection of the curved surface onto a vertical plane normal to R_h, and R_v acts through the centroid of the liquid volume.

Example 2.13

Figure 2.21 shows a view of a tank that contains kerosene. The bottom corners of the tank are rounded, with $r = 1$ m. The liquid depth is 2.2 m (from E to C), and the tank width (into the page) is 4 m. Determine the forces that act on the curved portion of the bottom labeled ABC.

Solution

The tank section ECBD is sketched in a profile view in Figure 2.22a. The forces acting on ABC are a horizontal force and a vertical force.

Figure 2.22b shows the pressure prism that results in a horizontal force R_f that acts on the vertical projection of the area, AC.

For kerosene, = 0.823(1 000) = 823 kg/m^3. The area of the vertical projection is wr where w is the width of the tank, 4 m, and the radius r is given as 1 m. Thus

$$A = wr = 4(1) = 4 \text{ m}^2$$

The distance to the centroid of the vertical projection is $(d - r) + r/2$:

$$z_c = (2.2 - 1) + 1/2 = 1.7 \text{ m}$$

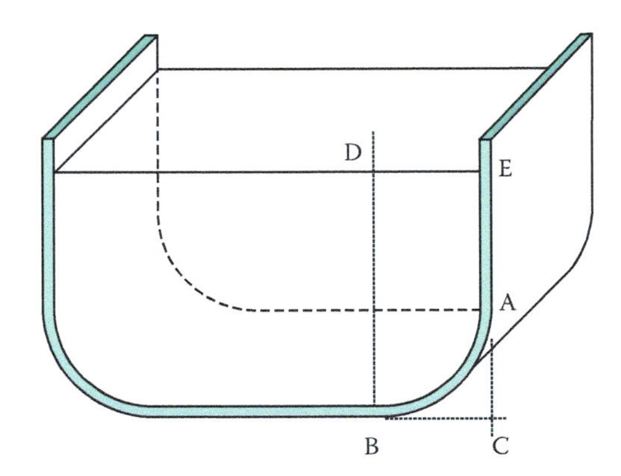

FIGURE 2.21 Kerosene tank with rounded corners.

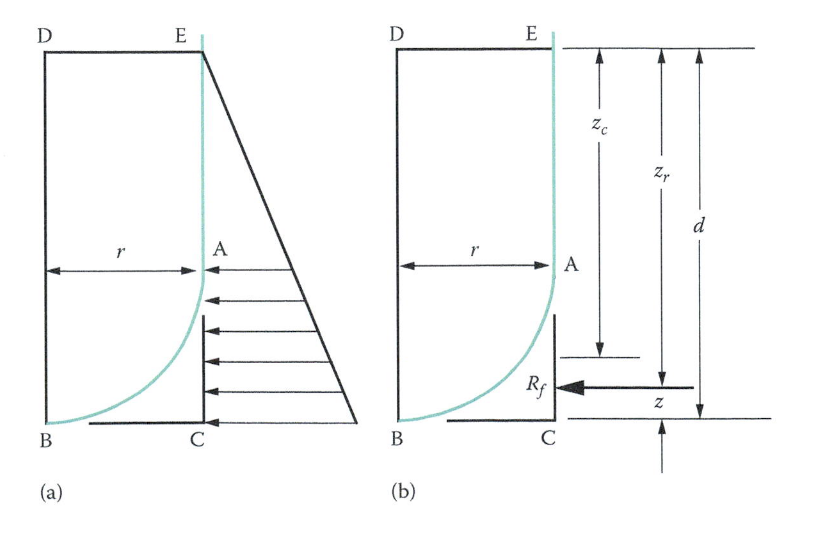

FIGURE 2.22 Tank showing the horizontal force acting on AC.

Therefore,

$$R_f = \rho g z_c A = 823(9.81)(1.7)(4) = 54\,900 \text{ N}$$

The second moment is

$$I_{xxc} = \frac{wr^3}{12} = \frac{4(1)^3}{12} = 0.333$$

The location of the force then is

$$z_r = z_c + \frac{I_{xxc}}{z_c A} = 1.7 + \frac{0.333}{1.7(4)} = 1.75 \text{ m}$$

The vertical force is shown in Figure 2.23. We break the region up into two areas, A_1 and A_2. The region labeled as A_1 is a parallelepiped of dimensions $(d - r)(r)(w)$. The volume of this area is

$$\forall = (2.2-1)(1)(4) = 4.8 \text{ m}^3$$

The force exerted by this volume is

$$R_{v1} = \rho \forall g = 823(4.8)(9.81) = 38\,753 \text{ N}$$

The location of this force, measured from BD is $x_{c1} = r/2 = 0.5$ m.

The second area A_2 has a volume of $\pi r^2 w/4$. So the force due to this volume is

$$R_{v2} = \rho \forall g = 823\pi(1)^2(4/4)(9.81) = 25\,364 \text{ N}$$

The location of this force is

$$x_{c2} = \frac{4r}{3\pi} = \frac{4(1)}{3\pi} = 0.424 \text{ m}$$

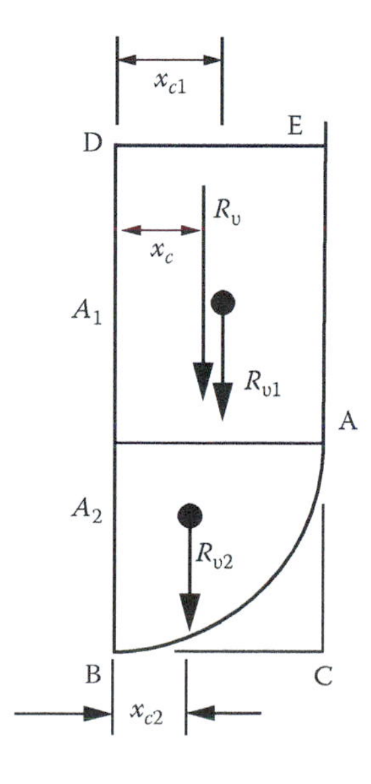

FIGURE 2.23 Tank showing the vertical forces acting on BC.

The resultant force is the sum of these two vertical forces:

$$R_v = 38\ 753 + 25\ 364 = 64\ 117\ \text{N}$$

The location of this force is found with

$$x_c = \frac{R_{v1}x_{c1} + R_{v2}x_{c2}}{R_{v1} + R_{v2}} = \frac{38\ 753(0.5) + 25\ 364(0.424)}{64\ 117}$$

$$x_c = 0.469\ \text{m}$$

The resultant force including horizontal and vertical components is

$$R = \sqrt{54\ 900^2 + 64\ 117^2} = 84\ 410\ \text{N}$$

This force acts at an angle of

$$\theta = \tan^{-1}\left(\frac{64\ 117}{54\ 900}\right) = 49.4°$$

In problems where a curved surface is immersed in a fluid, we may need an expression for the location of the centroid. For example, consider Figure 2.24a which is an ex-quarter circle area. It is desired to derive an expression for the distance to the centroid x_c. The region consists of a square area with an inscribed quarter circle removed. We can perform an analysis for the region and determine the location of x_c by integration. Alternatively, we can use information we have on existing cross sections and obtain a solution. We use this latter method.

The region may be divided into two areas whose centroid locations are known. These figures are shown graphically in Figure 2.24b and c. We first define a location to use as a reference, and this is labeled as point O.

The desired location of x_c is found with the following equation, in which all distances in the x direction are measured from point O:

$$x_c = \frac{A_1x_{c1} - A_2x_{c2}}{A_1 - A_2}$$

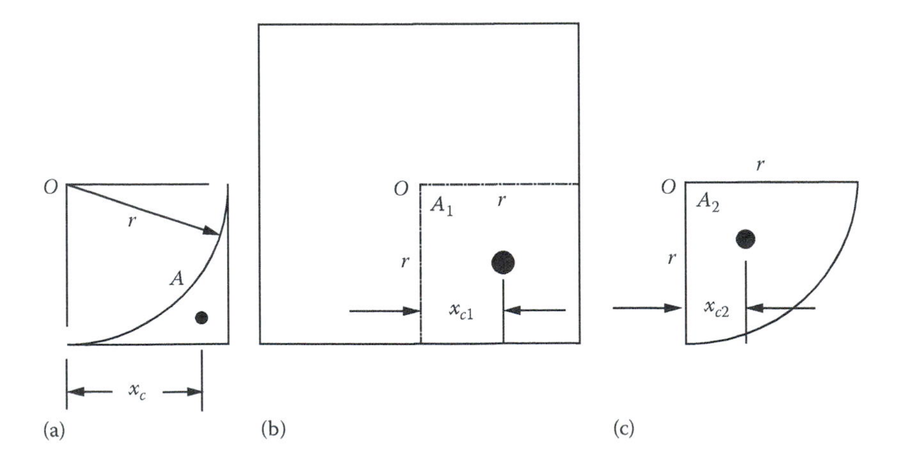

FIGURE 2.24 Ex-quarter circle showing location of centroid as a combination of two other geometries.

The terms in this equation are

$$A_1 = r^2 \quad A_2 = \frac{\pi r^2}{4}$$

(Appendix Table B.2)

$$x_{c1} = \frac{r}{2} \quad x_{c2} = \frac{4r}{3\pi}$$

Substituting,

$$x_c = \frac{r^2(r/2) - (\pi r^2/4)(4r/3\pi)}{r^2 - \pi r^2/4}$$

$$x_c = \frac{(r^3/2 - r^3/3)}{r^2 - \pi r^2/4}$$

or finally,

$$x_c = \frac{2r}{3(4-\pi)}$$

Example 2.14

Figure 2.25a shows a gate that is 4 ft wide (into the page) and has a curved cross section. When the liquid level gets too high, the moments due to liquid forces act to open the gate and allow some liquid to escape. For the dimensions shown, determine whether the liquid is deep enough to cause the gate to open. Take the liquid to be castor oil.

Solution

Figure 2.25b is a sketch of the gate with the equivalent forces and their locations. We now proceed to calculate the magnitude of the forces and to evaluate the moment these forces exert about the hinge. The force R_h is given by Equation 2.18:

$$R_h = \rho g z_c A_v$$

For castor oil, $\rho = 0.96(1.94)$ slug/ft^3 from Table A.5. The distance to the centroid of the vertical projected area is $z_c = 0.5$ ft. The vertical projected area becomes $A_v = 1(4) = 4$ ft^2. The horizontal force is then

$$R_h = 0.96(1.94 \text{ slug/ft}^3)(32.2 \text{ ft/s}^2)(0.5 \text{ ft})(4 \text{ ft}^2)$$

or

$$R_h = 120 \text{ lbf}$$

Now for A_v, which is a rectangle 4 ft wide × 1 ft tall,

$$I_{xcc} = \frac{4(1)^3}{12} = 0.333 \text{ ft}^4$$

The line of action of the horizontal force R_h is then

$$z_{rc} = z_c + \frac{I_{xcc}}{z_c A_v} = 0.5 + \frac{0.333}{0.5(4)}$$

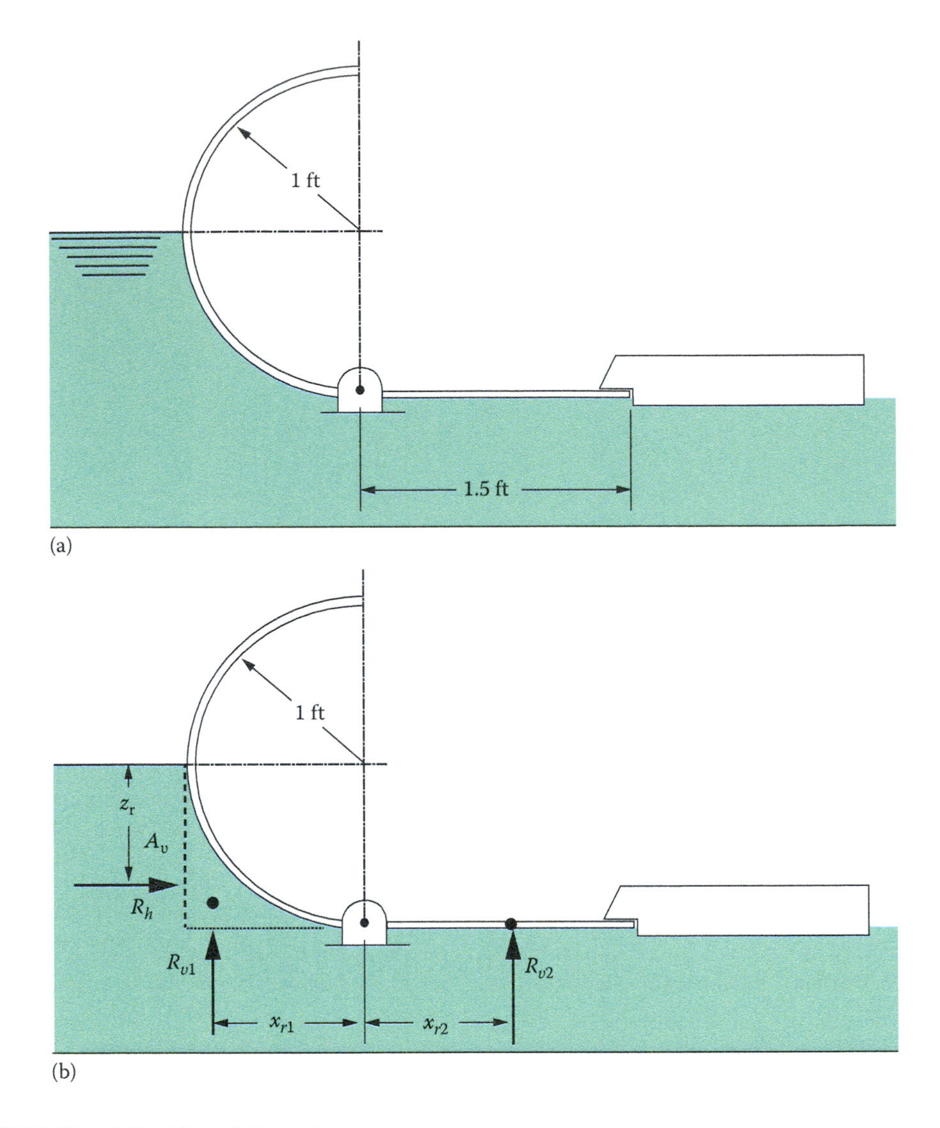

FIGURE 2.25 A liquid-retaining gate.

or

$$z_r = 0.667 \text{ ft from the free surface}$$

The vertical force acting on the curved part of the gate is given by

$$R_{v1} = \rho g \forall$$

where $\forall$ is the displaced volume of the liquid. Thus,

$$\forall = \frac{1}{4}\left(\frac{\pi D^2}{4}\right)b = \frac{1}{4}\left(\frac{\pi(2)^2}{4}\right)(4)$$

$$= 3.14 \text{ ft}^3$$

The vertical force is

$$R_{v1} = 0.96(1.94 \text{ slug/ft}^3)(32.2 \text{ ft/s}^2)(3.14 \text{ ft}^3) = 188 \text{ lbf}$$

The line of action of this vertical force is through the centroid of the ex-quarter circle:

$$x_{r1} = \frac{2r}{3(4-\pi)} = \frac{2(1)}{3(4-\pi)} = 0.776 \text{ ft.}$$

The third force to consider is also a vertical force due to the weight of the liquid. At a depth of 1 ft,

$$p = \rho g z = 0.96(1.94 \text{ slug/ft}^3)(32.2 \text{ ft/s}^2)(1 \text{ ft}) = 60 \text{ lbf/ft}^2$$

The equivalent force R_{v2} is the product of pressure and area over which the pressure acts:

$$R_{v2} = pA = (60 \text{ lbf/ft}^2)(1.5 \text{ ft})(4 \text{ ft})$$
$$= 360 \text{ lbf}$$

The line of action of this force is

$$x_{r2} = 0.75 \text{ ft to the right of the hinge}$$

To find out whether the gate tends to open, we sum moments about the hinge, assuming counterclockwise to be positive (an arbitrary decision). If our answer turns out negative, then we know that the forces exert moments that will open the gate. Thus,

$$\Sigma M = R_{v2}(x_{r2}) - R_{v1}(x_{r1}) - R_h(1 \text{ ft} - z_r)$$
$$= 360(0.75) - 188(0.776) - 120(1 - 0.667)$$

Solving, we get

$$\Sigma M = +84.2 \text{ ft lbf}$$

Because our answer is positive, we conclude that the gate will remain closed.

2.4 EQUILIBRIUM OF ACCELERATING FLUIDS

In the problems we have discussed thus far, forces due to pressure variations were simple to compute because the fluid was at rest and hence, there were no shear stresses between adjacent layers of fluid. A liquid transported at a uniform rate of acceleration is also motionless with respect to its container. Consequently, there are no shear stresses, and here too forces due to pressure variations are simple to calculate. Uniformly rotating fluids represent another case of interest. When a liquid is rotating at a constant angular velocity, it is acted upon by centripetal acceleration forces. Because the liquid is at rest with respect to its container, forces due to pressure variations are again simple to calculate.

To illustrate this phenomenon, consider the setup shown in Figure 2.26. A rectangular transparent box is attached to a model railroad flatcar that is free to traverse a track. A rope tied to the car is wrapped over a pulley and attached to a weight. As the weight falls, the car accelerates at a. Next we establish a coordinate system that moves with the car as shown in Figure 2.26. The positive x-axis is in the direction of motion of the car, the y-axis is directed inward to the plane of the paper, and the z-axis is upward. Applying Newton's law to the liquid in each of these directions gives the following pressure variations:

$$\frac{\partial p}{\partial x} = -\rho a \qquad (2.22a)$$

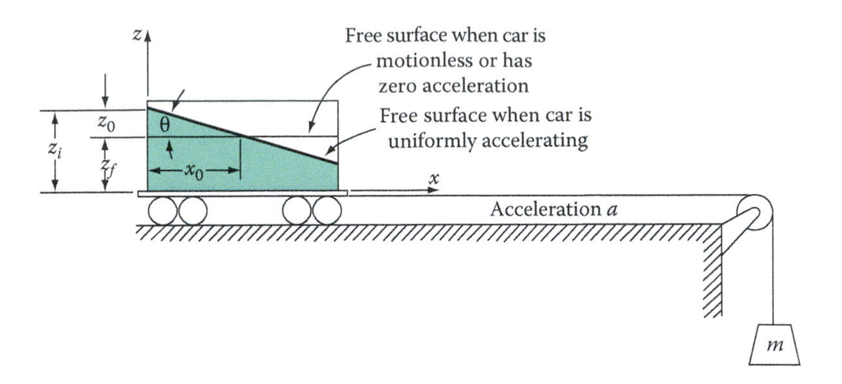

FIGURE 2.26 An experiment to investigate a uniformly accelerating liquid.

$$\frac{\partial p}{\partial y} = 0 \tag{2.22b}$$

$$\frac{\partial p}{\partial z} = -\rho g \tag{2.22c}$$

Integration of Equation 2.22a yields

$$p = -\rho a x + f_1(y, z) + C_1 \tag{2.23a}$$

where

f_1 is a function of y and z

C_1 is a constant of integration

Both naturally become zero when the partial derivative with respect to x is taken. Integration of Equation 2.22c yields

$$p = -\rho g z + f_2(x, y) + C_2 \tag{2.23b}$$

Taking the derivative of p in both equations with respect to y, we obtain

$$\frac{\partial p}{\partial y} = \frac{\partial f_1}{\partial y} \tag{2.24}$$

and

$$\frac{\partial p}{\partial y} = \frac{\partial f_2}{\partial y}$$

Inspection of Equation 2.22b leads to the conclusion that neither f_1 nor f_2 is a function of y. Equating Equations 2.23a and b, we get

$$-\rho a x + f_1(z) + C_1 = -\rho g z + f_2(x) + C_2$$

which leads to

$$f_1(z) = -\rho g z$$
$$f_2(x) = -\rho a x$$
$$C_1 = C_2$$

Therefore, the pressure in Equation 2.23 becomes

$$p = -\rho a x - \rho g z + C_1 \tag{2.25}$$

The constant C_1 is evaluated at the left edge of the tank where $x = 0$, $z = z_i$ = height of the liquid at the left wall, and $p = p_s$ = surface pressure. Substitution into Equation 2.25 gives

$$p_s = -\rho g z_i + C_1$$

or

$$C_1 = p_s + \rho g z_i$$

The pressure is then

$$p - p_s = \rho[g(z_i - z) - ax] \tag{2.26}$$

The equation of the free surface results when $p = p_s$ in Equation 2.26:

$$g(z_i - z) = ax$$

$$z = z_i - \frac{ax}{g} \tag{2.27}$$

This is a straight line with intercept z_i and slope $-a/g$.

Example 2.15

A lawn fertilizer company has a flatbed truck on which a rectangular, plastic tank is transported to various job sites. The tank is sketched in Figure 2.27. As indicated, the tank is 4 ft long, 3 ft tall, and 5 ft wide (into the plane of the page). The tank is filled to a depth of 2 ft with liquid fertilizer (specific gravity = 0.97). The truck can accelerate at a rate equal to $a = g/2$. Determine the equation of the free surface of the fertilizer at this rate.

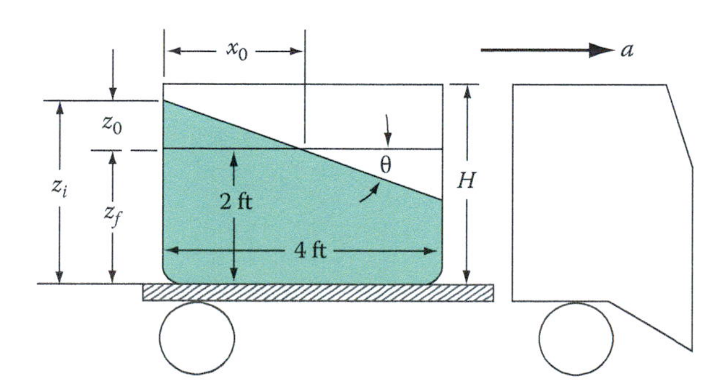

FIGURE 2.27 Accelerating truck with tank attached.

Solution

The surface during acceleration is shown in Figure 2.27. From the geometry, the slope of the line is

$$\tan\theta = \frac{z_0}{x_0} = \frac{a}{g} = \frac{g/2}{g} = 0.5$$

or

$$z_0 = 0.5x_0$$

With $x_0 = 2$ ft, $z_0 = 0.5(2) = 1$ ft. Also,

$$z_i = z_f + z_0 = 2 + 1 \text{ ft} = 3 \text{ ft}$$

The free surface then is given by

$$z = z_i - \left(\frac{a}{g}\right)x$$

or

$$z = 3 - 0.5x \text{ ft}$$

This equation is independent of fluid properties. Although the problem statement specified liquid fertilizer, the same equation would result for any liquid.

Consider next a liquid-filled cylindrical container rotating at a constant angular velocity (Figure 2.28). Cylindrical coordinates appear in the figure with the origin at the center of the bottom of the tank. The acceleration in the θ-direction is zero. The acceleration in the r-direction is $-r\omega^2$, where ω is the angular velocity. The acceleration in the z-direction is due to gravity. Applying Newton's law in the r- and z-directions to the forces acting on a fluid element yields

$$p\,dz(r\,d\theta) - \left(p + \frac{\partial p}{\partial r}dr\right)dz(r\,d\theta) = -\frac{r\omega^2}{g}\rho g\,dr(r\,d\theta)dz$$

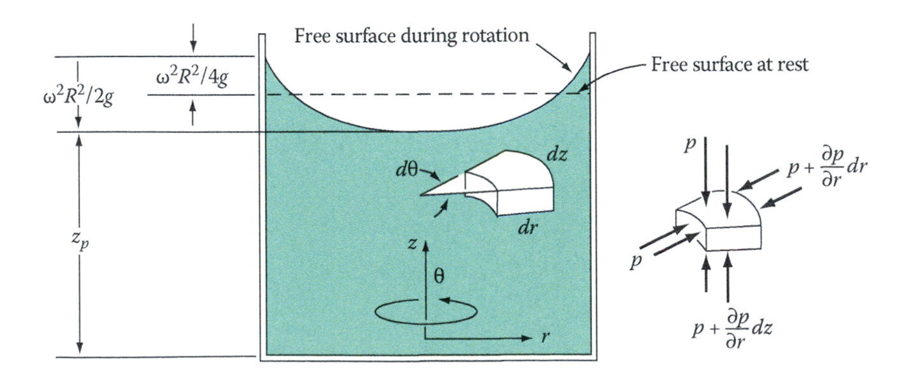

FIGURE 2.28 A rotating cylinder of radius R containing a liquid.

and

$$p\,dr(r\,d\theta) - \left(p + \frac{\partial p}{\partial z}dz\right)dr(r\,d\theta) = -\rho g(r\,d\theta)dr\,dz$$

These equations simplify to

$$\frac{\partial p}{\partial r} = \frac{r\omega^2}{g}\rho g$$

$$\frac{\partial p}{\partial z} = -\rho g$$

Integration brings

$$p = \frac{r^2\omega^2}{2g}\rho g + f_1(\theta, z) + C_1 \quad (2.28a)$$

$$p = -\rho g z + f_2(r, \theta) + C_2 \quad (2.28b)$$

where

f_1 and f_2 are functions of two variables
C_1 and C_2 are constants of integration

Differentiation of both equations with respect to θ yields

$$\frac{\partial p}{\partial \theta} = \frac{\partial f_1}{\partial \theta}$$

$$\frac{\partial p}{\partial \theta} = \frac{\partial f_2}{\partial \theta}$$

Because there is no acceleration in the θ-direction, both these equations equal zero. Therefore, neither f_1 nor f_2 is a function of θ. Equating Equations 2.28a and b, we obtain

$$\frac{r^2\omega^2}{2g}\rho g + f_1(z) + C_1 = -\rho g z + f_2(r) + C_2$$

after which we conclude that

$$f_1(z) = -\rho g z$$

$$f_2(r) = \frac{r^2\omega^2}{2g}\rho g$$

$$C_1 = C_2$$

Equation 2.28 now becomes

$$p = \frac{r^2\omega^2}{2g}\rho g - \rho g z + C_1 \tag{2.29}$$

At $r = 0$, we get $z = z_p$ and $p = p_s$ = surface pressure. By substitution, we obtain

$$p_s = -\rho g z_p + C_1$$

$$C_1 = ps + \rho g z_p$$

Thus, Equation 2.29 finally becomes

$$p - p_s = \rho g\left[\frac{r^2\omega^2}{2g} - (z - z_p)\right] \tag{2.30}$$

The equation of the free surface results if p is set equal to p_s:

$$z = \frac{r^2\omega^2}{2g} + z_p \tag{2.31}$$

which is the equation of a parabola. The liquid in the rotating cylinder forms a paraboloid of revolution. For this geometry, it can be shown that the distance between the bottom of the parabola and the top of the liquid at the wall is $R^2\omega^2/2g$ by setting $r = R$ in Equation 2.31. It can also be shown that the distance between the free surface before rotation to the bottom of the parabola is $R^2\omega^2/4g$.

Example 2.16

The rotating cylinder of Figure 2.28 is 30 cm in diameter and 42 cm high. It is filled to a depth of 35 cm with liquid. Determine the angular velocity required to spill liquid over the top.

Solution

Equation 2.31 is the free-surface equation:

$$z = \frac{r^2\omega^2}{2g} + z_p$$

If $r = R$, then $z = 0.42$ m. Thus,

$$0.42\,\text{m} = \frac{R^2\omega^2}{2g} + z_p$$

It is known that the free-surface height at rest is

$$0.35\,\text{m} = z_p + \frac{R^2\omega^2}{4g}$$

Combining these equations, we get

$$0.42\,\text{m} - \frac{R^2\omega^2}{2g} = 0.35\,\text{m} - \frac{R^2\omega^2}{4g}$$

or

$$0.07\,\text{m} = \frac{R^2\omega^2}{4g}$$

Solving for the angular velocity, we obtain

$$\omega^2 = \frac{4(0.07\,\text{m})(9.81\,\text{m/s}^2)}{(0.15\,\text{m}^2)} = 122$$

$$\omega = 11.0\,\text{rad/s} = 1.76\,\text{rev/s}$$

Therefore, liquid will spill over the top when

$$\omega > 11.0\ \text{rad/s}$$

2.5 FORCES ON SUBMERGED BODIES

When a body is submerged, the fluid exerts a net force on it in an upward direction. For example, a cube submerged in a fluid will have hydrostatic pressure acting on its lower surface and on its upper surface. The pressure exerted on its lower surface is greater than that exerted on its upper surface because pressure increases with depth. The difference in these pressures multiplied by the area over which they act is defined as the **buoyant force**, a force that acts in the direction opposite to that of the gravity force.

To illustrate this definition in more detail, consider the submerged object shown in Figure 2.29a. The object has dimensions $a \times b \times c$. Figure 2.29b shows the same object in a two-dimensional profile view. The pressure distributions acting on the upper and lower surfaces are found with

$$p_u = \rho g z_1$$

and

$$p_l = \rho g z_2$$

Forces are easier to work with than pressure distributions, and so we find it convenient to replace the pressures with equivalent forces. These forces are shown in Figure 2.29c and are determined using

$$R_u = p_u A = \rho g z_1(ac)$$

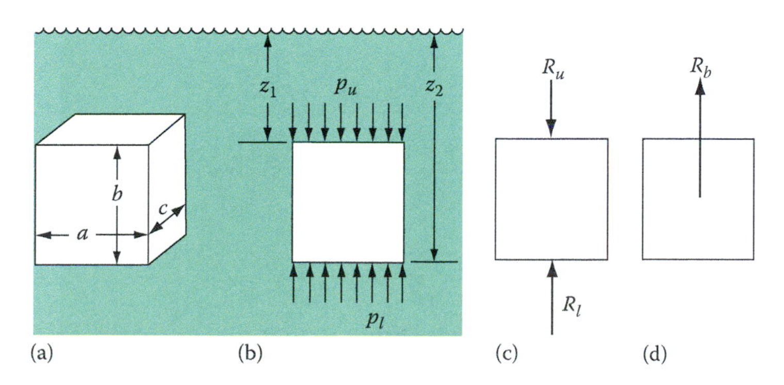

FIGURE 2.29 A submerged object and the hydrostatic forces acting on it.

and

$$R_l = p_l A = \rho g z_2(ac)$$

In order to simplify this development even further, we combine these two hydrostatic forces into one force that we have already defined as the **buoyant force** R_b:

$$R_b = R_l - R_u = p_l A - p_u A = \rho g z_2(ac) - \rho g z_1(ac)$$

or

$$R_b = \rho g(z_2 - z_1)(ac)$$

We see from the figure that $(z_2 - z_1) = b$, so that the buoyant force can be rewritten as

$$R_b = \rho g(abc) = \rho g \mathcal{V}$$

where the volume $\mathcal{V}$ of the object is (abc). The buoyant force therefore equals the weight of the volume of liquid displaced by the object.

Thus, we have illustrated the definition of the buoyant force as being the difference in hydrostatic forces (due to pressures) acting in the vertical direction on the object. Also, we have concluded that for the object of Figure 2.29, a consequence of our definition is that the buoyant force equals the weight of the liquid displaced by the object. To establish this fact with more mathematical rigor, however, we should work with an object of arbitrary shape to determine if the same conclusion can be drawn.

Consider an irregularly shaped submerged object as shown in Figure 2.30. A rectangular coordinate system is established with the z-direction positive downward. Choose a slice having a cross-sectional area dA and length $z_2 - z_1$. At 1 we have a pressure p_1, and at 2 we have p_2; both act over an area equal to dA. The buoyant force is due to this difference in pressure and is therefore equal to

$$dR_b = (p_2 - p_1)dA$$

acting upward. From hydrostatics, $p = \rho g z$. Substituting, we have

$$\begin{aligned} dR_b &= (\rho g z_2 - \rho g z_1)dA \\ &= \rho g(z_2 - z_1)dA \end{aligned}$$

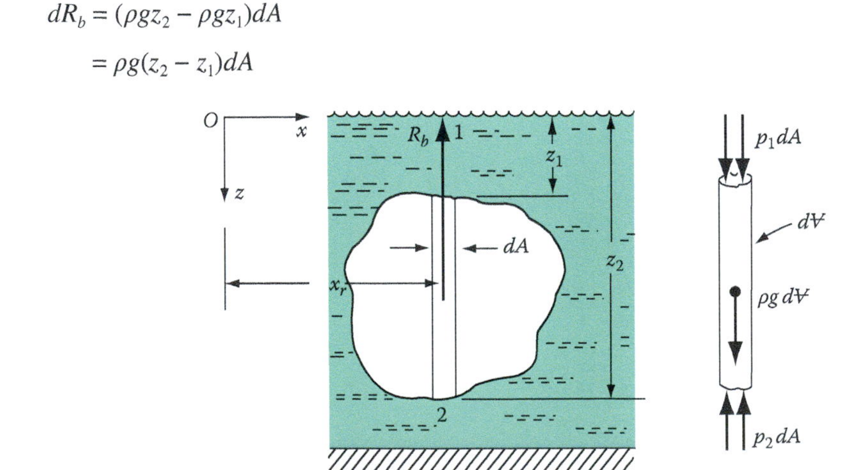

FIGURE 2.30 A submerged body.

But the volume of the element is $(z_2 - z_1)dA$, and our equation becomes

$$dR_b = \rho g \, d\mathcal{V}$$

Integrating over the entire volume gives the total vertical force:

$$R_b = \rho g \iiint d\mathcal{V} \tag{2.32}$$

Thus, the buoyant force equals the weight of the volume of fluid displaced. This concept is known as **Archimedes' principle**. (Recall that pressure does not vary with horizontal distance, so there are no unbalanced forces in the x- or y-direction.)

With reference to Figure 2.30, we can now evaluate the moment of the buoyant force about the origin:

$$R_b x_r = \rho g \iiint x \, d\mathcal{V}$$

Combining with Equation 2.32, we obtain an expression for the line of action of R_b:

$$x_r = \frac{\iiint x \, d\mathcal{V}}{\mathcal{V}} \tag{2.33}$$

Thus, the buoyant force acts through the centroid of the submerged volume: the **center of buoyancy**.

When the buoyant force exceeds the object's weight while submerged in a liquid, the object will float in the free surface. A portion of its volume will extend above the liquid surface, as illustrated in Figure 2.31. In this case,

$$dR_b = (\rho g z_2 - \rho_a g z_1) dA$$

where ρ_a is the air density or density of the fluid above the liquid. For most liquids,

$$\rho_a \ll \rho$$

and thus we can write

$$dR_b = \rho g z_2 dA = \rho g \, d\mathcal{V}_s$$

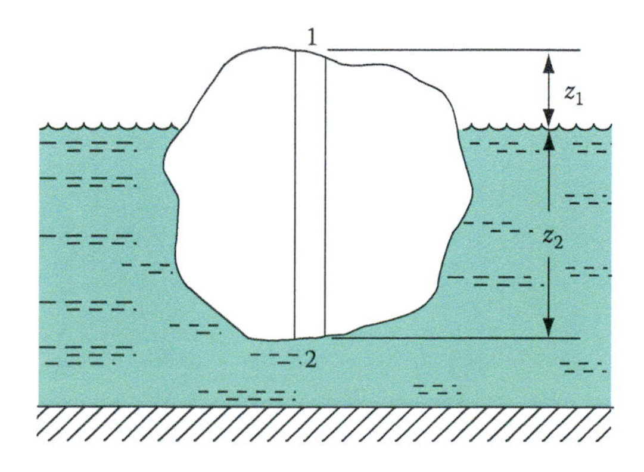

FIGURE 2.31 A floating body.

where $d\forall_s$ is the submerged volume. Integration gives

$$R_b = \rho g \iiint d\forall_s \tag{2.34}$$

Thus, the buoyant force exerted on a floating body equals the weight of the displaced volume of liquid. It can be shown that this force acts at the center of buoyancy of only the submerged volume.

Example 2.17

Figure 2.32 shows a 4 cm diameter cylinder floating in a basin of water, with 7 cm extending above the surface. If the water density is 1 000 kg/m^3, determine the density of the cylinder.

Solution

A free-body diagram of the cylinder includes gravity and buoyancy forces. The total weight of the object is supported by the buoyant force that acts on the submerged portion of the cylinder. Summing these forces, we get

$$mg = R_b$$

or

$$\rho_c \forall g = \rho \forall_s g$$

or

$$\rho_c = \frac{\rho \forall_s}{\forall}$$

where
ρ_c is the density of the cylinder
$\forall_s$ is the volume of the submerged portion of the cylinder

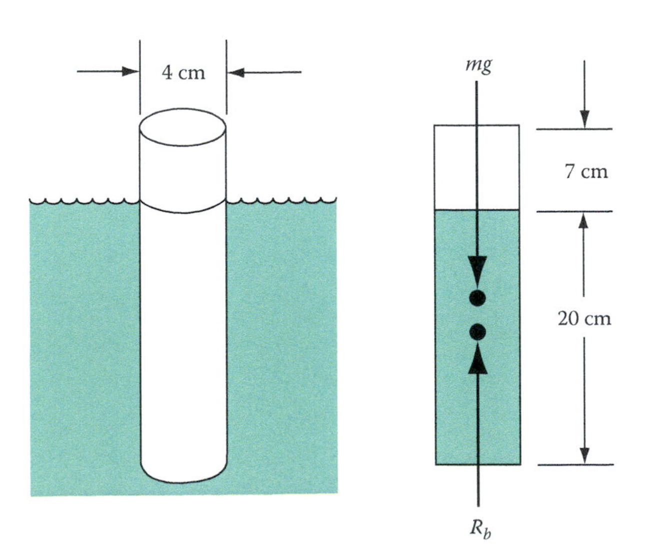

FIGURE 2.32 A cylinder floating in the surface of a liquid.

Thus,

$$\forall\!\!\!\!-\, = \left(\frac{\pi D^2}{4}\right)L$$

and

$$\forall\!\!\!\!-_s = \left(\frac{\pi D^2}{4}\right)L_s$$

where $L_s = 20$ cm as indicated in Figure 2.32. Substituting,

$$\rho_c = \frac{\rho(\pi D^2/4)L_s}{(\pi D^2/4)L} = \frac{\rho L_s}{L}$$

$$\rho_c = 1\,000\left(\frac{20}{27}\right)$$

or

$$\rho_c = 741 \text{ kg/m}^3$$

2.6 STABILITY OF SUBMERGED AND FLOATING BODIES

In this section, we consider two cases of stability: objects that are submerged and objects that are floating. An object is considered to be in stable equilibrium if it returns to its original position after being displaced slightly from that position. For the case of a submerged object, consider a gondola being carried in air by a balloon as sketched in Figure 2.33. Since the hot air in the balloon is lighter

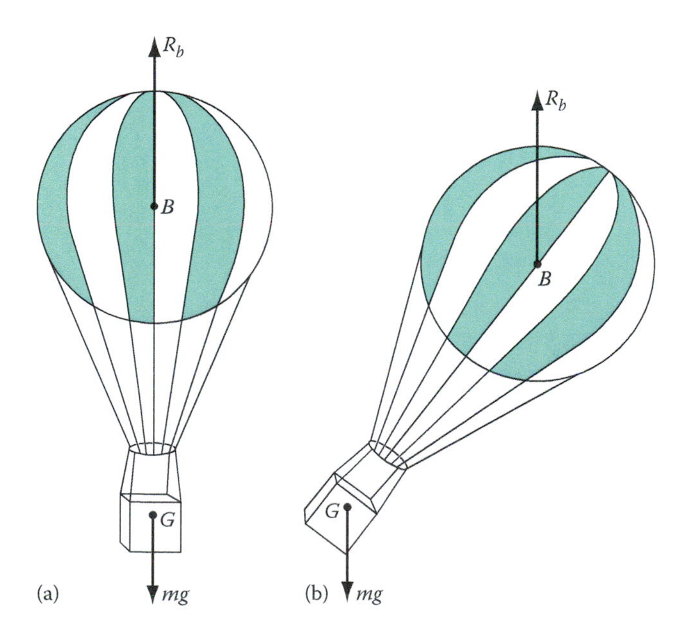

FIGURE 2.33 A balloon and basket submerged in air.

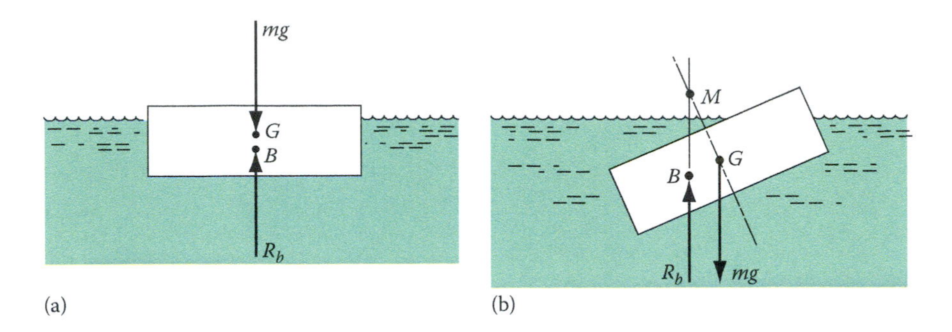

FIGURE 2.34 Stability of a floating body.

than ambient air at ground level, the balloon rises to a height at which atmospheric density nearly equals hot air density (accounting for the weight of the balloon). The buoyant force is shown to act through the center of buoyancy B. The weight of the system is shown to act at the center of gravity G. If the system is disturbed by, say, a gust of wind, then the buoyant force and the weight together form a moment that returns the balloon and basket to the equilibrium position. The system depicted in Figure 2.33a is then considered stable. For submerged bodies, we conclude that when the center of buoyancy lies directly above the center of gravity in a static position, the system is considered stable.

For the case of a body floating in a liquid surface, the stability requirements can be examined by using the body sketched in Figure 2.34. The buoyant force acts through B, and the weight acts through point G. Figure 2.34a depicts a stable situation; yet, unlike the submerged body of Figure 2.34, G is above B. Thus a somewhat different criterion is necessary for floating bodies.

First extend the line of action of R_b until it intersects the centerline of the body (Figure 2.34b). The point of intersection is called the **metacenter** M. The distance MG is known as the metacentric height. The force R_b acts through M. The object is stable if the metacenter lies above the center of gravity because in this case a restoring moment is set up when the object is disturbed from its equilibrium position.

Example 2.18

Figure 2.35a shows a bar of soap floating in a basin of water. Sketch its position for 0°, 30°, 60°, and 90° from the position of Figure 2.35a. Show the center of buoyancy (B) and the center of gravity (G), and comment on the stability.

Solution

Figure 2.35 shows all the positions with comments on the stability.

The preceding example shows how an intuitive analysis can be performed on a stability problem. In many cases, however, a mathematical approach is necessary. Consider the hull of a ship as sketched in Figure 2.36. A plan view is shown as well as a cross-sectional view at A–A. From the previous discussion, we know that there is a buoyant force acting at B and the weight of the entire ship acting at G. The weight of only the submerged volume $\forall_s$ also acts at B. When the ship is rolled (rotated about axis y–y), it becomes deflected through an angle θ known as the angle of heel. As shown in Figure 2.36c, the submerged volume changes geometry, and the center of buoyancy moves to B'. Correspondingly, a restoring moment is set up; this moment is given as $\Delta R_b x_0$. The weight of the submerged volume (equal to the weight of liquid displaced) now acts through B'. Summing moments about B, we have

$$\Delta R_b x_0 = \rho g \forall_s r \tag{2.35}$$

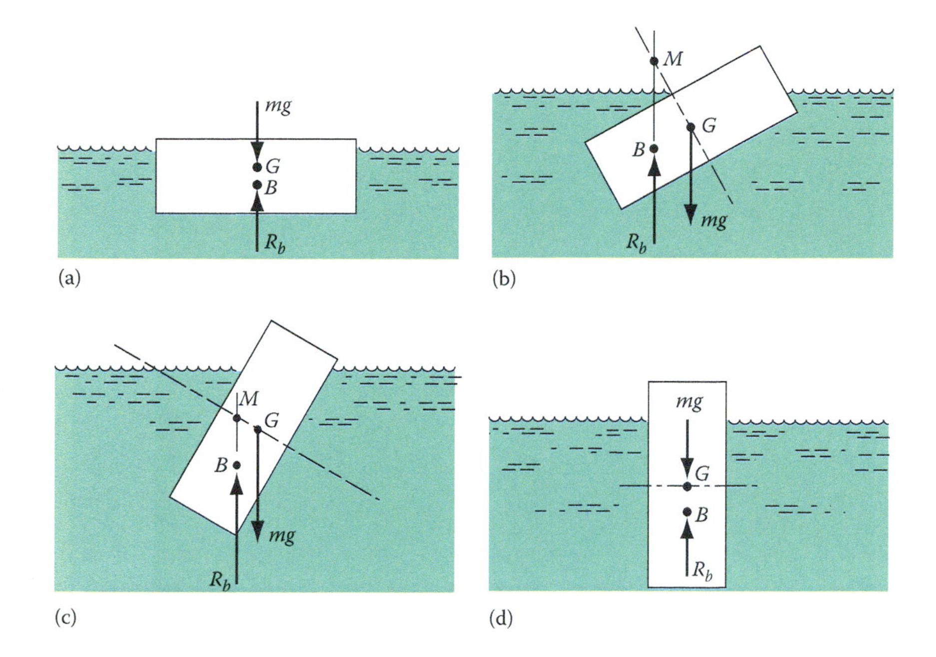

FIGURE 2.35 Solution to Example 2.18. (a) 0° stable, (b) 30° clockwise moment; return to 0°; M above G; stable, (c) 60° clockwise moment; return to 0°; M above G; stable, and (d) 90° unstable; any slight disturbance moves M above G giving a moment to move soap to 0°.

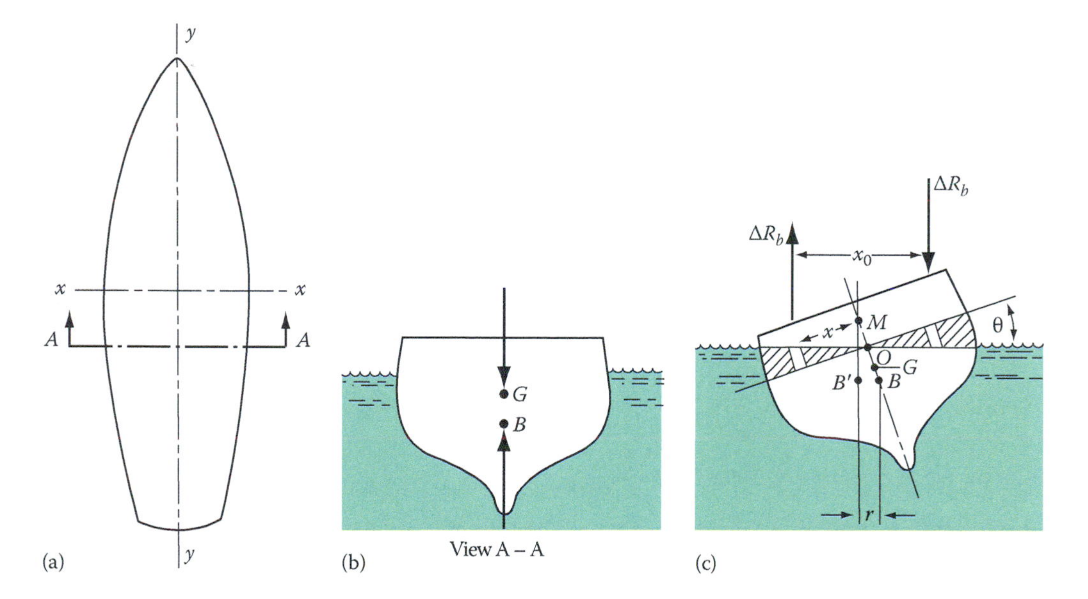

FIGURE 2.36 A sketch for calculating metacentric height.

The magnitude of $\Delta R_b x_0$ can be determined again by taking moments about point O, where the liquid surface intersects the cross section at the centerline. Choose an element of area dA a distance x from O. The volume of this element is $x \tan\theta\, dA$ or, for small θ, $x\theta\, dA$. The buoyant force due to it is ($\rho g x\theta\, dA$). The moment of the element about O is $\rho g x^2\theta\, dA$; therefore,

$$\Delta R_b x_0 = \iint \rho g x^2 \theta\, dA \tag{2.36}$$

Equating Equations 2.35 and 2.36 gives

$$\rho g\theta\iint x^2 dA = \rho g \forall_s r$$

The double integral is recognized as the moment of inertia about axis y–y of the ship. After simplification, we obtain

$$\theta I_{yyc} = \forall_s r \tag{2.37}$$

Now $r = MB \sin\theta = MB\theta$, for small θ. Combining this result with Equation 2.37, we have

$$\theta I_{yyc} = \forall_s MB\theta$$

or

$$MB = \frac{I_{yyc}}{\forall_s} \tag{2.38}$$

The metacentric height is found in terms of distance to be

$$MB = MG + GB$$

Thus, $MG = MB - GB$ or

$$MG = \frac{I_{yyc}}{\forall_s} - GB$$

The criteria for stability become

$$MG = \frac{I_{yyc}}{\forall_s} - GB > 0 \rightarrow \frac{I_{yyc}}{\forall_s} > GB \quad \text{(stable)} \tag{2.39a}$$

$$MG = \frac{I_{yyc}}{\forall_s} - GB = 0 \rightarrow \frac{I_{yyc}}{\forall_s} = GB \quad \text{(neutral)} \tag{2.39b}$$

$$MG = \frac{I_{yyc}}{\forall_s} - GB < 0 \rightarrow \frac{I_{yyc}}{\forall_s} < GB \quad \text{(unstable)} \tag{2.39c}$$

Example 2.19

It is known that a long log floats with its axis horizontal. A very short log, however, floats with its axis vertical. Determine the maximum length of a cylindrical walnut log 40 cm in diameter that will cause it to float with its axis vertical.

Solution

For walnut, the specific gravity ranges from 0.64 to 0.70 (Table A.7); we will use the average value of 0.67. Take the log as floating horizontally and about to tip to a vertical position. In this instance, then, I_{yyc} is the same for either position. We will calculate length L by using Equation 2.39b:

$$\frac{I_{yyc}}{\forall_s} = GB \quad \text{(neutral stability)}$$

Note that if L is larger than that calculated here, the log remains horizontal; if L is shorter, the log becomes vertical. From Table B.1 for a cylinder,

$$I_{yyc} = \frac{\pi D^4}{64} = \frac{\pi(0.4)^4}{64} = 1.26 \times 10^{-3}\,\text{m}^4$$

For a vertical log, G is $L/2$ from the bottom. The log has $s = 0.67$, and thus B is $0.67L/2$ from the bottom because the log is only 67% submerged, as shown in Figure 2.37. Hence,

$$GB = \frac{L}{2} - \frac{0.67L}{2} = 0.165L$$

With submerged volume given as

$$\forall = \frac{0.67\pi D^2 L}{4} = \frac{0.67\pi(0.4)^2 L}{4} = 8.42 \times 10^{-2} L\,\text{m}^{-3}$$

Equation 2.39b becomes, after substitution,

$$\frac{1.26 \times 10^{-3}}{8.42 \times 10^{-2} L} = 0.165L$$

or

$$L^2 = 0.0907$$

Solving, we get

$$L = 0.301\ \text{m} = 30.1\ \text{cm}$$

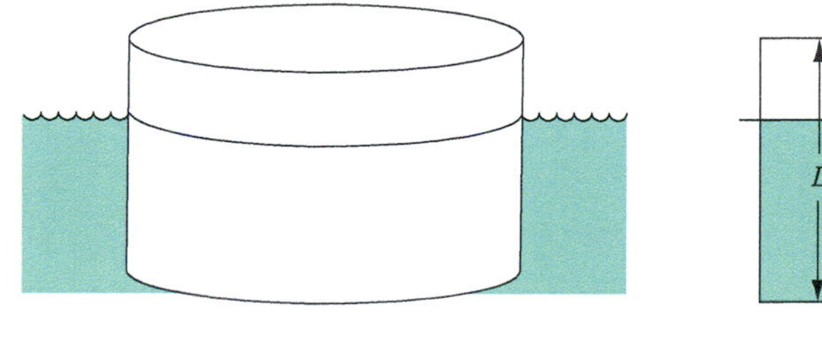

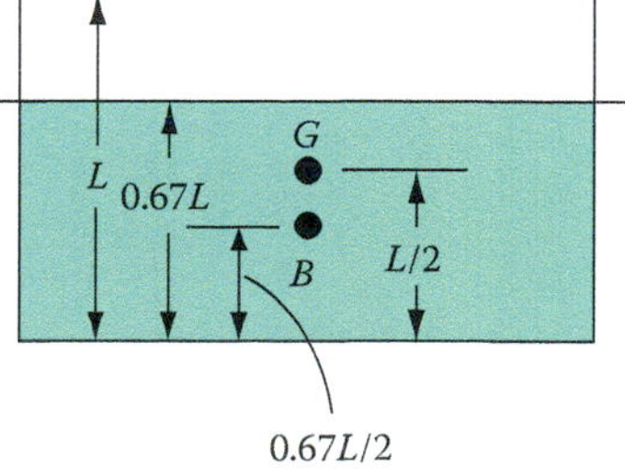

FIGURE 2.37 Floating log with axis vertical showing distances from bottom.

2.7 SUMMARY

In this chapter, we discussed pressure and its measurement by manometers and pressure gauges. We examined the pressure distributions existing in a static fluid in contact with submerged plane and curved surfaces. Moreover, we developed equations that are useful in finding forces and their locations that were equivalent to the pressure variations existing in static fluids. It was shown that accelerating liquid containers can be analyzed by methods of fluid statics. Finally, we derived expressions for predicting hydrostatic forces on submerged bodies and discussed the criterion of stability for submerged and for floating bodies.

INTERNET RESOURCES

1. Pressure can be measured with a pressure gauge as described in this chapter. Give a brief presentation on how pressure gauges are calibrated.
2. What is an inclined manometer, and when is it necessary to use one? Give a presentation on this topic.
3. Pressure transducers are electronic devices that are used to measure pressure and provide a digital output. Give a presentation on the types of pressure transducers that are available and on how accurate these devices are.

PROBLEMS

Hydrostatic Equation

2.1 Repeat the derivation in Section 2.1 for a three-dimensional prism to prove that pressure at a point is independent of direction.

2.2 A hydraulic jack is used to bend pipe as shown in Figure P2.2. What force F_2 is exerted on the pipe if F_1 is 450 N, assuming pressure is constant throughout the system.

2.3 A hydraulic elevator is lifted by a 12 in. diameter cylinder under which oil is pumped (Figure P2.3). Determine the output pressure of the pump as a function of z if the weight of the elevator and occupants is 1800 lbf. Take the specific gravity of the oil to be 0.86. The elevator cylinder does not contain oil.

2.4 A suited diver can dive to a depth at which the pressure is 1.8 kPa. Calculate the depth in seawater where this pressure exists.

2.5 A free diver (unsuited) can dive to a depth at which the pressure is about 70 psi. What is the corresponding depth in fresh water with $\rho = 1.94$ slug/ft^3?

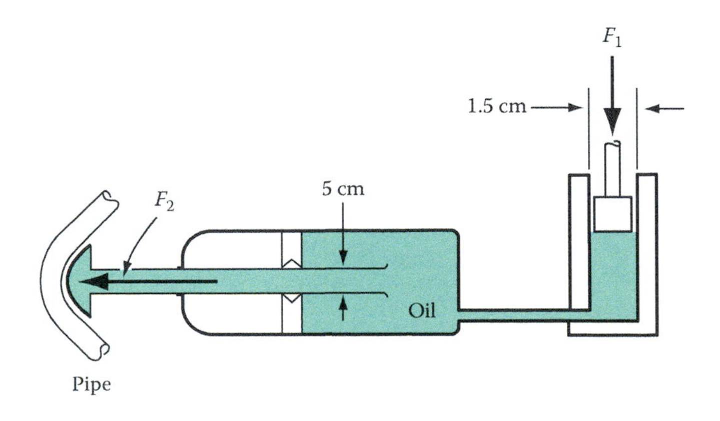

FIGURE P2.2

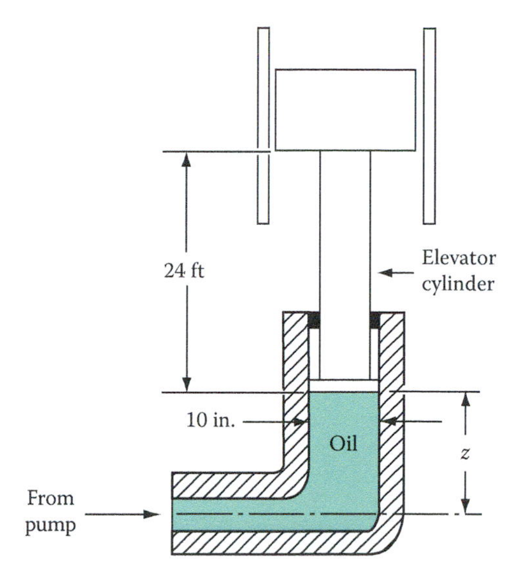

FIGURE P2.3

2.6 The deepest descent in seawater of a diving bell is 1370 ft. What is the pressure at this depth?

2.7 The deepest descent in seawater of a human-carrying vessel is 35,820 ft. What is the pressure at this depth?

2.8 A cylindrical tank is used to store methyl alcohol after its production and before it is pumped to its user. The tank is 35 ft in diameter and is filled to a depth of 10 ft. Calculate the pressure exerted at the bottom of the walls.

2.9 Linseed oil can be mixed with powders to produce paint. The oil is shipped via rail in tank cars that are 6 ft in diameter and 40 ft long. Calculate the pressure exerted at the bottom of a full tank.

2.10 The pressure at the bottom of an open top tank of ethyl alcohol is 17 psia. What is the depth of the liquid?

2.11 The output pressure of a pump is 3.5 psi. The pump is used to deliver propyl alcohol to a tank, as shown in Figure P2.11. To what depth can the tank be filled at this pump pressure, provided that inflow stops when pump pressure equals hydrostatic pressure in the tank? Ignore frictional losses.

2.12 Methyl alcohol and gasoline (assume octane) are mixed together in an open fuel tank. The methyl alcohol soon absorbs water from the atmosphere, and the water–alcohol mix separates from the gasoline, as shown in Figure P2.12. Find the pressure at the bottom of the tank wall.

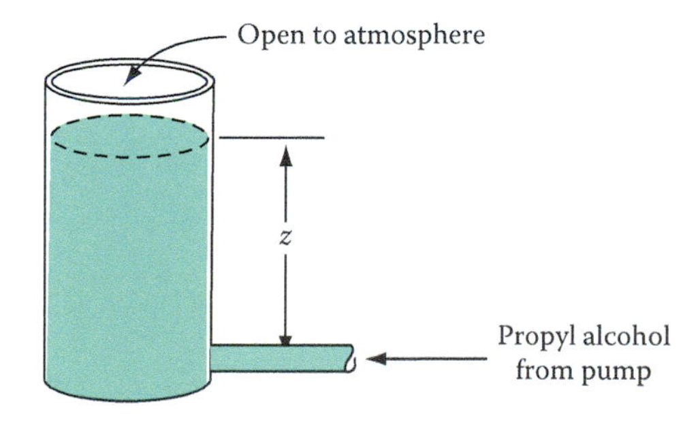

FIGURE P2.11

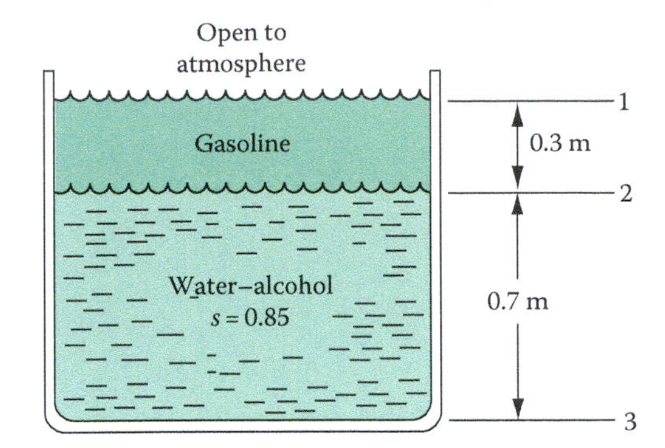

FIGURE P2.12

2.13 A coffee cup is 2 in. in diameter and filled to a depth of 3 in. with coffee (s = 1.03). Calculate the pressure difference between the top and the bottom of the glass sides.

2.14 A common drinking glass is 7 cm in diameter and filled to a depth of 10 cm with water. Calculate the pressure difference between the top and the bottom of the glass sides.

2.15 Figure P2.15 shows a tank containing three liquids. What is the expected reading on the pressure gauge?

2.16 A concrete dam is constructed as shown in Figure P2.16. When the water level on the left is 30 ft, determine the pressure at the bottom of the dam.

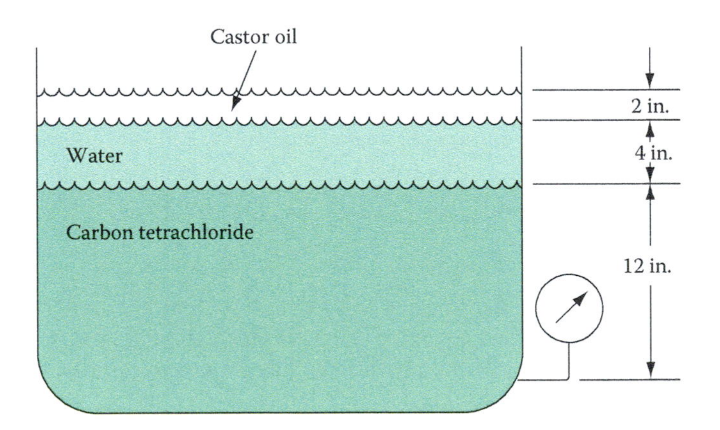

FIGURE P2.15

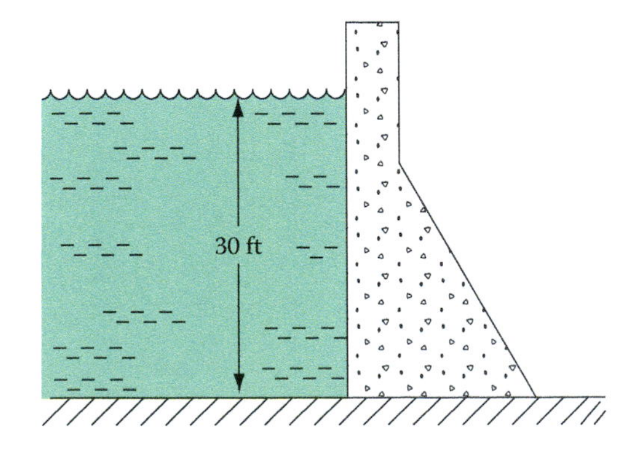

FIGURE P2.16

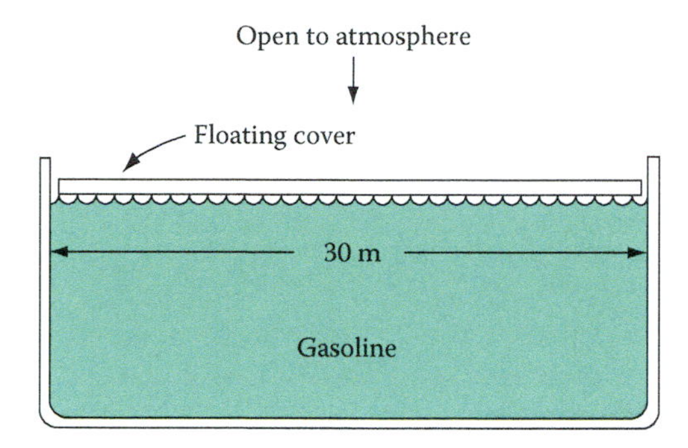

FIGURE P2.17

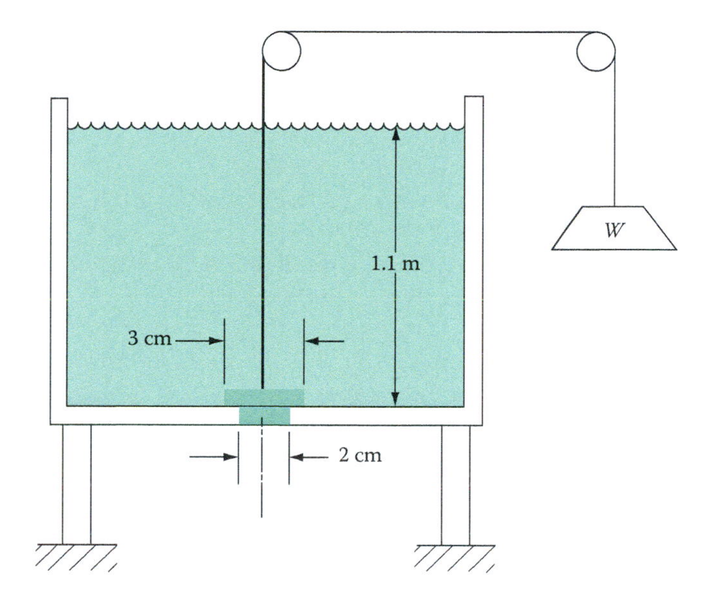

FIGURE P2.18

2.17 A floating cover is used in oil- and gasoline-storage tanks to keep moisture out (Figure P2.17). If the cover weighs 1 800 N and the depth of gasoline (assume octane) is 3 m, determine the pressure at the bottom of the tank wall.

2.18 Figure P2.18 shows a tank containing turpentine. The tank bottom has a plug with a cable attached that leads to a weight *W*. It is desired to have the plug close when the acetone depth is 1.1 m. Calculate the weight required to do this. Neglect the thickness of the plug.

Atmosphere

2.19 Calculate the atmospheric pressure at an elevation of 50,000 ft. Express the answer in kPa, psia, and psig.

2.20 At what atmospheric elevation is the pressure equal to 15 kPa if the temperature is –57°C? Assume a lapse rate of 6.5°C/km from sea level, where the temperature is taken to be 288 K.

2.21 Verify that the pressure variation in the stratosphere is given by

$$p = 22\ 500 \exp\left(\frac{11\ 000 - z}{6\ 315}\right)$$

where $T = -57°\text{C}$ = a constant, $z \geq 11\ 000$ m. Take $p = 22.5$ kPa at $z = 11\ 000$ m.

2.22 It was stated that in the troposphere the temperature is given by

$$T = T_o - \alpha h$$

where

T is the temperature from sea level to 11 000 m

T_o is 288 K (the temperature at sea level)

α is the lapse rate = 6.5°C/km

a. Starting with the differential form of the hydrostatic equation, substitute for ρ from the ideal gas law. For temperature, substitute the equation given here.

b. Separate the variables and integrate from p_1 = 101.3 kPa and z_1 = 0; to $p_2 = p$ and z_2 = 11 000 m.

c. Show that

$$\frac{p}{101.3} = (1 - 0.000\,022\,6z)^{5.26}$$

d. Plot p versus z (vertical axis) for $0 \le z \le 11\,000$ m.

e. Combine this plot with the result for $z \ge 11\,000$ m.

2.23 At what elevation is atmospheric pressure equal to 7 kN/m^2? Use the equation in Problem 2.21 or in Problem 2.22 as appropriate.

2.24 At what elevation is atmospheric pressure equal to 4 psia? Use the equation in Problem 2.21 or in Problem 2.22 as appropriate.

2.25 The elevation above sea level at Denver, CO, is 5283 ft, and the local atmospheric pressure there is 83.4 kPa. If the temperature there is 75°F, calculate the local air density and the barometric pressure in centimeters of mercury.

2.26 Select a city and use a reference (internet or otherwise) to determine its elevation above sea level. Use the appropriate equation to calculate the local atmospheric pressure there. Express the result in cm of mercury, and use a temperature of 25°C if necessary.

2.27 Determine the height of mercury in a barometer if atmospheric pressure is 12 psia.

2.28 At an elevation of 12 km, atmospheric pressure is 19 kN/m^2. Determine the pressure at an elevation of 20 km if the temperature at both locations is 57°C.

Manometry

2.29 Figure P2.29 shows a tank containing linseed oil and water. Attached to the tank 3 ft below the water–linseed oil interface is a mercury manometer. For the dimensions shown, determine the depth of the linseed oil.

2.30 A mercury manometer is used to measure pressure near the bottom of a tank containing acetone as shown in Figure P2.30. What is the depth d of the acetone if Δh is measured to be 4 in. of mercury and x is 2 in.?

2.31 Determine the pressure above the glycerine in Figure P2.31.

2.32 Determine the pressure of the water at the point where the manometer attaches. All dimensions are in centimeters (Figure P2.32).

2.33 Figure P2.33 shows a manometer and a gauge attached to the bottom of a tank of kerosene. The gauge reading is 3 psig. Determine the height z of mercury in the manometer.

2.34 A pump is a device that puts energy into a liquid in the form of pressure. The inlet side of a pump usually operates at less than atmospheric pressure. As shown in Figure P2.34, a manometer and a vacuum gauge are connected to the inlet side of a pump, and the vacuum gauge reads the equivalent of 34 kPa (absolute).

a. Express the gauge reading in psig.

b. Calculate the deflection when the manometer liquid is mercury.

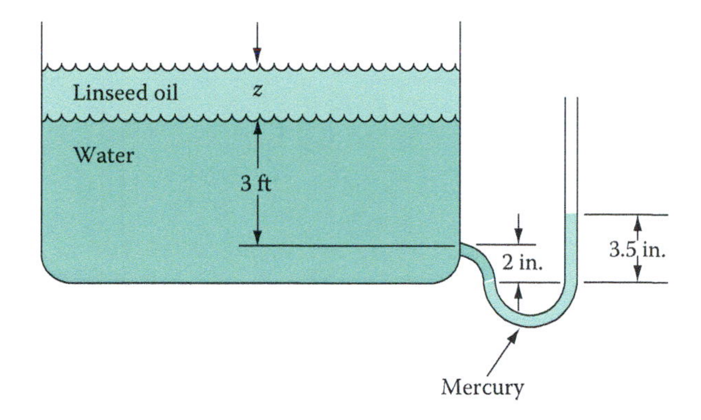

FIGURE P2.29

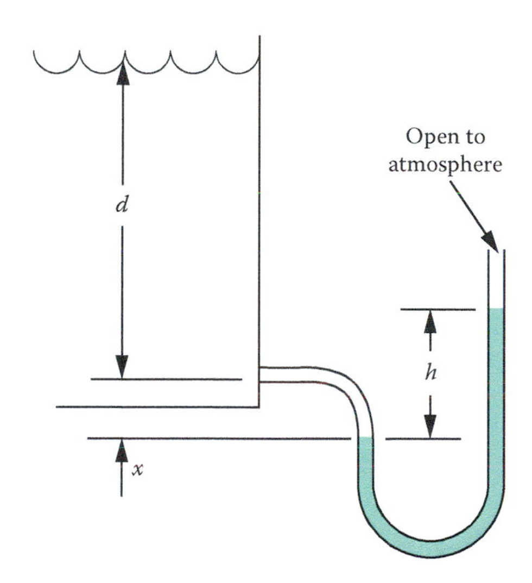

FIGURE P2.30

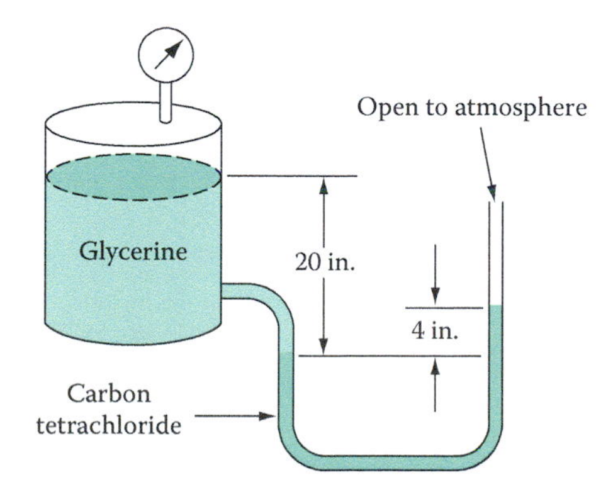

FIGURE P2.31

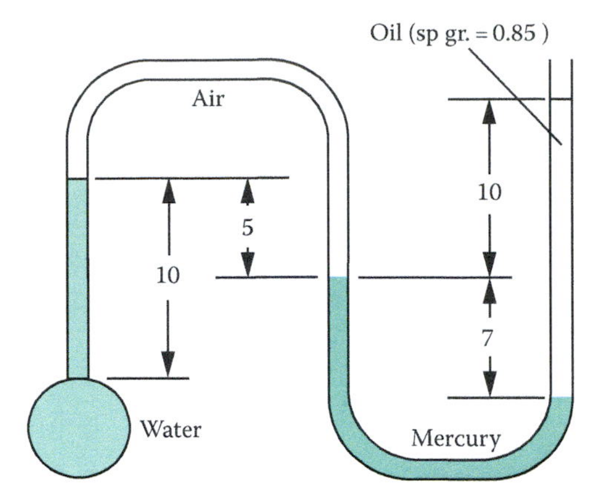

FIGURE P2.32

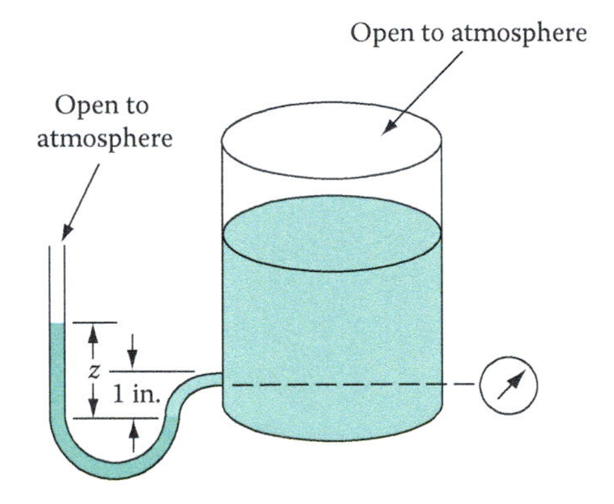

FIGURE P2.33

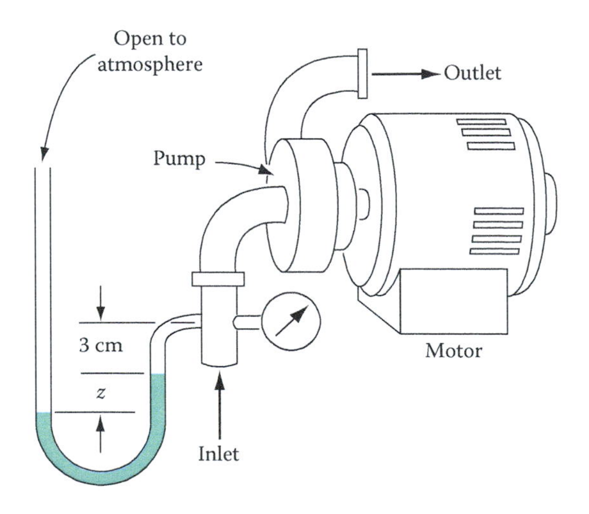

FIGURE P2.34

2.35 A manometer is used to measure pressure drop in a venturi meter as shown in Figure P2.35. Derive an equation for the pressure drop in terms of Δh. Make all measurements from the meter centerline, and ρ_2 is not negligible.

2.36 Two pipelines are connected with a manometer, as shown in Figure P2.36. Determine the pressure p_2 if the manometer deflection is 30 cm of water and p_1 is 55 kPa. The fluid in each pipe is air.

2.37 If turpentine pressure is 20 psi, find the water pressure for the system shown in Figure P2.37.

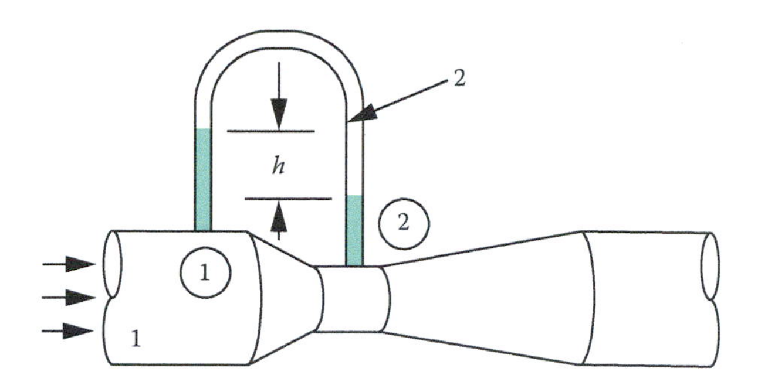

FIGURE P2.35

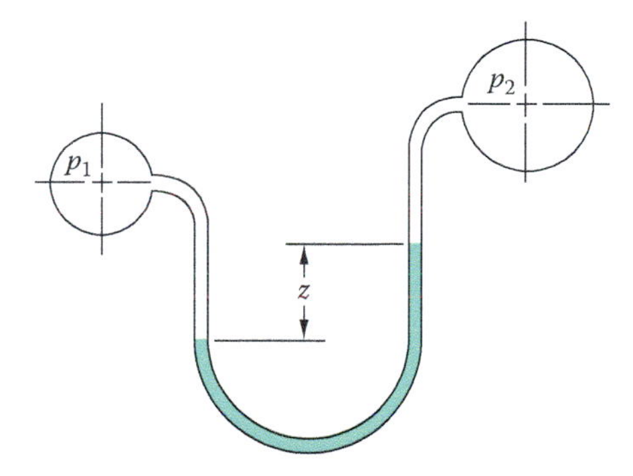

FIGURE P2.36

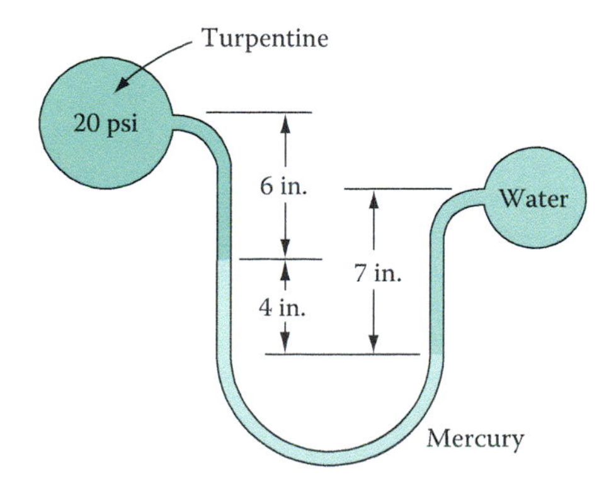

FIGURE P2.37

2.38 Determine the pressure difference between the linseed and castor oils of Figure P2.38. All dimensions are in inches.

2.39 Calculate the air pressure in the tank shown in Figure P2.39. Take atmospheric pressure to be 101.3 kPa.

2.40 The pressure of the air in Figure P2.40 is 10.7 kPa, and the manometer fluid is of unknown specific gravity s. Determine s for the deflections shown.

2.41 For the sketch of Figure P2.41 (all dimensions in inches), determine the pressure of the linseed oil if the glycerine pressure is 100 psig.

2.42 Find z in Figure P2.42 if $p_2 - p_1 = 2.5$ psig.

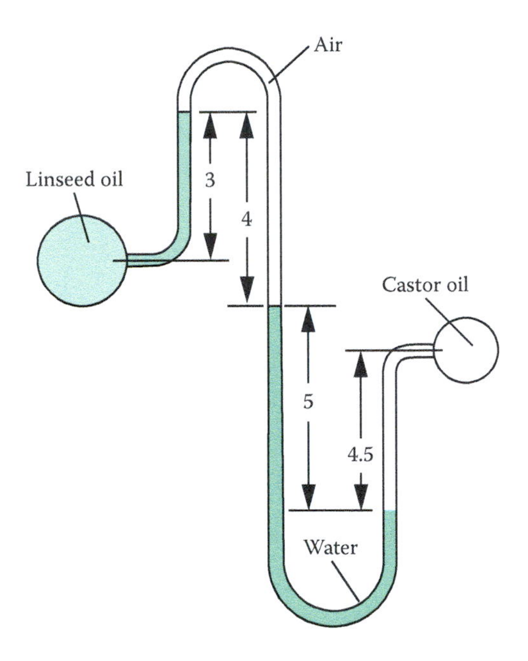

FIGURE P2.38

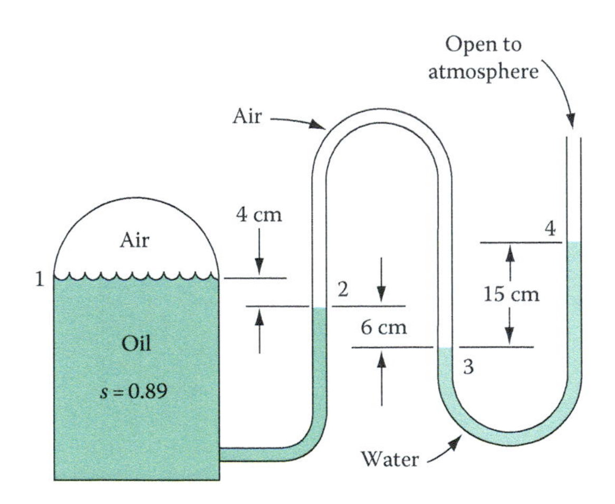

FIGURE P2.39

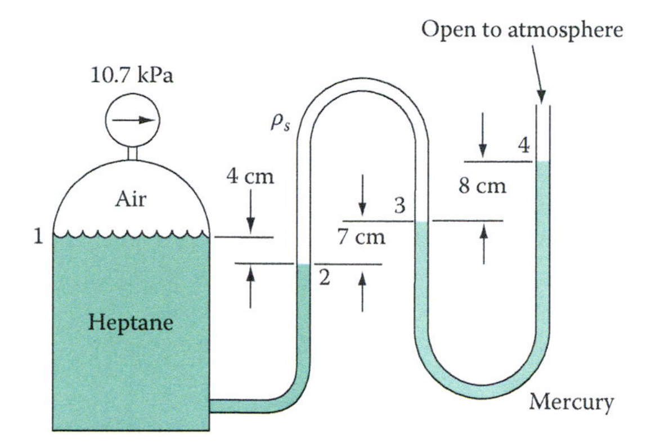

FIGURE P2.40

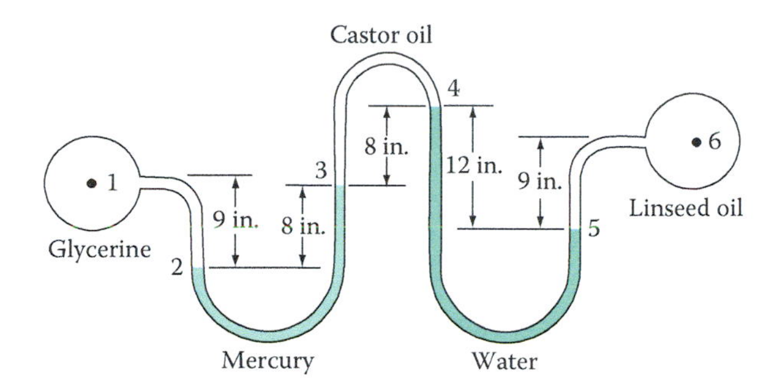

FIGURE P2.41

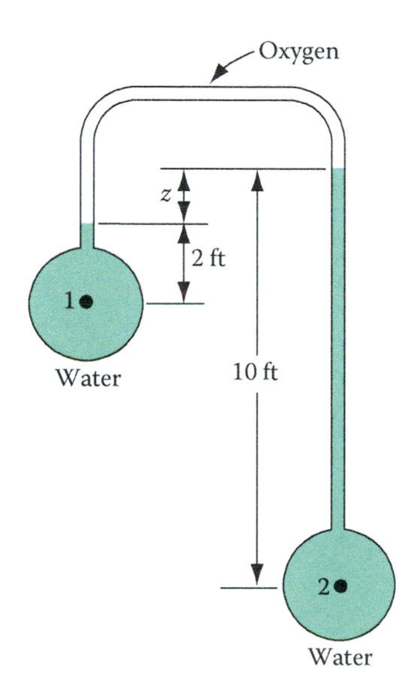

FIGURE P2.42

2.43 Two pipes containing air are connected with two manometers, as shown in Figure P2.43. One manometer contains glycerine, and the difference in liquid levels is 10 cm. The other manometer contains an unknown liquid, and it shows a difference in liquid levels of 15 cm. What is the density of the unknown liquid?

2.44 For the system of Figure P2.44, determine the pressure of the air in the tank.

2.45 An inverted U-tube manometer is connected to two sealed vessels, as shown in Figure P2.45. Heptane is in both vessels, and the manometer shows a difference in heptane levels of 10 cm. Calculate the pressure difference between the two vessels.

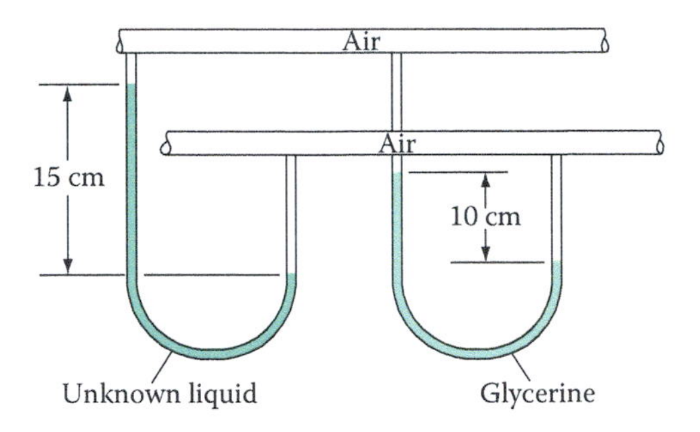

FIGURE P2.43

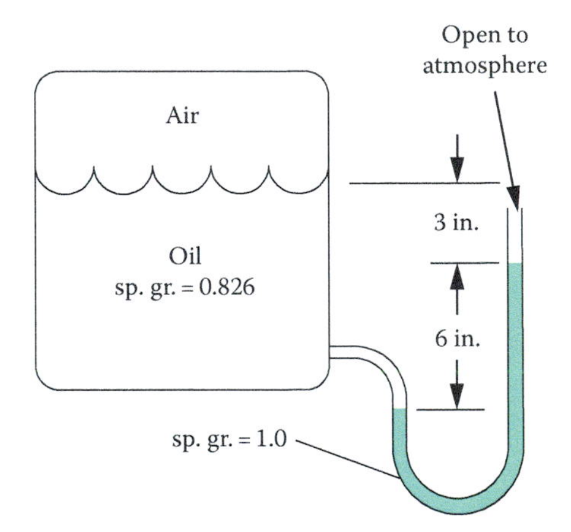

FIGURE P2.44

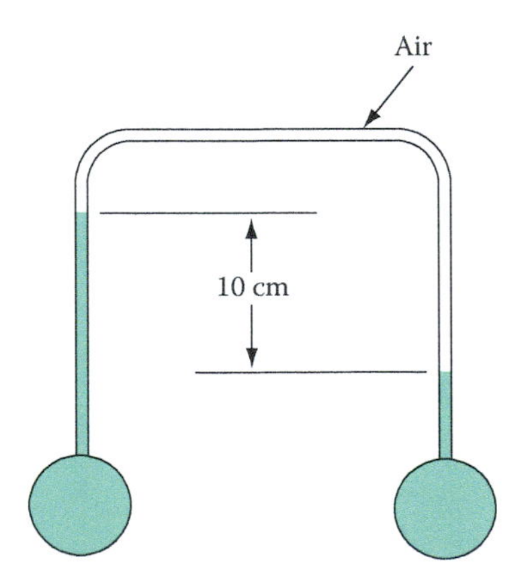

FIGURE P2.45

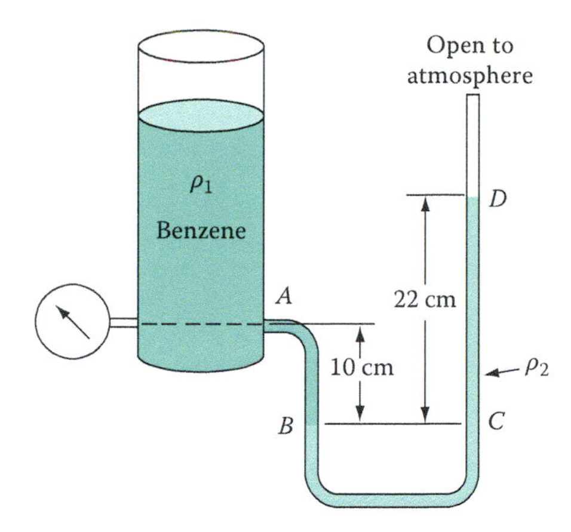

FIGURE P2.46

2.46 A manometer is used to measure pressure in a tank as a check against a simultaneous measurement with a gauge. The tank liquid is benzene; the manometer fluid is mercury. For the configuration shown in Figure P2.46, determine the gauge pressure at *A*.

Forces on Submerged Plane Surfaces

2.47 Write the equation for z_c and for z_r in terms of depth h for a rectangular gate that is completely in contact with liquid.

2.48 Write the equation for z_c and for z_r in terms of depth h for a circular gate that is completely in contact with liquid.

2.49 Write the equation for z_c and for z_r in terms of depth h for a triangular gate that is completely in contact with liquid.

2.50 Write the equation for z_c and for z_r in terms of depth h for a semicircular gate that is completely in contact with liquid.

2.51 Write the equation for z_c and for z_r in terms of depth h for an elliptical gate that is completely in contact with liquid.

2.52 Figure P2.52 shows a rectangular gate hinged at the bottom, and filled to a depth of $h = 0.3$ m with turpentine. The gate width into the page is 1 m. At the top of the gate is a stop, and the gate height H is 0.4 m. What is the magnitude of the force F_s at the stop?

2.53 Figure P2.53 is a sketch of an apparatus used in an experiment. The apparatus consists of one-fourth of a torus of rectangular cross section and made of clear plastic. The torus is attached to a lever arm, which is free to rotate (within limits) about a pivot point. The torus has inside and outside radii, R_i and R_o respectively, and it is constructed such that the center of these radii is at the pivot point of the lever arm. The torus is submerged in a liquid, and there will exist an unbalanced force that is exerted on the plane of dimensions $h \times w$. In order to bring the torus and lever arm back to their balanced position, a weight W must be added to the weight hanger. The force and its line of action can be found by summing moments about the pivot.

Following are data obtained on this apparatus, which shows the actual weight required to balance the lever arm while the torus is submerged. Substitute appropriately, and determine if the preceding equations accurately predict the weight W.

$L = 33.6$ cm $h = 10.2$ cm $w = 7.6$ cm
tank bottom to bottom of torus = 8.9 cm = d_2
pivot to tank bottom $d_1 = 31.8$ cm $W = 5.4$ N $z = 6.05$ cm

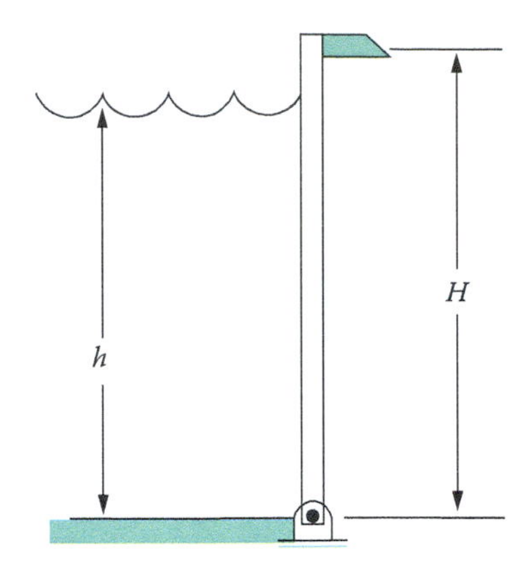

FIGURE P2.52

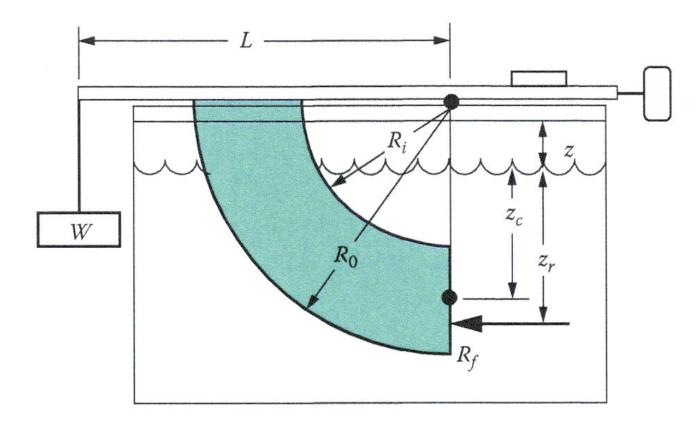

FIGURE P2.53

2.54 A hinged rectangular door 2 ft wide is free to rotate about point *O* but is held securely by a block at A. On the left side of the door is seawater filled to a depth of 3 ft. Find the restraining force at A. (See Figure P2.54.)

2.55 Figure P2.55 shows a rectangular retaining door holding water. Sketch the pressure prism and determine the resultant force on the door and its location.

2.56 A rectangular gate is used to separate carbon disulfide from liquid ether as shown in Figure P2.56. The gate is 0.4 m wide. Determine the height of ether that is required to balance the force exerted by the carbon disulfide such that the moment about *O* is zero.

2.57 A rectangular gate 1.5 ft wide is used as a partition to separate glycerine and water as shown in Figure P2.57. A stop is located on the floor of the water side of the gate. Calculate the force required to hold the door closed.

2.58 A rectangular gate is 9 ft high and 8 ft wide (into the page) and separates reservoirs of fresh water and oil. If the oil has a specific gravity of 0.85, determine the depth *h* required to make the reaction at B equal to 0. See Figure P2.58.

2.59 Figure P2.59 shows a rectangular gate holding water. Floating atop the water is a layer of oil. For the dimension shown,

a. Sketch the pressure prism for water only in contact with the door
b. Sketch an additional pressure distribution due to the oil
c. Determine the magnitudes of the resultant forces

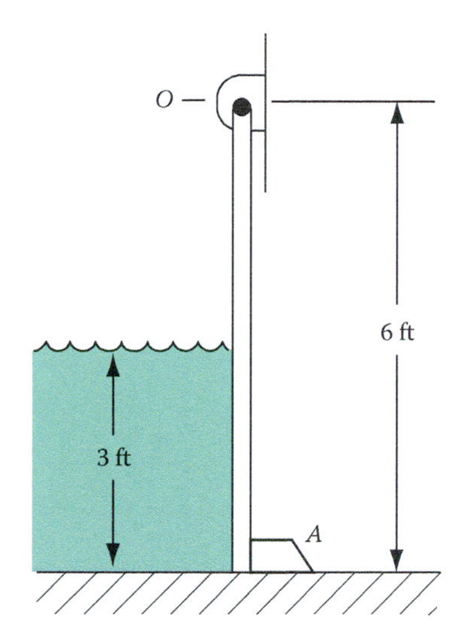

FIGURE P2.54

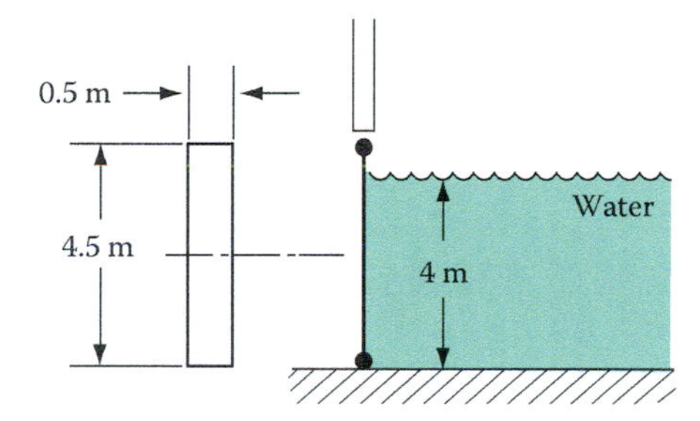

FIGURE P2.55

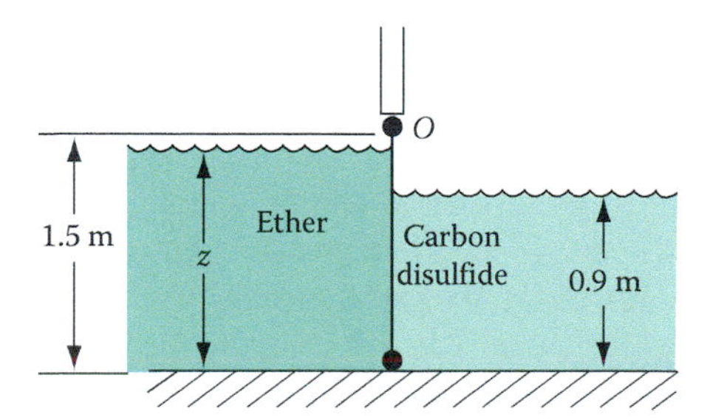

FIGURE P2.56

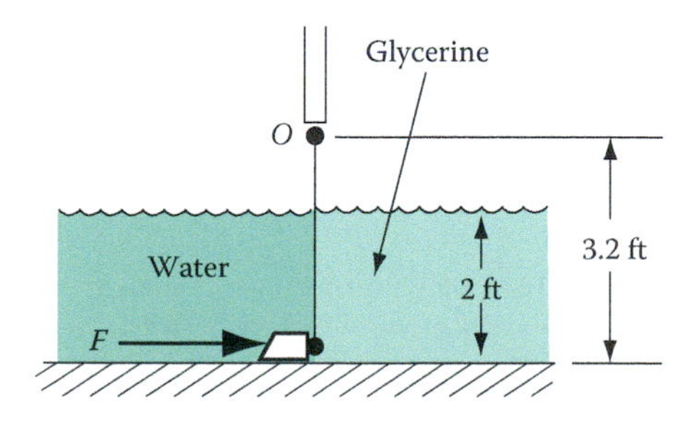

FIGURE P2.57

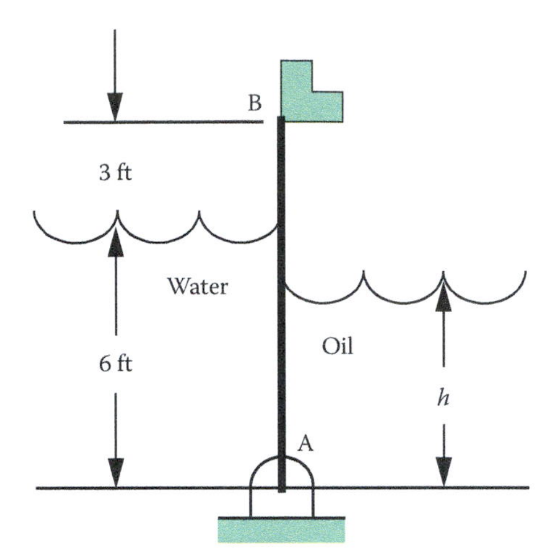

FIGURE P2.58

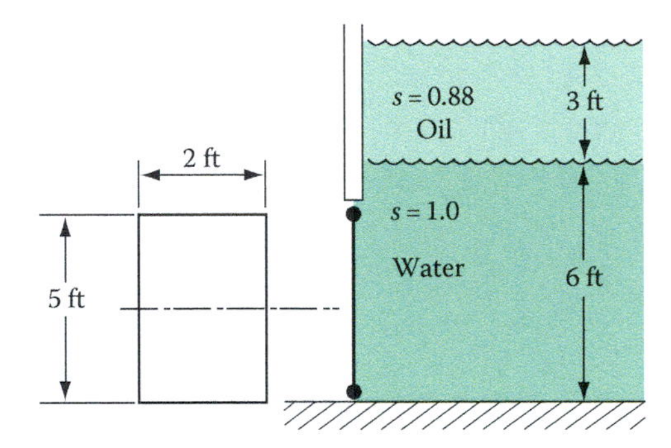

FIGURE P2.59

2.60 A rectangular gate is holding water that has a layer of kerosene over it. With both liquids in contact with the gate as shown in Figure P2.60:

a. Sketch the pressure prisms

b. Determine the magnitude of the resultant forces

Draw pressure profiles; separate profiles for each component

2.61 A rectangular gate 1.6 m wide is used as a partition as shown in Figure P2.61. One side has linseed oil filled to a depth of 10 m. The other has two fluids—castor oil over water. Determine the depth of castor oil required to keep the door stationary (no moment about point O).

2.62 Figure P2.62 shows a triangular retaining door holding acetone. Sketch the pressure prism and determine the resultant force on the door, and its location.

2.63 A porthole in the wall of a loaded ship is just below the surface of the water as shown in Figure P2.63. Sketch the pressure prism for the window, determine the magnitude of the resultant force, and find its location.

2.64 An aquarium is designed such that a viewing window is elliptical, as shown in Figure P2.64. Sketch the pressure prism for the window, determine the magnitude of the resultant force, and find its location.

2.65 A semicircular gate located in a wall is in contact with 3 ft of water, as shown in Figure P2.65. Sketch the pressure prism for the window, determine the magnitude of the resultant force, and find its location.

2.66 Figure P2.66 shows a 4 ft wide (into the page) gate that has an L-shaped cross section and is hinged at its corner. The gate is used to ensure that the water level does not get too high.

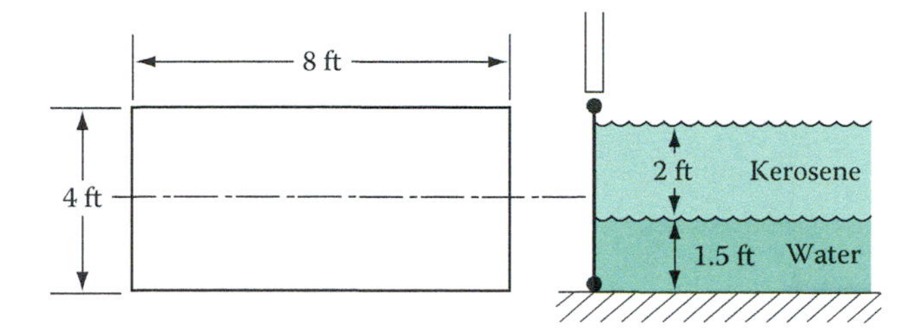

FIGURE P2.60

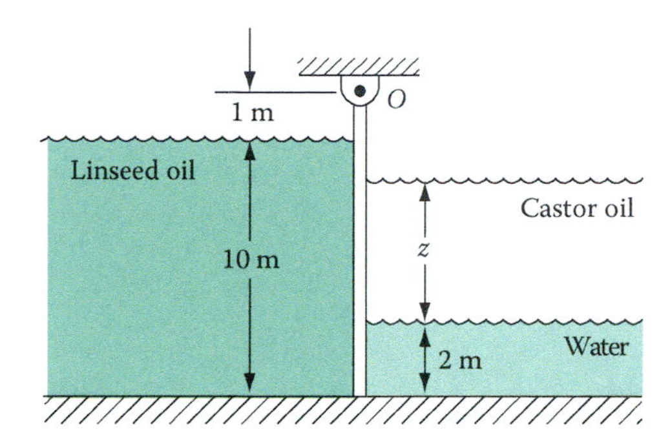

FIGURE P2.61

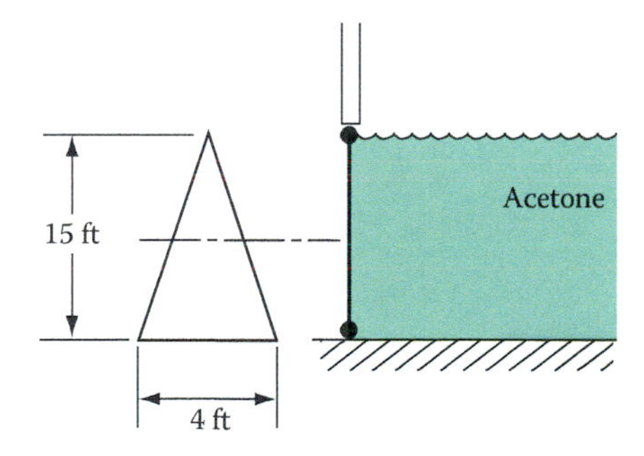

FIGURE P2.62

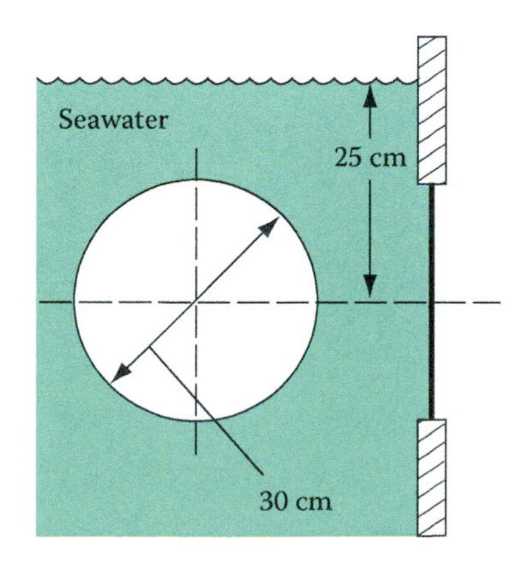

FIGURE P2.63

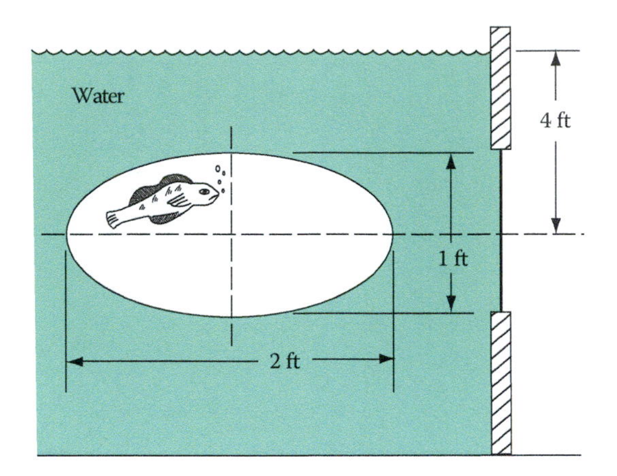

FIGURE P2.64

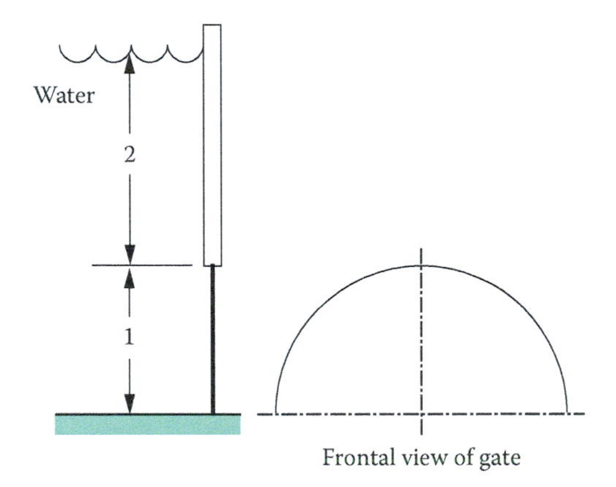

FIGURE P2.65

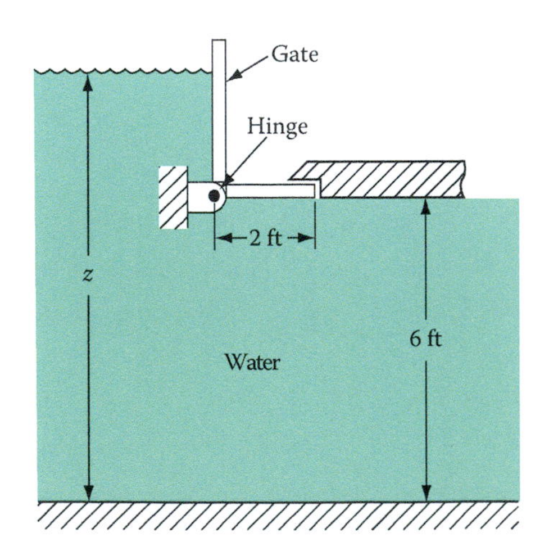

FIGURE P2.66

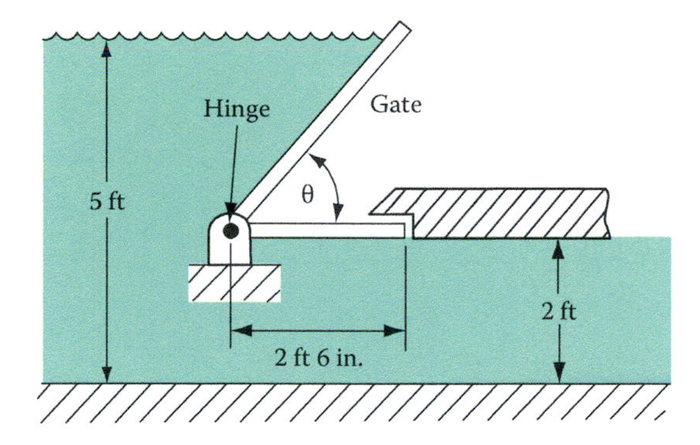

FIGURE P2.67

Once the level rises over a certain point, the fluid forces acting on the gate tend to open it and release some liquid. Determine the height z above which the gate tends to open.

2.67 Figure P2.67 shows a hinged gate used as a retainer for castor oil. The liquid depth to the horizontal portion of the gate is 2 ft, and the gate itself is to be designed so that the oil depth does not exceed 5 ft. When the depth is greater than 5 ft, the fluid forces act to open the gate, and some oil escapes through it. The gate is 1.2 ft wide (into the page). Determine the angle θ required for the gate to open when necessary.

2.68 A dam is constructed as in Figure P2.68. Determine the resultant force and its location acting on the inclined surface. Perform the calculations assuming a unit width into the page.

2.69 A tank having one inclined wall contains chloroform as shown in Figure P2.69. The inclined wall has a rectangular plug 20 cm wide × 40 cm high. Determine the force exerted on the plug if a pressure gauge at the tank bottom reads 17 kPa (gauge).

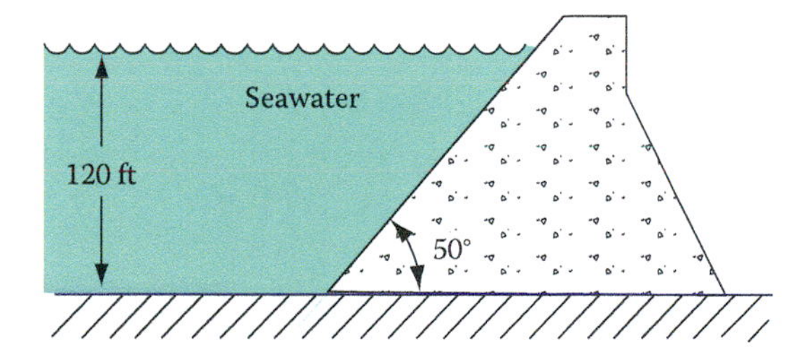

FIGURE P2.68

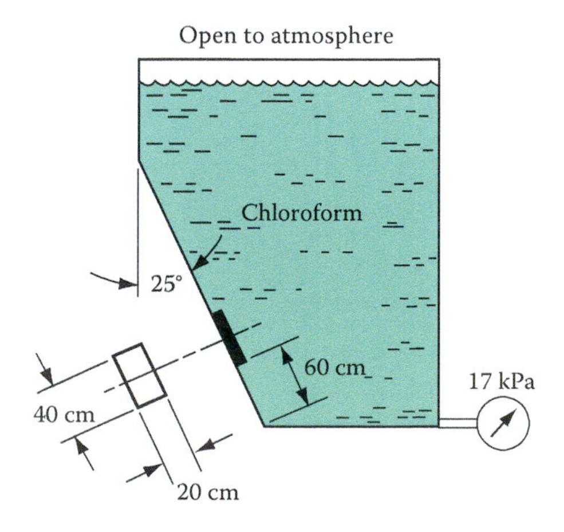

FIGURE P2.69

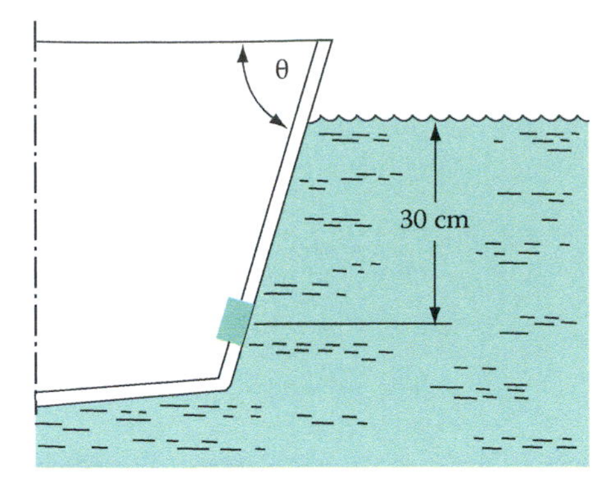

FIGURE P2.70

2.70 A small, narrow, flat-bottomed fishing boat is loaded down with two crew members and a catch of fish. The rear of the boat is a plane inclined at an angle θ of 75° with the horizontal. Thirty centimeters below the water surface is a hole in the boat plugged with a cork that is 8 cm in diameter. Determine the force and its location on the cork due to the saltwater if the liquid density is 1 025 kg/m³ (see Figure P2.70).

2.71 Figure P2.71 shows a tank that contains hexane. A circular gate in the slanted portion of one wall is 6 in. in diameter. Determine the resultant force acting on the gate and its location.

2.72 Figure P2.72 shows a rectangular gate hinged at the bottom, and filled to a depth of $h = 1$ m with castor oil. The gate width into the page is 0.5 m. At the top of the gate is a wire that goes over a pulley and is attached to a weight. What is the magnitude of the weight W required to hold the gate in position?
The angle θ is 45°, and the distance $H = 1.8$ m.

2.73 Figure P2.73 shows an inclined wall of a tank containing water. The tank wall contains a plug held in place by a weight. The liquid depth can be maintained constant in the tank, and the depth can be controlled by the amount of weight W that is used. For the conditions shown, determine the weight required to keep the plug in place.

2.74 A trough formed by two sides of wood is used to convey water. Every 0.8 m, a turnbuckle and wire are attached to support the sides. Calculate the tension in the wire using the data in Figure P2.74.

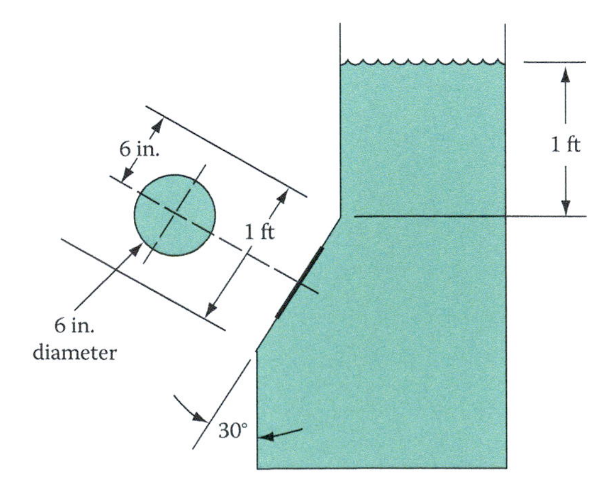

FIGURE P2.71

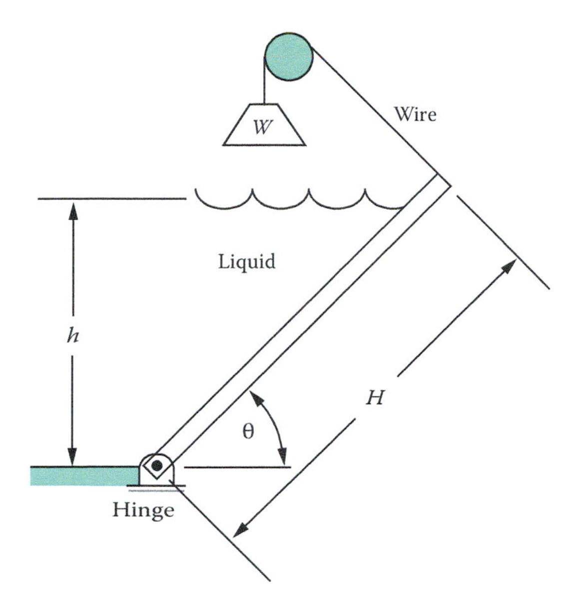

FIGURE P2.72

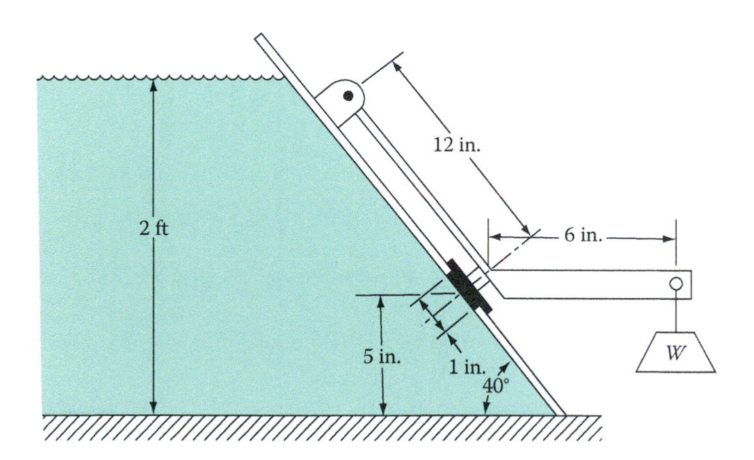

FIGURE P2.73

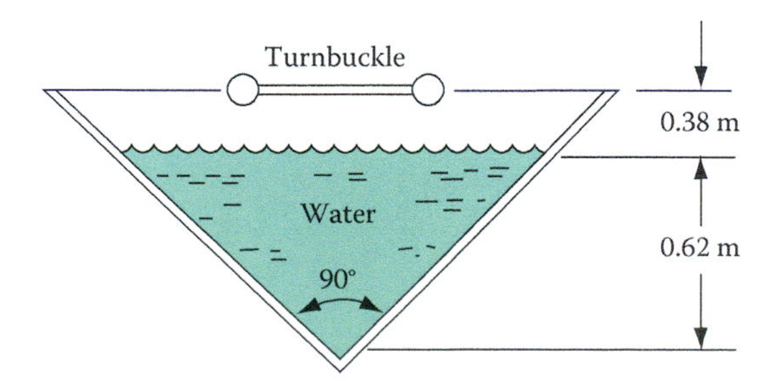

FIGURE P2.74

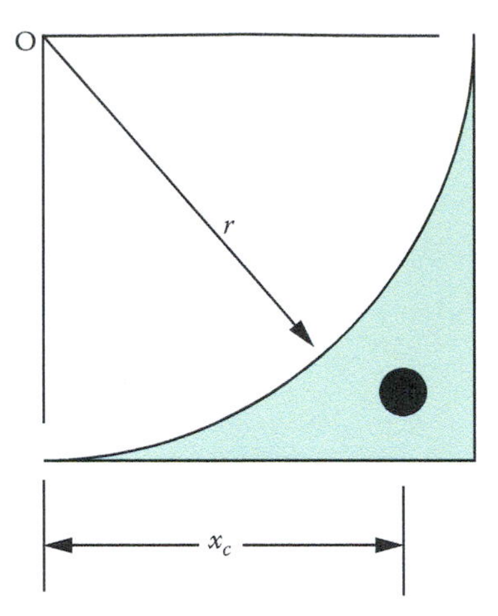

FIGURE P2.75

2.75 Shown in Figure P2.75 is an ex-quarter circle area, and it is desired to derive an expression for the distance to the centroid. The region consists of a square area with an inscribed quarter circle removed.

From calculus, the centroid of the region may be found with:

$$x_c = \frac{1}{A}\int_0^r \frac{1}{4}\left[f(x)\right]^2 dx$$

where

$$f(x) = (r^2 - x^2)^{1/2}$$

Substitute this function into the integral, perform the integration, and show that

$$x_c = \frac{2r}{3(4-\pi)}$$

Forces on Submerged Curved Surfaces

2.76 The reservoir of a tank truck is elliptical in cross section—4 ft high, 6 ft wide, and 7 ft long. Calculate the horizontal and vertical forces exerted on one of the lower quadrants of the tank when half-filled with gasoline (Figure P2.76). Take gasoline properties to be the same as those for octane.

2.77 Find the horizontal and vertical forces exerted on the portion labeled ABC of the elliptical tank truck of Figure P2.76 when it is half-filled with gasoline. Take gasoline properties to be the same as those for octane. The length of the tank is 7 ft.

2.78 The tank truck of Figure P2.76 has ends as sketched as in Figure P2.78. Determine the horizontal force acting on the end when the tank is filled with gasoline (assume the same properties as octane).

2.79 In many regions of the United States, home heating is effected by burning propane. A side view of a typical propane storage tank is sketched in Figure P2.79. Determine the forces exerted on quadrant DE for the case when the tank is filled to the top (point B). Tank length into the page is 2 m.

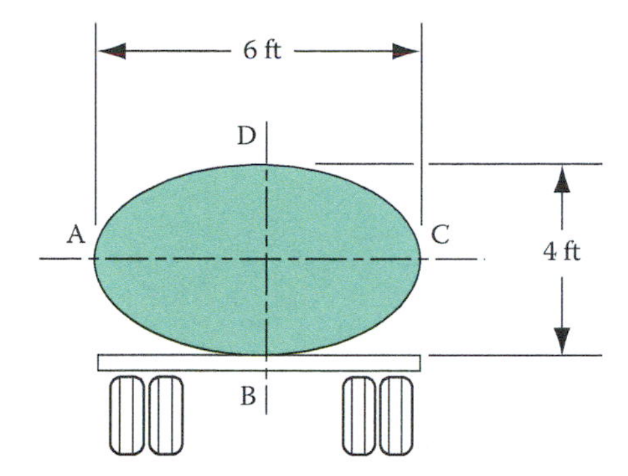

FIGURE P2.76

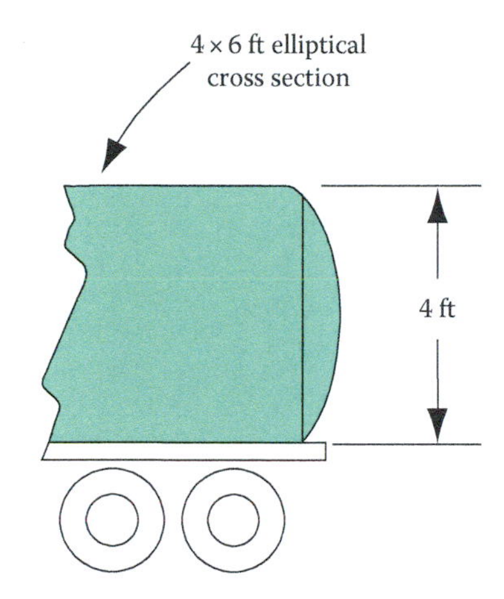

FIGURE P2.78

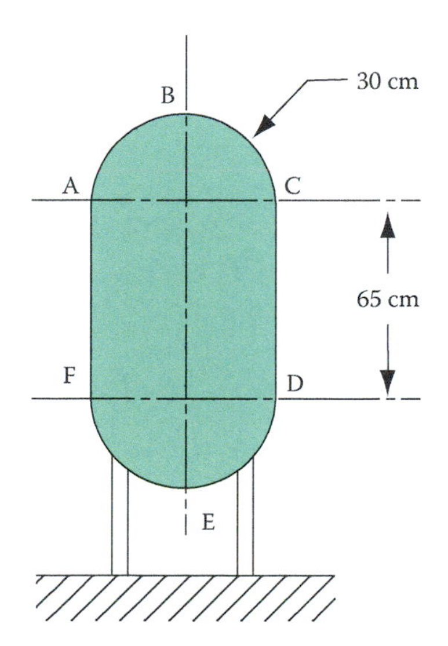

FIGURE P2.79

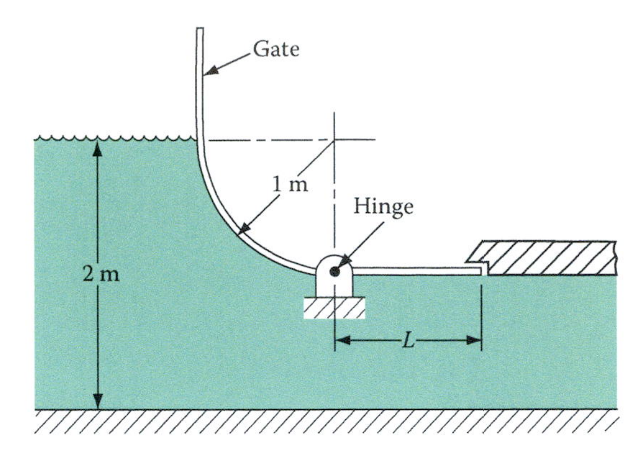

FIGURE P2.81

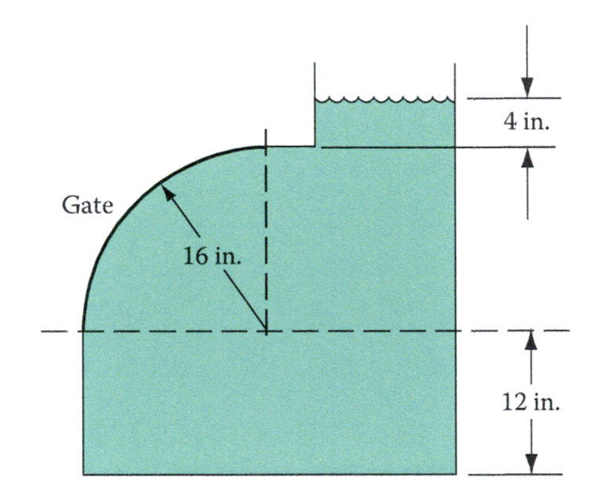

FIGURE P2.82

2.80 In many regions of the United States, home heating is effected by burning propane. A side view of a typical propane storage tank is sketched in Figure P2.79. Determine the forces exerted on the bottom portion labeled FED for the case when the tank is filled to the top (point B). Tank length into the page is 2 m.

2.81 Figure P2.81 shows a seawater-retaining gate with a curved portion. When the water level gets to a certain height, the fluid forces acting on the gate open it, and the seawater passes through. The gate is 0.7 m wide (into the page) and is to be designed so that the water depth does not exceed 2 m. Determine the length L of the straight portion of the gate required to do this.

2.82 Figure P2.82 shows the side view of a tank with a curved gate. The gate dimensions are 36 in. × 24 in. (into the page). The tank contains glycerine and is filled to a depth of 32 in. Determine the magnitude of the forces acting on the gate.

Equilibrium of Accelerating Liquids

2.83 At what acceleration must the car of Figure 2.25 be moving for the fluid to spill over the rear wall? The tank attached to the car is 6 in. long and 4 in. high. It is filled to a depth of 2 in. with linseed oil.

2.84 The car of Figure 2.25 is traveling at a constant velocity of 15 cm/s but goes around a curve of radius 56 cm. Viewing the car from the front, what is the equation of the surface? The tank attached to the car is 20 cm long and 10 cm high. It is filled to a depth of 5 cm with linseed oil.

2.85 The car of Figure 2.25 is decelerating uniformly at 12 g. Determine the free surface of the liquid. The car is 6 in. long and 4 in. high. It is filled to a depth of 2 in. with linseed oil.

2.86 A passenger in a car is holding a cup of coffee. The cup has an inside diameter of 7.8 cm and an internal height of 8 cm. The cup is filled with liquid to a height of 7 cm. The car accelerates uniformly from rest until it reaches 40 mph. What is the maximum acceleration rate that can be attained without spilling coffee over the top of the cup if the passenger holds the cup level?

2.87 At what rate of deceleration will the liquid level of Figure 2.26 just reach the top of the tank?

2.88 The liquid container of Figure 2.27 is rotating at 1.5 rev/s. Determine the shape of the liquid surface. The tank diameter is 1 ft; tank height is 18 in.; and the liquid depth when the tank is stationary is 6 in.

2.89 The liquid acceleration experiment depicted in Figure 2.25 is to be run with glycerin. The tank measures 6 in. long and 4 in. high. Glycerine is poured in until its height is 2 in. The weight is allowed to fall freely. Owing to wheel and pulley friction, the car accelerates at only $0.7g$. Determine the free surface of the glycerine.

Buoyancy

2.90 Derive an equation for the buoyant force exerted on a submerged cylinder with its axis vertical.

2.91 A 1.2 ft^3 block of aluminum is tied to a piece of cork, as shown in Figure P2.91. What volume of cork is required to keep the aluminum from sinking in castor oil?

2.92 What percentage of total volume of an ice cube will be submerged when the ice cube is floating in water?

2.93 A copper cylinder of diameter 4 cm and length 15 cm weighs only 14.5 N when submerged in liquid. Determine the liquid density.

2.94 A cylinder 8 cm in diameter is filled to a depth of 30 cm with liquid. A cylindrical piece of aluminum is submerged in the liquid, and the level rises 4 cm. The weight of the aluminum when submerged is 2.5 N. Determine the liquid density.

2.95 A submerged steel spherical buoy weighs 220 lbf out of water. A chain anchors the center of the buoy at a depth of 30 in. below the surface of the water (salt water of density 1.99 slug/ft^3). Calculate the tension in the chain (Figure P2.95).

2.96 A glass is filled to a depth d with linseed oil and allowed to float in a tank of water. To what depth can the glass be filled with linseed oil such that the tank water will just reach the brim of the glass? (See Figure P2.96.)

2.97 A cube of material is 4 in. on a side and floats at the interface between kerosene and water, as shown in Figure P2.97. Find the specific gravity of the material.

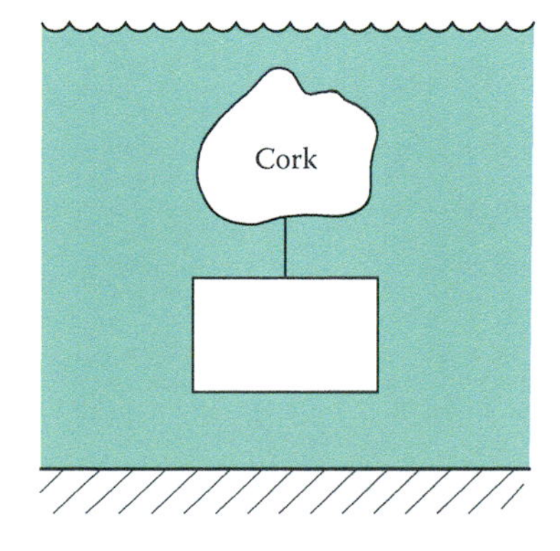

FIGURE P2.91

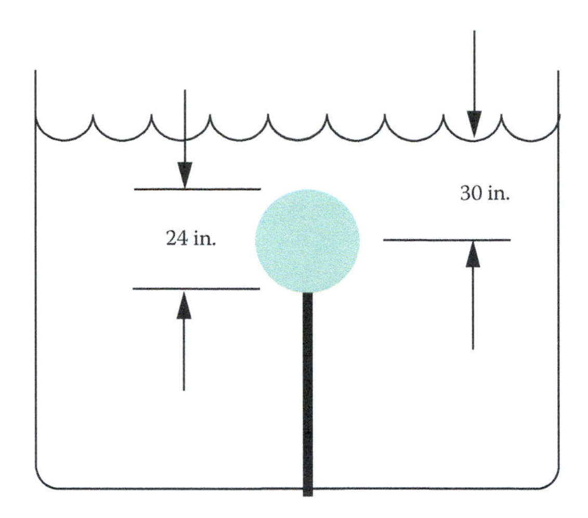

FIGURE P2.95

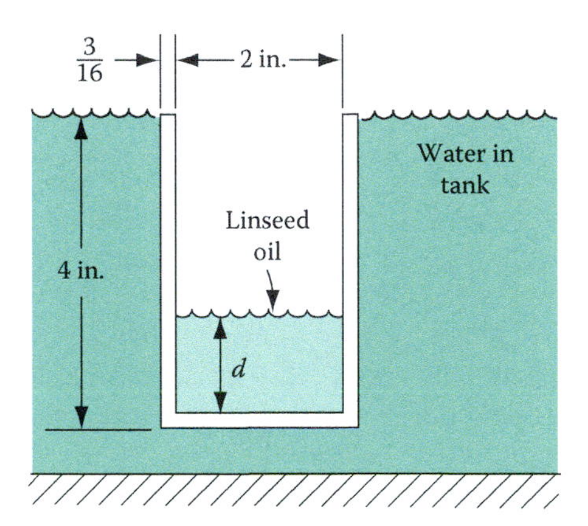

FIGURE P2.96

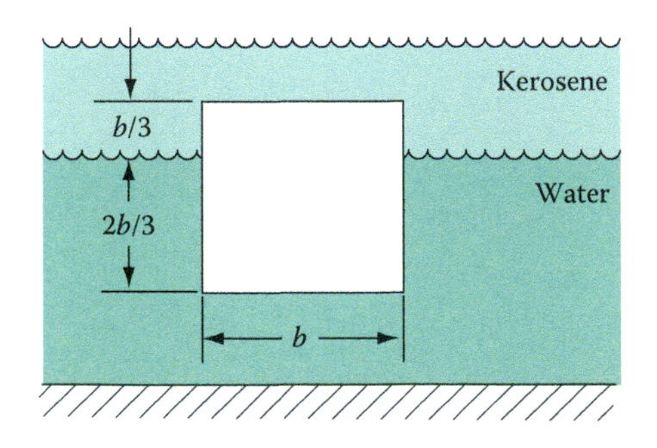

FIGURE P2.97

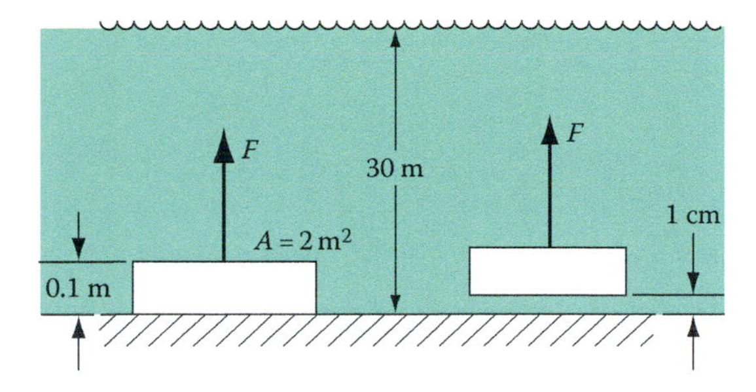

FIGURE P2.98

2.98 A block of tin of cross-sectional area 2 m^2 is resting on the flat bottom of a tank, as in Figure P2.98. No water can get beneath the block, which is only 0.1 m high. The water depth is 30 m.

a. Determine the force required to lift the block just off the bottom.

b. When the block has been raised just 1 cm and the water can get beneath it, determine the force required to lift the block further.

2.99 A flat barge carrying a load of dirt is sketched in Figure P2.99. The barge is 30 ft wide, 60 ft long, and 8 ft high. The barge is submerged 6.5 ft. Determine the weight of the dirt load if the weight of the barge is 400,000 lbf.

2.100 A bar of soap of dimensions 10 cm long, 5 cm wide, and 3 cm tall is floating in a basin of water with 8 mm extending above the surface. If the water density is 997 kg/m^3, determine the density of the soap (see Figure P2.100.).

2.101 Rework Example 2.19 for a balsa log, whose specific gravity is 0.125.

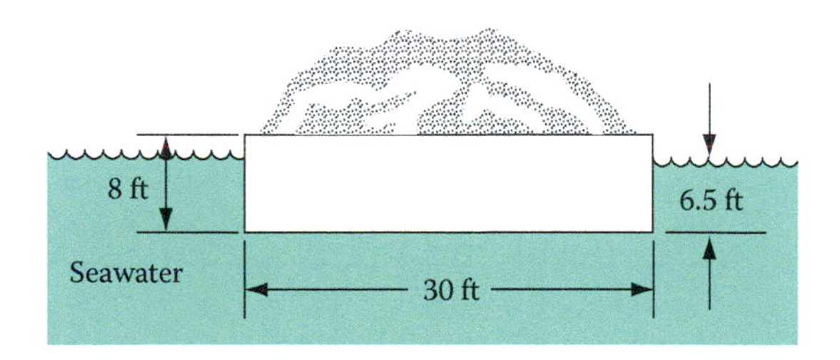

FIGURE P2.99

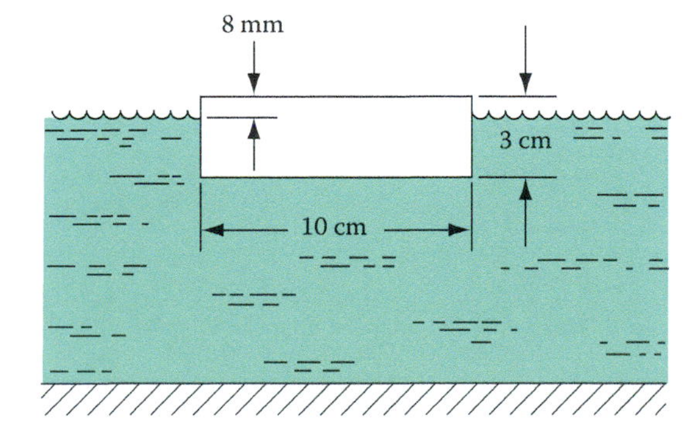

FIGURE P2.100

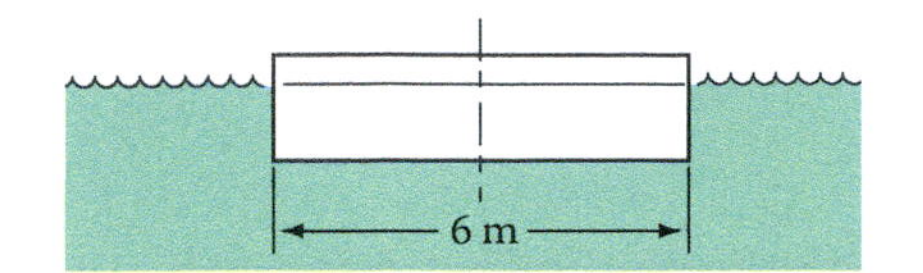

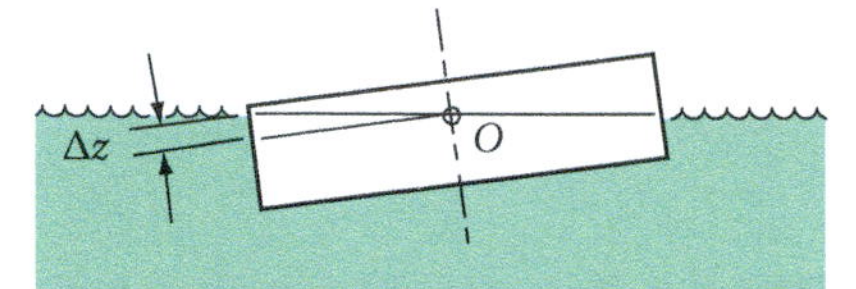

FIGURE P2.102

2.102 Figure P2.102 shows a prismatic body, such as a barge, floating in water. The body is 6 m wide and 18 m long, and it weighs 1.96 MN. Its center of gravity is located 30 cm above the water surface. Determine the metacentric height if Δz is 30 cm.

2.103 Figure P2.102 shows a prismatic body, such as a barge, floating in water. The body is 6 m wide and 18 m long, and it weighs 1.96 MN. Its center of gravity is located 30 cm above the water surface. Determine the restoring couple if Δz is 30 cm.

3 Basic Equations of Fluid Mechanics

The purposes of this chapter are to discuss the behavior of fluids while they are flowing, to present a unified mathematical approach to solving fluid flow problems, and to derive the basic equations of fluids mechanics. The continuity, momentum, and energy equations are derived for a general case. The Bernoulli equation is developed from the momentum and energy equations.

Before we begin a discussion of the equations, it is important to examine types of flows and how they can be characterized. **Closed-conduit flows** are completely enclosed by restraining solid surfaces; examples are flow through a pipe and flow between parallel plates where there is no free surface. **Open-channel flows** have one surface exposed to atmospheric pressure; examples are flow in a river and flow in a spillway. In **unbounded flows**, the fluid is not in contact with any solid surface; examples are the jet that issues from a household faucet and the jet from a can of spray paint.

3.1 KINEMATICS OF FLOW

In flow situations, solution of a problem often requires determination of a velocity. As we will see, however, velocity varies in the flow field. Moreover, the flow can be classified according to how the velocity varies. If the parameters of both the fluid and the flow are constant at any cross section normal to the flow, or if they are represented by average values over the cross section, the flow is said to be **one-dimensional**. Velocity distributions for one-dimensional flow are illustrated in Figure 3.1. Although flow velocity may change from point to point, it is constant at each location.

A flow is said to be **two-dimensional** if the fluid or the flow parameters have gradients in two directions. In Figure 3.2a—flow in a pipe—the velocity at any cross section is parabolic. The velocity is thus a function of the radial coordinate; a gradient exists. In addition, a pressure gradient exists in the axial direction that maintains the flow. That is, a pressure difference from inlet to outlet is imposed on the fluid that causes flow to occur. The flow is one-directional, but because we have both a velocity and a pressure gradient, it is two-dimensional. Another example of two-dimensional flow is given in Figure 3.2b. At the constant-area sections, the velocity is a function of one variable. At the convergent section, velocity is a function of two space variables. In addition, a pressure gradient exists that maintains the flow. Figure 3.2c gives another example of a one-directional, two-dimensional flow; gradients exist in two-dimensions.

It is possible to assume one-dimensional flow in many cases in which the flow is two-dimensional to simplify the calculations required to obtain a solution. An average constant velocity, for example, could be used in place of a parabolic profile, although the parabolic profile gives a better description of the flow of real fluids because velocity at a boundary must be zero relative to the surface (except in certain special cases such as rarefied gas flows).

A flow is said to be **three-dimensional** when the fluid velocity or flow parameters vary with respect to all three space variables. Gradients thus exist in three directions.

Flow is said to be **steady** when conditions do not vary with time or when fluctuating variations are small with respect to mean flow values, and the mean flow values do not vary with time. A constant flow of water in a pipe is steady because the mean liquid conditions (such as velocity and pressure) at one location do not change with time. If flow is not steady, it is called **unsteady**: mean flow conditions do change with time. An example would be the flow of water through a pipe during

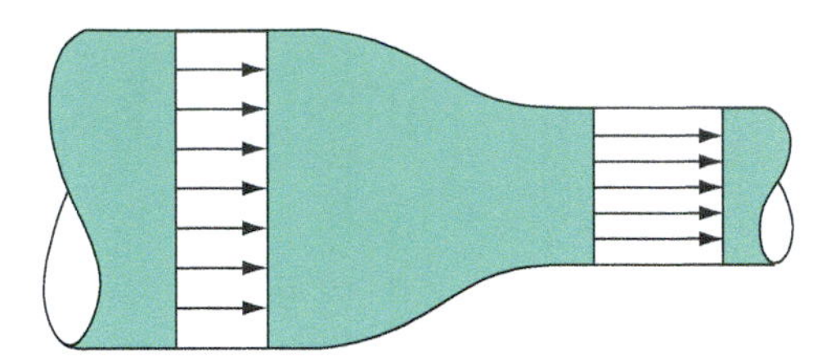

FIGURE 3.1 One-dimensional flow where velocity and pressure are uniform at any cross section.

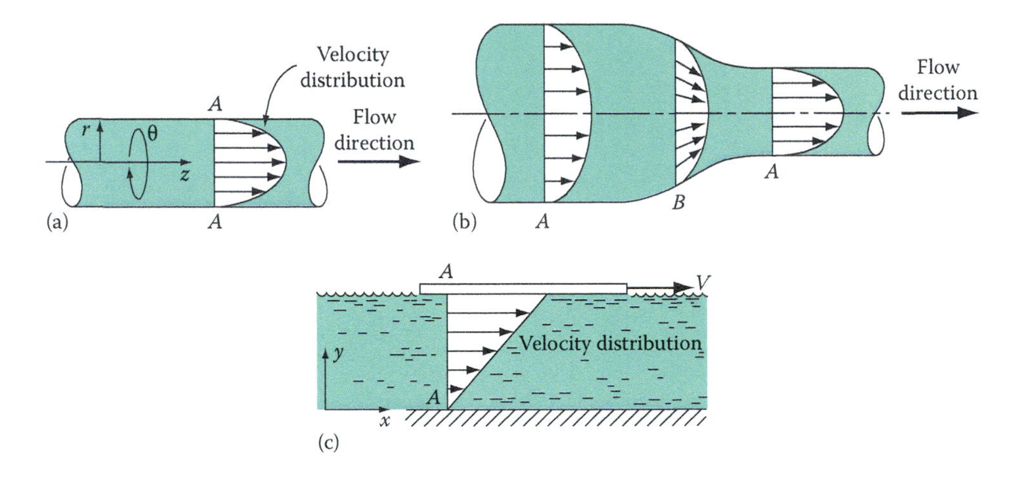

FIGURE 3.2 Two-dimensional flow.

the closing or opening of a valve. Emptying a sink creates an unsteady flow. In some unsteady flows, it may be permissible or even necessary to assume that steady flow exists to obtain a solution. The flow is then assumed to be **quasi-steady**.

Consider a 1/4 in. diameter tube used to siphon water from a tank 4 ft in diameter. The flow is unsteady; as the water level in the tank decreases, the velocity of the water in the tube decreases. Because the change in tank level is comparatively slow, however, the time dependence is not strong. Consequently, equations developed for steady flow could be applied to this unsteady problem at selected instants of time. An inaccuracy is involved in this procedure, however, and the question ultimately reduces to whether or not the magnitude of the error is acceptable. The final test of this approximation (or any mathematical model) is how well it describes the physical phenomena in question.

Velocity at a point has magnitude and direction. Velocity at a point is thus a vector quantity. It is often helpful to sketch the direction of velocity at each point in the flow field by using **streamlines**. Streamlines are tangent to the velocity vector throughout the flow field. The magnitude of the velocity at every point in the cross section is not specified—only direction. Furthermore, there is no flow across a streamline. Figure 3.3a shows a single streamline, and Figure 3.3b illustrates streamlines of flow in a diverging duct.

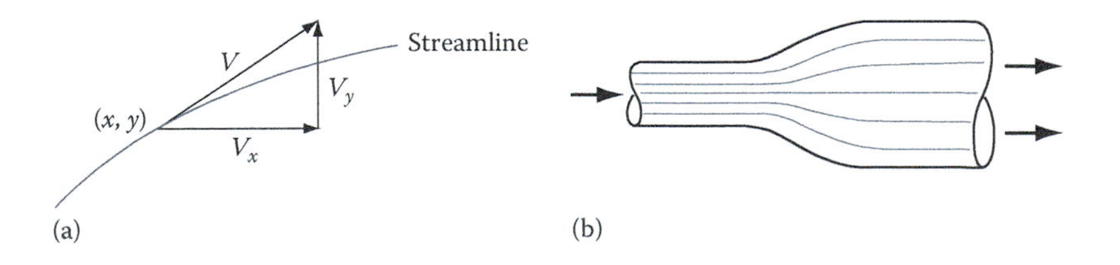

FIGURE 3.3 (a) Velocity vector at a point on a streamline and (b) streamlines in a diverging duct.

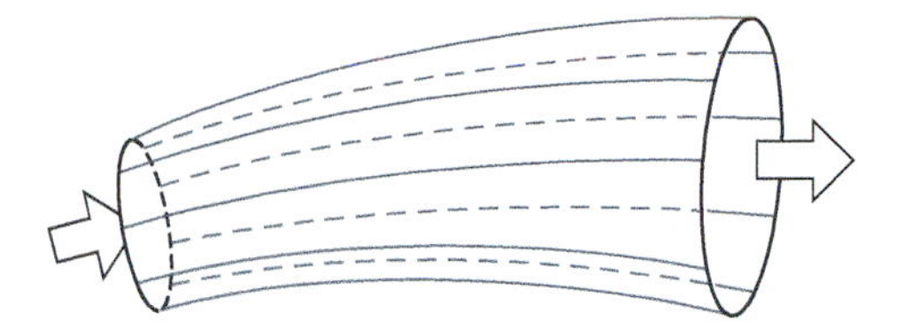

FIGURE 3.4 A streamtube formed by a bundle of streamlines.

A passageway for the fluid made up of a "bundle" of streamlines is called a **streamtube** (Figure 3.4). It is referred to as a passageway because its wall or boundary is made up of streamlines across which there is no flow. A pipe is an example of a streamtube.

Streaklines are yet another aid in visualizing flow direction. A **streakline** is defined as the locus of all fluid particles that have passed through a point. If we injected dye into the flow of a fluid at a single point, the dye would follow the path of the fluid particles passing through the point. A photograph of the flow pattern would then show a streak of dye: a streakline. If the flow varies with time, successive photographs show different streaklines. For steady flow, however, the flow does not vary with time; streaklines and streamlines are coincident. With these definitions and concepts of flow behavior, we can now proceed with the development of the equations of fluid mechanics.

3.2 CONTROL VOLUME APPROACH

As we saw in Section 3.1, there are many types of flow. We now address the question of how to determine the velocity in the flow field. There are two possible approaches: *the Lagrangian approach and the Eulerian approach.*

The **Lagrangian approach** is used in solid mechanics and involves describing the motion of a particle by position as a function of time. This approach might be used in describing the motion of an object falling under the action of gravity: $s = (1/2)gt^2$. At any time, distance from the body's original position is known. This approach is difficult to use in fluid mechanics, however, because a fluid is a continuous medium—a single fluid volume changes shape, and different particles within the fluid volume are traveling at different velocities. In other words, because of the nature of fluids, the Lagrangian approach is generally not a desirable method for analysis. There are some problems in fluid mechanics, however, where this approach can be used.

The **Eulerian** (or control volume) **approach** is preferred in fluid mechanics. In this method, we choose a region in the flow field for study. As an example, consider flow draining from a sink, as illustrated in Figure 3.5. A **control volume** is chosen about the region of study and is bounded by the dashed line called the **control surface**. Everything outside is called the **surroundings**. We can choose the control volume or shape of the control surface as necessary for convenience in solving the problem. In general, the shape of the control volume is selected such that fluid and flow properties can be evaluated at locations where mass crosses the control surface—or, if no mass enters or exits, where energy crosses the control surface. Furthermore, the control surface can move or change shape with time (as in Figure 3.5). The control volume is to fluids as the free-body diagram is to solids. The Eulerian approach lends itself nicely to the solution of fluid mechanics problems.

Our objective is to develop equations of fluid dynamics that, in effect, are conservation equations: the continuity equation (conservation of mass), the momentum equation (conservation of linear momentum), and the energy equation (conservation of energy). Each will be developed from a general conservation equation.

Let us define N as a flow quantity (mass, momentum, or energy) associated with a fluid volume or system of particles. Let n represent the flow quantity per unit mass. Thus,

$$N = \iiint n\rho \, d\forall \tag{3.1}$$

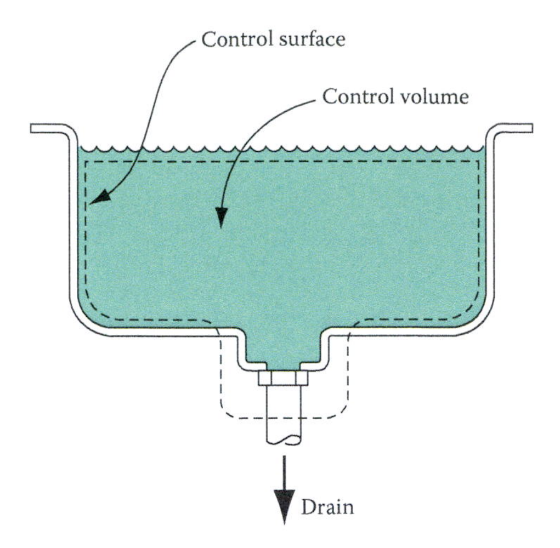

FIGURE 3.5 A control volume about the liquid in a sink.

Consider a system of particles at two different times: $\mathcal{V}_1$ at t_1 and $\mathcal{V}_2$ at t_2. Referring to Figure 3.6, we see that $\mathcal{V}_1$ is bounded by the solid line and $\mathcal{V}_2$ is bounded by the double line. Note that $\mathcal{V}_1$ consists of $\mathcal{V}_A$ and $\mathcal{V}_B$; $\mathcal{V}_2$ consists of $\mathcal{V}_B$ and $\mathcal{V}_C$. The control volume is bounded by the dashed line. The amount of flow quantity N contained in $\mathcal{V}_1$ is the amount in $\mathcal{V}_A$ and the amount in $\mathcal{V}_B$ at t_1 which is $N_{A_1} + N_{B_1}$. The amount of flow quantity contained in $\mathcal{V}_2$ is the amount in $\mathcal{V}_B$ and $\mathcal{V}_C$ at t_2, which is $N_{B_2} + N_{C_2}$. During the time interval, the change in N is, therefore,

$$\Delta N = \left(N_{B_2} + N_{C_2}\right) - \left(N_{A_1} + N_{B_1}\right)$$
$$= N_{B_2} - N_{B_1} + N_{C_2} - N_{A_1}$$

On a per-unit-time basis with $t_2 - t_1 = \Delta t$, we have

$$\frac{\Delta N}{\Delta t} = \frac{N_{B_2} - N_{B_1}}{\Delta t} + \frac{N_{C_2} - N_{A_1}}{\Delta t} \tag{3.2}$$

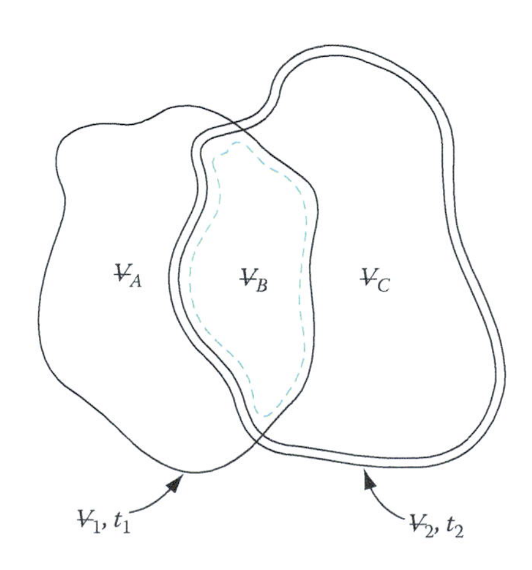

FIGURE 3.6 A fluid volume at two different times.

We now take the limit as Δt approaches zero to obtain an instantaneous time rate of change of flow quantity N:

$$\lim_{\Delta t\to 0}\frac{\Delta N}{\Delta t}=\lim_{\Delta t\to 0}\frac{N_{B_2}-N_{B_1}}{\Delta t}+\lim_{\Delta t\to 0}\frac{N_{C_2}-N_{A_1}}{\Delta t} \tag{3.3}$$

where

$$\lim_{\Delta t\to 0}\frac{\Delta N}{\Delta t}=\left.\frac{dN}{dt}\right|_{\text{system}}=\begin{cases}\text{instantaneous rate of}\\ \text{change of flow}\\ \text{quantity } N\end{cases}$$

$$\lim_{\Delta t\to 0}\frac{N_{B_2}-N_{B_1}}{\Delta t}=\left.\frac{\partial N}{\partial t}\right|_{\text{control volume}}=\begin{cases}\text{rate of accumulation (or}\\ \text{storage) of flow quantity}\\ N \text{ in control volume}\end{cases}$$

$$\lim_{\Delta t\to 0}\frac{N_{C_2}-N_{A_1}}{\Delta t}=\lim_{\Delta t\to 0}\frac{\partial N}{\Delta t}=\begin{cases}\text{net rate out (out minus in)}\\ \text{of flow quantity } N\\ \text{from control volume}\end{cases}$$

To obtain a specific limiting expression for this last term, consider a differential area dA through which fluid particles flow (Figure 3.7). The fluid velocity at dA is $\tilde{V}$, which has components normal and tangential to dA: V_n and V_t. The tangential velocity carries no fluid out of the control volume with it; all fluid leaving dA is in the V_n direction. During the time interval Δt, the mass of fluid crossing dA is

$$dm=\rho d\text{V}\!\!\!-$$

The differential volume of fluid $d\text{V}\!\!\!-$ is the product of cross-sectional area and height

$$d\text{V}\!\!\!-=(V_n\Delta t)dA$$

so that

$$dm=\rho(V_n\Delta t)dA$$

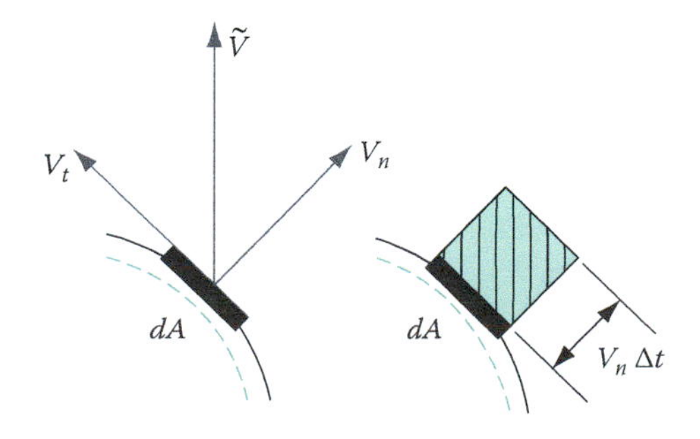

FIGURE 3.7 Flow through a differential area dA.

The amount of the quantity N being taken across the area dA is

$$\delta N = n\,dm$$

or

$$\delta N = n\rho V_n \Delta t\,dA.$$

Dividing by Δt gives

$$\frac{\delta N}{\Delta t} = n\rho V_n\,dA$$

and in the limit

$$\lim_{\Delta t \to 0} \frac{\delta N}{\Delta t} = \iint\limits_{\substack{\text{Control}\\ \text{Surface}}} n\rho V_n\,dA$$

By substitution into Equation 3.3, we have the **general conservation equation**:

$$\underbrace{\left.\frac{dN}{dt}\right|_{\text{S}}}_{\substack{\text{total}\\ \text{particles}}} = \underbrace{\left.\frac{\partial N}{\partial t}\right|_{\text{CV}}}_{\substack{\text{amount}\\ \text{stored}}} + \underbrace{\iint\limits_{\text{CS}} n\rho V_n\,dA}_{\substack{\text{net rate out}\\ \text{(out minus in)}}} \tag{3.4}$$

where
- S is the system
- CV is the control volume
- CS is the control surface

Equation 3.4 gives a relationship between the various quantities associated with a system of particles. In words, Equation 3.4 says

$$\begin{pmatrix}\text{instantaneous time}\\ \text{rate of change in } N\\ \text{for a system of particles}\end{pmatrix} = \begin{pmatrix}\text{instantaneous time}\\ \text{rate of accumulation of } N\\ \text{within the control volume}\end{pmatrix} + \begin{pmatrix}\text{amount of } N \text{ leaving}\\ \text{the control volume minus}\\ \text{the amount of } N \text{ entering}\end{pmatrix}$$

3.3 CONTINUITY EQUATION

The continuity equation is a statement of conservation of mass. The flow quantity N becomes m, and $n = m/m = 1$. Equation 3.4 becomes

$$\left.\frac{dm}{dt}\right|_{\text{S}} = \left.\frac{\partial m}{\partial t}\right|_{\text{CV}} + \iint\limits_{\text{CS}} \rho V_n\,dA$$

For a system of particles, mass is constant, because it can be neither created nor destroyed. Thus,

$$\left.\frac{\partial m}{\partial t}\right|_{CV} + \iint_{CS} \rho V_n \, dA = 0 \tag{3.5}$$

In other words, the mass entering the control volume equals mass stored plus mass leaving. For steady flow with no storage, the mass entering equals mass leaving. For steady flow, Equation 3.5 becomes

$$\iint_{CS} \rho V_n \, dA = 0 \tag{3.6}$$

For incompressible fluids (liquids), ρ is a constant, and we obtain the following for steady incompressible flow:

$$\iint_{CS} V_n \, dA = 0 \tag{3.7}$$

This equation states that the volume of fluid entering the control volume per unit time equals the amount leaving. Equation 3.7 also states that a velocity distribution normal to the control surface must be integrated over the cross-sectional area at exits and inlets to the control volume.

For many flow situations, however, a velocity distribution may not be known or derivable; in such cases, it is more convenient to use an average velocity at a cross section that is independent of area. (This assumes that we have one-dimensional flow or at least flow that can be treated as one-dimensional.) Thus, by definition,

$$\iint V_n \, dA = V \iint dA = VA$$

or

$$V = \frac{\iint V_n \, dA}{A} \tag{3.8}$$

in which V is the average velocity. The product $AV = Q$ is known as the volume flow rate with dimensions of L^3/T (m^3/s or ft^3/s). Also used in fluid mechanics is the term $\dot{m}$, defined as the mass **flow rate**, where $\dot{m} = \rho Q = \rho AV$. The mass flow rate has dimensions of M/T (kg/s or slug/s).

Example 3.1

Consider flow in a pipeline, as shown in Figure 3.8. Under certain flow conditions, the velocity distribution is given by:

$$V_z = 5\left(1 - \frac{r^2}{R^2}\right) \text{ ft/s}$$

where $2R$ is the pipe inside diameter. Determine the volume flow rate and the average velocity in terms of the radius. If the fluid is water, determine also the mass flow rate. Take the inside radius to be 3 in.

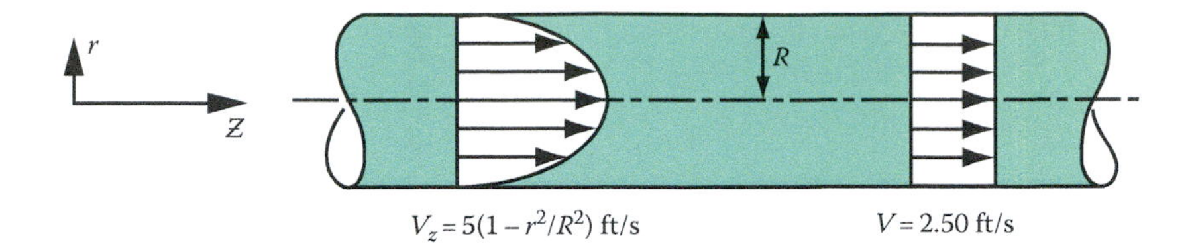

FIGURE 3.8 Flow in a pipeline.

Solution

The volume flow rate is the numerator of Equation 3.8:

$$Q = \iint V_n dA$$

in which we integrate from the centerline of the pipe to the inside radius (0 to R) and over the entire cross section (0 to 2π). The area dA is $r\,dr\,d\theta$. The velocity normal to the flow direction is V_z. Substituting,

$$Q = \int_0^{2\pi}\int_0^R 5\left(1 - \frac{r^2}{R^2}\right) r\,dr\,d\theta = 2\pi(5)\int_0^R \left(\frac{r - r^3}{R^2}\right) dr$$

$$Q = 10\pi\left(\frac{r^2}{2} - \frac{r^4}{4R^2}\right)\Bigg|_0^R = \frac{10\pi}{4}R^2 = \frac{10\pi}{4}\left(\frac{3}{12}\right)^2$$

or

$$Q = 0.491\ \text{ft}^3/\text{s}$$

The average velocity is the volume flow rate divided by the area:

$$V = \frac{Q}{A} = \frac{2.5\pi R^2}{\pi R^2}$$

$$V = 2.5\,\text{ft/s}$$

This result is shown in Figure 3.8. If the fluid is water, the mass flow rate is given by

$$\dot{m} = \rho Q = \rho A V = (1.94\ \text{slug/ft}^3)(0.491\ \text{ft}^3/\text{s})$$

$$\dot{m} = 0.952\ \text{slug/s}$$

Conceptually, we can replace the velocity distribution with an average velocity such that the volume flow rate remains unchanged (see Figure 3.8). In various practical problems, volume flow rate and area are known, but velocity distribution is not. In many cases, an average velocity is all that can be determined.

By using average velocity, the continuity equation becomes

$$\sum_{\text{in}} \rho A V = \sum_{\text{out}} \rho A V + \frac{\partial m}{\partial t}\bigg|_{\text{CV}} \tag{3.9}$$

For steady flow with no storage of mass,

$$\sum_{\text{in}} \rho AV = \sum_{\text{out}} \rho AV \tag{3.10}$$

Example 3.2

A pipeline with a 20 cm inside diameter is carrying liquid at a flow rate of 0.025 m^3/s. A reducer is placed in the line, and the outlet diameter is 15 cm. Determine the velocity at the beginning and end of the reducer.

Solution

Select the control volume as shown in Figure 3.9. Flow crosses the control surface at sections 1 and 2, where the streamlines are all parallel and flow properties are all known or can be determined. There is no mass stored. Applying Equation 3.6, we get

$$\iint_{\text{CS}} \rho V_n \, dA = 0 \quad (\text{mass flow out} - \text{mass flow in}) = 0$$

or

$$\iint_{2} \rho V_n \, dA - \iint_{1} \rho V_n \, dA = 0$$

With a liquid flowing, $\rho_1 = \rho_2$. Using the concept of average velocity, we have

$A_2V_2 = A_1V_1 = Q$ (a constant).

Now Q is given as 0.025 m^3/s. So at section 1,

$$V_1 = \frac{Q}{A_1} = \frac{0.025\,\text{m}^3/\text{s}}{(\pi/4)(0.20)^2\text{m}^2} = 0.31\text{ m/s}$$

At section 2,

$$V_2 = \frac{Q}{A_2} = \frac{0.025}{(\pi/4)(0.15)^2} = 1.4\text{ m/s}$$

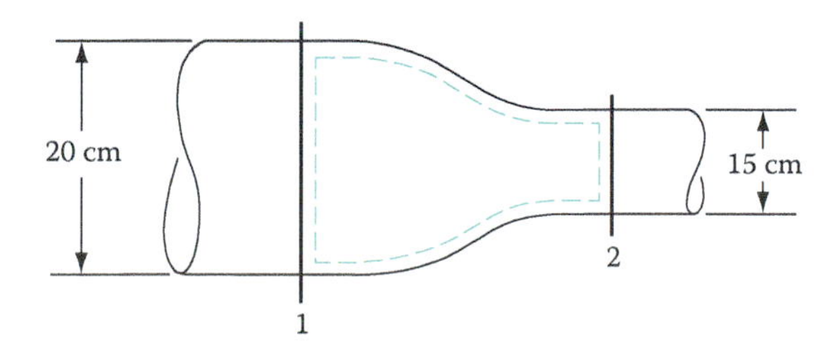

FIGURE 3.9 Sketch for Example 3.1.

Example 3.3

A "slop sink" typically may be found in a utility room. A sketch of such a sink is provided in Figure 3.10. The sink is rectangular in cross section, 40 cm × 48 cm, and contains water and liquid soap ($s = 0.98$) at a depth of 45 cm. A drain pipe at the bottom of the tank has an inside diameter of 3.8 cm, and the water outlet velocity there is a constant 1 m/s. Determine the time it takes to drain this tank from a depth of 40–15 cm.

Solution

We choose the control volume to be the volume of liquid in the tank at any time as indicated in the figure. The flow is unsteady, and the unsteady continuity equation applies:

$$\left.\frac{\partial m}{\partial t}\right|_{CV} + \iint_{CS} \rho V_n dA = 0$$

We now evaluate each term in the equation. The volume of fluid in the tank at any time is given by

$$\forall = \text{Area} \times \text{depth} = (0.40)(0.48)z = 0.192z \text{ m}^3$$

The density of the liquid is 980 kg/m³. The mass of liquid in the tank then is

$$m = \rho\forall = (980)(0.192z) = 188.2z \text{ kg}$$

The rate of change of mass in the tank then becomes

$$\frac{\partial m}{\partial t} = 188.2\frac{dz}{dt}$$

where the exact differential notation is used because depth varies only with time t. The second term in the continuity equation is evaluated for mass leaving and mass entering the control volume:

$$\iint_{CS} \rho V_n\, dA = \iint_{out} \rho V_n\, dA - \iint_{in} \rho V_n\, dA$$

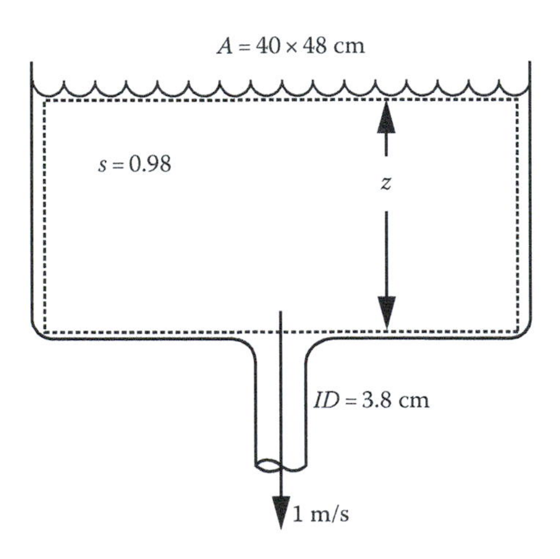

FIGURE 3.10 A utility room sink.

In this case, density is constant, and average velocity is given. With no mass entering, we have

$$\iint_{CS} \rho V_n dA = (\rho AV)_{out} - 0 = 980\frac{\pi(0.038)^2}{4}(1)$$

or

$$\iint_{out} \rho V_n dA = 1.11 \text{ kg/s}$$

Substituting into the continuity equation yields

$$188.2\frac{dz}{dt} + 1.11 = 0$$

Rearranging and separating variables,

$$dz = -5.91 \times 10^{-3}\, dt$$

$$\int_{0.4}^{0.15} dz = -5.91 \times 10^{-3} \int_0^t dt$$

$$0.15 - 0.4 = -5.91 \times 10^{-3} t$$

Solving,

$$t = 42 \text{ s}$$

This is a reasonable result for the conditions given. The exit velocity in a tank such as this varies with the depth of the liquid when gravity is the driving force. The reader may verify that density cancels from both terms in the continuity equation, and so the result is independent of fluid properties.

Example 3.4

In most metalworking shops, plate steel is cut by means of an oxyacetylene torch. Oxygen for cutting is supplied via tanks 30 cm in diameter × 1.3 m tall. These tanks are charged to an internal pressure of 14 MPa gauge. A valve 12.5 mm in diameter is located at the top of the tank. If the tank valve is opened fully, oxygen escapes at 2 m/s. Assuming that this exit velocity is constant, determine the tank pressure after 60 s. Take the temperature in the tank to be unchanging and equal to 25°C.

Solution

Select the control volume to include the entire tank contents and to cross just upstream of the valve perpendicular to the flow direction, as in Figure 3.11. Thus, flow leaves the control volume at the location where flow properties are known or can be determined.

The continuity equation is

$$\left.\frac{\partial m}{\partial t}\right|_{CV} + \iint_{CS} \rho V_n\, dA = 0$$

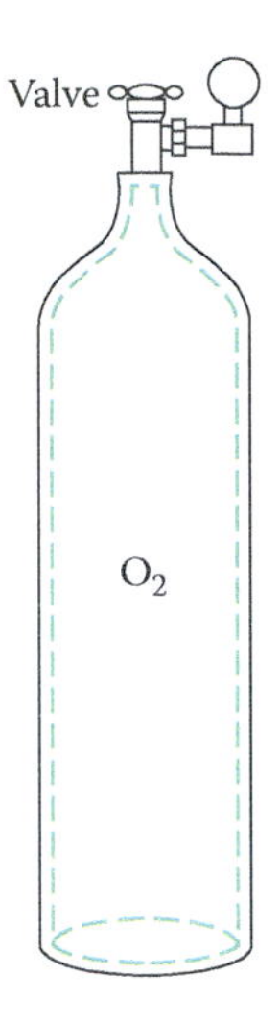

FIGURE 3.11 Sketch for Example 3.4.

To evaluate the first term, we find the mass in the tank at any time to be $m = \rho \forall$. Assuming an ideal gas, $\rho = p/RT$. For a constant-temperature process,

$$\left.\frac{\partial m}{\partial t}\right|_{CV} = \frac{\forall}{RT}\frac{dp}{dt}$$

Take the tank to be a right circular cylinder; thus,

$$\forall = \text{area} \times \text{height} = \frac{\pi(0.3)^2(1.3)}{4} = 0.091\,9\,\text{m}^3$$

From Table A.6 for oxygen, $R = 260$ J/(kg·K) and $T = 25 + 273 = 298$ K. We now have

$$\left.\frac{\partial m}{\partial t}\right|_{CV} = \frac{0.091\,9}{(260)(298)}\frac{dp}{dt} = 1.19 \times 10^{-6}\frac{dp}{dt}$$

The second term of the continuity equation, evaluated where mass crosses the control surface, becomes

$$\iint_{CS} \rho V_n \, dA = \rho A V\big|_{\text{outlet}}$$

$$= \frac{p}{RT}\frac{\pi D^2}{4}V$$

$$= \frac{p}{260(298)}\frac{\pi(0.012\,5)^2}{4}(2) = 3.16 \times 10^{-9}\,p$$

By substitution, we have

$$1.19 \times 10^{-6}\frac{dp}{dt} = -3.16 \times 10^{-9}\,p$$

After rearranging, we get

$$\int_{14\,000}^{p} \frac{dp}{p} = \int_{0}^{60} -2.66 \times 10^{-3}\, dt$$

Upon integration, we obtain

$$\ell\text{n} \frac{p}{14\,000} = 2.66 \times 10^{-3}(60) = -0.16$$

or

$$p = 14\,000\ \mathrm{e}^{-0.16}$$

Solving, we get

$$p = 11\,940 \text{ kPa}$$

3.4 MOMENTUM EQUATION

As we have seen, the continuity equation is a conservation of mass equation with which we can account for mass transfers across boundaries and mass storage within control volumes. Next, we will derive a conservation of momentum equation and use the same technique in applying it. There are expressions for linear momentum and for angular momentum. We will derive an equation for linear momentum. The angular momentum equation is discussed in Chapter 9.

3.4.1 Linear Momentum Equation

We will apply the general conservation equation: Let N be momentum mV, and n, which is N per unit mass, becomes V. By substitution into Equation 3.4, the general conservation equation, we obtain

$$\left.\frac{d}{dt}(mV)\right|_{\text{S}} = \left.\frac{\partial}{\partial t}(mV)\right|_{\text{CV}} + \iint_{\text{CS}} V \rho V_n \, dA \tag{3.11}$$

In addition, we have from Newton's law for the system

$$\sum F_x = \frac{d}{dt}(mV)_x$$

$$\sum F_y = \frac{d}{dt}(mV)_y$$

$$\sum F_z = \frac{d}{dt}(mV)_z$$

Combining these equations with Equation 3.11, we obtain a conservation of linear momentum equation for the three principal directions:

$$\sum F_x = \left.\frac{d}{dt}(mV)_x\right|_{\text{S}} = \left.\frac{\partial}{\partial t}(mV)_x\right|_{\text{CV}} + \iint_{\text{CS}} V_x \rho V_n \, dA \tag{3.12a}$$

$$\sum F_y = \frac{d}{dt}(mV)_y\bigg|_S = \frac{\partial}{\partial t}(mV)_y\bigg|_{CV} + \iint_{CS} V_y \rho V_n \, dA \tag{3.12b}$$

$$\sum F_z = \frac{d}{dt}(mV)_z\bigg|_S = \frac{\partial}{\partial t}(mV)_z\bigg|_{CV} + \iint_{CS} V_z \rho V_n \, dA \tag{3.12c}$$

Two velocities appear in the integrands of the last terms on the right-hand sides of the preceding equations. One of the velocities refers to a principal flow direction; for example, V_x. The other velocity, V_n, is normal to the control surface where mass crosses the boundary. These two velocities are not always equal. Because force and velocity have magnitude and direction, they are vectors. Thus, Equation 3.12 can be written in vector form as

$$\sum \mathbf{F} = \frac{d}{dt}(m\mathbf{V})\bigg|_S = \frac{\partial}{\partial t}(m\mathbf{V})\bigg|_{CV} + \iint_{CS} \mathbf{V}(\rho \mathbf{V} \cdot d\mathbf{A}) \tag{3.13}$$

This equation is a conservation of linear momentum equation. As required by Newton's law, the term $\sum \mathbf{F}$ represents all forces applied externally to the control volume. These include forces due to gravity, electric and magnetic fields, surface tension effects, pressure forces, and viscous forces (friction). The first term on the right-hand side represents the rate of storage of linear momentum in the control volume. The last term is a net rate out (out minus in) of linear momentum from the control volume. For steady one-dimensional flow, Equation 3.13 becomes, for any direction i,

$$\sum F_i = \iint_{CS} V_i \rho V_n \, dA$$

For one fluid stream entering and one leaving the control volume, we have the following for one-dimensional flow:

$$\begin{aligned}\sum F_i &= V_{2i}\rho_2 A_2 V_2 - V_{1i}\rho_1 A_1 V_1 \\ &= V_i \rho A V\big|_{out} - V_i \rho A V\big|_{in}\end{aligned}$$

or

$$\begin{aligned}\sum F_i &= \dot{m}(V_{2i} - V_{1i}) \\ &= \dot{m}(V_{out} - V_{in})_{i\,direction}\end{aligned} \tag{3.14}$$

The momentum equation can be used to set up a general equation for frictional effects existing within a pipe, for a certain type of phenomenon in open-channel flow, and for other kinds of applications-oriented problems. Because such applications will appear later in the text, we devote our study of the linear momentum equation here to some general one-dimensional problems for which Equation 3.14 applies.

Example 3.5

A gardener is squirting the side of a house with a hose. The nozzle produces a ½ in. diameter jet having a velocity of 5 ft/s. Determine the force exerted by the jet on the wall when the angle between the jet and the house is 90°.

Solution

A schematic of the jet in the vicinity of the wall is shown in Figure 3.12, along with the control volume selected. Also shown is the restraining force, F_z, which acts in the negative z-direction. For a smooth wall, the jet is divided upon impact into a thin sheet that travels in all directions, similar to a faucet jet striking a flat-bottomed sink. Flow crosses the control volume boundary at locations where liquid and flow properties can be determined. We have the continuity and momentum equations at our disposal. Recall that the continuity equation involves only the magnitude of each velocity, whereas the velocities in the momentum equation are dependent on the direction selected. Applying the one-dimensional continuity equation, we get

$$\rho_1 A_1 V_1 = \rho_2 A_2 V_2$$

Because the fluid is a liquid (incompressible), $\rho_1 = \rho_2$, and so

$$A_1 V_1 = A_2 V_2$$

We have not yet discussed frictional effects. However, in problems of this type, the effect of friction is to reduce the velocity of the liquid as it flows through the control volume. Assuming frictionless flow, we write

$$V_1 = V_2$$

Combining with the previous equation, we further conclude that $A_1 = A_2$.

The control volume has the external force acting directly on it. In this case, because of symmetry, we have forces only in the z-direction. We therefore write

$$\sum F_z = \dot{m}(V_{\text{out}} - V_{\text{in}})_z$$

The sum of the external forces acting on the control volume is the restraining force F_z acting in the negative z-direction, which equals the force exerted by the jet. Note that the velocity of flow out of the control volume in the z-direction is zero; the flow entering in the z-direction is denoted as V_1. By substitution,

$$\sum F_z = \rho_1 A_1 V_1 (0 - V_1) = -\rho_1 A_1 V_1^2$$

$$-F_z = -(1.94\,\text{slug/ft}^3)\frac{\pi}{4}\frac{(1/2)^2\,\text{ft}^2(5\,\text{ft/s})^2}{(12)^2}$$

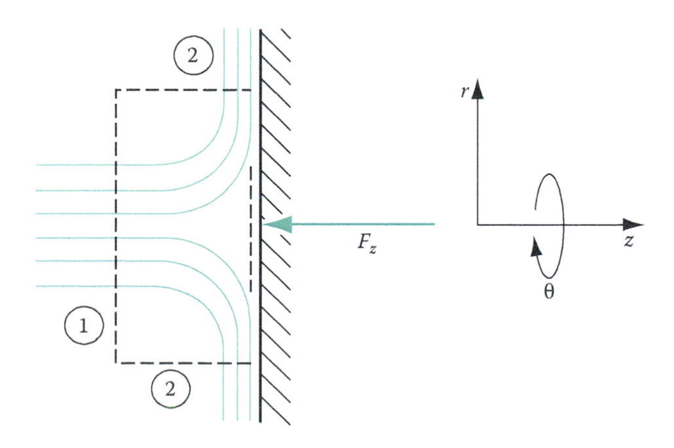

FIGURE 3.12 Sketch of the axisymmetric jet of Example 3.5.

Solving, we get

$$F_z = 0.066 \text{ lbf}$$

Note that the control volume was set up so that fluid crosses the control surface at a right angle.

Example 3.6 Laboratory Experiment: Impact of a Jet of Liquid

An experimental method of measuring the force exerted by a jet of liquid striking a stationary object involves the use of a specialized piece of equipment. Figure 3.13 is a schematic of the device used in such an experiment. The device consists of a catch basin within a sump tank. A submersible pump moves water from the sump tank, through a control valve, to the rotameter, and finally to the impact apparatus. After impacting an object, the water drains to the catch basin. On the side of the sump tank is a sight glass (not shown in Figure 3.13) showing the water depth in the catch basin.

The sump tank acts as a support for the table top which supports the impact apparatus. As shown in Figure 3.13, the impact apparatus contains a nozzle that produces a high-velocity jet of water. The jet is aimed at an object (such as a flat plate or hemisphere). The force exerted on the object causes the balance arm to which the plate is attached to deflect. A balancing weight is moved on the arm until the arm balances. A summation of moments about the pivot point of the arm allows for calculating the force exerted by the jet.

Method

The variables involved in this experiment are listed, and their measurements are described in the following text:

1. Volume rate of flow: measured with the rotameter.
2. Velocity of jet: obtained by dividing volume flow rate by jet area: $V = Q/A$. The jet is cylindrical in cross section.
3. Resultant force: found experimentally by summation of moments about the pivot point of the balance arm. The theoretical resultant force is found by use of an equation derived by applying the momentum equation to a control volume about the plate.

A hemispherically shaped object is placed in the path of the 1 cm diameter liquid jet as in Figure 3.14. The flow rate is measured as 24 L/min, and the balancing weight has a mass of 600 g. The lever arm from the pivot to where the object attaches is 15 cm. Determine the theoretical force exerted by the jet. If the balancing weight is moved to 9.6 cm, calculate the actual force.

Theoretical Force

The total force exerted by the jet equals the rate of change of momentum experienced by the jet after it impacts the object. See Figure 3.15 for a hemisphere.

The z direction is coincident with the direction of the incoming jet. The force needed to keep the hemisphere stationary is F, and it is in the negative z direction. The one-dimensional momentum equation applied to this figure is

$$-F = (\rho AV)(V_{zout} - V_{zin}) = (\rho AV)(-V - V) \quad \text{or} \quad F_z = 2\rho AV^2$$

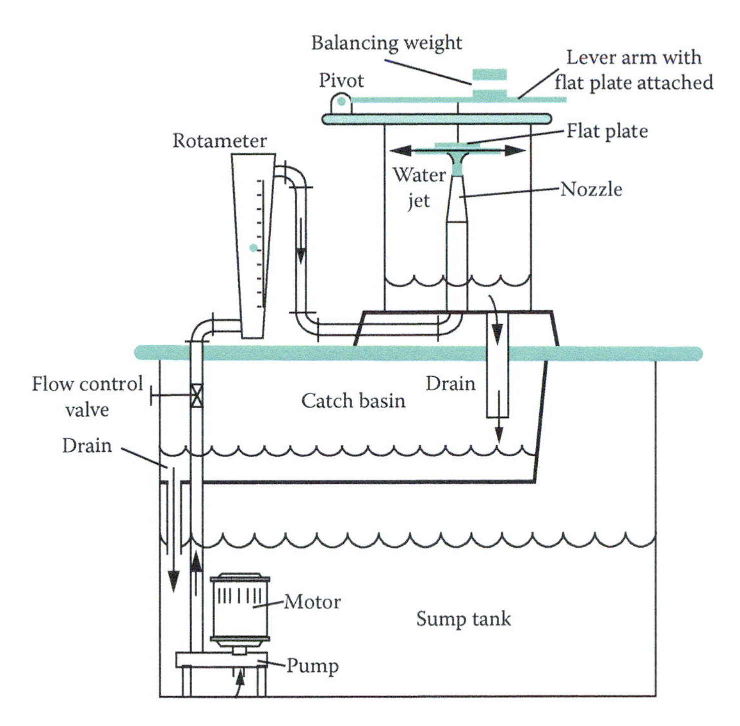

FIGURE 3.13 A schematic of the jet impact apparatus.

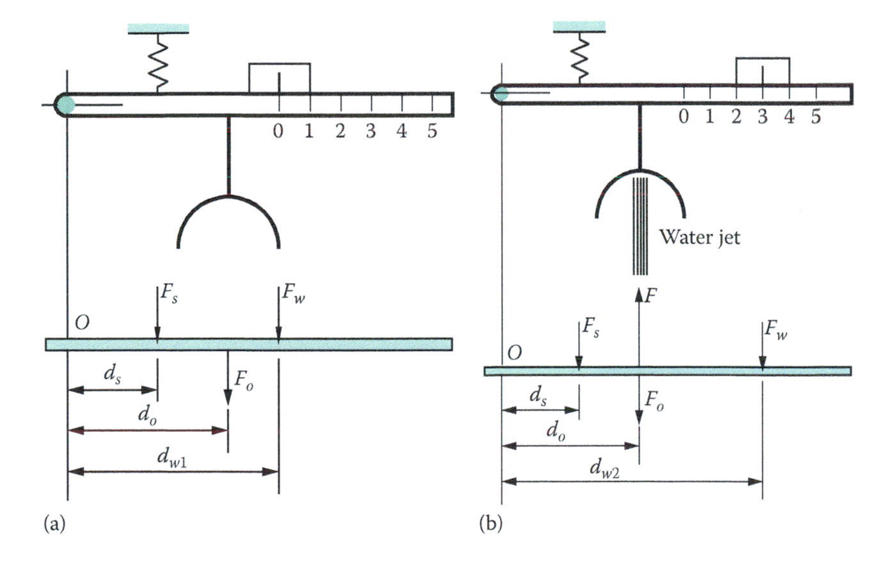

FIGURE 3.14 (a) Lever arm in zero position without any water flow and (b) lever arm in zero position when the water jet is on.

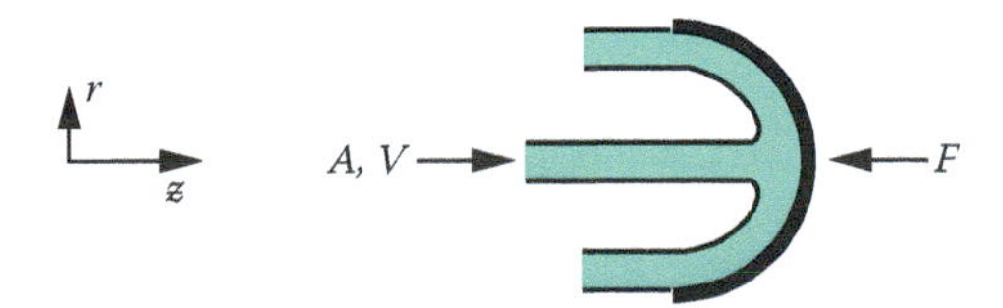

FIGURE 3.15 A jet of liquid impacting a hemisphere.

To determine the actual force, we sum moments about the hinge in Figure 3.14b, obtaining the following equation:

$$0 = -F_w d_{w1} - F d_0 + F_w d_{w2}$$

The force we are seeking is that exerted by the water jet F; rearranging gives

$$F d_0 = -F_w d_{w1} + F_w d_{w2} = F_w (d_{w2} - d_{w1})$$

or

$$F = \frac{F_w (d_{w2} - d_{w1})}{d_0}$$

From the given data, $m = 600$ g; $Q = 24$ 1/m; $d_0 = 15$ cm; $d_{w2} - d_{w1} = 9.6$ cm, and the jet diameter is 1 cm. Then

$$A = \frac{\pi(0.01)^2}{4} = 7.854 \times 10^{-5} \text{m}^2$$

$$Q = 24 \text{ 1/m} = 24(1 \times 10^{-3})/60 = 0.000\,4 \text{m}^3/\text{s}$$

$$V = \frac{Q}{A} = \frac{0.000\,4}{7.854 \times 10^{-5}} = 5.09 \text{m/s}$$

$$F_w = mg = 0.6(9.81) = 5.886 \text{N}$$

The theoretical force then is

$$F_z = 2\rho A V^2 = 2(1\,000)(7.854 \times 10^{-5})(5.09)^2 = 4.07 \text{ N}$$

The force obtained from the apparatus is

$$F = \frac{F_w (d_{w2} - d_{w1})}{d_0} = \frac{5.886(0.096)}{0.15}$$

$$F = 3.77 \text{ N}$$

The error is

$$\% \text{error} = \frac{4.07 - 3.77}{4.07} \times 100 = 7.37\%$$

Example 3.7

A vane in the shape of a flat plate is struck by a free jet. The vane has a velocity of V_v and is moving in the opposite direction of the jet (Figure 3.16a). Develop an expression for the force exerted on the vane.

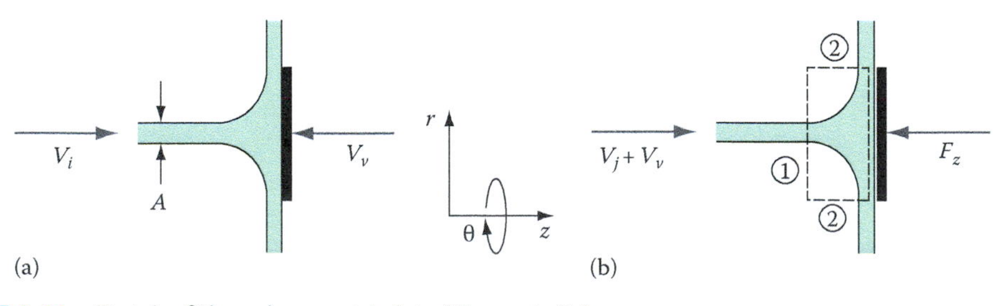

FIGURE 3.16 Sketch of the axisymmetric jet of Example 3.5.

Solution

Select a control volume as shown in the figure. Note that the control volume is moving with the vane and that the flow properties can be evaluated where fluid crosses the control surface. For convenience, let us impose a velocity of V_v acting to the right of each velocity currently in the figure. This arrangement will render the vane and the control volume motionless. The jet velocity now becomes $V_j + V_v$. The continuity equation for steady, one-dimensional flow of an incompressible fluid becomes

$$A_1 V_1 = A_2 V_2$$

where $V_1 = V_j + V_v$ and $V_1 = V_2$ for frictionless flow. The momentum equation for the z-direction is

$$\sum F_z = \dot{m}(V_{\text{out}} - V_{\text{in}})_z$$

The only external force acting on the control volume is F_z, the restraining force required to hold the vane stationary, which equals the force exerted by the jet. Thus,

$$-F_z = \dot{m}(0 - V_1) = \rho A(V_j + V_v)\,[0 - (V_j + V_v)]$$

$$F_z = \rho A(V_j + V_v)^2$$

If the vane were moving to the right instead, we would obtain

$$F_z = \rho A(V_j - V_v)^2$$

In this case, the force is zero when the vane is traveling at the jet velocity.

Example 3.8

Figure 3.17 shows a jet of water striking a curved vane, which turns the jet through an angle of 120°. The vane is moving to the right at a velocity of 1 m/s. The jet velocity is 2.5 m/s and its cross-sectional area is 0.03 m^2. Assuming no frictional losses between the jet and the surface, determine the reaction forces F_x and F_y.

Solution

Figure 3.17a shows the jet impacting the moving vane. Figure 3.17b shows the vane rendered stationary. Also shown are the reaction forces and the x–y axes. The control volume we select is moving with the vane itself and is set up so that fluid crosses the control surface at right angles. We use the relative velocity $V_j - V_v$, which is that of the jet with respect to the vane. For water, Table A.5 shows $\rho = 1\,000$ kg/m^3. The one-dimensional steady-flow continuity equation for incompressible flow is

$$A_1 V_1 = A_2 V_2$$

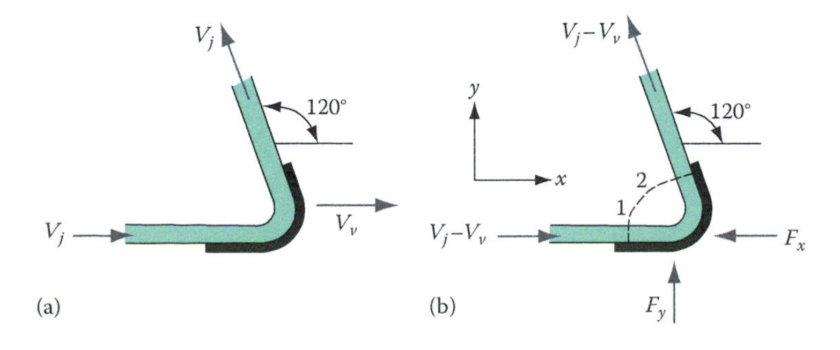

FIGURE 3.17 A liquid jet striking a moving vane.

where $V_1 = V_2 = V_j - V_v$ for frictionless flow. The momentum equation applied in the x-direction is

$$-F_x = \dot{m}(V_{\text{out rel}} - V_{\text{in rel}})_x$$

$$= \rho A(V_j - V_v)\,[-(V_j - V_v)\cos 60° - (V_j - V_v)]$$

$$F_x = \rho A(V_j - V_v)^2 (\cos 60° + 1)$$

Substituting,

$$F_x = 1\,000(0.03)(2.5 - 1)^2 (\cos 60° + 1)$$

Solving,

$$F_x = 101.3 \text{ N}$$

Applying the momentum equation in the y-direction, we write

$$F_y = \dot{m}(V_{\text{out rel}} - V_{\text{in rel}})_y$$

$$= \rho A(V_j - V_v)\,[(V_j - V_v)\sin 60° - 0]$$

$$F_y = \rho A(V_j - V_v)^2 \sin 60°$$

Substituting and solving gives

$$F_y = 1\,000(0.03)(2.5 - 1)^2 \sin 60°$$

$$F_y = 58.5 \text{ N}$$

Example 3.9

Figure 3.18 is a sketch of a jet pump. The flow of high pressure air in the smaller pipe is discharged into a larger pipe, and this entrains a secondary air stream into the region about the smaller pipe. We identify two sections in the pipeline. Section 1 is at the exit of the smaller pipe, and section 2 is a distance downstream where the two flows mix and the velocity profile is uniform. Moreover, at section 1, there are two inlet streams: 1a for the smaller pipe, and 1b for the larger pipe. Given the following data, determine the average pressure difference $(p_1 - p_2)$.

$D_1 = 1.25$ cm; $D_2 = 3.8$ cm; $V_{1a} = 20$ m/s; $V_{1b} = 10$ m/s.

Solution

Air density (assumed constant) $\rho = 1.19$ kg/m^3 (Table A.6)

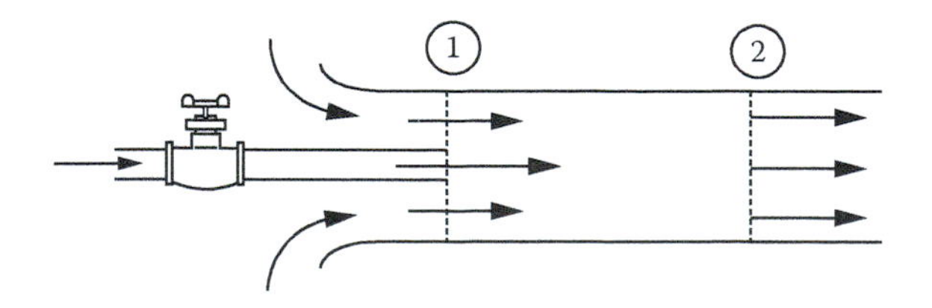

FIGURE 3.18 A jet pump.

The continuity equation applied between sections 1 and 2 is

$$A_{1a}V_{1a} + A_{1b}V_{1b} = A_2V_2$$

The areas are

$$A_{1a} = \frac{\pi(0.0125)^2}{4} = 1.23\times10^{-4}\,\text{m}^2$$
$$A_2 = 1.134\times10^{-3}\,\text{m}^2$$
$$A_{1b} = A_2 - A_{1a} = 1\times10^{-3}\,\text{m}^2$$

Substituting into the continuity equation, we get

$$1.23\times10^{-4}(20) + 1\times10^{-3}(10) = 1.134\times10^{-3}V_2$$

Solving,

$$V_2 = 11.1\ \text{m/s}$$

In problems of this type, we use a modified version of the momentum equation to solve for the pressure difference $p_1 - p_2$. The momentum at section 2 is $\rho A_2V_2^2$. The momentum of each stream at section 1 is $\rho A_{1a}V_{1a}^2$ and $\rho A_{1b}V_{2b}^2$. The momentum equation applied to sections 1 and 2 is

Sum of external force = change in momentum
section 2 minus section 1

$$A_2(p_1 - p_2) = \rho A_2V_2^2 - \rho A_{1a}V_{1a}^2 - \rho A_{1b}V_{2b}^2$$

Factoring out the density, and substituting,

$$1.134\times10^{-3}(\Delta p) = 1.19[1.134\times10^{-3}(11.1)^2 - 1.23\times10^{-4}(20)^2 - 1\times10^{-3}(10)^2]$$

Solving,

$$\Delta p = (p_1 - p_2) = -11.4\ \text{Pa}$$

Example 3.10

Figure 3.19 shows a copper reducing fitting located in a horizontal plane. Flow enters the fitting at section 1 with a velocity of 3 ft/s, goes through a 90° elbow and exits to atmospheric pressure. The inside diameter at section 1 is 3.86 in. and at section 2, the inside diameter is 1.96 in. The pressure at section 1 is measured as 15 psia. Determine the forces acting on the system: F_x and F_y. Assume frictionless flow and take the fluid to be water.

Solution

For water $\rho = 1.94\ \text{slug/ft}^3$ (Table A.5)
The diameters are given as

$$D_1 = 3.86\ \text{in.} = 0.322\ \text{ft} \quad D_2 = 1.96\ \text{in.} = 0.163\ \text{ft}^2$$

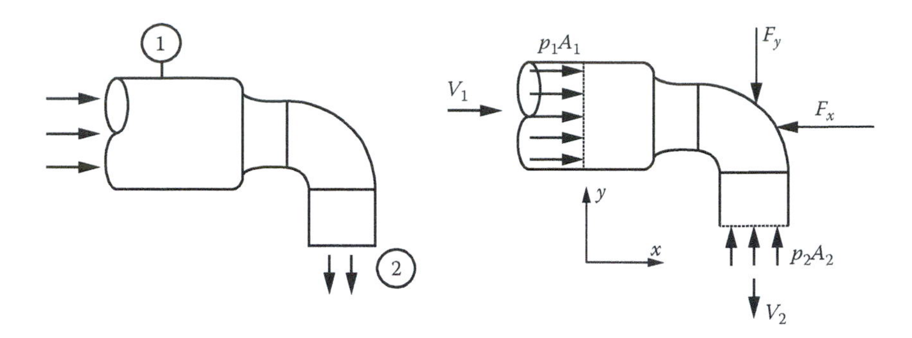

FIGURE 3.19 A reducing fitting and elbow.

The flow areas are

$$A_1 = \frac{\pi D_1^2}{4} = \frac{\pi(0.322)^2}{4} = 0.0813\,\text{ft}^2$$

$$A_2 = \frac{\pi D_2^2}{4} = \frac{\pi(0.163)^2}{4} = 0.0210\,\text{ft}^2$$

The pressure at section 1 is 15 lbf/in.^{2} = 2160 lbf/ft^2. The pressure at section 2 is atmospheric, 14.7 lbf/in.^{2} = 2117 lbf/ft^2. The velocity at section 1 is 3 ft/s.

The continuity equation applied between sections 1 and 2 is

$$A_1V_1 = A_2V_2$$

Substituting,

$$0.0813(3) = 0.0210V_2$$

$$V_2 = 11.6\text{ ft/s}$$

The second figure shows the forces acting on the elbow. These include the external forces required to hold the elbow in place (F_x and F_y), and the pressure x area forces. Also shown are the positive x and y directions. Summing forces in the positive x direction and setting them equal to the momentum change gives

$$-F_x + p_1A_1 = \rho A_1V_1\,(V_{x\text{out}} - V_{x\text{in}})$$

The term $V_{x\text{out}} = 0$. Rearranging and solving for the external force

$$-F_x + p_1A_1 = -\rho A_1V_1^2$$

$$F_x = p_1A_1 + \rho A_1V_1^2 = 2160(0.0813) + 1.94(0.0813)(3)^2$$

$$F_x = 177\text{ lbf}$$

Summing forces in the positive y direction and setting them equal to the momentum change gives

$$-F_y + p_2A_2 = \rho A_2V_2\,(V_{y\text{out}} - V_{y\text{in}})$$

The term $V_{yin} = 0$. Rearranging and solving for the external force

$$-F_y + p_2A_2 = \rho A_2V_2^2$$

$$F_y = p_2A_2 - \rho A_2V_2^2 = 2117(0.0210) - 1.94(0.0210)(11.6)^2$$

$$F_y = 38.8 \text{ lbf}$$

3.5 ENERGY EQUATION

In this section, we will develop an equation that expresses the conservation of energy. When energy crosses the boundary of a system, the energy in the system changes by an equal amount. So a decrease in the energy of a system equals the increase in energy of the surroundings and vice versa. Energy can cross a system boundary in the form of heat transfer (to or from the system), in the form of work (done by or on the system), or by mass entering or leaving. A process during which no heat is transferred to or from a system is known as an **adiabatic process**.

The law of conservation of energy states that for a system of particles,

$$E_2 - E_1 = \tilde{Q} - W'$$

or

$$dE = d\tilde{Q} - dW' \tag{3.15}$$

where

$E_2 - E_1$ is the change in energy experienced by the system
$\tilde{Q}$ is the heat added to the system
W' is work done by the system

The total energy is a property of the system and includes internal, kinetic, and potential energies:

$$E = U + KE + PE$$

On a per-unit-mass basis, we have

$$e = \frac{E}{m} = u + \frac{V^2}{2} + gz \tag{3.16}$$

Heat and work are not properties of the system, but they do represent energy transfers moving across the system boundaries in specific forms.

In applying the general conservation expression (Equation 3.4), $N = E$ and $n = e$. After substitution, we obtain

$$\left.\frac{dE}{dt}\right|_S = \left.\frac{\partial E}{\partial t}\right|_{CV} + \iint_{CS} e\rho V_n\, dA \tag{3.17}$$

which becomes

$$\left.\frac{dE}{dt}\right|_S = \frac{d\tilde{Q}}{dt} - \frac{dW'}{dt} = \left.\frac{\partial E}{\partial t}\right|_{CV} + \iint_{CS} e\rho V_n\, dA$$

or

$$\frac{d(\tilde{Q}-W')}{dt}=\left.\frac{\partial E}{\partial t}\right|_{CV}+\iint\limits_{CS}\left(u+\frac{V^2}{2}+gz\right)\rho V_n\,dA \tag{3.18}$$

The term W' consists of all forms of work crossing the boundary—shaft work, electric and magnetic work, viscous shear work, and flow work. The last of these forms represents work done by the fluid system in pushing mass into or out of the control volume. Thus,

$$\frac{dW'}{dt}=\frac{dW}{dt}+\frac{dW_f}{dt} \tag{3.19}$$

where dW_f/dt is the flow work per unit time. To develop an expression for dW_f/dt, consider an element of mass dm leaving a control volume through an area dA, as shown in Figure 3.20. The fluid in the control volume must work against the external pressure p to move dm out. With work = force × distance, we have

$$\left.\frac{dW_f}{dt}\right|_{out}=p\,dA\left.\frac{dL}{dt}\right|_{out}=\left.pV_n\,dA\right|_{out}=\left.\frac{p}{\rho}\rho V_n\,dA\right|_{out}$$

Similarly, for an element of mass entering, work is being done on the system.

Therefore,

$$\left.\frac{dW_f}{dt}\right|_{in}=-p\,dA\left.\frac{dL}{dt}\right|_{in}=-\left.\frac{p}{\rho}\rho V_n\,dA\right|_{in}$$

Combining equations, we obtain the flow work as

$$\left.\frac{dW_f}{dt}\right|_{S}=-\left.\frac{p}{\rho}\rho V_n dA\right|_{out}-\left.\frac{p}{\rho}\rho V_n\,dA\right|_{in}$$

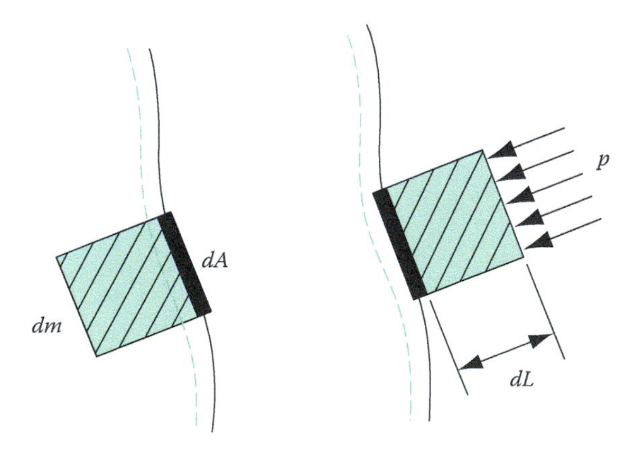

FIGURE 3.20 Sketch for the development of an expression for flow work W_f.

which can be written as

$$\left.\frac{dW_f}{dt}\right|_S = \iint_{CS} \frac{p}{\rho}\rho V_n dA \tag{3.20}$$

Substitution of Equations 3.19 and 3.20 into Equation 3.18 gives

$$\frac{d(\tilde{Q}-W)}{dt} - \iint_{CS} \frac{p}{\rho}\rho V_n\,dA = \left.\frac{\partial E}{\partial t}\right|_{CV} + \iint_{CS}\left(u+\frac{V^2}{2}+gz\right)\rho V_n\,dA$$

Combining the integral terms gives

$$\left.\frac{d}{dt}(\tilde{Q}-W)\right|_S = \left.\frac{\partial E}{\partial t}\right|_{CV} + \iint_{CS}\left(u+\frac{p}{\rho}+\frac{V^2}{2}+gz\right)\rho V_n dA$$

Enthalpy is a property defined as $h = u + p/\rho$; the conservation of energy equation now becomes

$$\left.\frac{d}{dt}(\tilde{Q}-W)\right|_S = \left.\frac{\partial E}{\partial t}\right|_{CV} + \iint_{CS}\left(h+\frac{V^2}{2}+gz\right)\rho V_n dA \tag{3.21}$$

A special form of the energy equation results for the case of steady, one-dimensional flow:

$$\left.\frac{d}{dt}(\tilde{Q}-W)\right|_S = \left[\left.\left(h+\frac{V^2}{2}+gz\right)\right|_{out} - \left.\left(h+\frac{V^2}{2}+gz\right)\right|_{in}\right]\rho AV \tag{3.22}$$

For many incompressible flows, $\tilde{Q}$ is usually assumed to be zero, and changes in internal energy are frequently negligible. Equation 3.22 then reduces to

$$-\frac{dW}{dt} = \dot{m}\left[\left.\left(\frac{p}{\rho}+\frac{V^2}{2}+gz\right)\right|_{out} - \left.\left(\frac{p}{\rho}+\frac{V^2}{2}+gz\right)\right|_{in}\right] \tag{3.23}$$

For compressible flows, however, the change in internal energy is significant.

We will deal primarily with energy conversions between various forms of thermal and mechanical energies. We will not consider energy transfers in nuclear or chemical reactions, nor will we examine work done as a result of magnetic, capillary, or electric effects. Equation 3.18 is generally known as the first law of thermodynamics, which, as we noted earlier, is a statement of the conservation of energy. For greater detail, the reader is referred to any text on classic engineering thermodynamics.

Example 3.11

A schematic of a garden fountain is given in Figure 3.21. A pump located beneath a water reservoir discharges a single jet vertically upward to a height of 6 ft above the reservoir surface. Determine the rate of work being done by the pump if the volume flow of liquid is 300 gpm. Neglect friction.

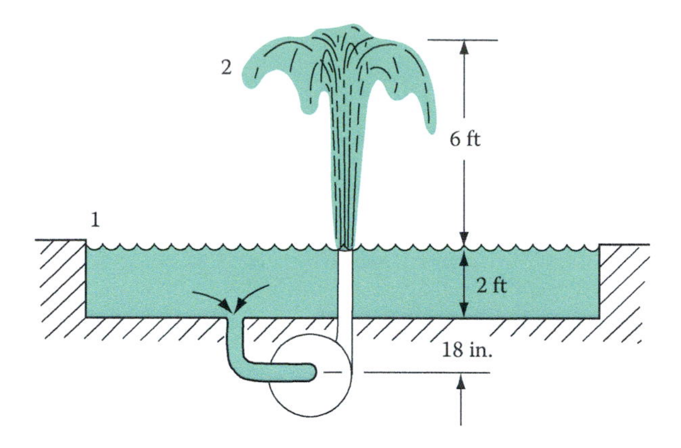

FIGURE 3.21 A garden fountain.

Solution

The control volume we select includes all the water in the reservoir, the pump and piping, and the amount discharged. We identify point 1 as the reservoir surface, whereas point 2 is at the top of the jet. Properties are known at both locations. Assuming negligible heat transfer, no change in internal energy, and steady, one-dimensional flow, Equation 3.23 applies:

$$-\frac{dW}{dt} = \dot{m}\left[\left(\frac{p_2}{\rho} + \frac{V_2^2}{2} + gz_2\right) - \left(\frac{p_1}{\rho} + \frac{V_1^2}{2} + gz_1\right)\right]$$

We next evaluate each term in the equation. The volume flow rate is given as

$$Q = 300 \text{ gal/min} \times 2.229 \times 10^{-3} = 0.67 \text{ ft}^3/\text{s}$$

The water density is 1.94 slug/ft^3, so the mass flow rate is

$$\dot{m} = \rho Q = (1.94 \text{ slug/ft}^3)(0.67 \text{ ft}^3/\text{s}) = 1.3 \text{ slug/s}$$

The pressure at sections 1 and 2 is equal to atmospheric pressure:

$$p_1 = p_2 = p_{atm}$$

The water velocity at the reservoir surface is zero; the water velocity at the top of the jet is also zero:

$$V_1 = V_2 = 0$$

If we let the reservoir surface be a datum from which to make vertical measurements, we have

$$z_1 = 0$$

and

$$z_2 = 6 \text{ ft}$$

Substituting these parameters into Equation 3.23 gives

$$-\frac{dW}{dt} = (1.3 \text{ slug/s})\left\{\left[0 + 0 + (32.2 \text{ ft/s}^2)(6 \text{ ft})\right] - (0 + 0 + 0)\right\}$$

Solving,

$$\frac{dW}{dt} = -251 \text{ ft}\cdot\text{lbf/s}$$

This power goes entirely into raising the water a distance of 6 ft. The negative sign indicates that work was done on the fluid. The horsepower is the customary unit for power in the English system. There are, by definition, 550 ft · lbf/s per horsepower. The pump power required is

$$\frac{dW}{dt} = -\frac{251 \text{ ft}\cdot\text{lbf/s}}{550 \text{ ft}\cdot\text{lbf/(s}\cdot\text{hp)}}$$

or

$$\frac{dW}{dt} = -0.46 \text{ hp}$$

Therefore, the pump must deliver 0.46 hp to the water in order to raise it by 6 ft.

Example 3.12

An amusement park water slide consists of a curving channel down which people slide on rubber mats. Water is pumped to the top of the slide, which is 60 ft high (see Figure 3.22). The inlet pipe to the pump is 6 in. in diameter; the outlet pipe to the slide is 4 in. in diameter. Determine the power required by the pump if the flow rate is 1.5 ft^3/s. Neglect friction.

Solution

We select section 1 as liquid surface in reservoir and section 2 as discharge from 4 in. pipe at the top of the slide. We apply the energy equation to these two sections, and include the pump input power. For no heat transfer, the energy equation is

$$-\frac{dW}{dt} = \dot{m}\left[\left(\frac{p_2}{\rho} + \frac{V_2^2}{2} + gz_2\right) - \left(\frac{p_1}{\rho} + \frac{V_1^2}{2} + gz_1\right)\right]$$

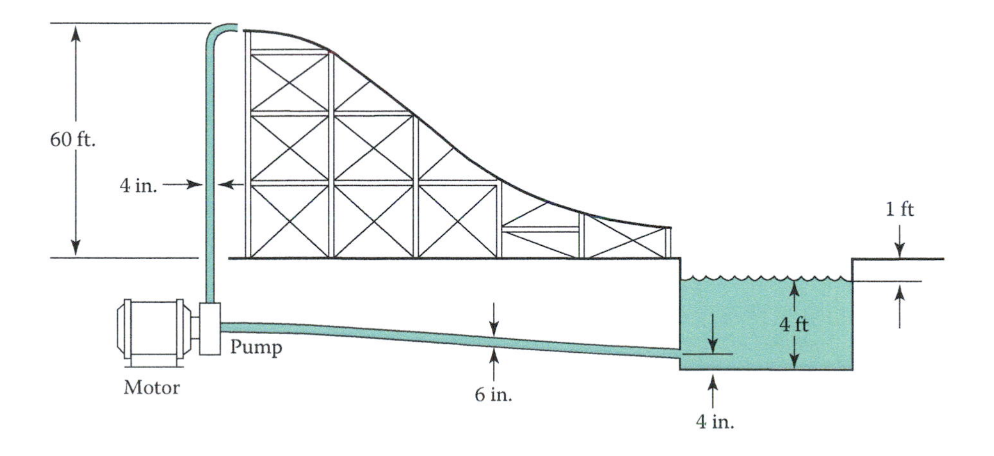

FIGURE 3.22 Amusement park water slide.

The pressure at both of these locations is atmospheric. The velocity of the free surface in the tank at the bottom of the slide is 0. Thus,

$$p_2 = p_1 = p_{atm}; \quad V_1 = 0.$$

The diameter of the outlet at the top of the slide is 4 in., so the area of the outlet pipe is

$$A_2 = \frac{\pi(4/12)^2}{4} = 0.0873\,\text{ft}^2$$

The velocity of water leaving the outlet pipe is

$$V_2 = \frac{Q}{A_2} = \frac{1}{0.0873} = 11.46\,\text{ft/s}$$

Using the tank bottom as a reference location, we have $z_1 = 4$ ft, and

$$z_2 = 4 + 1 + 60 = 65\text{ ft}$$

The mass flow rate of the water is

$$\dot{m} = \rho Q = 1.94(1.5) = 2.91\text{ slug/s}$$

Substituting into the energy equation gives

$$-\frac{dW}{dt} = 2.91\left[\left(0 + \frac{11.46^2}{2} + 32.2(65)\right) - (0 + 0 + 32.2(4))\right]$$

$$-\frac{dW}{dt} = 5907\,\text{ft}\cdot\text{lbf/s} = \frac{5907}{550} = 10.74\,\text{hp}$$

Note that the inlet pipe diameter was given but did not enter into the calculations. The answer is negative, which means that work was done on the fluid by the pump.

3.6 BERNOULLI EQUATION

The **Bernoulli equation** gives a relationship between pressure, velocity, and position or elevation in a flow field. Normally, these properties vary considerably in the flow, and the relationship between them if written in differential form is quite complex. The equations can be solved exactly only under very special conditions. Therefore, in most practical problems, it is often more convenient to make assumptions to simplify the descriptive equations. The Bernoulli equation is a simplification that has many applications in fluid mechanics. We will derive it in two ways. First, consider a flow tube bounded by streamlines in the flow field as illustrated in Figure 3.23. Recall that a streamline is everywhere tangent to the velocity vector and represents the path followed by a fluid particle in the stream; no flow crosses a streamline. Because the flow is steady and frictionless, viscous effects are neglected. Pressure and gravity are the only external forces acting. For this analysis, we will select a control volume, apply the momentum equation, and finally integrate the result along the streamtube. In the s-direction for the control volume, we have

$$\sum F_s = \iint_{\text{CS}} V_s \rho V_n \, dA$$

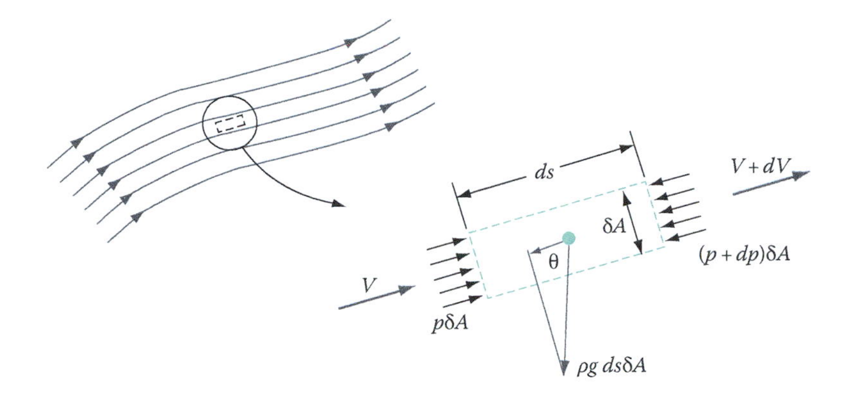

FIGURE 3.23 A differential control volume for the derivation of the Bernoulli equation.

Evaluating each term separately, we obtain

$$\sum F_s = p\delta A - (p + dp)\delta A - \rho g\, ds\, \delta A \cos\theta$$

where θ is the angle between the s-direction and gravity. Also,

$$\iint_{\text{CS}} V_s \rho V_n\, dA = (V + dV - V)\rho V\, \delta A$$

By substitution, we get

$$-dp\, \delta A - \rho g\, ds\, \delta A \cos\theta = \rho V\, dV\, \delta A$$

The term δA divides out; $ds \cos\theta$ is dz. Therefore, after simplification, we get

$$\frac{dp}{\rho} + VdV + gdz = 0$$

Integrating between points 1 and 2 along the streamtube gives

$$\int_1^2 \frac{dp}{\rho} + \frac{1}{2}(V_2^2 - V_1^2) + g(z_2 - z_1) = 0$$

For the special case of an incompressible fluid, density is constant (not a function of pressure), and the equation then becomes

$$\frac{p_2 - p_1}{\rho} + \frac{V_2^2 - V_1^2}{2} + g(z_2 - z_1) = 0 \tag{3.24}$$

or

$$\frac{p}{\rho} + \frac{V^2}{2} + gz = \text{a constant} \tag{3.25}$$

Equation 3.25 is the Bernoulli equation for steady, incompressible flow along a streamline with no friction (no viscous effects).

The Bernoulli equation can be developed from the one-dimensional energy equation (Equation 3.21):

$$\frac{d}{dt}(\tilde{Q}-W)\bigg|_S = \dot{m}\left[\left(h_2+\frac{V_2^2}{2}+gz_2\right)-\left(h_1+\frac{V_1^2}{2}+gz_1\right)\right]$$

Now let us briefly examine the conditions under which this equation becomes identical to the Bernoulli equation. For an incompressible flow with no shaft work, the energy equation reduces to

$$\frac{d\tilde{Q}}{dt}\bigg|_S = \dot{m}\left[\left(u_2+\frac{p_2}{\rho}+\frac{V_2^2}{2}+gz_2\right)-\left(u_1+\frac{p_1}{\rho}+\frac{V_1^2}{2}+gz_1\right)\right]$$

Now compare this equation to the Bernoulli equation for steady, frictionless, one-dimensional, incompressible flow (Equation 3.24). We conclude that for both equations to be identical, any change in internal energy of the fluid must equal the amount of heat transferred. It can be seen that for an incompressible flow with no work, no heat transfer, and no changes in internal energy, the energy equation and the Bernoulli equation derived from the momentum equation become identical. Thus, under certain flow conditions, the energy and momentum equations reduce to the same expression. Hence, the Bernoulli equation is referred to as the **mechanical energy equation**. For many flow problems, only the continuity and Bernoulli equations are required for a description of the flow.

Example 3.13

Figure 3.24 shows a vertically oriented flow passage, similar in shape to an automotive carburetor, through which air flows in the downward direction. Under certain flow conditions and carburetor dimensions, the pressure at the throat (section 2) is low enough such that fuel will be drawn upward from a tiny reservoir, and delivered to the air. The movement of air through the carburetor is induced by a low pressure at the bottom, and the velocity there is 0.5 m/s.

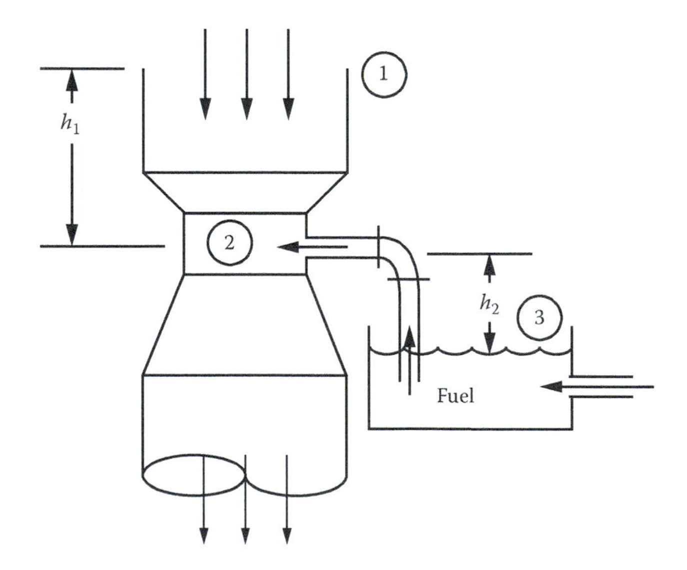

FIGURE 3.24 An automotive carburetor.

Determine the required pressure at the inlet (section 1) to bring fuel to the throat. Use the following dimensions: $D_1 = 3.6$ cm; $D_2 = 1.8$ cm; $h_1 = 4$ cm; $h_2 = 2$ cm. The fuel is gasoline (same properties as octane), and the air may be considered incompressible.

Solution: Fluid properties

Octane $\rho = 0.701(1\ 000)$ (Table A.5)

Air $\rho = 1.19\ \text{kg/m}^3$ (Table A.6)

The flow areas are

$$A_1 = \frac{\pi D_1^2}{4} = \frac{\pi(0.036)^2}{4} = 1.02 \times 10^{-3}\ \text{m}^2$$

$$A_2 = \frac{\pi D_2^2}{4} = \frac{\pi(0.018)^2}{4} = 2.54 \times 10^{-4}\ \text{m}^2$$

The hydrostatic equation is applied between sections 2 and 3. The pressure difference must be large enough to lift the fuel a distance of 2 cm. The hydrostatic equation is

$$p_2 + \rho g h_2 = p_3$$

With $p_3 = p_{atm} = 101\ 300$ Pa, we find

$$p_2 = 101\ 300 - 701(9.81)(0.02) = 101\ 162\ \text{Pa}$$

The continuity equation is applied at sections 1 and 2 to obtain a relationship between the velocities; V_2 was given as 0.5 m/s:

$$A_1 V_1 = A_2 V_2$$

$$1.02 \times 10^{-3} V_1 = 2.54 \times 10^{-4}(0.5)$$

$$\text{or} \quad V_1 = 0.125\ \text{m/s}$$

The pressure at section 1 is found by applying the Bernoulli equation between 1 and 2:

$$\frac{p_1}{\rho g} + \frac{V_1^2}{2g} + z_1 = \frac{p_2}{\rho g} + \frac{V_2^2}{2g} + z_2$$

If we take $z_2 = 0$, then $z_1 = 0.04$ m. Substituting,

$$\frac{p_1}{701(9.81)} + \frac{(0.125)^2}{2(9.81)} + 0.04 = \frac{101\ 162}{701(9.81)} + \frac{(0.5)^2}{2(9.81)} + 0$$

$$1.45 \times 10^{-4} p_1 + 7.96 \times 10^{-4} + 0.04 = 14.71 + 2.548 \times 10^{-2}$$

Solving

$$p_1 = 100\ 970\ \text{Pa} = 101\ \text{kPa}$$

Example 3.14

Consider the flow of water through a venturi meter, as shown in Figure 3.25. Liquid passing through encounters a contraction to the throat area, followed by a gradual expansion to the original diameter. A manometer is placed in the line to measure the pressure difference from the inlet to the throat. For the dimensions given, determine the volume flow rate of water through the meter.

Solution

We select the control volume so that liquid and flow properties are known or can be determined, as shown in the figure. Next, we assume one-dimensional, incompressible flow. The continuity and Bernoulli equations apply and are written from section 1 to 2:

$$Q = A_1V_1 = A_2V_2$$

and

$$\frac{p_1}{\rho} + \frac{V_1^2}{2} + gz_1 = \frac{p_2}{\rho} + \frac{V_2^2}{2} + gz_2$$

All measurements of height z are made from the centerline, so for a horizontal meter, $z_1 = z_2$. Rearranging the Bernoulli equation gives

$$\frac{p_1 - p_2}{\rho} = \frac{V_2^2 - V_1^2}{2}$$

From the continuity equation,

$$V_1 = \frac{Q}{A_1}$$

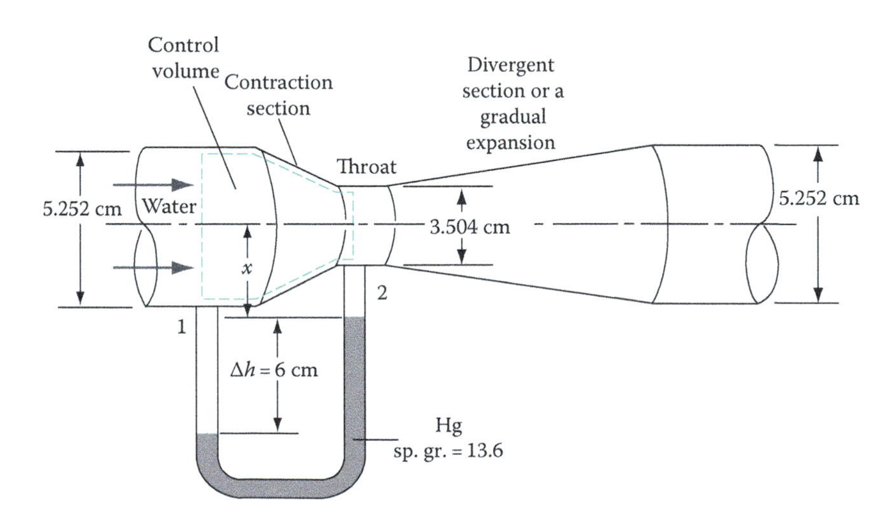

FIGURE 3.25 Flow of water through a venturi meter with pressure drop measured by a water-over-mercury U-tube manometer.

and

$$V_2 = \frac{Q}{A_2}$$

Combining these equations with the Bernoulli equation and simplifying gives

$$\frac{p_1 - p_2}{\rho} = \frac{1}{2}\left(\frac{Q^2}{A_2^2} - \frac{Q^2}{A_1^2}\right) \tag{i}$$

We now relate the manometer reading to the pressure difference. Assuming that the pressure varies negligibly across the pipe cross section (which is the case for a real flow situation), we select the pipe centerline as a reference location. Applying the hydrostatic equation from the centerline, we get

$$p_1 + \rho g x + \rho g\ \Delta h = p_2 + \rho g x + \rho_m g\ \Delta h$$

where
ρ is the flowing fluid
ρ_m is the manometer fluid

The term $\rho g x$ cancels from both sides of this equation, so that the distance x does not enter into the calculations. The pressure drop from 1 to 2 then is

$$p_2 - p_1 = \rho_m g\ \Delta h - \rho g\ \Delta h$$

or

$$\frac{p_2 - p_1}{\rho} = g\,\Delta h\left(\frac{\rho_m}{\rho} - 1\right)$$

Combining this result with Equation i yields

$$g\,\Delta h\left(\frac{\rho_m}{\rho} - 1\right) = \frac{Q^2}{2}\left(\frac{1}{A_2^2} - \frac{1}{A_1^2}\right)$$

Solving for volume flow rate, we get

$$Q = \sqrt{\frac{2g\,\Delta h(\rho_m/\rho - 1)}{\left(1/A_2^2\right) - \left(1/A_1^2\right)}}$$

We now evaluate all terms on the right-hand side:

$$A_1 = \frac{\pi D_1^2}{4} = \frac{\pi(0.052\,52)^2}{4} = 0.002\,17\,\text{m}^2$$

and

$$A_2 = \frac{\pi D_2^2}{4} = \frac{\pi(0.035\,04)^2}{4} = 0.000\,964\,\text{m}^2$$

From Table A.5 for mercury, $\rho = 13.6(1\,000)$ kg/m³; for water, $\rho = 1\,000$ kg/m³. Thus,

$$\frac{\rho_m}{\rho} = \frac{13.6(1\,000)}{1\,000} = 13.6$$

and

$$\Delta h = 0.06 \text{ m}$$

Substituting these quantities into the volume-flow-rate equation, we find

$$Q = \sqrt{\frac{2(9.81)(0.06)(13.6-1)}{(1/0.000\,964)^2 - (1/0.002\,17)^2}}$$

or

$$Q = 0.004\,16 \text{ m}^3\text{/s.}$$

Example 3.15

A siphon is set up to remove gasoline from a tank as illustrated in Figure 3.26. One-half-inch-inside-diameter flexible tubing is used. Estimate the exit velocity at the instant shown and the volume flow rate.

Solution

We will select a control volume such that point 1 is the velocity of the downward-moving liquid surface and point 2 is the end of the tube from which gasoline flows. Assuming quasi-steady, one-dimensional flow of an incompressible fluid, we can write the continuity and Bernoulli equations as

$$Q = A_1V_1 = A_2V_2$$

$$\frac{p_1}{\rho} + \frac{V_1^2}{2} + gz_1 = \frac{p_2}{\rho} + \frac{V_2^2}{2} + gz_2$$

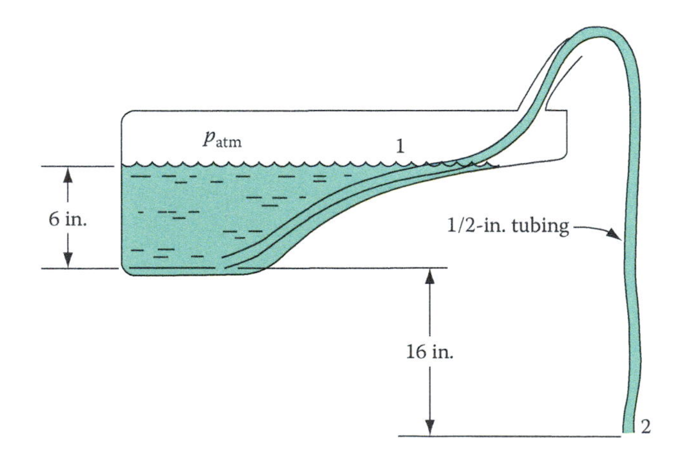

FIGURE 3.26 Siphoning gasoline from a tank.

The area of the reservoir at point 1 is much larger than that at point 2, so we regard V_1 in the Bernoulli equation as being negligible. Note also that when $p_1 = p_2 = p_{atm}$ the Bernoulli equation becomes

$$gz_1 - gz_2 = \frac{V_2^2}{2}$$

or

$$V_2 = \sqrt{2g(z_1 - z_2)}$$

This result is independent of liquid properties. Taking measurements from $z_2 = 0$, we have $z_1 = 22$ in. By substitution,

$$V_2 = \sqrt{2(32.2)\left(\frac{22}{12}\right)} = 10.86\,\text{ft/s}$$

The volume flow rate is

$$Q = A_2V_2 = \frac{\pi}{4}\frac{(0.5)^2}{(12)^2}10.86 = 0.015\,\text{ft}^3\text{/s}$$

This is an unsteady-flow problem because z_1 changes with time. In unsteady-flow problems, however, it is often necessary to assume a quasi-steady process wherein we can apply the steady-flow equations to any instant in time for the flow depicted. A more accurate description is obtained if a differential equation for z_1 is written and solved starting with the unsteady form of the continuity equation.

In Section 3.3, we solved a draining tank problem; specifically in Example 3.2, we had a tank that was being drained with a pipe. The exit velocity was constant. In a more realistic model, however, the fluid leaving a tank through a pipe at the bottom will have a velocity that varies with liquid height in the tank. We can formulate such problems now with the continuity and Bernoulli equations together.

Consider a tank of liquid with an attached pipe as indicated in Figure 3.27. Liquid drains from this tank and encounters friction within the pipe, which we assume is negligible. In the quasi-steady draining tank problems, we have considered thus far the liquid surface velocity in the tank is assumed negligible. In this analysis, however, it is our desire to determine how the depth of liquid in the tank varies with time, and account for the volume of liquid in the tank. We consider unsteady flow from the draining tank, and we wish to determine how long it will take for the tank to drain from an initial depth to some final depth.

As indicated in Figure 3.27, the tank is circular and has a diameter d. The pipe diameter is D, and the distance from the pipe exit to the free surface of the liquid in the tank is h. The distance from tank bottom to pipe exit is b, which is a constant. We identify the free surface of liquid in the tank as section 1, and the pipe exit as section 2. The equation relating areas and velocities at any time is

$$A_1V_1 = A_2V_2$$

At any time and for any depth, the velocity of the liquid surface at 1 (V_1) in terms of the velocity in the pipe is given by

$$V_1 = \frac{A_2}{A_1}V_2 \tag{3.26}$$

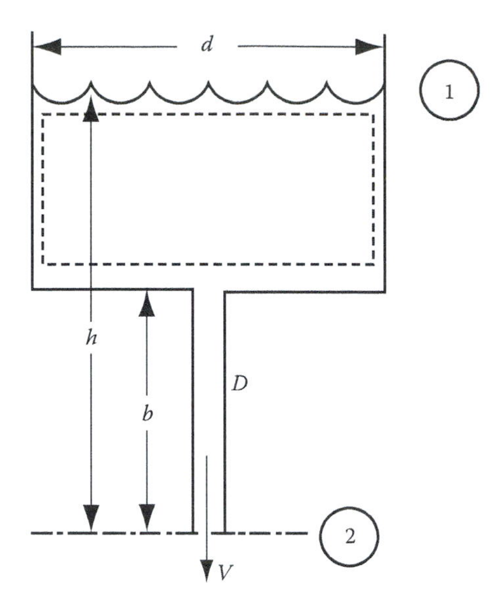

FIGURE 3.27 The unsteady draining tank problem.

The Bernoulli equation written from 1 (free surface in tank) to 2 (pipe exit) is

$$\frac{p_1}{\rho g} + \frac{V_1^2}{2g} + z_1 = \frac{p_2}{\rho g} + \frac{V_2^2}{2g} + z_2$$

Evaluating properties, we have

$$p_1 = p_2 = p_{\text{atm}} \quad z_2 = 0 \quad z_1 = h \quad V_2 = V$$

Simplifying Equation 3.26, and substituting these values, the Bernoulli equation becomes

$$\frac{V_1^2}{2g} + h = \frac{V_2^2}{2g}$$

or

$$\left(\frac{A_2}{A_1}\right)^2 \frac{V^2}{2g} + h = \frac{V^2}{2g}$$

which simplifies to

$$h = \frac{V^2}{2g}\left[1 - \left(\frac{A_2}{A_1}\right)^2\right]$$

Solving for velocity,

$$V = \left[\frac{2gh}{1-(A_2/A_1)^2}\right]^{1/2} \tag{3.27}$$

This is the equation of motion relating depth of liquid in the tank to efflux velocity. Note that if we assume the velocity at section 1 (V_1) is much smaller than the velocity in the exit pipe (V_2), the area ratio is negligible. Incorporating this assumption gives the familiar result: $V = \sqrt{2gh}$.

Next, we will write the unsteady continuity equation for the tank liquid:

$$0 = \frac{\partial m}{\partial t} + \iint_{CS} \rho V_n \, dA \tag{3.5}$$

where $\partial m/\partial t$ is the change of mass in the tank, and the integral term must be evaluated at locations where mass leaves the tank and where mass enters the tank. The mass of liquid in the tank at any time is

$$m = \rho \mathcal{V} = \rho \frac{\pi d^2}{4}(h-b) = \frac{\pi d^2}{4}\rho h - \frac{\pi d^2}{4}\rho b$$

Noting that b is a constant, the change of mass in the tank with respect to time is

$$\frac{\partial m}{\partial t} = \frac{\pi d^2}{4}\rho \frac{dh}{dt}$$

which is the first term on the right-hand side of Equation 3.5. The second term—the integral term—becomes

$$\iint_{CS} \rho V_n \, dA = \iint_{out} \rho V_n \, dA - \iint_{in} \rho V_n \, dA$$

With no fluid entering the tank, the preceding equation reduces to

$$\iint_{CS} \rho V_n dA = \rho \frac{\pi D^2}{4} V - 0$$

Substituting this equation and Equation 3.27 into Equation 3.5, we get

$$0 = \frac{\pi D^2}{4}\rho \frac{dh}{dt} + \rho \frac{\pi D^2}{4} V$$

or

$$0 = A_1 \frac{dh}{dt} + A_2 V$$

Rearranging,

$$\frac{dh}{dt} = -\frac{A_2}{A_1}V \tag{3.28}$$

For the draining tank problem, then, Equation 3.28 must be solved simultaneously with Equation 3.27. This will take into account the variation of liquid height and efflux velocity with time, as well as the fluid volume within the tank. A solution to Equation 3.28 can be obtained by direct integration. Note that Equations 3.27 and 3.28 are independent of fluid properties.

Substituting Equation 3.27 into Equation 3.28 gives

$$\frac{dh}{dt} = -\frac{A_2}{A_1}\left[\frac{2gh}{1-(A_2/A_1)^2}\right]^{1/2}$$

$$\frac{dh}{dt} = \left[\frac{2(A_2/A_1)^2 gh}{1-(A_2/A_1)^2}\right]^{1/2} \tag{3.29a}$$

All terms on the right-hand side of this equation except h are constants. The equation can thus be rewritten as

$$\frac{dh}{dt} = -C_1 h^{1/2} \tag{3.29b}$$

where

$$C_1 = \left[\frac{2(A_2/A_1)^2 g}{1-(A_2/A_1)^2}\right]^{1/2}$$

Rearranging Equation 3.29b, we have

$$h^{-1/2}\,dh = -C_1\,dt$$

We can integrate this expression from time = 0 (where $h = h_0$) to some future time t (where $h = h$):

$$\int_{h_0}^{h} h^{-1/2}dh = -\int_0^t C_1 dt$$

The result is

$$2\left(h^{1/2} - h_0^{1/2}\right) = -C_1 t$$

Solving for the height h at any time,

$$h = \left(h_0^{1/2} - \frac{C_1 t}{2}\right)^2 \tag{3.30}$$

This equation gives the variation of liquid height in the tank with depth. Substituting into Equation 3.27 gives the variation of velocity also with time. Fluid properties, specifically density, do not enter into the equations.

Example 3.16

The 20 cm diameter (=d) tank of Figure 3.27 contains water, and it is being drained by the attached piping system. The pipe itself has an inside diameter of 5.1 cm. Neglecting friction, determine the variation of velocity and height with time if the height h is allowed to vary from an initial value of 1 m–0.5 m.

Solution

For water, the density is $\rho = 1\ 000\ \text{kg/m}^3$. With the tank diameter of 0.2 m, we calculate

$$A_1 = \frac{\pi d^2}{4} = \frac{\pi(0.2)^2}{4} = 0.031\ 42\ \text{m}^2$$

The pipe area is

$$A_2 = \frac{\pi D^2}{4} = \frac{\pi(0.051)^2}{4} = 0.002\ \text{m}^2$$

The area ratio then is

$$\left(\frac{A_2}{A_1}\right)^2 = \left(\frac{0.002}{0.031\ 42}\right) = 0.0041$$

The constant then becomes

$$C_1 = \left[\frac{2(A_2/A_1)^2 g}{1-(A_2/A_1)^2}\right] = \left[\frac{2(0.004\ 1)(9.81)}{1-(0.004/1)}\right]^{1/2} = 0.284$$

Rearranging Equation 3.30 and solving for time, we have

$$t = -\frac{2\left(h^{1/2} - h_0^{1/2}\right)}{C_1}$$

Substituting $h = 0.5$ m, $h_0 = 1$ m, the time is

$$t = -\frac{2\left(0.5^{1/2} - 1^{1/2}\right)}{0.284}$$

$t = 2.06$ s

3.7 SUMMARY

In this chapter, we developed the basic equations of fluid mechanics by using the control volume approach. The continuity, momentum, and energy equations were written by beginning with the general conservation equation. The momentum and the energy equations both gave the Bernoulli equation for frictionless flow with no external heat transfer and no shaft work. Various examples were solved to illustrate the use of the expressions, but compressible flows have been reserved for Chapter 8 because a different mathematical formulation is necessary. Frictional or viscous forces were not included because they too will be covered in later chapters. The purpose of this chapter has been to present the basic equations and show how they are derived.

INTERNET RESOURCES

1. A number of problems in this chapter have dealt with a venturi meter. Locate manufacturers of venturi meters, and give a brief presentation on how they are constructed as well as how much they cost.
2. One of the problems in this chapter deals the force exerted on a single Pelton bucket? What is a Pelton bucket? How are they used, and where are some installations where they are located? (See Problem 3.31.)
3. Give a brief presentation on the Torricelli experiment described in this chapter (see Problem 3.51).
4. Provide a brief presentation on the Jet d'Eau found in Lake Geneva, Switzerland (see Problem 3.54).

PROBLEMS

Average Velocity

3.1 By integrating the expression

$$V_z = V_{max}\left(1 - \frac{r^2}{R^2}\right)$$

Over the cross-sectional area of a circular duct, show that the average velocity ½V_{max}.

3.2 Flow through a pipe under certain conditions has a velocity distribution given by

$$\frac{V_z}{V_{max}} = \left(1 - \frac{r}{R}\right)^{1/7}$$

Determine the average velocity in the pipe.

3.3 The velocity of flow in a pipe has a distribution given by

$$\frac{V_z}{V_{max}} = \left(1 - \frac{r}{R}\right)^{1/m}$$

In some cases, $m = 7$, or 8, or 9, depending on the volume flow rate through the pipe. Determine the average velocity in the pipe in terms of m.

3.4 Flow through a rectangular duct is illustrated in Figure P3.4. The cross section is assumed to be very wide compared to its height. Flow is in the z-direction. The velocity profile for such a duct is determined to be

$$V_z = V_{max}\left(\frac{h^2}{2\mu}\right)\left(\frac{1}{4} - \frac{y^2}{h^2}\right)$$

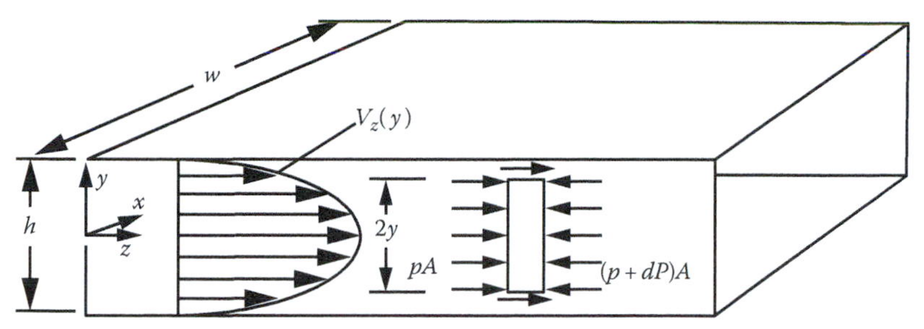

Laminar flow through a rectangular duct.

FIGURE P3.4

The volume flow rate is found by integrating the velocity V_z over the cross-sectional area:

$$Q = \int_0^w \int_{-h/2}^{+h/2} V_{max}\left(\frac{h^2}{2\mu}\right)\left(\frac{1}{4} - \frac{y^2}{h^2}\right) dydx$$

Perform this integration and show that the flow rate is given by

$$Q = \frac{h^3 w}{12\mu} V_{max}$$

The area of this duct is $A = hw$. Show that the average velocity in the duct is

$$V = \frac{h^2}{12\mu} V_{max}$$

3.5 Consider flow through the tubing arrangement illustrated in Figure P3.5. The velocity distribution in the 24 in. pipe is given by

$$V_z = 4\left(1 - \frac{r^2}{R^2}\right) \text{ft/s}$$

Calculate the volume flow rate through the system, and determine the average velocity at section 2. Is the result dependent on fluid properties?

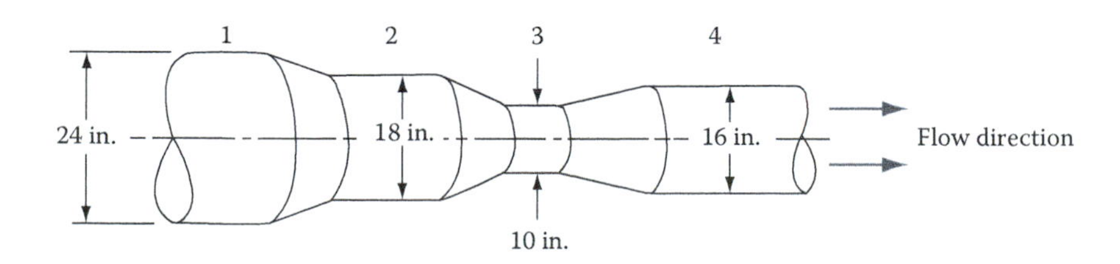

FIGURE P3.5

3.6 The velocity distribution in the 24 in. pipe of Figure P3.5 is given by

$$V_z = 0.8\left(1 - \frac{r^2}{R^2}\right) \text{ m/s}$$

Determine the volume flow rate through the system and the average velocity at section 3.

3.7 With regard to the piping system of Figure P3.5, express the average velocity at section 4 in terms of the average velocity at sections 1, 2, and 3.

3.8 Air is flowing through the configuration of Figure P3.5 at a mass flow rate of 0.003 slug/s. Assuming ideal gas behavior, calculate the velocity at section 1 if p_1 = 18 psia and T_1 = 520°R.

Continuity Equation: Steady Flow

3.9 Figure P3.9 shows a T-joint found in a water piping system. At section 1, the velocity is given by

$$V_z = 1\left(1 - \frac{r^2}{R^2}\right) \text{ ft/s}$$

The pipe radius R at section 1 is 3 in. At sections 2 and 3, the radius is 2 in. Measurements indicate that 2/3 of the total flow entering at section 1 leaves at section 2. Determine the average velocity at sections 2 and 3.

3.10 A long pipe has an inside diameter of 12 cm. Air enters the pipe at a pressure of 150 kPa and a temperature of 20°C. The mass flow rate of air is 0.05 kg/s. The pipe is heated by an external source, and the air thus becomes warmed as it travels through. At the exit, the air pressure is 140 kPa, and the temperature is 80°C. Is the velocity at the inlet equal to the velocity at the exit? Is the volume flow rate of the air at the inlet equal to the volume flow rate of the air at the outlet? Assume ideal gas behavior.

3.11 Freon-12 refrigerant enters a condenser (a heat exchanger) as vapor and leaves it as a liquid. Energy removed from the Freon-12 causes it to condense, and liquid leaves the exchanger at a velocity of 1 ft/s. The condenser is made of ¼ in. ID tubing, as indicated in Figure P3.11. Calculate the mass flow rate of Freon-12 entering the condenser.

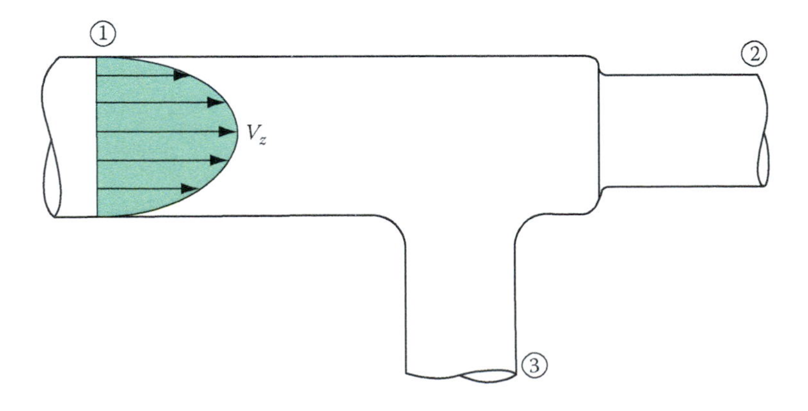

FIGURE P3.9

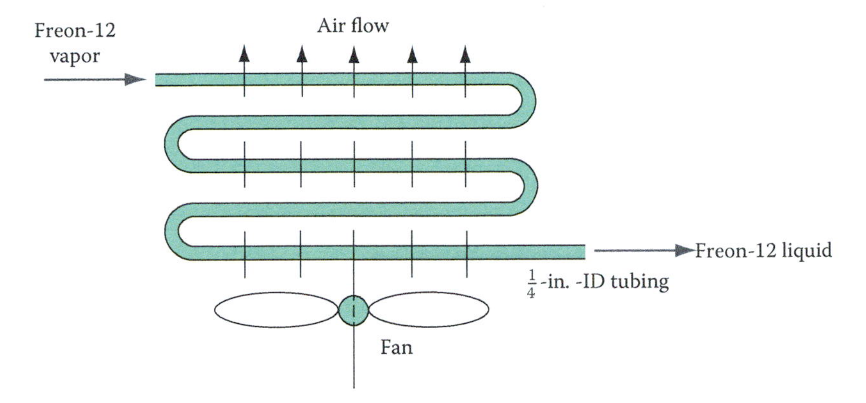

FIGURE P3.11

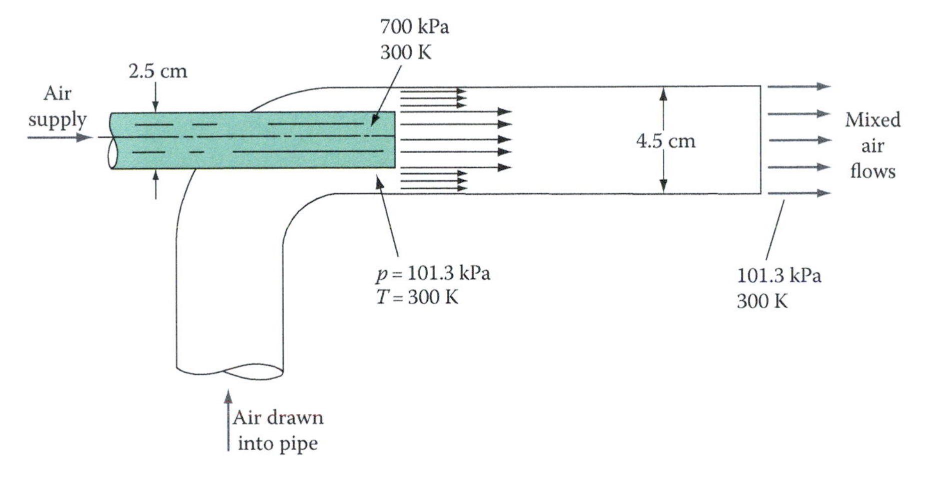

FIGURE P3.12

3.12 Figure P3.12 shows a jet pump that consists of a tube within a tube. Air is forced into the smaller tube at a pressure of 700 kPa, a temperature of 300 K, and a velocity of 3 m/s. This high-velocity jet draws atmospheric air into the larger tube at a pressure of 101.3 kPa, a temperature of 300 K, and a velocity of 8 m/s. Determine the velocity of the air drawn in by the jet.

3.13 Figure P3.13 shows a 30 m diameter fuel tank used to store heptane, which enters the tank through a 32 cm ID inlet line. The average velocity of flow in the inlet is 2 m/s. The initial depth of heptane in the tank is 1 m. Determine the time required for the heptane depth to reach 6 m.

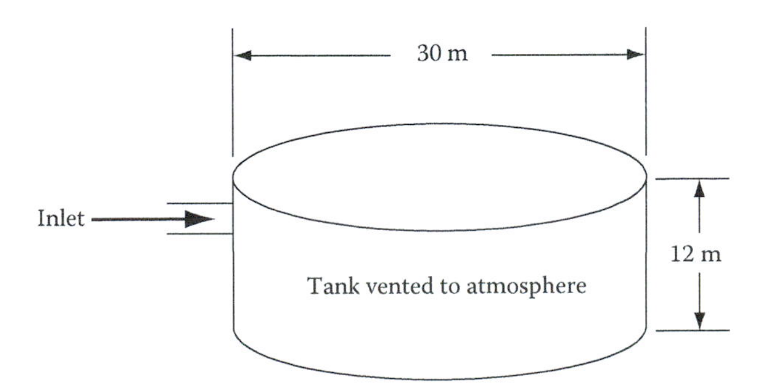

FIGURE P3.13

3.14 A 4 ft diameter tank containing solvent (acetone) is sketched in Figure P3.14. The solvent is drained from the bottom of the tank by a pump so that the velocity of flow in the outlet pipe is constant at 3 ft/s. If the outlet pipe has an inside diameter of 1 in., determine the time required to drain the tank from a depth of 3 ft to a depth of 6 in.

3.15 Figure P3.15 shows a diffuser that is part of a wind tunnel. Air enters a converging section (not shown), then passes through the test section (section 1) to the diffuser and to the fan (section 2). Sections 1 and 2 are both circular. The radius at section 1 is 1 ft and at section 2, the radius is 2 ft. For testing purposes, it is desired to have a velocity in the test section of 100 mph. Determine the velocity at section 2 and the flow rate if the air can be treated as incompressible.

3.16 A circular swimming pool 16 ft in diameter is being filled with two garden hoses, each having an inside diameter of 3/4 in. The velocities of flow from the hoses are 6 and 4 ft/s. Calculate the time required to fill the pool to a depth of 4.5 ft (see Figure P3.16). If we start filling the pool by 8:00 AM, will we be able to swim in a filled pool by 8:00 PM?

3.17 A tank with a volume of 3 ft^3 is charged to an internal pressure of 2000 psig with carbon dioxide. The gas is to provide an inert environment for a welding operation. The tank valve is regulated to allow 10 ft^3/h to escape. Assuming that the temperature in the tank remains constant at 65°F, estimate the tank pressure after 1 h of use. The escaping carbon dioxide is at 14.7 psia and 65°F.

3.18 An air compressor is used to pressurize an initially evacuated tank. The tank is 36 in. in diameter and 58 in. long. The supply line is 6 in. in diameter and conveys a flow of 5 ft/s. The air compressor's output pressure and temperature are constant at 50 psia and 90°F. The tank temperature of 70°F is also constant. Calculate the time required for the tank pressure to reach 15 psia.

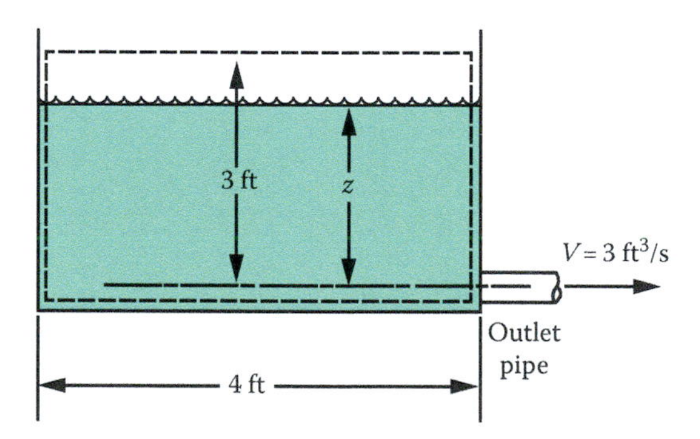

FIGURE P3.14

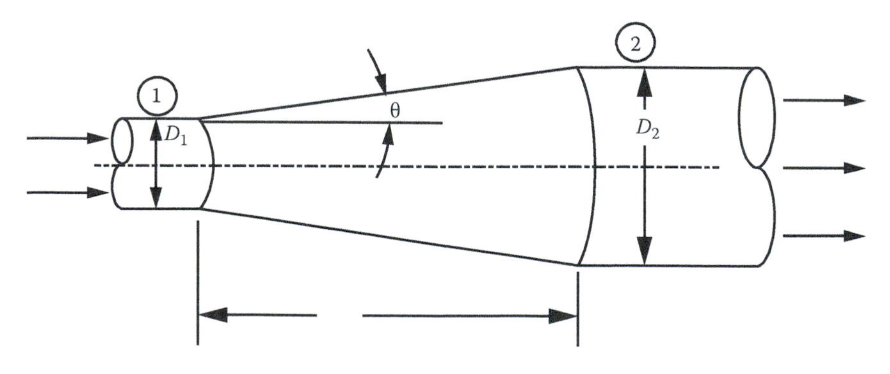

FIGURE P3.15

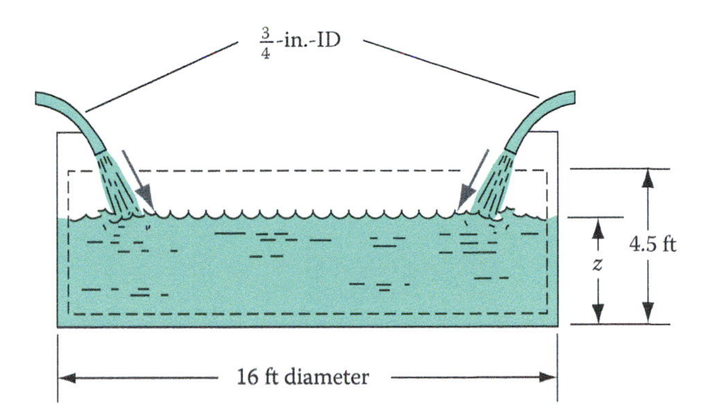

FIGURE P3.16

3.19 Consider a balloon that is being inflated with a fluid of density ρ. By completing the following steps, determine an equation for the change of radius of the balloon if its shape is spherical during inflation. Assume that the inlet flow rate to the balloon is a constant at 0.5 ft^3/s and that the balloon is inflated from an initial radius of 1 in. to a final radius of 12 in.

The radius of the balloon at any time is R, and ρ is the density of the fluid inside. During a time dt, the mass that has entered the balloon is $\rho AV\,dt$, where A is the area of the inlet and V is the velocity there. During the time interval, the balloon radius has increased to $R + dR$. The density of the air inside has changed to $\rho + d\rho$. The initial mass in the balloon of radius R is

$$m_t = \rho \forall = \rho \frac{4}{3}\pi R^3$$

After the time interval dt, the mass in the balloon is

$$m_{t+dt} = (\rho + d\rho)\frac{4}{3}\pi(R + dR)^3$$

The increase in mass of air in the balloon is given by

$$\frac{\Delta m}{\Delta t} = \frac{4}{3}\pi(R + dR)^3(\rho + d\rho) - \frac{4}{3}\rho\pi R^3$$

a. Expand the right-hand side, and by neglecting higher-order terms, show that the flow of fluid into the balloon is given by

$$\rho AV\,dt = \rho Q\,dt = \frac{4}{3}\pi R^3 d\rho + 4\rho\pi R^2 dR$$

b. If density is a constant, then $d\rho = 0$, and ρ cancels from the remaining terms. Verify that the result is

$$Q\,dt = 4\pi R^2 dR$$

c. By integrating this equation using limits of 0 to t for time, corresponding to a balloon radius increase from R_1 to R_2, show that the time required is given by

$$t = \frac{4\pi R_2^3}{3Q}\left(1 - \frac{R_1^3}{R_2^3}\right)$$

d. Substitute the given parameters into this equation and show that the time required to inflate the balloon from 1 to 12 in. is

$$t = 8.4 \text{ s}$$

3.20 In the previous problem, a balloon was inflated at a flow rate Q that was considered constant, and this could occur if the balloon is inflated from a tank of fluid. Suppose, however, that the inflation was done by a short winded individual and that the flow rate decreased linearly from some initial value to zero after so many seconds. The equation for this would be

$$Q = mt + b.$$

The constants m and b would be different for different individuals. As an example, the initial flow of air would be Q_i at $t = 0$. After t_f seconds or so (typically 5 s), the flow would be zero. Combining with the previous equation

$$Q_1 = m(0) + b \quad \text{so} \quad b = Q_1$$

and

$$0 = m(t_f) + Q_1 \quad \text{so} \quad m = \frac{-Q_i}{t_f}$$

a. Verify that the equation for flow rate becomes

$$\frac{Q}{Q_1} = \left(1 - \frac{t}{t_f}\right)$$

Following the derivation in the previous problem, we obtained

$$Q\,dt = 4\pi R^2 dR$$

Combine with the flow rate equation in part a here, and perform the integration over the limits 0 to t_f on time, and R_1 to R_2 on balloon radius. Show that

$$t_f = \frac{8\pi R_2^3}{3Q_i}\left(1 - \frac{R_1^3}{R_2^3}\right)$$

This equation may be used to find the final radius of the balloon after an inflation time of t_f.

Momentum Equation

3.21 A liquid jet issuing from a faucet impacts a flat bottom sink at a right angle. Just prior to impact, the jet velocity is 1 ft/s and the jet diameter is 3/8 in. Calculate the force exerted by the jet (Figure P3.21).

3.22 A jet of water impacts a flat plate that is attached to a lever arm. The lever arm is free to rotate about a hinge, and supports a weight. The velocity of the jet at impact with the plate is 5 m/s and the diameter is 7 cm. (a) Determine the force exerted on the flat plate by the jet. (b) If the lengths are $L_1 = 65$ cm, and $L_2 = 100$ cm, determine the magnitude of the weight that can be supported on the lever arm (Figure P3.22).

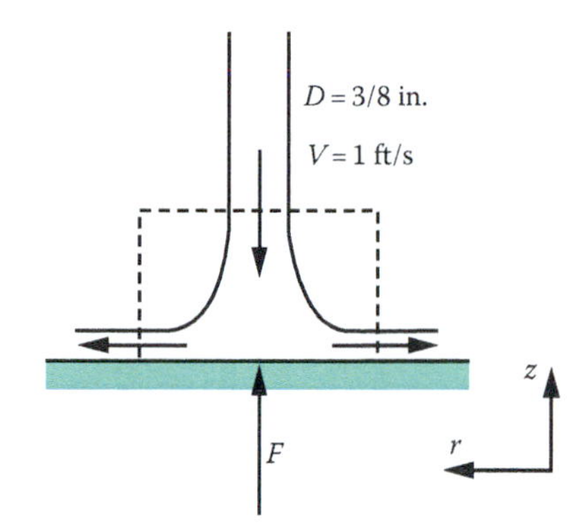

FIGURE P3.21

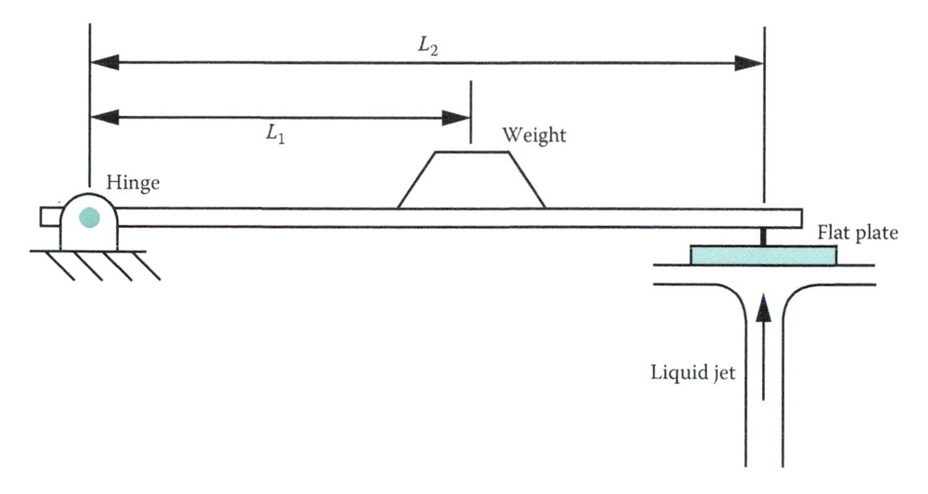

FIGURE P3.22

3.23 A two-dimensional jet in the form of a liquid sheet strikes a piece of angle iron as shown in Figure P3.23. Derive an expression for the force F.

3.24 The object in Figure P3.23 is moving to the right at half the jet velocity. Determine the force F.

3.25 A cylindrical jet impinges on a hemispherical object that is placed in the impact apparatus described in this chapter. For a flow rate of 22 1/min, the distance the 600 g weight has to move to balance the system is 78 mm. Determine the theoretical and actual forces exerted on the object.

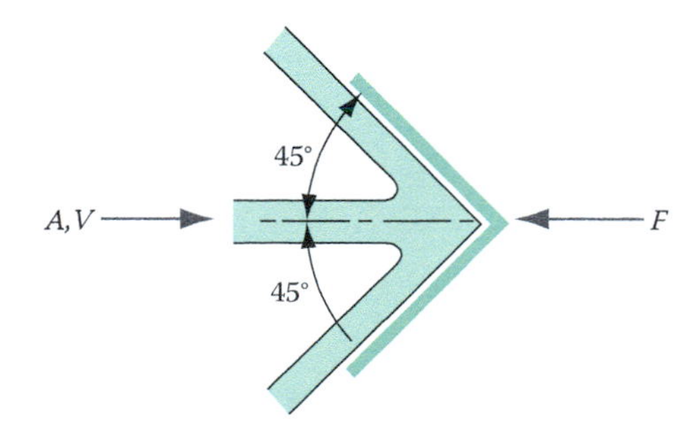

FIGURE P3.23

3.26 A cylindrical jet impinges on an object known as a Pelton bucket (see Figure P3.26). The jet impacts the bucket at the section labeled A–A. The bucket is placed in the impact apparatus described in this chapter. Show that the theoretical equation for the force exerted on the bucket is given by $F = \rho A_1 V_1^2 (1 - \cos\theta)$ (Figure P3.26).

3.27 A cylindrical jet impinges on an object known as a Pelton bucket (see the preceding problem). The jet impacts the bucket at the section labeled A–A. The bucket is placed in the impact apparatus described in this chapter. For a flow rate of 30 1/min, the distance the 600 g weight has to move to balance the system is 130 mm. Derive a theoretical equation for the force exerted on the bucket for an angle α of 165°. Compare the calculated force to that obtained with the impact apparatus.

3.28 A two-dimensional jet in the form of a liquid sheet impinges on a semicircular object as shown in Figure P3.28. Develop an expression for the restraining force.

3.29 The vane in Figure P3.28 moves to the right at a velocity equal to one-third the jet velocity. Determine the force F.

3.30 A two-dimensional jet in the form of a liquid sheet is turned through an angle of 30° by a curved vane as shown in Figure P3.30. Derive an expression for the restraining forces required to hold the vane stationary (Figure P3.30).

3.31 The forces F_x and F_y of Figure P3.31 are such that their magnitudes are related by $F_y = 3F_x$. Determine the angle θ through which the vane turns the liquid jet. The angle should vary over the range $0° \le \theta \le 90°$.

3.32 A two-dimensional jet in the form of a liquid sheet is turned through an angle of 110° by a curved vane, as shown in Figure P3.32. Derive an expression for the forces exerted on the object by the liquid.

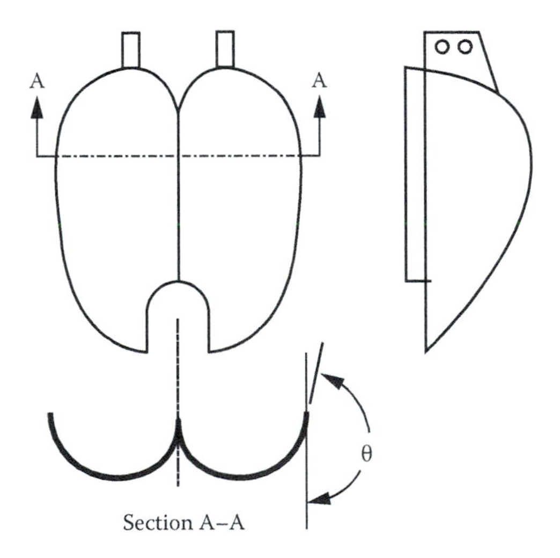

FIGURE P3.26

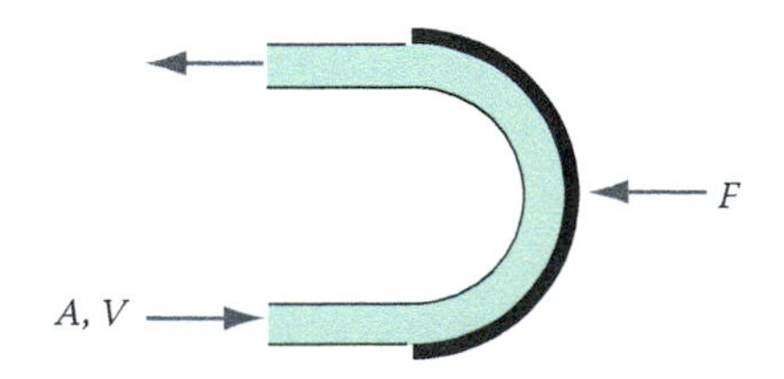

FIGURE P3.28

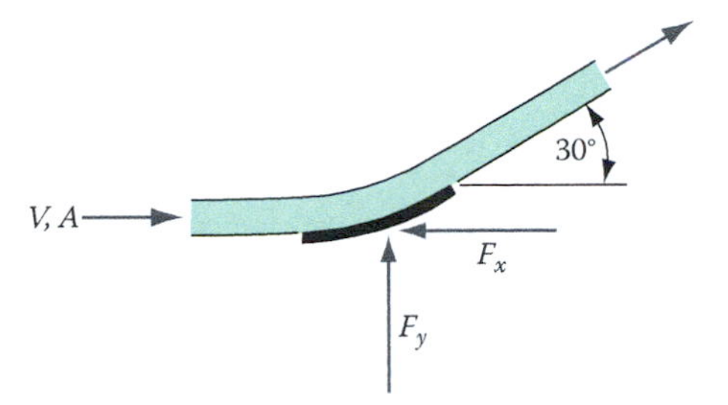

FIGURE P3.30

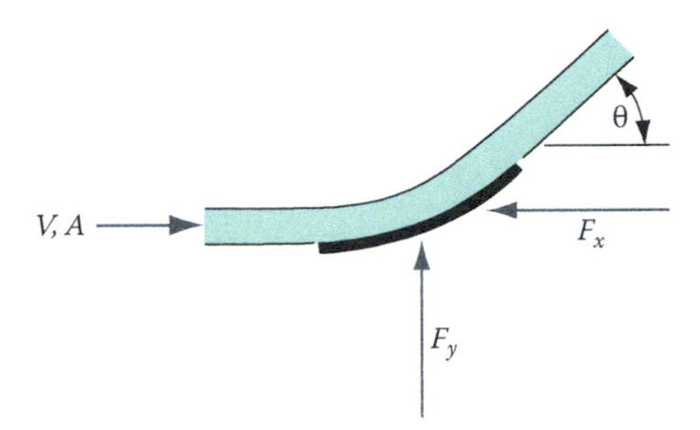

FIGURE P3.31

3.33 The forces F_x and F_y of Figure P3.32 are such that their magnitudes are related by $F_x = 3F_y$. Determine the angle through which the vane turns the liquid jet. The angle should range from 90° to 180°.

3.34 A liquid jet of velocity V and area A_1 strikes a vane that has a sharp-edged orifice, as shown in Figure P3.34. Liquid that leaves through the orifice has a volume flow rate equal to 1/2 that of the incoming jet. For frictionless flow, determine the force required to hold the plate stationary.

3.35 A liquid jet strikes a plate containing a sharp-edged orifice, as shown in Figure P3.35. The liquid jet velocity approaching the plate is V. For frictionless flow, determine the forces exerted on the plate by the liquid.

3.36 A liquid jet of velocity V and area A strikes a vane, as shown in Figure P3.36. The jet is deflected into two separate directions. Determine a relationship between the areas (inlet and both exits) if the magnitudes of the restraining forces are related by $F_x = 2F_y$ and the angle θ is 35°. Assume frictionless flow.

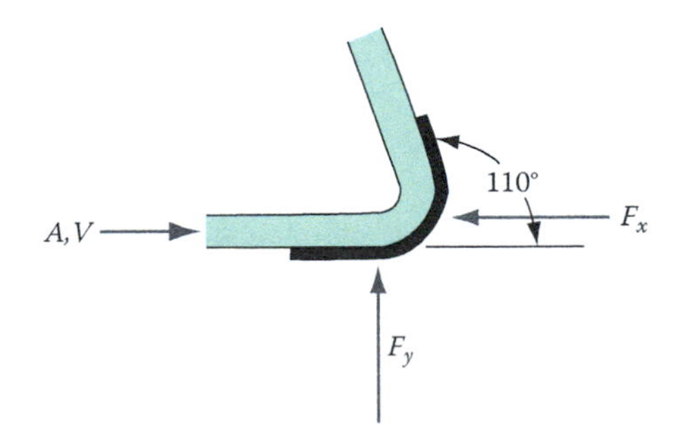

FIGURE P3.32

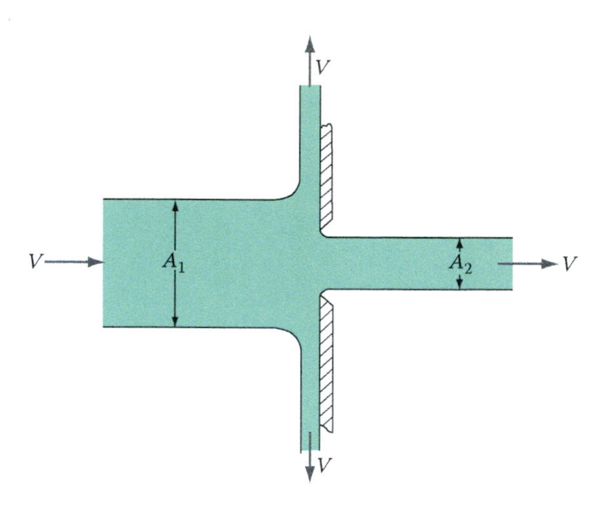

FIGURE P3.34

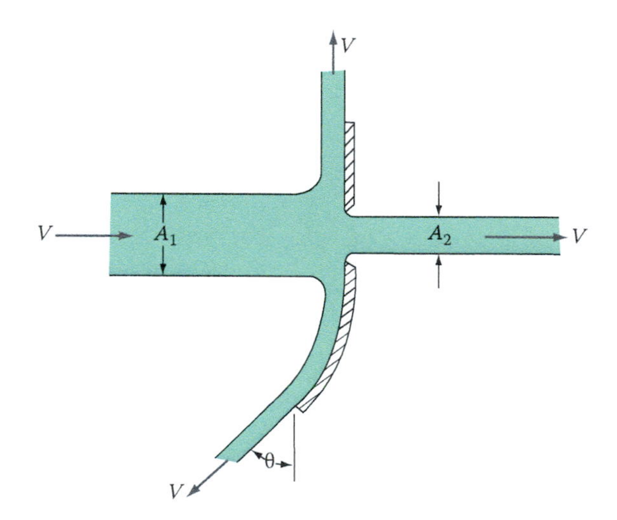

FIGURE P3.35

3.37 A water jet of velocity 12 ft/s and a cross-sectional area of 5 ft^2 strike a curved vane, as shown in Figure P3.37. The vane is moving at a velocity of 3 ft/s in the positive x-direction, and it deflects the jet through an angle θ of 60°. Assuming no frictional losses between the jet and the surface, determine the reaction forces.

3.38 Figure P3.38 shows a water jet of velocity V and area A that impacts a curved vane. Determine the forces exerted on the vane.

3.39 Figure P3.39 shows a copper reducing fitting located in a horizontal plane. Flow enters the fitting at section 1 with a velocity of V_1, goes through an elbow and exits to atmospheric pressure. The inside diameter at section 1 is D_1 and at section 2, the inside diameter is D_2. The pressure at section 1 is denoted as p_1. Determine the forces acting on the system: F_x and F_y. Assume frictionless flow.

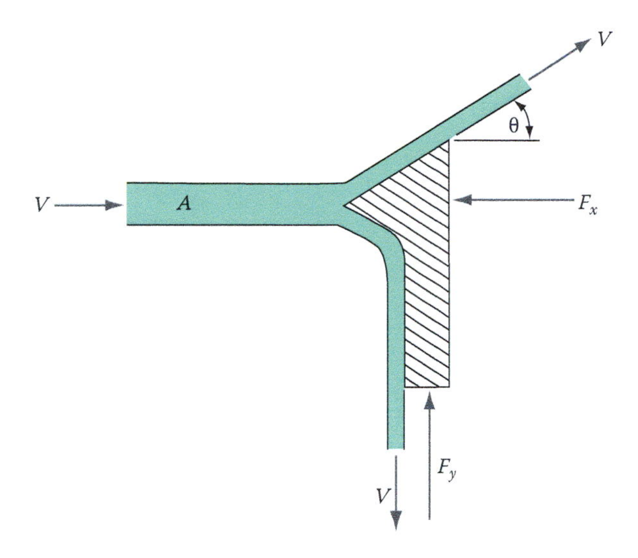

FIGURE P3.36

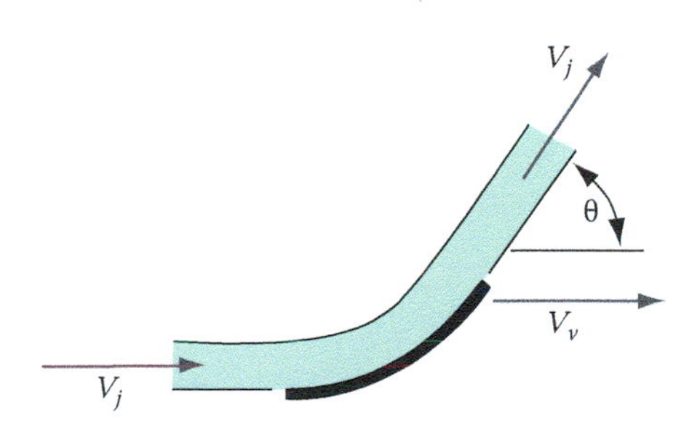

FIGURE P3.37

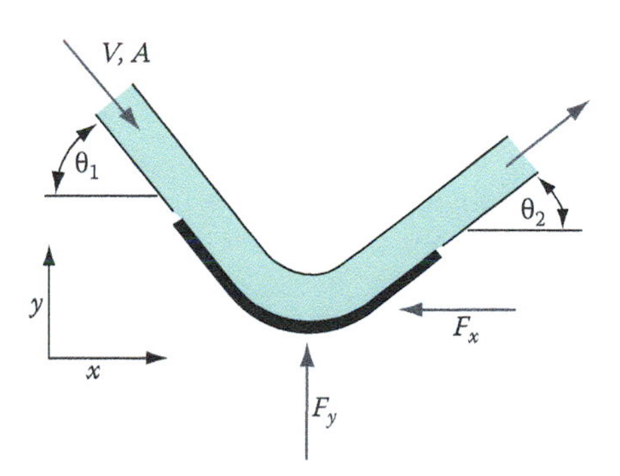

FIGURE P3.38

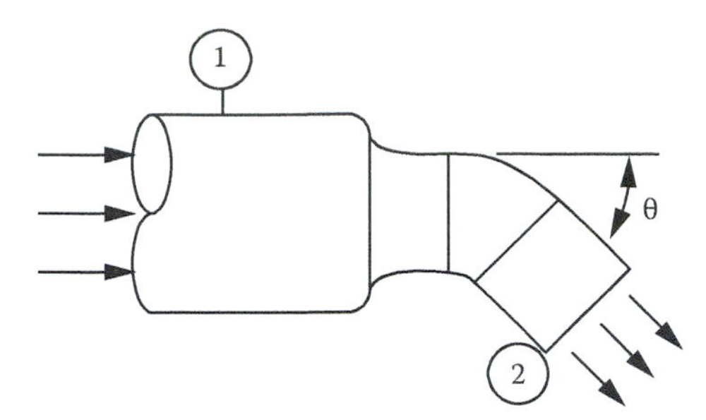

FIGURE P3.39

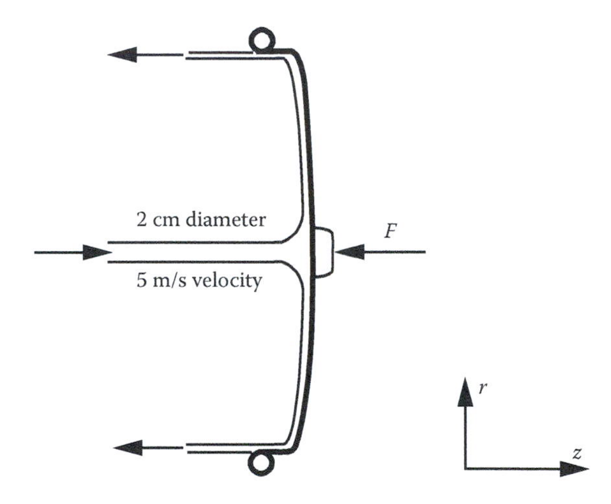

FIGURE P3.40

3.40 A garden hose is used to squirt water at someone who is protecting herself with a garbage can lid. Figure P3.40 shows the jet in the vicinity of the lid. Determine the restraining force F for the conditions shown.

Energy Equation

3.41 A pump is used in a piping system to move oil from one tank to another tank at a higher elevation, as indicated in Figure P3.41. The flow velocity in the piping system is 2 m/s and the oil has a specific gravity of 0.87. The pipe has an inside diameter of 10.23 cm. Determine the power required to pump the oil under these conditions.

3.42 A kerosene pump is illustrated in Figure P3.42. The flow rate is 0.015 m^3/s; inlet and outlet gage pressure readings are –8 and 200 kPa, respectively. Determine the required power input to the fluid as it flows through the pump.

3.43 Turbines convert the energy contained within a fluid into mechanical energy or shaft work. Turbines are often used in power plants with generators to produce electricity. One such installation is in a dam as shown in Figure P3.43. Water is permitted to flow through a passageway to the turbine, after which the water drains downstream. For the data given in Figure P3.43, determine the power available to the turbine when the discharge at the outlet is 30 m^3/s.

3.44 The Bernoulli equation was derived for an incompressible fluid. Derive a corresponding expression for an ideal gas at constant temperature.

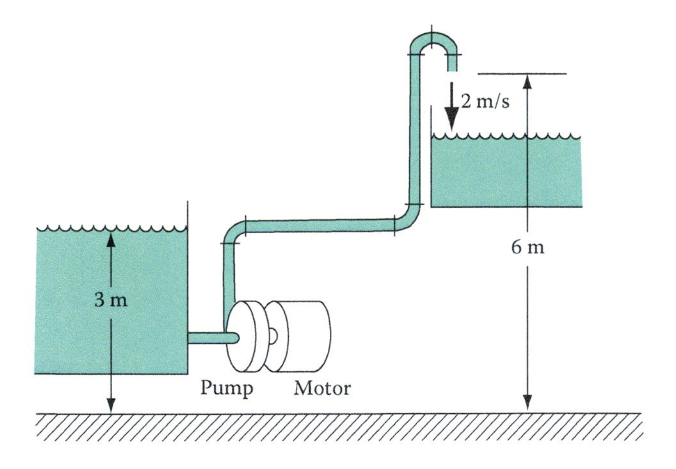

FIGURE P3.41

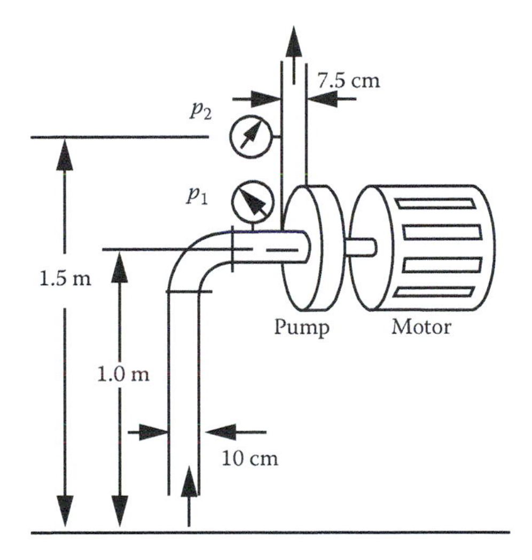

FIGURE P3.42

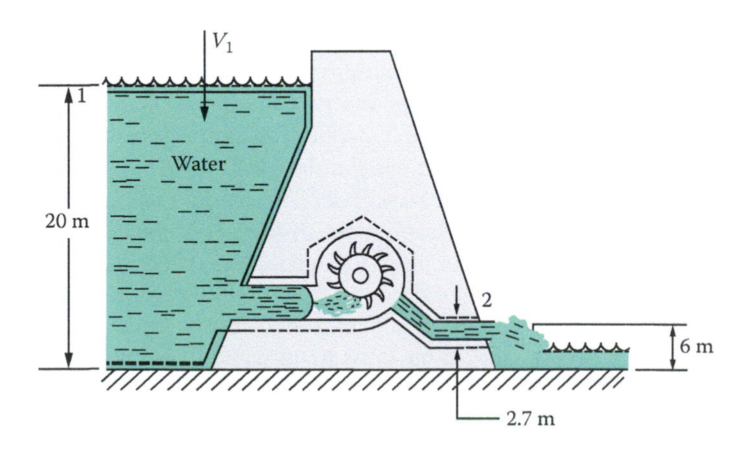

FIGURE P3.43

3.45 An axial flow window fan is used to ventilate a room. The fan is placed in a square housing 60 cm × 60 cm, which can be taken as the flow area. Typically, the difference in pressure across the fan is 0.08 cm of water. Determine the flow rate of air removed from the room if the fan power is 25 W. Take the room temperature to be 25°C. Neglect friction and assume the air to behave as an incompressible fluid (a good assumption at moderate temperatures and pressures).

3.46 An air turbine is used with a generator to generate electricity. Air at the turbine inlet is at 800 kPa and 25°C. The turbine discharges air to the atmosphere at a temperature of 11°C. Inlet and outlet air velocities are 100 and 2 m/s, respectively. Determine the work per unit mass delivered to the turbine from the air.

3.47 Diesel engines are used as motive power for pumping water. To recapture some of the energy contained in the diesel exhaust gases, they are to be run through a turbine. The exhaust gases enter the turbine in a line with a 12 in. inside diameter at a velocity of 10 ft/s with a temperature and pressure of 1200°F and 25 psia, respectively. At the turbine exit, the gases are at 600°F, 14.7 psia, and they have a velocity of 5 ft/s. Determine the work per unit mass delivered to the turbine and the horsepower. Take the properties of the exhaust gases to be the same as those for carbon dioxide and assume ideal gas behavior. (Enthalpy changes equal specific heat at constant pressure × temperature changes.) The process is considered to be adiabatic.

3.48 A steam turbine is used with an alternator in power plants for generation of electricity. Steam at 800 kPa and 400°C (enthalpy = 3 267.5 kJ/kg) traveling at 80 m/s enters the turbine from the boiler. The steam exits from the turbine; at a point 12 m below the inlet, the pressure is 10 kPa, the temperature is 100°C (enthalpy = 2 687.5 kJ/kg), and the velocity is 65 m/s. Determine the work per unit mass delivered to the turbine. (Note: These enthalpy values, taken from thermodynamics textbooks, are based on actual steam data.) Neglect heat-transfer effects.

3.49 Steam enters a turbine with a velocity of 30 m/s, with enthalpy of 3000 kJ/kg. At the outlet, the velocity is 100 m/s and enthalpy is 2600 kJ/kg. Assume a heat loss from the turbine of 0.60 kJ/s, with negligible changes in potential energy. Determine the power output of the turbine for a steady flow of 0.1 kg/s.

Bernoulli Equation

3.50 Figure P3.50 shows the exit of a spigot from which kerosene flows and impacts a flat surface. Section 1 has an inside diameter of 4 cm, and at section 2, the inside diameter is 2 cm. The pressure at sections 2 and 3 are both atmospheric (101.3 kPa). The distances h_1 and h_2 are 0.5 and 1 m, respectively. The pressure at section 1 is approximately 104 kPa. Determine the diameter of the jet at section 3.

3.51 Figure P3.51 depicts an interesting experiment to demonstrate what is known historically as the Torricelli law. A large diameter (~6 in.) clear plastic tube is mounted vertically and sealed at its bottom. At the top is a constant head reservoir with hose fittings attached. Water enters the reservoir and fills the vertical tube to a depth of H. Water that overflows the tube exits at the outlet, thereby providing a constant head in the tube. Pipes/valves are positioned at three locations on the vertical tube. One of the valves is opened and a water jet exits as depicted in the figure. If $z = 3$ ft and $H = 6$ ft, determine the distance x at the point where the jet impacts the surface. Take the jet exit diameter to be 1 in. and ignore frictional losses at the valve.

3.52 A water jet leaves a converging nozzle and is directed vertically upward as indicated in Figure P3.52. The diameter at section 1 is 3 cm, and at section 2, the diameter is 1 cm (i.e., jet diameter and nozzle exit diameter). If the pressure at section 1 is 25 kPa (gage), determine the height h_2 that the jet will attain. Take $h_1 = 20$ cm, and the pressure at section 2 to be atmospheric.

3.53 A household faucet produces a jet that gets smaller in diameter as it approaches the sink. At a certain flow, water coming out will fill an 8 oz glass in 8 s and produce a jet diameter at faucet exit of 1/2 in. If the sink bottom is 8 in. below the faucet exit, what is the jet diameter at impact?

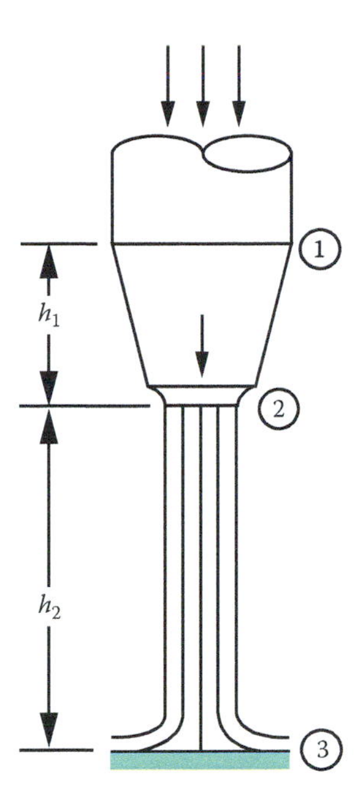

FIGURE P3.50

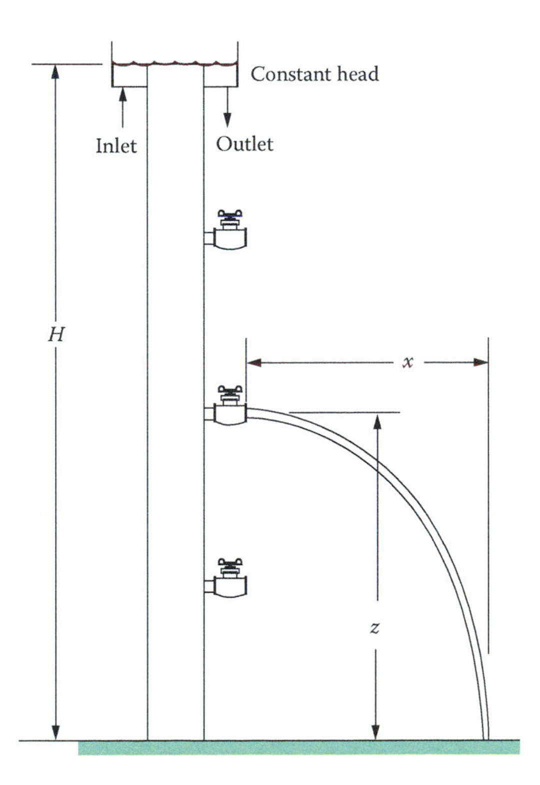

FIGURE P3.51

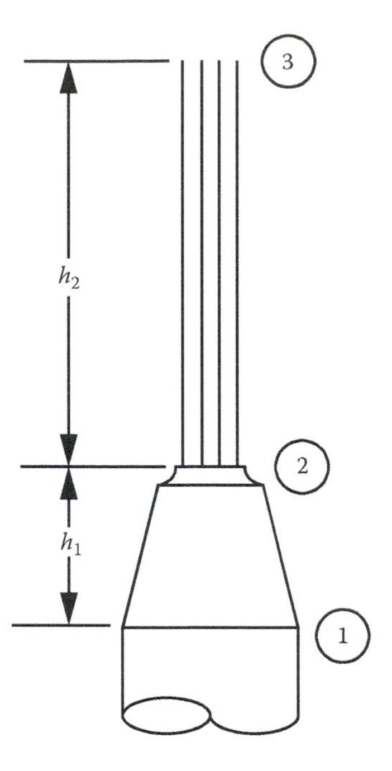

FIGURE P3.52

3.54 In Lake Geneva (Switzerland), there is a water jet (called the *Jet d'Eau*) that is discharged 130 m straight upward (measured from the surface of the lake). The exit of the discharge pipe is 20 cm in diameter. Calculate the velocity of the water as it leaves the pipe, and the volume flow rate of the water.

3.55 A carburetor of an automobile consists of a flow passage similar to that of a venturi meter. Air passes through the carburetor and draws in gasoline at the throat. Consider the venturi meter shown in the sketch. Air passes through and is accelerated to a high velocity at the throat. When the velocity increases, the pressure decreases to a point where water is drawn up into the throat through a small diameter tube. Determine the required flow air rate through the meter necessary to draw water up from the tank below. Treat the air as an incompressible fluid at room temperature conditions. Take the air pressure upstream of the meter to be 1.5 atm (see Figure P3.55).

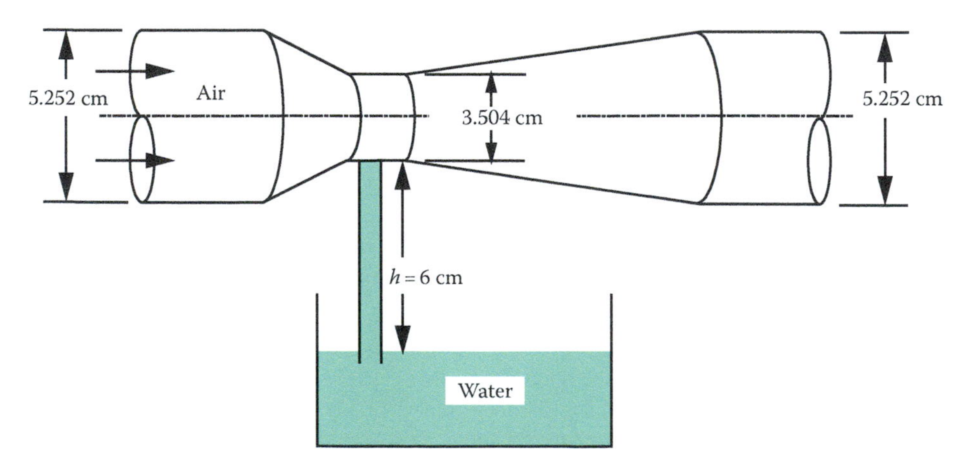

FIGURE P3.55

3.56 Air flows through a venturi meter as indicated in Figure P3.56. An air-over-water manometer is attached to the meter and when there is no air flow, the water level at the throat is 3 cm below the throat wall, as shown. Determine the volume flow rate of air required to raise the liquid level to the connection at the throat. Assume the air to be incompressible.

3.57 A venturi meter is a device placed in a pipeline and is calibrated to give the volume rate of liquid through it as a function of pressure drop (see Figure P3.57). For a flow rate of 4 ft³/s through the 16 × 8 in. meter shown, determine the reading Δh on the manometer. Assume that the liquid in the meter and in the manometer is water.

3.58 Linseed oil flows steadily through the system of Figure P3.58. A linseed-oil-over-water U-tube manometer between the 3 and 5 cm sections reads 6 cm. The oil is discharged through a 2 cm diameter exit. Calculate the exit velocity of the linseed oil.

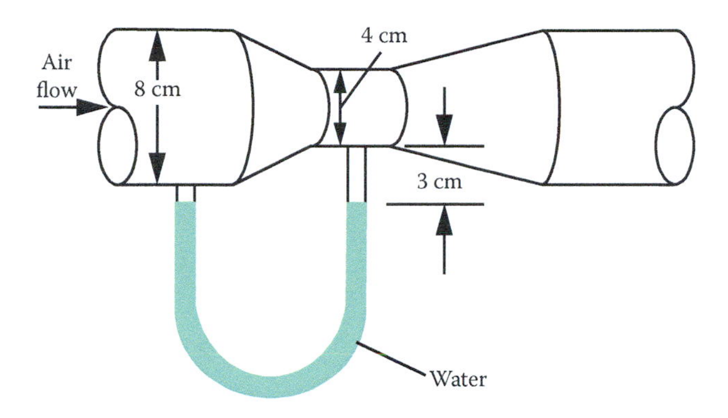

FIGURE P3.56

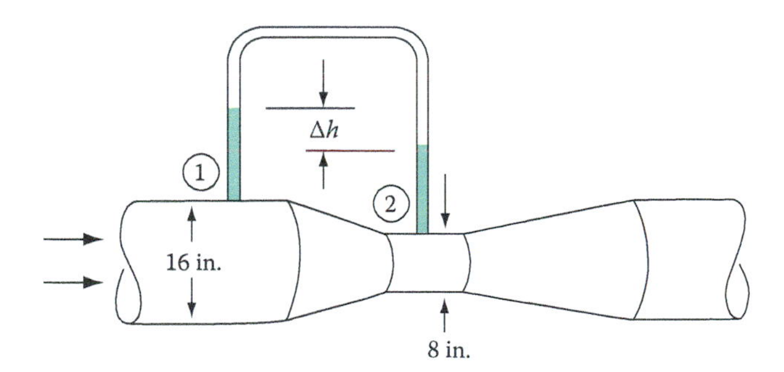

FIGURE P3.57

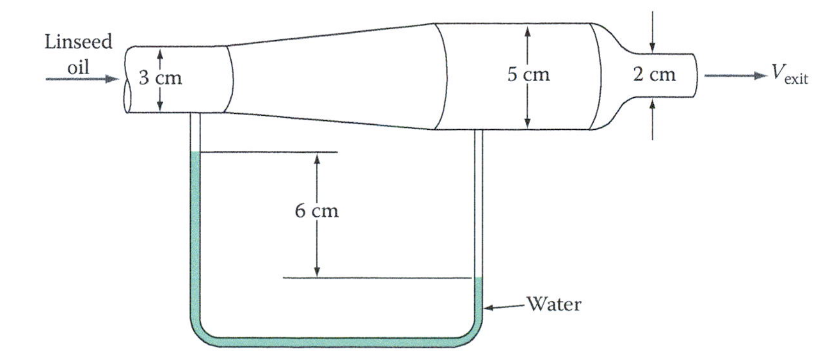

FIGURE P3.58

3.59 Water flows steadily through the system shown in Figure P3.59. The 4 in. section leads to a 3 in. throat, followed by a divergent section and, finally, a nozzle whose exit diameter is 2 in. An air-over-water manometer is connected between the 4 and 3 in. sections. Determine the expected reading Δh on the manometer if the velocity at the exit is 8 ft/s.

3.60 An open-top tank is discharging glycerine through an opening onto which an elbow has been placed, as in Figure P3.60. Determine the height of the glycerine jet, assuming frictionless flow and a tank diameter that is much greater than the exit pipe diameter.

3.61 Suppose the tank of Figure P3.61 is capped tightly and the space above the liquid surface is pressurized with compressed air. Determine the air pressure that would be required to make the liquid jet rise to a level equal to $3h$.

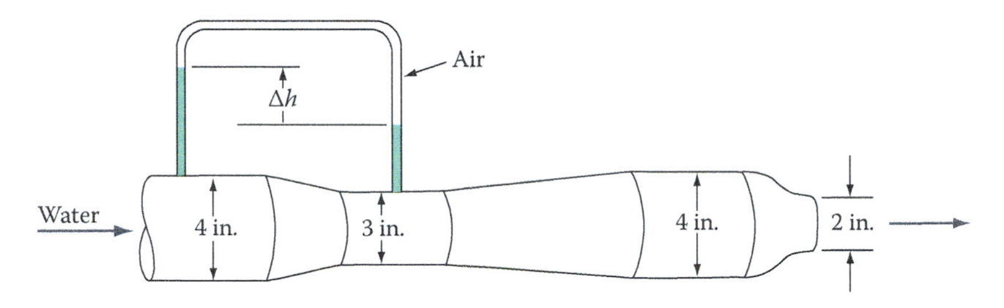

FIGURE P3.59

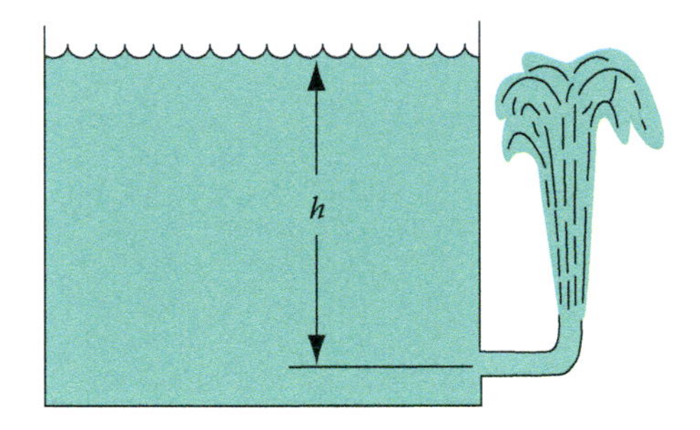

FIGURE P3.60

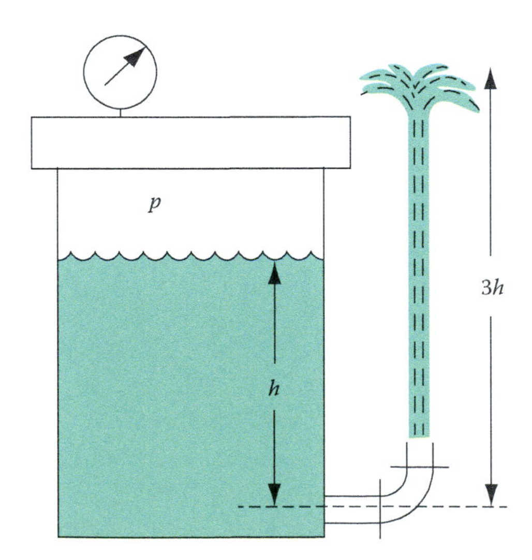

FIGURE P3.61

3.62 Water flows into a 1 m diameter tank at a rate of 0.006 m^3/s as illustrated in Figure P3.62. Water also leaves the tank through a 5 cm diameter hole near the bottom. At a certain height h, the efflux equals the influx. Determine the equilibrium height h.

3.63 Water flows into the tank of Figure P3.62 at a rate of 0.1 ft^3/s. Water also leaves the tank through a 1 in. diameter hole near the bottom. At a certain height h, the efflux and the influx are equal. Determine the equilibrium height h.

3.64 A siphon is used to drain a tank of water as shown in Figure P3.64. The siphon tube has an inside diameter of 3 in., and at the exit, there is a nozzle that discharges liquid in a 2 in. diameter jet. Assuming no losses in the system, determine the volume flow rate through the siphon. Calculate also the flow velocity in the 3 in. ID tube.

3.65 A siphon is used to drain a tank of water as shown in Figure P3.65. The siphon tube has an inside diameter of 3 in., and at the exit, there is a nozzle that discharges liquid in a 2 in. diameter jet. Assuming no losses in the system, determine the volume flow rate through the siphon. Calculate also the flow velocity in the 3 in. ID tube.

3.66 Figure P3.66 shows a water nozzle attached to a 3 in. inside diameter hose. The nozzle discharges a 1 in. outside diameter jet, and the pressure in the hose just upstream of the nozzle is 15 psig. Determine the volume flow rate through the nozzle if it is held horizontally.

3.67 A 3 cm water jet is discharged at 10 m/s at an angle of $\theta = 30°$ as shown in Figure P3.67. Determine the distance between the nozzle exit and the point of impact of the liquid jet.

3.68 A fire hydrant expelling water is shown schematically in Figure P3.68. The impact of the jet is measured to occur at $L = 6$ ft. Determine pressure within the hydrant if the nozzle is 18 in. above the ground. Take the kinetic energy within the center of the hydrant to be negligible.

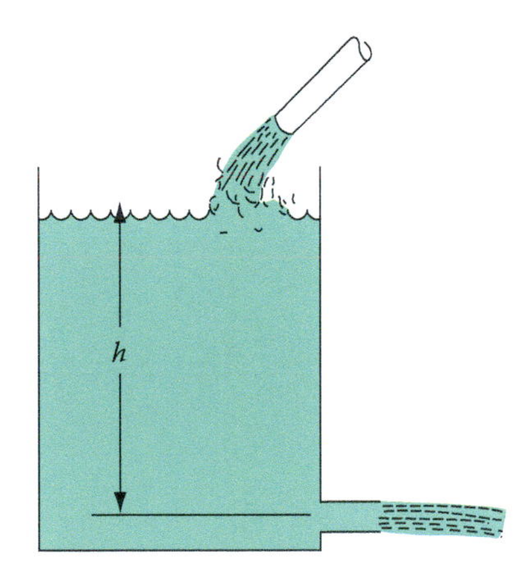

FIGURE P3.62

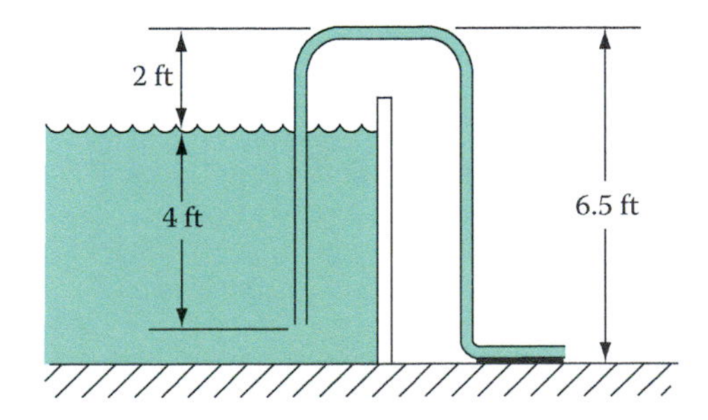

FIGURE P3.64

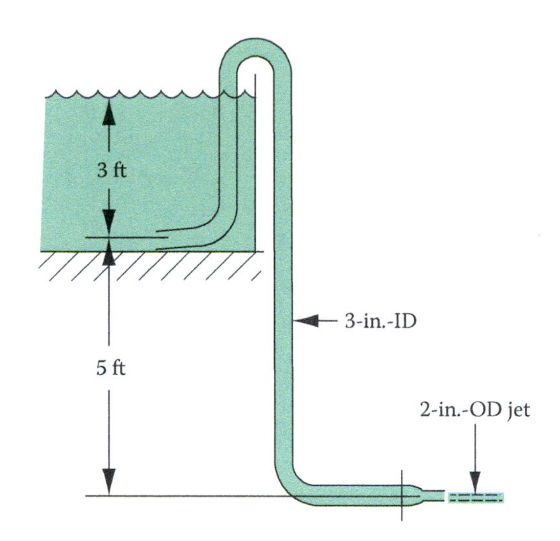

FIGURE P3.65

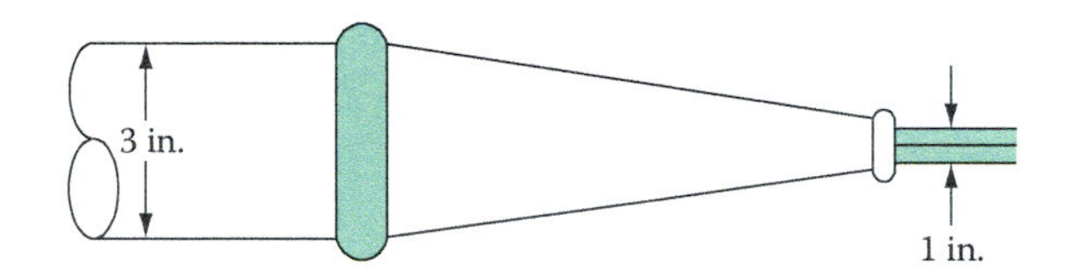

FIGURE P3.66

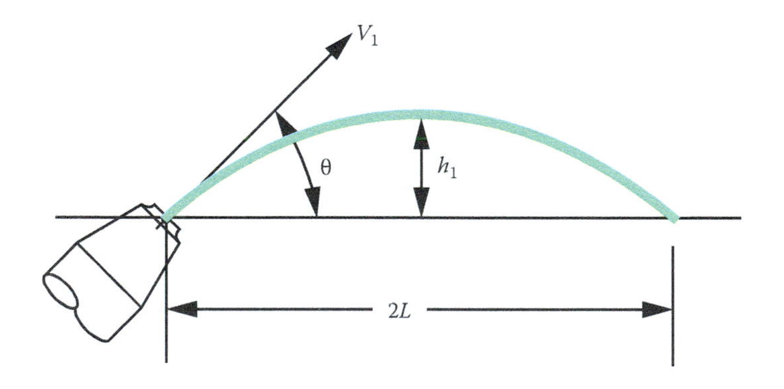

FIGURE P3.67

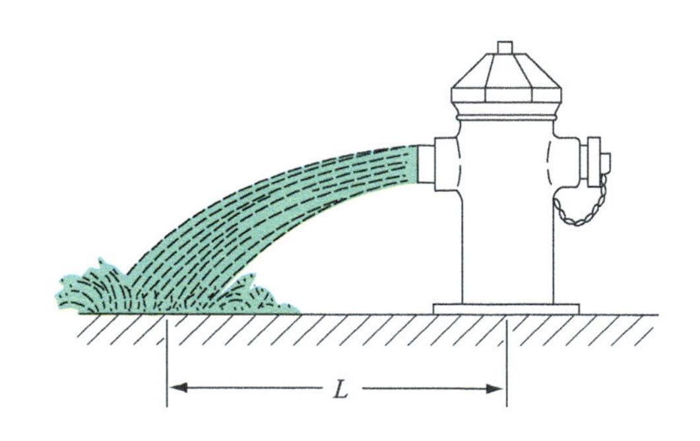

FIGURE P3.68

3.69 A hose whose outlet is inclined at an angle of 60° with the horizontal has an inside diameter of 4.3 in. A nozzle at the end has a diameter of 1.9 in. The maximum height attained by a jet of water from the nozzle is 3 ft above the nozzle exit. Determine the pressure inside the hose.

3.70 The tallest building in a small community is 60 ft tall. For fire protection, the fire marshal wants to be able to direct a water jet so that at the top of its trajectory, it will reach the roof of this building. If the water nozzle is to be located 30 ft from the building, determine the required water velocity at the nozzle exit. See Figure P3.70.

3.71 Consider water flow through an enlargement placed in a circular duct (Figure P3.71). A manometer is placed in the line and used to measure pressure difference across the expansion. For the dimension given, calculate the volume flow rate through the pipe. Take the manometer fluid to be mercury.

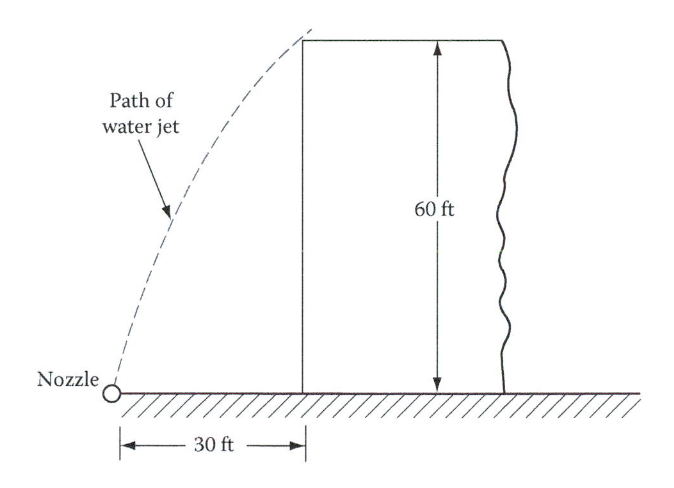

FIGURE P3.70

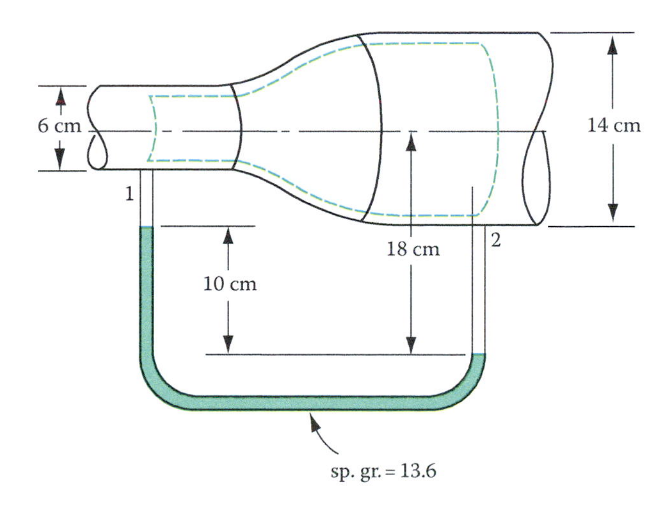

FIGURE P3.71

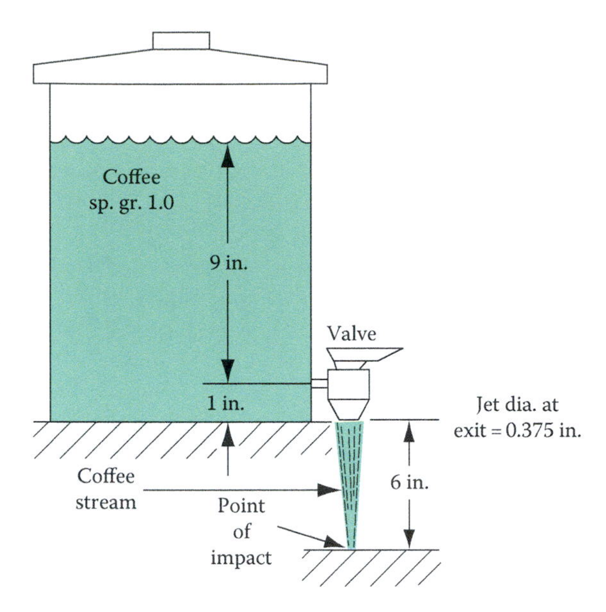

FIGURE P3.72

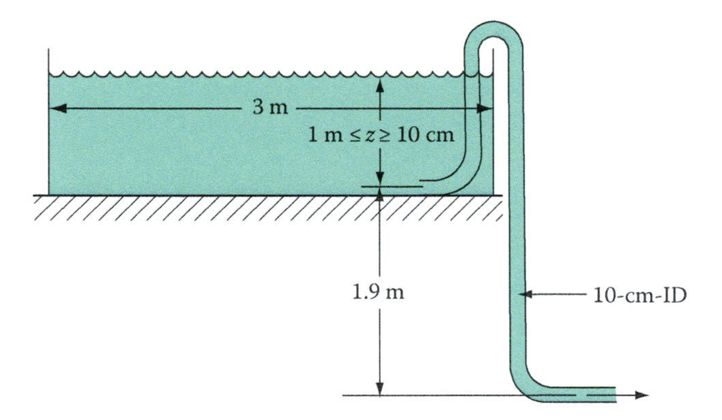

FIGURE P3.73

3.72 A pot contains coffee that is being drained through a valve. For the conditions shown in Figure P3.72, determine the time required to drain the pot from a depth of 10 to 2 in. The jet diameter is 0.375 in. at the valve exit.

3.73 Figure P3.73 shows a tank being drained by a 10 cm inside diameter siphon tube. Determine the time it takes to drain the tank from a depth of 1 m to a depth of 10 cm.

4 Dimensional Analysis and Dynamic Similitude

In the preceding chapter, we wrote equations that are basic to the study of fluid mechanics. We were able to apply them and solve them for many practical problems under special conditions. In a number of cases, the descriptive equations cannot be solved exactly, however. Consequently, we must formulate an alternative method of analysis. In this chapter, we will develop such a method by relying initially on the fact that equations must be dimensionally homogeneous. Further, we will define several dimensionless ratios that are commonly encountered in various flow situations. We will also see how these dimensionless ratios are used to advantage in correlating experimental data and in modeling problems.

Modeling problems involve testing a model of a prototype. Consider, for example, a company that wishes to manufacture an airplane. Under ideal conditions, a prototype might be built and tested in a wind tunnel in which a controlled airflow can be directed past the plane. Tests and measurements would be made on the stationary plane while it is in the wind tunnel. If the airplane is large, however, a sizable wind tunnel is needed with huge fans and a tremendous power requirement, since speeds in excess of 500 mi/h may be necessary. The cost of building a prototype aircraft and testing it in an appropriate facility can therefore be prohibitive. Alternatively, a scale model of the plane can be built and tested in a smaller tunnel. Measurements on the model can then be scaled up to predict performance of the prototype even though the prototype has not been built. Clearly, the cost is considerably less. Modeling wind flow past a building, for example, is an economical way of finding the effect of a hurricane on the structure. Thus, modeling problems have many useful applications. In this chapter, we will see how dimensionless ratios are used in modeling problems.

4.1 DIMENSIONAL HOMOGENEITY AND ANALYSIS

An equation must be dimensionally homogeneous. That is, all terms in the equation must have the same dimensions or units. To compare dimensions of various terms within an equation, it is convenient to first convert the dimensions into fundamental units. Suppose we are working within the British gravitational system, in which force (F), length (L), and time (T) are considered fundamental. Mass (M) is a derived unit defined as $M = F \cdot T^2/L$. Consider, for example, the Bernoulli equation, derived in Chapter 3:

$$\frac{p}{\rho} + \frac{V^2}{2} + gz = \text{a constant}$$

If we use the generalized notation, the dimensions of each term become

$$\frac{p}{\rho} = \frac{F}{L^2}\frac{L^3}{M} = \frac{F}{L^2}L^3\frac{L}{F \cdot T^2} = \frac{L^2}{T^2}$$

$$\frac{V^2}{2} = \frac{L^2}{T^2}$$

$$gz = \frac{L}{T^2}L = \frac{L^2}{T^2}$$

Thus, each term has dimensions of L^2/T^2 in a system in which F, L, and T dimensions are considered fundamental (British gravitational system). We could reformulate the dimensions of the Bernoulli equation in the system in which only mass (M), length (L), and time (T) are fundamental (SI system). Force is a derived unit defined as $F = M \cdot L/T^2$. Evaluating the Bernoulli equation in this system, we get

$$\frac{p}{\rho} = \frac{M \cdot L}{T^2} \frac{1}{L^2} \frac{L^3}{M} = \frac{L^2}{T^2}$$

$$\frac{V^2}{2} = \frac{L^2}{T^2}$$

$$gz = \frac{L}{T^2} L = \frac{L^2}{T^2}$$

In Chapter 1, we discussed the English engineering system, in which force, mass, length, and time are considered fundamental. In this system, equations are often written with a conversion factor because mass and force are both considered fundamental. The Bernoulli equation, for example, would be written as

$$\frac{p}{p} + \frac{V^2}{2g_c} + \frac{gz}{g_c} = \text{a constant}$$

where $g_c = 32.2\ \text{lbm} \cdot \text{ft/(lbf} \cdot \text{s}^2)$. Thus, g_c has dimensions of $M \cdot L/(F \cdot T^2)$. The dimensions of each term in the Bernoulli equation become

$$\frac{p}{\rho} = \frac{F}{L^2} \frac{L^3}{M} = \frac{F \cdot L}{M}$$

$$\frac{V^2}{2g_c} = \frac{L^2}{T^2} \frac{F \cdot T^2}{M \cdot L} = \frac{F \cdot L}{M}$$

$$\frac{gz}{g_c} = \frac{L}{T^2} \frac{F \cdot T^2}{M \cdot L} L = \frac{F \cdot L}{M}$$

Again each term of the equation has the same dimensions. The equation is thus dimensionally homogeneous. Table 4.1 provides dimensions of physical quantities that are commonly encountered in fluid mechanics.

4.1.1 Rayleigh Method

We now formulate the first technique of dimensional analysis—known as the **Rayleigh method—** by an example. Consider that we are trying to determine the force exerted on a flat plate by a jet of liquid. Without knowing what was derived in Chapter 3, we might speculate that this force is a function of liquid density (ρ), jet velocity (V), jet area (A), angle of attack between the jet and the plate (θ), and the distance between the nozzle and the plate (l). Thus, force (F) is an as yet undetermined function of these variables.

$$F = F(\rho, V, A, \theta, l) \tag{4.1}$$

TABLE 4.1
Dimensions and Units of Common Parameters in Fluid Mechanics

Symbol	Quantity	Dimensional System: F, L, T	M, L, T	F, M, L, T	SI	Units: British Gravitational or English
A	Area	L^2	L^2	L^2	m^2	ft^2
a	Acceleration	L/T^2	L/T^2	L/T^2	m/s^2	ft/s^2
a	Sonic velocity	L/T	L/T	L/T	m/s	ft/s
F	Force	F	$M \cdot L/T^2$	F	N = kg · m/s^2	lbf
L	Length	L	L	L	m	ft
m	Mass	$F \cdot T^2/L$	M	M	kg	slug or lbm
$\dot{m}$	Mass flow rate	$F \cdot T/L$	M/T	M/T	kg/s	slug/s or lbm/s
P, W	Power	$F \cdot L/T$	$M \cdot L^2/T^3$	$F \cdot L/T$	W = N · m/s	ft · lbf/s
p	Pressure	F/L^2	$M/(L \cdot T^2)$	F/L^2	Pa = N/m^2	lbf/ft^2
Q	Volume flow rate	L^3/T	L^3/T	L^3/T	m^3/s	ft^3/s
t	Time	T	T	T	s	s
$V \cdot U$	Velocity	L/T	L/T	L/T	m/s	ft/s
$\forall$	Volume	L^3	L^3	L^3	m^3	ft^3
W	Work	$F \cdot L$	$M \cdot L^2/T^2$	$F \cdot L$	J = N · m	lbf · ft
β	Coefficient of compressibility	L^2/F	$L \cdot T^2/M$	L^2/F	m^2/N	ft^2/lbf
θ	Angle (radians)	—	—	—	—	—
μ	Absolute viscosity	$F \cdot T/L^2$	$M/(L \cdot T)$	$F \cdot T/L^2$	N · s/m^2	lbf · s/ft^2
ν	Kinematic viscosity	L^2/T	L^2/T	L^2/T	m^2/s	ft^2/s
ρ	Density	$F \cdot T^2/L^4$	M/L^3	M/L^3	kg/m^3	slug/ft^3 or lbm/ft^3
σ	Surface tension	F/L	M/T^2	F/L	N/m	lbf/ft
τ	Shear stress	F/L^2	$M/L \cdot T^2$	F/L^2	Pa = N/m^2	lbf/ft^2
ω	Angular velocity	$1/T$	$1/T$	$1/T$	1/s	1/s

Next we assume that the functional dependence can be written as a product by separating the variables according to

$$F = C_1 \rho^{a1} V^{a2} A^{a3} \theta^{a4} l^{a5} \tag{4.2}$$

where the a_i exponents are unknown, as is the constant C_1. We now write Equation 4.2 with respective dimensions substituted for each term:

$$\frac{M \cdot L}{T^2} = C_1 \left(\frac{M}{L^3}\right)^{a1} \left(\frac{L}{T}\right)^{a2} \left(L^2\right)^{a3} (0)^{a4} (L)^{a5}$$

where we have arbitrarily elected to use the M, L, T system. However, either system is acceptable. The dimension of θ is in radians. Taking the natural logarithm of each quantity yields

$$\begin{aligned} ln M + ln L - 2 l n T = l n C_1 &+ (a_1 l n M - 3a_1 l n L) \\ &+ (a_2 l n L - a_2 l n T) + (2a_3 l n L) \\ &+ (a_4 l n 0) + (a_5 l n L) \end{aligned}$$

Since this equation must be dimensionally homogeneous, all coefficients of the time terms on both sides of the equation must be equal. Thus,

$$T\text{: } -2 = -a_2$$

For M and L, we have

$$M\text{: } 1 = a_1$$

$$L\text{: } 1 = -3a_1 + a_2 + 2a_3 + a_5$$

There are three equations and four unknowns. Solving simultaneously gives

$$a_2 = 2$$

$$a_1 = 1$$

$$a_3 = 1 - \frac{a_5}{2}$$

Substituting into Equation 4.2, we get

$$F = C_1 \rho V^2 \frac{A}{A^{a_5/2}} \theta^{a_4} l^{a_5}$$

Combining like terms yields

$$\frac{F}{\rho V^2 A} = C_1 \left(\frac{1}{A^{1/2}} \right)^{a_5} \theta^{a_4} \tag{4.3}$$

In functional form,

$$\frac{F}{\rho V^2 A} = F\left(\frac{l}{A^{1/2}}, \theta \right) \tag{4.4a}$$

As a check on this result, we see that each resulting term is dimensionless.

Note that in solving the algebraic equations simultaneously, a_5 was left as an unknown. If a_3 is left as unknown instead, the final form of the equation becomes

$$\frac{F}{\rho V^2 l^2} = F\left(\frac{A}{l^2}, \theta \right) \tag{4.4b}$$

Both forms, Equations 4.4a and 4.4b, are correct. In fact, multiplication of $F/\rho V^2 l^2$ by l^2/A in Equation 4.4b gives $F/V^2 A$ of Equation 4.4a. This result suggests that it is possible to manipulate the final result algebraically to obtain other forms with no loss in accuracy or generality. In other words, once the dimensionless terms or groups are determined, the result can be modified to obtain other dimensionless groups without starting over and solving for different exponents. As an example, consider Equation 4.4a:

$$\frac{F}{\rho V^2 A} = F\left(\frac{l}{A^{1/2}}, \theta \right) \tag{4.4a}$$

We can use any combination of groups to yield other groups. Thus,

$$\frac{F}{\rho V^2 A}\cdot\left(\frac{A^{1/2}}{l}\right)^2 = \frac{F}{\rho V^2 l^2}$$

The new term $F/\rho V^2 l^2$ can now be substituted in Equation 4.4a for either of the terms used to produce it to obtain a different relationship:

$$\frac{F}{\rho V^2 l^2} = F\left(\frac{l}{A^{1/2}},\theta\right) \tag{4.4c}$$

or

$$\frac{F}{\rho V^2 A} = F\left(\frac{F}{\rho V^2 l^2},\theta\right) \tag{4.4d}$$

Equations 4.4a through 4.4d are equally correct because in each equation, we have a dimensionally homogeneous relationship. In addition, as was done here, we can use the reciprocal of any term or the square of any term (or more than two terms) as long as our result is a relationship that is dimensionally homogeneous. Thus, dimensional analysis provides a means for us to express a functional relationship in terms of the variables of the problem at hand.

Returning to our original objective of predicting the force exerted on a flat plate by a jet of liquid, we would at this point have to make actual measurements on a jet impinging on a flat plate to determine values for the exponents and constants that are still unknown (e.g., C_1, a_4, and a_5 of Equation 4.3).

Thus, we began with a functional relationship in an equation (Equation 4.2) containing one unknown constant and five unknown exponents. By using dimensional analysis to reduce the number of unknowns to three, we are required to perform fewer experiments—and, moreover, the final form of the equation is known. At the outset, we speculated as to which variables were important. In performing experiments to evaluate exponents, we would learn whether the variables originally chosen were significant. If the final derived equation did not correlate the measurements satisfactorily, we would rederive an expression with dimensional analysis and include more quantities.

Example 4.1

An open-top tank is filled to a depth h with liquid of density ρ and viscosity μ. A hole of diameter D is drilled in the side of the tank, and a jet of liquid issues from it. Develop an expression for the efflux velocity V using dimensional analysis.

Solution

The problem does not state what variables to use in the analysis. We therefore rely on intuitive reasoning to determine which variables are significant. We hypothesize that the efflux velocity V depends on the liquid height h, the liquid properties ρ and μ, gravity g, and the hole diameter D. Using the M, L, T system, we write

$$V = V(h, D, \rho, \mu, g) = C_1 h^{a_1} D^{a_2} \rho^{a_3} \mu^{a_4} g^{a_5} \tag{i}$$

Substituting dimensions,

$$\frac{L}{T} = C_1 L^{a_1} L^{a_2}\left(\frac{M}{L^3}\right)^{a_3}\left(\frac{M}{L\cdot T}\right)^{a_4}\left(\frac{L}{T^2}\right)^{a_5}$$

For each dimension, the exponents may be written without the intermediate step of taking the logarithm. We have

$$M: 0 = a_3 + a_4$$
$$T: -1 = -a_4 - 2a_5$$
$$L: 1 = a_1 + a_2 - 3a_3 - a_4 + a_5$$

Solving the mass equation gives

$$\underline{a_3 = -a_4}$$

The time equation can be rearranged to get

$$\underline{a_5 = \frac{1}{2} - \frac{a_4}{2}}$$

Substituting these into the length equation gives

$$1 = a_1 + a_2 + 3a_4 - a_4 + \frac{1}{2} - \frac{a_4}{2}$$
$$\frac{1}{2} = a_1 + a_2 + \frac{3a_4}{2}$$

Solving,

$$\underline{a_1 = \frac{1}{2} - a_2 - \frac{3a_4}{2}}$$

Combining these exponents with Equation i of this example gives

$$V = C_1 \frac{h^{1/2}}{h^{a_2} h^{3a_4/2}} D^{a_2} \frac{1}{\rho^{a_4}} \mu^{a_4} \frac{g^{1/2}}{g^{a_4/2}}$$

Grouping terms with like exponents yields

$$\frac{V}{(gh)^{1/2}} = C_1 \left(\frac{D}{h}\right)^{a_2} \left(\frac{\mu}{\rho h^{3/2} g^{1/2}}\right)^{a_4}$$

In functional form,

$$\frac{V}{(gh)^{1/2}} = V\left(\left(\frac{D}{h}\right)\left(\frac{\mu}{\rho h^{3/2} g^{1/2}}\right)\right)$$

The original assumption of the variables thought important now has to be proven by experiment. If a low or no correlation (statistically) between the preceding variables is found to exist, the analysis must be repeated with more or with different terms to find a new functional dependence.

It is prudent in problems of the type in the preceding example to have methods of checking the results. As can be discerned from the solution, the final equation should contain only dimensionless terms or groups. In addition, we can determine how many dimensionless terms or groups to expect. Consider for example, the relationship written in Equation 4.1 for the force exerted on a flat plate by a jet of liquid:

$$F = F(\rho, V, A, \theta, l) \tag{4.1}$$

There are six variables in this equation (F, ρ, V, A, θ, and l). We used the M, L, T systems (three fundamental dimensions), which yielded three equations that related the a_i exponents. The number of dimensionless groups that result is

$$\begin{pmatrix}\text{Number of}\\ \text{variables}\end{pmatrix} - \begin{pmatrix}\text{Number of}\\ \text{fundamental}\\ \text{dimensions}\end{pmatrix} = \begin{pmatrix}\text{Number of}\\ \text{dimensionless}\\ \text{groups}\end{pmatrix}$$

For Equation 4.1, we would therefore expect to obtain

6 variables − 3 dimensions = 3 dimensionless groups

Equations 4.4a through 4.4d show this to be the case.

Likewise for the draining tank in Example 4.1, we wrote

$$V = V(h, D, \rho, \mu, g)$$

We used the M, L, T system and each dimension yielded an equation. We would therefore expect

6 variables − 3 dimensions = 3 dimensionless groups

4.1.2 Buckingham pi Method

Another approach to dimensional analysis is through the **Buckingham pi method**. In this method, the dimensionless ratios are called Π groups or Π parameters and could have been developed in a manner outlined by the following steps:

1. Select variables that are considered pertinent to the problem by using intuitive reasoning and write the relationship between the variables all on one side of the equation.
2. Choose repeating variables—that is, variables containing all m dimensions of the problem. It is convenient to select one variable that specifies scale (e.g., a length), another that specifies kinematic conditions (velocity), and one that is a characteristic of the fluid (such as density). For a system with three fundamental dimensions, ρ, V, and D are suitable. For a system with four fundamental dimensions, ρ, V, D, and g_c are satisfactory. The repeating variables chosen should be independent of each other.
3. Write Π groups in terms of unknown exponents.
4. Write the equations relating the exponents for each group and solve them.
5. Obtain the appropriate dimensionless parameters by substitution into the Π groups.

Solving a problem by following these five steps is illustrated in the next example.

Example 4.2

Liquid flows horizontally through a circular tube filled with sand grains that are spherical and of the same diameter. As liquid flows through this sand bed, the liquid experiences a pressure drop. The pressure drop $\Delta \rho$ is a function of average fluid velocity V, diameter D of the sand grains, spacing S between adjacent grains, liquid density ρ, and viscosity μ. Use dimensional analysis to determine an expression for the pressure drop in terms of these variables.

Solution

We solve using the F, M, L, T system:

Step 1: $f(\Delta \rho, V, D, S, \rho, \mu, g_c) = 0$.
Step 2: Repeating variables $\rightarrow \rho, V, D, g_c$.
Step 3: We are expecting three dimensionless groups or Π parameters (7 variables − 4 dimensions = 3 groups):

$$\Pi_1 = \rho^{a_1} V^{b_1} D^{c_1} g_c^{d_1} \Delta p^{e_1} = \left(\frac{M}{L^3}\right)^{a_1} \left(\frac{L}{T}\right)^{b_1} (L)^{c_1} \left(\frac{M \cdot L}{F \cdot T^2}\right)^{d_1} \left(\frac{F}{L^2}\right)^{e_1}$$

$$\Pi_2 = \rho^{a_2} V^{b_2} D^{c_2} g_c^{d_2} S^{e_2} = \left(\frac{M}{L^3}\right)^{a_2} \left(\frac{L}{T}\right)^{b_2} (L)^{c_2} \left(\frac{M \cdot L}{F \cdot T^2}\right)^{d_2} (L)^{e_2}$$

$$\Pi_3 = \rho^{a_3} V^{b_3} D^{c_3} g_c^{d_3} \mu^{e_3} = \left(\frac{M}{L^3}\right)^{a_3} \left(\frac{L}{T}\right)^{b_3} (L)^{c_3} \left(\frac{M \cdot L}{F \cdot T^2}\right)^{d_3} \left(\frac{F \cdot T}{L^2}\right)^{e_3}$$

Step 4:
Π_1 group:

$$M: 0 = a_1 + d_1$$

$$L: 0 = -3a_1 + b_1 + c_1 + d_1 - 2e_1$$

$$T: 0 = -b_1 - 2d_1$$

$$F: 0 = d_1 + e_1$$

Solving, we get $a_1 = -d_1$, $b_1 = -2d_1$, $c_1 = 0$, $e_1 = d_1$.

Π_2 group:

$$M: 0 = a_2 + d_2$$

$$L: 0 = -3a_2 + b_2 + c_2 + d_2 - e_2$$

$$T: 0 = -b_2 - 2d_2$$

$$F: 0 = d_2$$

Solving, we get $a_2 = 0$, $b_2 = 0$, $c_2 = -e_2$, $d_2 = 0$.

Π_3 group:

$$M: 0 = a_3 + d_3$$

$$L: 0 = -3a_3 + b_3 + c_3 + d_3 - 2e_3$$

$$T: 0 = -b_3 - 2d_3 + e_3$$

$$F: 0 = -d_3 + e_3$$

Solving, we get $a_3 = -d_3$, $b_3 = -d_3$, $c_3 = -d_3$, $e_3 = d_3$.

Step 5:

$$\Pi_1 = \left(\frac{\Delta p g_c}{\rho V^2}\right)^{d_1}$$

$$\Pi_2 = \left(\frac{D}{S}\right)^{c_2}$$

$$\Pi_3 = \left(\frac{\mu g_c}{\rho V D}\right)^{d_3}$$

We can now write the solution as

$$\frac{\Delta p g_c}{\rho V^2} = f\left(\frac{D}{S}, \frac{\mu g_c}{\rho V D}\right)$$

4.2 DIMENSIONLESS RATIOS

There are several main classes of problems in fluid mechanics: those that involve closed-conduit flows (such as pipe flow), those that involve flows with one surface exposed (such as flow in a river), those that involve flow past an object (such as an airplane in flight), and those that involve no contact with any surfaces (such as a spray). In this section, we will derive dimensionless ratios that are significant for these categories to serve as an introduction to each.

4.2.1 Flow in a Pipe or Conduit

Flow in a closed conduit results from a difference in pressure Δp from inlet to outlet. The pressure inside decreases linearly with length L and is affected by fluid properties and flow rate. If the velocity of flow is high with respect to the sonic velocity of the fluid medium, as with compressible flows, then sonic velocity too must be included. Thus, for a general pipe flow problem, we have

$$\Delta p = \Delta p(\rho, V, \mu, D, L, a) \tag{4.5a}$$

where a is introduced as the sonic velocity. Performing a dimensional analysis, we obtain

$$\frac{\Delta p}{\rho V^2} = \Delta p\left(\frac{\rho V D}{\mu}, \frac{V}{a}, \frac{L}{D}\right) \tag{4.5b}$$

The term on the left-hand side, when written $\Delta p/½\rho V^2$, is called the **pressure coefficient**; the ψ is added as a matter of custom to make the denominator a kinetic energy term. Next, $\rho VD/\mu$ is called the **Reynolds number** of the flow.

Finally, V/a is known as the **Mach number**, which is important in gas flows. It is interesting to manipulate these dimensionless ratios to discover that each is a ratio of forces. Suppose we examine the Reynolds number. The dimensions of numerator and denominator are

$$\rho V D = \frac{M}{L^3}\frac{L}{T}L = \frac{F \cdot T^2}{L^4}\frac{L}{T}L = \frac{F \cdot T}{L^2}$$

$$\mu = \frac{F \cdot T}{L^2}$$

The force in the numerator is an inertia force; the denominator is a viscous force. Thus, the Reynolds number is a ratio of inertia to viscous forces in the flow. Similarly, the pressure coefficient is a ratio of pressure forces to inertia forces. The Mach number is a ratio of dynamic force at the flow velocity to the dynamic force at the sonic velocity. Flow of incompressible fluids in conduits is treated in more detail in Chapter 5, and flow of compressible fluids in conduits is treated in more detail in Chapter 8.

4.2.2 Flow over Immersed Bodies

When an object flows through a fluid medium, forces are exerted on the object due to pressure variations in the flow field and due to skin friction or viscous effects along the surface. The force exerted in the direction of flow is called the **drag force** D_f. The flow acting perpendicular to the flow direction is the **lift force** L_f. Both act over an area. We therefore write

$$\Delta p = \Delta p(D_f, L_f, \rho, V, \mu, D) \tag{4.6a}$$

Solving, we obtain

$$\frac{\Delta p}{\rho V^2} = \Delta p\left(\frac{D_f}{\rho V^2 D^2}, \frac{L_f}{\rho V^2 D^2}, \frac{\rho V D}{\mu}\right) \tag{4.6b}$$

The term containing drag force when written as $D_f/½\rho V^2D^2$ is called the **drag coefficient** and represents the ratio of drag to inertia forces. The term $L_f/½\,\rho V^2D^2$ is called the **lift coefficient**. Flow over immersed bodies is treated in detail in Chapter 6.

4.2.3 Open-Channel Flow

Flow in an open channel such as a river or spillway is maintained by gravity forces analogous to pressure forces in pipe flow. Because pressure on the surface is atmospheric, pressure is not a significant parameter. Instead, differences in liquid depth at two points a distance L apart influence the flow. Therefore, we can write

$$\Delta z = \Delta z(V, g, L) \tag{4.7a}$$

Solving, we obtain

$$\frac{\Delta z g}{V^2} = \Delta z\left(\frac{V^2}{gL}\right) \tag{4.7b}$$

The term on the left-hand side, when written $\Delta zg/½V^2$, is called a **head coefficient**; V^2/gL (or in some texts $V/\sqrt{gL}$) is called the **Froude number**, which represents the ratio of inertia to gravity forces. Open-channel flows are treated in more detail in Chapter 7.

4.2.4 Unbounded Flows

Flows at gas–liquid or liquid–liquid interfaces where no solid surfaces are in contact are called unbounded flows. Examples include sprays and jets. In this case, pressure forces acting at an interface are balanced by surface tension forces that act over an area. Thus, we write

$$\Delta p = \Delta p(\rho, \sigma, L, V) \tag{4.8a}$$

TABLE 4.2
Some Common Dimensionless Ratios

Ratio	Name	Symbol	Force Ratio
$\rho VD/\mu$	Reynolds number	Re	Inertia/viscous
$\Delta p/½\rho V^2$	Pressure coefficient	C_p	Pressure/inertia
V^2/gL	Froude number	Fr	Inertia/gravity
$\rho V^2 L/\sigma$	Weber number	We	Inertia/surface tension
$D_f/½\rho V^2 D^2$	Drag coefficient	C_D	Drag/inertia
$L_f/½\rho V^2 D^2$	Lift coefficient	C_L	Lift/inertia
V/a	Mach number	M	Inertia at V/inertia at a
$F_T/½\rho V^2 D^2$	Thrust coefficient	C_T	Thrust/inertia

Solving yields

$$\frac{\Delta p}{\rho V^2} = \Delta \rho\left(\frac{\rho V^2 L}{\sigma}\right) \tag{4.8b}$$

where $\rho V^2 L/\sigma$ is known as the **Weber number** and represents the ratio of inertia to surface tension forces.

The previous discussion has introduced some basic concepts and dimensionless ratios encountered in fluid mechanics. They are by no means complete, however, as there are many more dimensionless ratios not discussed here. A summary of the common ratios appears in Table 4.2. Note that in this table, a number of ratios have $½\rho V^2$ (=kinetic energy) in the denominator. Often such ratios are simply called *coefficients*. Thus, drag force/kinetic energy may be called the drag coefficient.

4.3 DIMENSIONAL ANALYSIS BY INSPECTION

Table 4.2 contains dimensionless ratios that are commonly used in fluid mechanics. Therefore, when performing a dimensional analysis, we find it useful to express the result in terms of generally accepted ratios.

A table such as Table 4.2 can be used to advantage in performing a dimensional analysis. In Section 4.1, when an analysis was performed, the dimensions of each term were substituted into the equation, and exponents were equated. The equations were solved simultaneously to obtain values for the exponents. The values were then substituted back into the original equation, and like terms were collected to yield a functional relationship between dimensionless groups.

In this section, we will see that the technique of solving for exponents can be skipped with the aid of Table 4.2 if we remember that the objective is to find commonly recognized dimensionless groups. Suppose, for example, that we are analyzing a turbine that can be run with water or a liquid refrigerant as the working fluid. It is believed that the shaft torque T_s is a function of velocity (V), fluid density (ρ), viscosity (μ), angular velocity of the blades (ω), efficiency (η), and some characteristic dimension of the turbine blade geometry (D). We write

$$T_s = T_s\,(V, \rho, \mu, \omega, \eta, D) \tag{4.9}$$

In the F, L, T system, the dimensions of each term are written as

$$F \cdot L = T_s\,(L/T, F \cdot T^2/L^4, F \cdot T/L^2, 1/T, 0, L)$$

We have seven variables and three dimensions, so we are seeking 7 − 3 = 4 independent dimensionless groups. If we combine the geometry term D with rotational speed ω, we obtain dimensions of velocity V. Thus, $D\omega/V$ (or $V/D\omega$) is a dimensionless group. Efficiency η is already dimensionless. Density ρ can be combined with velocity V and geometry term D to obtain the dimensions of shaft torque T_s as follows:

$$\rho \quad V^2 \quad D^3 \rightarrow T$$

$$\frac{F \cdot T^2}{L^4} \cdot \frac{L^2}{T^2} \cdot L^3 = F \cdot L$$

Therefore, $T_s/\rho V^2 D^3$ (or $\rho V^2 D^3/T_s$) is our third dimensionless group. (Note that this term when divided by ψ could be called a torque coefficient = torque/kinetic energy.) Viscosity μ, velocity V, and geometry term D can be combined to give dimensions of torque as

$$\mu \quad V \quad D^2 \rightarrow T_s$$

$$\frac{F \cdot T}{L} \cdot \frac{L}{T} \cdot L^2 = F \cdot L$$

The ratio $T_s/\mu V D^2$ (or its reciprocal) is the fourth term. Note that the four ratios that we have written are all independent. That is, none of them can be obtained by manipulating the others together.

It is possible to put together another dimensionless group by using the variables of Equation 4.9. We see that the density ρ, velocity V, geometry term D, and viscosity μ can be combined to form a recognizable ratio: the Reynolds number $\rho V D/\mu$. This is a fifth group, but we are seeking only four *independent* groups. The Reynolds number can be obtained by multiplication of our third and fourth groups:

$$\frac{\rho V^2 D^3}{T_s} \frac{T_s}{\mu V D^2} = \frac{\rho V D}{\mu}$$

So even though we have five groups, they are not all independent. We can express our solution to this problem in terms of any of the four groups, but we want to give priority to the recognized groups without omitting terms that appear only once (efficiency η in this case).

The solution we select, then, is

$$\frac{T_s}{\frac{1}{2}\rho V^2 D^3} = T_s\left(\frac{V}{D\omega}, \eta, \frac{\rho V D}{\mu}\right) \tag{4.10}$$

Although $T_s/\mu V D^2$ can be used in place of $\rho V D/\mu$, the latter, being an easily recognizable quantity, is preferred.

Example 4.3

The designers of an all terrain vehicle decide not to use a propeller to move the vehicle through water. Instead, the rotation of the wheels is expected to provide the necessary thrust. The thrust or propulsive force F_T is determined to be a function of rotational speed of the wheels ω, diameter of the tires D, width of the tires W, and density ρ and viscosity μ of the liquid medium (muddy versus fresh water). Determine an expression for the propulsive force by using the inspection method.

Solution

The thrust can be written in functional form as:

$$F_T = f(\omega, D, W, \rho, \mu) = C_1 \omega^{a1} D^{a2} W^{a3} \rho^{a4} \mu^{a5}$$

Using the F, L, T, system we expect $6 - 3 = 3$ dimensionless groups. The dimensions of the variables are

$$F_T = F; \quad \omega = \frac{1}{T}; \quad D = L$$

$$W = L; \quad \rho = \frac{F \cdot T^2}{L^4}; \quad \mu = \frac{F \cdot T}{L^2}$$

For one of our groups, we examine the functional equation to determine if any obvious ratios stand out. The diameter and width have the same dimensions, so we identify D/W (or W/D) as one of our groups. If velocity appears as one of the variables, we can compose other recognizable groups. We do not have a velocity, but the dimension of ωD is L/T. The variables necessary for the Reynolds number can now be put together. The Reynolds number is defined as

$$\mathrm{Re} = \frac{\rho V D}{\mu}$$

In terms of the variables here,

$$\mathrm{Re} = \frac{\rho(\omega D)D}{\mu} = \frac{\rho \omega D^2}{\mu}$$

which is one of our groups. This group is known as the rotational Reynolds number.

We also recognize the force coefficient, defined as $F/(\frac{1}{2}\rho V^2 D^2)$. In terms of the variables here, the force coefficient becomes

$$\frac{F_T}{\frac{1}{2}\rho(\omega D)^2 D^2} = \frac{F_T}{\frac{1}{2}\rho \omega^2 D^4}$$

One (of many) solution(s) then is

$$\frac{F_T}{\frac{1}{2}\rho \omega^2 D^4} = F_T\left(\frac{\rho \omega D^2}{\mu}, \frac{D}{W}\right)$$

4.4 SIMILITUDE

In fluid mechanics, physical modeling depends on similitude, or similarity. That is, a prototype can be scaled down (or up), and a model can be built that represents the prototype. Measurements made on the model can then be scaled up (or down) to predict what will occur with the prototype. In many cases, a model is much less costly to build and easier to take data from than a prototype. Before a scale model can be constructed, however, certain rules of similitude must be observed. Because the prototype and model must be geometrically and dynamically similar, we will discuss the rules of similarity first and then apply them to various problems.

4.4.1 Geometric Similarity

Geometric similarity between two configurations is achieved when they are of different sizes but otherwise appear identical. Consider two cylinders of different diameters and lengths as shown in Figure 4.1. The prototype has diameter D and length L; the model has diameter D_M and length L_M. The ratios of diameters and of lengths are

$$\frac{D_M}{D} = \lambda_1$$

$$\frac{L_M}{L} = \lambda_2$$

We require that for geometric similarity, $\lambda_1 = \lambda_2$; in other words, we require that every linear dimension on the two must be related by the same scale ratio.

As another example, consider two square tubes of different sizes (see Figure 4.2). For geometric similarity, we require that

$$\frac{L_M}{L} = \frac{S_M}{S} = \lambda \tag{4.11}$$

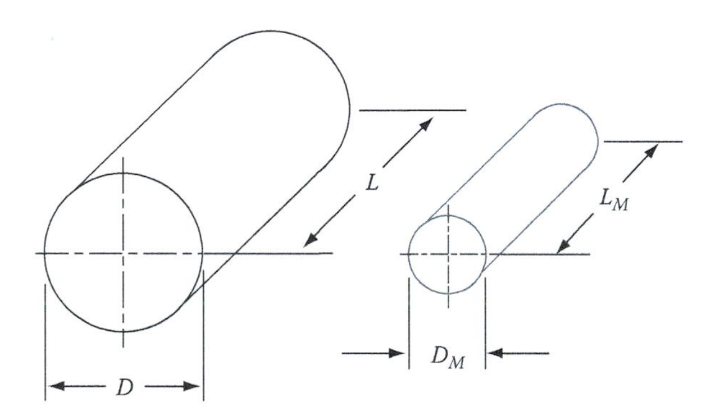

FIGURE 4.1 Similar cylinders.

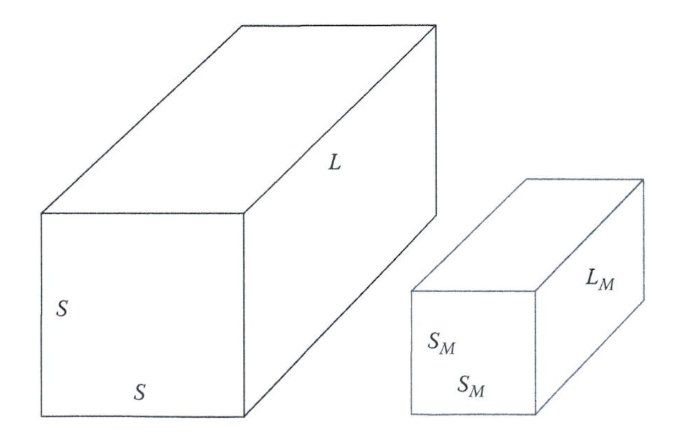

FIGURE 4.2 Similar square tubes.

4.4.2 Dynamic Similarity

Dynamic similarity is the second condition that must be met before two flow situations can be considered similar. The dynamic parameters must be related in a certain manner. Just as Equation 4.11 gives a scale ratio for length, we must have corresponding expressions for force, mass, and time. These are

$$l_M = \lambda l$$

$$F_M = \eta F \tag{4.12a}$$

$$m_M = \zeta m \tag{4.12b}$$

$$t_M = \tau t \tag{4.12c}$$

respectively, where the subscript M refers to the model.

From these expressions, we can obtain scale ratios for any dynamic or kinematic parameter and for fluid properties:

$$\begin{aligned} &\text{Velocity:} && V_M = \frac{\lambda}{\tau} V \\ &\text{Acceleration:} && a_M = \frac{\lambda}{\tau^2} a \\ &\text{Density:} && \rho_M = \frac{\zeta}{\lambda^3} \rho \\ &\text{Pressure:} && p_M = \frac{\eta}{\lambda^2} p \end{aligned} \tag{4.12d}$$

Whether we are working with a model or a prototype, Newton's law applies:

$$F = ma \tag{4.13a}$$

and

$$F_M = m_M a_M \tag{4.13b}$$

By substitution from Equations 4.11 to 4.12, Equation 4.13b becomes

$$\eta F = \zeta m \frac{\lambda}{\tau^2} a$$

Dividing by Equation 4.13a gives

$$\eta = \frac{\zeta \lambda}{\tau^2}$$

or

$$1 = \frac{\zeta \lambda}{\tau^2 \eta}$$

The form of this equation will be adjusted slightly to obtain an expression that is more convenient to use in fluid mechanics. Multiplying numerator and denominator by λ^3 yields

$$1 = \frac{\zeta\lambda^4}{\tau^2\lambda^3\eta}$$

Separating these terms appropriately gives

$$1 = \frac{\zeta}{\lambda^3}\frac{\lambda^2}{\tau^2}\frac{\lambda^2}{\eta}$$

or

$$1 = \frac{(\rho_M/\rho)(V_M^2/V^2)(l_M^2/l^2)}{F_M/F}$$

This equation can be rewritten as

$$\left.\frac{F}{\rho V^2 l^2}\right|_M = \frac{F}{\rho V^2 l^2} \tag{4.14}$$

Equation 4.14 expresses the condition for dynamic similarity between model and prototype; the value of the dimensionless parameter $F/\rho V^2 l^2$ must be the same in model and in prototype at geometrically similar locations. The force F includes gravity forces, electric forces, pressure forces, magnetic forces, and surface tension forces.

Let us examine a case in which pressure forces are important. The force due to pressure is $F = \Delta p l^2$. Substitution into Equation 4.14 gives

$$\left.\frac{\Delta p l^2}{\rho V^2 l^2}\right|_M = \frac{\Delta p l^2}{\rho V^2 l^2}$$

or, after multiplication by 2,

$$\left.\frac{\Delta p}{\frac{1}{2}\rho V^2}\right|_M = \frac{\Delta p}{\frac{1}{2}\rho V^2} \tag{4.15}$$

Thus, for dynamic similarity in problems in which pressure forces are important, the pressure coefficient in model and prototype must be identical at geometrically similar locations.

Let us next examine viscous forces. For Newtonian fluids,

$$\tau = \mu\frac{dV}{dy}$$

and

$$F = \tau l^2 = \mu\frac{dV}{dy}l^2$$

We can write

$$F = \mu \frac{\Delta V}{\Delta l} l^2$$

or

$$F \propto \mu V l$$

Using Equation 4.14, we have

$$\left.\frac{\mu V l}{\rho V^2 l^2}\right|_M = \frac{\mu V l}{\rho V^2 l^2}$$

or, inverting, we get

$$\left.\frac{\rho V l}{\mu}\right|_M = \frac{\rho V l}{\mu} \tag{4.16}$$

Thus, when viscous or frictional effects are important, the Reynolds number between model and prototype must be the same.

Let us now examine a case in which gravity forces are important. The force due to gravity is $F = mg$, which is proportional to $\rho l^3 g$. Substituting into Equation 4.14 yields

$$\left.\frac{\rho l^3 g}{\rho V^2 l^2}\right|_M = \frac{\rho l^3 g}{p V^2 l^2}$$

or, inverting,

$$\left.\frac{V^2}{gl}\right|_M = \frac{V^2}{gl} = \text{Froude number} \tag{4.17}$$

Thus, the Froude number in both model and prototype must be identical at geometrically similar locations for dynamic similarity in problems in which gravity forces are important.

Finally, consider surface tension forces. The force due to surface tension is $F = \sigma l$. Combining with Equation 4.14, we obtain

$$\left.\frac{\sigma l}{\rho V^2 l^2}\right|_M = \frac{\sigma l}{p V^2 l^2}$$

or, inverting,

$$\left.\frac{\rho V^2 l}{\sigma}\right|_M = \frac{\rho V^2 l}{\sigma} \tag{4.18}$$

Thus, in problems in which surface tension forces are important, the Weber numbers must be identical in model and prototype. The importance of using the widely recognized ratios in fluid mechanics is now evident.

4.4.3 Modeling

We have seen that geometric and dynamic similarity must be ensured if a prototype is to be modeled properly. Furthermore, these criteria imply that various dimensionless ratios between model and prototype must be identical. Use of these dimensionless ratios in modeling problems is best illustrated by example.

Example 4.4

A capillary tube has an 8-mm inside diameter through which liquid fluorine refrigerant R-11 flows at a rate of 0.03 cm^3/s. The tube is to be used as a throttling device in an air conditioning unit. A model of this flow is constructed by using a pipe of 2 cm inside diameter and water as the fluid medium.

a. What is the required velocity in the model for dynamic similarity?
b. When dynamic similarity is reached, the pressure drop in the model is measured as 40 Pa.

What is the corresponding pressure drop in the capillary tube?

Solution

a. In this case, the model is much larger than the prototype. For dynamic similarity between the two, the Reynolds numbers must be identical. Thus,

$$\left.\frac{\rho VD}{\mu}\right|_M = \frac{\rho VD}{\mu}$$

or

$$V_M = V\left(\frac{D}{D_M}\right)\left(\frac{\mu_M}{\mu}\right)\left(\frac{\rho}{\rho_M}\right)$$

In the prototype,

$$A = \frac{\pi(0.008)^2}{4} = 5.03\times10^{-5}\ \text{m}^2$$

so, from continuity,

$$V = \frac{Q}{A} = \frac{0.03\times10^{-4}}{5.03\times10^{-5}} = 0.059\,6\ \text{m/s}$$

From Table A.5,

$$\frac{\mu_M}{\mu} = \frac{0.89\times10^{-3}}{0.42\times10^{-3}}$$

and

$$\frac{\rho_M}{\rho} = \frac{1}{1.48}$$

By substitution,

$$V_M = 0.059\,6\left(\frac{0.008}{0.02}\right)\left(\frac{0.89}{0.42}\right)(1.48)$$

$$V_M = 0.075 \text{ m/s}$$

b. In this case, the pressure coefficient in model and prototype must be the same. Thus,

$$\left.\frac{\Delta p}{\rho V^2}\right|_M = \frac{\Delta p}{\rho V^2}$$

or

$$\Delta p = \Delta p_M\left(\frac{\rho}{\rho_M}\right)\left(\frac{V^2}{V_M^2}\right)$$

By substitution, we get

$$\Delta p = (40 \text{ Pa})\left(\frac{1.48}{1}\right)\left(\frac{0.059\,6}{0.075}\right)^2$$

$$\Delta p = 37.4 \text{ Pa}$$

Example 4.5

The characteristics of a ship 50 ft in length are to be studied with a 5-ft-long model. The ship velocity is 12 knots. What is the required velocity of the model for dynamic similarity? Measurements on the model indicate a drag force of 5 lbf. What is the expected drag on the prototype? Assume water to be the fluid in both cases and neglect viscous effects.

Solution

The flow involves an open-channel type of geometry, so dynamic similarity is achieved for equal Froude numbers. Thus,

$$\left.\frac{V^2}{gl}\right|_M = \frac{V^2}{gl}$$

or

$$V_M^2 = V^2\frac{l_M}{l}$$

The scale ratio is given as

$$\lambda = \frac{l_M}{l} = \frac{5 \text{ ft}}{50 \text{ ft}} = \frac{1}{10}$$

With the ship velocity given as 12 knots, the corresponding model velocity for dynamic similarity is

$$V_M^2 = (12)^2 \frac{1}{10} = 14.4$$

or

$$V_M = 3.8 \text{ knots}$$

To relate drag measurements between model and prototype, we use the drag coefficient. We have, from Table 4.2,

$$\left.\frac{D_f}{\frac{1}{2}\rho V^2 D^2}\right|_M = \frac{D_f}{\frac{1}{2}\rho V^2 D^2}$$

Since the liquid is the same in both cases, we can solve for the drag force exerted on the prototype to get

$$D_f = D_{fM} \frac{V^2}{V_M^2} \frac{D^2}{D_M^2}$$

All parameters are known. Substituting gives

$$D_f = 5\left(\frac{12}{3.8}\right)^2 \left(\frac{10}{1}\right)^2$$

$$D_f = 5000 \text{ lbf}$$

Thus, if dynamic similarity conditions are met, the drag force on the prototype will be 5000 lbf.

Example 4.6

Octane is used as a fuel for a certain engine and is sprayed into air at the engine intake. The spray nozzle orifice is 0.122 in. in diameter. The average velocity of the droplet–air mix is 10 ft/s, and the octane concentration is small enough that the density of the mix is about equal to that of air. The system is to be modeled with an orifice that is 0.25 in. in diameter spraying water in air. Determine the average velocity of the water–air mix for dynamic similarity between the two.

Solution

In this case, the Weber numbers must be identical:

$$\left.\frac{\rho V^2 l}{\sigma}\right|_M = \frac{\rho V^2 l}{\sigma}$$

Taking the densities in both cases as being equal gives

$$V_M^2 = V^2 \frac{l}{l_M} \frac{\sigma_M}{\sigma}$$

Surface tension values are given in Table A.5; by substitution, we get

$$V_M^2 = (10)^2\left(\frac{0.122}{0.25}\right)\left(\frac{72}{21.14}\right) = 1.66$$

Solving, we get

$$V_M = 12.9 \text{ ft/s}$$

Three devices that are commonly used in model studies are a wind tunnel (Figure 4.3), a water tunnel (Figure 4.4), and a tow tank. The top view of a wind tunnel is given schematically in Figure 4.3. As indicated, the motor and fan assembly circulates air through the system. The ductwork is designed so that losses due to friction are minimized by the use of turning vanes and gradually diverging ducts. Flow at the test section must be as smooth and as uniform as possible. Models of automobiles, trucks, airplanes, groups of houses, buildings, and the like can be placed in the test section to determine the effects of speed or winds such as hurricanes. Data so obtained can then be scaled up to predict similar effects on the prototype.

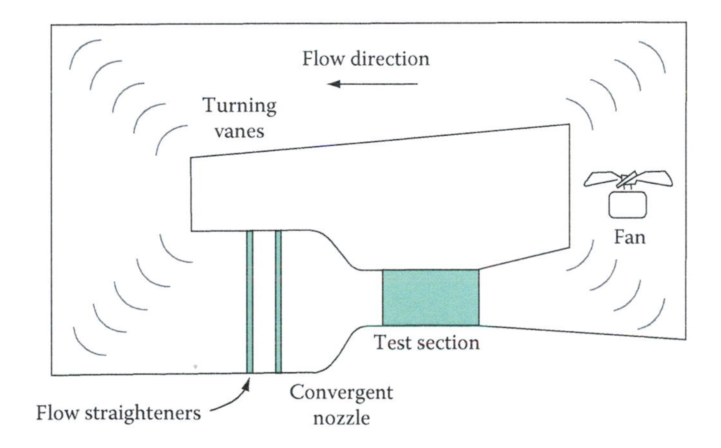

FIGURE 4.3 Schematic of a closed-loop wind tunnel.

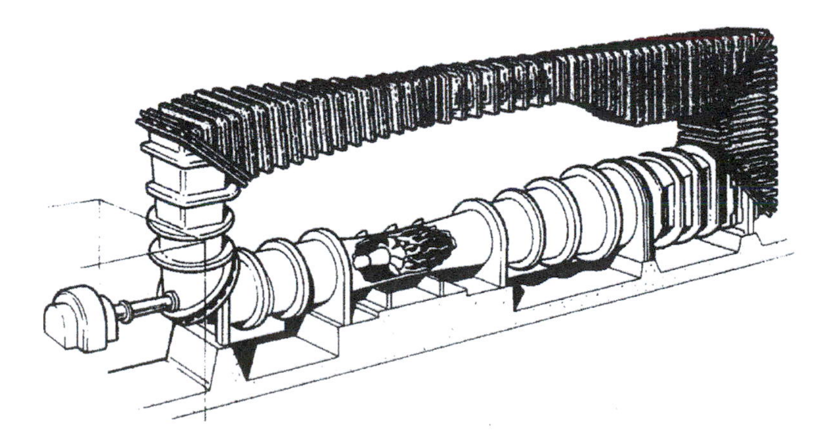

FIGURE 4.4 Schematic of a closed-loop water tunnel.

A tow tank consists of a huge reservoir of water, usually rectangular in cross section, that is used extensively for testing ship hull designs. Once a design is made, a scale model is built and tested in the basin for drag and other forces. The appropriately instrumented model is usually pulled along the water surface to obtain data.

It is not always possible to assure dynamic similarity between model and prototype, however. In the tow tank, for example, it is nearly impossible to satisfy both Froude and Reynolds number similarity. Despite this shortcoming, modeling remains a viable alternative to testing a full-scale prototype.

4.5 CORRELATION OF EXPERIMENTAL DATA

This section presents data from an actual experiment and then correlates them using dimensionless groups.

Consider a tube with an inside diameter of 0.545 in. through which water is made to flow. The tube is placed in a horizontal position, and pressure taps on it are located at 5, 3, and 2 ft apart. The pressure taps are connected to manometers that give the pressure drop in terms of a head loss as water flows through the tube. As can be discerned from the data, head loss increases with flow rate. The test setup is shown in Figure 4.5.

For various settings of the flow control valve, the flow rate is measured with the venturi meter, and the corresponding head loss in the tube is measured with inverted U-tube manometers. The head loss is due entirety to friction between the water and the tube inside surface. Data taken directly are given in columns 1, 2, and 3 of Table 4.3.

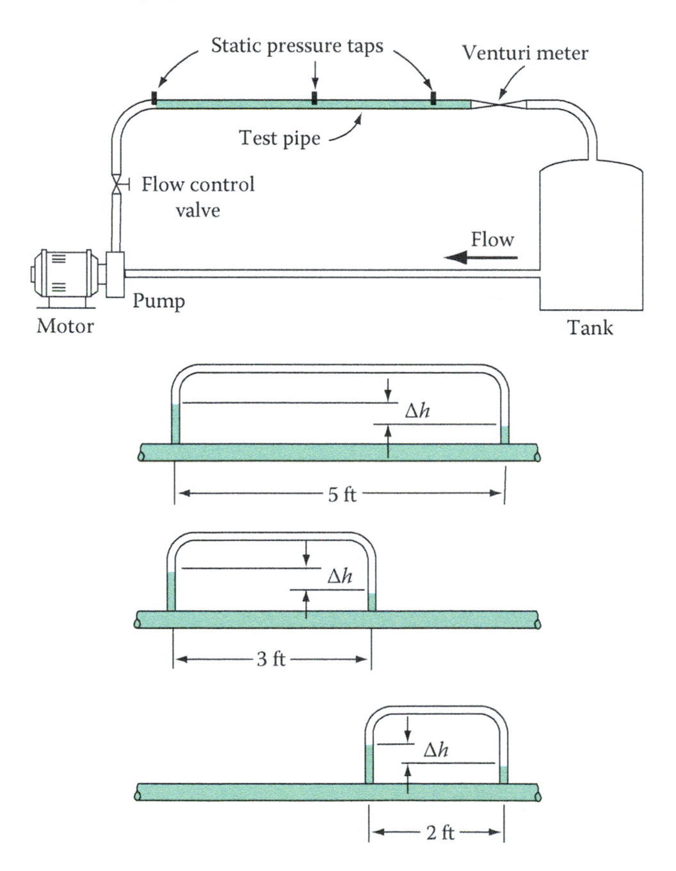

FIGURE 4.5 Measurement of head loss for water flowing through a tube.

TABLE 4.3

Raw and Reduced Data for the Pressure Drop Experiment

$D = 0.545$ in. $= 0.0454$ ft; $A = 0.00162$ ft²; $\rho = 1.94$ slug/ft³

$\mu = 1.90 \times 10^{-5}$ lbf·s/ft² $\nu = \mu/\rho = 3.15 \times 10^{-4}$ ft²/s

Group I Data Point	$L = 5$ ft Q (gpm)	Δh (in.)	V (ft/s)	Re	L/D	$\Delta h/L$
1	1.5	2.5	2.06	9 568	110.1	0.04
2	1.85	3.69	2.54	11 800		0.06
3	2	4.38	2.75	12 757		0.07
4	2.45	6.25	3.37	15 627		0.10
5	2.8	8.88	3.85	17 860		0.15
6	3.3	11.75	4.54	21 049		0.20
7	3.5	13.13	4.81	22 325		0.22
8	3.9	16.13	5.36	24 876		0.27
9	4.1	17.81	5.64	26 152		0.30
Group II Data Point	**$L = 3$ ft Q (gpm)**	**Δh (in.)**	**V (ft/s)**	**Re**	**L/D**	**$\Delta h/L$**
11	1.5	1.5	2.06	9 568	66.1	0.04
12	1.85	2.2	2.54	11 800		0.06
13	2	2.63	2.75	12 757		0.07
14	2.45	3.8	3.37	15 627		0.11
15	2.8	5.33	3.85	17 860		0.15
16	3.3	7	4.54	21 049		0.19
17	3.5	7.88	4.81	22 325		0.22
18	3.9	9.7	5.36	24 876		0.27
19	4.1	10.7	5.64	26 152		0.30
Group III Data Point	**$L = 2$ ft Q (gpm)**	**Δh (in.)**	**V (ft/s)**	**Re**	**L/D**	**$\Delta h/L$**
21	1.5	1	2.06	9 568	44.0	0.04
22	1.85	1.48	2.54	11 800		0.06
23	2	1.75	2.75	12 757		0.07
24	2.45	2.5	3.37	15 627		0.10
25	2.8	3.56	3.85	17 860		0.15
26	3.3	4.7	4.54	21 049		0.20
27	3.5	5.25	4.81	22 325		0.22
28	3.9	6.45	5.36	24 876		0.27
29	4.1	7.13	5.64	26 152		0.30

We wish to correlate the data using dimensionless groups obtained by methods described in this chapter. The head loss is predicted to be a function of

$$\Delta h = \Delta h(V, \rho, \mu, D, L) \tag{4.19}$$

where

Δh is the head loss in the pipe (related to pressure drop by $\Delta h = \Delta p/\rho g$)
V is the average velocity of water in the tube
ρ and μ is the density and viscosity of the water, respectively
D is the inside diameter of the tube
L is the length between pressure taps (5 ft, 3 ft, or 2 ft)

Using the MLT system and the inspection method, we write

Variable	Dimension	Definition
Δh	L	Head loss, ft
V	L/T	Velocity = Q/A, ft/s
ρ	M/L^3	Density = 1.94 slug/ft³
μ	$M/(L \cdot T)$	Viscosity = 1.9×10^{-5} lbf·s/ft²
D	L	Inside diameter 0.545 in. = 0.0454 ft
L	L	5, 3, or 2 ft

We have six variables and three dimensions, so we are seeking three groups. For flow of a fluid in a duct, the viscosity is important, so the Reynolds number will be one of the groups: Re = $\rho VD/\mu$. Other obvious results are $\Delta h/L$, and D/L. We could use $\Delta h/L$ or $\Delta h/D$. Because D is a constant in these data, $\Delta h/L$ is a better choice. (This conclusion is based on experience. If we are as yet inexperienced, we would select either $\Delta h/L$ or $\Delta h/D$, and proceed. The validity of the resulting relationship would be determined by the value of a correlation coefficient.) The dimensionless correlation then is

$$\frac{\Delta h}{L} = \Delta h\left(\frac{\rho VD}{\mu}, \frac{L}{D}\right)$$

We can now begin to reduce the data. The inside area of the tube is

$$A = \frac{\pi D^2}{4} = \frac{\pi(0.545/12)^2}{4} = 0.00162 \text{ ft}^2$$

The velocity can now be calculated. A sample calculation for the first row of the data is

$$Q = (1.50 \text{ gal/min})\left(\frac{1}{448.77}\right) = 0.00334 \text{ ft}^3/\text{s}$$

$$V = \frac{Q}{A} = \frac{0.00334 \text{ ft}^3/\text{s}}{0.00162 \text{ ft}^2} = 2.06 \text{ ft/s}$$

$$\text{Re} = \frac{\rho VD}{\mu} = \frac{(1.94 \text{ slug/ft}^3)(2.06 \text{ ft/s})(0.054 \text{ ft})}{1.9 \times 10^{-5} \text{ lbf}\cdot\text{s/ft}^2} = 9568$$

and

$$\frac{\Delta h}{L} = \frac{(2.5/12) \text{ ft}}{5 \text{ ft}} = 0.0417$$

These values are shown in the first row of the table. The table also shows results of calculations made for all the data points. A graph of the raw data is provided in Figure 4.6. A graph of the dimensionless groups obtained here is provided in Figure 4.7. L/D is a constant in each group and is not included in this analysis.

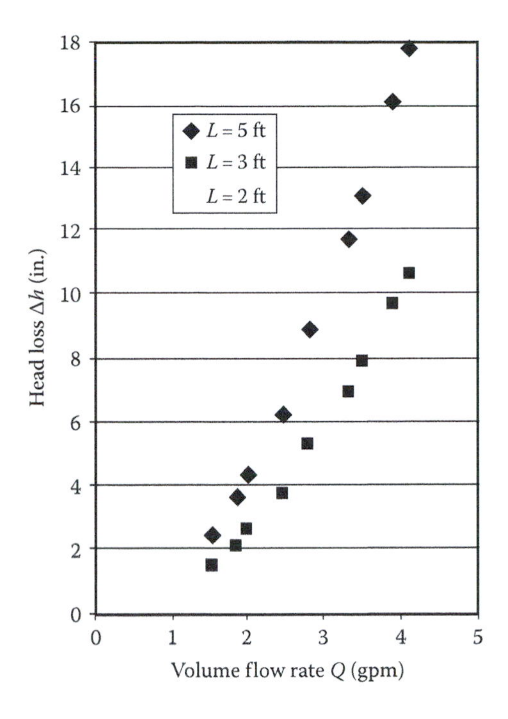

FIGURE 4.6 Graph of the raw data.

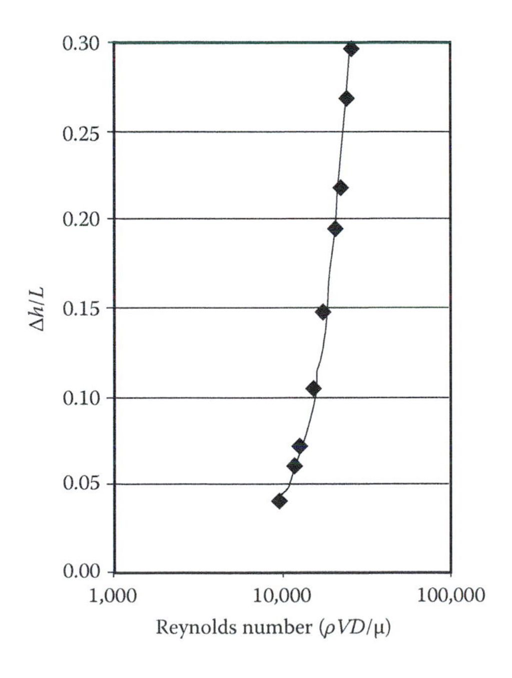

FIGURE 4.7 Graph of the reduced data using dimensionless groups.

We can now obtain an equation for the line of Figure 4.7. Using a spreadsheet, the equation is

$$\frac{\Delta h}{L} = 6 \times 10^{-10} (\text{Re})^{1.9686}$$

with a correlation coefficient of $R^2 = 0.9989$.

The raw data graphed as three curves in Figure 4.6 all condense onto one line in Figure 4.7. The correlation coefficient is very high. One obvious advantage of using dimensionless groups is that we can use Figure 4.7 to obtain the head loss for pressure taps that are 4 ft apart, or any distance between 2 and 5 ft. The advantages of dimensional analysis in data reduction are evident.

4.6 SUMMARY

In this chapter, we used dimensional analysis to develop nondimensional ratios that are useful for correlating data. We also derived various dimensionless ratios such as Reynolds, Froude, and Weber numbers and examined the requirements of geometric and dynamic similarity. Similitude and modeling were discussed, as were dimensionless ratios that are important in various flow situations.

INTERNET RESOURCES

1. Find a sketch and a design of an open loop subsonic wind tunnel, and provide a brief presentation on its design.
2. Find a sketch and a design of a closed-loop subsonic wind tunnel, and provide a brief presentation on its design.
3. Find a sketch and a design of a tow tank used in evaluating ship hulls, and provide a brief presentation on its design.

PROBLEMS

Dimensions of Terms and Equations

4.1 An airplane wing traveling through air experiences a frictional effect known as the drag force, which may be expressed in dimensionless form as a drag coefficient. The drag coefficient C_D is defined as

$$C_D = \frac{D_f}{\frac{1}{2}\rho V^2 A}$$

where

D_f is the drag force
ρ is the density of air
V is the wing velocity
A is the plan form area of the wing

Show that the drag coefficient is dimensionless.

4.2 Another form of the Bernoulli equation is obtained by dividing by g, yielding

$$\frac{p}{\rho g} + \frac{V^2}{2g} + z = \text{a constant}$$

What is the dimension of each term?

4.3 The equation for volume flow rate for laminar flow of a Newtonian fluid in a circular duct is given by

$$Q = \frac{\pi R^4}{8\mu}\left(\frac{\Delta p}{L}\right)$$

where
- Q is the volume flow rate
- $\Delta p/L$ is the pressure drop per unit length
- R is the pipe radius
- μ is the absolute viscosity of the fluid

Determine if this equation is dimensionally consistent.

4.4 The velocity profile for laminar flow in a circular tube is given by

$$V_z = \frac{\Delta p}{L}\frac{R^2}{4\mu}\left[1-\left(\frac{r}{R}\right)^2\right]$$

where
- $\Delta p/L$ is the pressure drop per unit length
- R is the tube radius
- μ is the absolute viscosity of the fluid

Show that the right-hand side of this equation has the correct dimensions.

4.5 The equation for volume flow rate for laminar flow of a Newtonian liquid down an incline is given by

$$Q = \frac{gh^3b}{3v}\sin\theta$$

where
- Q is the volume flow rate of liquid
- h is the depth of liquid
- b is the width of channel down which liquid flows
- v is the kinematic viscosity of liquid
- $\sin\theta$ is the slope of the incline (dimensionless)
- g is the acceleration due to gravity

Determine if this equation has the proper dimensions.

4.6 The equation for velocity for laminar flow of a Newtonian fluid down an incline is given by

$$V_x = \frac{\rho g}{\mu}\frac{h^2}{2}\left(\frac{2z}{h}-\frac{z^2}{h^2}\right)\sin\theta$$

where
- ρ is the density of liquid
- μ is the absolute viscosity of liquid
- g is the acceleration due to gravity
- h is the depth of liquid
- $\sin\theta$ is the slope of incline (dimensionless)

Determine the dimensions of this equation.

4.7 The steady-state flow energy equation from Chapter 3 can be written as

$$\frac{d(Q-W)}{dt} = \dot{m}\left[\left(h_2 + \frac{V_2^2}{2} + gz_2\right) - \left(h_1 + \frac{V_1^2}{2} + gz_1\right)\right]$$

What are the dimensions of each term?

Dimensional Analysis: Four- and Five-Variable Problems

4.8 A weir is an obstruction placed in the flow of a liquid in an open channel. The liquid height upstream of the weir can be used to determine the volume flow rate over the weir. Assuming that the volume flow Q is a function of upstream height h, gravity g, and channel width b, develop an expression for Q by using dimensional analysis.

4.9 Sonic velocity in a gas is assumed to be a function of gas density ρ, pressure p, and dynamic viscosity μ. Determine a relationship using dimensional analysis.

4.10 For flow over a flat plate, the flow velocity V in the vicinity of the surface varies with the wall shear stress τ_ω, distance from the wall y, and fluid properties ρ and μ. Use dimensional analysis to determine an expression relating these variables.

4.11 The height h a liquid attains inside a partly submerged capillary tube is a function of surface tension σ, tube radius R, gravity g, and liquid density ρ. Determine an expression for the height by using dimensional analysis.

4.12 In a falling-sphere viscometer, spheres are dropped through a liquid, and their terminal velocity is measured. The liquid viscosity is then determined. Perform a dimensional analysis for the viscometer assuming that viscosity μ is a function of sphere diameter D and mass m, local acceleration due to gravity g, and liquid density ρ.

4.13 The power $\tilde{P}$ delivered by an internal combustion engine is a function of the mass flow rate of air consumed, the energy contained in the fuel E, the rotational speed of the engine ω, and the torque T_s developed. Determine an expression for power by using dimensional analysis.

4.14 A rocket expels high-velocity gases as it travels through the atmosphere. The thrust F_T exerted depends on exit pressure of the gases p_e, ambient pressure p_a, gas velocity at exit V, and exit area A. Use dimensional analysis to develop an expression for the thrust.

4.15 A specially made centrifugal pump is designed to move a highly viscous liquid. The impeller of this pump is merely a flat disk that rotates within the fluid. The torque T required to rotate the disk is a function of disk diameter D, rotational speed ω, density ρ, and viscosity μ of the liquid. Determine a relationship among these variables.

4.16 The main bearing on an engine crankshaft is lubricated with engine oil. The torque due to friction T_o depends on shaft diameter D, rotational speed ω, lubricant viscosity μ, and density ρ. Use dimensional analysis to derive an expression for torque.

Dimensional Analysis: Six-Variable Problems

4.17 Figure P4.17 is a sketch of a conically shaped element in a viscometer, which is used to measure the viscosity of oil. The element is rotated at a specific rotational speed, and the torque is measured. Appropriately calibrating the device will provide a relationship between the viscosity and the pertinent variables. The torque T required to rotate the cone in a liquid is a function of the angular velocity ω, the clearance δ, the half angle θ, the cone height H, and the viscosity μ of the oil. Determine a relationship among these variables.

4.18 A surface explosion sends a sound wave through air. The wave travels at the sonic velocity. The pressure ratio across the wave p_2/p_1 is a function of the energy released by the

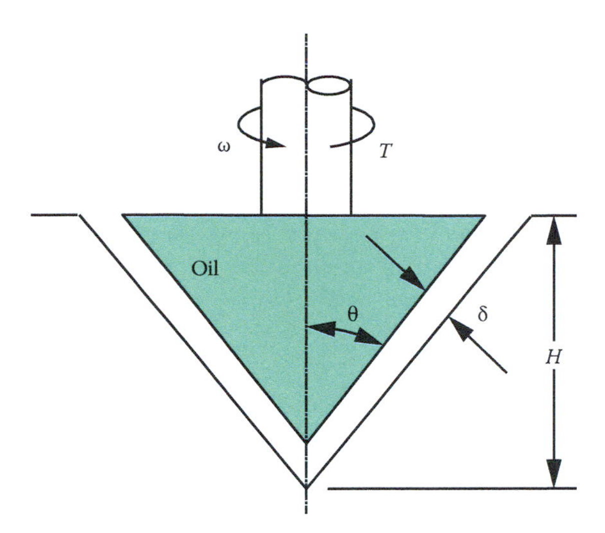

FIGURE P4.17 A conically shape element in a viscometer.

explosive E, distance from blast location d, sonic velocity in the air a, time t since the explosion, and density ρ of the air. Use dimensional analysis to develop an expression for the pressure ratio.

4.19 The friction factor f is a dimensionless quantity used in pipe flow problems as an aid in calculating pressure drop. The friction factor is a function of fluid properties density ρ and viscosity μ, of average velocity V, of pipe diameter D, and of the roughness of the pipe wall. The wall roughness is traditionally represented by a single term ε having the dimension of length. Determine a relationship between these variables to predict friction factor.

4.20 An open-top tank is filled to a depth h with liquid of density ρ and viscosity μ. A hole of diameter D is drilled in the side of the tank, and a jet of liquid issues from it. Assuming that the efflux velocity V is a function of these parameters and also of gravity g, develop an expression for V using dimensional analysis, and the F, M, L, T system. Compare the results to those obtained in Example 4.1.

4.21 A pneumatic spray nozzle is a two-fluid device that directs high-velocity air at a liquid jet to break the liquid into a spray. The average droplet diameter D of the spray depends on the mass flow rate of air, the volume flow of liquid Q, the density of each fluid ρ_a and ρ_1, and the difference in velocity of the two streams ΔV. Use dimensional analysis to derive a relationship for D.

4.22 The efficiency of a pump is assumed to be a function of discharge flow rate Q, pressure increase Δp, pipe diameter D, and fluid properties density ρ and viscosity μ. Use dimensional analysis to find an expression for pump efficiency.

4.23 Power input to a pump depends on flow rate Q, pressure rise Δp (outlet minus inlet), liquid density ρ, efficiency η, and impeller diameter D. Determine an expression for power by using dimensional analysis.

4.24 Vanes located on the periphery of a rotor are struck by a jet of liquid, causing the rotor to rotate. The torque measured at the rotor shaft T_o depends on the rotational speed of the rotor ω, the jet velocity V, the jet area A, the radial distance R from shaft center to point of impact of the jet, and the liquid density ρ. Use dimensional analysis to determine an expression for the torque.

4.25 The power required to drive a fan dW/dt is a function of fluid properties ρ and μ, fan blade diameter D, mass flow rate, and rotational speed ω. Use dimensional analysis to determine an expression for power.

4.26 A water turbine is located in a dam. The shaft torque T_s delivered to its blades is a function of the water flow rate Q through the system, total water head H upstream of the dam, liquid density ρ, angular velocity of the blades ω, and efficiency η. Determine an expression for torque.

4.27 A two-nozzle water sprinkler discharges water while the sprinkler head itself rotates. The angular velocity ω depends on water density ρ, volume flow of water Q, exit velocity through the nozzles V, nozzle exit area A, and torque T_s to overcome bearing friction. Use dimensional analysis to develop an expression for the angular velocity.

4.28 A jet-propelled boat takes in water at its bow, passes it through pumps, and discharges it through ducts at the stern. The propulsive force F_p is a function of the velocity of the boat V, water velocity at the exit duct V_e, exit pressure p_e, pumping power $\tilde{P}$ and liquid density ρ. Use dimensional analysis to develop an expression for the propulsive force.

4.29 The designers of an all-terrain vehicle decide not to use a propeller to move the vehicle through water. Instead, the rotation of the wheels is expected to provide the necessary thrust. The thrust or propulsive force F_T is determined to be a function of rotational speed of the wheels ω, diameter of the tires D, width of the tires W, and density and viscosity ρ and μ of the liquid medium (muddy water vs. fresh water). Determine an expression for the propulsive force by using the inspection method. Include g_c and compare the results to those obtained in Example 4.3.

4.30 Paint is modeled as a Bingham plastic that may be described by the following equation:

$$\tau = \tau_o + \mu_o \frac{dV}{dy}$$

where

τ_o is its initial yield stress
μ_o is its viscosity

The sharpened end of a nail is dipped into a can of paint. When removed, a droplet of paint whose volume is $\forall$ will drain and fall from the tip of the nail. The volume of the droplet that forms is a function of the paint density ρ, initial yield stress τ_o, viscosity μ_o, surface tension σ, and the nail diameter D. Determine a functional relationship among these variables.

4.31 In a spray drying operation, eggs in liquid form are sprayed into a heated drying chamber to produce powdered eggs. Moisture from the liquid droplets is released to the hotter environment through the surface area of the drops. The amount of moisture released in terms of mass of liquid per unit time is a function of volume flow rate through the nozzle Q, liquid density ρ, surface tension σ, average droplet diameter D, and residence time t that the droplets are suspended before they settle. Use dimensional analysis to derive an expression for the moisture released.

4.32 The amount of time t that clothes must stay in a washing machine for cleaning is a function of the height of the agitator h, the frequency of oscillation or its reciprocal ω, the mass of water in the tank m, the mass of soap used m_s, and the tank volume V. Use dimensional analysis to determine an expression for the time.

4.33 A sphere falling in air will ultimately reach a terminal velocity. The terminal velocity V is a function of the sphere diameter D, the density of air ρ, the kinematic viscosity of air ν, the frontal area of the sphere A, and the drag force D_f exerted on the sphere. Using dimensional analysis, find an expression to relate these variables.

Dimensional Analysis: Seven- and Eight-Variable Problems

4.34 A thin sheet of liquid flows down a roof. The liquid sheet velocity is a function of density ρ, slope angle θ, viscosity μ, surface tension σ, roof length l, and gravity g. Develop an expression for the velocity using the F, M, L, T system with g_c.

4.35 A viscous liquid is placed in the annulus between concentric cylinders. The outer cylinder is fixed, while the inner cylinder rotates. Determine an expression for torque T_o required to rotate the inner cylinder if it depends on the constant rotational speed ω, the liquid properties ρ and μ, the cylinder diameters D_2 and D_1, and the inner cylinder length L.

4.36 When a vertical flat surface is sprayed with paint, the paint will run if too thick a coat is applied. A thick coat, however, is desirable in many cases to make the surface coating glossy. An optimum thickness t is thought to exist, which is a function of droplet diameter D of the spray, initial yield stress τ_o, viscosity μ_o, surface tension σ and density ρ of the paint, and flow rate Q. Use dimensional analysis to develop an expression for the thickness.

4.37 The lift force on a wing L_f is a function of its chord length c, velocity V, maximum thickness t, angle of attack α, fluid density ρ, fluid viscosity μ, and sonic velocity a. Determine a dimensionless relationship among these variables.

4.38 A jet engine takes in air at atmospheric pressure, compresses it, and uses it in a combustion process to generate high-temperature gases. These gases are then passed through a turbine to provide power for compressing the inlet air. After passing through the turbine, the gases are exhausted to the atmosphere to deliver thrust. The thrust F_T developed is a function of inlet air density ρ_a and pressure p_a, pressure after the compressor p_c, energy content in the fuel E, exhaust gas pressure p_e, and velocity V. Use dimensional analysis to develop an expression for the thrust.

4.39 A tank is filled with liquid being pumped through a pipe into the tank bottom. The height h that the liquid can reach within the tank is a function of the pump output pressure p, atmospheric pressure p_{atm}, the pipe diameter D, the liquid velocity in the pipe V, and the liquid properties ρ and μ:

$$h = h(p, p_{atm}, D, V, \rho, \mu)$$

Determine a functional relationship to predict the liquid height h.

4.40 A venturi meter is shown schematically at several places within the text. It is desired to derive an equation to predict the volume flow rate through the meter. Using dimensional analysis, start with the following equation and derive the appropriate, descriptive dimensionless groups:

$$Q = Q(g, \Delta h, D_1, D_2, \rho, \mu)$$

where
- D_1 is the upstream diameter
- D_2 is the diameter at the throat
- ρ is the density of the fluid
- μ is the viscosity of the fluid
- Δh is the pressure drop

4.41 It is desired to conduct some tests on a butterfly valve. A butterfly valve is merely a disk, with a shaft attached, that is located within a pipeline. The disk outside diameter and the inside diameter of the pipeline are approximately equal. Rotating the shaft will cause the disk to rotate and block the flow area with an adequate seal. We want to determine the electric motor power required to automatically close the valve. The motor power $\tilde{P}$ required is a function of the maximum torque encountered T_o, the desired rotational speed ω, the mass of the disk m, the average velocity of the flow V, the disk diameter D, and density ρ and viscosity μ of the fluid. Using dimensional analysis, develop an expression to relate these variables.

Miscellaneous Problems

Only the dependent variable is given in the following problems, and you are required to select the appropriate independent variables based on your experience and intuition. Include enough independent variables so that at least three dimensionless groups result.

4.42 Liquid lubricant is purchased in pressurized spray cans. When the button is depressed, a spray of lubricant is produced in a conical configuration. The average diameter of the droplets produced is a function of a number of variables. Use dimensional analysis to find an expression for average droplet diameter.

4.43 The drag force exerted by air friction on a moving automobile must be overcome by power produced by the engine. Use dimensional analysis to formulate an expression for power.

4.44 The wings of a flying airplane experience a lift force L_f great enough to elevate the plane. Determine the dimensionless ratio (or ratios) that affects (or affect) the lift force.

4.45 An upright funnel is plugged and then filled with liquid. When the plug is removed, liquid issues from the end and empties the funnel after time t has elapsed. Use dimensional analysis to develop an expression for t.

4.46 A fan is a device used to increase the energy of air by raising its pressure. The increase in pressure experienced by the air is a function of a number of variables. Use dimensional analysis to determine an expression for the pressure rise.

4.47 A pump is a device used to increase energy in a liquid by raising its pressure. A compressor is such a device made for gases. When gases are compressed, however, both temperature and pressure increase. Thus, many compressors have coolers that reduce the compressed air temperature to that of the ambient temperature. Given such an arrangement, develop an expression to predict the power required for compressing a gas. Neglect heat-transfer effects.

4.48 Liquid flows horizontally through a circular tube filled with sand grains that are spherical and of the same diameter. As liquid flows through this sand bed, the liquid experiences a pressure drop. Use dimensional analysis to determine an expression for the pressure drop in terms of the pertinent variables.

4.49 Paint (a Bingham fluid) is applied with a spray gun by forcing it under high pressure through a small orifice. The average diameter of the droplets produced is known to be a function of many variables. Use dimensional analysis to find an expression for average droplet diameter D.

4.50 Fins are used by a swimmer to increase the swimmer's forward velocity. The propulsive force is increased because of the fins. Use dimensional analysis to develop a functional equation to relate the propulsive force to the pertinent variables.

Modeling Problems

4.51 Acetone flows through a tube at a volume flow rate of 1 ft^3/s. If water is used in the tube instead of acetone, what volume flow rate is required for complete dynamic similarity?

4.52 Water flows in a pipe with an inside diameter of 12 in. The average velocity of the water is 1 ft/s. For dynamically similar flow, determine the average flow velocity for air at 170°F and 18 psig in a 6 in. pipe.

4.53 Various control systems use pneumatic devices to direct the desired operations. These devices require that air be piped to them. In one design, the pneumatic line diameter is 1/2 in. and the air flow rate is 0.03 $in.^3/s$. If such a line is to be modeled with water flowing through a 6 in. ID line, determine the average water velocity required for dynamic similarity.

4.54 Gasoline (assume octane) flows at a rate of 4 in./s in a 3/8 in. diameter fuel line of an automobile. The flow is modeled with water flowing in a 1 in. ID pipe. What is the corresponding water velocity in the water pipe? Assuming that the pressure drop in the water pipe is 4.5 psi, determine the pressure drop in the fuel line.

4.55 Benzene flows at a rate of 0.1 ft^3/s in a 2 in. diameter pipe. The pressure drop measured over a stretch of pipe is to be determined by making suitable measurements on a model. The model is a 1 in. pipe carrying water. Determine the volume flow rate of water required for dynamic similarity. If the pressure drop in the water line is 4 psi, predict the corresponding pressure drop in the benzene line.

4.56 A fan is to be purchased to provide adequate ventilation for workers in an underground mine. The fan is to move air at 0.05 m^3/s through a mine shaft 2.5 m high by 2 m wide. A model of the duct 25 cm by 20 cm has been constructed. If water is used as the fluid medium in the model, determine the required flow rate for dynamic similarity. For the characteristic length of each duct, use $D = 4$ area/perimeter.

4.57 An airplane wing of chord length 0.8 m travels through air at 5°C and 75 kPa. The wing velocity is 120 m/s. A one-tenth scale model of the wing is tested in a wind tunnel at 25°C and 101 kPa. For dynamic similarity between the two, determine the velocity required in the tunnel.

4.58 You are asked to determine the drag force exerted on an automobile at a speed of 60 mph and an air temperature of 62°F. A one-fifth scale model is to be tested in a water tunnel at 70°F. Determine the required water velocity for dynamic similarity between the full-size car and the model. If the drag force on the model is measured to be 10 lbf, determine the expected drag force on the automobile.

4.59 The drag characteristics of a blimp traveling at 15 ft/s are to be studied by modeling. The prototype is 25 ft in diameter and 200 ft long. The model is one-twelfth scale and will be tested underwater. What velocity must the model have for dynamic similarity between the two? If the drag on the model is 3.5 lbf, what is the corresponding drag for the prototype? Neglect buoyant effects.

4.60 A ship 104 m in length is tested by a model that is only 12 m long. The ship velocity is 40 km/h. What model velocity is required for dynamic similarity if the liquid used for each is the same? If the drag on the model is 8 N, what drag is expected for the prototype? Neglect viscous effects.

4.61 Swim fins are tested by a diver in a water tunnel. The fins are 60 cm long and develop a thrust of 10 N when the relative velocity between the water and the user is 1.25 m/s. The manufacturer has decided to market a geometrically similar fin for children that is 30 cm long. What is the expected thrust if test conditions are dynamically similar to those on the full-size fin?

4.62 A ship 300 ft long travels through water at 15 knots. A one-fiftieth scale model is to be tested in a towing basin containing a liquid mixture of water and a thinning agent. Determine the kinematic viscosity of the mixture if the Reynolds number and the Froude number are the same for both model and prototype. Also calculate the required velocity of the model.

4.63 Tests on drag force exerted on tractor-trailer trucks are to be made. For actual conditions, a truck travels at 55 mph and pulls a trailer that is about 40 ft long. It is desired to model this situation with a 5-ft-long model in a wind tunnel. What is the required flow velocity in the wind tunnel for dynamically similar conditions between the model and the prototype? Suppose that the 5-ft-long model could be submerged in water that is circulated in a water tunnel. What is the required water velocity for dynamically similar conditions between model and prototype?

4.64 A propeller of a ship is to be modeled. The model has a diameter of 10 cm and it exerts a propulsive force of 150 N when the flow velocity is 0.5 m/s. The prototype is 2 m in diameter. What is the velocity and the propulsive force for the prototype that correspond to what was obtained with the model? The fluid is water in both cases.

4.65 A yacht is 21 ft long and a model of it is only 1 ft long. The yacht can travel at 10 knots. What velocity of the model is required for dynamic similarity between the two if water is the fluid in both cases? What velocity is required if glycerine is used for tests on the model?

4.66 The finish on the hull of a ship has an influence on the power required to move the ship through water. Tests are made by coating an aluminum plate with the surface finish of interest and then dragging the plate through water. Consider a painted aluminum plate 2 m long that is dragged at 5 knots through water. The measured drag force is 10 N. It is proposed to use the same finish on a ship hull that is 8 m long. Determine the velocity of the prototype required for dynamic similarity between the ship and the plate. Estimate the drag force exerted on the ship using the drag force obtained on the plate. Water is the fluid in both cases.

4.67 Pump performance data are sometimes correlated by using several dimensionless terms, two of which are

$$\text{Volumetric flow coefficient} = \frac{Q}{\omega D^3}$$

$$\text{Energy transfer coefficient} = \frac{g\Delta H}{\omega^2 D^2}$$

where

Q is the volume flow rate through the pump
ω is the rotational speed of pump
D is the characteristic length associated with the pump, usually the dimension of an internal component of the pump
g is the acceleration due to gravity
ΔH is the energy added to fluid in the form of an increase in head, expressed as a length dimension

Actual performance data on a centrifugal pump are as follows: rotational speed = 3500 rpm; total head = energy added = 80 ft of water; volume flow rate = 50 gpm; characteristic dimension = 5⅛ in.; fluid = water. It is desired to operate this pump at 1750 rpm and to change the characteristic dimension to 4⅝ in. Determine how the new configuration will affect the flow rate and the total head/energy added to the fluid.

4.68 Assume that the same fluid is used in model and prototype for any system. Determine, then, the ratio of velocities—model to prototype—if the criterion for similarity is (a) Froude number, (b) Reynolds number, and (c) Weber number.

4.69 Kerosene flows through a 2 in. ID pipe at 4 ft/s. This flow is modeled using a 6 in. ID pipe with water as the working fluid.

a. What is the corresponding water velocity in the model for dynamic similarity between model and prototype?
b. Suppose that by mistake the Froude number is used rather than the correct dimensionless group. What is the corresponding water velocity in the model when the Froude number is the same in the model and the prototype?

4.70 Acetone is a solvent that is used as a degreaser. The acetone is sprayed through a nozzle onto a surface that needs cleaning. It is believed that the cleaning action is most effective with a nozzle having an orifice diameter of 0.25 cm and discharge velocity (at the nozzle exit) of 1 m/s. A similar cleaning system is to be tested using carbon tetrachloride and a nozzle whose orifice diameter is 0.30 cm.

a. What is the required discharge velocity of the carbon tetrachloride if the Reynolds number is used to assure similarity?
b. Suppose that the Froude number is used instead. What would the corresponding discharge velocity be in this case?

c. What if the Weber number is used? What is the corresponding discharge velocity of the carbon tetrachloride in this case?
d. Which is the correct dimensionless group to use for this problem?

CORRELATION OF EXPERIMENTAL DATA

4.71 Consider a tube with an inside diameter D through which water flows. The tube is placed in a horizontal position, and pressure taps on it are located 5 ft apart. The pressure taps are connected to a manometer that gives the pressure drop in terms of a head loss as water flows through the pipe. Head loss increases with flow rate. The test setup is shown in Figure 4.5, and the data obtained are provided in the following table. Determine an equation to fit the data, and predict $\Delta h/L$ in terms of the Reynolds number. ($L = 5$ ft; $D = 0.785$ in.).

Data pt.	Q (gpm)	Δh (in.)
1	1.5	0.5
2	1.85	0.5
3	2.0	0.75
4	2.45	0.93
5	2.8	1.00
6	3.3	1.75
7	3.5	1.81
8	3.9	2.31
9	4.1	2.63

4.72 Consider a tube with an inside diameter D through which water flows. The tube is placed in a horizontal position, and pressure taps on it are located 5 ft apart. The pressure taps are connected to a manometer that gives the pressure drop in terms of a head loss as water flows through the pipe. Head loss increases with flow rate. The test setup is shown in Figure 4.5, and the data obtained are provided in the following table. Determine an equation to fit the data, and predict $\Delta h/L$ in terms of the Reynolds number. ($L = 5$ ft; $D = 1.025$ in.).

Data pt.	Q (gpm)	Δh (in.)
1	1.5	0.063
2	1.85	0.13
3	2.0	0.19
4	2.45	0.31
5	2.8	0.44
6	3.3	0.63
7	3.5	0.69
8	3.9	0.75
9	4.1	0.88

4.73 A hemisphere is placed in a wind tunnel and oriented as shown in the following figure; the air ($T = 24$°C) moves past the hemisphere at a uniform and controllable velocity. The air exerts a drag force on the hemisphere and its mounting stand equal to W. When the drag of the mounting stand is subtracted from the raw numbers, drag force on the hemisphere D_f versus velocity V data are obtained. The experiment is performed with three different-sized hemispheres: 1, 2, and 3 in. diameters; the results are as given in the following table. Graph the drag coefficient as a function of the Reynolds number.

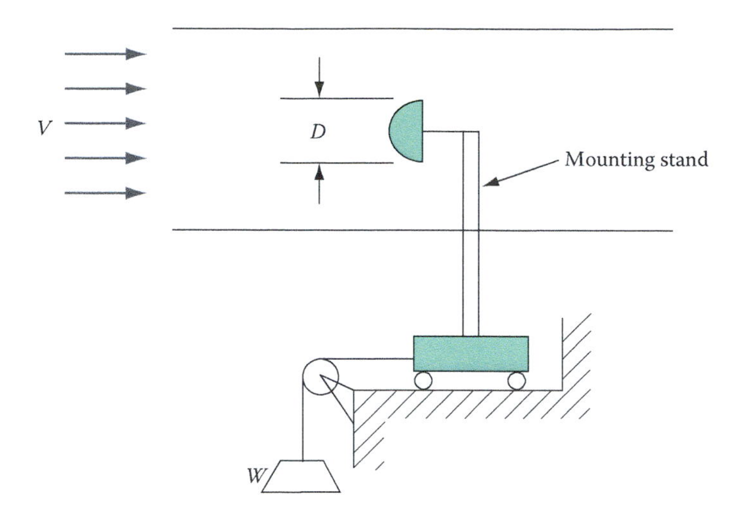

D = 1 in.		D = 2 in.		D = 3 in.	
D_f (ibf)	V (mph)	D_f (ibf)	V (mph)	D_f (ibf)	V (mph)
0.038	45.1	0.072	44.3	0.155	44.0
0.037	39.6	0.072	41.1	0.141	40.9
0.029	30.9	0.043	30.6	0.083	30.3
0.019	20.0	0.026	20.0	0.040	20.0
0.011	10.0	0.014	10.0	0.020	10.0

Data obtained by Mary Franck and Darryl Wilson.

4.74 Data in the accompanying table are for a water pump operating at ω = 1750 rpm. (These are actual pump performance data obtained from a catalog.) The first column is of volume flow rate in gpm. The second column is of a change in head ΔH when the pump is operated with an 8 in. (=characteristic length, D) diameter impeller. The parameter ΔH is defined as

$$\Delta H = \left\{ \left(\frac{p}{\rho g} + \frac{V^2}{2g} + z \right)\Bigg|_{\substack{\text{pump}\\\text{outlet}}} - \left(\frac{p}{\rho g} + \frac{V^2}{2g} + z \right)\Bigg|_{\substack{\text{pump}\\\text{inlet}}} \right\}$$

The third and fourth columns are also of ΔH but for different impeller sizes.

	Fluid = Water	ω = 1750 rpm	ΔH in ft of Water
Q, gpm	ΔH (D = 8 in.)	ΔH (D = 7 ½ in.)	ΔH (D = 7 in.)
80	67	59	48
160	65	56	45
240	58	50	40
320	50	39	28
400	35	20	—

a. Construct a graph of ΔH (vertical axis) versus Q for this pump.
b. From a dimensional analysis performed for a pump, the following groups can be derived:

$$\frac{g\Delta H}{\omega^2 D^2} = \text{energy transfer coefficient}$$

$$\frac{Q}{\omega D^3} = \text{volume flow coefficient}$$

and

$$\frac{dW/dt}{\rho\omega^3 D^5} = \frac{Qg\Delta H}{\omega^3 D^5} = \text{power coefficient}$$

Construct a graph of the energy transfer coefficient (vertical axis) versus the volume flow coefficient.
c. Construct a graph of the power coefficient (vertical axis) versus the volume flow coefficient.

5 Flow in Closed Conduits

Flow in closed conduits is a very important part of the study of fluid mechanics—primarily because examples are commonplace. Water for domestic use is distributed to all parts of the household in pipes; sewers and drains carry wastewater away. Crude oil is pumped from well to refinery in pipes. Natural gas is brought to the user via pipes. Heated air is distributed to all parts of a house in circular and rectangular ducts. Many other examples of conduit flow can be found in everyday life. In this chapter, we will examine the variables that are important in mathematically describing such flows.

The purpose of this chapter is to describe laminar and turbulent flow phenomena, to determine the effect of viscosity, to present the equations of motion for conduit flow, and to analyze various piping systems. Because pumps are commonly used to convey liquid in pipes, a brief survey of pumps is also presented.

5.1 LAMINAR AND TURBULENT FLOWS

Experiments with flow in pipes demonstrated that two different flow regimes exist—*laminar* and *turbulent*. When laminar flow exists in a system, the fluid flows in smooth layers called *laminae*. A fluid particle in one layer stays in that layer. The layers of fluid slide by one another without apparent eddies or swirls. Turbulent flow, on the other hand, exists at much higher flow rates in the system. In this case, eddies and vortices mix the fluid by moving particles tortuously about the cross section.

The existence of two types of flow is easily visualized by examining results of experiments performed by Osborne Reynolds. His apparatus is shown schematically in Figure 5.1a. A transparent tube is attached to a constant-head tank with water as the liquid medium. The opposite end of the tube has a valve to control the flow rate. Dyed water is injected into the water at the tube inlet, and the resulting flow pattern is observed. For low rates of flow, something similar to Figure 5.1b results. The dye pattern is regular and forms a single line like a thread. There is no lateral mixing in any part of the tube, and the flow follows parallel streamlines. This type of flow is called **laminar**, or **viscous, flow**.

As the flow rate of water is increased beyond a certain point, the dye is observed not to follow a straight threadlike line but to disperse. The dyed water mixes thoroughly with the pipe water as shown in Figure 5.1c as a result of erratic fluid behavior in the pipe. This type of flow is called **turbulent flow**. The Reynolds number at the point of transition between laminar and turbulent flows is called the **critical Reynolds number**.

This experiment can be repeated with several pipes of different diameter. A dimensional analysis, when performed and combined with data, shows that the criterion for distinguishing between these flows is the Reynolds number:

$$\mathrm{Re} = \frac{\rho V D}{\mu} = \frac{VD}{\nu} \tag{5.1}$$

where

V is the average velocity of the flow
D is the inside diameter of the tube

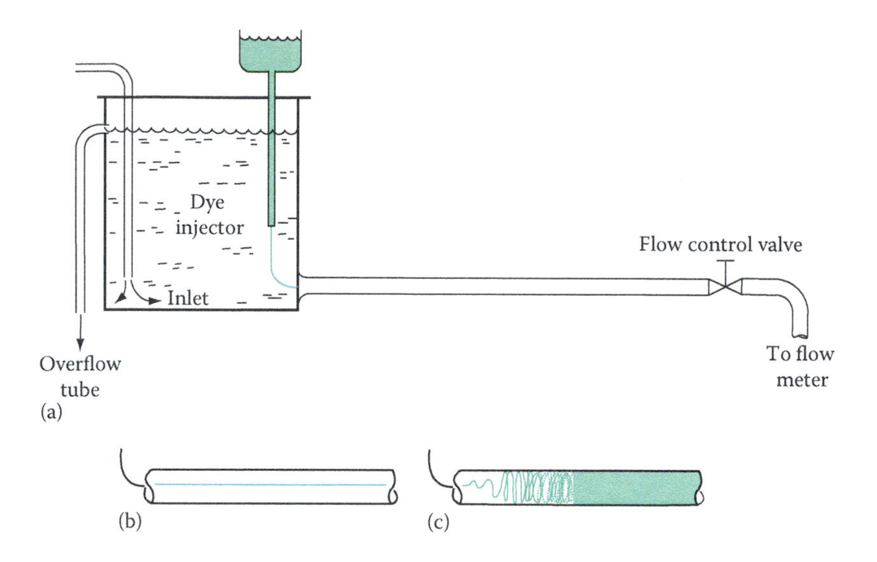

FIGURE 5.1 An experiment for visualizing laminar and turbulent flows.

For straight circular pipes, the flow is always laminar for a Reynolds number less than about 2 100. The flow is usually turbulent for Reynolds numbers over 4 000. For the transition regime in between, the flow can be either laminar or turbulent, depending upon details of the apparatus that cannot always be predicted or controlled. For our work, we will sometimes need to have an exact value for the Reynolds number at transition. We will arbitrarily choose this value to be 2 100.

A distinction must be made between laminar and turbulent flows, because the velocity distribution within a duct is different for each. Figure 5.2a shows a coordinate system for flow in a tube, for example, with corresponding velocities. As indicated, there can be three different *instantaneous* velocity components in a conduit—one for each of the three principal directions. Furthermore, each of these velocities can be dependent upon at most three space variables and one time variable. If the flow in the tube is laminar, we have only one nonzero instantaneous velocity: V_z. Moreover, V_z is a function of only the radial coordinate r, and the

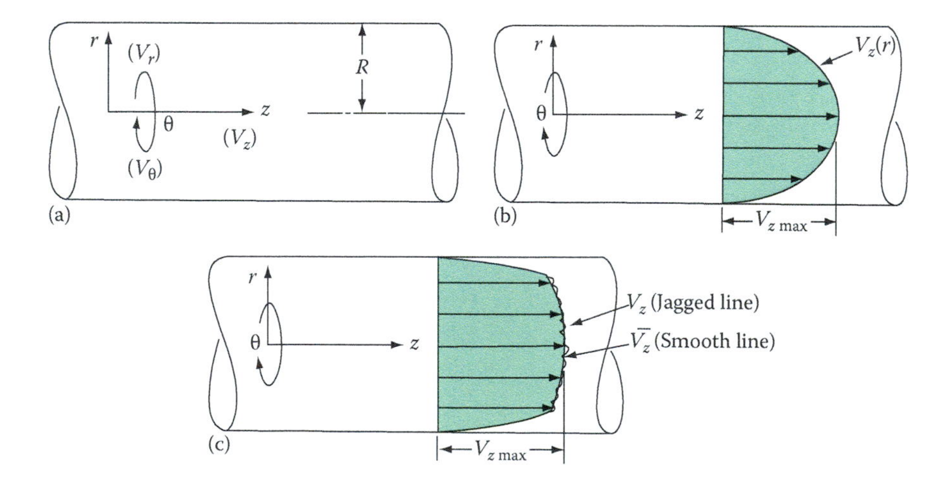

FIGURE 5.2 (a) Cylindrical coordinates and velocity directions in a tube, (b) axial velocity distribution for laminar flow in a tube, and (c) axial velocity distribution for turbulent flow in a tube.

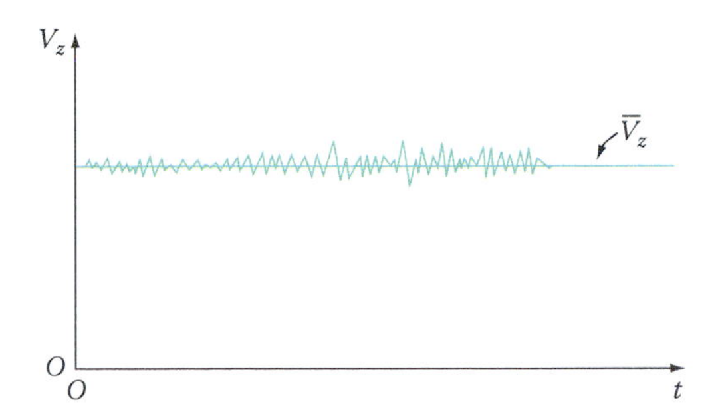

FIGURE 5.3 Variation of centerline axial velocity with respect to time (turbulent flow).

velocity distribution is parabolic as shown in Figure 5.2b. An equation for this distribution can be derived from the equation of motion performed later in this chapter.

If the flow is turbulent, all three instantaneous velocities V_r, V_θ, and V_z are nonzero. Moreover, each of these velocities is a function of all three space variables and of time. An equation for velocity (V_r, V_θ, or V_z) is not derivable from the equation of motion. Therefore, to envision the axial velocity, for instance, we must rely on experimental data. Figure 5.2c shows the axial instantaneous velocity V_z for turbulent flow in a tube.

The instantaneous velocity V_z fluctuates randomly about the mean axial velocity $\overline{V}_z$. We would need a very sensitive measuring instrument to obtain V_z but a relatively insensitive instrument to obtain $\overline{V}_z$. If the velocity profile in the turbulent flow of Figure 5.2c were measured at another instant in time, a different instantaneous profile (V_z) would result. It would, however, fluctuate about the same mean axial velocity $\overline{V}_z$. Thus, the mean axial velocity would be obtained by removing the time dependence (by integration if the function were known) from the instantaneous velocity V_z. The time dependence can be further illustrated by looking at data in the axial or z-direction at a point (say the tube centerline) as a function of time (see Figure 5.3).

As indicated in the preceding discussion, the axial velocity V_z in turbulent flow fluctuates about some mean velocity. In general, the fluctuations are small in magnitude, but they cause slower-moving particles in one region of the pipe cross section to exchange position with faster-moving particles in another region. This is in contrast to what happens in laminar flow, in which a fluid particle in one layer stays in that layer. The fluctuations in turbulent flow are responsible for a mixing effect that manifests itself in a more evened-out velocity profile than that for laminar flow. Also, these fluctuations cause the mixing of the injected water with the tube water in the Reynolds experiment of Figure 5.1c. For both laminar and turbulent flows, maximum velocity in the axial direction, $V_{z\,max}$, occurs at the centerline of the duct or conduit. These comments concerning laminar and turbulent flows are summarized in Table 5.1.

The Reynolds number (Equation 5.1) is used to distinguish between laminar and turbulent flows. The velocity in the Reynolds number expression is the average velocity V. In principle, the average velocity is obtained by integrating the equation for instantaneous velocity V_z over the cross-sectional area and dividing the result by the area. This procedure is correct for laminar, transition, or turbulent flow. If there is no equation available, then experimental means are necessary to find the average velocity. In the simplest case, we measure volume flow rate (for an incompressible fluid) and divide by cross-sectional area: $V = Q/A$.

5.2 EFFECT OF VISCOSITY

Table 5.1 shows the velocity distribution for laminar and turbulent flows. In both cases, the velocity at the wall is zero. This phenomenon, called the **nonslip condition**, is due to viscosity. In laminar flow, in which inertia or momentum of the fluid is small, the viscous effect is able

TABLE 5.1
Comparison of Laminar and Turbulent Flows

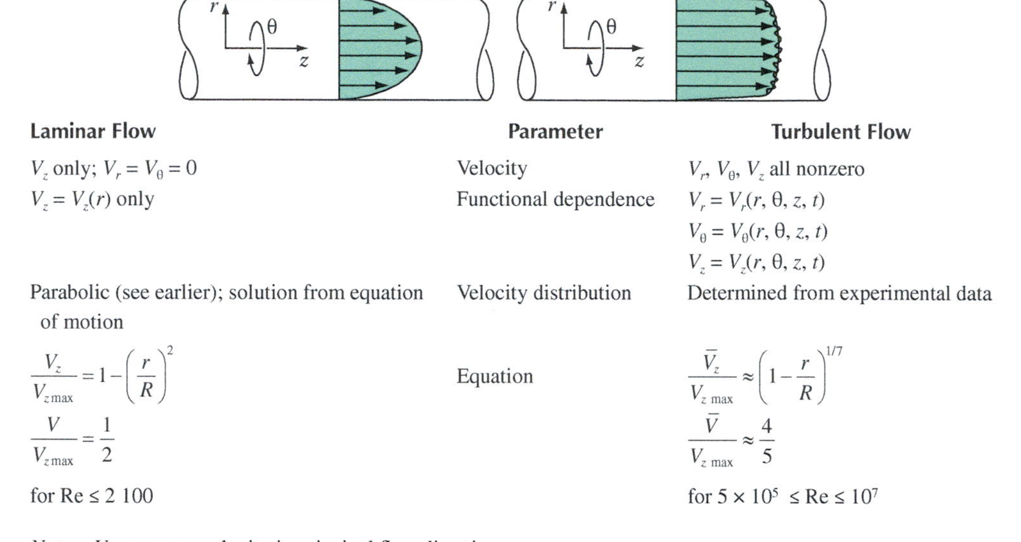

Laminar Flow	Parameter	Turbulent Flow
V_z only; $V_r = V_\theta = 0$	Velocity	V_r, V_θ, V_z all nonzero
$V_z = V_z(r)$ only	Functional dependence	$V_r = V_r(r, \theta, z, t)$ $V_\theta = V_\theta(r, \theta, z, t)$ $V_z = V_z(r, \theta, z, t)$
Parabolic (see earlier); solution from equation of motion	Velocity distribution	Determined from experimental data
$\dfrac{V_z}{V_{z\max}} = 1 - \left(\dfrac{r}{R}\right)^2$ $\dfrac{V}{V_{z\max}} = \dfrac{1}{2}$ for Re ≤ 2 100	Equation	$\dfrac{\overline{V}_z}{V_{z\max}} \approx \left(1 - \dfrac{r}{R}\right)^{1/7}$ $\dfrac{\overline{V}}{V_{z\max}} \approx \dfrac{4}{5}$ for $5 \times 10^5 \le \text{Re} \le 10^7$

Note: V = average velocity in principal flow direction.

to penetrate farther into the cross section from the wall than it can in turbulent flow. Stated another way, in turbulent flow, in which the viscous effects are small, the momentum or inertia of the flow is able to penetrate farther outward toward the wall from the centerline than it can in laminar flow. This penetration of momentum or inertia is called the **momentum transport phenomenon**.

Consider the case of a fluid between two parallel plates, with the upper plate moving as shown in Figure 5.4. The upper plate has fluid adhering to it owing to friction. The plate exerts a shear stress on the particles in layer A. This layer in turn exerts a shear stress on layer B and so on. It is the x-component of velocity in each layer that causes this shear stress to be propagated in the negative z-direction; that is, the A layer pulls the B layer along, and so forth. As this shear stress approaches the stationary wall, movement is retarded by the effect of zero velocity at the bottom propagating upward; that is, the E layer retards the D layer, and so on. The momentum (mass·velocity) of the plate in the x-direction is transported in the z-direction. The resultant effect on velocity is the distribution sketched in Figure 5.4.

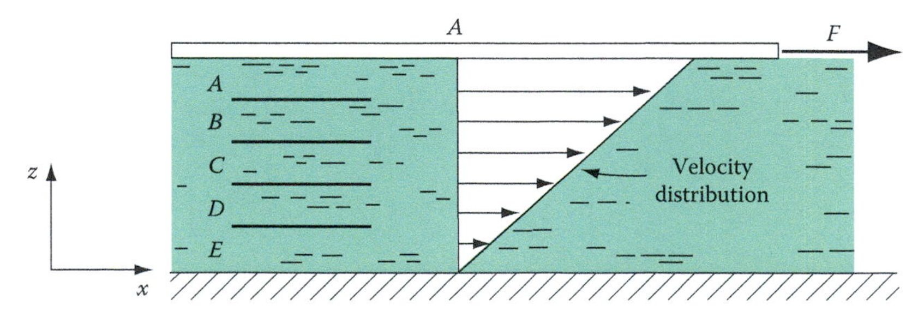

FIGURE 5.4 Layers of fluid flow between parallel plates.

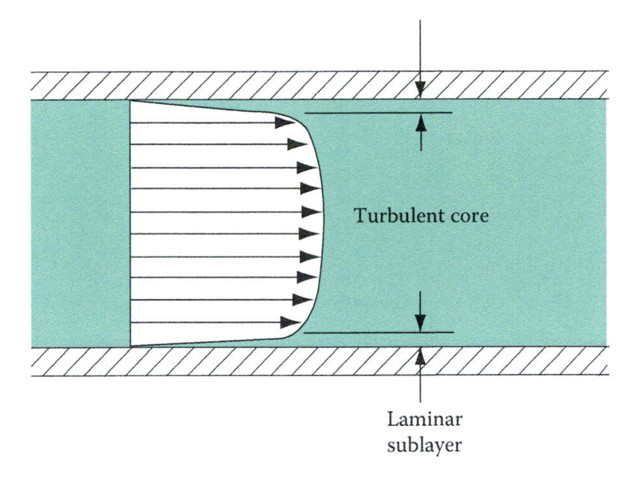

FIGURE 5.5 Laminar sublayer in turbulent pipe flow.

As we saw earlier, in turbulent flow, the velocity at a stationary wall is zero. Near the wall, then, there must be a region of flow that is laminar. This region is called the **laminar sublayer**, and the flow in the remainder of the cross section is called the **turbulent core**. Both regions are illustrated for flow in a pipe in Figure 5.5.

5.2.1 Entrance Effects

Another effect of viscosity is evident at the entrance to a pipe, as illustrated in Figure 5.6. Flow is uniform at the entrance; but as the fluid travels downstream, the effect of zero wall velocity propagates throughout the cross section. The flow is divided into a viscous region and a core region. Particles in the core do not sense that a wall is present. Eventually, the core disappears, and the velocity distribution becomes fully developed. A fully developed profile does not change with further increases in length. Mathematically, we write

$$\frac{d(V_z)}{dz} = 0 \quad \text{(fully developed laminar flow)}$$

$$\frac{d(\bar{V}_z)}{dz} = 0 \quad \text{(fully developed turbulent flow)}$$

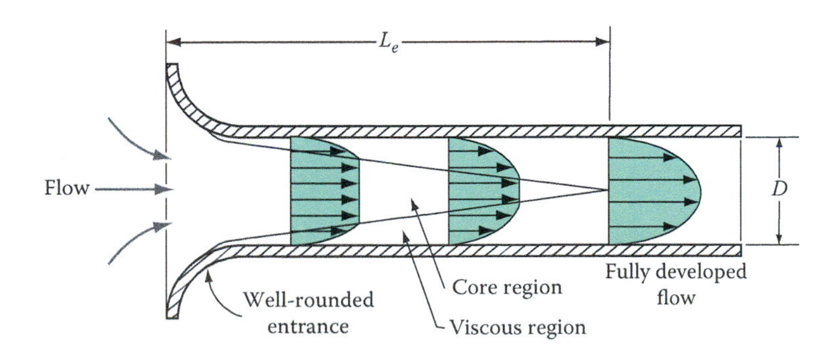

FIGURE 5.6 Laminar flow near a pipe inlet.

The distance L_e (Figure 5.6) is called the **entrance length**, and its magnitude is dependent upon the forces of inertia and viscosity. It has been determined from numerous experimental and analytical investigations that the entrance length can be estimated with

$$L_e = 0.06D(\text{Re}) \quad \text{(laminar flow)} \tag{5.2a}$$

$$L_e = 4.4D(\text{Re})^{1/6} \quad \text{(turbulent flow)} \tag{5.2b}$$

where $\text{Re} = \rho VD/\mu = VD/\nu$. For laminar flow, we see that the entrance length varies directly with the Reynolds number. For the largest Reynolds number encountered in laminar flow of 2 100, Equation 5.2a predicts

$$L_e = 0.06D(2\,100) = 126D$$

Thus, 126 diameters is the maximum length that would be required for fully developed conditions to exist in laminar flow.

For turbulent flow, the entrance length varies with the one-sixth power of the Reynolds number. Conceptually, there is no upper limit for the Reynolds number in turbulent flow, but in many engineering applications, $10^4 < \text{Re} < 10^6$. Over this range, we calculate, with Equation 5.2b:

$$20 < \frac{L_e}{D} < 44$$

So in turbulent flow, the entrance length values are considerably less than the 126 diameters required at a Reynolds number of 2 100. The reason that a shorter length is required in turbulent flow is the mixing action. For abrupt or sharp-edged entrances, additional turbulence is created at the inlet. The effect is to further decrease the inlet length required for fully developed flow to exist.

Example 5.1

Castor oil is flowing from a tank through a well-rounded entrance into a pipe whose inside diameter is 5.3 cm. The flow rate is 0.01 m^3/s. Is the flow laminar or turbulent? What is the entrance length required for the flow to become fully developed?

Solution

Castor oil $\rho = 960$ kg/m^3, $\mu = 650 \times 10^{-3}$ N·s/m^2 (Table A.5)

The flow area is

$$A = \frac{\pi D^2}{4} = \frac{\pi(0.053\text{ m})^2}{4} = 0.002\,206\text{ m}^2$$

The average velocity in the pipe is

$$V = \frac{Q}{A} = \frac{0.01\text{ m}^3/\text{s}}{0.002\,206\text{ m}^2} = 4.5\text{ m/s}$$

The Reynolds number becomes

$$\text{Re} = \frac{\rho VD}{\mu} = \frac{(960\text{ kg/m}^3)(4.5\text{ m/s})(0.053\text{ m})}{650\times 10^{-3}}$$

$$\text{Re} = 352$$

The flow is laminar

The entrance length for fully developed flow is

$$L_e = 0.06D\,\text{Re} = 0.06(0.053\,\text{m})(352)$$

$$L_e = 1.12\,\text{m}$$

Thus, the velocity profile remains virtually unchanged after a distance of 1.12 m has been reached.

5.3 PIPE DIMENSIONS AND SPECIFICATIONS

This section presents various standards regarding pipe specifications. Pipes are specified according to what is called a **nominal diameter**—for example, ⅛-nominal or 1½-nominal.* The nominal diameter does not necessarily indicate the exact inside or outside diameter of the pipe, however. Another specification is the **pipe schedule**. A schedule 40 pipe is considered standard in 6-nominal and lower pipe sizes. A schedule 80 pipe has a thicker wall than a schedule 40, so schedule 80 in the lower sizes is designated XS for extra strong. Table C.1 gives the dimensions of numerous pipe sizes from ⅛-nominal to 40-nominal for wrought steel and wrought iron pipe.

A proper specification for a pipe would thus include a nominal diameter and a schedule. For example, a 1-nominal, schedule 40 pipe would have an outside diameter of 3.34 cm (1.315 in.) and an inside diameter of 2.664 cm (0.0874 ft), where these dimensions are obtained from Table C.1. The nominal diameter specifies the outside diameter of the pipe, whereas the schedule specifies the wall thickness and hence the inside diameter. Thus, all 1-nominal pipe, regardless of schedule, has the same outside diameter.

During the early days of pipe manufacturing, the nominal diameter was equal to the inside diameter of the pipe. However, improvements in the strengths of materials have meant that pipe-wall thickness could be decreased with little or no decrease in the ability of the pipe to perform satisfactorily. The options were to increase the inside diameter or to decrease the outside diameter. A decision was made based on the need to use fittings (elbows, T-joints, valves, and so on) with pipes. Fittings attach to the outside wall, so it seemed prudent to keep the outside diameter the same, decreasing the wall thickness required by enlarging the inside diameter. Hence, for 12-nominal and smaller pipe sizes, the nominal and actual diameters are not equal. For 14-nominal and larger, the nominal diameter is equal to the outside diameter.

Pipes are attached together in various ways. Pipe ends can be threaded, for example, and the number of threads per length (usually per inch) is standardized. Standard pipe threads are tapered (Figure 5.7). Before attachment, pipe threads are wrapped with special tape or coated with a viscous compound to ensure a fluid-tight connection. Threaded connections are common with cast iron,

FIGURE 5.7 A threaded pipe end showing a taper.

* It is standard industrial practice to specify nominal pipe sizes with an inch dimension (e.g., 1 ½ in. nominal). The nominal dimension is considered to be a name for the size only. For convenience, the "inch" is dropped from the designation in this book because the reference may be to an English or to an SI measurement. For further discussion, refer to ASTM E 380.76, p. 6, Section 3.4.3.1, and to ASME Guide SI-1, p. 9, para 8.

FIGURE 5.8 A flanged pipe connection.

wrought iron, and wrought steel pipe. Alternatively, if the metal is weldable, pipes can be welded together to form a fluid-tight connection. Welding is common in the larger pipe sizes.

Pipe ends may also have flanges attached. Flanges are made in various sizes. Before two flanges are bolted together to form a secure connection, a rubber or gasket-type material is placed between them to ensure a fluid-tight joint. A flanged pipe joint is shown in Figure 5.8.

Another type of pipe material is polyvinyl chloride (PVC) plastic pipe, which can be threaded or joined together with an adhesive. Plastic pipe can be specified in the same manner as wrought steel pipe.

Another conventional circular conduit for conveying fluid is called **tubing**. Copper tubing is used extensively in plumbing applications. It is important to realize that there is a difference between pipe and tubing. The most obvious difference is that tubing has a much thinner wall than does pipe. Also, there is an entirely different way to specify tubing. For copper water tubing, there are three types—K, L, and M—and a number of standard sizes. Type K is used for underground service and general plumping. Type L is used for interior plumbing. Type M is for soldered fittings only. Dimensions of copper tubing for ¼-standard to 12-standard are provided in Table C.2.

A proper specification for tubing would include a standard size and type. For example, a 1-standard, type K copper tube has an outside diameter of 2.858 cm (1.125 in.) and an inside diameter of 2.528 cm (0.08292 ft), from Table C.2. The standard dimension fixes the outside diameter, whereas the type specifies the wall thickness directly or the inside diameter indirectly.

Another type of tubing in common use is refrigeration tubing. It is specified in the same way as the copper tubing just discussed and is also usually made of copper. The difference is that refrigeration tubing is ductile (in fact, it is easily bent by hand), but copper water tubing is quite rigid.

Copper tubing ends can be flared with a tool to make them suitable for joining, as illustrated in Figure 5.9a. Another joining method is by the use of a compression fitting (Figure 5.9b). The tube is inserted through a ring that is part of the fitting. As the fitting nut is tightened, the ring is compressed, causing the copper tube to expand tightly against the inside wall of the fitting. Still another joining method is soldering. The tube is inserted into a fitting and the two are soldered together—commonly called *sweating*.

5.3.1 Equivalent Diameters for Noncircular Ducts

Noncircular conduits can be found in many fluid-conveying systems. For example, sheet metal is bent appropriately into a rectangular cross section for use in heating or air conditioning ducts,

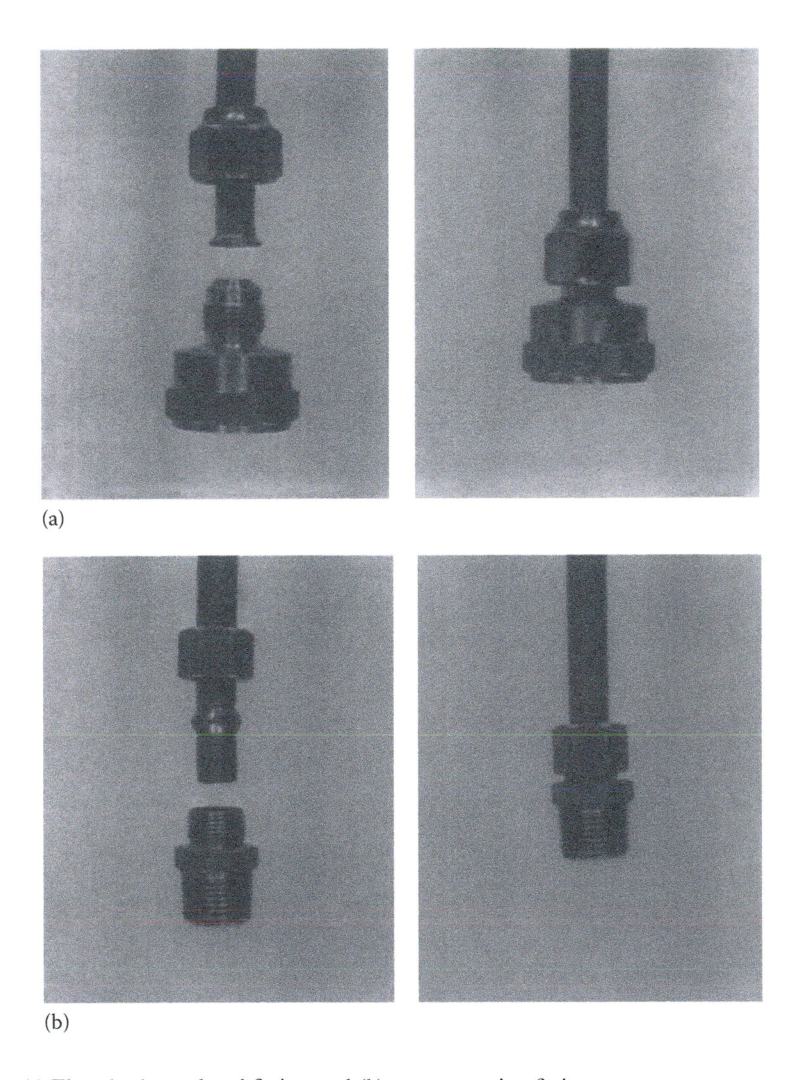

FIGURE 5.9 (a) Flared tube end and fitting and (b) a compression fitting.

gutters, and downspouts. In double-pipe heat exchangers, one tube is placed within a larger tube, and the fluid-flow section between the tubes is an annulus. Recall that the definition of Reynolds number ($VD/\nu = \rho VD/\mu$) contains a dimension or characteristic length D that is used to represent the flow area. The question then arises as to what dimension to select for rectangular, square, annular, or any other noncircular cross section. We examine two different choices that have been proposed for modeling flows in noncircular conduits.

The first characteristic dimension that we discuss is called the **hydraulic diameter**, D_h. The hydraulic diameter is defined as

$$D_h = \frac{4 \times \text{cross-sectional area of fluid flow}}{\text{perimeter of duct in conduct with fluid}} \tag{5.3}$$

For a circular cross section flowing full, $A = \pi D^2/4$ and $P = \pi D$. Thus,

$$D_h = \frac{\pi D^2}{\pi D} = D = \text{diameter of pipe}$$

For a square duct flowing full, $A = s^2$ and $P = 4s$. Therefore,

$$D_h = \frac{4s^2}{4s} = s = \text{length of one side}$$

Hydraulic diameter can also be determined when the duct is only half full. For a circular duct flowing half full, $A = \pi D^2/8$ and $P = \pi D/2$. Then

$$D_h = \frac{\pi D^2/2}{\pi D/2} = D = \text{diameter of pipe}$$

The hydraulic diameter arises from applying the momentum equation to flow in a duct, which is illustrated in the next section.

The second characteristic dimension we discuss (used primarily in open-channel flow modeling) is called the **hydraulic radius**, R_h, defined as

$$R_h = \frac{\text{cross-sectional area of fluid flow}}{\text{perimeter of duct in contact with fluid}}$$

For a circular cross section flowing full, the hydraulic radius is

$$R_h = \frac{\pi D^2/4}{\pi D} = \frac{D}{4}$$

For a circular duct, we would like to use the diameter D (not $D/4$) as the characteristic length, and so, the hydraulic diameter D_h is preferred over the hydraulic radius in closed-conduit flows.

In this chapter, we will adopt the traditional approach and use the hydraulic diameter for closed-conduit flow modeling. It is necessary to be able to use a hydraulic diameter, because it is best to have only one characteristic dimension to represent the duct shape of an irregular cross section. Equation 5.3 is a general expression that applies to any cross section of flow.

5.4 EQUATION OF MOTION

Consider flow in a pipe of constant diameter where pressure is measured at two different points a distance L apart (see Figure 5.10). Assume one-dimensional, steady, uniform flow of an incompressible fluid; assume also that there is no heat transfer and no shaft work being done. If as a first approximation, we further assume frictionless flow, the momentum and energy equations reduce to the same expression—the Bernoulli equation. Applying Bernoulli's equation, we obtain

$$\frac{p_1}{\rho g} + \frac{V_1^2}{2g} + z_1 = \frac{p_2}{\rho g} + \frac{V_2^2}{2g} + z_2$$

For a horizontal pipe, $z_1 = z_2$. From continuity, $A_1V_1 = A_2V_2$. Because $D_1 = D_2$, then $A_1 = A_2$; therefore, $V_1 = V_2$. The Bernoulli equation reduces to

$$p_1 = p_2$$

This result is not an accurate description of the situation, however. For flow to be maintained in the direction indicated in Figure 5.10, p_1 must be greater than p_2 in an amount sufficient to overcome

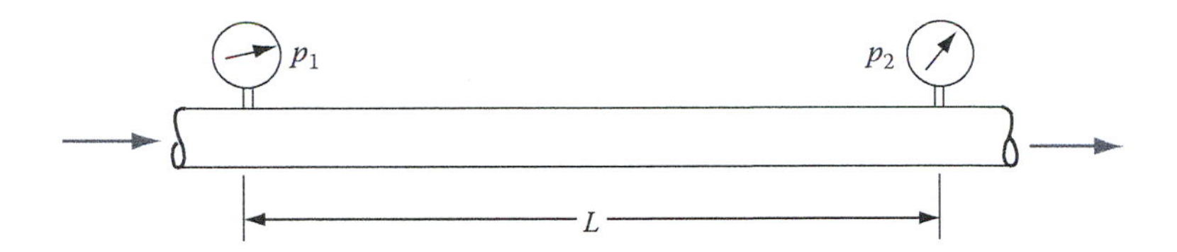

FIGURE 5.10 Flow in a constant-diameter pipe.

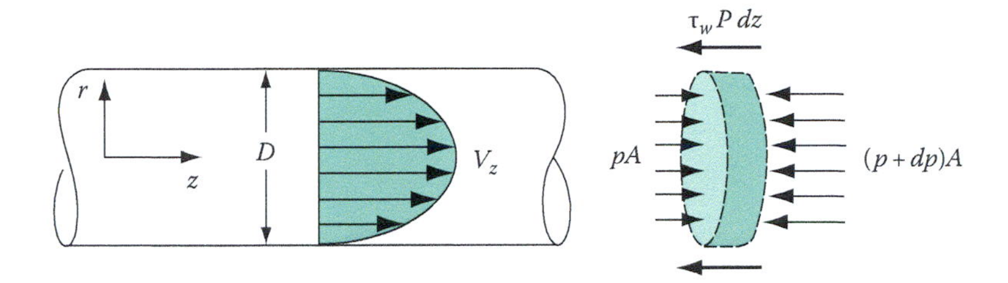

FIGURE 5.11 Control volume of a system: flow in a duct.

friction between the fluid and the pipe wall. To be able to apply Bernoulli's equation, we must first modify the equation by adding a friction term to the right-hand side. We next derive an equation for the friction term for fully developed, steady, one-dimensional flow of an incompressible fluid.

Consider flow in a pipe, as shown in Figure 5.11. A control volume that extends to the wall (where the friction force acts) is selected for analysis. Note that a circular cross section is illustrated, but the results are general until we substitute specific equations for the geometry of the cross section. The forces acting on the control volume are pressure normal to the surface and shear stress acting at the wall. The momentum equation is

$$\sum F_z = \iint V_z \rho V_n \, dA$$

Since the flow out of the control volume equals the flow in, the right-hand side of this equation is zero. The sum of the forces is

$$pA - \tau_w P\,dz - (p + dp)A = 0$$

where

A is the cross-sectional area

$P\,dz$ is the surface area (perimeter times length) over which the wall shear τ_w acts

The equation reduces to

$$\tau_w P\,dz + A\,dp = 0$$

Rearranging and solving for pressure drop, we get

$$dp = -\tau_w \frac{P\,dz}{A}$$

Multiplying numerator and denominator by 4 gives

$$dp = -\tau_w \frac{4P\,dz}{4A}$$

Recalling the definition of hydraulic diameter D_h (=4*A*/*P*), the preceding equation becomes

$$\frac{dp}{dz} = -\frac{4\tau_w}{D_h} \tag{5.4}$$

We have thus expressed the pressure drop per unit length of conduit in terms of the wall shear and the hydraulic diameter. The equation was modified to obtain the hydraulic diameter, which is our preferred choice for the characteristic dimension. Equation 5.4 is a general expression for any cross section, because at this point, we have yet to specify whether the conduit is circular, square, or annular, for instance. We next introduce a friction factor *f*, which is customarily defined as the ratio of friction forces to inertia forces*:

$$f = \frac{4\tau_w}{\frac{1}{2}\rho V^2} \tag{5.5}$$

By substitution into Equation 5.4, we obtain

$$dp = -\frac{\rho V^2}{2}\frac{f\,dz}{D_h} \tag{5.6}$$

Integrating this expression from point 1 to point 2 a distance *L* apart in the conduit yields

$$\int_{p_1}^{p_2} dp = -\frac{\rho V^2}{2}\int_0^L \frac{f\,dz}{D_h}$$

Assuming that the friction factor *f* is constant gives

$$p_2 - p_1 = -\frac{\rho V^2}{2}\frac{fL}{D_h} \tag{5.7}$$

where the negative sign indicates that pressure decreases with increasing *z* in the flow direction due to friction. With this equation, the Bernoulli equation can be rewritten as

$$\frac{p_1}{\rho g} + \frac{V_1^2}{2g} + z_1 = \frac{p_2}{\rho g} + \frac{V_2^2}{2g} + z_2 + \sum \frac{fL}{D_h}\frac{V^2}{2g} \tag{5.8}$$

which now takes wall friction into account. The summation indicates that if several pipes of different diameter are connected in series, the total friction drop is due to the combined effect of them all. The frictional effect is manifested in either a heat loss from the fluid or a gain in internal energy of the fluid.

---

* The friction factor defined in Equation 5.5 is known as the Darcy–Weisbach friction factor. The Fanning friction factor, *f*′, is used in some texts and is defined as

$$f' = \frac{\tau_w}{\frac{1}{2}\rho V^2}$$

The Darcy–Weisbach equation is preferred here because the 4 in Equation 5.4 becomes included in *f*, which yields a simpler formulation for Equation 5.6 and other equations that follow.

Example 5.2

A 2-nominal pipe is inclined at an angle of 30° with the horizontal and conveys 0.001 m^3/s of methyl alcohol uphill. Determine the pressure drop in the pipe if it is 7 m long (see Figure 5.12). Take the friction factor f to be 0.02.

Solution

From Table A.5, we read for methyl alcohol

$$\rho = 0.789(1\,000)\ \text{kg/m}^3$$

From Table C.1, we find for 2-nominal, schedule 40 pipe (because a schedule was not specified, we assume the standard):

$$D = 5.252\ \text{cm}, \quad A = 21.66\ \text{cm}^2$$

The continuity equation applied to the pipe is

$$Q = A_1V_1 = A_2V_2$$

and because diameter is constant, $A_1 = A_2$. Therefore, $V_1 = V_2$.

The Bernoulli equation with the friction term applies:

$$\frac{p_1}{\rho g} + \frac{V_1^2}{2g} + z_1 = \frac{p_2}{\rho g} + \frac{V_2^2}{2g} + z_2 + \sum \frac{fL}{D_h}\frac{V^2}{2g}$$

If z_1 is our reference point, then $z_1 = 0$ and $z_2 = 3.5$ m. The average velocity is

$$V = \frac{Q}{A} = \frac{0.001}{21.66 \times 10^{-4}} = 0.462\ \text{m/s}$$

Rearranging the Bernoulli equation, we obtain, for a circular duct with $D_h = D$,

$$p_1 - p_2 = \frac{\rho}{2}\left(V_2^2 - V_1^2\right) + \rho g(z_2 - z_1) + \frac{fL}{D}\frac{\rho V^2}{2}$$

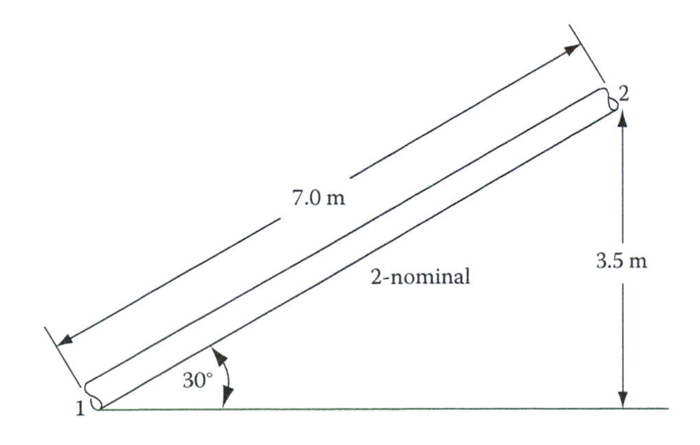

FIGURE 5.12 Sketch for Example 5.2.

Substituting,

$$p_1 - p_2 = 0 + 0.789(1\,000)(9.81)(3.5 - 0) + \frac{0.02(7)}{0.052\,52}\frac{0.789(1\,000)(0.462)^2}{2}$$

$$= 27\,090 + 224.5 = 27\,315 \text{ Pa}$$

or

$$p_1 - p_2 = 27.1 \text{ kPa}$$

5.5 FRICTION FACTOR AND PIPE ROUGHNESS

Section 5.4 introduced the concept of friction and a friction factor. In this section, we will discuss this concept further for two flow regimes—laminar and turbulent flows. Because the friction factor is influenced by the fluid velocity, we begin with a velocity distribution for circular pipe flow and later extend the results to other cross sections.

Consider laminar flow in a circular pipe as shown in Figure 5.13. Cylindrical coordinates are chosen with z as the axial direction. A control volume that does not extend to the solid surface is selected as shown, and in applying the momentum equation, we obtain

$$\sum F_z = \iint_{\text{CS}} V_z \rho V_n \, dA$$

Because the pipe is of constant diameter, the right-hand side of this equation is zero. Summing forces acting on the control volume yields

$$pA + \tau \, dA_p - (p + dp)A = 0$$

Simplifying,

$$\tau \, dA_p - A \, dp = 0 \tag{5.9}$$

For a circular conduit, $A = \pi r^2$, and the area over which the shear stress acts is $dA_p = 2\pi r \, dz$. After substitution and simplification, Equation 5.9 becomes

$$\tau(2\pi r \, dz) - \pi r^2 \, dp = 0$$

$$\frac{dp}{dz} = \frac{2\tau}{r} \tag{5.10}$$

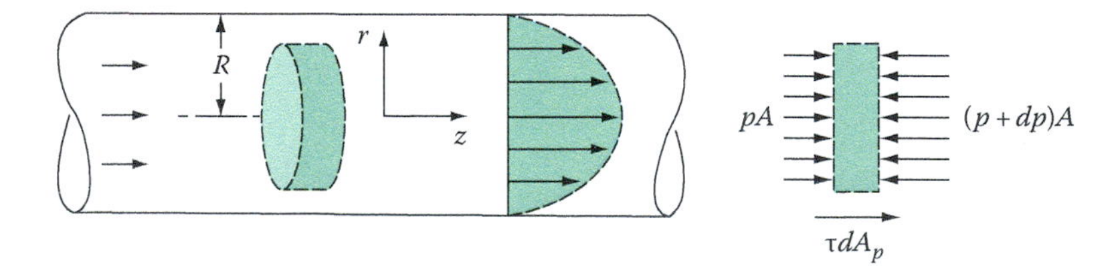

FIGURE 5.13 Laminar flow in a circular pipe.

Equation 5.10 is valid for any circular cross section conveying any fluid. For a Newtonian fluid,

$$\tau = \mu \frac{dV_z}{dr}$$

and, by substitution,

$$\frac{dp}{dz} = \frac{2}{r}\mu \frac{dV_z}{dr} \tag{5.11}$$

Now *dp*/*dz* is a pressure drop per length of pipe; pressure does not vary across the cross section, and thus, p is a function of neither r nor θ, only z. The right-hand side of the equation is a function of r alone. Because r and z are independent variables, the entire equation must equal a constant and must therefore be directly integrable to find the velocity V_z. Rearranging Equation 5.11 yields

$$\frac{dV_z}{dr} = \left(\frac{dp}{dz}\right)\frac{r}{2\mu}$$

Integrating gives

$$V_z = \left(\frac{dp}{dz}\right)\frac{r^2}{4\mu} + C_1 \tag{5.12}$$

where C_1 is a constant of integration. At the boundary $r = R$, the velocity $V_z = 0$. Substituting, we obtain

$$0 = \left(\frac{dp}{dz}\right)\frac{R^2}{4\mu} + C_1$$

or

$$C_1 = -\left(\frac{dp}{dz}\right)\frac{R^2}{4\mu}$$

The velocity distribution now becomes

$$V_z = \left(\frac{dp}{dz}\right)\frac{1}{4\mu}(r^2 - R^2)$$

or

$$V_z = \left(-\frac{dp}{dz}\right)\frac{R^2}{4\mu}\left[1 - \left(\frac{r}{R}\right)^2\right] \tag{5.13}$$

Note that *dp*/*dz* is negative because *dp* becomes smaller as *dz* becomes larger. That is, the pressure decreases with increasing length, and the term (–*dp*/*dz*) is actually positive.

The average velocity is obtained from

$$V = \frac{\iint V_z \, dA}{A} = \frac{(-R^2/4\mu)(dp/dz)\int_0^{2\pi}\int_0^R (1 - r^2/R^2) r \, dr \, d\theta}{\pi R^2}$$

or

$$V = \frac{R^2}{8\mu}\left(-\frac{dp}{dz}\right) \tag{5.14}$$

In Section 5.4, Equation 5.6 was developed to express the friction factor in terms of the average velocity as

$$dp = -\frac{\rho V^2}{2}\frac{f \, dz}{D_h}$$

which becomes, for a circular duct of radius R,

$$\frac{dp}{dz} = -\frac{\rho V^2}{2}\frac{f}{2R} \tag{5.15}$$

Combining with Equation 5.14 gives

$$V = -\frac{R^2}{8\mu}\left(-\frac{\rho V^2}{2}\right)\frac{f}{2R}$$

Solving for the friction factor, we get

$$f = \frac{32\mu}{\rho V R} = \frac{64\mu}{\rho V D}$$

or

$$f = \frac{64}{\text{Re}} \tag{5.16}$$

for laminar flow in circular conduits.

For turbulent flow, a rigorous mathematical derivation of an accurate velocity distribution for pipes is not possible. Consequently, we must resort to intuitive reasoning, experimental data, and dimensional analysis to develop a means for finding friction factor f. We know that frictional effects are greatest at the wall, where a shear stress acts owing to viscosity. An examination of flow near a wall would show that a great number of protuberances exist; each protuberance may contribute to the frictional effect. Furthermore, some of these protuberances extend beyond the laminar sublayer into the turbulent core as shown in Figure 5.14. We conclude, then, that wall roughness influences the friction factor. Other parameters of significance are the fluid density and viscosity, average velocity of flow, and pipe diameter. Thus,

$$f = f(\rho, V, D, \mu, \varepsilon)$$

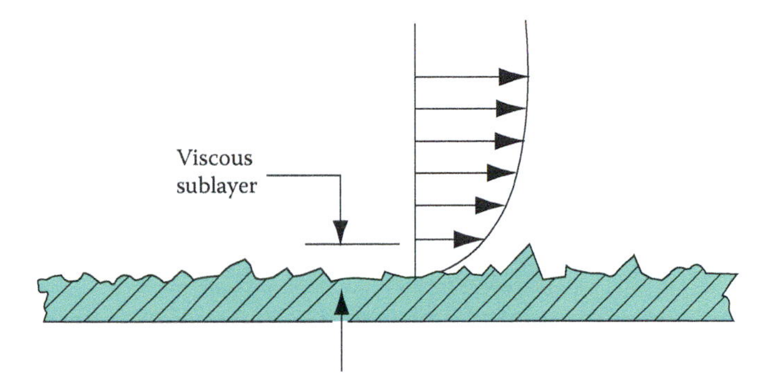

FIGURE 5.14 Turbulent flow in the vicinity of a wall.

where ε is a characteristic dimension of the wall roughness or, more simply, an average height of the protuberances. Performing a dimensional analysis, we obtain

$$f = f\left(\frac{\rho VD}{\mu}, \frac{\varepsilon}{D}\right) \tag{5.17}$$

Many investigators have performed experiments to obtain data on pipe friction. In some experiments, pipe walls were coated with sand particles of constant height. Thus, a control for ε was established, and today, commercially available pipe is assigned an equivalent sand roughness for calculation purposes. A plot of the data to predict f with the Reynolds number Re and the relative roughness ε/D as independent variables is called a **Moody diagram** (Figure 5.15). The Reynolds number ranges from 600 to 100 000 000; the friction factor ranges from 0.008 to 0.1.

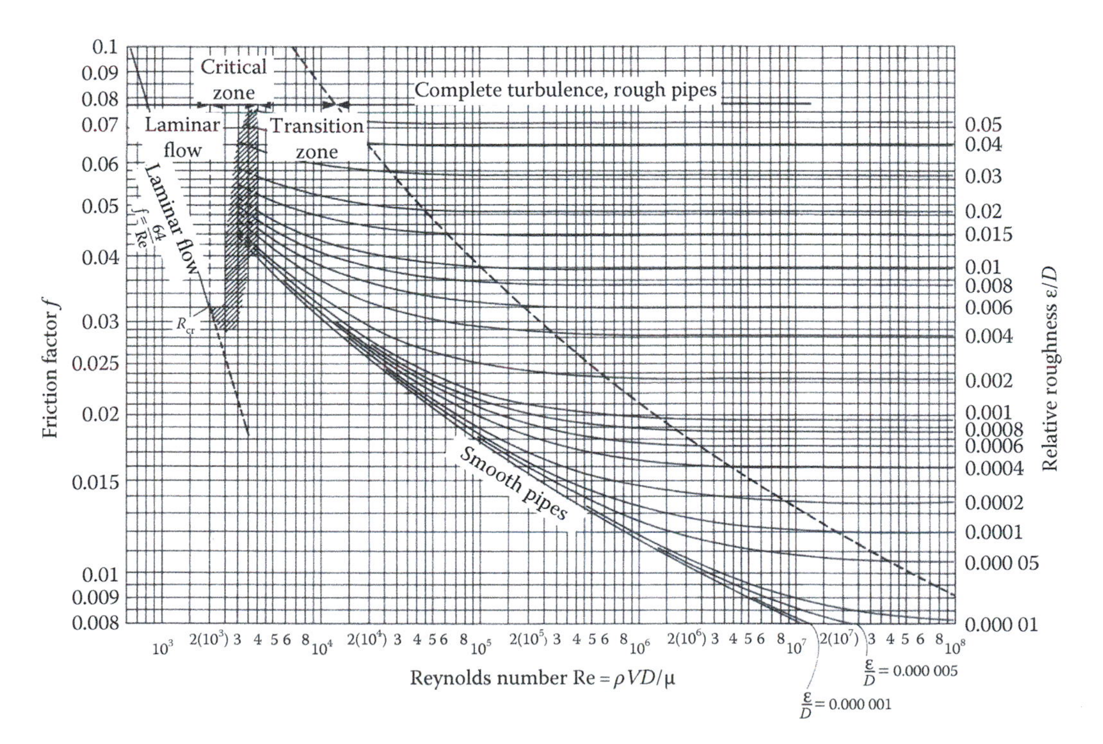

FIGURE 5.15 Friction factors for pipe flow (Moody diagram). (Adapted from Moody, L.F., *Trans. ASME*, 68, 672, 1944. With permission.)

TABLE 5.2
Values of ε for Various Materials

Pipe Material	ε (ft)	ε (cm)
Riveted steel	0.003–0.03	0.09–0.9
Concrete	0.001–0.01	0.03–0.3
Wood stave	0.0006–0.003	0.018–0.09
Cast iron	0.00085	0.025
Galvanized surface	0.0005	0.015
Asphalted cast iron	0.0004	0.012
Commercial steel or wrought iron	0.00015	0.0046
Drawn tubing	0.000005	0.00015

In the laminar range, a plot of friction factor for circular pipes is given. For laminar flow, the friction factor depends only on Reynolds number and is not affected by wall roughness. Beyond this region, a critical zone exists wherein it is difficult to predict the friction factor, because the flow can be either laminar or turbulent. Beyond the critical zone is a smooth-to-rough transition regime for rough pipes. In this region, the friction factor depends on both the relative roughness and the Reynolds number. Transition extends as far as the dashed line, beyond which a completely rough regime exists for rough pipes. In the turbulent region, the Reynolds number no longer influences the friction factor, as in the laminar and transition zones. For complete turbulence, the friction factor depends only on relative roughness.

It is necessary to have information on ε to find *f*. Typical values of ε for various types of material are given in Table 5.2. These data are for new, clean pipe. After years of service, minerals dissolved in the flowing fluid may deposit on the conduit wall, or the wall itself may corrode. Both processes affect ε and decrease the inside diameter. The net effect is to increase friction and reduce the volume-carrying capacity of the pipe. In water pipes for home use, for example, calcium deposits are common. In the petroleum industry, pipes carrying crude oil are cleaned frequently to remove deposits and prevent buildup. Buildup on the tube wall may have a substantial effect on ε.

5.5.1 Flow through Pipes of Noncircular Cross Sections

Flow through noncircular cross sections is somewhat involved, and caution must be exercised when determining the friction factor *f*. Consider the case of fully developed laminar flow through a rectangular duct, as shown in Figure 5.16. The flow is assumed to be two-dimensional—that is, the width w is very large in comparison to the height h. Applying the momentum equation gives

$$\sum F_z = \iint_{CS} V_z \rho V_n \, dA$$

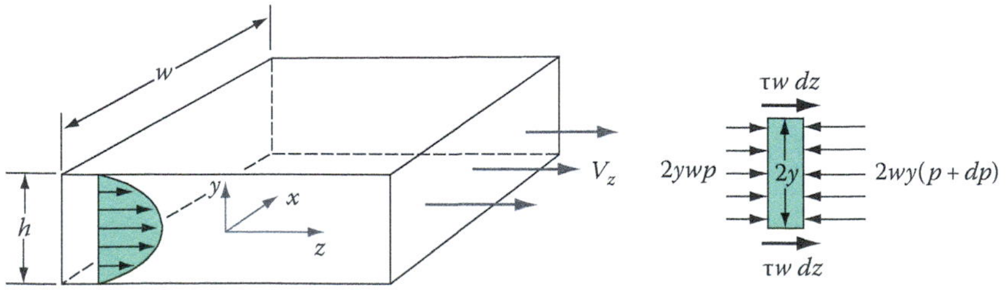

FIGURE 5.16 Laminar flow through a rectangular duct.

Because the cross section is constant, the right-hand side of this equation is zero. Summing forces acting on the control volume gives

$$\sum F_z = 0 = 2wyp + \tau w\, dz - 2wy(p + dp)$$

or

$$\frac{dp}{dz} = \frac{\tau}{y} \tag{5.18}$$

For a Newtonian fluid,

$$\tau = \mu \frac{dV_z}{dy}$$

By substitution into Equation 5.18, we have

$$\frac{y}{\mu}\frac{dp}{dz} = \frac{dV_z}{dy}$$

Integrating gives an expression for V_z:

$$V_z = \frac{y^2}{2\mu}\frac{dp}{dz} + C_1$$

where C_1 is a constant. Now at $y = \pm h/2$, we find that $V_z = 0$. Therefore,

$$0 = \frac{h^2}{8\mu}\frac{dp}{dz} + C_1$$

$$C_1 = \left(-\frac{dp}{dz}\right)\frac{h^2}{8\mu}$$

The velocity becomes

$$V_z = \frac{h^2}{2\mu}\left(-\frac{dp}{dz}\right)\left(\frac{1}{4} - \frac{y^2}{h^2}\right) \tag{5.19}$$

The average velocity is found from

$$V = \frac{\int_{-h/2}^{h/2} V_z\, dy}{h}$$

$$V = \frac{h^2}{12\mu}\left(-\frac{dp}{dz}\right) \tag{5.20}$$

The friction factor defined previously is obtained from Equation 5.6 as

$$dp = -\frac{\rho V^2}{2}\frac{f\,dz}{D_h}$$

For this rectangular duct, the hydraulic diameter is

$$D_h = \frac{4A}{P} = \frac{4(hw)}{2h+2w} = \frac{4hw}{2w} = 2h \tag{5.21}$$

because $w \gg h$. Combining these equations with Equation 5.20 gives

$$V = \frac{h^2}{12\mu}\left(\frac{\rho V^2}{2}\frac{f}{2h}\right)$$

Solving for f gives

$$f = \frac{48\mu}{\rho V h} = \frac{96\mu}{\rho V(2h)}$$

or

$$f = \frac{96}{\text{Re}} \tag{5.22}$$

for two-dimensional fully developed laminar flow through a rectangular duct with the Reynolds number based on the length $2h$. This result is for an infinitely wide rectangle; results are available for other ratios of h/w, as shown in Figure 5.17. The product fRe ranges from 96 for a two-dimensional flow to 56.91 for a square. Thus, for laminar flow, the surface roughness does not affect f. For turbulent flow in a rectangular or square duct, the Moody diagram can be applied if the hydraulic diameter is used in expressions for relative roughness and Reynolds number.

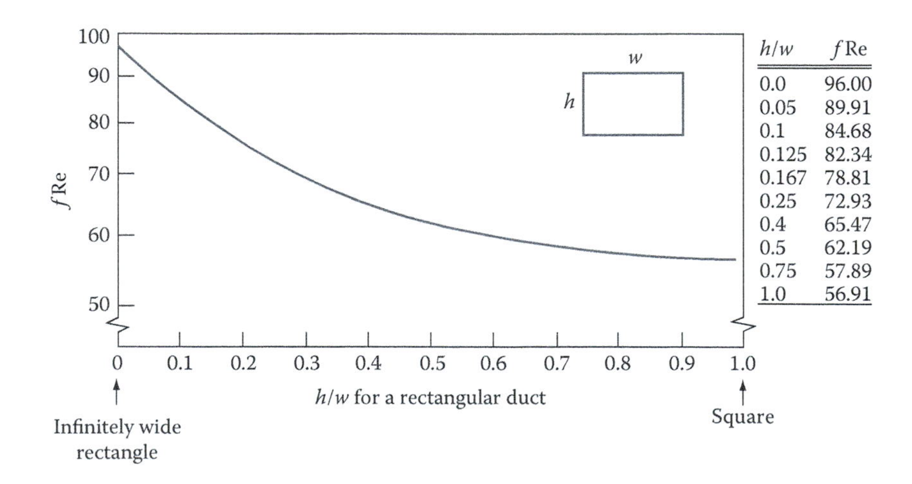

h/w	fRe
0.0	96.00
0.05	89.91
0.1	84.68
0.125	82.34
0.167	78.81
0.25	72.93
0.4	65.47
0.5	62.19
0.75	57.89
1.0	56.91

FIGURE 5.17 Friction factor for laminar flow in a rectangular duct.

5.5.2 Flow through an Annulus

The third type of flow that we will consider is flow in an annulus. Care must be exercised, however, in identifying the dimensions of the annulus. Figure 5.18 shows a cross-sectional view of a concentric annular duct. The annular flow area is bounded by the inside surface of the outer duct (radius R_1) and the outside surface of the inner duct (R_2). We define the ratio of these diameters as

$$\kappa = \frac{R_2}{R_1}$$

and we conclude that $0 < \kappa < 1$.

For fully developed laminar flow in a concentric annular duct, the control volume is taken to be a cylindrical shell, as also shown in Figure 5.18. Applying the momentum equation yields

$$\sum F_z = \iint_{\text{CS}} V_z \rho V_n \, dA$$

Because the cross section is constant, the right-hand side of this equation is zero. Summing forces acting on the control volume gives

$$\sum F_z = 0 = pA + (\tau + d\tau)dA_{P_1} - \tau \, dA_{P_2} - (p + dp)A$$

The shell area A is $2\pi r \, dr$. The surface areas dA_{P_i} are

$$dA_{P_1} = 2\pi(r + dr)dz$$

$$dA_{P_2} = 2\pi r \, dz$$

By substitution, the momentum equation becomes

$$(\tau + d\tau)(2\pi)(r + dr)dz - \tau(2\pi r)dz - 2\pi r \, dr \, dp = 0$$

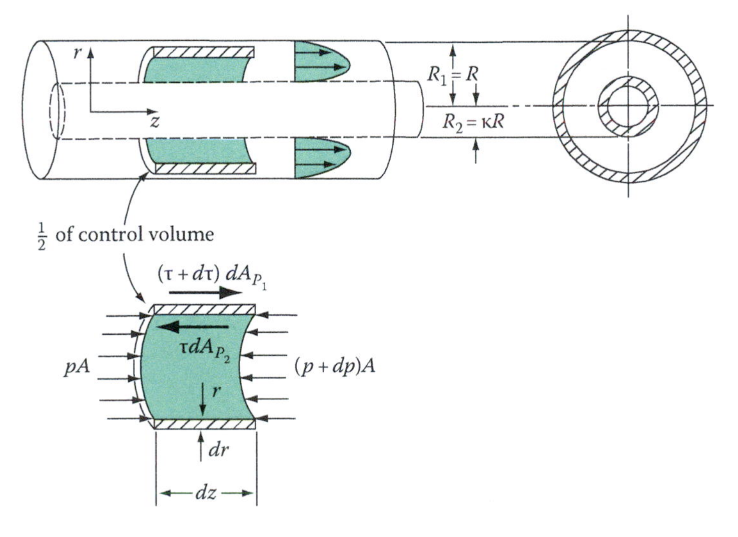

FIGURE 5.18 Laminar flow in an annulus.

Simplifying, we get

$$\tau\, dr + r\, d\tau + dr\, d\tau = r\, dr \frac{dp}{dz}$$

The term $dr\, d\tau$ is small in comparison to the others and can be omitted. The equation becomes

$$\frac{\tau}{r} + \frac{d\tau}{dr} = \frac{dp}{dz}$$

For a Newtonian fluid,

$$\tau = \mu \frac{dV_z}{dr}$$

By substitution, we obtain

$$\frac{\mu}{r}\frac{dV_z}{dr} + \frac{d}{dr}\left(\frac{\mu\, dV_z}{dr}\right) = \frac{dp}{dz}$$

or

$$\frac{1}{r}\frac{d}{dr}\left(r\frac{dV_z}{dr}\right) = \frac{1}{\mu}\frac{dp}{dz}$$

After multiplication by r and integration with respect to r, we have

$$r\frac{dV_z}{dr} = \frac{r^2}{2\mu}\frac{dp}{dz} + C_1$$

Division by r and integration once again with respect to r yields

$$V_z = \frac{r^2}{4\mu}\frac{dp}{dz} + C_1 \ln(r) + C_2 \tag{5.23}$$

We have two boundary conditions:

1. $r = R$; $V_z = 0$
2. $r = \kappa R$; $V_z = 0$ $(0 < \kappa < 1)$

Applying the first, we have

$$0 = \frac{R^2}{4\mu}\frac{dp}{dz} + C_1 \ln(R) + C_2$$

or

$$C_2 = -\frac{R^2}{4\mu}\frac{dp}{dz} - C_1 \ln(R) \tag{5.24a}$$

Applying the second boundary condition yields

$$0 = \frac{\kappa^2 R^2}{4\mu}\frac{dp}{dz} + C_1 \ln(\kappa R) + C_2$$

or

$$C_2 = -\frac{\kappa^2 R^2}{4\mu}\frac{dp}{dz} - C_1 \ln(\kappa R) \tag{5.24b}$$

Equating these expressions for C_2 gives

$$C_1 = \frac{1}{\ln(\kappa)}\frac{R^2}{4\mu}\frac{dp}{dz}(1-\kappa^2)$$

$$C_2 = -\frac{R^2}{4\mu}\frac{dp}{dz}\left(1 + \frac{1-\kappa^2}{\ln(\kappa)}\ln(R)\right)$$

Substituting into Equation 5.23 and simplifying, we get

$$V_z = \frac{R^2}{4\mu}\left(-\frac{dp}{dz}\right)\left[1 - \frac{r^2}{R^2} - \frac{1-\kappa^2}{\ln(\kappa)}\ln\left(\frac{r}{R}\right)\right] \tag{5.25}$$

The average velocity is found by integration:

$$V = \frac{\int_{\kappa R}^{R} V_z 2\pi r\, dr}{\pi(R^2 - \kappa^2 R^2)}$$

Solving, we get

$$V = \frac{R^2}{8\mu}\left(-\frac{dp}{dz}\right)\left((1+\kappa^2) + \frac{1-\kappa^2}{\ln(\kappa)}\right) \tag{5.26}$$

The hydraulic diameter for an annular flow area is

$$D_h = \frac{4A}{P} = \frac{4\pi(R^2 - \kappa^2 R^2)}{2\pi R + 2\pi\kappa R} = 2(R - \kappa R)$$

or

$$D_h = 2R(1 - \kappa) \tag{5.27}$$

Equation 5.6 gives the friction factor in terms of the pressure drop as

$$\frac{dp}{dz} = -\frac{\rho V^2}{2}\frac{f}{D_h}$$

Combining with Equations 5.27 and 5.26 yields

$$\frac{1}{f} = \frac{\text{Re}}{64}\left(\frac{1+\kappa^2}{(1-\kappa^2)} + \frac{1+\kappa}{(1-\kappa)\ln(\kappa)}\right) \tag{5.28}$$

for fully developed laminar flow in a concentric annulus. For turbulent flow, the results for circular tubes are applicable when hydraulic diameter is used in expressions for relative roughness and Reynolds number and when κ is less than or equal to 0.75 (based on experimental results). The friction factor for concentric annular ducts is about 6%–8% greater than that for a smooth circular tube in the range of Reynolds numbers from 15 000 to 150 000.

Summarizing, we have considered flow in three types of cross sections—circular, rectangular, and annular. We have stated the friction factor for each of these shapes in both laminar and turbulent flows. The results are provided in Table 5.3.

5.5.3 Miscellaneous Geometries

In heat-exchanger applications, it is not uncommon to find a tube that has been manufactured with internally spiraling fins. The added internal fins increase the surface area, and the heat-transfer rate is increased over that obtained from using an unfinned or plain tube. The friction factor for internally finned tubes is given in Problem 5.74 of the exercises.

In a number of industrial fluid-handling processes, coil-shaped tubes are used. The friction factor for such an arrangement is given in Problem 5.75.

5.6 SIMPLE PIPING SYSTEMS

Simple piping systems are those that contain no fittings, such as elbows or tee joints. The only head loss in such systems is due to the friction at the wall. In this section, we consider examples that illustrate the use of the Moody chart.

In simple piping systems, six variables enter the problem:

L is the pipe length
D is the pipe diameter or hydraulic diameter of conduit
$\nu = (\mu/\rho)$ = kinematic viscosity of fluid
ε is the wall roughness
Q is the volume flow rate
$h_f = (V^2/2g)(fL/D)$ = head loss due to friction

Of these variables, L, ν, and ε are generally known. Thus, three types of problems are commonly encountered:

1. Given L, ν, ε, Q, and D, find h_f.
2. Given L, ν, ε, D, and h_f, find Q.
3. Given L, ν, ε, h_f, and Q, find D.

The solution technique for each of these problem types is illustrated by the following examples.

TABLE 5.3

Summary of Friction Factor Determinations for Three Common Cross Sections

	D_h	Re	V_z (Laminar)	V (Laminar)	f (Laminar)	f (Turbulent)
Circular	D	$\dfrac{\rho VD}{\mu}$	$V_Z = \left(-\dfrac{dp}{dz}\right)\dfrac{R^2}{4\mu}\left[1-\left(\dfrac{r}{R}\right)^2\right]$	$V = \dfrac{R^2}{8\mu}\left(-\dfrac{dp}{dz}\right)$	$\dfrac{64}{\text{Re}}$	Moody diagram (Figure 5.15)
Rectangular 2-D flow ($w \gg h$)	$2h$	$\dfrac{\rho V(2h)}{\mu}$	$V_Z = \dfrac{h^2}{2\mu}\left(-\dfrac{dp}{dz}\right)\left(\dfrac{1}{4}-\dfrac{y^2}{h^2}\right)$	$V = \dfrac{h^2}{12\mu}\left(-\dfrac{dp}{dz}\right)$	$\dfrac{96}{\text{Re}}$	Moody diagram (Figure 5.15)
Rectangular general	$\dfrac{2hw}{h+w}$	$\dfrac{2\rho V}{\mu}\left(\dfrac{hw}{h+w}\right)$	—	—	Figure 5.17	Moody diagram (Figure 5.15)
Concentric annular duct	$2R(1-\kappa)$	$\dfrac{\rho V(2R)}{\mu}(1-\kappa)$	$V_Z = \dfrac{R^2}{4\mu}\left(-\dfrac{dp}{4\mu}\right) \times \left[1-\dfrac{r^2}{R^2}-\dfrac{1-\kappa^2}{\ell n(\kappa)}\ell n\left(\dfrac{r}{R}\right)\right]$	$V = \dfrac{R^2}{8\mu}\left(-\dfrac{dp}{dz}\right)\left((1+\kappa^2)+\dfrac{1-\kappa^2}{\ell n(\kappa)}\right)$	$\dfrac{1}{f} = \dfrac{\text{Re}}{64}\left(\dfrac{1+\kappa^2}{1-\kappa^2}+\dfrac{1+\kappa}{(1-\kappa)\ell n(\kappa)}\right)$	Moody diagram (Figure 5.15) only if $\kappa \le 0.75$ (with small error)

where $\kappa = \dfrac{\text{outer diameter of inner conduit}}{\text{inner diameter of outer conduit}}$

Example 5.3

Castor oil flows through 100 m of 4-nominal, schedule 40 wrought iron pipe. Determine the pressure drop experienced by the liquid if the volume flow rate is 0.01 m^3/s.

Solution

We know the volume flow rate, Q = 0.01 m^3/s, and the pipe length, L = 100 m. From the appendix tables,

Castor oil $\rho = 0.960(1\,000)$ kg/m^3 [Table A.5]

$\mu = 650 \times 10^{-3}$ N·s/m^2

4-nom, sch 40 $D = 10.23$ cm [Table C.1]

$A = 82.19 \times 10^{-4}$ m^2

From Table 5.2, we find, for wrought iron,

$\varepsilon = 0.004\,6$ cm

The continuity equation for this problem is

$$Q = A_1V_1 = A_2V_2$$

where the subscript 1 refers to inlet conditions, and the subscript 2 refers to conditions at the end of the pipe. Because diameter is constant, $A_1 = A_2$ and so $V_1 = V_2$. The Bernoulli equation with the friction factor term applies:

$$\frac{p_1}{\rho g} + \frac{V_1^2}{2g} + z_1 = \frac{p_2}{\rho g} + \frac{V_2^2}{2g} + z_2 + \sum \frac{fL}{D_h}\frac{V^2}{2g}$$

Evaluating known terms, we have

$V_1 = V_2$ (from continuity)

$z_1 = z_2$ (for a horizontal pipe)

The Bernoulli equation reduces to

$$p_1 - p_2 = \frac{fL}{D}\frac{\rho V^2}{2}$$

By definition, and after substitution,

$$V = \frac{Q}{A} = \frac{0.01}{82.19 \times 10^{-4}} = 1.22 \text{ m/s}$$

So,

$$\text{Re} = \frac{\rho VD}{\mu} = \frac{960(1.22)(0.1023)}{650 \times 10^{-3}} = 184$$

We thus have laminar flow, because the Reynolds number is less than 2 100. The friction factor for laminar flow of a Newtonian fluid in a circular duct is (from Table 5.3)

$$f = \frac{64}{\text{Re}} = \frac{64}{184}$$

or

$$f = 0.347$$

Substituting into the Bernoulli equation, we get

$$p_1 - p_2 = \frac{0.347(100)}{0.1023}\frac{960(1.22)^2}{2}$$

or $p_1 - p_2 = 2.42 \times 10^5\ \text{N/m}^2 = 242\ \text{kPa}$

Another way to express this result is in terms of a head loss, defined as

$$\Delta h = \frac{p_1 - p_2}{\rho g} = \frac{2.42 \times 10^5}{960(9.81)}$$

$$\Delta h = 25.7\ \text{m of castor oil}$$

Note that for laminar flow, the value of the roughness ε does not enter into the problem.

Example 5.4

Decane flows through a 4-nominal, schedule 40 pipe that is 100 m long at a flow rate of 0.01 m^3/s. Calculate the corresponding pressure drop if the pipe is made of commercial steel.

Solution

From the appendix tables and from Table 5.2, we read

Decane	$\rho = 0.728(1\,000)\ \text{kg/m}^3$ $\mu = 0.859 \times 10^{-3}\ \text{N}\cdot\text{s/m}^2$	[Table A.5]
4-nom, sch 40	$D = 10.23\ \text{cm}$ $A = 82.19 \times 10^{-4}\ \text{m}^2$	[Table C.1]
Commercial steel	$\varepsilon = 0.0046\ \text{cm}$	[Table 5.2]

The continuity equation for this problem is

$$Q = A_1V_1 = A_2V_2$$

Because diameter is constant, $A_1 = A_2$ and so $V_1 = V_2$. The Bernoulli equation with the friction factor term applies:

$$\frac{p_1}{\rho g} + \frac{V_1^2}{2g} + z_1 = \frac{p_2}{\rho g} + \frac{V_2^2}{2g} + z_2 + \sum \frac{fL}{D_h}\frac{V^2}{2g}$$

Evaluating known terms, we write

$V_1 = V_2$ (from continuity)

$z_1 = z_2$ (for a horizontal pipe)

The Bernoulli equation reduces to

$$p_1 - p_2 = \frac{fL}{D}\frac{\rho V^2}{2}$$

The average velocity is

$$V = \frac{Q}{A} = \frac{0.01}{82.19 \times 10^{-4}} = 1.22 \text{ m/s}$$

Therefore,

$$\text{Re} = \frac{\rho VD}{\mu} = \frac{0.728(1\,000)(1.22)(0.1023)}{0.859 \times 10^{-3}} = 1.05 \times 10^5$$

The flow is thus turbulent. The friction factor is found from Figure 5.14:

$$\text{Re} = 1.05 \times 10^5$$

$$\frac{\varepsilon}{D} = \frac{0.004\,6}{10.23} = 0.000\,45$$

Substituting into the Bernoulli equation gives

$$p_1 - p_2 = \frac{0.021(100)}{0.4801} \frac{0.728(1.94)(3.08)^2}{2} = \frac{0.021(100)}{0.1023} \frac{0.728(1\,000)(1.22)^2}{2}$$

Solving,

$$p_1 - p_2 = 1.11 \times 10^4 \text{ Pa} = 11.1 \text{ kPa}$$

The previous two examples are identical except for the fluids used. The pressure drop for the castor oil was found to be 242 kPa. For decane flowing in the same size pipe with the same velocity, the pressure drop was found as 11.1 kPa. The difference is due to the viscosities of the two fluids. The castor oil is highly viscous, resulting in a large frictional effect and a correspondingly larger pressure drop. The flow of the castor oil is laminar while that of the decane is turbulent.

Example 5.5

Benzene flows through a 2-nominal, schedule 40 wrought iron pipe. The pressure drop measured at points 350 m apart is 34 kPa. Determine the flow rate through the pipe.

Solution

The method and calculations in the preceding example, where pressure drop is unknown, are quite straightforward. In this example, volume flow rate Q is the unknown; therefore, velocity V and friction factor f are also unknown. A trial-and-error solution method will be required to solve this problem, but the technique is simple. From property and data tables, we determine the following:

Benzene	$\rho = 0.876(1\,000)$ kg/m^3 $\mu = 0.601 \times 10^{-3}$ N·s/m^2	[Table A.5]
12-nom, sch 80	$D = 5.252$ cm $A = 21.66$ cm^2	[Table C.1]
Wrought iron	$\varepsilon = 0.004\,6$ cm	[Table 5.2]

The continuity equation is

$$Q = A_1V_1 = A_2V_2$$

Because diameter is constant, $A_1 = A_2$ and so $V_1 = V_2$. The Bernoulli equation with the friction factor term applies:

$$\frac{p_1}{\rho g} + \frac{V_1^2}{2g} + z_1 = \frac{p_2}{\rho g} + \frac{V_2^2}{2g} + z_2 + \sum \frac{fL}{D_h}\frac{V^2}{2g}$$

Evaluating properties,

$V_1 = V_2$ (from continuity)

$z_1 = z_2$ (for a horizontal pipe)

$L = 350$ m (given)

$p_1 - p_2 = 34$ kPa (given)

The Bernoulli equation becomes

$$p_1 - p_2 = \frac{fL}{D}\frac{\rho V^2}{2}$$

In problems of this type, where volume flow rate Q is the unknown, it is convenient to solve the Bernoulli equation for velocity:

$$V = \sqrt{\frac{2D(p_1 - p_2)}{\rho fL}}$$

Trial and error is necessary because velocity is unknown, but it is needed to calculate the Reynolds number, which in turn is needed to determine the friction factor. Substituting yields

$$V = \sqrt{\frac{2(0.052\ 52)(34\ 000)}{876f(350)}}$$

or

$$V = \frac{0.108}{\sqrt{f}}$$

The Reynolds number is

$$\text{Re} = \frac{\rho VD}{\mu} = \frac{876V(0.052\ 52)}{0.601 \times 10^{-3}}$$
$$= 7.66 \times 10^4\ V$$

In addition, we have

$$\frac{\varepsilon}{D} = \frac{0.004\ 6\ \text{cm}}{5.252} = 0.000\ 88$$

With reference to the Moody diagram, Figure 5.14, we know that our operating point is somewhere along the $\varepsilon/D = 0.000\ 88$ line. As our first estimate, we assume a value for the friction factor that corresponds to the fully turbulent value for this line. Thus, we have the following first trial:

$f = 0.019$ (fully turbulent value for $\varepsilon/D = 0.000\ 88$)

Then

$$V = \frac{0.108}{\sqrt{f}} = \frac{0.108}{0.019} = 0.784 \text{ m/s} \quad \text{(from Equation i)}$$

$$\text{Re} = 7.66 \times 10^4 (0.784) \quad \text{(from Equation ii)}$$

Thus,

$$\left.\begin{aligned} \text{Re} &= 6 \times 10^4 \\ \frac{\varepsilon}{D} &= 0.000\,88 \end{aligned}\right\} f = 0.023 \quad \text{(from Figure 5.14)}$$

For the second trial,

$$f = 0.023 \quad V = \frac{0.108}{\sqrt{0.023}} = 0.712 \text{ m/s}$$

Flow in closed conduits

$$\left.\begin{aligned} \text{Re} &= 7.66 \times 10^4 (0.712) = 5.45 \times 10^4 \\ \frac{\varepsilon}{D} &= 0.000\,88 \end{aligned}\right\} f \approx 0.023 \quad \text{(from Figure 5.14)}$$

which agrees with our assumed value. So

$$V = 0.712 \text{ m/s}$$

$$Q = AV = (21.66 \times 10^{-4})(0.712)$$

or $Q = 0.001\,5 \text{ m}^3/\text{s}$

The method converges very rapidly. Seldom are more than two trials necessary.

Example 5.6

Glycerine flows through a 2-standard, type K copper tube that is 20 ft long. The head loss Δh over this length is 1 ft of glycerine. Calculate the flow rate through the pipe.

Solution

Length and head loss are given in this problem, and volume flow rate is unknown. We proceed as in the last example. From property and data tables, we obtain the following values:

Glycerine	$\rho = 1.263(1.94)$ slug/ft^3 $\mu = 1983 \times 10^{-5}$ lbf·s/ft^2	[Table A.5]
2-std, type K	$D = 0.1633$ ft $A = 0.02093$ ft^2	[Table C.2]
Drawn tubing	$\varepsilon = 0.000005$ ft	[Table 5.2]

The continuity equation is

$$Q = A_1 V_1 = A_2 V_2$$

Because diameter is constant, $A_1 = A_2$ and so $V_1 = V_2$. The Bernoulli equation with the friction factor term applies:

$$\frac{p_1}{\rho g} + \frac{V_1^2}{2g} + z_1 = \frac{p_2}{\rho g} + \frac{V_2^2}{2g} + z_2 + \sum \frac{fL}{D_h}\frac{V^2}{2g}$$

Evaluating properties,

$V_1 = V_2$ (from continuity)

$z_1 = z_2$ (for a horizontal pipe)

$L = 20$ ft (given)

$\Delta h = \dfrac{(p_1 - p_2)}{\rho g} = 1$ ft (given)

The Bernoulli equation becomes

$$\frac{p_1 - p_2}{\rho g} = \Delta h = \frac{fL}{D}\frac{V^2}{2g}$$

Solving for velocity V in terms of head loss Δh gives

$$V = \sqrt{\frac{2gD\Delta h}{fL}} = \sqrt{\frac{2(32.2)(0.1633)(1)}{20f}}$$

or

$$V = \frac{0.725}{\sqrt{f}} \tag{i}$$

The Reynolds number is

$$\text{Re} = \frac{\rho VD}{\mu} = \frac{1.263(1.94)V(0.1633)}{1983 \times 10^{-5}}$$
$$= 20.18V \tag{ii}$$

Also,

$$\frac{\varepsilon}{D} = \frac{0.000005}{0.1633} = 0.00003$$

At this point, we see that the coefficient of velocity V in the Reynolds number equation is quite small. We conclude that the flow is probably laminar. So as our first trial, we substitute for the friction factor:

$$f = \frac{64}{\text{Re}} = \frac{64\mu}{\rho VD}$$

which is valid for laminar flow of a Newtonian fluid in a circular duct. Squaring Equation i and substituting for the friction factor gives

$$V^2 = \frac{0.526}{f} = \frac{0.526}{64}\frac{\rho VD}{\mu}$$

Simplifying and substituting,

$$V = 0.0082\frac{\rho D}{\mu} = \frac{0.0082(1.263)(1.94)(0.1633)}{1983 \times 10^{-5}}$$

$$= 0.166 \text{ ft/s}$$

As a check on the laminar flow assumption,

$$\text{Re} = 20.18V = 20.18(0.166) = 3.34 < 2100$$

So the flow is laminar. The volume flow rate is

$$Q = AV = 0.02093(0.166)$$

$$Q = 0.0035 \text{ ft}^3\text{/s} = 1.6 \text{ gpm}$$

Example 5.7

A riveted steel pipeline is used to convey 2 ft^3/s of gasoline at a distance of 800 ft. The available pump can overcome a pressure drop of 40 psi. Select a suitable pipe size for the installation. Assume that gasoline properties are the same as those for octane.

Solution

Diameter is unknown in this problem, and it is unlikely that we can find a diameter that will satisfy both the given flow rate and the pressure drop. The flow rate is usually (but not always) the more important parameter.

Because the diameter D is unknown, the friction factor f and velocity V are also unknown, and again a trial-and-error solution method is required. From property and data tables, we determine the following:

Octane $\rho = 0.701(1.94)$ slug/ft^3
$\mu = 1.07 \times 10^{-5}$ lbf·s/ft^2 [Table A.5]

Riveted steel $\varepsilon = \dfrac{0.003 + 0.03}{2} = 0.0165$ ft [Table C.1]

(Table 5.2 shows that the roughness ε for riveted steel varies from 0.003 to 0.03 ft. For our purposes, the average value of 0.0165 ft is satisfactory.) The continuity equation is

$$Q = A_1V_1 = A_2V_2$$

Because diameter is constant, $A_1 = A_2$ and so $V_1 = V_2$. The Bernoulli equation with the friction factor term applies:

$$\frac{p_1}{\rho g} + \frac{V_1^2}{2g} + z_1 = \frac{p_2}{\rho g} + \frac{V_2^2}{2g} + z_2 + \sum \frac{fL}{D_h}\frac{V^2}{2g}$$

Evaluating properties,

$V_1 = V_2$ (from continuity)

$z_1 = z_2$ (for a horizontal pipe)

$L = 800$ ft (given)

$$p_1 - p_2 = (40 \text{ lbf/in.}^2)(144 \text{ in.}^2/\text{ft}^2) \qquad \text{(given)}$$
$$= 5760 \text{ lbf/ft}^2$$

The Bernoulli equation becomes

$$p_1 - p_2 = \frac{fL}{D}\frac{V^2}{2}$$

The right-hand side of this equation contains a velocity term. We were given volume flow rate, and so, it is convenient at this point to substitute for the velocity:

$$V = \frac{Q}{A} = \frac{4Q}{\pi D^2}$$

The Bernoulli equation now becomes

$$p_1 - p_2 = \frac{fL}{D}\frac{\rho}{2}\frac{16Q^2}{\pi^2 D^4}$$

Rearranging and solving for diameter gives

$$D = \sqrt[5]{\frac{8\rho fLQ^2}{\pi^2(p_1 - p_2)}}$$

Substituting all known quantities and simplifying, we get

$$D = \sqrt[5]{\frac{8(0.701)(1.94)f(800)(2)^2}{\pi^2(5760)}}$$
$$= 0.907 f^{1/5} \qquad \text{(i)}$$

Likewise, the Reynolds number can be expressed in terms of volume flow rate:

$$\text{Re} = \frac{\rho VD}{\mu} = \frac{\rho D}{\mu}\frac{4Q}{\pi D^2} = \frac{4\rho Q}{\pi D\mu}$$

Substituting,

$$\text{Re} = \frac{4(0.701)(1.94)(2)}{\pi D(1.07\times 10^{-5})}$$
$$= \frac{3.24\times 10^5}{D} \qquad \text{(ii)}$$

As a first trial, assume $f = 0.02$ (a completely random selection); then

$$D = 0.907f^{1/5} = 0.907(0.02)^{1/5} = 0.415 \text{ ft}$$

Also,

$$\left.\begin{aligned} &\text{Re} = 7.8\times 10^5 \quad \text{(Equation ii)} \\ &\frac{\varepsilon}{D} = \frac{0.0165}{0.415} = 0.0398 \end{aligned}\right\} f = 0.065 \quad \text{(from Figure 5.14)}$$

For the second trial:

$$f = 0.065, \quad D = 0.525 \text{ ft}$$

$$\left.\begin{aligned} &\text{Re} = 6.2\times 10^5 \\ &\frac{\varepsilon}{D} = \frac{0.0165}{0.525} = 0.0314 \end{aligned}\right\} f = 0.057 \quad \text{(from Figure 5.14)}$$

For the third trial:

$$f = 0.057, \quad D = 0.511 \text{ ft}$$

$$\left.\begin{aligned} &\text{Re} = 6.3\times 10^5 \\ &\frac{\varepsilon}{D} = \frac{0.0165}{0.511} = 0.032 \end{aligned}\right\} f \approx 0.057 \quad \text{(from Figure 5.14)}$$

The calculations indicate that diameter $D = 0.511$ ft. This value does not appear explicitly in Table C.1, however, because pipe and tubing are sized in discreet values. So, we choose the next larger size. Anything smaller will exceed the pressure drop. Furthermore, nothing special was specified in the problem statement, so in addition, we choose the standard schedule within the size we select. We see that 6-nominal schedule 40 is too small. The next larger size is the one that we select for our solution:

8-nominal schedule 40
$D = 0.6651$ ft

5.7 MINOR LOSSES

The losses due to flow through valves or fittings in a pipeline are known as **minor losses**. As fluid flows through a fitting, it may encounter an abrupt change in area or flow direction. Such abrupt changes cause pressure drops in the fluid flow. The flow pattern existing at a pipe inlet, for example, is not fully developed, and a pressure loss results that takes energy away from the flow. In a **globe valve**, the fluid is forced through a very small opening as it travels through, which results in a pressure loss. This loss is customarily expressed in terms of a loss coefficient K:

$$\Delta p = -K\frac{\rho V^2}{2} \tag{5.29}$$

For any piping system, the overall loss in pressure due to the combined effect of all fittings is found by summing the loss coefficients.

It is interesting to compare the preceding equation for minor loss to that for the friction loss. Both have the same form; for pipe friction, we have

$$\Delta p = -\frac{fL}{D}\frac{\rho V^2}{2} = \begin{pmatrix}\text{friction}\\ \text{loss term}\end{pmatrix}\cdot\begin{pmatrix}\text{kinetic energy}\\ \text{of the flow}\end{pmatrix}$$

Similarly, for the minor losses,

$$\Delta p = -\sum K\frac{\rho V^2}{2} = \begin{pmatrix}\text{sum of loss}\\ \text{coefficients}\end{pmatrix}\cdot\begin{pmatrix}\text{kinetic energy}\\ \text{of the flow}\end{pmatrix}$$

For many fittings, the loss coefficient K must be measured; in a few cases, however, it can be determined analytically.

An alternative method of accounting for minor losses involves the use of what is called an **equivalent length**. Conceptually, the loss due to a fitting is equal to the loss due to a certain length of pipe in the same pipeline. So, a fitting could be removed and replaced with additional pipe. We write

$$K = \frac{fL_{eq}}{D}$$

where L_{eq} is the length of pipe required. For further discussion, see Problem 5.58.

In the design of pipelines, energy loss due to friction is dominant for pipe lengths of 100 ft (30.5 m) or greater. For shorter lengths, losses at elbows, valves, tee joints, and the like may be equal to or greater than the frictional losses. These losses, calculated in the form of Equation 5.29, can be incorporated into Bernoulli's equation with friction. The result is

$$\frac{p_1}{\rho g} + \frac{V_1^2}{2g} + z_1 = \frac{p_2}{\rho g} + \frac{V_2^2}{2g} + z_2 + \sum \frac{fL}{D_h}\frac{V^2}{2g} + \sum K \frac{V^2}{2g} \quad (5.30)$$

which is called the **modified Bernoulli equation**. Minor losses due to fittings are merely added together. Thus, it is important to have information on K, the loss coefficient for various fittings, as will be discussed in the following paragraphs.

Consider flow through an elbow as shown in Figure 5.19. As the fluid is turned, the fluid tends to separate from the inner wall. The separation region consists mostly of swirls and eddies; so, a portion of the cross section is not being used to convey fluid in the primary flow direction. The result is a pressure drop in the fitting. Table 5.4 gives the loss coefficient for various elbows, both flanged and threaded. Since similar losses occur in tee joints, loss coefficients for these fittings are also given in Table 5.4.

Next consider flow through a contraction, as illustrated in Figure 5.20. At section 1, the flow begins to separate so that the main stream can negotiate the corner. At section 2, the flow area has reduced to a minimum, called the *vena contracta*. At section 3, the flow again fills the pipe. The separation between sections 1 and 3 causes a pressure drop. Empirical results for the loss coefficient in a reducing bushing are given in Table 5.4.

Flow through a sudden enlargement is shown in Figure 5.21. At section 1, the flow encounters the enlargement and separates. At section 2, the flow again fills the cross section. By applying the momentum equation to sections 1 and 2, we obtain

$$\sum F_z = \iint_{CS} V_z \rho V_n \, dA$$

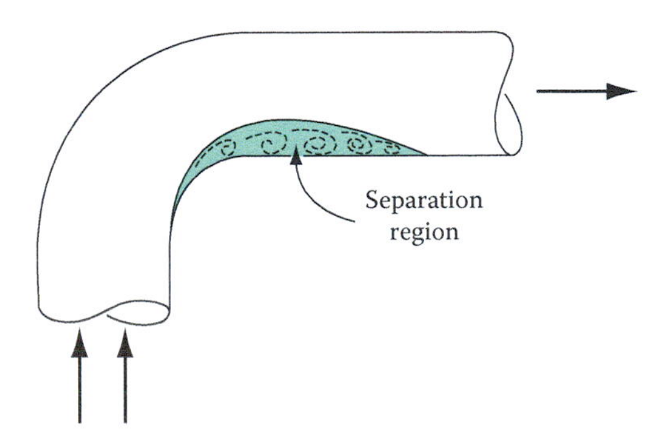

FIGURE 5.19 Flow through an elbow.

TABLE 5.4
Loss Coefficients for Various Fittings

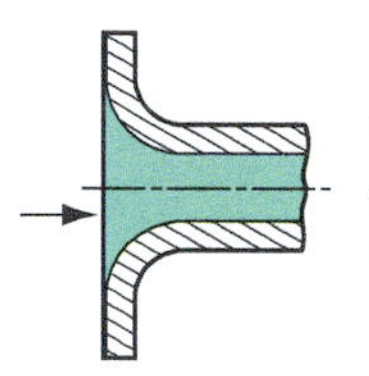

Well-rounded entrance or bell-mouth inlet
$K = 0.05$

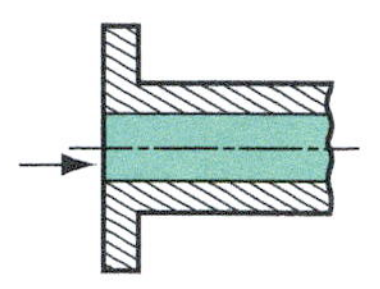

Square-edged inlet
$K = 0.5$

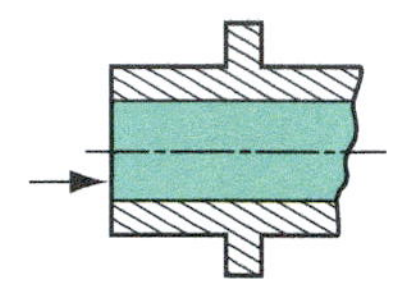

Re-entrant inlet or inward-projecting pipe
$K = 1.0$

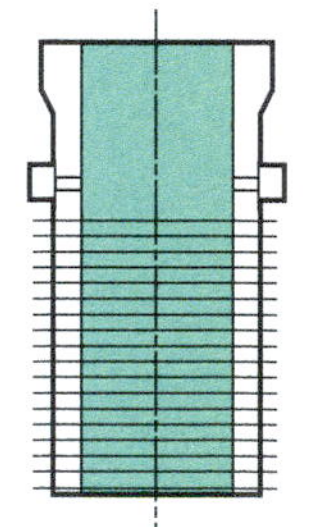

Basket stainer
$K = 1.3$

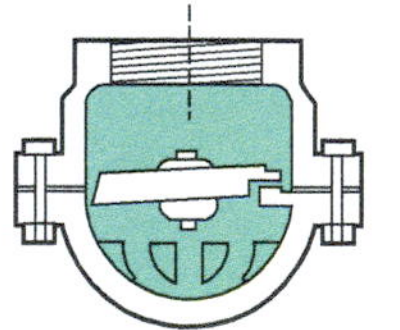

Foot valve
$K = 0.8$

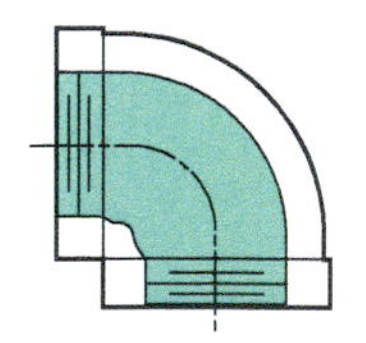

Threaded 90° elbow
$K = 1.4$ (regular)
$K = 0.75$ (long radius)

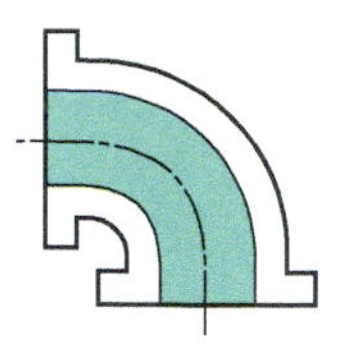

Flanged 90° elbow
$K = 0.31$ (regular)
$K = 0.22$ (long radius)

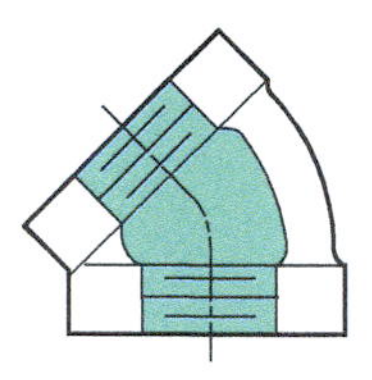

Threaded 45° elbow
$K = 0.35$ (regular)

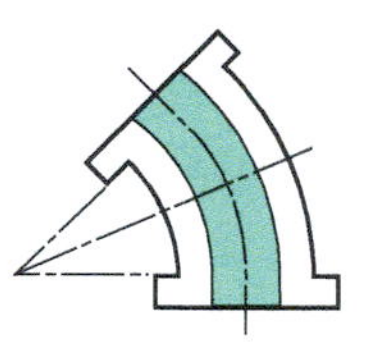

Flanged 45° elbow
$K = 0.17$ (long radius)

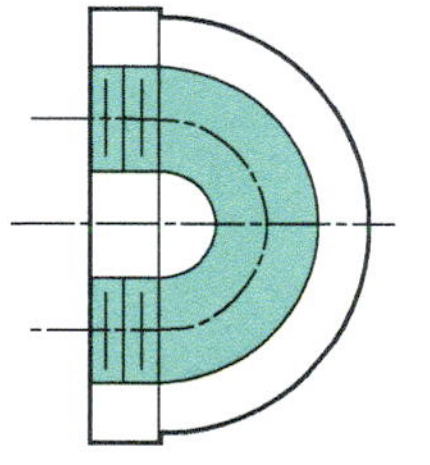

Threaded return bend
$K = 1.5$ (regular)

Flanged return bend
$K = 0.30$ (regular)
$K = 0.20$ (long radius)

(Continued)

TABLE 5.4 (*Continued*)

Loss Coefficients for Various Fittings

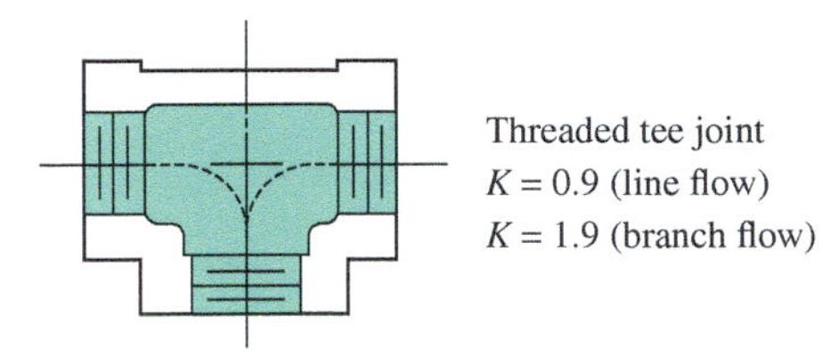

Threaded tee joint
$K = 0.9$ (line flow)
$K = 1.9$ (branch flow)

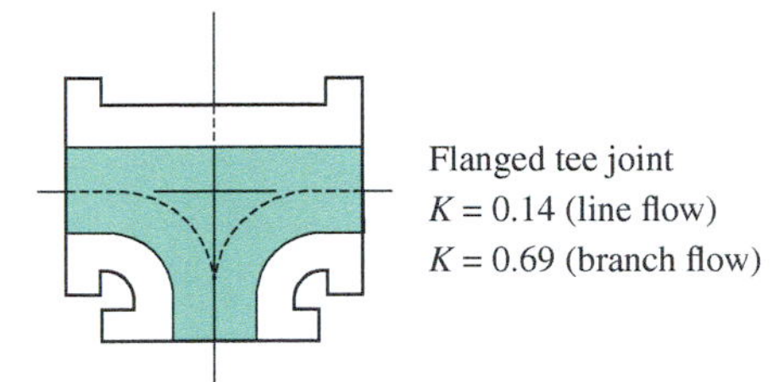

Flanged tee joint
$K = 0.14$ (line flow)
$K = 0.69$ (branch flow)

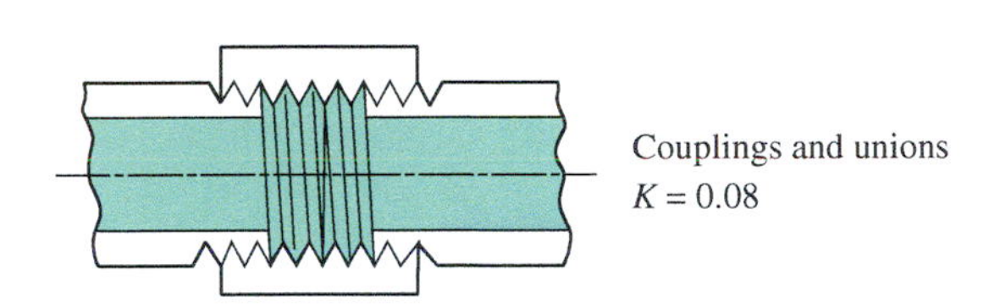

Couplings and unions
$K = 0.08$

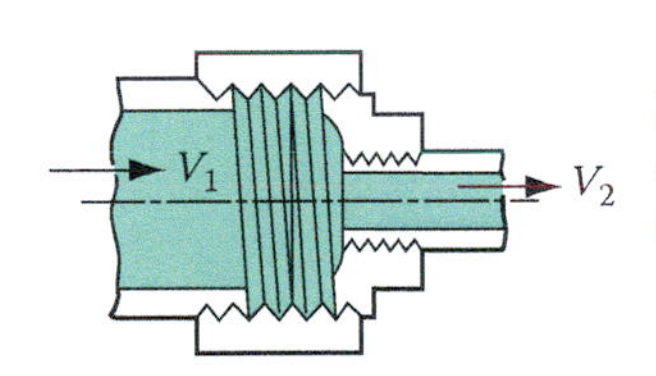

Reducing bushing and coupling (sudden contraction)

$$h_m = K\frac{V_2^2}{2g}$$

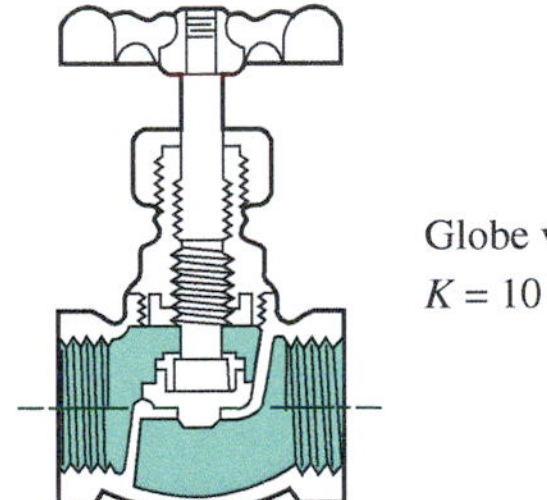

Globe valve fully open
$K = 10$

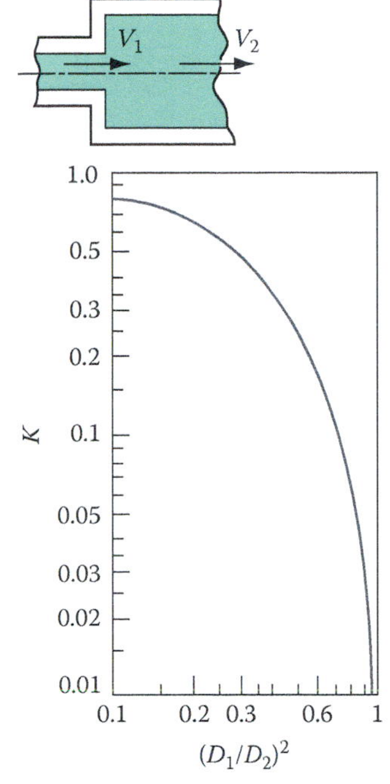

$$h_m = \left(1 - \frac{D_2^2}{D_2^2}\right)^2 \frac{V_1^2}{2g} = \left(\frac{D_2^2}{D_1^2} - 1\right)^2 \frac{V_2^2}{2g}$$

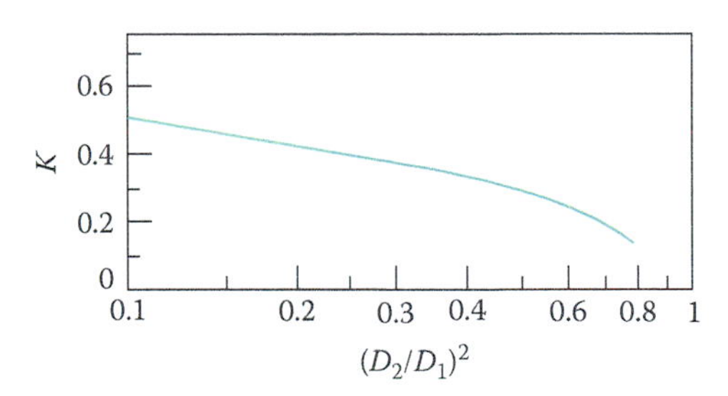

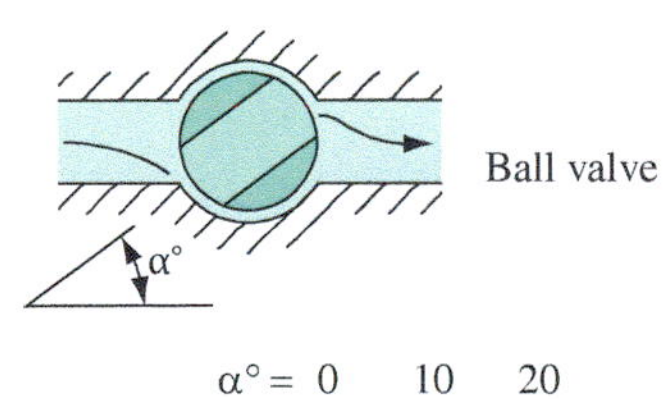

α° = 0	10	20
K = 0.05	0.29	1.56

α° = 30	40	50	60	70	80
K = 5.47	17.3	25.6	206	485	∞

(*Continued*)

TABLE 5.4 *(Continued)*

Loss Coefficients for Various Fittings

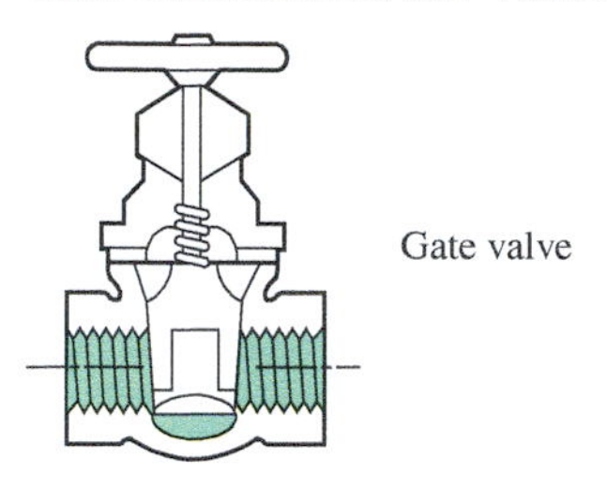

Fraction closed =	0	1/4	3/8	1/2
$K =$	0.15	0.26	0.81	2.06

Fraction closed =	5/8	3/4	7/8
$K =$	5.52	17.0	97.8

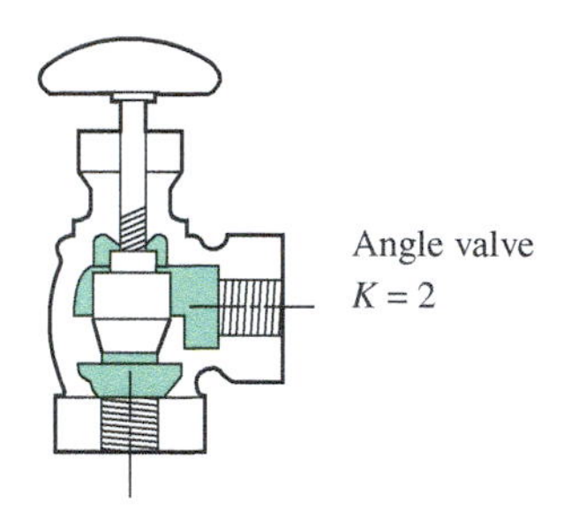

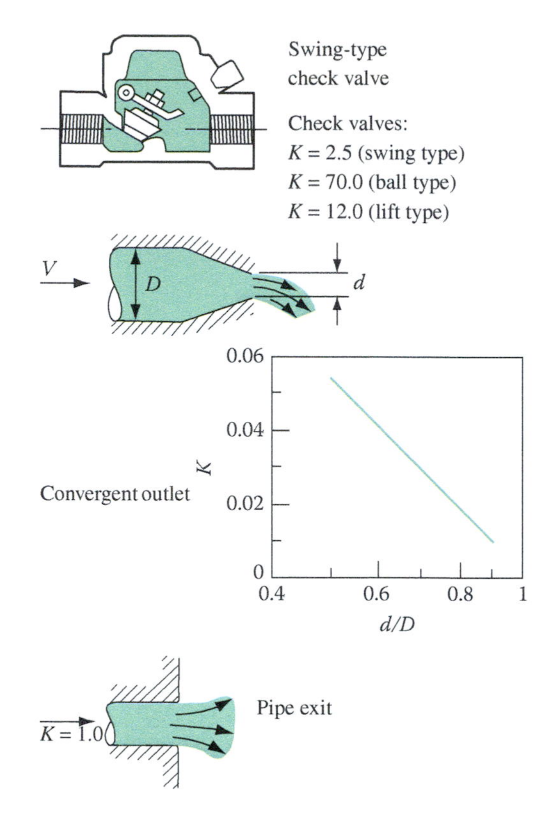

Source: Courtesy of Hydraulic Institute, Parsippany, NJ. www.Pumps.org.

Note: Many fittings have losses that vary with nominal diameter D_n: K for threaded fittings is taken at $D_n = 1$ in.; K for flanged fittings is taken at $D_n = 4$ in. With plastic pipe, fittings can be sleeved and cemented. With copper tubing, fittings can be soldered together. Both can be taken as equivalent to threaded fittings.

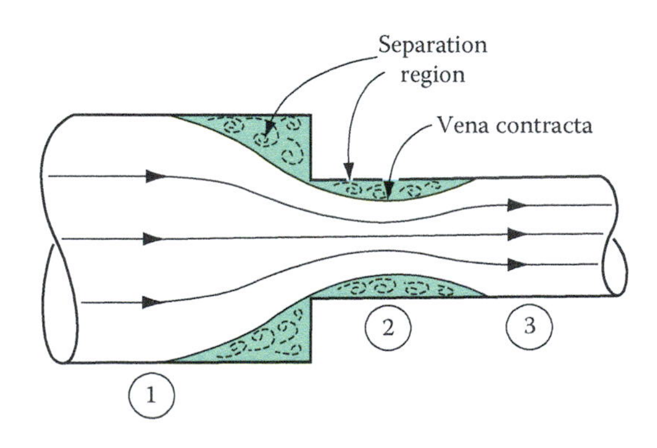

FIGURE 5.20 Flow through a contraction (such as a reducing bushing).

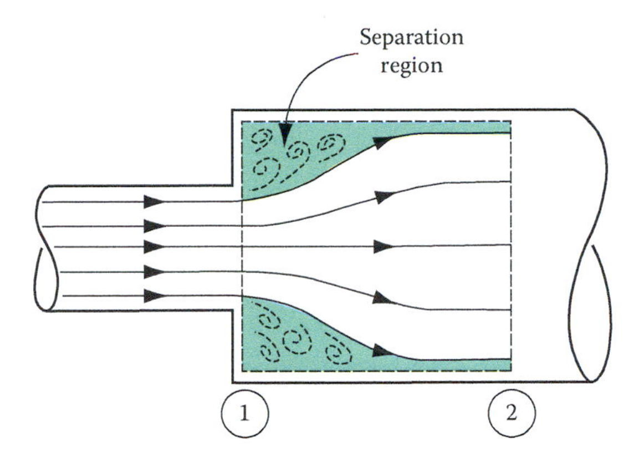

FIGURE 5.21 Flow through a sudden enlargement.

or

$$p_1A_2 - p_2A_2 = \rho A_2V_2(V_2 - V_1)$$

From continuity, $A_1V_1 = A_2V_2$. By substitution and after simplification, we have

$$\frac{p_1 - p_2}{\rho g} = \frac{V_2^2}{g}\left(1 - \frac{A_2}{A_1}\right) \tag{5.31}$$

By applying the modified Bernoulli equation to sections 1 and 2, we get

$$\frac{p_1}{\rho g} + \frac{V_1^2}{2g} + z_1 = \frac{p_2}{\rho g} + \frac{V_2^2}{2g} + z_2 + \sum \frac{fL}{D}\frac{V^2}{2g} + \sum K\frac{V^2}{2g}$$

or

$$\frac{p_1 - p_2}{\rho g} = \frac{V_2^2 - V_1^2}{2g} + K\frac{V_1^2}{2g} \tag{5.32}$$

Combining Equations 5.31 and 5.32 yields

$$\begin{aligned}\frac{V_2^2}{g}\left(1 - \frac{A_2}{A_1}\right) &= \frac{V_2^2}{2g}\left(1 - \frac{V_1^2}{V_2^2}\right) + K\frac{V_1^2}{2g} \\ &= \frac{V_2^2}{2g}\left(1 - \frac{A_2^2}{A_1^2}\right) + K\frac{A_2^2}{A_1^2}\frac{V_2^2}{2g}\end{aligned}$$

Rearranging and simplifying, we get

$$K = 1 - \frac{2A_1}{A_2} + \frac{A_1^2}{A_2^2}$$

or

$$K = \left(1 - \frac{A_1}{A_2}\right)^2 \quad (5.33)$$

The corresponding minor loss is

$$h_m = \frac{V_1^2}{2g}\left(1 - \frac{A_1}{A_2}\right)^2 = \left(\frac{A_2}{A_1} - 1\right)^2 \frac{V_2^2}{2g} \quad (5.34)$$

where the continuity equation has been used. Table 5.4 plots K versus A_1/A_2 for use with Equation 5.34. Note that as A_2 approaches infinity, $K \rightarrow 1$, which is for an exit from a pipe into a reservoir (also given in Table 5.4).

Loss coefficients for valves are usually measured. Table 5.4 gives K for a globe valve, a gate valve, an angle valve, a ball valve, and check valves.

The values of the loss coefficient for most fittings are sensitive to nominal diameter D_n. In fact, values of K for many of the fittings in Table 5.4 are available for diameters ranging from 0.3 to 4 in. for threaded fittings and from 1 to 20 in. for flanged fittings. The values of K given in Table 5.4 correspond to $D_n = 1$ for threaded fittings and $D_n = 4$ for flanged fittings. These values are satisfactory for design.

Difficulties may arise in trying to evaluate minor losses for a given piping system. Before solving a number of piping problems, let us review the control volume concept and how to apply it using the modified Bernoulli equation (Equation 5.30). First we decide where the control volume is to be located, and we then identify those places where mass crosses the control surface. The modified Bernoulli equation contains pressure p, velocity V, and height z terms, which apply *only* at locations where mass crosses the boundary—nothing inside or outside of the control volume can be described or accounted for by these terms. The friction factor term fL/D and the minor loss coefficient term ΣK account for what happens inside (and not outside) of the control volume. To illustrate these concepts, consider Figure 5.22. Five drawings of the same pipe are shown, a different control volume having been selected for each case. As indicated in Figure 5.22a, the control volume extends from the free surface labeled "1" to the free surface labeled "2" and includes all the liquid in between. The modified Bernoulli equation is written, and the terms are evaluated. The pressures p_1 and p_2 are equal to atmospheric pressure. The velocities at the free surfaces are negligible. The heights z_1 and z_2 are measured from the same reference line, which could be at any location, although here we arbitrarily select the bottom of the lower tank. The friction factor f applies only to the pipe itself, as does the length L and the inside diameter D. The minor losses include all fittings encountered by a fluid particle in going from 1 to 2. In this case, we include an entrance, a valve, two elbows, and an exit loss. The coefficient of the friction and minor loss terms is $V^2/2g$, where V is the velocity associated with these terms—that is, the velocity in the pipe itself.

In Figure 5.22b, the control volume extends from the free surface at 1 to a section downstream of the pipe exit where the exit kinetic energy of the flow has dissipated so that the velocity V_2 is zero. The pressures p_1 and p_2 are equal to atmospheric pressure. The velocity V_1 is negligible, and the heights are measured from a reference plane. The friction factor term fL/D applies to the pipe. The minor loss term includes an entrance, a valve, two elbows, and an exit loss. The kinetic energy of the fluid within the pipe is $V^2/2g$. The resultant equation is the same as in Figure 5.22a. Thus, if we say that the outlet pressure p_2 equals atmospheric pressure in the open-ended pipe of Figure 5.22b, then the corresponding velocity V_2 is zero, and we would include an exit loss in our minor losses. Note that if $V_2 = 0$, the area A_2 is infinite, so the product Q is not zero.

In Figure 5.22c, the control volume extends from the free surface at point 1 to some point 2 downstream. The pressure p_1 is equal to atmospheric pressure. The pressure p_2 is unknown. The velocity $V_2 = V$ is the velocity in the pipe, and V_1 is negligible in comparison to V_2. The heights z_1 and z_2 are again measured from the same plane. The friction term fL/D applies to only the pipe. The minor

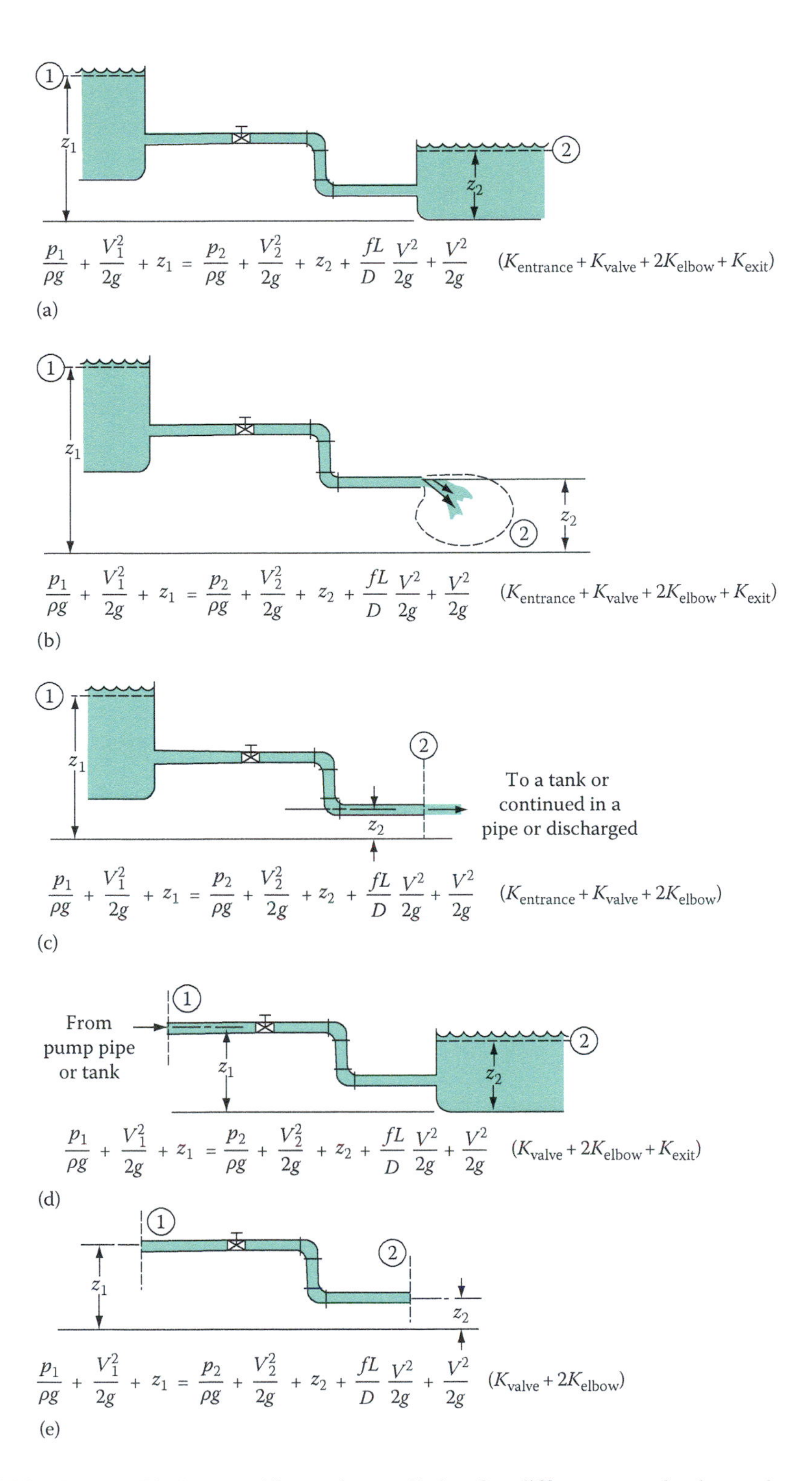

FIGURE 5.22 The modified Bernoulli equation applied to five different control volumes drawn about the same piping system. (Suggested by Professor Ernest Blattner.)

losses that we include here are an entrance, a valve, and two elbows. We do not include an exit loss in this case because the actual loss itself would not be realized until after the exit is encountered, which is outside the control volume where none of the terms in the modified Bernoulli equation apply. The kinetic energy within the pipe is based on velocity V ($=V_2$).

In Figure 5.22d, the control volume extends from some location labeled 1 to the free surface of the liquid in a tank. The pressure p_1 is unknown, but p_2 equals atmospheric pressure. The velocity V_1 is the velocity in the pipe, and V_2 is negligible. The heights z_1 and z_2 are measured from the same datum. The friction term fL/D applies only to the pipe. The minor losses include a valve, two elbows, and an exit loss. An entrance loss is not included in this case. The kinetic energy of the flow within the pipe is based on the velocity V ($=V_1$).

In Figure 5.22e, the control volume includes the liquid contained in the pipe between points 1 and 2. The pressures p_1 and p_2 are not known. The velocities V_1 and V_2 are nonzero but equal, so the kinetic energy terms containing them in the modified Bernoulli equation cancel each other. The heights z_1 and z_2 are measured from the same plane. The friction term fL/D applies to the pipe. The minor losses that we consider include a valve and two elbows. The kinetic energy of the flow is based on velocity V, which equals V_1 and V_2.

We are now able to solve piping problems that are more complex than those in Section 5.6. The types of problems, however, are the same—where h_f is unknown, where Q is unknown, or where D is unknown.

Example 5.8

In a processing plant, turpentine is piped from tanks to cans that are to be sealed and sold to retail outlets. A portion of the pipeline is sketched in Figure 5.23. There are 60 m of 12-nominal pipe and 22 m of 8-nominal pipe. All elbows are standard and flanged, and the line is made of schedule 80 wrought iron pipe. Determine the pressure drop $p_1 - p_2$ if the volume rate of flow is 0.05 m³/s.

Solution

The control volume we select includes all the liquid in the piping system from gauge to gauge. We use the property and data tables:

Turpentine	$\rho = 870$ kg/m³ $\mu = 1.375 \times 10^{-3}$ N·s/m²	[Table A.5]
12-nom, sch 80	$D_{12} = 28.89$ cm $A_{12} = 655.50$ cm²	[Table C.1]
8-nom, sch 80	$D_8 = 19.37$ cm $A_8 = 294.70$ cm²	[Table C.1]
Wrought iron	$\varepsilon = 0.004\ 6$ cm	[Table 5.2]

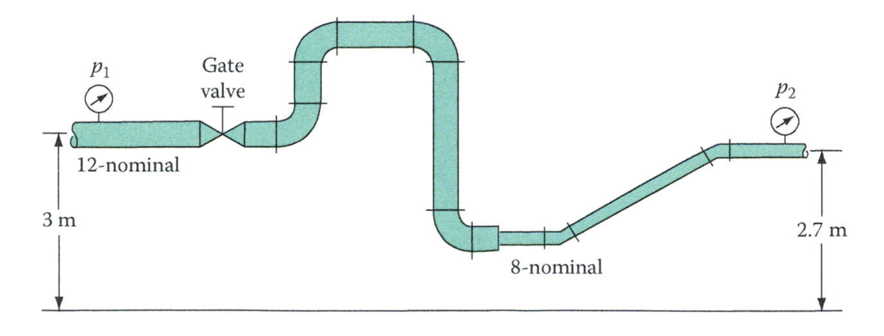

FIGURE 5.23 A pipeline for Example 5.8.

Next, we write the modified Bernoulli equation from sections 1 to 2:

$$\frac{p_1}{\rho g}+\frac{V_1^2}{2g}+z_1=\frac{p_2}{\rho g}+\frac{V_2^2}{2g}+z_2+\left(\frac{fL}{D}+\sum K\right)\frac{V^2}{2g}\bigg|_{12\text{-pipe}}+\left(\frac{fL}{D}+\sum K\right)\frac{V^2}{2g}\bigg|_{8\text{-pipe}}$$

The flow velocities are

$$V_{12}=\frac{Q}{A_{12}}=\frac{0.05}{0.065\,550}=0.763\text{ m/s}$$

$$V_8=\frac{Q}{A_8}=\frac{0.05}{0.029\,470}=1.70\text{ m/s}$$

Thus,

$$\text{Re}_{12}=\frac{\rho VD}{\mu}=\frac{870(0.763)(0.289)}{1.375\times10^{-3}}=1.39\times10^5$$

$$\text{Re}_8=\frac{870(1.70)(0.194)}{1.375\times10^{-3}}=2.08\times10^5$$

Also,

$$\frac{\varepsilon}{D_{12}}=\frac{0.004\,6}{28.89}=0.000\,16$$

$$\frac{\varepsilon}{D_8}=\frac{0.004\,6}{19.37}=0.000\,24$$

From the Moody diagram,

$$f_{12}=0.018$$

$$f_8=0.017\,5$$

The minor losses in the 12-pipe are from one gate valve and four standard elbows:

$$\Sigma K|_{12}=0.15+4(0.31)=1.39$$

The minor losses in the 8-pipe area are from the contraction $D_8^2/D_{12}^2=0.45$ and two 45° elbows:

$$\Sigma K|_8=0.29+2(0.17)=0.63$$

After rearranging the Bernoulli equation and substituting, we obtain

$$\begin{aligned}\frac{p_1-p_2}{\rho g}&=\frac{(1.70)^2}{2(9.81)}-\frac{(0.763)^2}{2(9.81)}+(2.7-3)+\left(\frac{0.018(60)}{0.289}+1.39\right)\\&\quad\times\frac{(0.763)^2}{2(9.81)}+\left(\frac{0.017\,5(22)}{0.194}+0.63\right)\frac{(1.70)^2}{2(9.81)}\\&=0.147-0.029\,6+(-0.3)+0.152+0.384\\&=0.353\text{ m of turpentine}\end{aligned}$$

or

$$p_1 - p_2 = 0.353(870)(9.81) = 3\ 010\ \text{N/m}^2$$

$$p_1 - p_2 = 3.01\ \text{kPa}$$

Example 5.9

A water tank is fitted with a drain and outlet pipe, as sketched in Figure 5.24. The system has 82 ft of commercial steel pipe of 1½-nominal diameter. Determine the flow rate through the pipe. All fittings are threaded and regular.

Solution

We select as our control volume all the fluid in the tank and in the piping system. Section 1 is the free surface of the tank liquid, and section 2 is as indicated in the figure. The area $A_2 \to \infty$, which means that $V_2 \rightharpoonup 0$, $P_2 = P_{atm}$, and we include an exit loss in the sum of minor losses. From the property tables.

Water	$\rho = 1.94$ slug/ft^3 $\mu = 1.9 \times 10^{-5}$ lbf·s/ft^2	[Table A.5]
1½-nom, sch 40	$D = 0.1342$ ft $A = 0.01414$ ft^2	[Table C.1]
Commercial steel	$\varepsilon = 0.00015$	[Table 5.2]
Minor losses	$\sum K = K_{entrance} + K_{return\ bend} + K_{elbow} + K_{globe\ valve} + K_{exit}$ $\sum K = 0.5 + 1.5 + 1.4 + 10 + 1.0 = 14.4$	[Table 5.4]

Write the Bernoulli equation from section 1 (the free surface of the liquid in the tank) to section 2:

$$\frac{p_1}{\rho g} + \frac{V_1^2}{2g} + z_1 = \frac{p_2}{\rho g} + \frac{V_2^2}{2g} + z_2 + \sum \frac{fL}{D_h}\frac{V^2}{2g} + \sum K \frac{V^2}{2g}$$

The figure shows that $p_1 = p_2 = p_{atm}$ and that if $z_2 = 0$, then $z_1 = 23$ ft. The reservoir surface velocity V_1 is negligible in comparison to the velocity in the pipe, and $V_2 = 0$. Because all the piping is of the same diameter, the equation becomes

$$z_1 = \frac{V^2}{2g}\left(\frac{fL}{D} + \sum K\right)$$

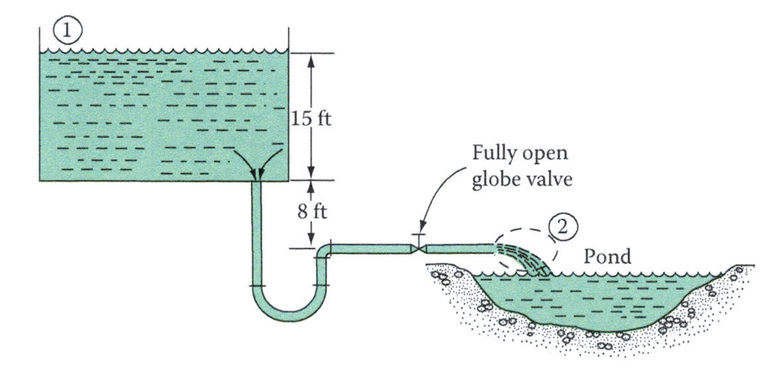

FIGURE 5.24 Piping system for Example 5.9.

By substitution,

$$23(2)(32.2) = V^2\left(\frac{f(82)}{0.1324} + 14.4\right)$$

or

$$V = \left(\frac{1480}{14.4 + 611f}\right)^{1/2}$$

Because V and f are unknown, a trial-and-error solution is required. As a first trial, assume $f = 0.03$; then

$$V = 6.72 \text{ ft/s}$$

$$\left.\begin{aligned} \text{Re} &= \frac{\rho VD}{\mu} = \frac{1.94(6.72)(0.1342)}{1.9\times10^{-5}} = 9.21\times10^4 \\ \frac{\varepsilon}{D} &= \frac{0.00015}{0.1342} = 0.0011 \end{aligned}\right\}\text{Moody diagram: } f = 0.02$$

As a second trial assume $f = 0.02$; then

$$V = 7.45 \text{ ft/s}$$

$$\left.\begin{aligned} \text{Re} &= 1.02\times10^5 \\ \frac{\varepsilon}{D} &= 0.0011 \end{aligned}\right\}\text{Moody diagram: } f = 0.02 \quad \text{(close enough)}$$

Using $f = 0.02$, $V = 7.5$ ft/s, and the continuity equation, we find

$$Q = AV = 0.01414(7.5)$$

$$Q = 0.106 \text{ ft}^3\text{/s} = 6.4 \text{ ft}^3\text{/min}$$

Example 5.10

Methyl alcohol is used in a processing plant where a flow rate of 0.4 m^3/min of the liquid must be supplied. The available liquid pump can supply this flow rate only if the pressure drop in the supply line is less than 15 m head of water. The pipeline is made up of 50 m of soldered drawn copper tubing and follows the path shown in Figure 5.25. Determine the minimum size of tubing required.

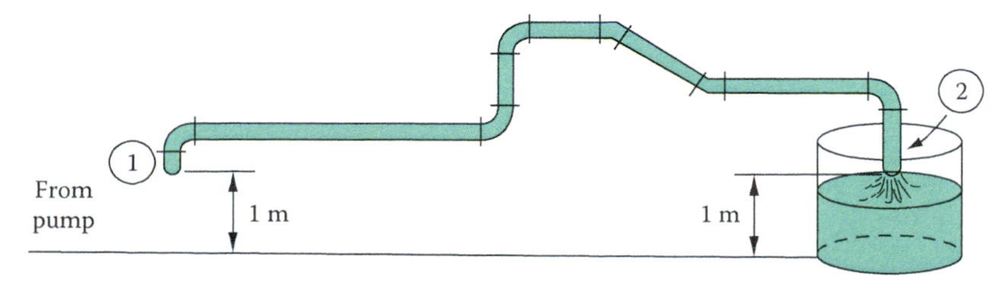

FIGURE 5.25 Sketch for Example 5.10.

Solution

The control volume we select extends from the inlet at section 1 to the outlet at section 2. The area at 2 is large enough so that the liquid pressure after the exit is atmospheric. We use the property and data tables to find

Water	$\rho = 1\,000$ kg/m^3	[Table A.5]
Methyl alcohol	$\rho = 789$ kg/m^3 $\mu = 0.56 \times 10^{-3}$ N·s/m^2	[Table A.5]
Drawn tubing	$\varepsilon = 0.000\,15$ cm	[Table 5.2]
Minor losses	Soldered fittings, as stated in a footnote of Table 5.4, may be assumed to have the same loss as threaded fittings:	

$$\sum K = 4K_{90°\text{ elbow}} + 2K_{45°\text{ elbow}} + K_{\text{exit}}$$

$$= 4(1.4) + 2(0.35) + 1 = 7.3 \qquad \text{[Table 5.4]}$$

Write the Bernoulli equation from 1 to 2 as

$$\frac{p_1}{\rho g} + \frac{V_1^2}{2g} + z_1 = \frac{p_2}{\rho g} + \frac{V_2^2}{2g} + z_2 + \sum \frac{fL}{D}\frac{V^2}{2g} + \sum K \frac{V^2}{2g}$$

The pressure drop is given in terms of meters of water:

$$\frac{p_1 - p_2}{\rho g} = 15 \text{ m of water}$$

where ρ is for water. To convert the pressure drop to meters of alcohol, multiply by the ratio of densities:

$$\frac{p_1 - p_2}{\rho g} = \frac{15(1\,000)}{789} = 19.0 \text{ m of methyl alcohol}$$

The velocity at point 1 is the same as that at point 2 because the area at these points is the same. Moreover, $z_1 = z_2$. Thus, the Bernoulli equation becomes

$$\frac{p_1 - p_2}{\rho g} = \frac{V^2}{2g}\left(\frac{fL}{D} + \sum K\right)$$

From continuity, we have $V = 4Q/\pi D^2$. By substitution into the Bernoulli equation, we get

$$19.0 = \frac{16Q^2}{2g\pi^2 D^4}\left(\frac{f(50)}{D} + 7.3\right)$$

With $Q = 0.4$ m^3/min $= 0.006\,7$ m^3/s, the equation becomes

$$\frac{19.0(2)(9.81)\pi^2}{16(0.006\,7)^2} = \frac{f(50)}{D^5} + \frac{7.3}{D^4}$$

or

$$f = 1.025 \times 10^5 D^5 - 0.146D \qquad \text{(i)}$$

A trial-and-error approach is required. In this case, because of the form of the equation, it is easier to assume a diameter. As a first trial, assume

$$D = 0.05 \text{ m}, \quad f = 1.025\times10^5(0.05)^5 - 0.146(0.05) = 0.025$$

$$A = \frac{\pi D^2}{4} = \frac{\pi(0.05)^2}{4} = 0.00196 \text{ m}^2$$

$$V = \frac{Q}{A} = \frac{0.006\,7}{0.001\,96} = 3.41 \text{ m/s}$$

$$\text{Re} = \frac{\rho VD}{\mu} = \frac{789(3.41)(0.05)}{0.56\times10^{-3}}$$

$$\left.\begin{array}{l} \text{Re} = 2.4\times10^5 \\ \dfrac{\varepsilon}{D} = \dfrac{0.000\,15 \text{ cm}}{5 \text{ cm}} = 0.000\,03 \end{array}\right\} f = 0.015 \quad \text{(from Figure 5.14)}$$

The diameter for this value of the friction factor is our second assumed value. Successive trials with Equation i give the following values:

D	f
0.05	0.025
0.04	0.005 4
0.045	0.013
0.046	0.015 3

Continuing, we have the following as the next trial:

$$D = 0.046, \quad f = 0.015, \quad A = \frac{\pi(0.046)^2}{4} = 0.001\,66 \text{ m}^2$$

$$V = \frac{0.006\,7}{0.001\,66} = 4.03 \text{ m/s}; \quad \text{Re} = \frac{789(4.03)(0.046)}{0.56\times10^{-3}}$$

$$\left.\begin{array}{l} \text{Re} = 2.6\times10^5 \\ \dfrac{\varepsilon}{D} = \dfrac{0.000\,15 \text{ cm}}{4.6 \text{ cm}} = 0.000\,033 \end{array}\right\} f \approx 0.015 \quad \text{(close enough; from Figure 5.14)}$$

The solution suggests that $D = 4.6$ cm. Referring to Table C.2 for type M copper tubing, we seek a diameter equal to or greater than 4.6 cm. Anything smaller will reduce the volume flow rate; anything larger will work, but in order to not incur an unnecessary expense, we select the next larger diameter. Thus, we specify

2-standard type M
$D = 5.102$ cm

Example 5.11 Laboratory Experiment—Determination of Friction Factor

Figure 5.26 is a schematic of a pipe flow test rig. The apparatus contains a sump tank, which is used as a water reservoir from which one or two centrifugal pumps discharge water to the pipe circuit. The circuit itself consists of six different diameter lines and a return line, all made of schedule 80 PVC pipe. The circuit contains ball valves for directing and regulating the flow, and can be used

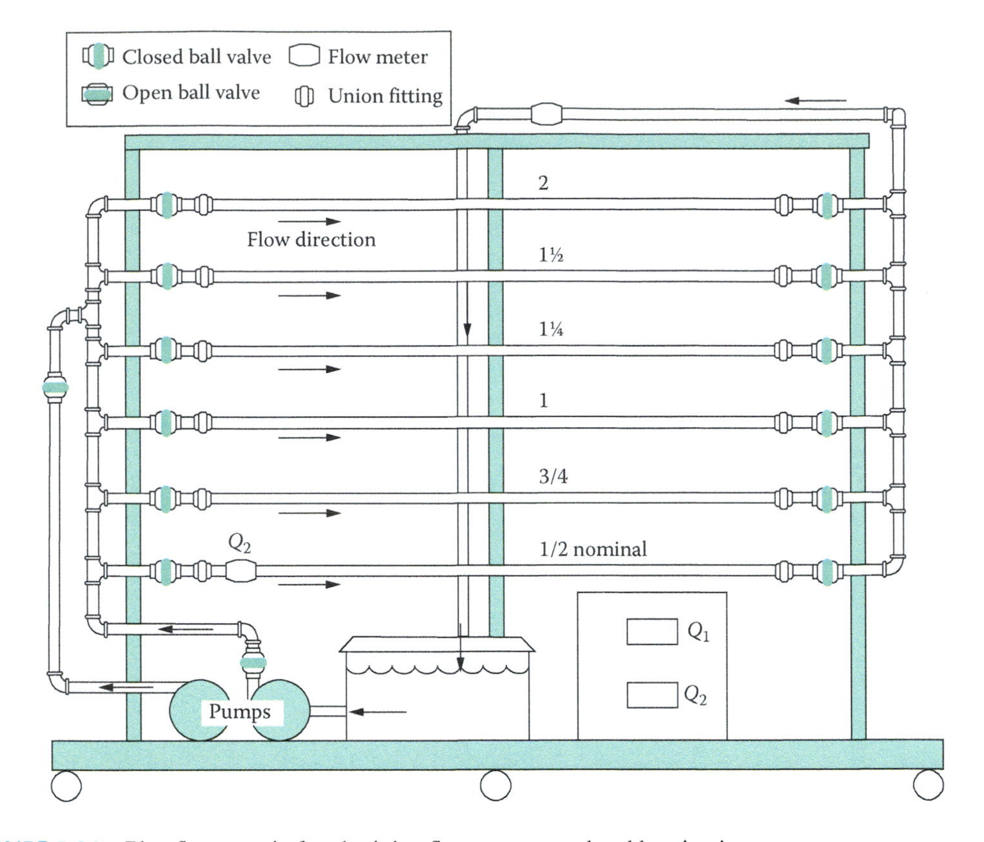

FIGURE 5.26 Pipe flow test rig for obtaining flow rate versus head loss in pipes.

to make up various series and parallel piping combinations. The circuit has provision for measuring pressure loss through the use of static pressure taps and a manometer board (not shown).

The 6 lines are ½, ¾, 1, 1¼, 1½, and 2 in. schedule 80 pipe. The topmost line is a return line, and it is made of 2 in. pipe as well. The apparatus contains two flow meters. The ½ line contains a flow meter, which is used only for that pipe. The other flow meter in the return line is for all other flows through the system. Because the circuit contains flow meters, the measured pressure losses can be obtained as a function of flow rate.

The appropriate valves on the apparatus are opened/closed to obtain the desired flow path. The valve closest to the pump(s) on the downstream side is used to vary the volume flow rate. With the pump on, pressure drops and the actual volume flow rates are recorded. The valve closest to the pump is used to change the volume flow rate. Again the pressure drops and flow rates are recorded.

This procedure is repeated for many different volume flow rates and corresponding pressure drops.

Equations

With head loss data, the friction factor can be calculated with

$$f = \frac{2g\Delta h}{V^2(L/D)}$$

It is customary to graph the friction factor as a function of the Reynolds number:

$$\text{Re} = \frac{VD}{\nu}$$

The f vs Re graph (the Moody Diagram) is traditionally drawn on a log-log grid. The graph also contains the roughness coefficient ε/D. For this experiment, the roughness factor ε is that for smooth walled tubing.

The flow of water was directed through the 1-nominal pipe, and the following (raw) data were obtained:

$$Q = 22.1 \text{ gpm}, \quad \Delta h = 12.5 \text{ in.}, \quad L = 35 \text{ in.}$$

The head loss Δh as measured with an inverted U-tube manometer, and the distance between pressure taps is 35 in. Determine the friction factor.

Solution

From the property tables,

Water	$\rho = 1.94 \text{ slug/ft}^3$	$\mu = 1.9 \times 10^{-5} \text{lbf} \cdot \text{s/ft}^2$	[Table A.5]
1 nom sch 40	$D = 0.07975$ ft	$A = 0.005 \text{ ft}^2$	[Table C.1]
Roughness	ε = drawn: tubing, assumed ≈ 0		

The raw data may be converted as follows:

$$Q = 22.1 \text{ gpm} = 22.1(0.002228) = 0.0492 \text{ ft}^3\text{/s}$$

$$\Delta h = 12.5 \text{ in.} = 1.042 \text{ ft}$$

$$L = 35 \text{ in.} = 2.92 \text{ ft}$$

The average velocity is

$$V = \frac{Q}{A} = \frac{0.0492}{0.005} = 9.85 \text{ ft/s}$$

The friction factor becomes

$$f = \frac{2g\Delta h}{V^2(L/D)} = \frac{2(32.2)(1.042)}{(9.85)^2(2.92/0.07975)}$$

$$f = 0.019$$

The Reynolds number is

$$\text{Re} = \frac{\rho VD}{\mu} = \frac{1.94(9.85)(0.07975)}{1.9 \times 10^{-5}}$$

$$\text{Re} = 8.02 \times 10^4$$

In the process of constructing a Moody diagram, we would have one data point.

5.8 PIPES IN PARALLEL

Complex pipe friction problems sometimes involve fluid flow in a parallel piping system. In pipeline design, a common method of increasing the capacity of a line is by looping or laying a pipeline parallel to the main line, as shown in Figure 5.27. The pressure drop from A to B along either path (A_2 or A_3) is the same. Usually, the diameters of all the pipes are known; the problem

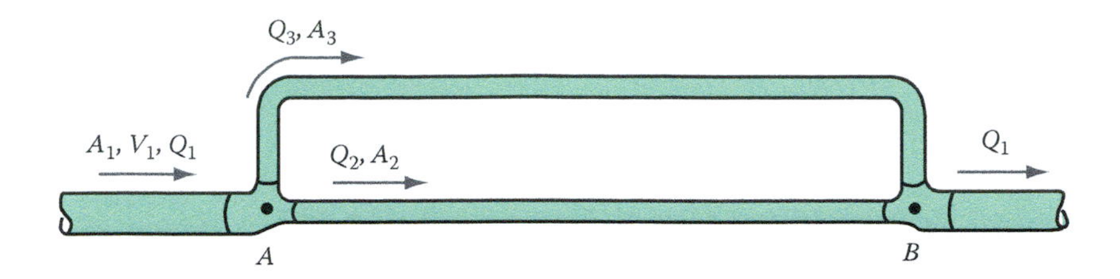

FIGURE 5.27 Pipes in parallel.

involves finding Q_2 and Q_3. The solution technique requires use of the Moody diagram and trial and error. The technique is best illustrated by an example.

Example 5.12

For the piping system of Figure 5.27, the distance from A to B is 4500 ft, and the main line is made of 10-nominal, schedule 40 wrought iron pipe. The attached loop is 8-nominal, schedule 40 wrought iron pipe. The flow Q_1 is 0.3 ft³/s of water. Determine the flow rate in both branches.

Solution

Minor losses are usually neglected in this type of problem. We use the property and data tables to find

Water	$\rho = 1.94$ slug/ft³ $\mu = 1.9 \times 10^{-5}$ lbf · s/ft²	[Table A.5]
10-nom, sch 40	$D_2 = 0.8350$ ft $A_2 = 0.5476$ ft²	[Table C.1]
8-nom, sch 40	$D_3 = 0.6651$ ft $A_3 = 0.3474$ ft²	[Table C.1]
Wrought iron	$\varepsilon = 0.00015$ ft	[Table 5.2]

(The subscripts 2 and 3 refer to the labels in Figure 5.27.) From the continuity equation,

$$Q_1 = Q_2 + Q_3 = 0.3 \text{ ft}^3/\text{s}$$

Also $p_A - p_B$ along Q_2 must be equal to $p_A - p_B$ along Q_3. Thus,

$$(p_A - p_B)|_2 = (p_A - p_B)|_3$$

or

$$\frac{f_2 L_2}{D_2}\frac{V_2^2}{2g} = \frac{f_3 L_3}{D_3}\frac{V_3^2}{2g}$$

With $L_2 = L_3$, the equation reduces to

$$\frac{f_2 V_2^2}{D_2} = \frac{f_3 V_3^2}{D_3}$$

In terms of volume flow rate,

$$\frac{f_2}{D_2}\frac{16 Q_2^2}{\pi^2 D_2^4} = \frac{f_3}{D_3}\frac{16 Q_3^2}{\pi^2 D_3^4}$$

or

$$\frac{f_2 Q_2^2}{D_2^5} = \frac{f_3 Q_3^2}{D_3^5}$$

By substitution, the equation now becomes

$$2.46\ f_2 Q_2^2 = 7.68\ f_3 Q_3^2$$

or

$$Q_2 = Q_3 \sqrt{\frac{3.11 f_3}{f_2}}$$

From Table 5.2, $\varepsilon = 0.000\ 15$; hence,

$$\frac{\varepsilon}{D_2} = 0.000180$$

$$\frac{\varepsilon}{D_2} = 0.000226$$

The Reynolds number is

$$\mathrm{Re} = \frac{\rho V D}{\mu} = \frac{\rho D}{\mu} \frac{4Q}{\pi D^2} = \frac{4\rho Q}{\pi D \mu}$$

For each line,

$$\mathrm{Re}_2 = \frac{4(1.94)Q_2}{\pi(0.8350)(1.9 \times 10^{-5})} = 1.58 \times 10^5 Q_2$$

$$\mathrm{Re}_3 = 1.99 \times 10^5 Q_3$$

From these calculations, our working equations are

$$Q_2 + Q_3 = Q = 0.3\ \mathrm{ft^3/s}$$

$$\mathrm{Re}_2 = 1.58 \times 10^5 Q_2 \qquad \frac{\varepsilon}{D_2} = 0.000180$$

$$\mathrm{Re}_3 = 1.99 \times 10^5 Q_3 \qquad \frac{\varepsilon}{D_3} = 0.000226$$

$$Q_2 = Q_3 \sqrt{\frac{3.119 f_3}{f_2}}$$

Because $D_2 > D_3$ we expect that $Q_2 > Q_3$.

As a first trial assume that $Q_2 = ½Q = 0.15\ \mathrm{ft^3/s}$; then

$$Q_3 = 0.30 - 0.15 = 0.15$$

$$\mathrm{Re}_2 = 2.37 \times 10^4$$

$$\mathrm{Re}_3 = 2.99 \times 10^4$$

From the Moody diagram

$$f_2 = 0.0255$$

$$f_3 = 0.0241$$

and by substitution into the working equation, we get

$$Q_2 = 0.258\ \text{ft}^3/\text{s}$$

If this value is used for the second trial, the next iteration gives $Q_2 = 0.0898$. The third iteration gives $Q_2 = 0.33$, which shows that successive trials lead to divergence. A more successful method involves taking as a second assumed value the average of the initially assumed Q_2 and the calculated Q_2. For the second iteration, therefore, assume that

$$Q_2 = \frac{0.15 + 0.26}{2} = 0.20$$

Then

$$Q_3 = 0.1$$

$$\text{Re}_2 = 3.17 \times 10^4$$

$$\text{Re}_3 = 1.99 \times 10^4$$

$$f_2 = 0.024$$

$$f_3 = 0.0265$$

and, by equation,

$$Q_2 = 0.18$$

For the next trial, assume that

$$Q_2 = \frac{0.18 + 0.20}{2} = 0.19$$

Then

$$Q_3 = 0.11$$

$$\text{Re}_2 = 3.01 \times 10^4$$

$$\text{Re}_3 = 2.19 \times 10^4$$

$$f_2 = 0.024$$

$$f_3 = 0.026$$

and, by equation,

$$Q_2 = 0.20$$

which is close enough to the assumed value. Thus, the solution is

$$Q_2 = 0.19 \text{ ft}^3/\text{s} \quad \text{and} \quad Q_3 = 0.11 \text{ ft}^3/\text{s}$$

An alternative iteration scheme involves assuming choices for the friction factors where fully developed turbulent flow values are taken as first estimates.

All pipe friction problems in this chapter have been solved by using the Moody diagram. Although reference to a figure is highly impractical if one is using a computer for the calculations, the computer is well suited for these problems, especially the trial-and-error type. Thus, an equation fit of the Moody diagram is helpful.

Many equations have been written, and a sampling is provided here. These equations are easily entered on a programmable calculator or in a computer program. They apply for turbulent flow conditions.

Chen equation:

$$\frac{1}{\sqrt{f}} = -2.0\log\left\{\frac{\varepsilon}{3.706\,5D} - \frac{5.042\,2}{\text{Re}} \times \log\left[\frac{1}{2.825\,7}\left(\frac{\varepsilon}{D}\right)^{1.109\,8} + \frac{5.850\,6}{\text{Re}^{0.898\,1}}\right]\right\} \tag{5.35}$$

The Haaland equation

$$f = \left\{-0.782\ \ln\left[\frac{6.9}{\text{Re}}\right] + \left(\frac{\varepsilon}{3.7D}\right)^{1.11}\right\}^{-2} \tag{5.36}$$

The Swamee–Jain equation

$$f = \frac{0.250}{\{\log[(\varepsilon/3.7D) + (5.74/\text{Re}^{0.9})]\}^2} \tag{5.37}$$

If ε/D and Re are known, then f can be determined, just as with the Moody diagram.

5.9 PUMPS AND PIPING SYSTEMS

A driving force is required to make fluid flow from one point to another in a closed conduit. Where elevation differences exist, this force may be gravity, but the driving force is usually supplied by pumps (for liquids) or blowers (for gases). These devices add to the mechanical energy of the fluid in the form of velocity, pressure, and/or elevation increases. Common methods of effecting the increase are by positive displacement or centrifugal action. In this section, we discuss pumps in general and learn how to calculate pumping power requirements. Techniques for determining fan power requirements are similar to those for pumps.

With **positive-displacement pumps**, a finite volume of liquid is drawn into a chamber and then forced out under high pressure. In a **reciprocating positive-displacement pump**, the chamber is formed by a cylinder-piston arrangement. As the piston is withdrawn from the cylinder, liquid is drawn in through the inlet. On the return stroke, the liquid is forced out the outlet tube. One-way valves are placed in the liquid lines to control flow direction. In a **rotary positive-displacement pump**, the chamber moves from inlet to discharge and back. In a **rotary gear pump**, the space between intermeshing, rotating gears carries liquid from inlet to discharge (see Figure 5.28). Positive-displacement pumps can be used to deliver very high pressures, but pulsations in the flow

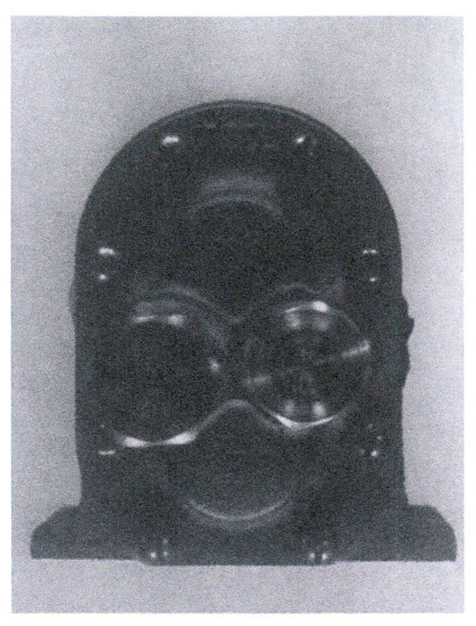

FIGURE 5.28 A positive-displacement pump. Liquid enters on one side of housing, moves along the housing periphery between gear teeth, and exits on the other side of the housing.

are also present. At constant speed, the flow capacity of these pumps is a constant. In general, the discharge rate is directly dependent on the speed and independent of the type of liquid. For gases, rotary blowers and compressors are available that work on the same principles.

Centrifugal pumps are more commonly used than positive-displacement pumps in process conditions. **Centrifugal pumps** add energy to the fluid by means of a rotating impeller within a casing. Figure 5.29 shows a frontal view of an impeller and the casing cover. A centrifugal pump is shown schematically in Figure 5.30. The liquid enters the pump in the axial direction (with respect to the motion of the impeller). It enters the rotating eye of the impeller and moves radially through the channels between the vanes. At the periphery, the liquid exits through the discharge pipe. The rotating impeller imparts a high-velocity head to the liquid, a portion of which is converted to pressure head after the liquid leaves the impeller. Pumping power requirements are determined with the energy equation, in combination with the modified Bernoulli equation.

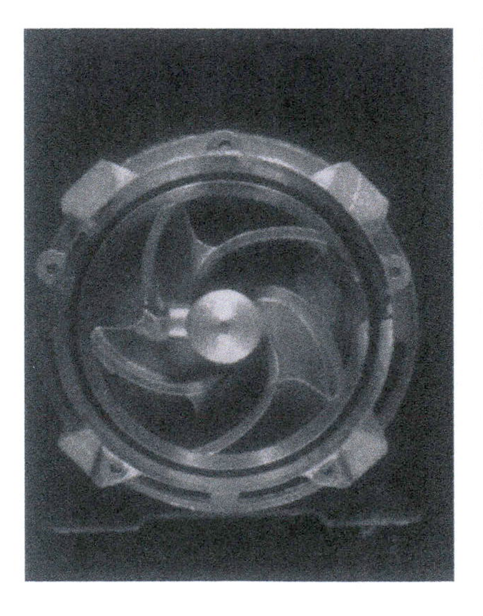

FIGURE 5.29 Centrifugal pump impeller and casing.

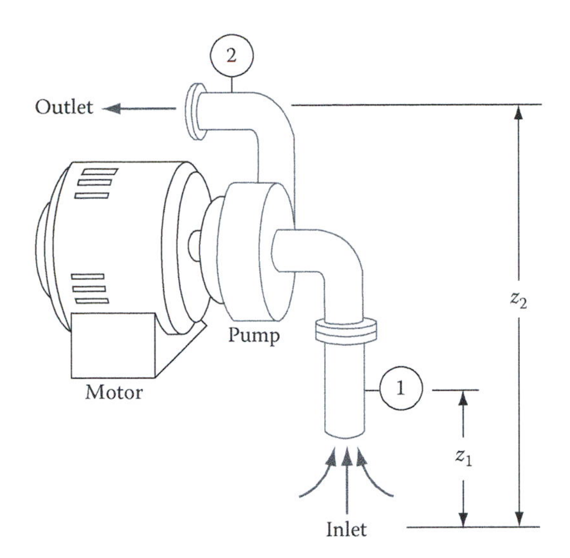

FIGURE 5.30 Schematic of a centrifugal pump.

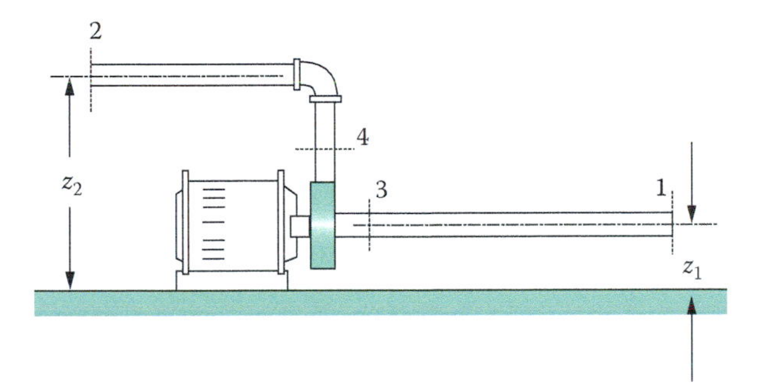

FIGURE 5.31 Pipe/pump combination.

Consider the pump and piping system shown in Figure 5.31. We select section 1 to be upstream of the pump at the pipe inlet, and section 2 downstream of the pump at the pipe outlet. Section 3 is just upstream of the pump, and section 4 is just downstream of the pump. The modified Bernoulli equation (with friction) applied to the inlet line (sections 1–3) is

$$\frac{p_1}{\rho g}+\frac{V_1^2}{2g}+z_1=\frac{p_3}{\rho g}+\frac{V_3^2}{2g}+z_3+\left(\sum\frac{fLV^2}{D_h 2g}+\sum K\frac{V^2}{2g}\right)_{\text{inlet line}}$$

The energy equation written about the pump, sections 3–4, is

$$\frac{p_3}{\rho g}+\frac{V_3^2}{2g}+z_3=\frac{p_4}{\rho g}+\frac{V_4^2}{2g}+z_4+\frac{1}{\dot{m}g}\frac{dW}{dt}$$

The modified Bernoulli equation for the pump outlet line, sections 4–2, is

$$\frac{p_4}{\rho g}+\frac{V_4^2}{2g}+z_4=\frac{p_2}{\rho g}+\frac{V_2^2}{2g}+z_2+\left(\sum\frac{fLV^2}{D_h 2g}+\sum K\frac{V^2}{2g}\right)_{\text{outlet line}}$$

Adding the preceding equations gives for the pump and piping system combination:

$$\frac{p_1}{\rho g}+\frac{V_1^2}{2g}+z_1=\frac{p_2}{\rho g}+\frac{V_2^2}{2g}+z_2+\left(\sum\frac{fLV^2}{D_h 2g}+\sum K\frac{V^2}{2g}\right)_{\text{inlet line}}$$
$$+\left(\sum\frac{fLV^2}{D_h 2g}+\sum K\frac{V^2}{2g}\right)_{\text{outlet line}}+\frac{1}{\dot{m}g}\frac{dW}{dt} \tag{5.38}$$

This is the most general equation for a pump and pipeline combination.

For many **pumps**, the inlet line is larger in diameter than the outlet line. A smaller inlet line will have a larger pressure drop, and if the pressure at the pump inlet (section 3) is too small, the fluid may boil (at room temperature) inside the pump. This phenomenon is referred to as **cavitation**. The vapor bubbles make their way through the impeller, and eventually collapse. This has an erosive effect on the impeller and on the pump housing. Cavitation is to be avoided, and using a larger inlet line helps with this objective. Pump manufacturers usually provide information on when a pump will cavitate.

Many pumps, however, are made with equal inlet and outlet lines. These can be used if cavitation will not be a problem. In such cases, Equation 5.38 applied to the system can be simplified, as indicated in the following example.

Example 5.13

A house is located near a freshwater lake. The homeowner decides to install a pump near the lake to deliver 25 gpm of water to a tank adjacent to the house. The water can then be used for lavatory facilities or sprinkling the lawn. For the system sketched in Figure 5.32, determine the pump power required.

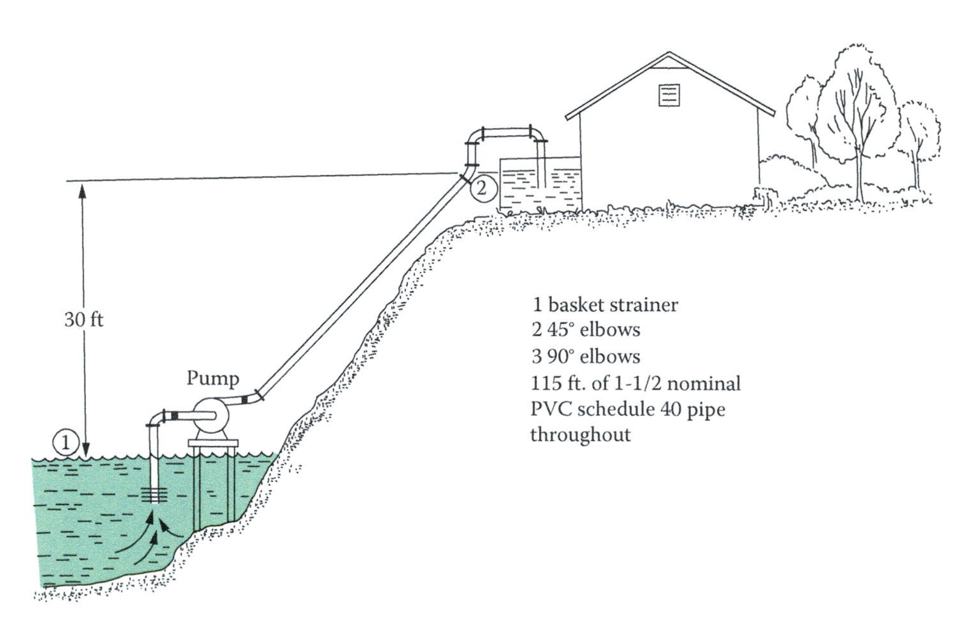

FIGURE 5.32 Pumping system for Example 5.13.

Solution

The control volume we select includes all liquid in the piping system and in the pump. The control volume extends from the free surface of the lake (section 1) to the free surface of the receiver tank (section 2). From the property and data tables, we determine the following:

Water $\rho = 1.94$ slug/ft^3 [Table A.5]
$\mu = 1.9 \times 10^{-5}$ lbf·s/ft^2

1½-nom, sch 40 $D = 0.125$ ft [Table C.1]
$A = 0.01227$ ft^2

Smooth pipe $\varepsilon \approx 0$

Minor losses

$$\sum K = K_{\text{basket strainer}} + 2K_{45^\circ \text{ elbow}} + 3K_{90^\circ \text{ elbow}} + K_{\text{exit}}$$
$$= 1.3 + 2(0.35) + 3(1.4) + 1.0$$
$$= 7.2 \qquad \text{[Table 5.4]}$$

Equation 5.38, for equal inlet and outlet lines, becomes:

$$\frac{p_1}{\rho g} + \frac{V_1^2}{2g} + z_1 = \frac{p_2}{\rho g} + \frac{V_2^2}{2g} + z_2 + \sum \frac{fL}{D}\frac{V^2}{2g} + \sum K\frac{V^2}{2g} + \frac{1}{\dot{m}}\frac{dW}{dt}\frac{1}{g}$$

Thus, only inlet and outlet conditions must be known. At sections 1 and 2 (both are reservoir surfaces), velocity is negligible. With section 1 as the reference,

$$-\frac{1}{\dot{m}}\frac{dW}{dt} = z_2 g + \frac{V^2}{2}\left(\frac{fL}{D} + \sum K\right)$$

Thus, the pump power goes into lifting the fluid 30 ft, and in overcoming friction.

The volume and flow mass required are

$$Q = 25 \text{ gpm} \times 0.002228 = 0.0557 \text{ ft}^3/\text{s}$$

$$\dot{m} = \rho Q = 0.108 \text{ slug/s}$$

The velocity is

$$V = \frac{Q}{A} = \frac{0.0557}{0.01227} = 4.54 \text{ ft/s}$$

with which we obtain

$$\text{Re} = \frac{1.94(4.54)(0.125)}{1.9 \times 10^{-5}} = 5.8 \times 10^4$$

It is permissible to use the "smooth pipe" curve on the Moody diagram for PVC pipe. At Re = 5.8×10^4, we read $f = 0.020\,5$.

By substitution into the equation, we have

$$-\frac{dW}{dt}=0.108\left[30(32.2)+\frac{(4.52)^2}{2}\left(\frac{0.0205(115)}{0.125}+7.2\right)\right]$$

$$-\frac{dW}{dt}=0.108(966+268)=133\ \text{ft}\cdot\text{lbf/s}$$

By definition, 1 hp = 550 ft · lbf/s; therefore,

$$\frac{dW}{dt}=-0.242\ \text{hp}$$

The negative sign indicates power transferred to the fluid. This value represents the power that must be added to the fluid. The power input to the pump shaft must be greater, however, to overcome losses. It is considered a good design practice to place a strainer on the submerged pipe inlet of Figure 5.31. Also, to prevent backflow through the system due to a siphoning effect, it is necessary to install a one-way valve in the flow line.

5.10 SUMMARY

This chapter presented the basic design procedures for piping systems. Laminar and turbulent flow regimes were defined. The friction factor and the relative roughness were both discussed, and their importance in pipe friction calculations was illustrated. Losses due to fittings were also discussed. Finally, pumping requirements for existing systems were calculated by application of the energy equation.

It should be pointed out that all the example problems in this chapter dealt with circular pipe, although other cross sections were discussed. The methods of this chapter can easily be extended to noncircular cross sections by using the concept of hydraulic diameter. Such problems are included in the exercises that follow.

INTERNET RESOURCES

1. Locate an animation of how a gear pump works, and give a brief presentation about it. Where would one use a gear pump?
2. Give a presentation on something called a propeller pump. How does fluid flow through the impeller of such a pump?

PROBLEMS

Laminar versus Turbulent Flow

5.1 Methyl alcohol flows in a tube with an inside diameter of 6 in. at a rate of 1 in.^{3}/s. Is the flow laminar or turbulent?

5.2 Castor oil is flowing at 15 ft/s in a pipe with an inside diameter of 4 in. Is the flow laminar or turbulent?

5.3 Air at atmospheric pressure and 25°C flows in a pipe having an inside diameter of 8 cm. What is the maximum average velocity permissible for the flow to be considered laminar?

5.4 Water flows in a 2 in. ID pipe at an average velocity of 1 in./s. Is the flow laminar?

Hydraulic Diameter

5.5 A rectangular duct has internal dimensions of 4 × 1.5 in. Calculate the hydraulic diameter.

5.6 A rectangular duct is 1 cm tall and 10 cm wide. What is its hydraulic diameter?

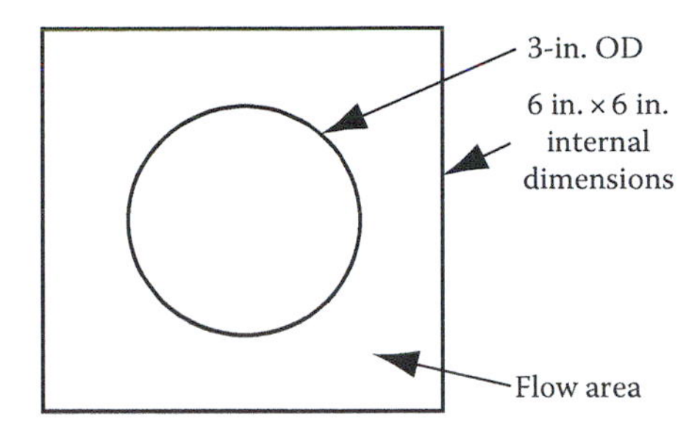

FIGURE P5.12

5.7 An annular duct is formed by an outer tube with an inside diameter of 10 cm and an inner tube with an outside diameter of 5 cm. Calculate the hydraulic diameter.

5.8 A concentric circular annulus conveys benzene at 0.02 m^3/s. The annulus is made up of two tubes. The inside diameter of the outer tube is 20 cm, and the outside diameter of the inner tube is 10 cm. Determine the hydraulic diameter and the Reynolds number.

5.9 Oxygen at 25°C and atmospheric pressure is conveyed at 0.3 m^3/s through a 30 × 10 cm rectangular duct. Is the flow laminar or turbulent?

5.10 A rectangular duct has inside dimensions of 12 × 18 in. What is the hydraulic diameter of the duct? If 30 ft^3/min of air at 71°F flows through the duct, determine the Reynolds number.

5.11 A circular annulus is made up of two concentric tubes. The inner tube has an outside diameter of 15 cm, whereas the outer tube has an inside diameter of 40 cm. The annulus will convey carbon tetrachloride under laminar flow conditions. Calculate the maximum permissible average velocity in the conduit.

5.12 A conduit that conveys air is rectangular, having dimensions of 6 × 6 in. Due to space limitations, a circular tube having an outside diameter of 3 in. has to be mounted within this duct, and it has been decided to mount the tube directly in the center (see Figure P5.12). Determine the hydraulic diameter of the duct flow area.

Entrance Length

5.13 Do the entrance length equations for laminar and for turbulent flow predict the same entrance length when the Reynolds number is 2 100? Should they? Why or why not?

5.14 Water at room temperature flows through a well-rounded entrance to a 10-cm-ID pipe. Calculate the velocity for a Reynolds number of 2 100. Determine also the entrance length required for the flow to become fully developed also at a Reynolds number of 2 100.

5.15 Propylene glycol flows from a tank into a 2-nominal, schedule 80 pipe. Measurements indicate that the flow becomes fully developed 50 diameters downstream. Determine the volume flow rate through the pipe in cubic meters per second if laminar conditions exist.

5.16 Propylene glycol flows from a tank into a 2-nominal, schedule 80 pipe. Measurements indicate that the flow becomes fully developed 18 diameters downstream. Determine the volume flow rate through the pipe in cubic meters per second if turbulent conditions exist.

5.17 Ethyl alcohol flows into a 6 cm ID pipe through a well-rounded entrance. After 20 cm of pipe length, measurements taken in the pipe show that the flow is fully developed. Determine the volume flow rate in the pipe, assuming laminar flow exists.

5.18 Helium at 25°C and 110 kPa (absolute) flows into a copper tube (1-1/4-standard, type M) that is 10 m long. Is the flow fully developed at tube end if the helium mass flow rate is 0.02 kg/s?

5.19 Turpentine flows at 0.03 ft^3/s from a tank into a 1-nominal, schedule 40 pipe. Determine the length required for the flow to become fully developed.

Piping Systems: Pressure Drop Unknown

(*indicates that minor losses are to be included)

5.20 Kerosene flows at a flow rate of 8 gpm through a ¾ standard type K drawn tubing, which is laid out horizontally and is 140 ft long. Determine the pressure drop.

5.21 Water flows at a flow rate of 5.0 L/s through a 2-nominal schedule 40 commercial steel pipe. It is laid out horizontally and is 15 m long. Calculate the pressure drop.

5.22 Chloroform flows at a flow rate of 15 gpm through a 1 standard type M drawn tubing, which is laid out horizontally and is 75 ft long. Determine the pressure drop.

5.23 Acetone flows at a flow rate of 1.0 L/s through a 1 standard type K drawn tubing. It is laid out horizontally and is 35 m long. Calculate the pressure drop.

5.24 Propylene flows at a flow rate of 250 gpm through a 4-nominal schedule 40 commercial steel pipe, which is laid out horizontally and is 405 ft long. Determine the pressure drop.

5.25 Ethylene glycol flows at a flow rate of 1 gpm through a ⅛-nominal schedule 40 wrought iron pipe, which is laid out horizontally and is 25 ft long. Determine the pressure drop.

5.26 Carbon tetrachloride flows at a flow rate of 1.0 L/s through a 1 standard type M drawn tubing. It is laid out horizontally and is 25 m long. Calculate the pressure drop.

5.27 A main duct of an air-conditioning system is a rectangular conduit of dimension 1 × 0.5 m. It conveys air at a flow rate of 1.1 kg/s. The conduit wall is galvanized sheet metal. Calculate the pressure drop over the 16 m length of conduit.

5.28 A double pipe (or concentric tube) heat exchanger consists of two tubes or pipes. One tube is placed within the other, creating two flow passages. One of these is the inner tube through which a fluid passes. The other flow passage is an annulus through which a second fluid passes.

Figure P5.28 is a sketch of six double pipe heat exchangers connected in series. Two fluids pass through the system. One fluid passes through the inner tube of all six exchangers, and a second fluid passes through the annulus of all six exchangers. The double pipe heat exchanger shown is made of 2 standard copper tube placed within a 4 standard copper tube, both type M. Ethylene glycol flows through the annulus at an average velocity of 2.5 m/s. The six exchangers of the figure are connected in series such that the annulus length is 10 m. Determine the pressure drop experienced by the ethylene glycol. Neglect changes in potential energy.

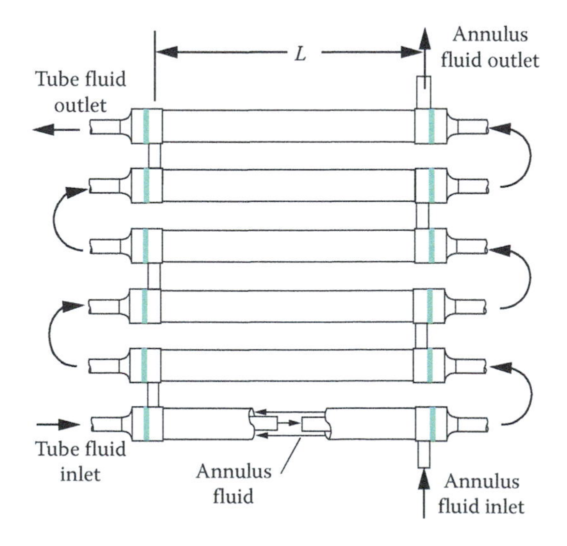

FIGURE P5.28

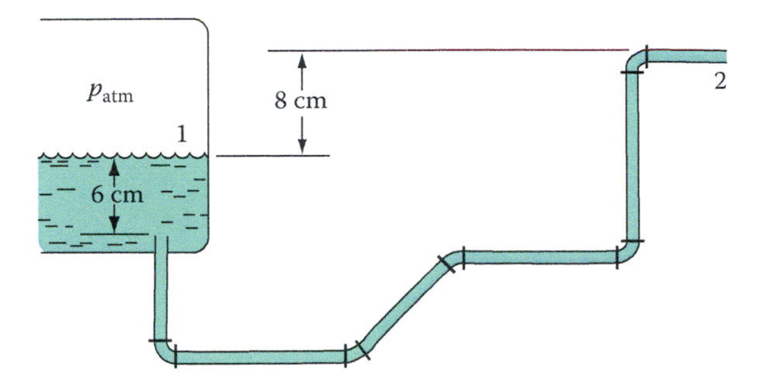

FIGURE P5.29

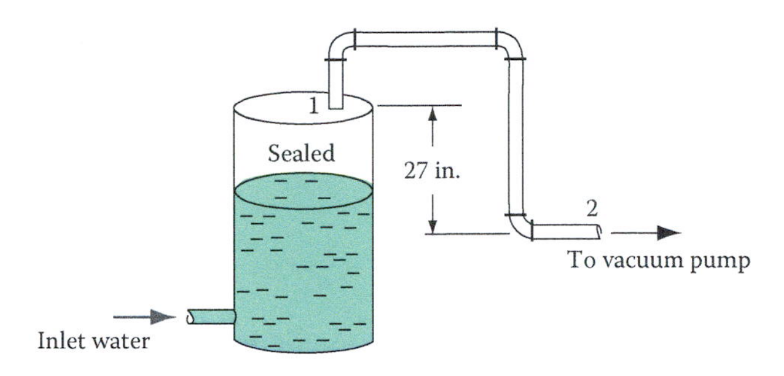

FIGURE P5.30

5.29* The fuel line in an automobile is made of drawn copper tubing, ⅜-standard, type M. A schematic of a fuel line from tank to fuel pump is shown in Figure P5.29. For a fuel flow of 10 cm^3/s, determine the pressure drop in this portion of the line (1–2). The system includes 3.3 m of tubing. Take the fuel to have the same properties as those of octane. Include the minor losses.

5.30* A vacuum pump is used to lower the existing pressure above a tank of water. As air is removed, water from a sump tank fills the main tank, as shown schematically in Figure P5.30. The inlet flow of water is 600 ft^3/h. If the air pipe is schedule 40, 2-nominal wrought iron, determine the pressure drop in the piping system (1–2). All fittings are regular and threaded, and the total pipe length is 71 ft. Neglect compressibility effects, and include minor losses.

5.31 An underground salt dome can be used as a storage volume for gasoline. Access to the salt-dome is established with several circular annuli. Consider one annulus and take the saltdome to be filled with gasoline, as sketched in Figure P5.31. The outer pipe is schedule 160, 24-nominal; the inner pipe is schedule 80, 12-nominal. Both pipes are made of cast iron. When salt water is pumped into the pipe, gasoline is forced upward through the annulus. For a saltwater flow rate of 10 ft^3/s, determine the pressure drop of the gasoline in the annulus. Take the depth to be 1,500 ft and neglect entrance and exit losses. Take the viscosity of salt water to be equal to that of plain water and the properties of gasoline to be the same as those of octane.

Piping Systems: Volume Flow Rate Unknown

(*indicates that minor losses are included)

5.32 A 12-nominal, schedule 80 wrought iron pipe is inclined at an angle of 10° with the horizontal and conveys chloroform downhill (Figure P5.32). If the allowable pressure drop in the pipe is 2 psi and the pipe length is 100 ft, determine the volume carrying capacity in the pipe.

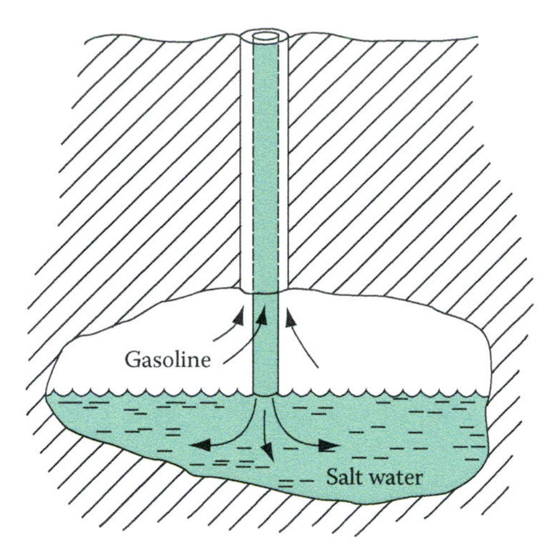

FIGURE P5.31

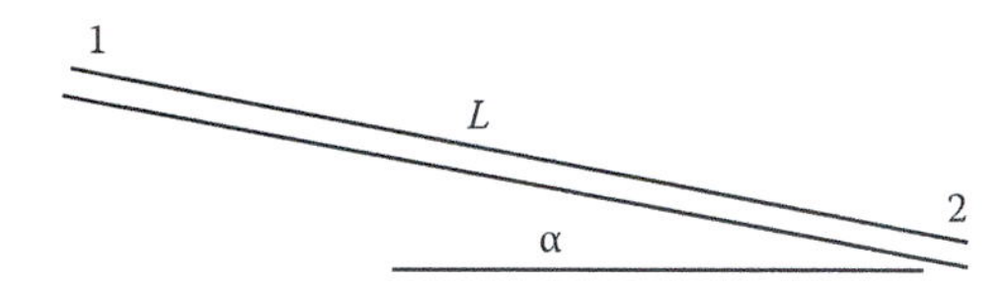

FIGURE P5.32

5.33 Ether flows through a horizontally laid, 4-nominal schedule 40 wrought iron pipe. The pressure drop measured at points 280 m apart is 100 kPa. Determine the volume flow rate through the pipe.

5.34 Propyl alcohol flows in a 4 standard type L drawn tubing that is 20 m long. The pressure drop over this length is 15. Determine the volume flow rate under these conditions.

5.35 Molasses flows through a schedule 80, 4-nominal stainless steel pipe to a bottling machine in a production plant. The pipe is 250 ft long, and the pressure drop is 12 psia. Determine the volume flow rate. Take the properties of molasses to be the same as those of glycerine and the pipe wall to be smooth. The molasses is at room temperature.

5.36 Octane flows in a 1-nominal schedule 40 wrought iron pipe that is 35 m long. The pressure drop over this length is 100 kPa. Determine the volume flow rate under these conditions.

5.37 A refinery plant separates crude oil into various components. One constituent produced is heptane, which is conveyed through a schedule 30, 14-nominal cast-iron pipe that is 150 ft long. The pressure drop in the pipe is 0.75 psi. Determine the volume flow rate of heptane in the pipe.

5.38 A rectangular conduit of dimension 5 × 7 in. conveys hydrogen. The conduit wall is asphalt coated and is 25 ft long. The hydrogen compressor provides enough power to overcome a pressure drop of 0.01 in. of water. Determine the mass flow of hydrogen.

5.39 Benzene flows through a 12-nominal schedule 80 wrought iron pipe. The pressure drop measured at points 1200 ft apart is 5 psi. Determine the volume flow rate through the pipe.

5.40 Hexane flows in a 1 standard type K drawn tubing that is 10 m long. The pressure drop over this length is 50. Determine the volume flow rate under these conditions.

5.41 An annular flow passage is formed by placing a 1 standard type M copper tube within a 3 standard type M copper tube. The annulus is 2 m long and conveys glycerine. The pressure drop over the 2 m length is 19 kPa. Determine the volume flow rate of glycerine.

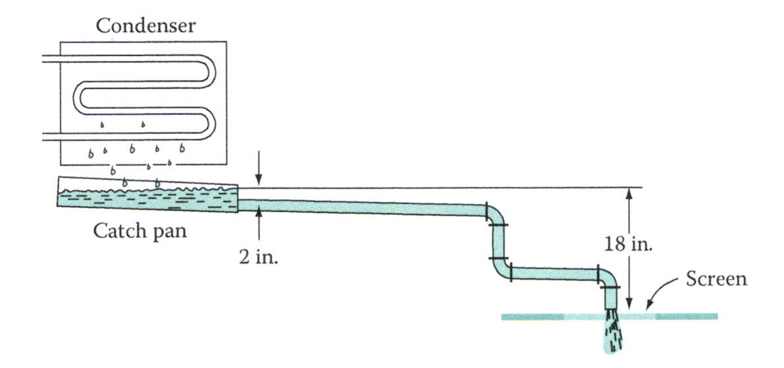

FIGURE P5.42

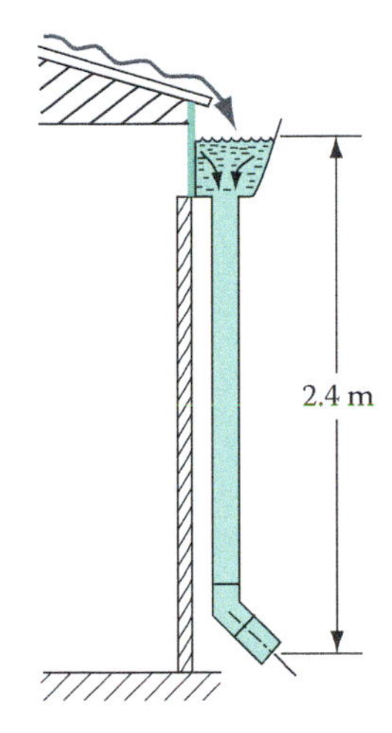

FIGURE P5.43

5.42* The evaporator of a household air conditioner is located in the attic. Water at 40°F condenses from the air when the air is cooled by moving past the coil. The condensed water falls downward into a catch pan. A pipe of ½-nominal schedule 40 PVC pipe is attached to the pan and conveys to water to a roof vent located in the soffit. A schematic of the piping system is shown in Figure P5.42. There are 8 ft of pipe and 3 standard elbows. The water discharge system is completely gravity fed. Determine the volume flow rate through the pipe for the case shown.

5.43* Gutters and downspouts are attached to roofs to drain rainwater. Both circular and rectangular downspouts are conventionally used. Figure P5.43 is a schematic of a downspout containing 2.2 m of galvanized ductwork and one 45° elbow. Determine the volume carrying capacity if the duct cross section is round with an inside diameter of 13.1 cm.

5.44* Repeat Problem 5.43 but take the duct cross section to be a 7 × 10 cm rectangle.

5.45* Repeat Problem 5.43 for a duct that is square in cross section: 8 × 8 cm.

Piping Systems: Diameter Unknown

(*indicates that minor losses are included)

5.46 Linseed oil is often used as a wood finish and can be purchased in half-gallon containers. On the assembly line, one machine can fill and cap 20 containers per minute. The linseed oil tank is located 36 ft from a machine. The oil is pumped from the tank to the machine;

the pump outlet pressure is 35 psig, and the machine requires 15 psig pressure for optimum operation. Select a suitable diameter for a pipeline if commercial steel is to be used.

5.47 A nozzle is used to provide a water spray for keeping dust from escaping from a dirt pile into the atmosphere. Just before entering the nozzle, the fluid must have a pressure of 60 psi at a flow rate of 12 ft^3/min. The water source is a city water main in which the pressure is maintained at 70 psi. The distance from the main to the proposed nozzle location is 65 ft. Select a suitable pipe size for the installation if galvanized iron pipe is all that is available.

5.48 A diesel engine is used as a power source for a generator as part of an electrical backup system for a remotely located manufacturing plant. The diesel requires 0.01 m^3/s of kerosene. A kerosene tank is located 10 m from the engine. The tank pressure is maintained at a constant 200 kPa, and the engine fuel injectors require that the kerosene be delivered at 115 kPa or less. Drawn copper tubing will be used for the fuel line; select an appropriate diameter.

5.49 A riveted steel pipeline is used to convey 0.5 m^3/s of gasoline (assume properties the same as octane) at a distance of 40 km. The available pump can overcome a friction loss of 300 kPa. Select a suitable pipe size for the installation.

5.50* Ethylene glycol is used in a heat recovery system where exhausted warm air passes over soldered copper tubing with fins attached. A schematic of the piping system is shown in Figure P5.50. The system must be able to convey 0.002 m^3/s of ethylene glycol at an allowable pressure drop (1–2) of 50 kPa. Select a suitable tubing diameter. The total length of tubing is 12 m.

5.51* Overhead sprinklers are commonly used in buildings for fire protection. If a fire starts, sufficient heat is produced to melt a metal tab that opens the sprinkler head. One such installation is shown schematically in Figure P5.51. Water is supplied at the main at a pressure of 75 psig. The nozzle requires a pressure of 60 psig at a flow rate of 15 gpm for effective coverage. Assuming that 39 ft of wrought iron pipe with threaded fittings is to be used in the installation, select a suitable pipe diameter.

5.52 An annulus is to be made of two type M copper tubes that are 2.5 m long. The outer tube size is limited by space and is 2 standard, while the size of the inner tube must be selected. The fluid is turpentine, and the available pump can overcome a pressure loss of 30 kPa while delivering 0.01 m^3/s of liquid. Select a suitable inner tube size.

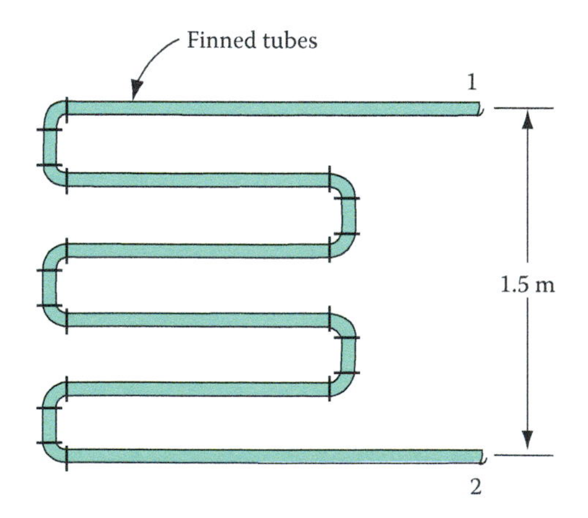

FIGURE P5.50

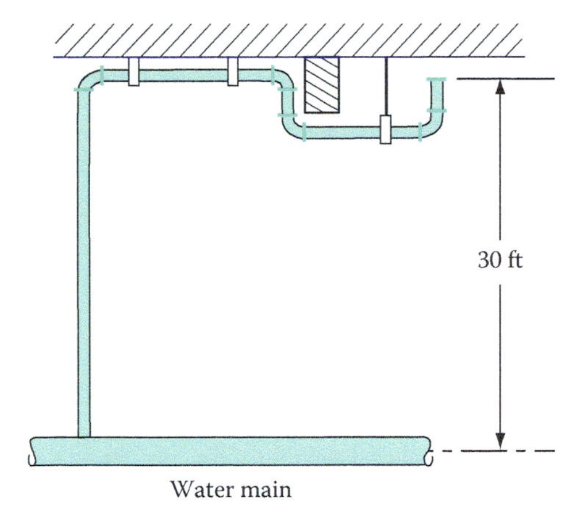

FIGURE P5.51

Surface Roughness

5.53 The owners of an above ground pool wish to empty it by using a 7.1 cm ID plastic hose as a siphon (Figure P5.53). The 12 m long flexible hose is placed over the pool wall, and the siphon is started. The volume flow rate of water through the hose is 0.008 m^3/s. Determine the friction factor f and an equivalent roughness ε. The tube configuration shows a return bend and an elbow (assume same minor losses as flanged fittings).

5.54 Ethylene glycol is used in an experiment to determine the roughness of a pipe material. The liquid is pumped through the pipe at a flow rate of 0.01 m^3/s. The measured pressure drop is 3.8 kPa over a 3 m length. The pipe itself is 3-nominal schedule 40. Determine the roughness ε.

5.55 Water is used to determine the roughness ε of a new pipe material. The pipe is 2-nominal schedule 40, and it is 21 ft long. The water is pumped through at a rate of 50 gpm, and the measured pressure loss over 18 ft is 1.0 psi. Determine the roughness ε.

5.56 Table 5.2 gives values of surface roughness for new, clean conduit surfaces. Conduit walls become rougher with time, however, owing to corrosion, incrustations, and deposition of materials (usually minerals) on the pipe wall. The time it takes for the wall surface to

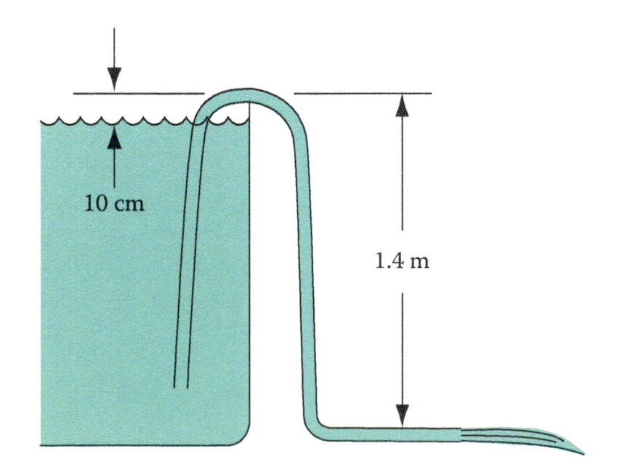

FIGURE P5.53

roughen depends greatly on the fluid conveyed. It has been found that the absolute roughness ε increases linearly with time according to

$$\varepsilon_t = \varepsilon + kt$$

where
- ε is the absolute roughness of a new surface
- k is the a constant to be determined from tests
- t is the time in years*

A 4-nominal cast iron pipe conveys water at a rate of 0.015 m^3/s. Determine the pressure drop per kilometer of new pipe. Next, estimate the pressure drop per kilometer of the pipe after 35 years if the flow velocity remains unchanged. Every 9 years the absolute roughness of the wall doubles.

5.57 A 16-nominal cast iron pipe conveys carbon disulfide at a rate of 9 ft^3/s. The pipe itself is 1100 ft long. Determine the pressure drop in the pipe when new if it is laid horizontally.

Previous experiments with this fluid-pipe combination have shown that $k = 0.00007$ ft/year. Assuming the pressure drop in the pipe is the same as when new, determine the volume flow rate of liquid through the pipe after 10 years have elapsed (see the preceding problem).

Equivalent Length

5.58 As seen by working the more complex pipe friction problems, minor losses can consume significant amounts of energy in the form of a pressure loss. Also, the inclusion of minor losses in the Modified Bernoulli Equation can make the trial-and-error type problems tedious. Efforts have therefore been made to represent minor losses using what is known as **equivalent length.** Recall from the modified Bernoulli equation the friction and minor loss terms:

$$\frac{fL}{D_h}\frac{V^2}{2g} + \sum K\frac{V^2}{2g} = \left(\frac{fL}{D_h} + \sum K\right)\frac{V^2}{2g}$$

The concept of equivalent length allows us to replace the minor loss coefficient term with

$$\sum K = \frac{fL_{eq}}{D_h}$$

where f is the friction factor that applies to the pipe, D_h is the characteristic length, and L_{eq} is the equivalent length of all the fittings. Physically what we are doing is calculating the length of pipe (of same material, size, and schedule) we can "replace" the fittings with to obtain the same pressure drop. Calculate the equivalent length of the fittings for the following data: $\Sigma K = 14.4$; $f = 0.034$; and $D = 0.1342$ ft.

5.59 Repeat the previous problem for the following data: $\Sigma K = 7.2$; $f = 0.0205$; and $D = 0.125$ ft.

* Colebrook, C. F. and C. M. White, The reduction of carrying capacity of pipes with age, *Journal of the Institute of Civil Engineers* (London), 1937.

Parallel Piping Systems

5.60 Two pipes connected in parallel convey acetone from one location to another. The pipes are made of 24-nominal schedule 20 and 12-nominal schedule 40, both cast iron. For a volume flow rate of 0.25 m³/s, determine the volume flow rate in each pipe.

5.61 An automatic sprinkler system for a narrow plot of lawn is sketched in Figure P5.61. Water is supplied by the main such that p_1 = 400 kPa (gage) and Q_1 is variable. The sprinkler pipeline is made of PVC pipe, schedule 40. For a wide open ball valve, determine the flow delivered to each sprinkler head. Do not neglect minor losses. Details regarding fluid paths and notes are as follows:

From 1 to the T-joint at 2:	1½ nominal; pipe length is 6.5 m
From 2 to the sprinkler head at 3:	1½ nominal; pipe length is 0.3 m
From 2 to the sprinkler head at 4:	1 nominal; pipe length is 8.3 m
Fittings:	K for each sprinkler head is 50, which includes the exit loss; T-joint at 2; reducing bushing and elbow between 2 and 4; ball valve.

5.62 A home air conditioning system contains a fan that forces air over cooling coils (condenser). The cooled air (T = 7°C) goes to a plenum chamber to which three circular ducts are attached as shown in Figure P5.62. The chamber pressure is 108 kPa, and the ducts are of lengths 2, 2.6, and 3.8 m. Ducts all have a 30 cm inside diameter and are made of galvanized sheet metal. Neglect minor losses and determine the volume flow in each if the air flow into the plenum chamber is 20 m³/min and each duct exhausts into rooms at 101.3 kPa pressure. Assume constant temperature.

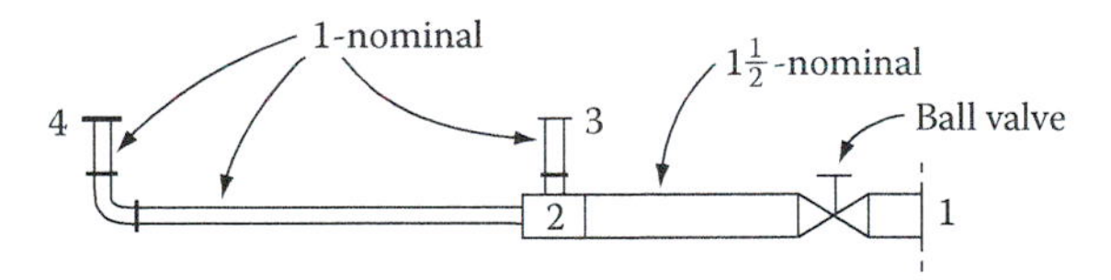

FIGURE P5.61

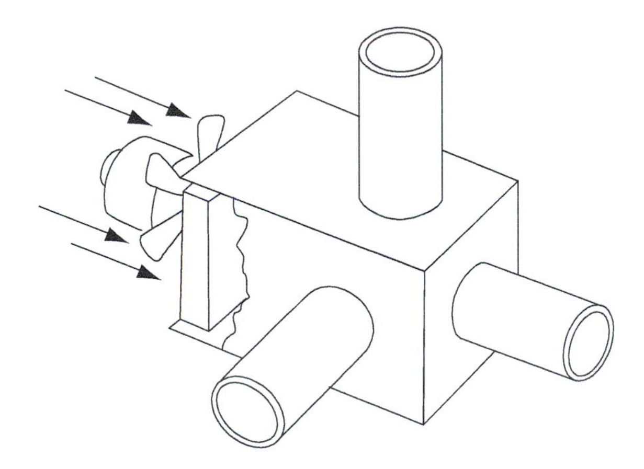

FIGURE P5.62

Pumps and Piping Systems

5.63 The homeowners of Example 5.13 have decided that they would rather install a smaller pump to deliver 15 gpm rather than 25 gpm.

Determine the pump power required for the smaller pump.

5.64 Octane is to be pumped in a piping system that is routed from storage tanks to the main pump by a number of smaller pumps. One such arrangement is sketched in Figure P5.64. This pump must supply 0.4 m^3/s of octane to the main pump. All fittings are flanged; the pipe is wrought iron, schedule 80, 24 nominal, with $L = 65$ m. The absolute pressure at section 2 is 282.5 kPa. Determine the power required to be transferred to the liquid. Assuming an overall pump-motor efficiency of 65%, determine the input power required by the motor.

5.65 In a dairy products processing plant, milk* (ρ = 1 030 kg/m^3, $\mu = 2.12 \times 10^{-3}$ N·s/m^2) is pumped through a piping system from a tank to a container packaging machine. The pump and piping are all stainless steel (smooth walled) arranged as shown in Figure P5.65.

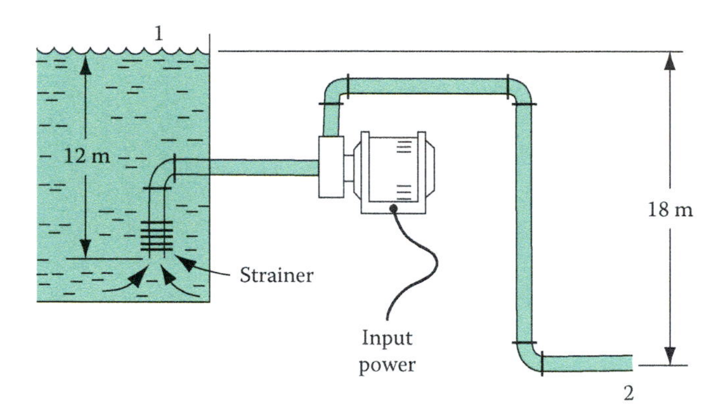

FIGURE P5.64

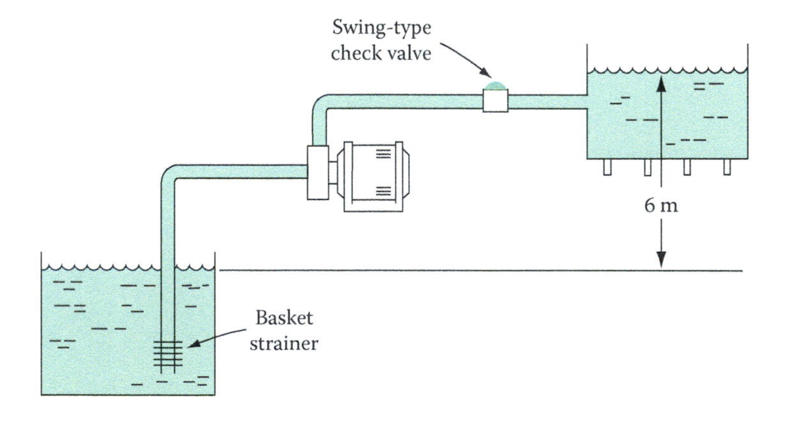

FIGURE P5.65

* Geonkoplis, C. G. *Transport Processes and Unit Operations*. Boston, MA: Allyn & Bacon, 1978, p. 629, App. Table A.4-2.

The pump inlet line (4-nominal schedule 40 pipe) is 2 m long. The pump outlet line (3½-nominal schedule 40 pipe) is 15 m long. All fittings are flanged, and the flow rate through the system is 0.015 m^3/s. Determine the electrical power input to the pump if the pump-motor efficiency is 70%.

5.66 Since the early 1900s, the Grand Canyon has become a popular place for tourists to visit. The lack of potable water became a major, hindrance to visitors, however, as permanent natural sources of water are rare or nonexistent on the rims. In 1928, the Utah Parks Company constructed a 12,500 ft pipeline of 3½-nominal pipe. Water is piped up to the North Rim from Roaring Springs, in Roaring Springs Canyon, a vertical distance of almost 3900 ft (1 200 m). Water flows through the pipe at 95 gpm into one of two 2,000,000 gallon storage tanks. Water can then be taken from the tanks and used for showers, consumption, irrigation, etc. Given this information, calculate the pump power required for this installation. Assume that inlet and outlet pressures are equal, that minor losses can be neglected and that the pipe is made of wrought iron. (Much of the data for this problem were obtained from a National Park Service tourist publication.)

5.67 Decorative water fountains consist of a water reservoir and underground piping to and from a pump. One such system is shown schematically in Figure P5.67. The inlet line to the pump is 20 m of 16-nominal PVC pipe. The outlet line from the pump is 18 m of 12-nominal PVC pipe. The outlet line leads to the bottom of an annular flow line. The expansion there has a loss value of $K = 2$ based on the kinetic energy in the 12-nominal line. The annular flow passage has a length $L = 180$ cm, and is bounded by $D_o = 30$ cm, and $D_i = 20$ cm. It too is made of PVC material. There is a negligible loss at the exit of the annulus; the pressure at that location is atmospheric. What is the pump power required for the flow configuration shown? If the pump-motor combination has an efficiency 92%, determine the electrical power requirements. If electricity costs \$0.05/kW·h, determine the cost of running the fountain for 8 h.

5.68 Derive Equation 5.25 in detail starting with Equation 5.23.

5.69 Derive Equation 5.28 in detail starting with Equation 5.25.

5.70 Show that Equation 5.28 reduces to Equation 5.16 when $\kappa = 0$.

5.71 Select five Reynolds numbers and ε/D values at random from the Moody and obtain the corresponding friction factors. Substitute the Reynolds numbers and the ε/D values into any of the pipe friction equations and calculate f. Compare your results to the values read from the Moody diagram.

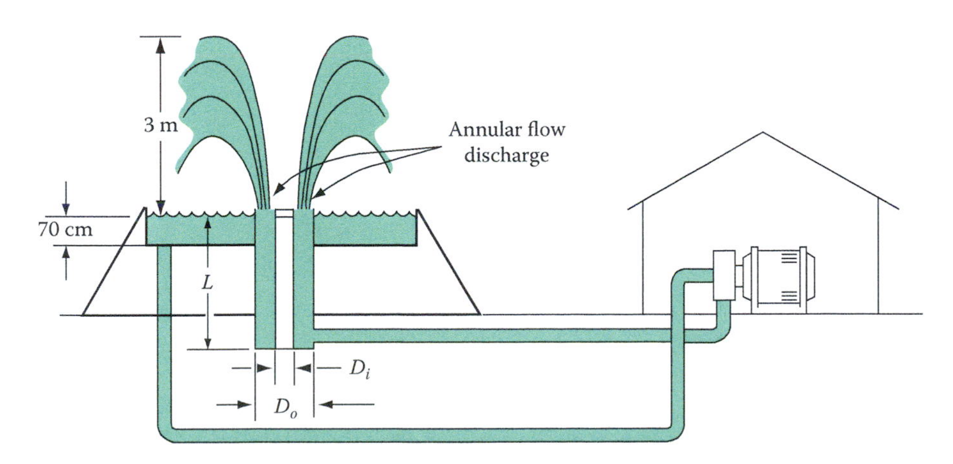

FIGURE P5.67

5.72 Consider the graph and table of $f\text{Re}$ versus h/w for a rectangular duct, Figure 5.17. Using the tabulated values, show that the equation of the line is

$$f\text{Re} = 96 - 95(h/w) + 56(h/w)^2$$

5.73 Consider the graph and table of $f\text{Re}$ versus h/w for a rectangular duct, Figure 5.17. Using the tabulated values, construct a graph of hydraulic diameter D_h versus h/w.

5.74 Figure P5.74 shows an end view of an internally finned tube. The spiraled fins are an integral part of the tube wall. The friction factor for such a configuration is given by

$$f = \frac{0.046}{\text{Re}^{0.2}} \frac{D_e}{D_i} \sec^{3/4} \theta$$

where

θ is the helix angle measured from tube axis
D_i is the diameter to base of fins
D_e is the effective diameter based on actual cross-sectional area (tube area minus fin area)
$\text{Re} = VD_h/\nu > 2\ 100$ (turbulent flow)
D_h is the hydraulic diameter = 4 × cross sectional area/perimeter

A finned tube has the same basic dimensions of 1 standard type K drawn tubing with internal fins attached. The fins are rectangular in cross section (from an end view perpendicular to the axis of the tube). They are equal to the wall thickness in height, and they are double the wall thickness in width. The tube contains equally spaced fins, and the helix angle is 30° from the tube axis. The tube conveys water at a flow rate of 0.002 m³/s.

(a) Determine the friction factor and the pressure drop per unit length.
(b) Determine the friction factor and pressure drop per unit length if no fins are attached to the drawn tubing wall and the flow rate is the same.

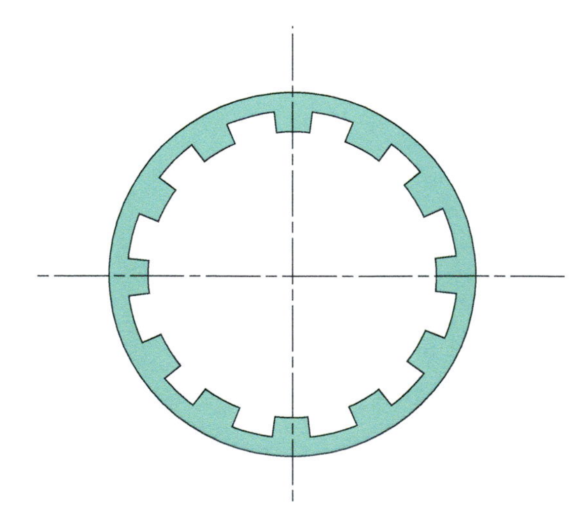

FIGURE P5.74

5.75 The friction factor for flow in a coiled tube is somewhat different from that for a straight tube. Correlations based on experimental work have been derived for laminar and turbulent flows in a coiled tube. These are as follows:

Laminar flow:

$$\frac{f_{\text{coiled}}}{f_{\text{straight}}} = \left\{1 - \left[1 - \left(\frac{5.8}{D_e}\right)^{0.45}\right]^{2.22}\right\}^{-1} \quad \text{valid for } 5.8 < D_e < 1\,000$$

Turbulent flow:

$$\frac{f_{\text{coiled}}}{f_{\text{straight}}} = \left[\text{Re}\left(\frac{D}{D_c}\right)^2\right]^{0.05} \quad \text{valid if} \left[\text{Re}\left(\frac{D}{D_c}\right)^2\right] > 6$$

where

f_{coiled} is the friction factor for the coiled tube
f_{straight} is the friction factor if the coiled tube were straight
D_e is the Dean number = $(\text{Re}/2)(D/D_c)^{1/2}$
D is the pipe inside diameter
D_c is the diameter of coil measured to centerline
$\text{Re} = VD/\nu$

Also for flow in a curved tube, the critical Reynolds number above which fully turbulent flow exists in the duct is given by

$$\text{Re}_{\text{cr}} = 2 \times 10^4 \left(\frac{D}{D_c}\right)^{0.32}$$

The flow is thus turbulent if $\text{Re} > \text{Re}_{\text{cr}}$.

Two enterprising chemical engineering students have decided to produce homemade beer for their own consumption. In the final phase of the production process, beer is siphoned from a five gallon container through a plastic hose that leads to a coiled copper tube as shown in Figure P5.75. The copper tube is located within a water tank that is used to cool the beer. The tube has a straight length that extends to the bottom of the water tank and then coils upward making loops until it reaches the top of the water tank. The beer properties change somewhat during the time that it passes through the water tank, but for our purposes, we can assume them to be essentially constant. They are: specific gravity = 1.0042 and $\mu = 1.4 \times 10^{-5}$ lbf · s/ft^2. For the following data, determine the volume flow rate of beer through the tubing:

$L = 22$ ft (total for the copper tubing)

$D_c = 8$ in.

$D = ¼$ in. ID drawn copper tubing

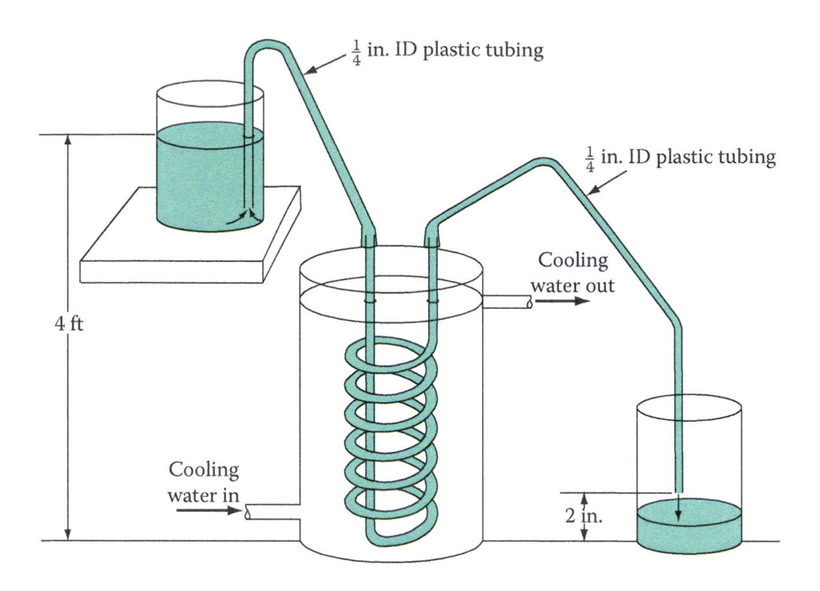

FIGURE P5.75

Number of coils = 6½

There are 10 ft of ¼ in. ID plastic tubing (which has the same roughness factor as drawn copper tubing). Neglect minor losses. Does coiling the tube have a significant effect?

Experimental Results

5.76 An apparatus used to obtain pressure drop due to friction versus flow rate was described in this chapter. Following are results obtained with that apparatus. Reduce the data and determine friction factor and Reynolds number.

2-nominal schedule 80 PVC pipe
$L = 2$ ft; fluid is water

Trial	Q (GPM)	Δh (ft)
1	7	0.02
2	13.3	0.06
3	14	0.06
4	19.4	0.08
5	22.7	0.1
6	27	0.1
7	31.1	0.12
8	33.2	0.14
9	40.5	0.23

5.77 Repeat Problem 5.76 for the following data:

1-nominal schedule 80 PVC pipe
$L = 35$ in.; fluid is water

Trial	Q (GPM)	Δh (ft)
1	5.02	0.15
2	9.35	0.34
3	14.9	0.65
4	17.4	0.83
5	20.5	1.04
6	24.6	1.29
7	27.1	1.52
8	29.3	1.78
9	38.3	2.82

System Curve

5.78 A **system curve** for a piping system is defined as a graph of any flow variable as a function of volume flow rate Q (horizontal axis). A system curve can be drawn for any piping system.

Water flows through 1000 ft of 2 nominal schedule 40 galvanized pipe. Construct a system curve of pressure drop versus flow rate over a velocity range of 4–10 ft/s.

6 Flow over Immersed Bodies

In this chapter, we examine the forces that are exerted on a body moving through a fluid. Consider, for example, a wing or airfoil moving at a velocity V through air. For convenience, and in keeping with the control volume approach, we impose a velocity on the system equal to $-V$, thus rendering the airfoil stationary while air moves past the wing. The air velocity is given the symbol U, or U_∞, to denote this transformation. The forces of interest are the forces act parallel and perpendicular to the principal airspeed direction. These forces are that the drag and lift forces as illustrated in Figure 6.1.

The purpose of this chapter is to describe the effects that cause lift and drag forces to be exerted on immersed bodies—namely, boundary layer growth and separation. Data or drag of variously shaped bodies are presented, and, finally, the combined effects of lift and drag on different airfoils are discussed.

6.1 FLOW PAST A FLAT PLATE

In this section, we examine flow past a flat plate, with particular interest in the fluid velocity in the vicinity of the plate surface. Consider uniform flow passing over a flat plate that is aligned parallel to the flow direction. The leading edge of the plate is sharpened so that flow past the upper surface (our region of interest) encounters no interference that a blunt leading edge would provide. Upstream of the plate, the flow is uniform with a velocity equal to U_∞. The flow is laminar near the leading edge of the plate. Somewhere downstream, the flow becomes turbulent. The region between the laminar and turbulent zones is a transition zone. See Figure 6.2.

Along the plate surface, the velocity is zero due to the nonslip condition. The velocity profile increases from zero at the wall to the free stream value or nearly so at a certain vertical distance from the plate. This vertical distance, called the **boundary layer thickness** δ, is shown schematically in Figure 6.2. Also shown in the figure is a depiction of the laminar and turbulent velocity profiles. Figure 6.3 provides a comparison of these profiles, and as can be seen, the laminar profile near the wall is more positively sloped than the turbulent profile. There is considerably more mixing action in the turbulent case, which tends to distribute the kinetic energy of the flow more evenly.

6.1.1 Boundary Layer Growth

As seen in Figure 6.2, the boundary layer thickness δ increases with distance along the plate. Moreover, there are three definite zones: the laminar zone near the leading edge, the transition zone, and the turbulent zone downstream. Note that a laminar sublayer exists in the turbulent zone because the velocity along the wall must be zero. Also shown in the figure is a **displacement thickness** δ*, which refers to the displacement of the external flow due to the presence of the plate, which in turn has caused a boundary layer to form. The flow must be displaced an amount δ* to satisfy the continuity equation. Upstream the velocity is U_∞, and the flow area for an undisturbed height is the product of H and a unit width into the page. The continuity equation applied to the flow is

$$\text{Area} \times \text{Velocity upstream} = \text{Area} \times \text{Velocity anywhere along the plate}$$

or

$$\underbrace{U_\infty H}_{\text{in free stream}} = \underbrace{U_\infty \delta^*}_{\text{displacement thickness}} + \underbrace{U_\infty (H-\delta)}_{\text{between } H \text{ and boundary layer}} + \underbrace{\int_0^\delta V_x \, dy}_{\text{within boundary layer}} \tag{6.1}$$

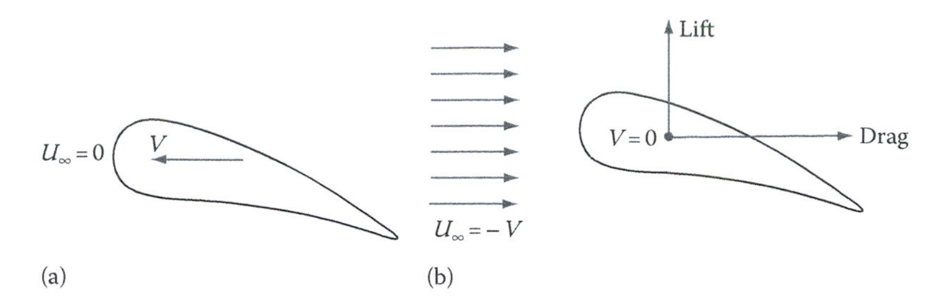

FIGURE 6.1 Forces acting on an airfoil. (a) Airfoil moving in air and (b) control volume with airfoil stationary.

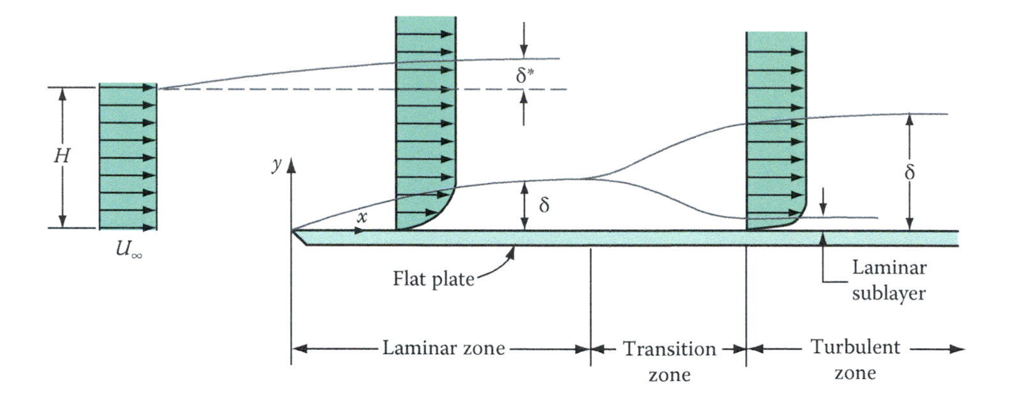

FIGURE 6.2 Growth of the boundary layer along a flat plate. The vertical scale is greatly exaggerated.

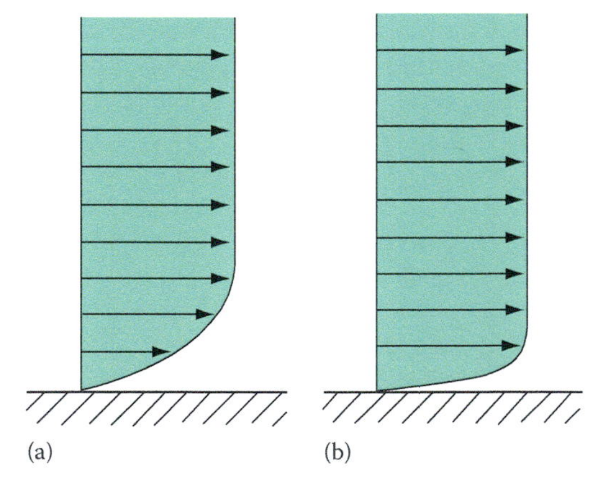

FIGURE 6.3 Laminar (a) and turbulent (b) boundary-layer velocity profiles.

There is a drag force acting on the plate that is due to skin friction between the fluid and the plate. In the laminar zone, the wall shear stress decreases with distance. In the turbulent zone, the wall shear stress is greater than that in the laminar zone and likewise decreases with distance, but much more slowly than that in the laminar zone. Both contribute to the force that tends to move the plate in the direction of flow.

Details of the flow in the boundary layer are derived in Chapter 13. For purposes of illustration, the results of the analysis are given here. It can be shown that for the laminar boundary layer, the boundary layer thickness for flow over a flat plate is

$$\delta = \frac{5.0x}{\sqrt{\mathrm{Re}_x}} \tag{6.2}$$

where

$\mathrm{Re}_x = \rho U_\infty x/\mu$

x is the distance along the plate

In addition, the displacement thickness is

$$\delta^* = \frac{1.73x}{\sqrt{\mathrm{Re}_x}} \tag{6.3}$$

and the drag force exerted on the plate in the direction of flow is given by

$$D_f = \frac{0.664 b\rho U_\infty^2 x}{\sqrt{\mathrm{Re}_x}} \left(\text{laminar flow}\right) \tag{6.4}$$

where b is the width of the flat plate.

Experiments have shown that transition between the laminar and turbulent boundary layers occurs at a Reynolds number of approximately 5×10^5. Furthermore, if the boundary layer is turbulent, it can be shown that the boundary layer thickness is

$$\delta = \frac{0.37x}{(\mathrm{Re}_x)^{1/5}} \tag{6.5}$$

The displacement thickness is

$$\delta^* = \frac{0.046x}{(\mathrm{Re}_x)^{1/5}} \tag{6.6}$$

and the drag force is

$$D_f = \frac{0.036\rho U_\infty^2 bx}{\left(\mathrm{Re}_x\right)^{1/5}} \left(\text{turbulent flow}\right) \tag{6.7}$$

These equations are valid if it is assumed that the turbulent boundary layer begins at the leading edge of the plate. Although this is not actually the case, desired results can be obtained by using experimental results as demonstrated in Example 6.1.

It is customary to express data on drag as drag coefficient versus Reynolds number, where both are defined as

$$C_D = \frac{D_f}{\frac{1}{2}\rho U_\infty^2 A} \tag{6.8}$$

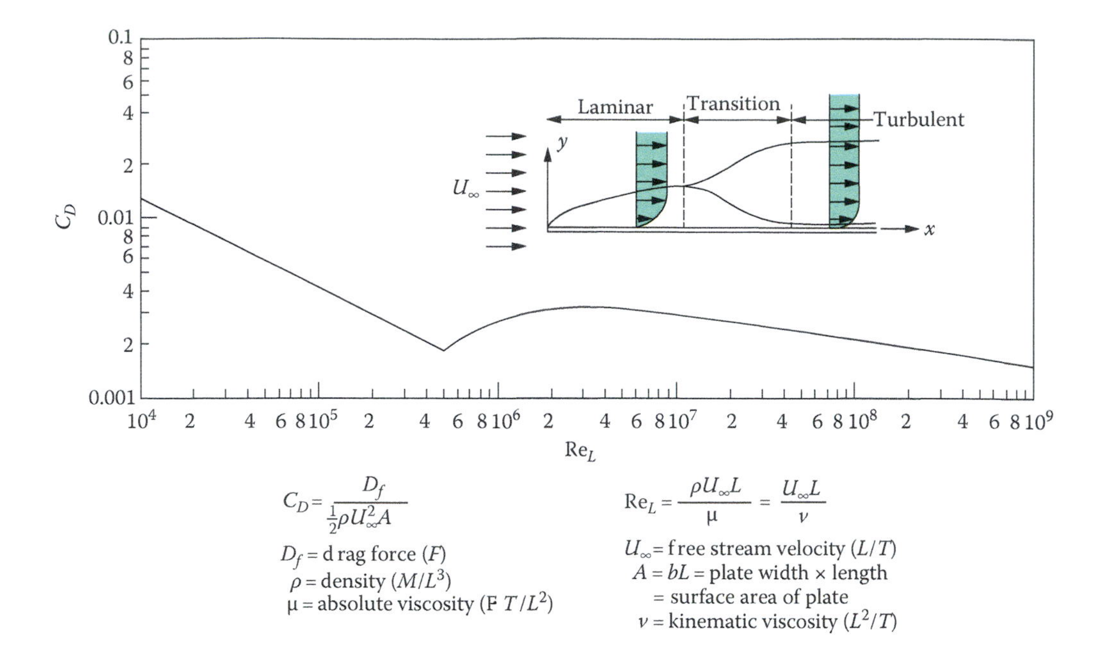

FIGURE 6.4 Drag coefficient versus Reynolds number at the end of a surface for flow over a flat plate.

and

$$Re = \frac{\rho U_\infty D}{\mu} \tag{6.9}$$

where

- D_f is the drag force
- ρ is the density of the fluid
- μ is the viscosity of the fluid
- U_∞ is the free-stream velocity
- A is the projected frontal area or some other characteristic area of the body
- D is the characteristic dimension of the object

For flow over a flat plate, we can express Equations 6.4 and 6.6 in these and calculate drag coefficients and Reynolds numbers. When these results are supplemented with experimental data, the graph of Figure 6.4 results. The drag coefficient varies from 0.001 to 0.1, while the Reynolds number ranges from 10^4 to 10^9. Note that the characteristic length used in the Reynolds number of Figure 6.4 is the length of the plate. It can be seen that the drag coefficient reaches a low point at transition and again at high Reynolds numbers.

Example 6.1

A 10 m/s wind at a temperature of 22°C blows over the flat plate of Figure 6.2. The plate is 12 m long and 6 m wide.

a. Sketch the boundary layer growth for both laminar and turbulent regimes.
b. Determine the force acting on the plate due to the fluid in contact with it.

Solution

a. The laminar boundary layer is given by

$$\delta = \frac{5.0x}{\sqrt{\mathrm{Re}_x}} = \frac{5.0x}{(\rho U_\infty x/\mu)^{1/2}}$$

or

$$\delta = 5.0\sqrt{\frac{\mu x}{\rho U_\infty}}$$

From Table A.3, for air at 22°C, ρ = 1.197 kg/m^3 and μ = 18.22 × 10^{-6} N · s/m^2. By substitution, we get

$$\delta = 5.0\sqrt{\frac{\left(18.22\times10^{-6}\right)(x)}{1.197(10)}}$$

$$= 6.17\times10^{-3}\sqrt{x} \qquad \text{(i)}$$

up to transition. The distance along the wall where transition occurs, x_{critical}, or x_{cr}, is found from

$$\mathrm{Re}_x = 5\times10^5 = \frac{\rho U_\infty x_{\mathrm{cr}}}{\mu}$$

so

$$x_{\mathrm{cr}} = \frac{5\times10^5\left(18.22\times10^{-6}\right)}{1.197(10)} = 0.76\ \mathrm{m}$$

For turbulent flow,

$$\delta = \frac{0.37x}{(\mathrm{Re}_x)^{1/5}} = \frac{0.37x}{\left(\rho U_\infty x/\mu\right)^{1/5}}$$

$$= 0.37x^{4/5}\left[\frac{18.22\times10^{-6}}{1.97(10)}\right]^{1/5}$$

or

$$\delta = 0.0254x^{4/5} \quad \text{for } x \ge 0.76\ \mathrm{m} \qquad \text{(ii)}$$

A plot of Equations i and ii is given in Figure 6.5.

b. The drag force is found by applying Equations 6.4 and 6.7. However, it is easier to use Figure 6.4.

The Reynolds number at the plate end is calculated as

Re_L = 7.9 × 10^6

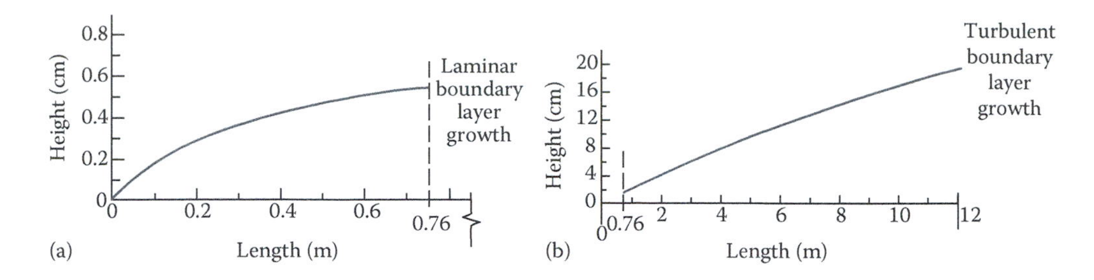

FIGURE 6.5 Plot of (a) laminar and (b) turbulent boundary layer growths on the plate of Example 6.1.

At this value, Figure 6.4 shows

$$C_D = \frac{D_f}{\frac{1}{2}\rho U_\infty^2 A} = 0.003$$

The drag force is found to be

$$D_f = 0.003\left(\frac{1}{2}\right)(1.197)\left(10^2\right)(12)$$

or

$$D_f = 12.9\ \text{N}$$

We see that the laminar boundary layer thickness grows to just less than 0.8 cm and begins transition within approximately 1 m from the leading edge. The turbulent boundary layer thickness grows to nearly 20 cm over a 12 m length. The combined effect is to exert a drag force on the surface of 12.9 N, most of which is due to the turbulent boundary layer (i.e., turbulent flow characteristics).

6.1.2 Separation

We have discussed the formation of a boundary layer in flow over a flat plate. Now we examine fluid behavior in flow over a curved surface. Consider flow past a curved boundary like that illustrated in Figure 6.6. Again the surface is stationary, and the free-stream velocity is U_∞. Point *A* at the nose is where the velocity normal to the surface is zero. This point is referred to as a **stagnation point**, and the pressure measured at *A* is termed **stagnation pressure**. The boundary layer begins its growth from here. At *B* and *C*, the boundary layer has experienced a growth that is intuitively predictable. Over the rear portion of the surface, starting at point *D*, the pressure increases with distance. The fluid particles are slowed down in the boundary layer. The decelerating effect is due to the positive or adverse pressure gradient that has developed. If the decrease in kinetic energy is great, a region of flow reversal may form. The velocity distribution changes as depicted at points *E* and *F*. The velocity at the wall is zero owing to viscosity. At point *D*, where separation begins, $dV_x/dy = 0$ at the wall surface. The region of flow reversal is called the **separation region** because the forward flow has been separated from the boundary by the adverse pressure gradient $dp/dx > 0$. Vortices, swirls, and, in general, reversed flows occur in the separation region. Moreover, the pressure in the separation region is nearly equal to the pressure at point *D*. The location of point *D* has a strong effect on the drag and lift forces exerted. This effect is discussed in more detail in the following sections.

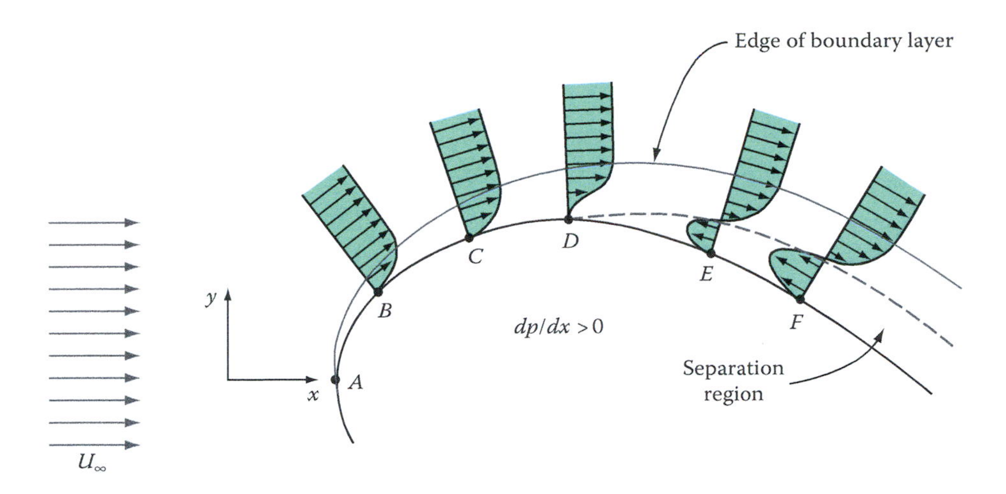

FIGURE 6.6 Flow past a curved surface.

6.2 FLOW PAST VARIOUS TWO-DIMENSIONAL BODIES

The drag force exerted on various types of bodies is presented in this section. The first body to be discussed is an infinitely long cylinder immersed in a uniform flow of velocity U_∞. The velocity profile and pressure distribution about the cylinder are important. From observations of the flow pattern around the cylinder, streamlines can be sketched as in Figure 6.7. Point A is the stagnation point, and B is the point of separation. Beyond point B and immediately downstream of the cylinder is the separation region (or wake). Separation occurs because of the adverse pressure gradient that forms. Because the flow cannot easily negotiate the turn past B, the main flow separates from the cylinder boundary. The complex nature of the flow makes it impossible to determine the velocity distribution at every point on the cylinder. To determine the pressure distribution, therefore, experimental means are necessary.

Suppose that a hollow cylinder drilled every 10° with a static pressure tap is placed in a uniform flow (such as that produced in a wind tunnel). Each tap is connected to a manometer. The arrangement shown in Figure 6.8 would give the pressure distribution on the surface of the cylinder. Only half the cylinder is used because the distribution is symmetric. Experiment shows that at the stagnation point (A), pressure is greater than the reference pressure (atmospheric pressure) because the velocity at this point is zero. The pressure measured at this point is the stagnation pressure. The pressure decreases with the angle θ until at 70° (or thereabouts), the minimum pressure corresponding to maximum velocity is reached. Beyond this, at 120°, separation occurs because the flow cannot negotiate the abrupt boundary change. In the separation region, the

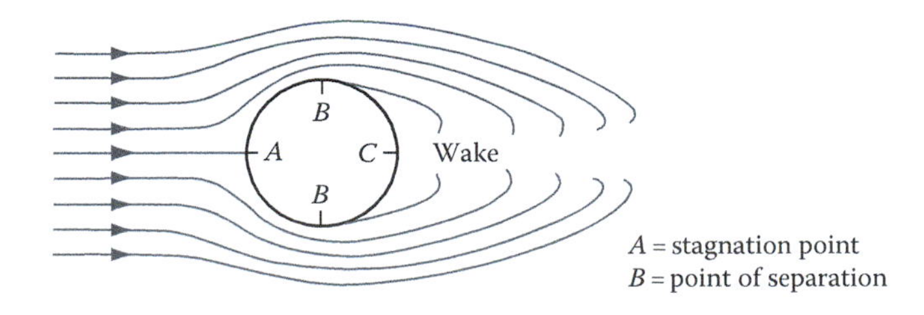

FIGURE 6.7 Uniform flow past an infinitely long cylinder.

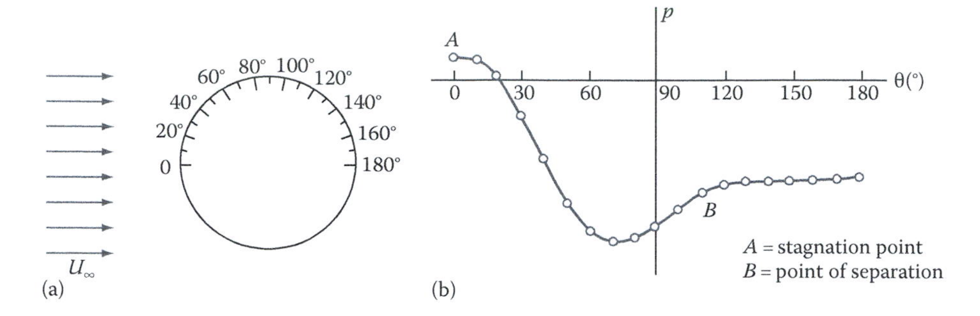

FIGURE 6.8 An experimental arrangement to determine pressure distribution on the surface of a cylinder immersed in a uniform flow: (a) each pressure tap is connected to a manometer and (b) representative results.

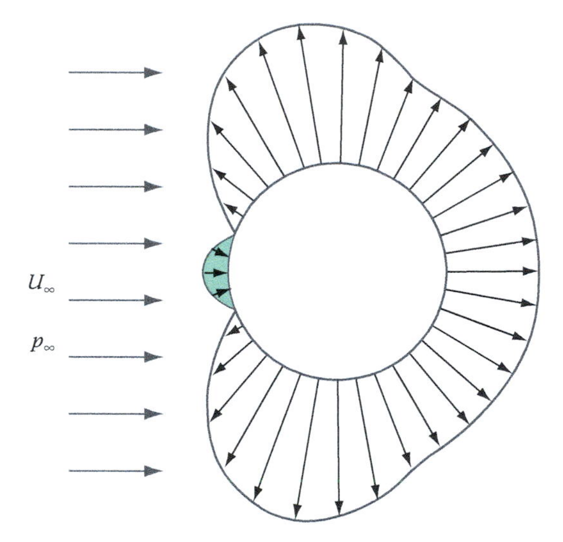

FIGURE 6.9 Polar plot of pressure distribution on the surface of a cylinder immersed in a uniform flow.

pressure on the surface of the cylinder is a constant and very nearly equals that at 120°. For a more meaningful representation, a typical pressure distribution is sketched on a polar coordinate grid in Figure 6.9. The cylinder surface is drawn as a circle, and the pressure at any point along the surface is represented as a line with an arrowhead. The length of any line is proportional to the magnitude of the pressure at its location. The direction (signified by the arrowhead) indicates whether the pressure at a location is greater or less than the free-stream pressure p_∞. A line with an arrow pointing at the cylinder surface means that the pressure at that location is greater than p_∞, and the opposite is also true.

From Figure 6.9, it is apparent that the pressure distribution over the front half of the cylinder is different from that over the rear half. This difference in pressure, which acts over the projected frontal area of the cylinder, results in a net force acting on the cylinder in the direction of flow. This net force due to pressure differences is called **pressure drag** or **form drag**. Form drag combined with skin friction drag gives the total drag exerted on the cylinder.

The form drag force exerted on a cylinder can be calculated from a pressure distribution obtained experimentally by integration over the cylinder surface. Figure 6.10 gives the direction

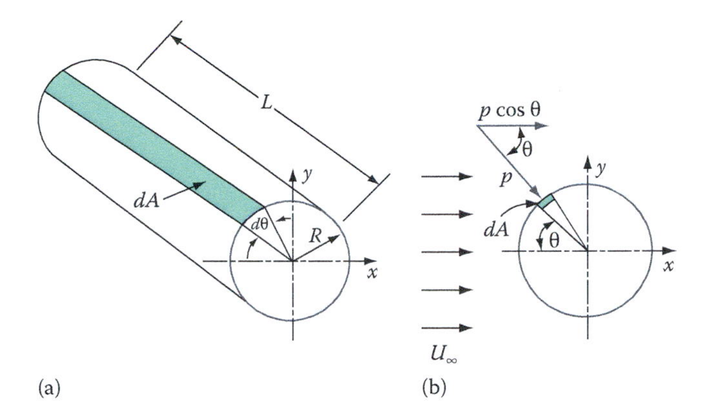

FIGURE 6.10 (a) Definition sketch of the area dA on a cylinder and (b) pressure p and its component, $p \cos \theta$, in the main flow direction.

of the pressure force component that contributes to the drag. By integration of experimental data, we obtain

$$D_f = \int_0^{2\pi}\int_0^{L} p\cos\theta \, dA$$

The form drag can be obtained from integration of the pressure distribution about the surface of the cylinder:

$$D_f = RL\int_0^{2\pi} p\cos\theta \, d\theta = DL\int_0^{\pi} (\Delta p + p_\infty)\cos\theta \, d\theta$$

where $D = 2R$ has been substituted, and symmetry assumed. Continuing,

$$D_f = DL\int_0^{\pi} \Delta p\cos\theta \, d\theta + DL\int_0^{\pi} p_\infty \cos\theta \, d\theta$$

The second integral is equal to zero. Therefore,

$$D_f = DL\int_0^{\pi} \Delta p\cos\theta \, d\theta$$

How these equations are used to predict drag is demonstrated in the following example.

Example 6.2

The accompanying table gives pressure versus θ for a 3.75 in.-diameter cylinder that is 10.5 in. high and immersed in a uniform flow of 53.5 ft/s. The pressure data collected are Δp ($= p - p_\infty$) versus θ. The ambient pressure $p_\infty = 14.7$ psi. Obtain and estimate of the form drag by integrating the pressure distribution around the surface of the cylinder.

θ°	Δp, psf	Δp cos θ
0	1.07	1.07
10	1.29	1.27
20	0.210	0.197
30	−1.80	−1.56
40	−4.72	−3.62
50	−8.15	−5.24
60	−11.6	−5.80
70	−14.4	−4.92
80	−16.1	−2.79
90	−15.9	0.00
100	−14.8	2.57
110	−13.7	4.70
120	−10.7	5.36
130	−9.87	6.34
140	−9.44	7.23
150	−9.44	8.18
160	−9.44	8.87
170	−9.44	9.30
180	−9.44	9.44

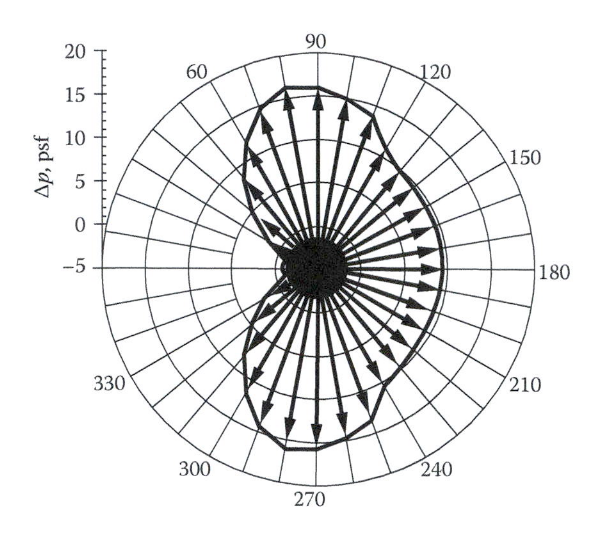

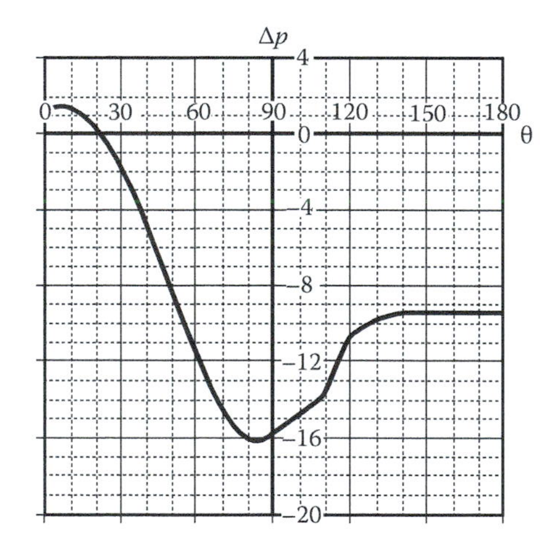

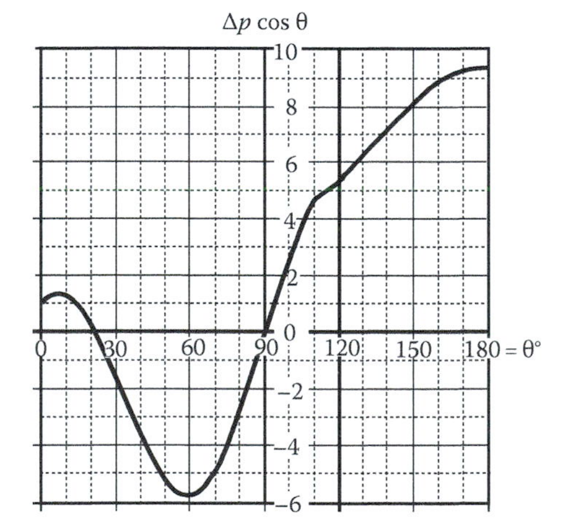

$$L = 10.5\,\text{in.}\quad D = 3.75\,\text{in.}\quad U_\infty = 53.5\,\text{ft/s}$$

$$\text{Properties}\quad \rho = 0.00232\,\text{slug/ft}^3\quad \mu = 0.3801\times10^{-6}\ \text{lbf}\cdot\text{s/ft}^2$$

a. Sketch Δp versus θ on polar coordinate paper.
b. Sketch Δp versus θ on rectangular coordinate paper.
c. Sketch Δp cos θ versus θ on rectangular coordinate paper. Determine the area under curve noting which portions are positive and which are negative. Calculate the form drag force.
d. Calculate the drag coefficient and the Reynolds number.

Solution

The first two columns of the following table are the raw data obtained from the apparatus. The third column is the product of $\Delta p \cos \theta$, also as a function of θ.

The following figures are the graphs called for in parts a, b, and c of the problem statement.

The area under the third graph can be obtained by counting the number of squares on $\Delta p \cos \theta$ versus θ graph = + 2 − 24 + 56 = 34 (an estimate). Each square has an area equal to (1 psf) (0.175 rad). Thus,

$$\int_0^{\pi} \Delta p \cos\theta \; d\theta = 34(1)(0.175) = 5.9\,\text{psf} = \frac{D_f}{DL}$$

$$(\text{c})\; D_f = 5.9(3.75/12)(10.5/12) = 1.62\,\text{lbf} = D_f$$

$$(\text{d})\; C_D = \frac{D_f}{\frac{1}{2}\rho U_\infty 2(DL)} = \frac{5.9}{\frac{1}{2}(0.00232)(53.5)2} = 1.78$$

$$\text{Re} = \frac{\rho U_\infty D}{\mu} = \frac{0.00232(53.5)(3.75/12)}{0.3801\times 10^{-6}} = 1.0\times 10^5$$

These values are in the same range as the curve in the text, but do not match exactly.

For another illustration of the concepts of skin friction drag and form drag, examine two cases of flow past a flat plate (Figure 6.11). In one case, the plate is aligned parallel to the flow, and there is no separation from the boundary. Here the total drag on the plate is due to skin friction. In the second case, the plate is normal to the flow. Separation occurs, a wake forms, and the drag exerted on the plate is due primarily to the pressure difference from front to back.

The difference between a laminar and a turbulent velocity profile has a significant effect on where separation occurs, and on the drag force exerted. The turbulent profile can offer more resistance to an adverse pressure gradient. For a turbulent boundary layer, we therefore expect that separation will occur farther downstream than that for a laminar boundary layer. To see how the location of the point of separation affects flow past a cylinder, refer to Figure 6.12. In the laminar case, the boundary layer remains laminar to the point of separation. In the turbulent case, the laminar boundary layer experiences a transition to a turbulent boundary layer; as a result, the point of separation is moved farther downstream on the cylinder surface, and the form drag is thus reduced. A common technique of inducing transition is to roughen the surface of the object. A familiar example for a sphere is the surface of a golf ball.

Figure 6.13 is a graph of drag coefficient as a function of Reynolds number for a cylinder having a large length-to-diameter ratio. The drag coefficient decreases steadily from 60 to 1 as the Reynolds number increases from 10^{-1} to 10^3. For a Reynolds number in the range 10^3–10^4, C_D remains

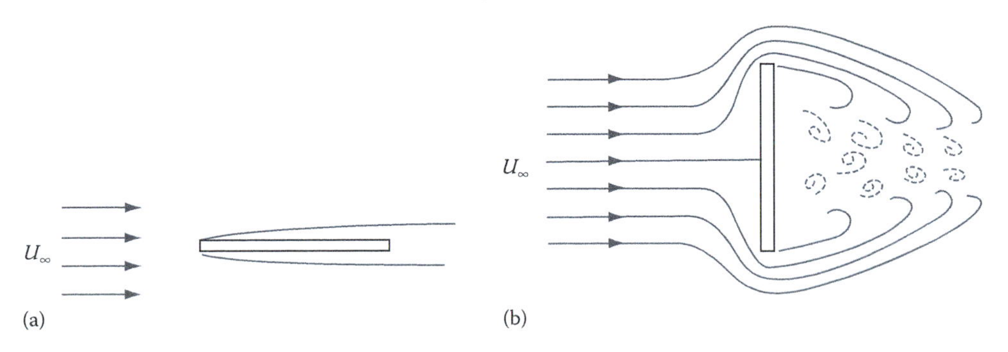

FIGURE 6.11 Two cases of flow past a flat plate. (a) Plate aligned with flow direction and (b) plate normal to flow direction.

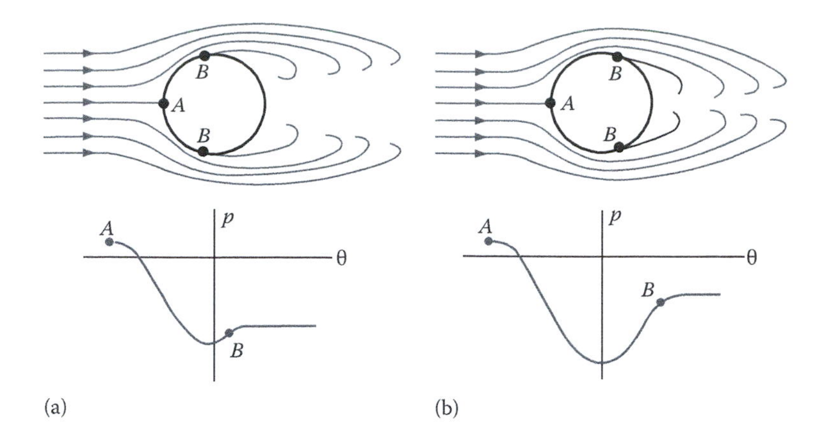

FIGURE 6.12 A comparison of (a) laminar and (b) turbulent separation effects.

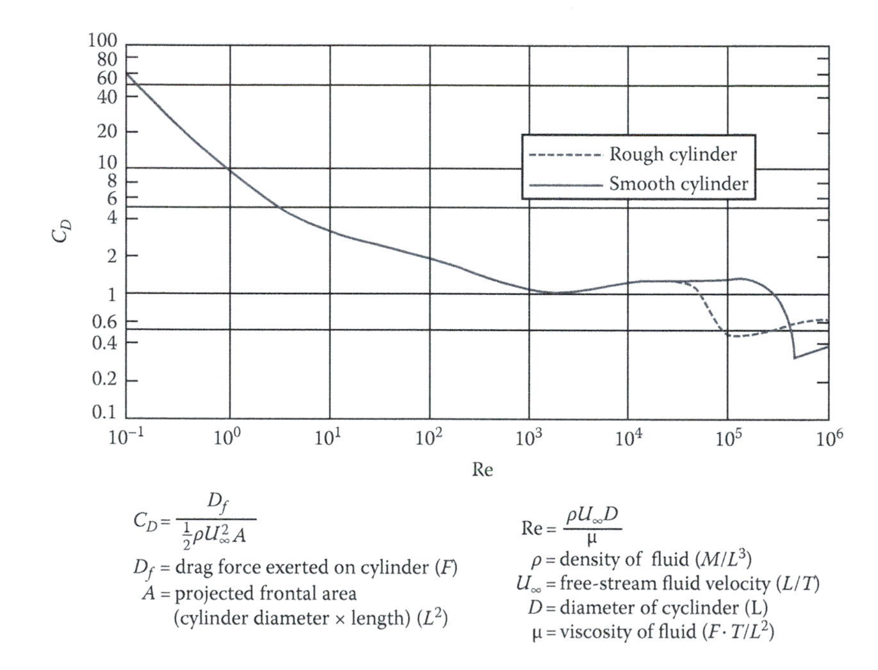

FIGURE 6.13 Drag coefficient as a function of Reynolds number for long circular cylinders. (Adapted from Schlichting, H., *Boundary Layer Theory*, 7th ed., McGraw-Hill, Inc., New York, 1979. With permission.)

approximately constant. At a Reynolds number greater than 4×10^4, a roughened cylinder exhibits a different drag coefficient curve than does a smooth cylinder. The entire plot can be obtained from experimental results measured with the setup shown in Figure 6.14.

Example 6.3

A flagpole consists of two telescoping cylinders. The smaller cylinder is made of 3½ nominal, schedule 40 galvanized pipe, which is 6 ft long. This pipe just fits into 4-nominal, schedule 40 pipe (also galvanized). The pipes are welded together end to end to form a flagpole that is to extend 12 ft above ground. Local safety requirements stipulate that the flagpole must

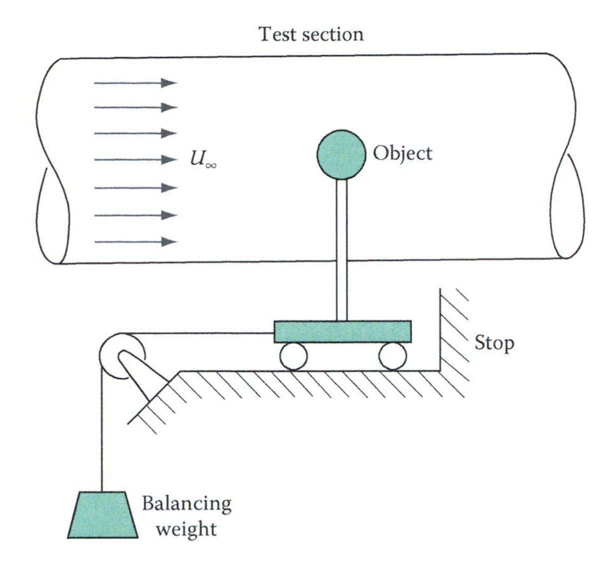

FIGURE 6.14 An experimental setup for measuring drag on an object immersed in a uniform flow of velocity U_∞.

withstand 100 mi/h winds. Before deciding on how deep to bury the lower portion of the pole, it is necessary to calculate forces due to the wind.

a. Determine the total drag force exerted on the installed flagpole for a wind speed of 80 mi/h.
b. Determine the moment at the base of the pole exerted by these forces. Assume a temperature of 71°F.

Solution

We can determine the drag forces acting on the flagpole by using Figure 6.13. We use the property tables to find

Air at 71°F	$\rho = 0.00232\ \text{slug/ft}^3$	
	$\mu = 0.3801 \times 10^{-6}\ \text{lbf} \cdot \text{s/ft}^2$	Table A.3
3½ - nom, sch 40	OD = 4 in. = 0.333 ft	Table C.1
4 - nom, sch 40	OD = 4.5 in. = 0.375 ft	Table C.1

The flow velocity is

$$U_\infty = \frac{100\ \text{mi}}{1\ \text{h}} \frac{5280\ \text{ft}}{1\ \text{mi}} \frac{1\ \text{h}}{3600\ \text{s}} = 147\ \text{ft/s}$$

The Reynolds number of the flow past the $3\frac{1}{2}$-nominal pipe is

$$\text{Re}_3 = \frac{\rho U_\infty D}{\mu} = \frac{0.00232(147)(0.333)}{0.3801 \times 10^{-6}} = 2.99 \times 10^5$$

Similarly, for the 4-nominal pipe,

$$\text{Re}_4 = \frac{\rho U_\infty D}{\mu} = \frac{0.00232(147)(0.375)}{0.3801 \times 10^{-6}} = 3.37 \times 10^5$$

For these Reynolds numbers, Figure 6.13 gives

$$C_{D3} = 1.2$$

$$C_{D4} = 1.1$$

where the smooth cylinder curve was selected for a galvanized surface. (It is difficult to read the figure with great accuracy.) By definition,

$$C_D = \frac{D_f}{\frac{1}{2}\rho U_\infty^2 A}$$

which is rearranged to solve for the drag force:

$$D_f = C_D \frac{1}{2}\rho U_\infty^2 A$$

The area A is the projected frontal area. A cylinder viewed from the direction of the approach flow appears as a rectangle whose area is length times diameter. For the pipes of this example,

$$A_3 = LD|_3 = 6(0.333) = 2.0 \text{ ft}^2$$

$$A_4 = LD|_4 = 6(0.375) = 2.25 \text{ ft}^2$$

a. Substituting, we find the drag force for each as

$$D_{f3} = \frac{1.2}{2}(0.00232)(147)^2(2.0)$$

$$D_{f3} = 60.2 \text{ lbf}$$

and

$$D_{f4} = \frac{1.1}{2}(0.00232)(147)^2(2.25)$$

$$D_{f4} = 62.0 \text{ lbf}$$

b. The moment exerted at the base will equal the sum of force times distance for both pipes. We take each force to act at the centroid of the projected frontal area. So the force exerted on the $3\frac{1}{2}$-nominal pipe acts at a distance of 9 ft (= 6 + 6/2) from the base. The force exerted on the 4-nominal pipe acts at a distance of 3 ft (= 6/2) from the base. The moment, then, is

$$\sum M_{\text{base}} = 60.2(9) + 62.0(3)$$

$$\sum M_{\text{base}} = 728 \text{ ft lbf}$$

Drag coefficient versus Reynolds number is available for other two-dimensional shapes; these data are provided in Figure 6.15.

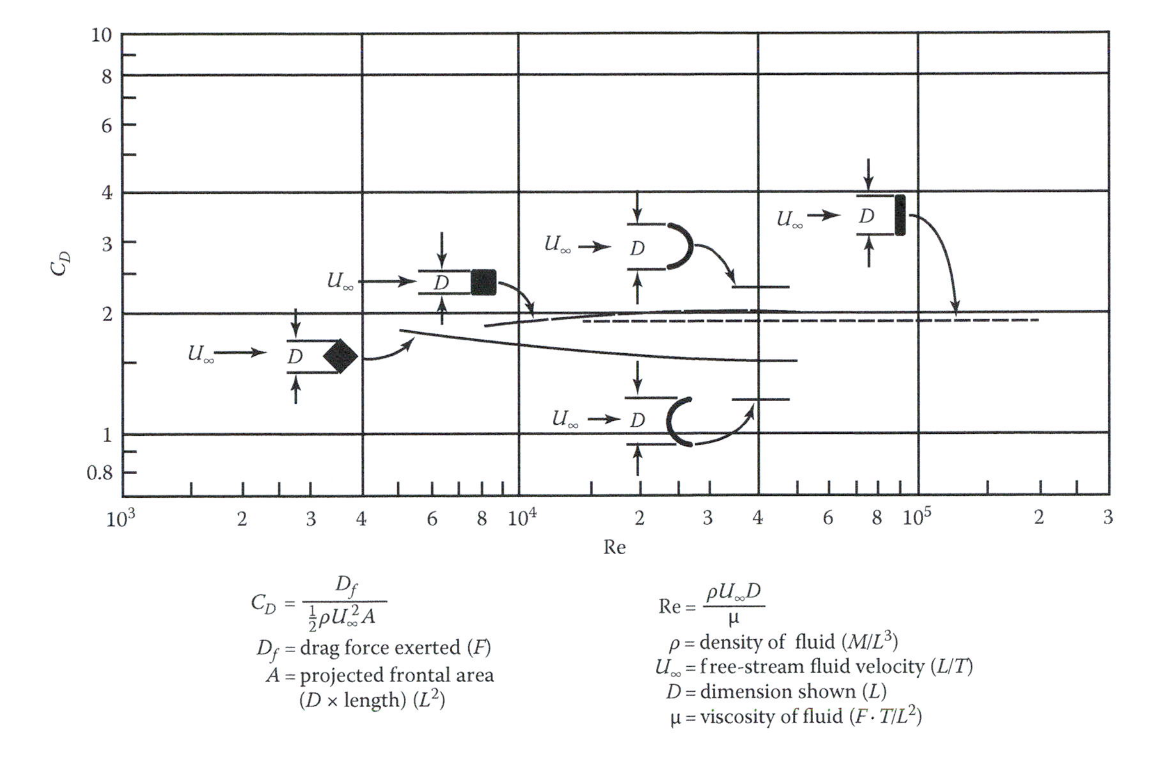

FIGURE 6.15 Drag coefficient of various two-dimensional bodies. (Data from several sources; see references at end of this text.)

Example 6.4

A 6 m-tall utility pole used for a streetlight is square in cross section. At its base, the pole is 15 × 15 cm, and at the top it is 7 × 7 cm. The cross section tapers linearly from base to top. The wind velocity past the pole is 30 m/s. Determine the drag force exerted on the pole for both square cylinder configurations illustrated in Figure 6.15. Take the air temperature to be 25°C.

Solution

The pole is tapered, but we can estimate the force acting on the entire pole by using dimensions at its midpoint. We use Table A.3 to find

$$\text{Air at } 26.85°\text{C} \quad \rho = 1.177\ \text{kg/m}^3 \quad \mu = 18.46 \times 10^{-6}\ \text{N}\cdot\text{s/m}^2$$

At the midpoint of the pole, its dimension is 11 × 11 cm [= (15 + 7)/2]. We perform the calculations for two configurations that have these dimensions. The first is for $D_1 = 0.11$ m. The second is for $D_2 = 0.11/\cos 45° = 0.156$ m. The Reynolds numbers are calculated as

$$\mathrm{Re}_1 = \frac{\rho U_\infty D_1}{\mu} = \frac{1.177(30)(0.11)}{18.46 \times 10^{-6}} = 2.1 \times 10^5$$

$$\mathrm{Re}_2 = \frac{1.177(30)(0.156)}{18.46 \times 10^{-6}} = 3.0 \times 10^5$$

From Figure 6.15, we read

$$C_{D1} = 2$$
$$C_{D2} = 1.5$$

By definition,

$$C_D = \frac{D_f}{\frac{1}{2}\rho U_\infty^2 A}$$

which is rearranged to solve for the drag force:

$$D_f = C_D \frac{1}{2}\rho U_\infty^2 A$$

For the first configuration, the area is

$$A_1 = 0.11(6) = 0.66 \text{ m}^2$$

Likewise,

$$A_2 = 0.156(6) = 0.93 \text{ m}^3$$

The drag forces are calculated as

$$D_{f1} = \frac{2}{2}(1.177)(30)^2(0.66)$$

$$D_{f1} = 700 \text{ N}$$

and

$$D_{f2} = \frac{1.5}{2}(1.177)(30)^2(0.93)$$

$$D_{f2} = 740 \text{ N}$$

6.3 FLOW PAST VARIOUS THREE-DIMENSIONAL BODIES

This section presents data on drag coefficients for several three-dimensional bodies. A curve for a sphere is given in Figure 6.16. Note that the drag coefficient decreases from more than 200 to 0.4 over the Reynolds number range of 10^{-1} to 10^4. The drag coefficient is nearly constant for Reynolds numbers from 10^3 to 2×10^5, but there is a sudden drop in drag coefficient beyond 2×10^5. An analytic solution has been derived by Stokes for flow about a sphere. The analysis shows that for Re < 1,

$$C_D = \frac{24}{\text{Re}} \tag{6.10}$$

where $\text{Re} = \rho U_\infty D/\mu$.

A second graph of drag data on spheres is presented in Figure 6.17. The vertical axis of this graph, drag coefficient C_D, is identical to that of Figure 6.16. The horizontal axis, however, is of a different parameter: $\text{Re}\sqrt{C_D}$. The curve of Figure 6.17 extends from about 2 on the horizontal axis to about 4×10^5 and looks similar to that of Figure 6.16.

Figures 6.16 and 6.17 are useful when making calculations on finding what is known as **terminal velocity** for spheres. When a body travels steadily through a fluid (acceleration is zero) and the

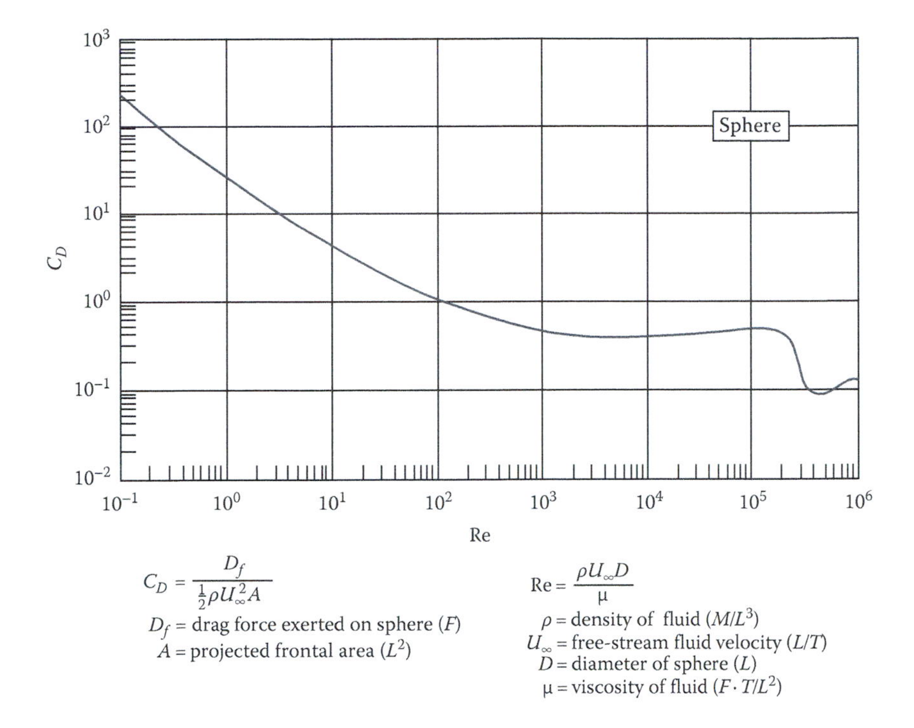

FIGURE 6.16 Drag coefficient as a function of Reynolds number for spheres. (Adapted from Schlichting, H., *Boundary Layer Theory*, 7th ed., McGraw-Hill, Inc., New York, 1979. With permission.)

forces acting are in equilibrium, the body is said to be traveling at a constant terminal velocity. One familiar example of this phenomenon is a parachutist. There are other examples: a sphere, such as a ball bearing falling through liquid; pulverized coal particles suspended in an airstream that is traveling in a pipe to a boiler; and liquid fuel particles sprayed into the airstream of an engine carburetor. The terminal velocity of a particle can be determined from the data presented in this chapter. In the case of a sphere, finding its terminal velocity using Figure 6.16 involves a trial-and-error calculation method. Trial and error is avoided when using Figure 6.17. Both graphs are merely different forms of the same data.

Consider a sphere of known properties falling through a fluid at terminal velocity. The forces acting on the sphere are due to drag, buoyancy, and gravity, as shown in Figure 6.18. Applying Newton's law, we have

$$\sum F = ma = 0$$

because the sphere is not accelerating. By substitution, we get

$$mg - \rho g \forall - \frac{1}{2} C_D \rho V^2 A = 0$$

Now the sphere volume is $\forall = \pi D^3/6$, and the projected frontal area is $A = \pi D^2/4$. If the sphere density ρ_s is known, the preceding equation can be rewritten as

$$\rho_s g \frac{\pi D^3}{6} - \rho g \frac{\pi D^3}{6} - \frac{1}{2} C_D \rho V^2 \frac{\pi D^2}{4} = 0 \qquad (6.11a)$$

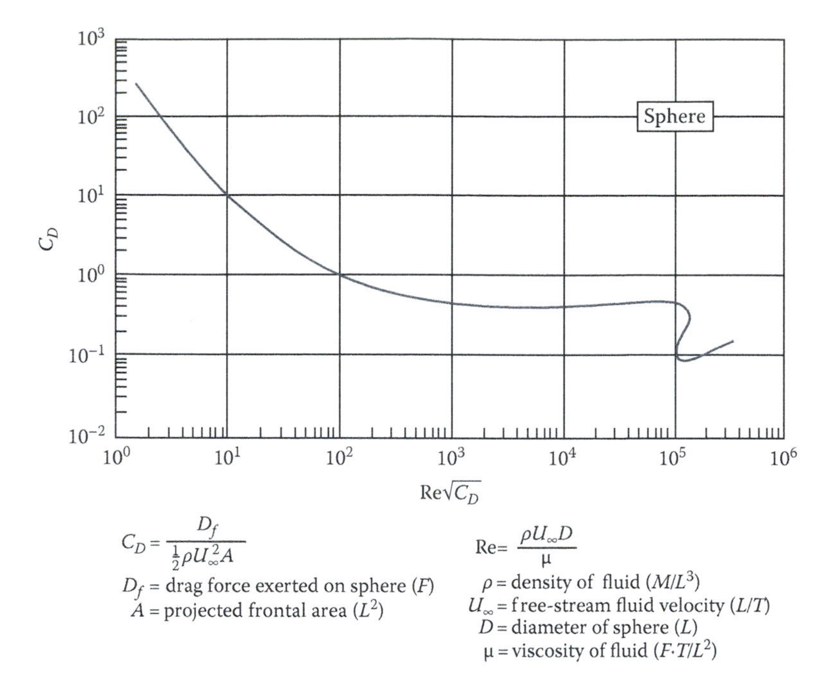

FIGURE 6.17 Drag coefficient as a function of $\mathrm{Re}\sqrt{C_D}$ for spheres. (Note on how this figure was developed: an enlarged copy of Figure 6.16 was scanned into a computer. The drawing was scaled and digitized to yield tabular data C_D versus Re. The data were then manipulated appropriately to obtain $\mathrm{Re}\sqrt{C_D}$, which was graphed versus C_D.)

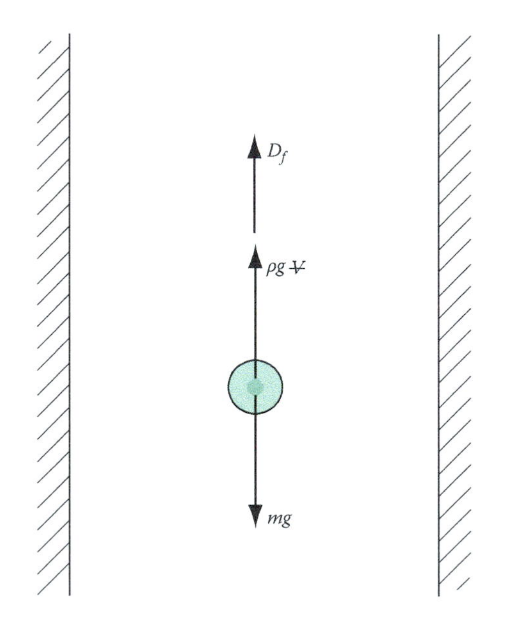

FIGURE 6.18 Forces acting on a sphere falling at terminal velocity in a fluid.

Solving for the terminal velocity, we obtain

$$V = \left[\frac{4}{3}\frac{gD}{C_D}\left(\frac{\rho_s}{\rho} - 1\right)\right]^{1/2} \tag{6.11b}$$

The difficulty in using this equation is that the drag coefficient C_D depends on the terminal velocity, which is not known. The technique for finding velocity is illustrated in the following example.

Example 6.5

A 1.3 cm-diameter marble is falling through a tank of water at 25°C. Calculate the terminal velocity of the marble, assuming that it has the same properties as glass.

Solution

From Table A.7, for glass, $s = 2.6$ (the average value). From Table A.5, for water, $\rho = 1\ 000$ kg/m^3 and $\mu = 0.89 \times 10^{-3}$ N·s/m^2. By substitution into Equation 6.11b, we get

$$V = \left[\frac{4}{3}\frac{gD}{C_D}\left(\frac{\rho_s}{\rho} - 1\right)\right]^{1/2}$$

$$= \left\{\frac{4}{3}\frac{(9.81)(0.013)}{C_D}\left[\frac{2.6(1\ 000)}{1\ 000} - 1\right]\right\}^{1/2}$$

$$= \frac{0.521\ 6}{\sqrt{C_D}} \tag{i}$$

Also,

$$\text{Re} = \frac{\rho VD}{\mu} = \frac{1\ 000(V)(0.013)}{0.89 \times 10^{-3}} = 1.46 \times 10^4\ \text{V}$$

Assume $C_D = 1.0$; then $V = 0.522$ m/s and Re = 7.6 × 10^3. From Figure 6.16 at Re = 7.6 × 10^3, we find $C_D = 0.4$, which is our second trial value. Therefore, assume that $C_D = 0.4$; then $V = 0.825$ m/s and Re = 1.21 × 10^4. From Figure 6.16 at Re = 1.21 × 10^4, we find $C_D = 0.4$, which checks with the assumed value. Thus, the terminal velocity of the marble is

$$V = 0.825 \text{ m/s}$$

The trial-and-error process can be avoided by using Figure 6.17. We begin in the usual way and derive Equation i:

$$V\sqrt{C_D} = 0.521\ 6 \tag{i}$$

Next, we multiply both sides by $\rho D/\mu$ to obtain

$$\frac{\rho VD}{\mu}\sqrt{C_D} = 0.5216\frac{1\ 000(0.013)}{0.89 \times 10^{-3}}$$

or

$$\text{Re}\sqrt{C_D} = 7.62 \times 10^3$$

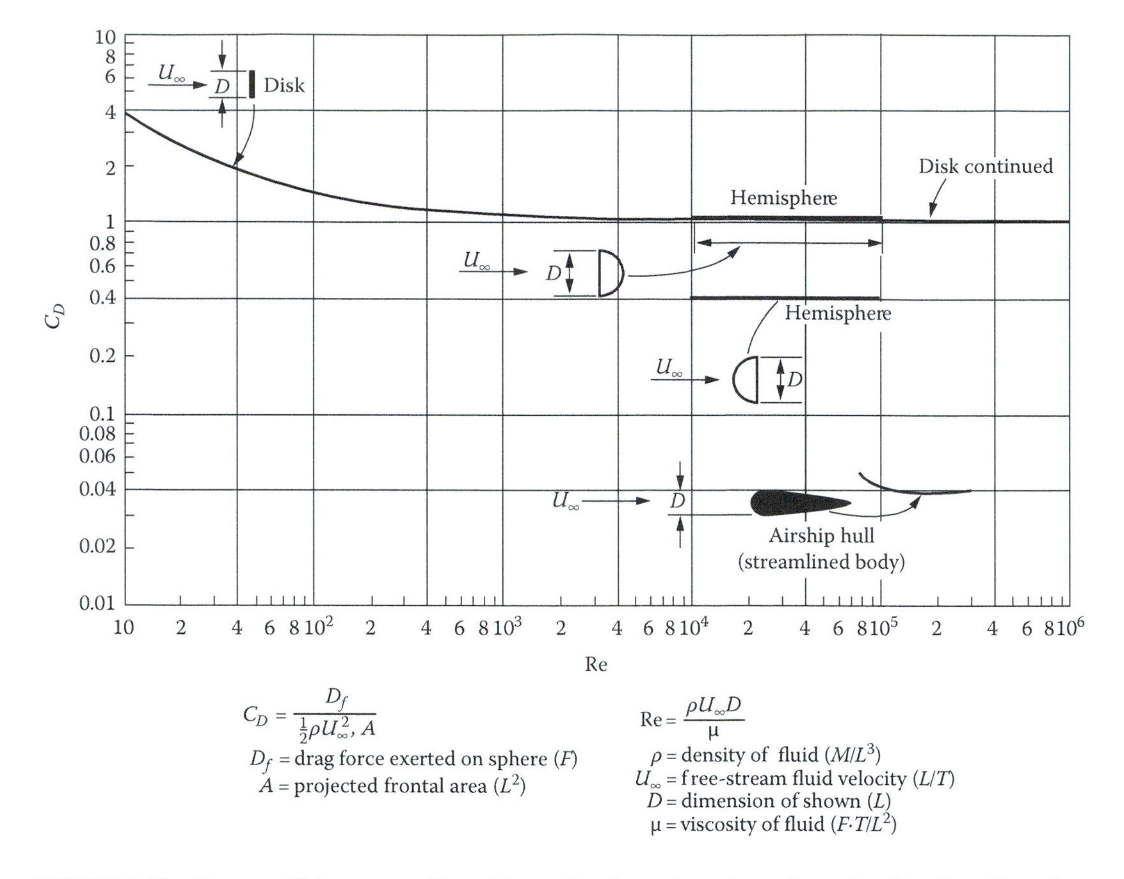

FIGURE 6.19 Drag coefficient versus Reynolds number for various three-dimensional bodies. (Data from various sources; see references at end of text.)

At this value, Figure 6.17 gives

$$C_D = 0.4$$

which leads to the same result obtained after several trials with Figure 6.16.

Curves of drag coefficient versus Reynolds number for various three-dimensional bodies are provided in Figure 6.19. Data for a disk, two hemispheres, and a streamlined body of revolution are shown.

6.4 APPLICATIONS TO GROUND VEHICLES

As mentioned in Section 6.2, the total drag force exerted on an object immersed in a uniform flow is made of two components: skin friction drag and form (or pressure) drag. In this section, we discuss how these two components affect ground vehicles. Specifically, we shall examine bicycle–rider combinations, automobiles, and tractor–trailer trucks.

First, it is necessary to define the concept of *streamlining*. A ground vehicle requires power to move it over land. A portion of this power goes to overcoming the rolling resistance offered by friction between the tires (or wheels) and the road, and by friction in bearings or other surfaces moving with respect to each other. Another portion of the power required goes to overcoming the drag encountered by the vehicle as it moves through air. With ground vehicles, it will in general not be possible to significantly reduce the skin friction drag. We can, however, modify the shape of the vehicle so that the form or pressure drag is reduced. When the shape is modified and the pressure drag is reduced, the object is said to be **streamlined**.

Consider, for example, the streamlined body of revolution in Figure 6.19. The thickness of the rear half of the object gradually decreases with length. Because the flow no longer has an abrupt turn to make, separation occurs over only a small portion of the trailing edge. Thus, form drag has been minimized, and most of the drag exerted on the streamlined shape is skin friction drag. Note that a typical airship hull is shaped like the streamlined body of revolution of Figure 6.19.

6.4.1 Bicycle–Rider Combinations

The design of a conventional bicycle remained almost unchanged for nearly a century. Aerodynamically, a bicycle–rider combination is not considered to be streamlined. A tremendous wake exists behind the bicycle and especially the rider, which contributes to a substantial form or pressure drag. Designers have long recognized that air resistance is a significant factor, but constraints have been placed on design that have prevented streamlining of the vehicle. Until 1900, the crouched position commonly used with downturned handlebars was an accepted means of reducing air resistance. The multiple-rider bicycle was placed ahead of a single rider to reduce the wind resistance encountered by the single rider. After 1900, a streamlined enclosure for bicycle riders, resembling a dirigible, was patented. Human-powered vehicles set speed records with this enclosure and others like it. As a result, streamlined bicycles have been pedaled to speeds in excess of 60 mi/h (27 m/s). (The record for an unaided bicycle–rider combination is 43.45 mi/h (19.42 m/s), set by a world-class racing cyclist.) Also in an effort to reduce wind resistance, a recumbent bicycle, in which the rider pedaled while in a reclined position, was built and later streamlined.

Figure 6.20 shows some of the early designs of aerodynamic devices for bicycles. Note that in all cases, there is a gradual reduction in the thickness of the trailing edge, the objective being to reduce the size of the wake behind the vehicle. Figure 6.21 shows a few of the more conventional designs.

As we have seen, a rider must overcome rolling resistance and air resistance to move the vehicle. Neglecting rolling resistance for the moment, we know that the drag force increases with the square of the velocity. Power is proportional to the product of drag force and velocity, or, in other words, power is proportional to the cube of the velocity. Riding at 20 mi/h (8.9 m/s), for example, requires eight times the power required to ride at 10 mi/h (4.5 m/s). Studies have shown that a well-trained athlete can produce 1 hp (750 W) for about 30 s and about 0.4 hp (300 W) for about 8 h. A healthy individual can sustain 1 hp (750 W) of output for about 12 s and roughly 0.1 hp (75 W) for 8 h.

Table 6.1 shows aerodynamic data for a number of bicycle–rider combinations. Drag coefficient and rolling resistance data are given for a speed of 20 mi/h (8.9 m/s). Several trends can be noticed from the data. First, in comparing the "arms straight" (no. 2) position to the "fully crouched" (no. 3) position, we see that the frontal area, the drag coefficient, and the drag force are all smaller when the rider is in the crouched position. Second, when the components—brake calipers, crankset, rims, and so on—are streamlined, there is only a small reduction in the drag force (compare no. 3 to no. 4) over the unstreamlined case. Third, the technique of riding in the wake of another rider—called drafting, no. 7—provides a decrease in the drag force (compare no. 3 to no. 7). Fourth, aerodynamic devices (nos. 5, 8, and 9) yield substantial reductions in drag force over the unstreamlined rider (no. 1). Other trends can be seen as well.

Figure 6.22 gives data on drag coefficient versus Reynolds number for bicycle–rider combinations with and without streamlining. Streamlining in this case is achieved with a wing-shaped shell that fits over the rider. From a top view, the fairing appears to be like the streamlined body of revolution in Figure 6.19.

Once the drag force (or the total resisting force, which equals drag + rolling resistance) is known, the power required to maintain speed is given by

$$\frac{dW}{dt} = D_f U_\infty \tag{6.12}$$

Note that in Table 6.1 data are given both for drag force and for rolling resistance.

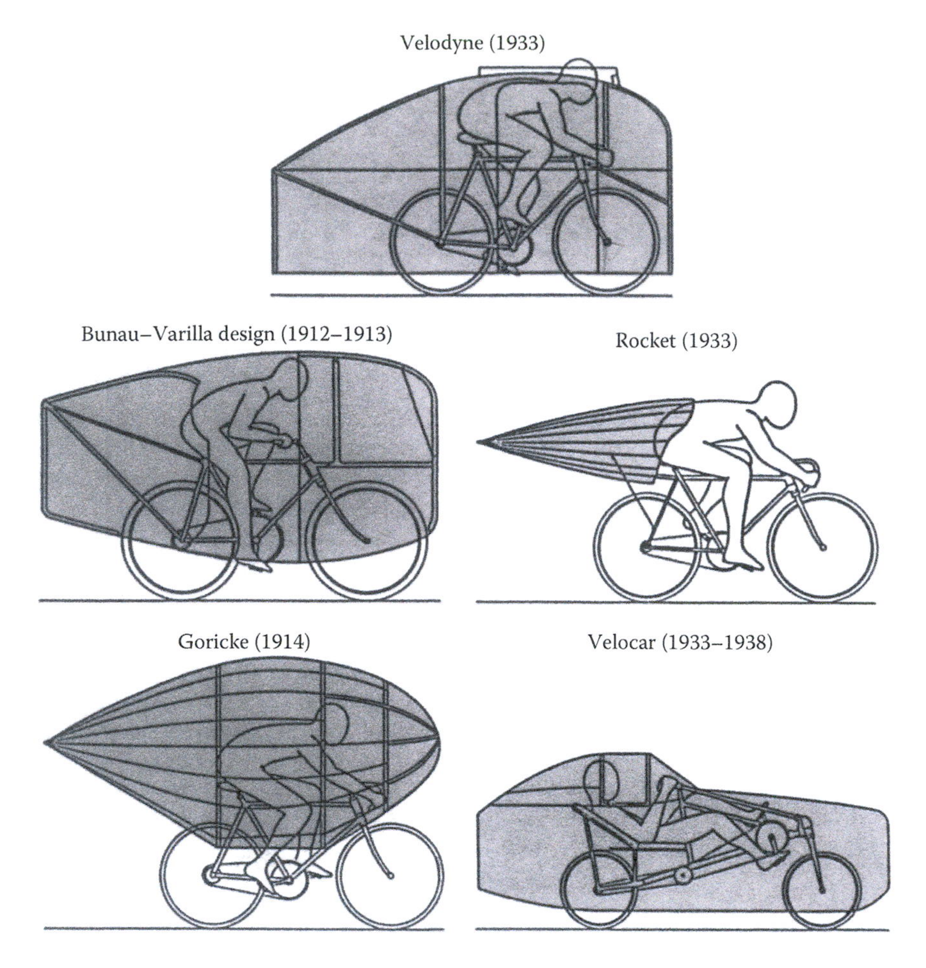

FIGURE 6.20 Early designs of aerodynamic devices for human-powered vehicles. Dates indicate the year of invention or the year when speed records were set. (Reprinted from Gross, A.C. et al., *Sci. Am.*, 249, 142, December 1983. With permission.)

Example 6.6

A cyclist rides at 20 mi/h, and 20% of her output goes into overcoming rolling resistance. What power is required to maintain this speed if the "not-streamlined" curve of Figure 6.22 applies?

Solution

Figure 6.22 was prepared by using data listed with the figure. We assume the rider of this example has similar body features. Our result must, therefore, be an estimate at best. The rider velocity is

$$U_\infty = \frac{15\,\text{mi}}{1\,\text{h}} \cdot \frac{5280}{3600} = 22.0\,\text{ft/s}$$

The Reynolds number for use with Figure 6.22 is

$$\text{Re} = \frac{U_\infty D}{\nu} = \frac{22.0(5.42)}{1.69 \times 10^{-4}} = 7.1 \times 10^5$$

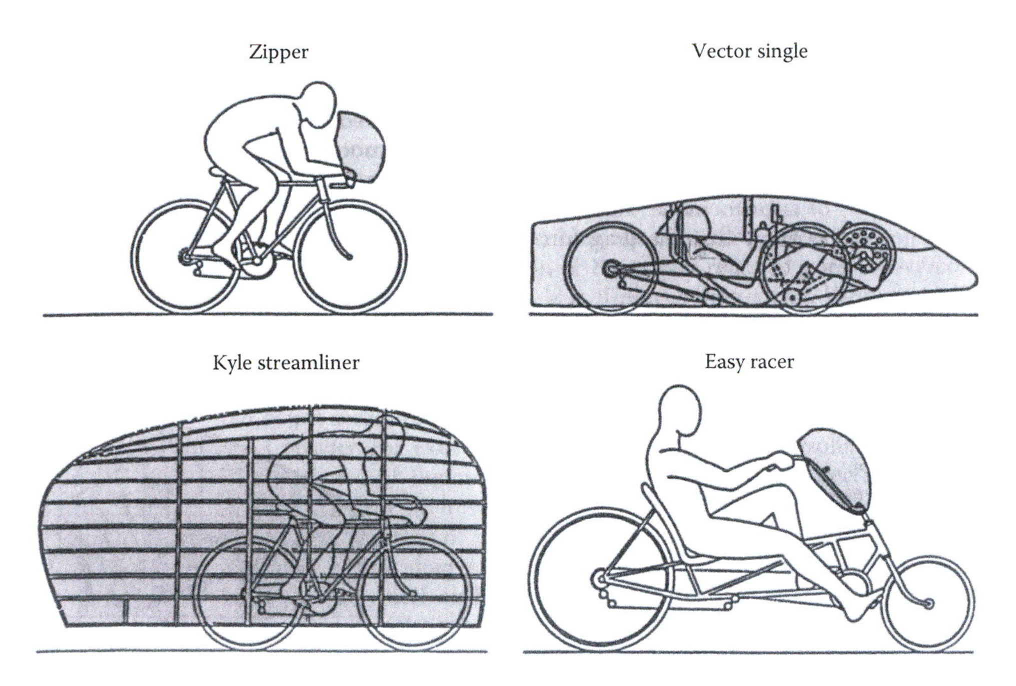

FIGURE 6.21 Conventional designs of aerodynamic devices for human-powered vehicles. (Reprinted from Gross, A.C. et al., *Sci. Am.*, 249, 142, December 1983. With permission.)

The drag coefficient is read as

$$C_D = \frac{D_f}{\frac{1}{2}\rho U_\infty^2 A} \approx 0.95$$

The drag force becomes

$$D_f = \frac{0.95}{2}(0.0022)(22.0)^2(5.87) = 2.97\,\text{lbf}$$

The total force the rider must overcome includes the 2.97 lbf drag force plus 20% of the total for rolling resistance. Thus,

$$F_{\text{total}} = 2.97 + 0.2F_{\text{total}}$$

or

$$0.80F_{\text{total}} = 2.97$$

Solving,

$$F_{\text{total}} = 3.71\ \text{lbf}$$

The power required is force times velocity:

$$\frac{dW}{dt} = 3.71(22.0) = 81.7\,\text{ft}\cdot\text{lbf/s}$$

TABLE 6.1

Aerodynamic and Rolling Resistance Data for Several Bicycle–Rider Configurations Traveling at a Speed of 20 mi/h (8.9 m/s)

Configuration		Drag Force, D_f		Rolling Resistance		Drag Coefficient	Frontal Area, A	
		lbg	N	lbg	N	$C_D = \dfrac{Df}{\frac{1}{2}\rho U_\infty^2 A}$	ft²	m²
European upright commuter	40 lb bike, 160 lb rider, tires: 27 in. diameter, 90 psi	6.14	27.3	1.20	5.34	1.1	5.5	0.51
Touring (arms straight)	25 lb bike, 160 lb rider, tires: 27 in. diameter, 90 psi	4.40	19.6	0.33	3.69	1.0	4.3	0.40
Racing (fully crouched)	20 lb bike, 160 lb rider, tires: 27 in. diameter, 105 psi	3.48	15.5	0.54	2.4	0.88	3.9	0.36
Aerodynamic components (fully crouched)	20 lb bike, 160 lb rider, tires: 27 in., diameter, 105 psi	3.17	14.5	0.54	2.4	0.83	3.9	0.36
Partial fairing (zipper, crouched)	21 lb bike, 160 lb rider, tires: 27 in. diameter, 105 psi	2.97	13.2	0.54	2.4	0.70	4.1	0.38
Recumbent (Easy Racer)	27 lb bike, 160 lb rider, tires: 20 in. front, 27 in. rear, 90 psi	2.97	13.2	0.94	4.2	0.77	3.8	0.35
Drafting (closely following another bicycle)	20 lb bike, 160 lb rider, tires: 27 in. diameter, 105 psi	1.94	8.63	0.54	2.4	0.50	3.9	0.36
Blue bell (two wheels one rider)	40 lb bike, 160 lb rider, tires: 20 in. front, 27 in. rear, 105 psi	0.61	2.7	0.8	4	0.12	5.0	0.46
Vector single (three wheels)	68 lb bike, 160 lb rider, tires: 24 in. front, 27 in. rear	0.51	2.3	1.02	4.54	0.11	4.56	0.424

Source: Modified from Gross, A.C. et al., *Sci. Am.*, 249(6), 142, December 1983.

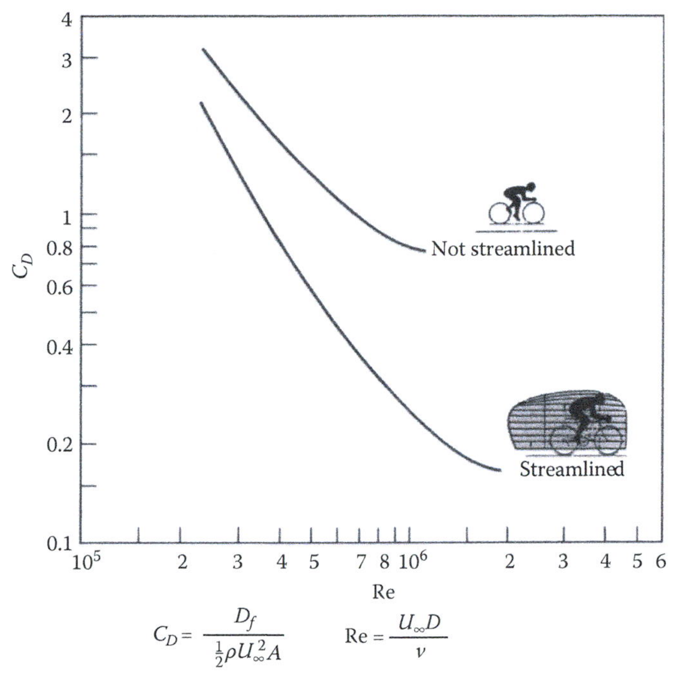

A = projected frontal area; both cases = 5.87 ft^2 = 0.545 m^2
D = height of each rider = 5.42 ft = 1.65 m
ρ = 0.002 2 slug/ft^3 = 1.13 kg/m^3
$\nu = 1.69 \times 10^{-4}$ ft^2/s = 1.54×10^{-5} m^2/s

FIGURE 6.22 Drag coefficient versus Reynolds number for streamlined and unstreamlined bicycle–rider combinations. (Data from *ASME*, 2(1), I.)

With 1 hp = 550 ft·lbf/s, we calculate

$$\frac{dW}{dt} = \frac{81.7}{550}$$

or

$$\frac{dW}{dt} = 0.15 \text{ hp}$$

6.4.2 Automobiles

For ground vehicles, the power required to propel an object through a fluid depends on the drag force, as we have seen. In the case of an automobile traveling at constant velocity, the power required to maintain speed is the sum of the power required to overcome rolling resistance between the tires and the road plus the power required to overcome aerodynamic drag. At low speeds, the predominant resistance to motion is rolling resistance. At high speeds, aerodynamic drag predominates as the resisting force.

Figure 6.23 shows a pressure profile sketched on a side view of an automobile. As with the cylinder of Figure 6.8, data are represented as arrows, and the length of each line is proportional to the magnitude of the pressure at each location. An arrow pointing away from the vehicle signifies

FIGURE 6.23 Pressure profile on an automobile.

a pressure that is less than the free-stream pressure, and the converse is also true (somewhat different than in Figure 6.8). The pressure profile shows that there is a difference in pressure from the front of the vehicle to the rear. The horizontal component of each pressure, multiplied by the area over which it acts, gives the drag force. The area of importance is the projected frontal area for the vehicle.

Table 6.2 gives data useful for estimating automobile drag coefficients. The drag coefficient for a particular body style is calculated with

$$C_D = 0.16 + 0.009\,5 \sum_i C_{Di} \tag{6.13}$$

where values of C_{Di} are obtained and summed from each of eight categories listed. It can be seen from the sketches that squared ends, where the flow must make abrupt turns, contribute greatly to the overall drag of the vehicle.

Example 6.7

Fluid Mechanics Laboratory Experiment: Measurement of Drag on Model Vehicles

Introduction: Drag force measurements on various bodies can be obtained using a subsonic wind tunnel, which can be found in most laboratories. Making measurements of drag force versus velocity using spheres, hemispheres, disks, and flat plates are classical experiments. Drag force versus velocity measurements on ground vehicles, such as bicycles, trucks and automobiles, seem to be more compelling. The difficulty with using automobile models is in finding the frontal area needed in the equation for drag coefficient. Previous studies indicate that a frontal area for a vehicle of interest is not easily obtained; however, current data on drag force or drag coefficient are widely available on the internet. Results of various studies indicate that the drag (and drag coefficient) vary with speed.

Apparatus: Figure 6.24 is a sketch of the subsonic wind tunnel used to obtain drag versus velocity data. An automobile model is placed in the test section as indicated in Figure 6.25. The automobile model is secured with a fishing line that extends outward from the front of the tunnel. The line then goes over three pulleys, and is attached to a spring scale mounted on top of the test section. Air moving past the model exerts a drag force that is measured directly with the spring scale. A differential pressure meter is attached to a static pressure tap located in the top of the test section. The meter provides a reading of pressure difference between that in the test section and atmospheric pressure. Alternatively, an inclined manometer can be used to find the pressure difference.

TABLE 6.2
Estimates of Automobile Drag Coefficient

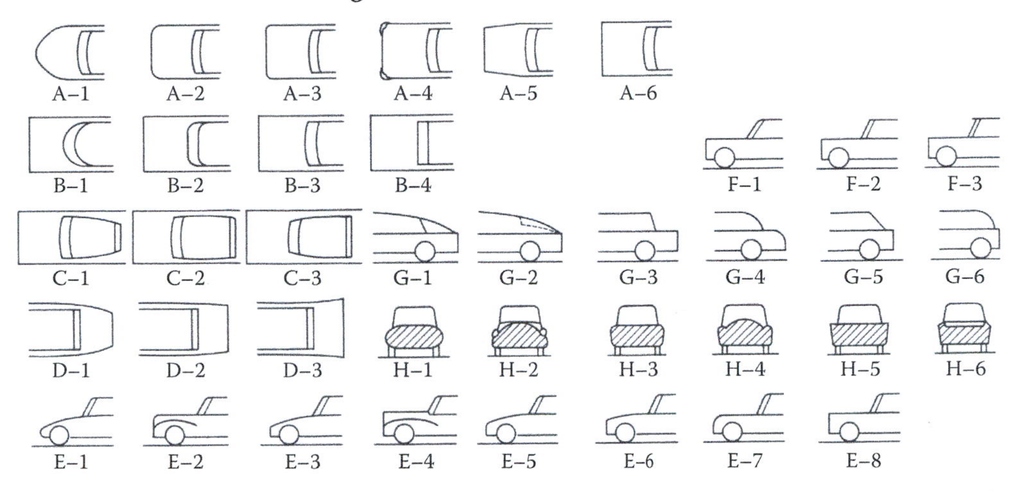

A. Plan view, front end
A–1 Approximately semi-circular (1)
A–2 Well-rounded outer quarters (2)
A–3 Rounded corners without protuberances (3)
A–4 Rounded corners with protuberances (4)
A–5 Squared tapering-in corners (5)
A–6 Squared constant-width front (6)

B. Plan view, windshield
B–1 Full wraparound (approximately semicircular) (1)
B–2 Wraparound ends (2)
B–3 Bowed (3)
B–4 Flat (4)

C. Plan view, roof
C–1 Well- or medium-tapered to rear (1)
C–2 Tapering to front and rear (maximum width at BC post) or approximately constant width (2)
C–3 Tapering to front (max. width at rear) (3)

D. Plan view, lower rear end
D–1 Well- or medium-tapered to rear (1)
D–2 Small taper to rear or constant width (2)
D–3 Outward taper (or flared-out fins) (3)

E. Side elevation, front end
E–1 Low, rounded front, sloping up (1)
E–2 High, tapered, rounded hood (1)
E–3 Low, squared front, sloping up (2)
E–4 High, tapered, squared hood (2)
E–5 Medium-height, rounded front, sloping up (3)
E–6 Medium-height, squared front, sloping up (4)
E–7 High, rounded front, with horizontal hood (4)
E–8 High, squared front, with horizontal hood (5)

F. Side elevation, windshield peak
F–1 Rounded (1)
F–2 Squared (including flanges or gutters) (2)
F–3 Forward-projecting peak (3)

G. Side elevation, rear roof/trunk
G–1 Fastback (roofline continuous to tail) (1)
G–2 Semi fastback (with discontinuity in line to tail) (2)
G–3 Squared roof with trunk rear edge squared (3)
G–4 Rounded roof with rounded trunk (4)
G–5 Squared roof with short or no trunk (4)
G–6 Rounded roof with short or no trunk (5)

H. Front elevation, cowl and fender cross section at windshield
H–1 Flush hood and fenders, well-rounded body sides (1)
H–2 High cowl, low fenders (2)
H–3 Hood flush with rounded-top fenders (3)
H–4 High cowl with rounded-top fenders (3)
H–5 Hood flush with square-edged fenders (4)
H–6 Depressed hood with high squared-edged fenders (5)

Source: Reprinted from Bolz, R.E. and Tuve, G.L. (eds.), *CRC Handbook of Table for Applied Engineering Science*, 2nd ed., CRC Press, Cleveland, OH, 1973. With permission.
Note: Drag rating values in parentheses are for use in Equation 6.13.

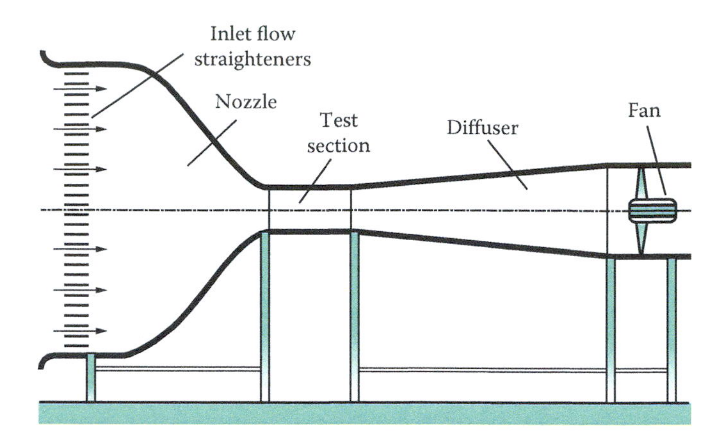

FIGURE 6.24 Subsonic wind tunnel used to make drag force measurement.

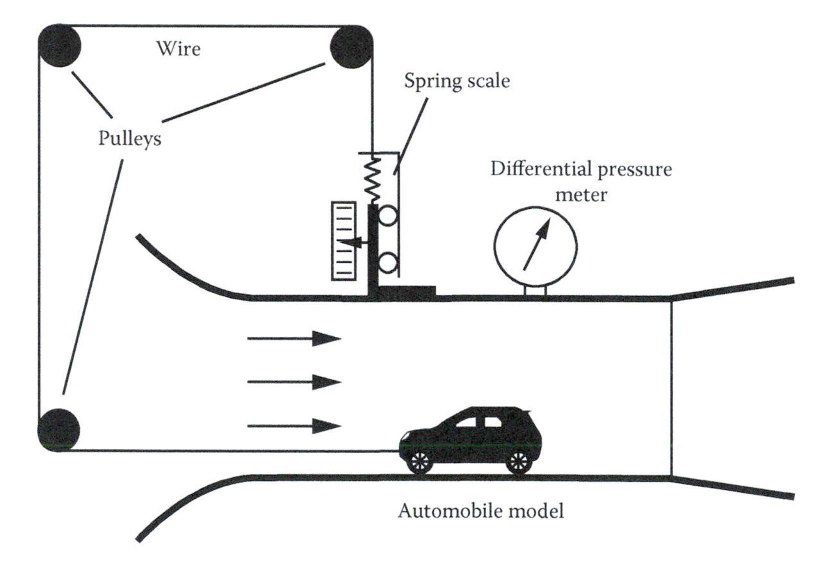

FIGURE 6.25 Sketch of the wind tunnel test section with instrumentation.

Analysis: A reading on the differential pressure meter is used with the Bernoulli equation to calculate the air speed in the test section. We identify section 1 as being far upstream of the tunnel inlet, and section 2 is at the test section where the differential pressure meter is attached (Figure 6.26). We apply the Bernoulli equation to these sections:

$$\frac{p_1}{pg} + \frac{V_1^2}{2g} + z_1 = \frac{p_2}{pg} + \frac{V_2^2}{2g} + z_2 \tag{6.14}$$

where

p_1 is the pressure far upstream $(= p_{atm})$
V_1 is the velocity upstream $(= 0)$
p_2 is pressure
V_2 is velocity, respectively, at the test section

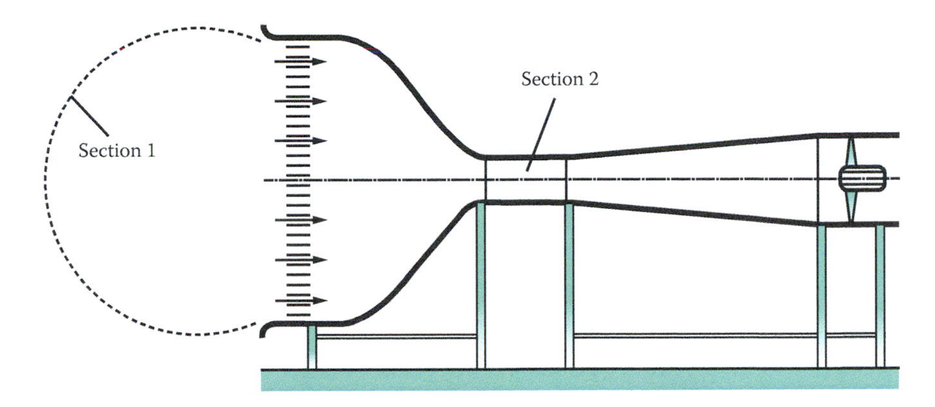

FIGURE 6.26 Sketch of the wind tunnel with sections defined.

The area section 1 approaches infinity, so the velocity there is zero and the pressure there is atmospheric. Changes in potential energy are neglected. Equation 6.14 reduces to

$$\frac{p_1}{pg} - \frac{p_2}{pg} = \frac{V_2^{\,2}}{2g}$$

From the hydrostatic equation,

$$\frac{p_1}{pg} - \frac{p_2}{pg} = \Delta h$$

Combining the preceding equations yields an expression for the velocity at the test section:

$$V_2 = \sqrt{2g\Delta h} \tag{6.15}$$

The drag force D_f is measured directly with the spring scale. The drag coefficient is defined as

$$C_d = \frac{2D_f}{\rho V_2^{\,2} A}$$

The difficulty encountered in automobile vehicle tests is in finding the frontal area to use in Equation 6.3. Often what is reported in the literature is the product of drag coefficient and area. Various mathematical packages can be used to determine area. Alternatively, however, the models used were photographed from the front in a well lit setting, and the photographs were uploaded into photo editing software. An outline of the vehicle was carefully traced and the background removed. The software measured the pixel count of the outlined area, and the frontal area was obtained. The Reynolds number is given by $Re = VD/\nu$, in which the characteristic dimension D is taken to be the bumper to bumper length of the vehicle.

FIGURE 6.27 Frontal view of the Ford F-350 model truck used in this experiment.

Results: The model vehicle used in this study is a Ford F-350 truck, scaled 1.31. The frontal area is 5.23 in.^{2} The width × length of this model is 3.15 in. × 3.18 in. The drag force velocity measurements are shown in the first two columns of the following table:

$p_t - p$ psi	Drag Force oz (N)	Δh in m	V m/s (mph)	C_D	Re
0.032	0.250 (0.070)	19.1	19.4 (43.4)	0.09	3.06×10^5
0.042	0.375 (0.104)	25.1	22.2 (49.7)	0.11	3.51×10^5
0.051	0.500 (0.139)	30.5	24.5 (54.8)	0.12	3.87×10^5
0.063	1.250 (0.348)	37.7	27.2 (60.8)	0.24	4.30×10^5
0.072	1.500 (0.417)	43.0	29.1 (65.1)	0.25	4.60×10^5
0.089	2.375 (0.660)	53.2	32.3 (72.0)	0.32	5.11×10^5
0.101	3.125 (0.869)	60.4	34.4 (77.0)	0.37	5.44×10^5
0.117	4.000 (1.112)	69.9	37.1 (83.0)	0.41	5.86×10^5
0.134	4.875 (1.355)	80.1	39.7 (88.8)	0.43	6.27×10^5

The raw data were obtained using instruments calibrated in US customary (or Engineering) units, and the results are tabulated. Equation 6.15 was used to convert the pressure change reading to a velocity in m/s. Drag coefficient and Reynolds number were also calculated and the results are given in the last two columns of the table. Figure 6.27 shows a frontal view of the vehicle, while Figure 6.28a and b is the drag-velocity and drag coefficient-Reynolds number graphs, respectively.

6.4.3 Tractor–Trailer Trucks*

Tractor–trailer trucks are used primarily for transporting goods from one location to another. The actual transportation process itself adds no tangible value to the product being shipped, although the purchaser must necessarily pay for the service. It is, therefore, important to reduce this cost in any feasible manner. One way of cutting transportation costs is to reduce the fuel consumption of the shipping vehicle, which in turn can be done by reducing the power needed to overcome drag

* This discussion is based on a report, courtesy of Professor Colin H. Marks, entitled *A Study of Aerodynamic Methods for Improving Truck Fuel Economy*, prepared by F.T. Buckley Jr., C. H. Marks, and W. H. Easton Jr. of the University of Maryland, for the National Science Foundation.

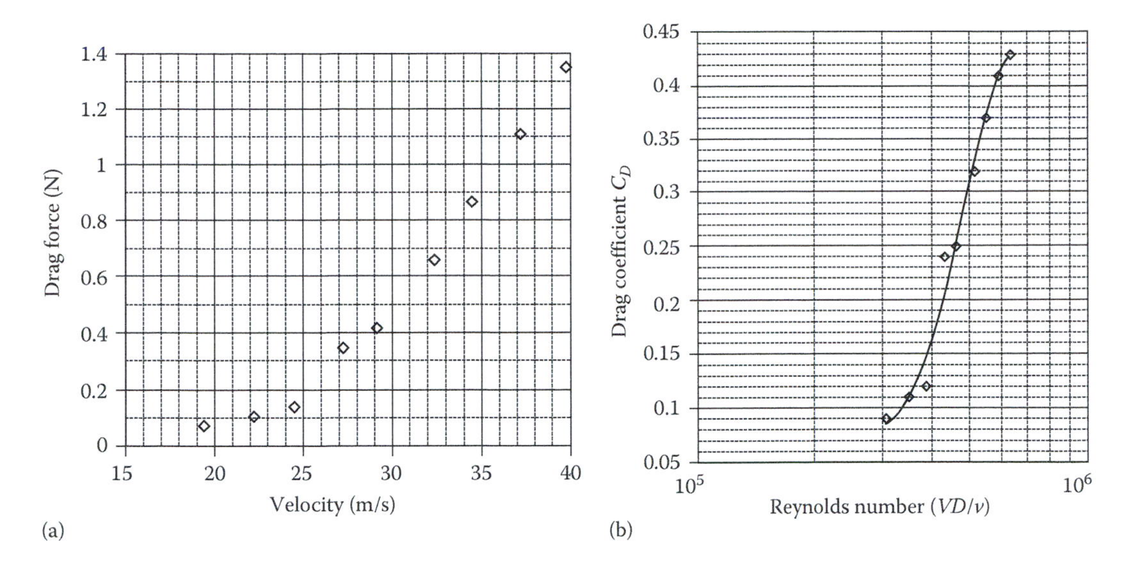

FIGURE 6.28 Graphs of the results of this experiment. Note that the Reynolds number axis is logarithmic.

forces. Thus, streamlining can become a significant cost-effective technique for reducing the price we have to pay for desired goods.

The vehicle that we consider in this discussion is a cab-over-engine tractor–trailer combination. Figure 6.29 shows plan and profile views of the vehicle as well as streamlines of airflow about the vehicle. Neglecting the wake behind the vehicle, we shall focus on the interaction between the cab and the trailer. As shown in the figure, there are a number of locations where the flow has separated from the vehicle surface. Figure 6.29 shows the vehicle moving through still air.

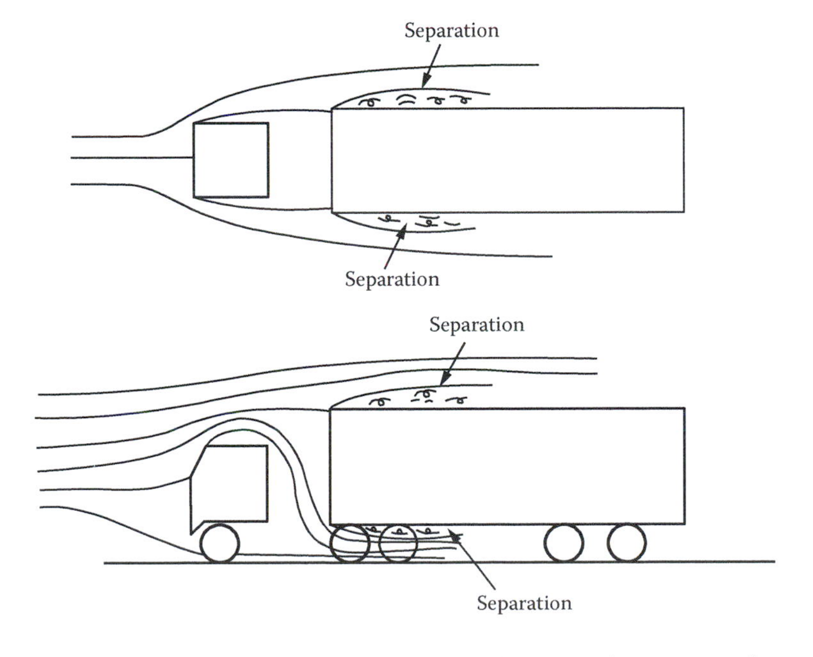

FIGURE 6.29 Plan and profile views of streamlines about a cab-over-engine tractor–trailer combination.

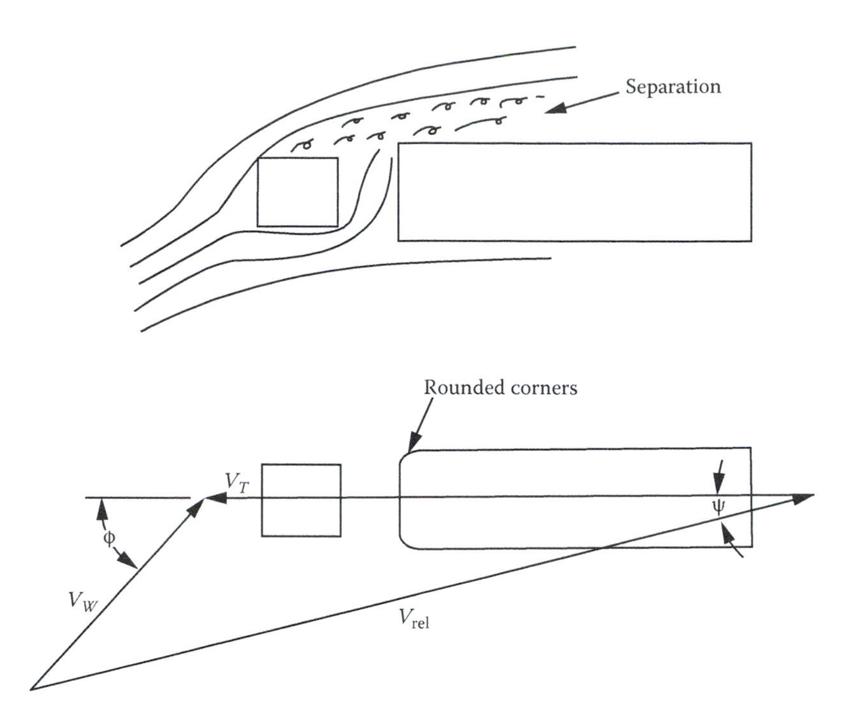

FIGURE 6.30 Wind velocity diagram for analysis of flow about a truck.

Figure 6.30 shows the flow pattern that results if there is a cross wind. Note the substantial separation region on the downstream side of the truck. In both figures, it appears that the flow patterns contain regions that contribute greatly to the drag exerted on the vehicle.

Figure 6.30 also shows a definition sketch of the velocities and angles associated with the analysis. The velocity vector labeled V_T is the forward velocity of the vehicle. The vector labeled V_W is the wind velocity, which makes an angle φ with the axis of the truck, and the angle ψ is defined as the yaw angle.

Experiments with a number of designs of aerodynamic devices have yielded useful results that lead to substantial savings in fuel costs. One small but important design change involves rounding the front corners on the trailer of the vehicle (see Figure 6.31). This change tends to reduce the separation region along the front sides of the trailer (see top portion of Figure 6.29). A second drag reduction method involves the addition of what is known as a **fairing**. Figure 6.32 shows a fairing placed on the roof of the cab. The fairing illustrated is as long and wide as the cab roof, and it extends to the height of the trailer. It contains no abrupt changes in curvature that might produce separation. Of the fairing designs currently in use or available, the one illustrated here gives the greatest reduction in drag for the cab-over-engine vehicle. A third drag reduction technique involves the use of a gap seal, as illustrated in Figure 6.31. A **gap seal** is a flat vertical piece of material that extends from the bottom of the trailer and cab to the top of the trailer and fairing. The gap seal is most effective when there is a crosswind. Recent designs of tractor trailers do not use gap seals, however. Instead, the sides of the cab are extended back and the effect is to reduce the amount of crosswind that passes through.

In the study upon which this discussion is based, a ⅛-scale model of a truck was constructed and tested in a wind tunnel. Data were recorded at Reynolds numbers of 10^6 (using the trailer width as the characteristic length) to simulate full-scale vehicles traveling at highway speeds. Underbody details, air horns, and cab roof lights were not included in the model because they have a negligible effect on the drag coefficient. The exhaust stack (see top portion of Figure 6.31) was found to have the greatest effect at a yaw angle of 0°, but its effect is not considered here.

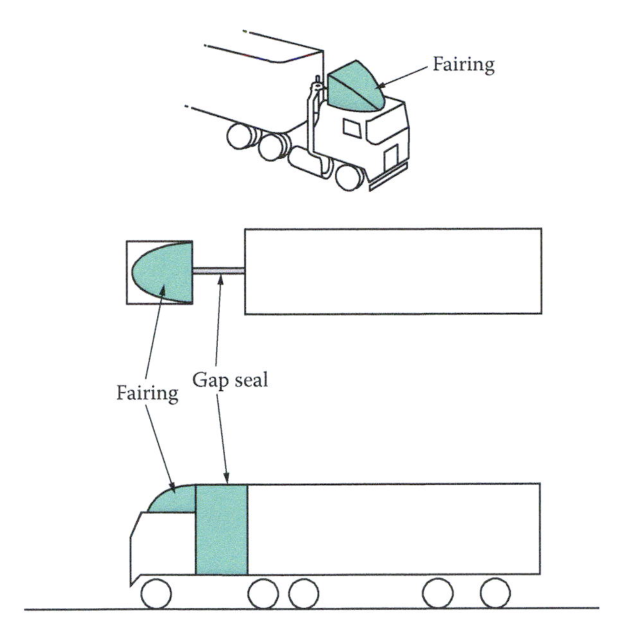

FIGURE 6.31 A fairing and gap seal attached to a tractor–trailer truck.

TABLE 6.3

Wind-Averaged Drag Coefficient $\bar{C}_D$ for a Number of Tractor–Trailer Configurations

Model Description	Basic Configuration	Seal (Figure 6.31)	With Fairing and Gap Seal (Figure 6.26)
Square-edged trailer	0.96	0.76	0.70
Round-edged trailer	0.89	0.73	0.68

Conditions:

$$\text{Re} = \frac{V_T D}{\nu} = 10^6 \text{ where } D = \text{trailer width} = 8\text{ ft} = 2.44\text{ m (full scale)}$$

$$\bar{C}_D = \frac{D_f}{\frac{1}{2}\rho V_T^2 A} \text{ where A = trailer width above ground} \times \text{width}$$

$$= 108\text{ ft}^2 = 10.0\text{ m}^2 \text{ (full scale)}$$

The problem of reducing drag on tractor–trailer trucks is in finding the effects of various aerodynamic devices. Experimental means are necessary in such evaluations, and results are expressed in terms of a percent reduction in drag rather than in absolute values. For example, Table 6.3 provides results of measurements taken using several modifications to existing vehicles.

Example 6.8

A truck having a square-edged trailer like that shown in Figure 6.31 travels about 100,000 mi/year (highway driving) and averages 5 mi/gal of fuel. The owner purchases a fairing and attaches it to the cab. By how much are the fuel costs reduced in 1 year if the owner pays \$5.00/gal of fuel?

Solution

Without any fairing, the owner's fuel cost is

$$\text{Fuel cost} = (100{,}000\ \text{mi/year})(1\ \text{gal/5 mi})(\$5.00/\text{gal})$$
$$= \$100{,}000/\text{year}$$

By attaching a fairing, the drag coefficient is reduced by 21% (calculated as (0.96 – 0.76)/0.96 with values from Table 6.3). The mileage increase of the vehicle is 5 × 0.21 = 1.05 mi/gal. In other words, with the fairing, the owner can expect to get 6.05 mi/gal. The fuel cost becomes

$$\text{Fuel cost with fairing} = (100{,}000\ \text{mi/year})(1\ \text{gal/6.05 mi}) \times (\$5.00/\text{gal})$$
$$= \$82{,}645/\text{year}$$

The savings in fuel costs become \$100,000 – \$82,645 or

Savings = \$17,355/year

6.5 LIFT ON AIRFOILS

In the preceding sections, we discussed the effect of drag on various bodies. In this section, we examine both lift and drag forces on airfoils. As illustrated in Figure 6.32, the geometric dimensions of importance are the wingspan b, the length or chord of the wing c, and the maximum thickness D. A common definition associated with airfoil geometry is **aspect ratio**: $AR = b/c$. The dynamic features of importance include the free-stream velocity, the lift and drag forces, and the angle of attack α of the wing. These features are also shown in Figure 6.32.

From observations of flow patterns about airfoils at various angles of attack, streamlines can be sketched as in Figure 6.33. It is known that as the angle of attack α is increased, fluid moving over the top surface must accelerate to keep up with the slower-moving fluid on the bottom surface. Moreover, the faster-moving fluid with the higher kinetic energy must have a correspondingly lower pressure. The difference between the pressure acting on the lower surface and that acting on the upper surface results in a net lifting force on the foil. The lift force, like the drag force, is usually expressed in terms of a lift coefficient:

$$C_L = \frac{L_f}{\frac{1}{2}\rho U_\infty^2 A} \tag{6.16}$$

where A is the planform area (the area seen in the plan view) perpendicular to the chord (a constant for the wing).

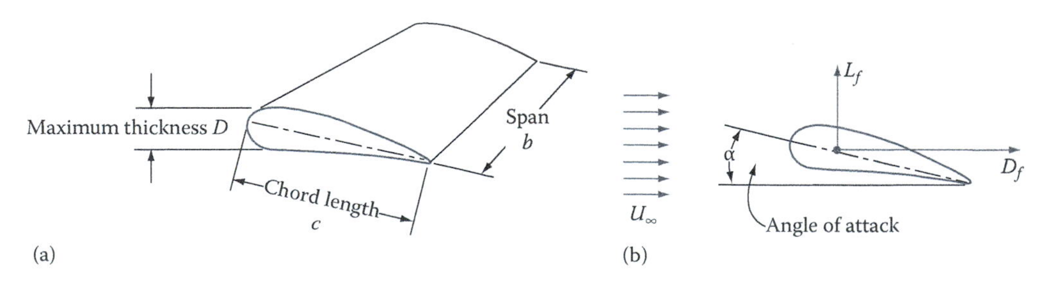

FIGURE 6.32 Geometric and dynamic parameters of airfoils.

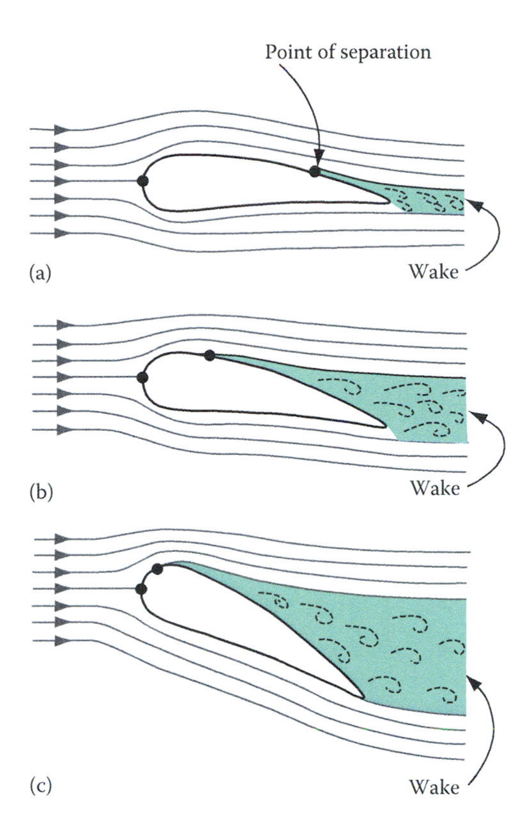

FIGURE 6.33 Streamlines of flow about an airfoil at various angles of attack.

As the angle of attack is increased, the point of separation on the upper surface moves toward the leading edge of the foil, as in Figure 6.33. The pressure in the wake is nearly equal to that at the separation point. When the separation point is almost to the leading edge, the entire upper surface is in the wake; the flow has separated from the entire upper surface. For this condition, the pressure on the upper surface is approximately the same as that upstream with the net effect that a decrease in lift is noticed at this angle of attack. This condition is called **stall**; its location is shown in Figure 6.34, a typical plot of C_L versus α. Figure 6.35 gives a measured pressure distribution acting on an airfoil immersed in a uniform flow.

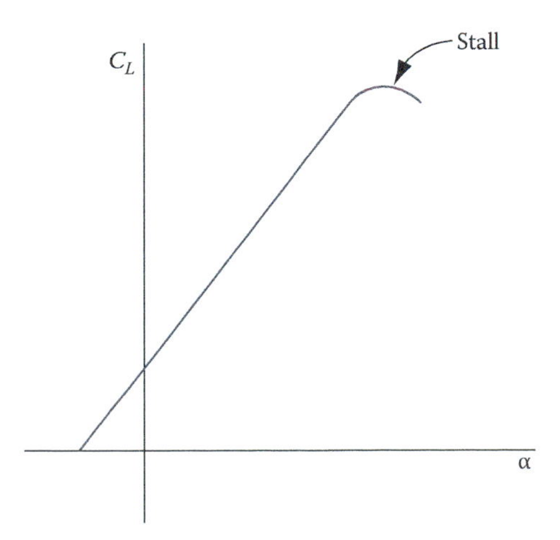

FIGURE 6.34 A typical plot of lift coefficient versus angle of attack for an airfoil.

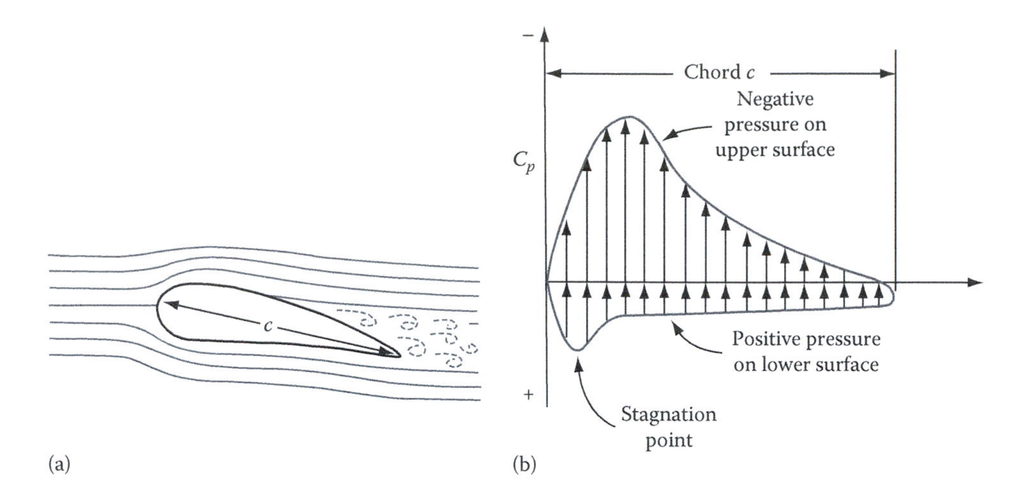

FIGURE 6.35 Pressure coefficient versus chord length for an airfoil. (a) Streamlines and (b) pressure distribution.

Abundant data on lift and drag of various airfoils have been obtained by wind tunnel testing. The results of such tests are typically presented as plots of lift and drag coefficients versus angle of attack. Efficiency of an airfoil, however, is more appropriately defined as a function of the ratio of lift to drag or C_L/C_D. To consolidate the results of a series of tests on one airfoil at a given free-stream velocity, a polar diagram is helpful. A **polar diagram** is a plot of C_L versus C_D with various angles of attack represented by different points on the curves. The ratio C_L/C_D at any point is the slope of the line from the origin to that point; the maximum value of the ratio is the line that passes through the origin tangent to the curve. The stall point is easily discernible. A polar diagram for the Clark Y airfoil is given in Figure 6.36. The data were obtained with a rectangular airfoil (36 ft span by 6 ft chord). An important reference is that which corresponds to zero lift—in this case, –5.6°. In general, this point corresponds to minimum drag.

Example 6.9

The cruising speed of an airplane with a Clark Y airfoil cross section is 150 mi/h. The plane flies at an altitude of 20,000 ft (p_∞ = 6.8 psi, T_∞ = 450°R). Using the data of Figure 6.32, determine the lift and drag forces for an angle of attack of 8.0°. Calculate the horsepower required to overcome drag.

Solution

The forces are found from

$$L_f = \frac{1}{2} C_L \rho V^2 A$$

$$D_f = \frac{1}{2} C_D \rho V^2 A$$

From Figure 6.32, at α = 8°, we find that C_L = 1.0 and C_D = 0.065. From the ideal gas law,

$$\rho = \frac{p}{RT} = \frac{6.8(144)}{1710(450)} = 0.0013\,\text{slug/ft}^3$$

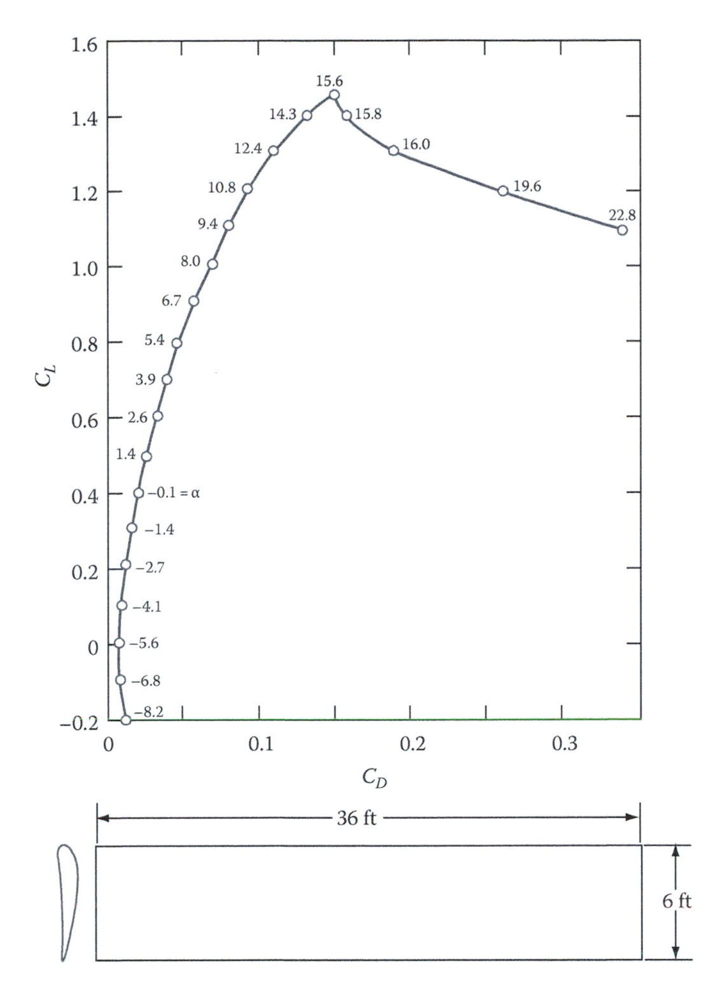

FIGURE 6.36 A polar diagram for the Clark Y airfoil. (Data from Silverstein, A., NACA Report 502, 1934, p. 15.)

The velocity is

$$V = \frac{150(5280)}{6300} = 220 \text{ ft/s}$$

The area used in wing calculations is the planform area, 6 × 36 = 216 ft². Substituting,

$$L_f = \frac{1}{2}(1.0)(0.0013)(220)^2(216)$$

$$L_f = 6769 \text{ lbf}$$

Similarly

$$D_f = \frac{1}{2}(0.065)(0.0013)(220)^2(216)$$

$$D_f = 442 \text{ lbf}$$

The power required to overcome drag is

$$\frac{dW}{dt} = D_f V = 442(220) = 97\ 240 \text{ ft} \cdot \text{lbf/s}$$

With 1 hp = 550 ft·lbf/s,

$$\frac{dW}{dt} = \frac{97\ 240}{550} = 177 \text{ hp}$$

Thus far in our discussion of flow past wings, we have discussed lift and drag forces exerted on a wing in a two-dimensional analysis. There is a three-dimensional effect known as a **wing vortex**, which is illustrated in Figure 6.37.

The air in the wake of a wing is set in motion, and at the wing tip this motion is recognized as a *vortex*. We normally refer to this phenomenon as a vortex being *shed* from the wing. The rotational movement of the air is said to occur about an imaginary line called a **vortex filament**. At the trailing edge of the wing, we have what is referred to as a **vortex sheet**, which "rolls up" into a pair of discrete vortices. The distance between the rolled-up vortices is less than the wing span. In a fluid that has zero viscosity, the vortices continue to infinity. In a real fluid, however, the vortices become unstable and break up due to viscous effects.

The vorticity shed from a wing is usually greatest near the wing tip. Air pressure below the wing is greater than that above it. The pressure difference causes air to flow around the tip from bottom to top, as indicated in Figure 6.37. There is a decrease in pressure near the center of the vortex, which leads to a corresponding decrease in temperature. If the humidity is high, the temperature drop causes vapor to condense, which makes the helical flow of a trailing vortex clearly visible. At takeoff and landing of an aircraft, detractable wing flaps are used as an aid to increase the lift, and similar vortices are shed at the tips of these flaps.

The trailing vortices of a moving wing cause what is known as a **downwash** in the wake of the wing (Figure 6.37). The downwash affects a large region or volume of air that extends well above

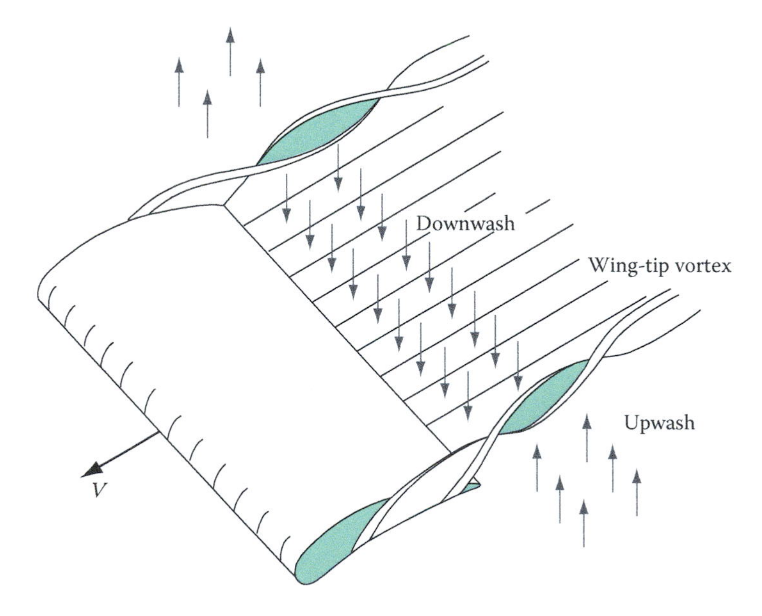

FIGURE 6.37 Three-dimensional flow effects about a wing showing helical motion of the wingtip vortices, induced downwash in the wake of the wing, and induced upwash outboard of the wing.

and below the aircraft. One effect is that when the plane flies close to the ground, some of the downwash is suppressed, leading to an increase in lift. This increase is manifested in a tendency of the plane to "float" during the time that it is landing and when its height is less than one-third to one-half of the wing span.

An interesting phenomenon occurs when an airplane flies into the trailing vortex shed by another airplane. The vortex causes a lift increase on one end of the wing of the trailing craft and a lift decrease on the other end. This results in a rolling moment exerted on the aircraft, which is a dangerous condition, especially if the trailing airplane is a much smaller craft. Thus the Federal Aviation Administration specifies air traffic rules, which, among other things, require that a minimum separation distance be maintained between flying aircraft. Although the vortices eventually break up, they exist for a long enough time to cause a limitation on the capacity of airport runways.

An airplane in the downwash of another will experience what is called **induced drag**. The trailing airplane is essentially pushed down, and the flow of air past the wing is affected. Figure 6.38 shows how this phenomenon is modeled. Figure 6.38a shows flow past a wing and the angle of attack α. Figure 6.38b shows a velocity diagram with the air velocity U_∞, the downwash velocity vector, and the relative air velocity U_{rel}. The angle between the air velocity U_∞ and the relative air velocity U_{rel} is denoted as β. The angle between the chord c of the wing and the relative air velocity U_{rel} is the effective angle of attack α_{eff} $(= \alpha - \beta)$. For a real wing traveling into a downwash, the lift and the drag are both affected. The change in drag is called induced drag. With regard to operation of the aircraft, the engines must exert additional power so that additional lift will overcome the downwash forces. Alternatively, the aircraft must change its orientation so that the angle of attack is increased, again so that additional lift will overcome downwash forces.

There are other practical applications of this three-dimensional effect. One occurs in the case of a fighter aircraft that is approaching a tanker aircraft from the rear. The fighter flies into the downwash of the tanker and, to avoid a dangerous situation, the engine thrust of the fighter must be markedly increased.

Another practical example deals with the **upwash** beyond the tip of the wing, or outboard of a lifting wing, as shown in Figure 6.37. Downwash causes downward forces to be exerted on a trailing craft, whereas upwash causes upward forces to be exerted. With regard to the power (= drag force × velocity) required, downwash forces cause an increase in drag to be exerted on a trailing craft; an increase in power is thus required to restore the aircraft to its prior condition. Upwash forces cause an increase in lift, and a corresponding decrease in drag and power will be realized. Birds flying in a V-formation make use of upwash forces. Figure 6.39 shows a frontal view of a flying bird with

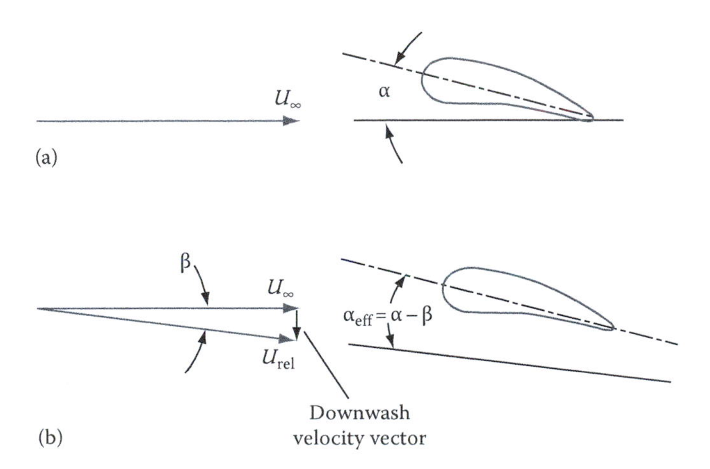

FIGURE 6.38 Flow past a wing showing how angle of attack is affected by downwash.

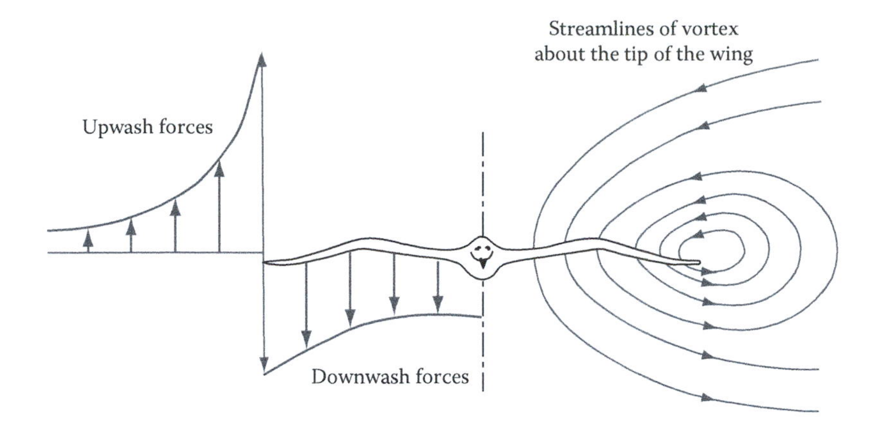

FIGURE 6.39 Flow field about a lifting wing of a bird as viewed in the downstream direction. (From Lissaman, P.B.S. and Shollenberger, C.A., *Science*, 168(3934), 1003, May 1970. With permission.)

wings extended. As indicated, there is a downwash in the wake of the wing with arrows pointing downward. Outboard of the wing tip, there is an upwash. Also shown are streamlines of flow about the wing. When in a V-formation, each bird (except the lead bird) flies in the upwash of its neighbors, with less lifting power needed.*

Figure 6.40 shows a graph of drag ratio versus spacing for a V-formation of birds. The vertical axis is a ratio of the ideal drag of an entire formation to the sum of the individual drags of the birds

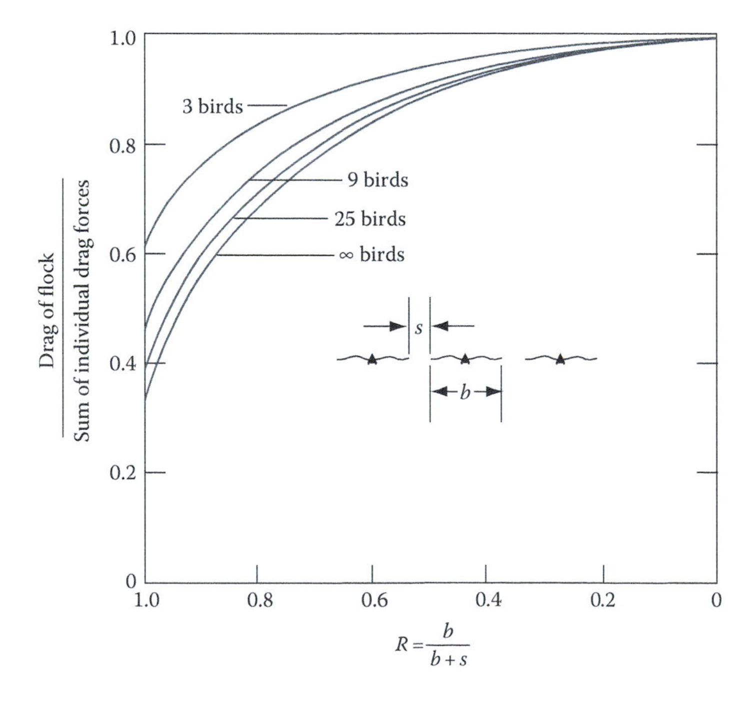

FIGURE 6.40 Drag reduction by V-formation flight. (From Lissaman, P.B.S. and Shollenberger, C.A., *Science*, 168(3934), 1003, May 1970. With permission.)

* These results were obtained from an analysis that was performed to model the flight formation of birds: P. B. S. Lissaman and C. A. Shollenberger, Formation flight of birds, *Science*, 168(3934): 1003–1005, May 1970.

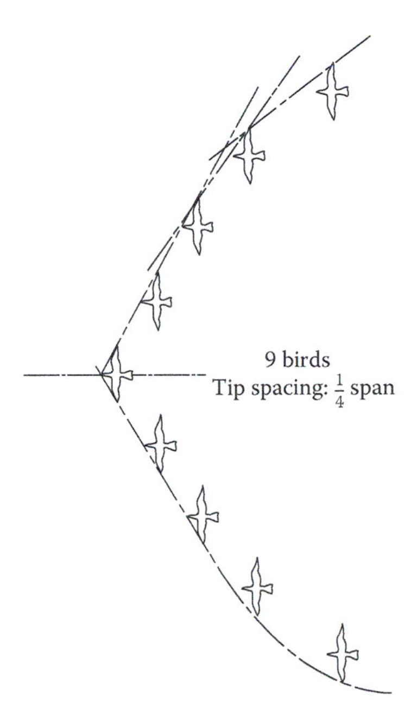

FIGURE 6.41 Optimum V-formation for nine birds flying at a tip spacing equal to one-fourth of the span. (From Lissaman, P.B.S. and Shollenberger, C.A., *Science*, 168(3934), 1003, May 1970. With permission.)

flying solo. The horizontal axis is a ratio of the wing span, b, to the wing span plus spacing, $b + s$, as defined in the diagram. The graph shows that a minimum drag ratio occurs when the spacing s equals 0 ($R = 1$). In addition, there is little difference between the performance of 25 birds and an infinite number of birds.

The ideal formation is not exactly a V, but instead is a formation that is more swept back at the tips and less at the apex. Such a formation is indicated in Figure 6.41, which shows the plan view of a nine-bird formation with a tip spacing s equal to one-fourth of the span b. Note that the outermost birds are a little further behind their neighbors and that this lag distance changes as we approach the apex.

6.6 SUMMARY

In this chapter, we examined the concepts of boundary layer growth and separation. We saw how these phenomena affect forces that are exerted on variously shaped objects immersed in a uniform flow. The drag and lift coefficients were defined, and data on drag and lift forces were expressed in terms of these coefficients.

INTERNET RESOURCES

1. Recent research work on drag reduction for tractor–trailer trucks has yielded new devices. What are these devices, and how much of a reduction in drag does each one provide? How much in annual fuel savings?
2. Obtain data on drag coefficient of vehicles and compare those values to ones predicted in this chapter.

3. Many wing shapes have been tested and catalogued by NACA. What is NACA, and what is the numbering system used to denote various wing shapes?
4. What are the devices used in airplanes to diminish the effect of upwash/downwash?
5. What is the coast down procedure used to obtain drag data on ground vehicles?

PROBLEMS

6.1 A 10 m/s wind at 22°C flows past a flat plate that is 12 m long and 6 m wide (same as Example 6.1). Calculate the displacement thickness for the plate, and plot δ^* versus x.

6.2 Air at 22°C flows past a flat plate that is 10 m long and 4 m wide. What is the velocity U_∞ required for the boundary layer to remain laminar over the entire plate? What is its thickness at the end of the plate?

6.3 Air at 22°C flows past a flat plate that is 8 m long and 4 m wide. The velocity U_∞ is such that the boundary layer remains laminar over the entire plate; with transition occurring at the end, determine the drag force exerted on the plate. Plot the boundary layer thickness versus x.

6.4 Air at 22°C and a velocity of 12 mph flows past a flat plate that is 10 m long and 4 m wide. Determine the drag force exerted on the plate.

6.5 A sailboat has a rudder that is 6 in. wide and extends into water a distance of 18 in. At 5 knots, what is the drag force exerted on the rudder?

6.6 A 6 m/s wind blows past a billboard measuring 10 m wide by 4 m high. The billboard is aligned with the wind direction as shown in the plan view of Figure P6.6. Calculate the force exerted on the billboard ($T = 22$°C).

6.7 Air at 20°C and 101.3 kPa (absolute) flows over a flat plate at a speed of 15 m/s. How thick is the boundary layer at a distance of 1 m from the front edge of the plate? What is the Reynolds number at that point?

6.8 A square duct in an air conditioning system is used to convey 50°F air to a plenum chamber. The duct is 3 ft on a side and 9 ft long. Estimate the skin friction drag exerted on the inside walls if the average air velocity in the duct is 4 ft/s.

6.9 A boxcar is pulled at a velocity of 18 m/s along a straight stretch of rail. The boxcar is 16.8 m long, and the sides are 2.44 m high. Assuming that the sides are perfectly flat, estimate the skin friction drag force exerted on them.

Flow Past Two-Dimensional Objects

6.10 The following data give pressure versus θ for a 3.75 in.-diameter cylinder that is 10.5 in. high and immersed in a uniform flow of 36.5 mi/h. The pressure data collected are Δp $(= p - p_\infty)$, and the ambient pressure $p_\infty = 14.7$ psi. The form drag is obtainable from integration of the pressure distribution around the surface of the cylinder.

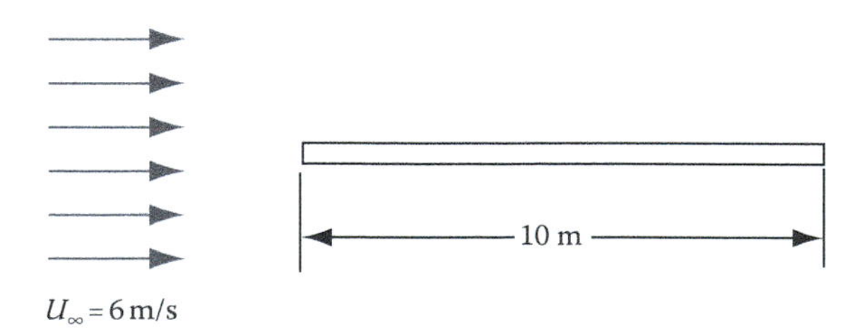

FIGURE P6.6

$L = 10.5\,\text{in.}$ $D = 3.75\,\text{in.}$ $U_\infty = 43.3\,\text{ft/s}$

Properties $\rho = 0.00232\,\text{slug/ft}^3$ $\mu = 0.3801 \times 10^{-6}\ \text{lbf}\cdot\text{s/ft}^2$

a. Sketch Δp versus θ on polar coordinate paper.
b. Sketch Δp versus θ on rectangular coordinate paper.
c. Sketch $\Delta p \cos\theta$ versus θ on rectangular coordinate paper. Determine the area under curve noting which portions are positive and which are negative. Calculate the form drag force.
d. Calculate the drag coefficient and the Reynolds number.

θ°	**Δp, psf**
0	0.900
10	0.820
20	−0.0860
30	−1.63
40	−3.48
50	−5.58
60	−7.99
70	−9.58
80	−10.3
90	−9.88
100	−9.45
110	−8.81
120	−7.52
130	−7.09
140	−6.87
150	−6.87
160	−6.87
170	−6.87
180	−6.87

6.11 The following data give pressure versus θ for a 9.53 cm diameter cylinder that is 26.7 cm high and immersed in a uniform flow of 10.3 m/s. The pressure data collected are Δp $(= p - p_\infty)$, and the ambient pressure $p_\infty = 101.3$ kPa. The form drag is obtainable from integration of the pressure distribution around the surface of the cylinder.

$L = 26.7\,\text{cm}$ $D = 9.53\,\text{cm}$ $U_\infty = 10.3\,\text{m/s}$

Properties $\rho = 1.197\,\text{kg/m}^3$ $\mu = 18.22 \times 10^{-6}\ \text{N}\cdot\text{s/m}^2$

a. Sketch Δp versus θ on polar coordinate paper.
b. Sketch Δp versus θ on rectangular coordinate paper.
c. Sketch $\Delta p \cos\theta$ versus θ on rectangular coordinate paper. Determine the negative. Calculate the form drag force.
d. Calculate the drag coefficient and the Reynolds number.

θ°	Δp, psf
0	22.6
10	20.6
20	−2.06
30	−41.1
40	−103
50	−162
60	−226
70	−269
80	−290
90	−278
100	−267
110	−247
120	−236
130	−232
140	−226
150	−226
160	−226
170	−226
180	−226

6.12 The following data give pressure versus θ for a 3.75 in.-diameter cylinder that is 10.5 in. high and immersed in a uniform flow of 22 ft/s. The pressure data collected are Δp $(= p - p_\infty)$, and the ambient pressure $p_\infty = 14.7$ psi. The form drag is obtainable from integration of the pressure distribution around the surface of the cylinder.

$L = 10.5\,\text{in.}$ $D = 3.75\,\text{in.}$ $U_\infty = 22\,\text{ft/s}$

Properties $\rho = 0.00232\,\text{slug/ft}^3$ $\mu = 0.3801\times 10^{-6}\,\text{lbf}\cdot\text{s/ft}^2$

a. Sketch Δp versus θ on polar coordinate paper.
b. Sketch Δp versus θ on rectangular coordinate paper.
c. Sketch Δp cos θ versus θ on rectangular coordinate paper. Determine the area under curve noting which portions are positive and which are negative. Calculate the form drag force.
d. Calculate the drag coefficient and the Reynolds number.

θ°	Δp, psf
0	0.086
10	0.043
20	−0.13
30	−0.43
40	−1.12
50	−1.72
60	−2.223
70	−2.58
80	−2.58
90	−2.49
100	−2.49
110	−2.49
120	−2.49

θ°	Δp, psf
130	−2.49
140	−2.49
150	−2.49
160	−2.49
170	−2.45
180	−2.40

6.13 A cylindrical smokestack is 26 m high and has an average diameter of 3.5 m. For a wind speed past the smokestack of 1.5 m/s, determine the drag force exerted ($T = 7°C$).

6.14 A flagpole consists of two sections and each is 6 m long. The lower section has a diameter of 15 cm and the upper section has a diameter of 12.5 cm. Determine the bending moment exerted at the base of the pole if no flag is flying and if the wind velocity is 50 km/h. Assume smooth cylinder surfaces.

6.15 An automobile aerial consists of three telescoping sections (6, 4, and 2 mm in diameter). When fully extended, each section is 23 cm long. For an automobile speed of 75 km/h, estimate the force exerted on the aerial and the moment exerted on its base ($T = 27°C$).

6.16 An electrical wire on a utility pole is 1 cm in diameter and 20 m long. If a 4 m/s wind passes, determine the drag force exerted on the wire. Ignore any translational motion of the wire.

6.17 Two cylinders are attached to a balance arm, as illustrated in Figure P6.17. The cylinder on the left has a roughened surface; otherwise, they are identical. A uniform flow of air is directed upward past both of them. At what velocity will the balance arm tilt clockwise? At what velocity will it tilt counterclockwise? Take the diameter of each cylinder to be 15 cm.

6.18 A Savonius rotor is a type of windmill (Figure P6.18). For a uniform airspeed of 3 ft/s directed at a stationary rotor, estimate the torque exerted about the shaft. (The data of this chapter will not yield an accurate result because here the half cylinders are attached together.)

6.19 Two square rods of equal length are immersed in a uniform flow of oxygen, as illustrated in Figure P6.19. The oxygen velocity is 5 m/s. If D_1 is 10 cm, determine D_2, assuming that the drag force exerted on each is the same.

6.20 Consider airflow past a square cylinder 10 cm on a side and 6 m long as shown in Figure P6.20a. The free-stream velocity is 4.5 m/s.

a. Use Figure 6.15 to find the drag force exerted.
b. Find the drag force for the cross section consisting of three flat plates pieced together as in Figure P6.20b. Compare the results.

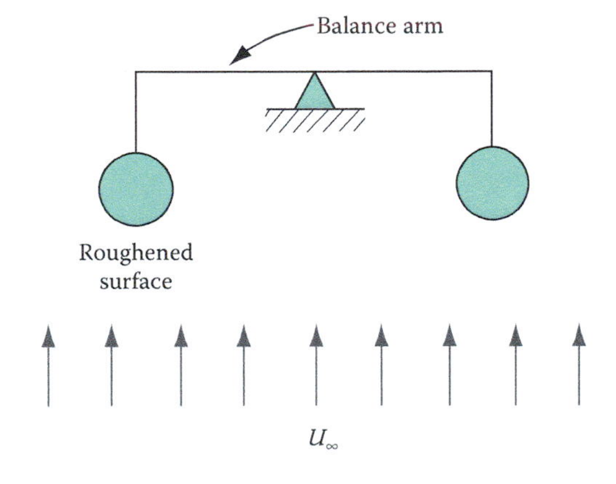

FIGURE P6.17

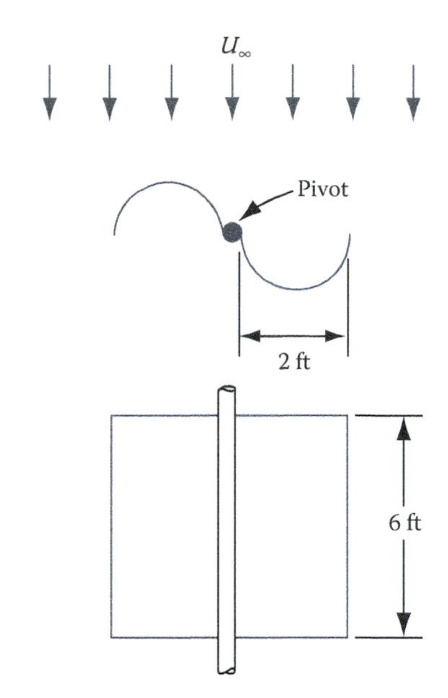

FIGURE P6.18

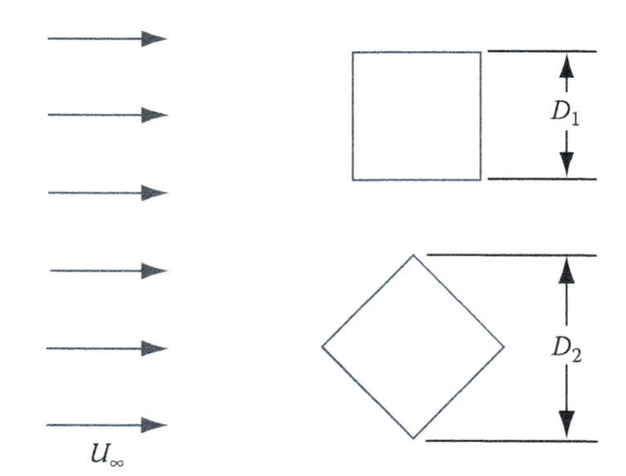

FIGURE P6.19

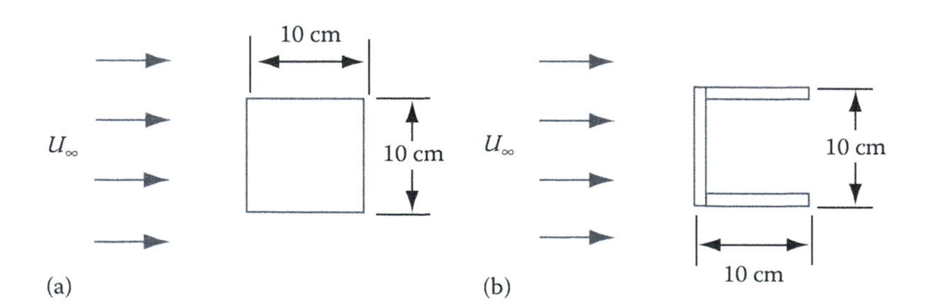

FIGURE P6.20

Flow Past a Sphere

6.21 A stainless steel sphere of diameter 2 cm falls through castor oil. Calculate the terminal velocity of the sphere.

6.22 Paraffin in the form of a sphere is released in a tank of water. Determine the terminal velocity of the sphere if its diameter is 3 cm.

6.23 A 3 cm diameter steel sphere is dropped into a tank of glycerine. Calculate its terminal velocity.

6.24 A simple trick performed with a table tennis ball and a vacuum cleaner exit hose is sketched in the Figure P6.24. The exhaust air causes the ball to remain suspended at a certain height above the exit. If a typical table tennis ball weighs 3.08 g and is 3.81 cm in diameter, determine the air velocity required to perform the trick. Take the exhaust air temperature to be 37°C.

6.25 A steel sphere is dropped into a tank of glycerine. Calculate the terminal velocity of the sphere if its diameter is 3 mm.

6.26 The force balance equation for a sphere falling through a liquid at terminal velocity is

$$mg - pg\forall - \frac{1}{2}C_D\rho V^2 A = 0$$

The sphere volume is $\pi D^3/6$, and its area is $\pi D^2/4$. When combined with the preceding equation, we get

$$V = \left[\frac{4}{3}\frac{gD}{C_D}\left(\frac{\rho_s}{\rho} - 1\right)\right]^{1/2}$$

But when the Reynolds number of the sphere is less than 1, the drag coefficient is given by

$$C_D = \frac{24}{\text{Re}}$$

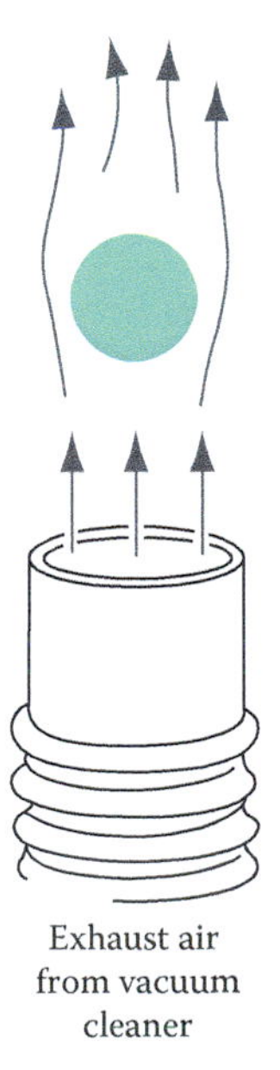

FIGURE P6.24

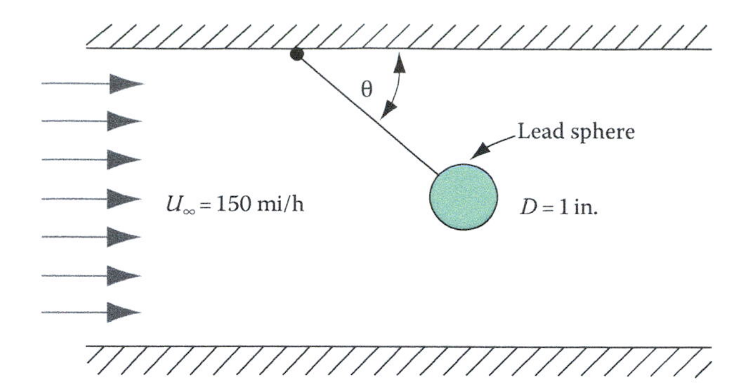

FIGURE P6.27

Show that when Re < 1, the terminal velocity is

$$V = \frac{\rho g D^2}{18\mu}\left(\frac{\rho_s}{\rho} - 1\right)$$

6.27 A 1 in. diameter lead ball is suspended by a thin piece of thread in the test section of a wind tunnel as indicated in the figure. The uniform speed past the sphere is 150 mph. Determine the resultant angle that the thread makes with the horizontal (Figure P6.27). Take $T = 62°F$.

6.28 A brass sphere that is 1.2 cm in diameter falls through a tank of turpentine. Determine the terminal velocity of the sphere.

6.29 If a solid sphere made of brass is dropped into a tank of glycerine, its terminal velocity is measured as 30 mm/s. Determine the sphere's diameter.

6.30 A gasoline (assume octane) fuel droplet 500 μm in diameter is injected into an airstream at the intake of an engine. The air velocity is a uniform 2 m/s downward. Determine the final velocity of the droplet ($T = 22°C$).

Flow Past Three-Dimensional Objects

6.31 A hemisphere whose diameter is 10 in. is supported in a uniform airflow of 12 ft/s as indicated in the Figure P6.31. What diameter sphere will have the same drag exerted on it by the air at the same uniform flow velocity?

6.32 Repeat Problem 6.31 but turn hemisphere around (Figure P6.32).

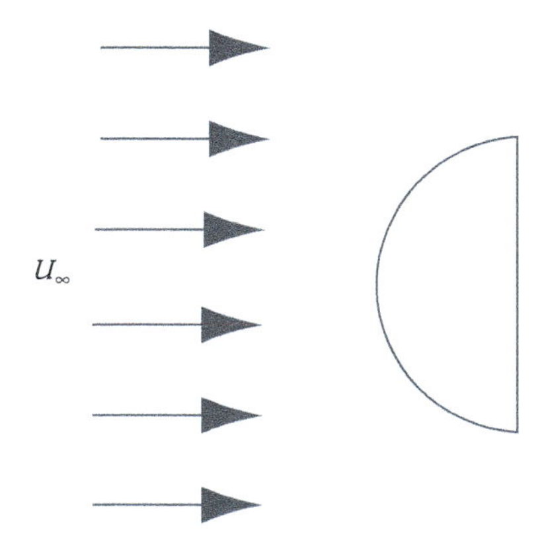

FIGURE P6.31

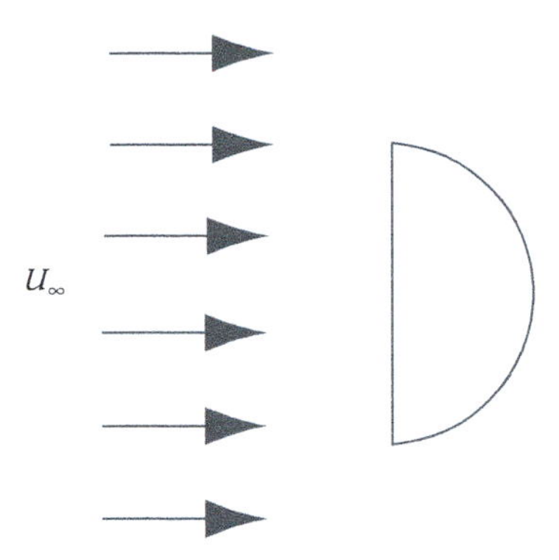

FIGURE P6.32

6.33 A motorist wishes to install a rear view mirror on the front fender of his car. The mirror housing is a hemisphere of diameter 6 in.; the convex side faces forward. Determine the additional drag on the automobile at 20 mph if two of these mirrors are installed.

6.34 A disk and a hemisphere are attached to a balance arm, as shown in the Figure P6.34. Both are subjected to a uniform upward air velocity of 15 m/s. The hemisphere is 0.06 m in diameter. If the scale is to balance, what is the required diameter of the disk?

6.35 Repeat the Problem 6.34 but replace the hemisphere with a streamlined body having $D = 20$ cm.

6.36 A four-cup wind speed indicator or anemometer is shown in a plan view in Figure P6.36. The bearings on which the assemblage turns are rusty, and the torque required to start the cups rotating is high. Determine the torque exerted if a velocity U_∞ of 12 mph is required to start rotation. Ignore the effects of the cups labeled A and the connecting rods ($T = 62°$F).

6.37 An ornamental iron gate consists of six vertical rods welded at their ends to horizontal crosspieces. Each vertical rod is 1/2 × 1/2 in. square, and 5 ft 6 in. long; the horizontal top and bottom pieces are 3/4 in. diameter cylinders 42 in. long. The square rods are placed as illustrated in Figure P6.37. For a uniform wind speed of 20 mi/h, estimate the total force exerted on the gate. ($T = 71°$F.)

6.38 A 12 in.-diameter telephone pole 30 ft tall has four crosspiece wire hangers attached as in Figure P6.38. The crosspieces are 2 in. square in cross section and 4 ft long; they are separated by a distance of 1 ft. Estimate the moment exerted about the base of the pole in a uniform wind of velocity 25 mi/h. Assuming that the surface of the pole affects the results (smooth versus rough), perform calculations for both cases.

6.39 A billboard measures 12 m wide by 6 m tall. It is supported by two vertical posts which are 10 cm in diameter (Figure P6.39). The posts are set in the ground, and they hold the billboard 10 m above the ground. Estimate the moment exerted about the posts at ground level if the wind velocity is a uniform 0.5 m/s.

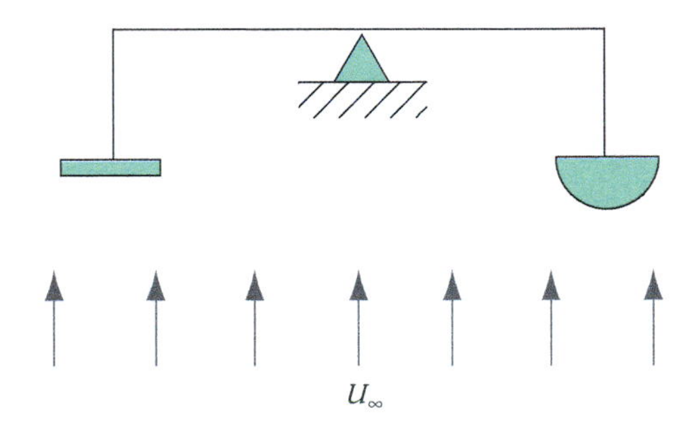

FIGURE P6.34

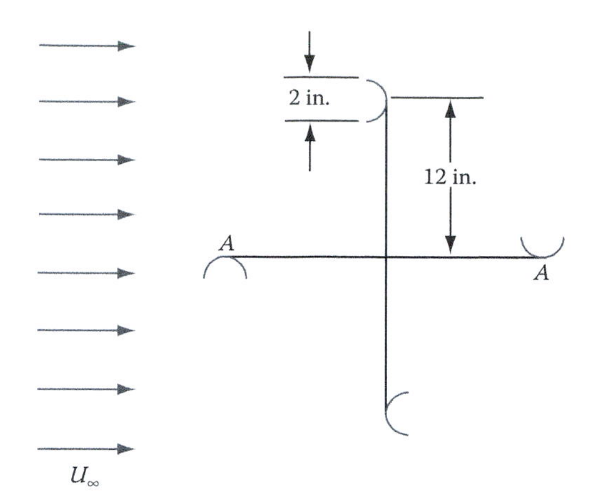

FIGURE P6.36

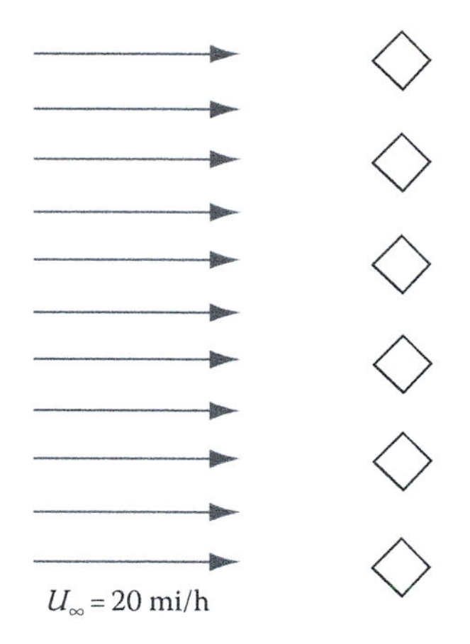

FIGURE P6.37

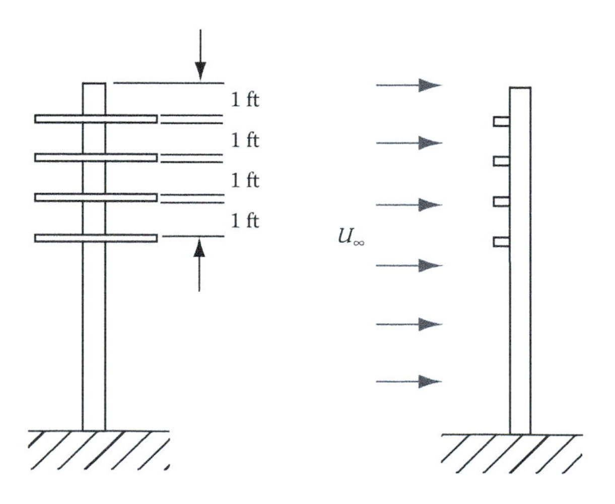

FIGURE P6.38

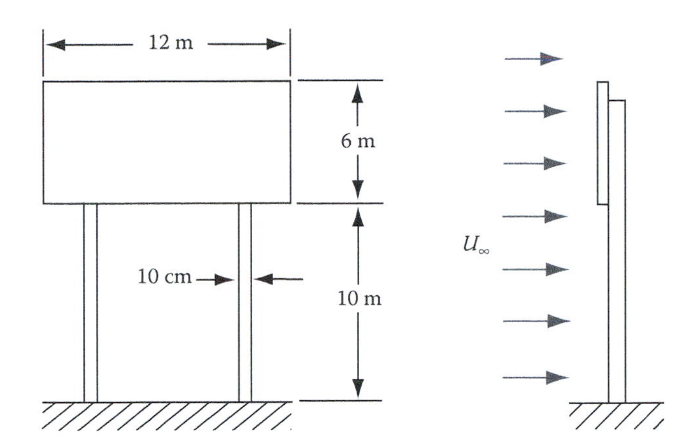

FIGURE P6.39

Flow Past Ground Vehicles: Cyclists

6.40 Calculate the power required to overcome drag for a cyclist riding at 10 m/s in a touring position (no. 2 of Table 6.1). Is it enough to power a 100 W light bulb?

6.41 Calculate the power required to overcome drag for a cyclist riding at 8.9 m/s in the fully crouched position (Table 6.1).

6.42 Calculate the power required to overcome drag for a cyclist riding at 8.9 m/s on recumbent bicycle (Table 6.1).

6.43 a. Determine the power required for a cyclist to ride at 20 mi/h in the racing position (no. 3 of Table 6.1)

b. If the cyclist maintains this speed and drafts another cyclist (no. 7), what power is required?

6.44 Which vehicle requires more power to overcome drag at 20 mi/h: the vector single of Table 6.1 or the streamlined vehicle of Figure 6.22? Calculate the power for both vehicles.

6.45 A cyclist rides at 7 m/s. At this velocity, 18% of her total power input goes into overcoming rolling resistance. If the cyclist uses this same input to power the streamlined vehicle of Figure 6.22, what is the expected velocity, again assuming that 18% of the input goes into overcoming rolling resistance?

6.46 An athlete who is in fairly good condition can maintain an output equivalent to 0.1 hp for 1 h on a bicycle. Determine how far this rider can travel after 1 h if only 90% of the rider's output goes to overcoming drag. (Assume that Figure 6.22 applies.)

6.47 A recreational cyclist has an output equivalent to 0.08 hp for 1 h on a bicycle. This cyclist fervently studies the concept of streamlining and uses a fairing. Determine how far this rider can travel after 1 h if 90% of this rider's output goes to overcoming drag. (Assume that Figure 6.22 applies.)

Flow Past Ground Vehicles: Automobiles

6.48 According to a popular automotive magazine, the drag coefficient of a 1986 Honda Accord LXi is 0.32, and its frontal area is 53 in. × 66 in. wide. Using a drag coefficient of 0.32, calculate the power required for the vehicle to maintain a speed of 70 mi/h.

6.49 According to the Internet, the drag coefficient of a 2014 Volkswagen XL1 is approximately 0.188 if its frontal area is 2.6 m^2. Calculate the power required for the vehicle to maintain a speed of 30 m/s.

6.50 According to the internet, the drag coefficient of a Tesla Model S is approximately 0.24 if its frontal area is 2.5 m^2. Calculate the power required for the vehicle to maintain a speed of 30 m/s.

6.51 According to a popular automotive magazine, the drag coefficient of a 1986 Mazda RX-7 is 0.29, and its frontal area is 126.5 cm 160 cm wide. Locate such a vehicle and determine whether Equation 6.13 predicts the same result. Using a drag coefficient of 0.29, calculate the power required for the vehicle to maintain a speed of 24.6 m/s.

6.52 The first two columns of the following table are data obtained of drag force versus velocity for a model of a 1957 Chevy. Dimensions of the vehicle are also provided. Graph these data with velocity on the horizontal axis. Calculate drag coefficient and Reynolds number, and graph drag coefficient versus Reynolds number (horizontal axis).

Frontal area = 5.73 in.^{2} (= 0.003 7 m^2); width = 3.15 in. (= 0.08 m)

Length = 10.75 in. (= 0.273 m)

V m/s (mph)	Chevrolet N
19.4 (43.4)	0.209
22.2 (49.7)	0.313
24.5 (54.8)	0.417
27.2 (60.8)	0.487
29.1 (65.1)	0.660
32.3 (72.0)	0.834
34.4 (77.0)	1.077
37.1 (83.0)	1.251
39.7 (88.8)	1.460

6.53 The first two columns of the following table are data obtained of drag force versus velocity for a model of a 1970 Dodge Challenger R/T. Dimensions of the vehicle are also provided. Graph these data with velocity on the horizontal axis. Calculate drag coefficient and Reynolds number, and graph drag coefficient versus Reynolds number (horizontal axis).

Frontal area = 4.35 in.^{2} (= 0.002 8 m^2); width = 3.18 in. (= 0.08 m)

Length = 11.25 in. (= 0.286 m)

V m/s (mph)	Dodge N
19.4 (43.4)	0.070
22.2 (49.7)	0.070
24.5 (54.8)	0.104
27.2 (60.8)	0.209
29.1 (65.1)	0.278
32.3 (72.0)	0.626
34.4 (77.0)	0.834
37.1 (83.0)	0.973
39.7 (88.8)	1.112

Flow Past Ground Vehicles: Trucks

6.54 If the square-edged trailer of Example 6.7 were replaced with a round-edged trailer, how would the savings in fuel costs be affected? Use the data for the square-edged, unstreamlined trailer as a basis for comparison.

6.55 A truck having a square-edged trailer like that shown in Figure 6.27 travels 160 000 km/year and averages 2 000 km/(m^3 of fuel). A fairing and a gap seal are then attached to the vehicle. Calculate the power required to overcome drag at 24.6 m/s for both configurations and compare the results.

6.56 A truck having a round-edged trailer like that shown in Figure 6.27 travels 80,000 mi/year and averages 5.1 mi/(gal of fuel). Diesel fuel costs $1.16/gal.

a. Calculate the yearly cost of fuel to run the vehicle.

b. Calculate the yearly cost of fuel to run the vehicle if it has a fairing attached.

c. Calculate the yearly cost of fuel to run the vehicle if it has a fairing and a gap seal attached.

6.57 Calculate the horsepower required to overcome drag for a tractor–trailer combination where the trailer has rounded front edges for the following cases: (a) basic configuration, (b) with a fairing attached, and (c) with a fairing and gap seal attached. Take the speed in all cases to be 65 mi/h.

Flow Past Airfoils/Aircraft

6.58 A Delta airlines publication reports speed and thrust ratings for various aircraft. A B747-400 has 4 Pratt & Whitney turbofan engines, each generating 56,000 lbf of thrust. The velocity is 561 mph. How much power does this aircraft have?

6.59 A Delta airlines publication reports speed and thrust ratings for various aircraft. An MD90 has two engines, and each generates 125 kN of thrust. At a velocity of 819 km/h, how much power does this aircraft have?

6.60 The data given in the accompanying Figure P6.60 are for a NACA 23012 airfoil.

a. Construct a polar diagram for the airfoil.

b. Determine the angle of attack for maximum C_L/C_D (Data from NACA Report 524.)

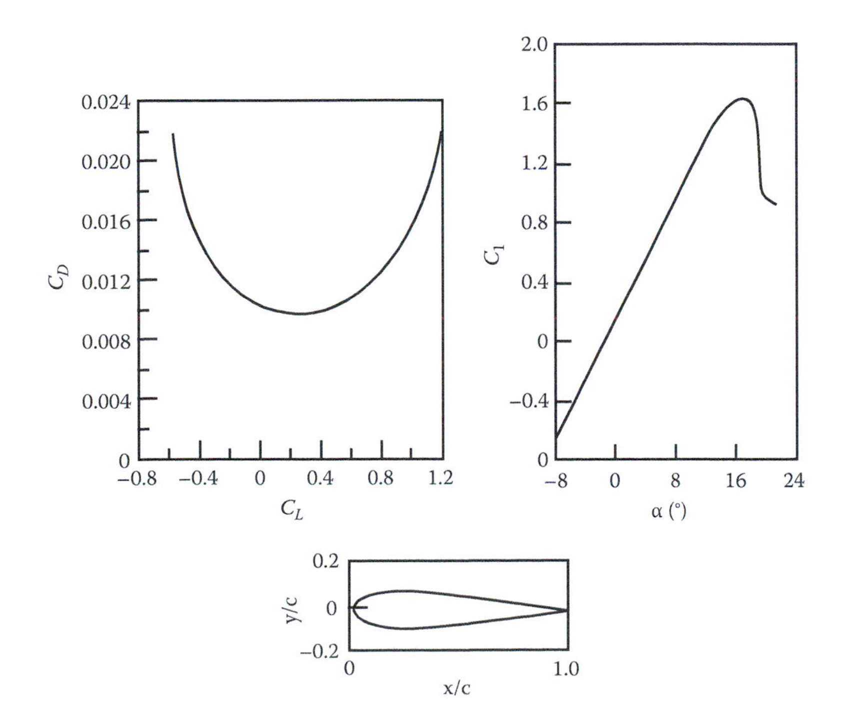

FIGURE P6.60

6.61 An airplane with a NACA 23012 airfoil cruises at 138 m/s at an altitude of 6,100 m ($p_\infty = 46.5$ kPa, $T_\infty = 250$ K). The airfoil has an aspect ratio of 10 with a span of 36 m. Using the data of the previous problem, determine the lift and drag forces. Then determine the horsepower required to overcome drag. Take the angle of attack to be 8°.

6.62 The following data were obtained on a NACA 2412 airfoil. Plot the polar diagram for the airfoil. (Data from NACA Technical Note 404.)

α	C_D	C_L
−3.4	0.0096	−0.178
−1.7	0.0092	−0.026
−0.4	0.0090	+0.133
+1.1	0.0090	+0.288
+2.6	0.0095	+0.439
+5.6	0.0110	+0.744
+8.7	0.0143	+1.049
+11.8	0.0201	+1.328
+13.4	0.0261	+1.457
+15.0	0.0352	+1.566
+15.8	0.0422	+1.589
+19.8	0.2269	+1.307
+26.8	0.4189	+1.003

6.63 Airfoils have a maximum lift coefficient corresponding to a certain angle of attack. When operated beyond this angle of attack, the wing stalls, as indicated in Figure 6.34 in general and in Figure 6.36 for the Clark Y airfoil. At the stall condition, the wing is supporting the dead weight W of the aircraft. Thus, at stall we have

$$L_f = W = \left(C_{L\max}\right)\frac{\rho V_{\text{stall}}^2 A}{2}$$

The stall velocity found in this equation is the absolute minimum landing speed for an aircraft. A modified version of the North American P-51D Mustang Fighter plane of World War II is the speed record holder for a propeller-driven plane (499 mi/h).

Consider a Mustang racer having a wing area of 233 ft^2 and a weight of 9,500 lbf. Its landing speed is designed to be 1.25 times the stall speed. At stall, the lift coefficient of the wing is 1.65. Calculate the stall velocity of the aircraft and its minimum landing speed.

6.64 A Clark Y airfoil is on an aircraft whose stall speed is 100 mi/h. Determine the weight of the aircraft if the wing area is 216 ft^2. (See Problem 6.63.)

6.65 An aircraft has a Clark Y airfoil and is flying at 130 knots, with an angle of attack of the wing of 8°. The plane is following another aircraft such that the downwash vector is represented by a velocity of 8 knots. Determine

a. The effective angle of attack
b. The change in the lift coefficient
c. The change in the drag coefficient
d. The ratio lift coefficient/drag coefficient for both angles

6.66 Drag coefficient versus Reynolds number for a streamlined and an unstreamlined cyclist was provided in this chapter. Use the not-streamlined cyclist curve and

a. Read values from the graph and prepare a table of C_D and corresponding Re
b. Use the data below the graph with your table to produce values of free-stream velocity U_∞ and drag force D_f

c. Make calculations of power for each velocity and drag force
d. Construct a graph of power dW/dt (vertical axis) versus velocity U_∞. Let U_1 vary from 5 to 25 mi/h

6.67 Drag coefficient versus Reynolds number for a streamlined and an unstreamlined cyclist was provided in this chapter. Use the streamlined cyclist curve, and
a. Read values from the graph and prepare a table of C_D and corresponding Re
b. Use the data below the graph with your table to produce values of free-stream velocity U_∞, and drag force D_f
c. Make calculations of power for each velocity and drag force
d. Construct a graph of power dW/dt (vertical axis) versus velocity U_∞. Let U_1 vary from 5 to 40 mi/h

6.68 A 3 in.-diameter sphere was immersed in a uniform flow of air at room temperature. The sphere has a smooth surface; as a result of measurements made, the following data were reported:

Smooth Surface Sphere

Free-Stream Velocity, U_∞ (ft/s)	Drag Force, D_f (lbf)
19.8	0.025
30.3	0.043
38.8	0.045
45.1	0.083
48.9	0.102
55.2	0.129
59.5	0.151
64.2	0.173
69.2	0.201
77.2	0.232

Use these data to produce a graph of drag coefficient versus Reynolds number and compare the results to those in Figure 6.16. (Data obtained by Lance Lane.)

6.69 A 3 in.-diameter sphere was immersed in a uniform flow of air at room temperature. The sphere has a roughened surface and as a result of measurements made, the following data were reported:

Rough Surface Sphere

Free-Stream Velocity, U_∞ (ft/s)	Drag Force, D_f (lbf)
19.8	0.025
25.6	0.027
34.3	0.026
38.8	0.039
47.5	0.057
51.8	0.069
58.4	0.084
63.2	0.105
67.7	0.126
81.0	0.192

Use these data to produce a graph of drag coefficient versus Reynolds number and compare the results to those in Figure 6.16. (Data obtained by Lance Lane.)

7 Flow in Open Channels

The flow of liquid in a conduit may be either pipe flow or open-channel flow. The feature that differentiates these two types is that open-channel flow must have a free surface subject to atmospheric pressure or a constant pressure—examples include flow in a river, a roadside gutter, an irrigation canal, or a submerged culvert. There are many examples of open-channel flow, and in this chapter, we will examine the variables that are useful in mathematically modeling such flows. Topics of discussion will include types of open-channel flow, properties of open channels, momentum and energy considerations, critical flow, gradually varied flow, and rapidly varied flow.

7.1 TYPES OF OPEN-CHANNEL FLOWS

Open-channel flows can be classified into many categories according to how the depth of the flow changes with respect to time and position:

- *Steady flow* exists if the volume flow rate does not change during the interval of interest.
- *Unsteady flow* exists if the volume flow rate does change with time. (Floods and surges are examples.)
- *Nonuniform flow* exists in a channel if liquid enters or leaves along the course of the flow. This type is also known as spatially varied or discontinuous flow. (Flow in roadside gutters, main drainage canals, flood surges, and feeding channels of irrigation systems are examples.)
- *Uniform flow* exists if the depth is uniform (i.e., unchanging) at every section of the channel. Uniform flow occurs when the flow is steady.
- *Steady uniform flow* exists if the depth of flow does not change during the time interval under consideration.
- *Unsteady uniform flow* exists if the water surface fluctuates while remaining parallel to the channel bottom—a practically impossible condition.
- *Varied flow* exists if the depth changes with increasing distance along the channel.
- *Rapidly varied flow* exists if the liquid depth changes abruptly over a relatively short channel length. (Hydraulic jump and hydraulic drop are examples.) Otherwise, the flow is gradually varied.

These descriptions are schematically summarized in Figure 7.1. In this chapter, we are concerned only with steady uniform flow and varied flow.

The forces of importance in open-channel flow are due to inertia, viscosity, and gravity. The ratio of the inertia force to the viscous force is the *Reynolds number*; accordingly, the flow can be further classified as laminar, transitional, or turbulent. Experiments have shown that the flow is always turbulent when the Reynolds number exceeds 1 000* and that most open-channel flows are turbulent.

The ratio of inertia to gravity forces is known as the *Froude number* given by $\mathrm{Fr} = V^2/gl$. Accordingly, the flow can be further classified as critical, subcritical, or supercritical. When the Froude number of the flow equals 1, the flow is *critical*. If the Froude number is less than 1, the flow

* Based on one-fourth of the hydraulic diameter (as defined in Chapter 5) for characteristic length; see Section 7.2.

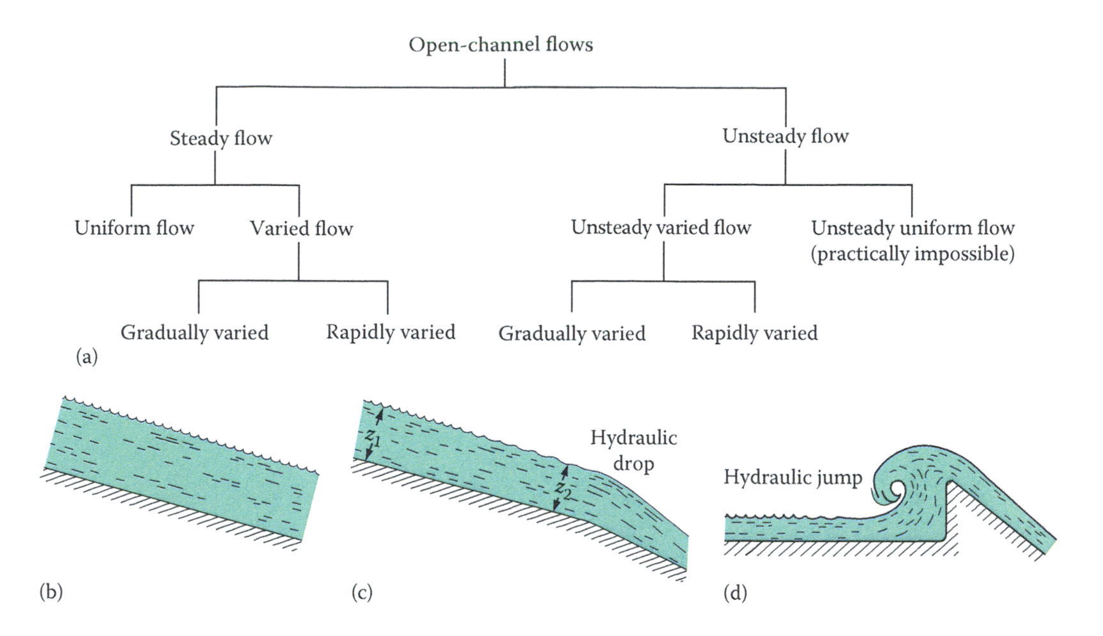

FIGURE 7.1 Types of open-channel flows. (a) Summary of types, (b) uniform flow, (c) gradually varied flow, and (d) rapidly varied flow.

is *subcritical*—it has a low velocity and can be described as tranquil and streaming. When the Froude number is greater than 1, the flow is *supercritical*—it has a high velocity and can be described as rapid and torrential.

7.2 OPEN-CHANNEL GEOMETRY FACTORS

An open-channel conduit can be either natural or artificial. Natural channels include brooks, streams, and rivers—in general, any watercourse that exists naturally. Artificial open-channel conduits include canals, flumes, and culverts. The analysis of artificial channels produces results that are applicable to natural conditions.

In this section, we will consider geometric factors of four different types of prismatic channels—that is, channels with unvarying cross sections and constant bottom slopes. These cross sections are rectangular, triangular, trapezoidal, and circular. A rectangular section is the shape produced by common building materials such as brick and timber. A triangular section is usually constructed for small ditches and roadside gutters. The rectangular and triangular sections are both special cases of the trapezoidal section. The circular section is conventionally used for sewers and culverts of small to intermediate size. Each of these cross sections is shown schematically in the first column of Table 7.1.

The geometric factors for each cross section given in Table 7.1 are defined as follows:

- *Depth of flow* z is the vertical distance from the channel bottom to the liquid surface.
- *Top width* b_t is the width of the channel section at the free surface.
- *Flow area* A is the cross-sectional area of flow perpendicular to the flow direction.
- *Wetted perimeter* P is the length of the channel cross section in contact with liquid.
- *Hydraulic radius* R_h is the ratio of area to wetted perimeter:

$$R_h = \frac{A}{P} \tag{7.1}$$

TABLE 7.1

Geometric Elements of Various Channel Sections

Section	Area, A	Wetted Perimeter, P	Hydraulic Radius, R_h	Top Width, b_t	Hydraulic Depth, z_m	Section Factor, Z	Depth to Centroid, z_c
b_t, z, b	bz	$b+2z$	$\frac{bz}{b+2z}$	b	z	$bz^{1.5}$	$\frac{z}{2}$
b_t, 1, m, z, b	$(b+mz)z$	$b+2z\sqrt{1+m^2}$	$\frac{(b+mz)z}{b+2z\sqrt{1+m^2}}$	$b+2mz$	$\frac{(b+mz)z}{b+2mz}$	$\frac{[(b+mz)z]^{1.5}}{\sqrt{b+2mz}}$	$\frac{z(3b+2mz)}{6(b+mz)}$
b_t, 1, m, z	mz^2	$2z\sqrt{1+m^2}$	$\frac{mz}{2\sqrt{1+m^2}}$	$2mz$	$\frac{z}{2}$	$\frac{\sqrt{2}}{2}mz^{2.5}$	$\frac{z}{3}$
b_t, α, z, D	$(\alpha-\sin\alpha)\frac{D^2}{8}$	$\frac{\alpha D}{2}$	$\frac{D}{4}\left(1-\frac{\sin\alpha}{\alpha}\right)$	$(\sin\frac{1}{2}\alpha)D$ or $2\sqrt{z(D-z)}$	$\frac{D}{8}\left(\frac{\alpha-\sin\alpha}{\sin(1/2)\alpha}\right)$	$\frac{\sqrt{2}}{32}\frac{(\alpha-\sin\alpha)^{0.5}}{(\sin(1/2)\alpha)^{0.5}}D^{2.5}$	—

Source: Adapted from Chow, V.T., *Open-Channel Hydraulics*, McGraw-Hill, Inc., New York, 1959.

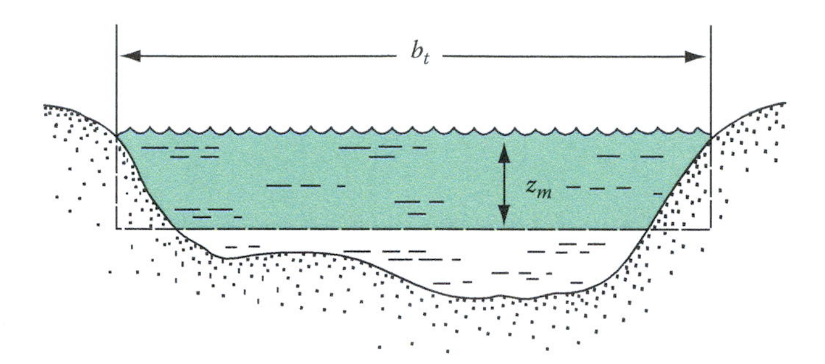

FIGURE 7.2 Equivalent or mean depth.

This definition seems to be inconsistent with the definition of hydraulic diameter in Chapter 5:

$$D_h = \frac{4A}{P}$$

because for a pipe, $R_h = \pi R^2/2\pi R = R/2$ and $D_h = 4\pi R^2/2\pi R = 2R$. However, most hydraulics (open-channel flow) texts use Equation 7.1 because of its convenience. Consider an open channel of rectangular cross section flowing with liquid of depth z. The width is b. The flow area thus becomes bz, and the wetted perimeter is $b + 2z$. For very wide channels, $b \gg 2z$, and the hydraulic radius becomes

$$R_h = \frac{bz}{b+2z} \approx \frac{bz}{b} = z$$

where z is the depth of flow.

- *Mean depth* z_m is the ratio of water area to top width:

$$z_m = \frac{A}{b_t} \tag{7.2}$$

  This expression, useful in modeling wide natural channels, allows us to calculate the depth of an equivalent rectangular channel. For the wide channel of Figure 7.2, an equivalent rectangular channel would have depth z_m.
- *Section factor* Z (used in making critical flow calculations) is the product of water area and square root of hydraulic depth:

$$Z = A\sqrt{z_m} = A\sqrt{\frac{A}{b_t}} \tag{7.3}$$

- *Depth to centroid* z_c is the distance from the free surface to the centroid of the flow area.

Table 7.1 provides definitions for each of these factors for four different channel sections.

7.3 ENERGY CONSIDERATIONS IN OPEN-CHANNEL FLOWS

The equations of motion for flow in open channels are the continuity, momentum, and energy equations. The momentum equation is particularly useful for flows with friction, which are primarily uniform, and rapidly varied flows (discussed in Section 7.7.2). The energy or Bernoulli equation is applicable to steady one-dimensional frictionless flows or gradually varied flows where area changes predominate over frictional effects. We illustrate the use of the energy equation in solving frictionless open-channel flow problems in this section.

7.3.1 Flow under a Sluice Gate

A sluice gate is a vertical gate used in spillways to retain the flow. A sketch of a partially raised gate is shown in Figure 7.3. For frictionless flow under a partially raised gate in a rectangular channel, we can write the following equations:

$$\text{Continuity: } Q = AV = bz_1V_1 = bz_2V_2$$
$$\text{Energy: } \frac{p_1}{\rho g} + \frac{V_1^2}{2g} + z_1 = \frac{p_2}{\rho g} + \frac{V_2^2}{2g} + z_2 \tag{7.4}$$

where b is the width of the channel. Now, $p_1 = p_2 = p_{\text{atm}}$; and with $V_1 \ll V_2$, the energy equation becomes

$$z_1 = \frac{V_2^2}{2g} + z_2 \tag{7.5}$$

Combining with Equation 7.4 and rearranging give

$$1 = \frac{Q^2}{2b^2z_2^2gz_1} + \frac{z_2}{z_1}$$

or

$$2\left(1 - \frac{z_2}{z_1}\right) = \frac{Q^2}{b^2z_1z_2^2g} \tag{7.6}$$

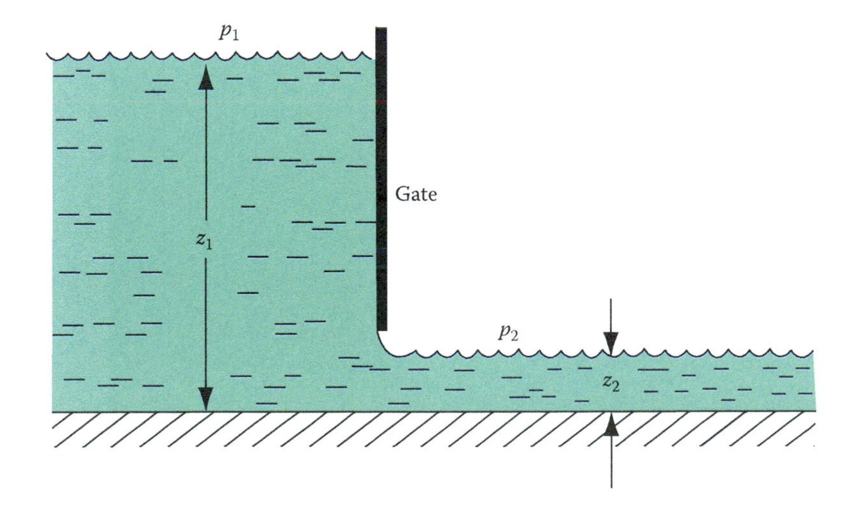

FIGURE 7.3 Flow under a sluice gate.

To remove z_2 from the denominator, multiply Equation 7.6 by $(z_2/z_1)^2$:

$$2\left(\frac{z_2}{z_1}\right)^2\left(1-\frac{z_2}{z_1}\right)=\frac{Q^2}{b^2 z_1^3 g} \tag{7.7}$$

This equation is the frictionless sluice gate equation for a rectangular channel. Note that when the gate is touching the bottom of the channel, $z_2 = 0$. When the gate is withdrawn from the liquid, $z_1 = z_2$. Thus, z_2/z_1 varies from 0 to 1. The right-hand side of Equation 7.7 is the upstream Froude number. A rectilinear plot of the sluice gate equation is given in Figure 7.4a. Point A represents a condition of maximum flow determined by differentiating Equation 7.7 and setting the result equal to zero:

$$\frac{d(Q^2/b^2 z_1^3 g)}{d(z_2/z_1)}=2\left(\frac{z_2}{z_1}\right)^2(-1)+2(2)\left(\frac{z_2}{z_1}\right)\left(1-\frac{z_2}{z_1}\right)=0$$

or

$$2-3\left(\frac{z_2}{z_1}\right)=0$$

with solution

$$\frac{z_2}{z_1}=\frac{2}{3} \tag{7.8}$$

Thus, maximum flow occurs at this ratio of liquid depths. Substituting Equation 7.8 back into Equation 7.7 yields

$$2\left(\frac{2}{3}\right)^2\left(1-\frac{2}{3}\right)=\frac{Q_{max}^2}{b^2 z_1^3 g}$$

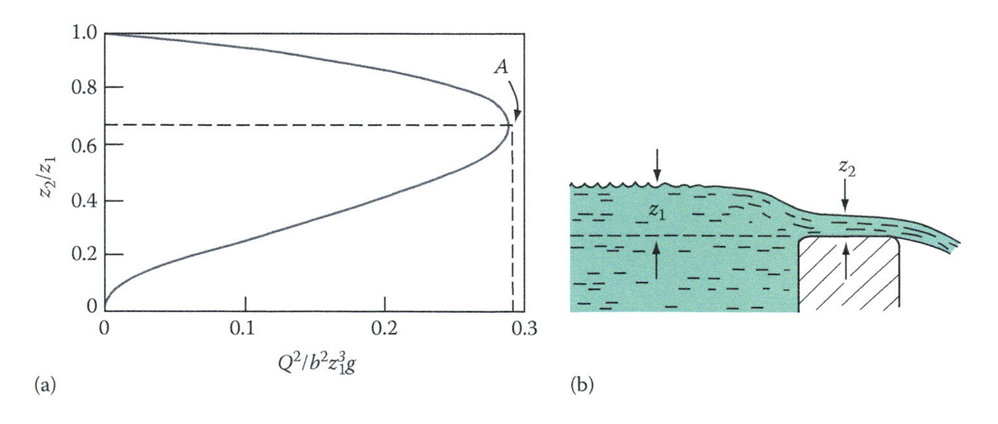

FIGURE 7.4 (a) A plot of the sluice gate equation for a rectangular channel and (b) flow over the crest of a dam.

from which we obtain the upstream Froude number at maximum flow:

$$\frac{Q_{\max}^2}{b^2 z_1^3 g} = \frac{8}{27} \tag{7.9}$$

The downstream Froude number is

$$\frac{Q_{\max}^2}{b^2 z_2^3 g} = \frac{Q_{\max}^2}{b^2 z_1^3 g}\left(\frac{z_1}{z_2}\right)^3 = \frac{8}{27}\left(\frac{3}{2}\right)^3$$

or

$$\frac{Q_{\max}^2}{b^2 z_2^3 g} = 1 \tag{7.10}$$

Thus, at maximum flow conditions, the flow upstream of the sluice gate is subcritical, whereas downstream the flow is critical.

Data taken from a sluice gate under laboratory conditions agree well with the curve of Figure 7.4a for frictionless flow over the lower portion of the curve—that is, for $z_2/z_1 < (2/3)$. Above this point, the assumption $V_1 \ll V_2$ is no longer valid.

The equation and results obtained for a sluice gate are applicable to flow over a crest of a dam shown in Figure 7.4b, Let z_1 be the upstream height and z_2 be the height directly over the dam. The flow rate curve that results is given in Figure 7.4a.

7.3.2 Flow through a Venturi Flume

Consider a channel of rectangular cross section where the width gradually changes with distance (Figure 7.5). We apply Bernoulli's equation to any streamline of the flow; but for convenience,

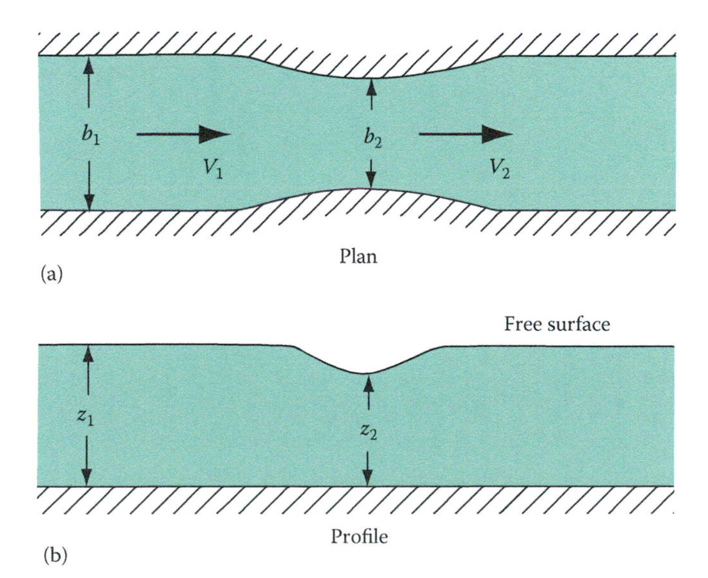

FIGURE 7.5 Flow through a rectangular venturi flume.

we choose the free surface where pressure is a constant. Applied to points 1 and 2, Bernoulli's equation is

$$\frac{p_1}{\rho g} + \frac{V_1^2}{2g} + z_1 = \frac{p_2}{\rho g} + \frac{V_2^2}{2g} + z_2$$

With $p_1 = p_2$ and $Q = b_1 z_1 V_1 = b_2 z_2 V_2$, after simplification and substitution, we have

$$\frac{Q^2}{2b_1^2 z_1^2 g} + z_1 = \frac{Q^2}{2b_2^2 z_2^2 g} + z_2$$

or

$$\frac{Q^2}{2g}\left(\frac{1}{b_1^2 z_1^2} - \frac{1}{b_2^2 z_2^2}\right) = z_2 - z_1$$

Solving for flow rate, we obtain

$$Q = \left[\frac{2g(z_1 - z_2)}{1/b_2^2 z_2^2 - 1/b_1^2 z_1^2}\right]^{1/2} \tag{7.11}$$

Thus, volume flow rate in a rectangular channel can be measured by inserting a venturi flume and measuring liquid depths at the minimum width and upstream sections.

When Bernoulli's equation was applied in the preceding problems, the following term appeared:

$$E = \frac{V^2}{2g} + z \tag{7.12}$$

The sum of kinetic and potential energies is labeled E and defined as the *specific energy* of the flow. Let us examine this equation in general. From continuity, $V = Q/A$; after substitution, we have

$$E = \frac{Q^2}{2gA^2} + z \tag{7.13}$$

and thus for a given rate of flow, E depends only on z. If the depth of flow is plotted as a function of specific energy for a constant area and flow rate, the resulting graph is referred to as a *specific-energy diagram* (Figure 7.6). Two intersecting curves result for any constant Q line, such as BC and AC in Figure 7.6. Curve BC asymptotically approaches a 45° line that passes through the origin. (A 45° line corresponds to channels of zero or very small slopes.) Curve AC asymptotically approaches the horizontal axis.

There are several points of interest in the diagram. First, the point at which minimum specific energy exists, E_{min}, corresponds to what is called critical flow; the flow depth is thus called critical depth. The Froude number is unity at the critical depth z_{cr}. To investigate this point, we first differentiate the expression for E:

$$E = \frac{Q^2}{2gA^2} + z$$

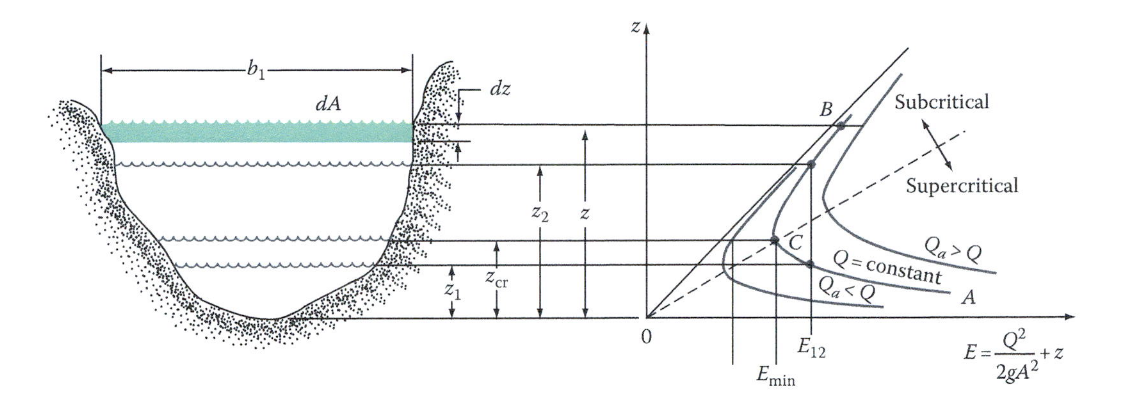

FIGURE 7.6 Specific-energy diagram and corresponding depths for various conditions.

With Q equal to a constant,

$$\frac{dE}{dz} = -\frac{Q^2}{gA^3}\frac{dA}{dz} + 1 = -\frac{V^2}{gA}\frac{dA}{dz} + 1 \tag{7.14}$$

At the free surface,

$$dA = b_t dz$$

and, by definition, the mean depth is

$$z_m = \frac{A}{b_t}$$

By substitution into Equation 7.14, we obtain

$$\frac{dE}{dz} = -\frac{V^2 b_t}{gA} + 1 = -\frac{V^2}{gz_m} + 1$$

At the minimum, $dE/dz = 0$. Thus, at critical flow,

$$\frac{V^2}{gz_m} = \text{Fr} = 1 \tag{7.15}$$

So at the critical state, $z_m = z_{cr}$—that is, the critical depth equals the mean flow depth.

Figure 7.6 shows that two depths correspond to any value of specific energy greater than the minimum; for example, z_1 and z_2 correspond to E_{12}. The depths z_1 and z_2 are called alternate depths. If the depth is greater than the critical depth, the velocity will be less than the critical flow velocity, and subcritical flow therefore exists. If the depth of flow is less than z_{cr}, the flow is supercritical because the velocity will be greater than the critical flow velocity. Thus, by definition, z_1 is the depth of a supercritical flow, whereas z_2 is the depth of a subcritical flow. Generalizing, we conclude that

$$\begin{aligned} &V = \sqrt{gz_{cr}} \quad \text{Fr} = 1 \quad \text{(critical flow)} \\ &V < \sqrt{gz_{cr}} \quad \text{Fr} < 1 \quad \text{(subcritical flow)} \\ &V > \sqrt{gz_{cr}} \quad \text{Fr} > 1 \quad \text{(subcritical flow)} \end{aligned} \tag{7.16}$$

which is true for any of the prismatic channels sketched in Table 7.1.

Example 7.1

Consider a rectangular channel of width 8 m.

a. Plot a family of specific-energy lines for $Q = 0$, 10, 20, 40, and 80 m³/s.
b. Draw the locus of critical depth points.
c. Plot Q as a function of critical depth.

Solution

For a rectangular channel, $A = bz$; therefore,

$$E = \frac{Q^2}{2gA^2} + z$$

becomes

$$E = \frac{Q^2}{2gb^2z^2} + z = \frac{Q^2}{2(9.81)(8)^2z^2} + z$$

$$= \frac{Q^2}{1\,260z^2} + z$$

The resulting equations are

$$Q = 0 \quad E = z$$

$$Q = 10 \quad E = \frac{1}{12.6z^2} + z$$

$$Q = 20 \quad E = \frac{1}{3.14z^2} + z$$

$$Q = 40 \quad E = \frac{1}{0.785z^2} + z$$

$$Q = 80 \quad E = \frac{1}{0.196z^2} + z$$

A plot of these equations appears in Figure 7.7a along with the locus of critical depth points. A plot of Q versus z_{cr} is given in Figure 7.7b. For other than rectangular cross sections, area either is derived or can be obtained from Table 7.1.

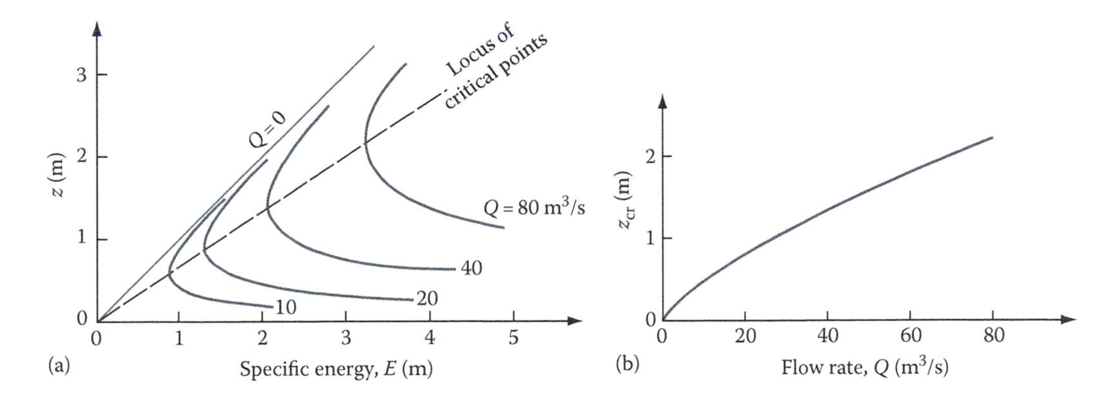

FIGURE 7.7 Solution curves for Example 7.1.

7.4 CRITICAL FLOW CALCULATIONS

It has been shown that at critical flow, $z_m = z_{cr}$; therefore,

$$V_{cr} = \sqrt{gz_{cr}} = \sqrt{gz_m}$$

or

$$\frac{V_{cr}^2}{g} = z_m \tag{7.17a}$$

from which we have

$$\frac{V_{cr}^2}{2g} = \frac{z_m}{2} \tag{7.17b}$$

where the notation V_{cr} has been introduced to denote the velocity corresponding to critical depth or critical velocity. Thus, at critical flow in any channel, half the mean depth equals the kinetic energy of the flow. The section factor Z has been previously defined as

$$Z = A\sqrt{z_m}$$

Combining this definition with $Q_{cr} = AV_{cr}$ and Equation 7.17, we obtain the following at critical flow:

$$Z = \frac{Q_{cr}}{\sqrt{g}} \tag{7.18}$$

Because Z is a function of depth, Equation 7.18 implies that only one depth is possible for critical flow. Similarly, for a fixed depth, there is only one flow rate, Q_{cr}, that corresponds to critical flow.

Example 7.2

Determine the critical depth and corresponding critical velocity of the rectangular channel of Example 7.1 (channel width $b = 8$ m) for a volume flow rates ranging from 0 to 80 m^3/s.

Solution

From Table 7.1 for a rectangular channel, $A = bz_{cr}$. Also, at critical flow, $z_{cr} = z_m$. We write

$$V_{cr} = \frac{Q}{A} = \frac{Q}{bz_{cr}}$$

Substituting into Equation 7.17a gives

$$\frac{V_{cr}{}^2}{g} = z_m = z_{cr}$$

Combining these equations, we get

$$\frac{Q^2}{b^2 z_{cr}^2 g} = z_{cr}$$

or

$$z_{cr}^3 = \frac{Q^2}{b^2 g}$$

$$z_{cr} = \left(\frac{Q^2}{b^2 g}\right)^{1/3}$$

Also,

$$V_{cr} = \frac{Q}{bz_{cr}}$$

For a channel width of $b = 8$ m, and with $g = 9.81$ m/s^2 the preceding equations give the results provided in the following table. A graph of Q versus z_{cr} is provided in Example 7.1.

Q (m^3/s)	z_{cr} (m)	$V_{cr} = Q/(bz_{cr})$ (m/s)
0	0.00	0
10	0.54	2.31
20	0.86	2.91
40	1.37	3.66
80	2.17	4.61

A sample calculation for $Q = 40$ m^3/s is as follows:

$$V_{cr} = \frac{Q}{A} = \frac{Q}{bz_{cr}} = \frac{40}{8z_{cr}}$$

$$z_{cr} = \frac{Q^2}{b^2 z_{cr}^2 g} = \frac{25}{9.81 z_{cr}^2}$$

$$z_{cr}^3 = 2.55$$

$$z_{cr} = 1.37\,\text{m}$$

$$V_{cr} = \frac{Q}{bz_{cr}} = \frac{40}{8z_{cr}} = 3.66\,\text{m/s}$$

This result can be obtained by using Equation 7.18 where the section factor for a rectangular channel is found from Table 7.1:

$$Z = \frac{Q_{cr}}{\sqrt{g}}$$

Substituting, we get

$$bz_{cr}^{1.5} = \frac{Q_{cr}}{\sqrt{g}}$$

For $Q = 40$ m^3/s and $b = 8$ m, we substitute to find

$$z_{cr}^{1.5} = \frac{40}{8\sqrt{9.81}} = 1.60$$

or

$$z_{cr} = 1.37 \text{ m}$$

Example 7.3

A trapezoidal channel with a bottom width of 15 ft has sides with slopes $1/m = 2$. For a flow rate of 60 ft^3/s, determine the critical depth.

Solution

From Table 7.1, the section factor for a trapezoidal channel is

$$Z = \frac{[(b+mz)z]^{1.5}}{(b+2mz)^{0.5}}$$

With $b = 15$ ft, $m = ½$, and $Q_{cr} = 60$ ft^3/s, substitution into Equation 7.18 gives the following at critical flow:

$$\frac{[(15+0.5z_{cr})z_{cr}]^{1.5}}{(15+z_{cr})^{0.5}} = \frac{60}{\sqrt{32.2}} = 10.57 \qquad \text{(i)}$$

By trial and error, we find

$$z_{cr} = 0.785 \text{ ft}$$

Note that the left hand side of Equation i is a constant for a given channel. To repeat the calculations for other flow rates involves changing 60 ft^3/s in Equation i to the desired value. Results for other flow rates are provided in the following table:

Q (ft^3/s)	**z_{cr} (ft)**
0	0.000
10	0.239
20	0.379
40	0.600
60	0.785
80	0.949
100	1.10

The "goal seek" feature in a spreadsheet makes such calculations easy to perform, although tedious.

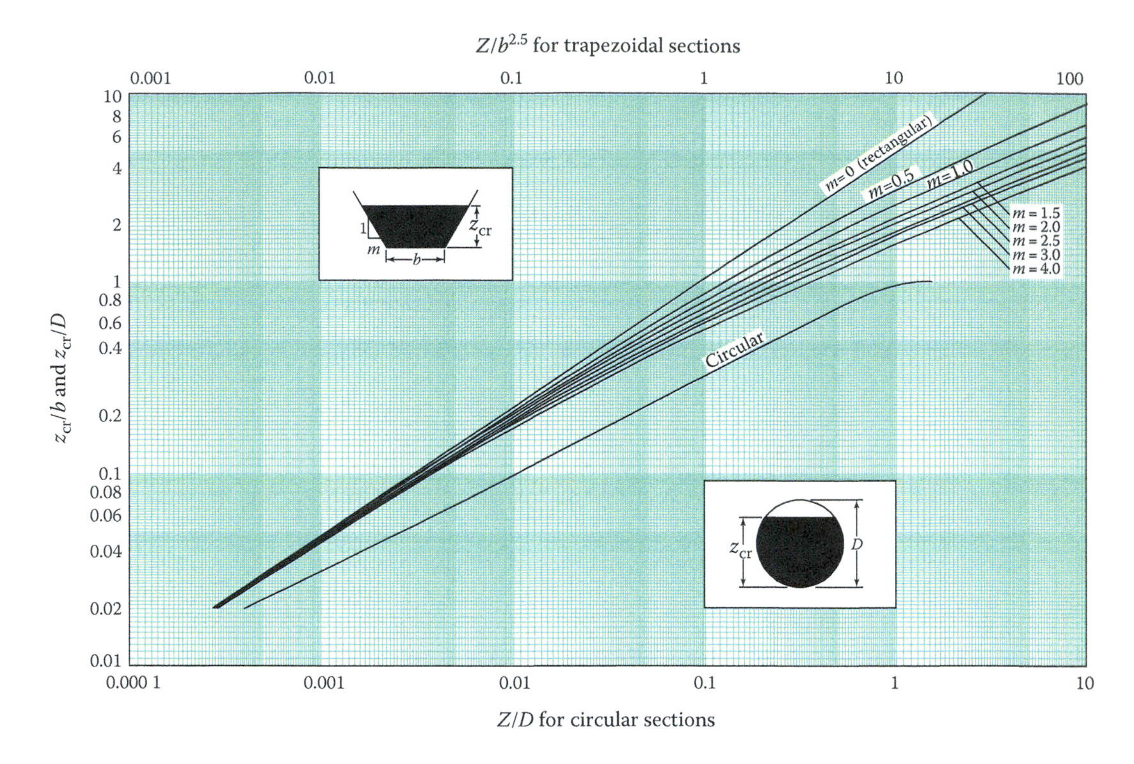

FIGURE 7.8 Curves for determining critical depth in trapezoidal and circular channels. (From Chow, V.T., *Open-Channel Hydraulics*, McGraw-Hill, Inc., New York, 1959, p. 65. With permission.)

Owing to the variety of channel widths, slopes, and volume flow rates requiring trial-and-error solutions for critical depth determinations, design calculations can be quite time consuming.

Consequently, a chart has been prepared for critical depth computations (Figure 7.8). Use of the chart is straightforward, as is illustrated in the following example.

Example 7.4

Rework Example 7.3 using Figure 7.8; $m = 1/2$, $b = 15$ ft, $Q = 60$ ft³/s.

Solution

To use the chart, we first find

$$\frac{Z}{b^{2.5}} = \frac{Q_{cr}}{b^{2.5}g^{0.5}} = \frac{60}{15^{2.5}\,32.2^{0.5}} = 0.012$$

With m given as ½, enter the chart at the horizontal axis for $Z/b^{2.5} = 0.012$ to the $m = ½$ line; find

$$\frac{z_{cr}}{b} = 0.052$$

Solving, we obtain

$$z_{cr} = 0.052(15)$$

$$z_{cr} = 0.78 \text{ ft}$$

7.5 EQUATIONS FOR UNIFORM OPEN-CHANNEL FLOWS

In this section, we will develop equations for laminar and for turbulent open-channel flow. The conditions are that the liquid depth and flow rate are constant at any cross section; the water surface and the channel bottom are parallel. Up to now, we have considered the open channels of study to be horizontal, whereas in fact there must be at least a very mild slope in the channel to sustain the flow. The forces of importance are friction (due to viscosity) and gravity. Pressure forces are not as significant. For open-channel flow to exist, gravity forces must overcome the friction or viscous forces, and this requires a nonzero slope.

7.5.1 Laminar Open-Channel Flow

Let us first consider laminar flow down an incline. This type of flow occurs in a thin sheet of liquid flow on, for example, a long, sloping roof. Frictional forces are significant. The case we will examine is flow down a very wide incline that is essentially two-dimensional. Figure 7.9 illustrates the situation—the x-direction is along the surface, the z-direction is normal to it, and the y-direction is into the plane of the page. For laminar flow, only the velocity in the x-direction, V_x, is nonzero. Moreover, we expect V_x to be a function of only z. An expected-velocity profile (V_x versus z) is shown in the figure.

Figure 7.10a shows a control volume consisting of an infinitesimal volume element with x-directed forces of pressure, gravity, and shear acting on it. By applying the momentum equation, we obtain

$$\sum F_x = ma = 0$$

because the acceleration is zero. By substitution,

$$\sum F_x = -\tau\, dx\, dy + (\tau + d\tau) dx\, dy + \rho g\, dx\, dy\, dz \sin\theta + p\, dy\, dz - \left(p + \frac{\partial p}{\partial x} dx \right) dy\, dz = 0$$

Simplifying yields

$$\frac{d\tau}{dz} + \rho g \sin\theta - \frac{\partial p}{\partial x} = 0 \tag{7.19}$$

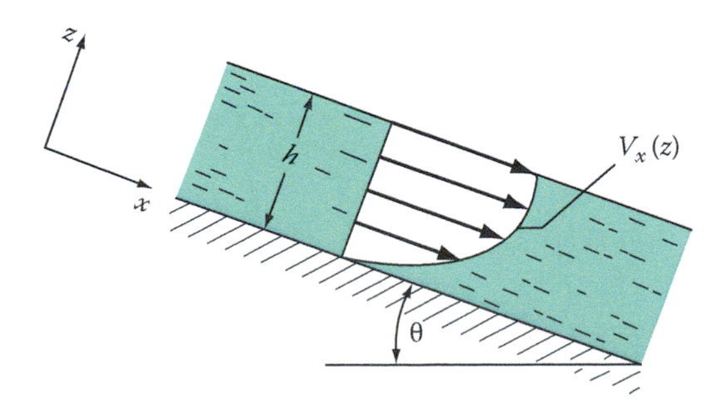

FIGURE 7.9 Laminar uniform open-channel flow.

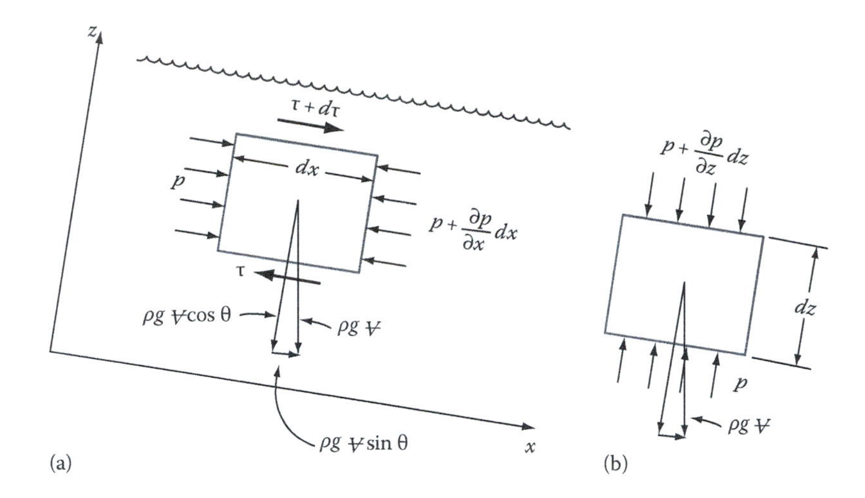

FIGURE 7.10 Control volume with forces acting on it for uniform open-channel flow.

To evaluate p and $\partial p/\partial x$, we now write the momentum equation for the z-direction. Referring to Figure 7.10b, we have, for a nonaccelerating control volume,

$$\sum F_z = 0$$

or

$$\sum F_z = p\,dx\,dy - \left(p + \frac{\partial p}{\partial z}dz\right)dx\,dy + \rho g\,dx\,dy\,dz\cos\theta = 0$$

Simplifying, we get

$$\frac{\partial p}{\partial z} = \rho g\cos\theta \tag{7.20}$$

Integrating with respect to z gives

$$p = \rho g z\cos\theta + f_1(x) + C_1 \tag{7.21}$$

where
$f_1(x)$ is an unknown function of x
C_1 is a constant of integration

To evaluate $f_1(x)$ and C_1, we must ensure that Equation 7.21 satisfies the boundary condition

At $z = h$, $p = p_{atm}$ = a constant

Applying the boundary condition to Equation 7.21 brings

$$p_{atm} = \rho g h\cos\theta + f_1(x) + C_1$$

We therefore conclude that

$$f_1(x) = 0$$

and

$$C_1 = p_{atm} - \rho gh \cos\theta$$

Substitution into Equation 7.21 gives

$$p - p_{atm} = \rho gz\left(1 - \frac{h}{z}\right)\cos\theta \tag{7.22}$$

which is the hydrostatic equation. Differentiating with respect to x, we get

$$\frac{\partial p}{\partial x} = 0$$

Thus, the x-directed momentum Equation 7.19 becomes

$$\frac{d\tau}{dz} = \rho g\sin\theta$$

Integrating with respect to z yields

$$\tau = -\rho gz \sin\theta + C_2 \tag{7.23a}$$

This equation must satisfy the boundary condition

At $z = h$, $\tau = 0$ = shear stress exerted on surface of liquid by air

Applying the boundary condition to Equation 7.23a, we obtain

$$C_2 = \rho gh \sin\theta$$

and therefore,

$$\tau = -\rho g(z - h)\sin\theta \tag{7.23b}$$

For laminar flow in the x-direction, Newton's law of viscosity is

$$\tau = \mu\frac{dV_x}{dz}$$

Combining with Equation 7.23b and rearranging give

$$\frac{dV_x}{dz} = \frac{\rho g}{\mu}(h - z)\sin\theta$$

After integrating with respect to z, we have

$$V_x = \frac{\rho g}{\mu}\left(hz - \frac{z^2}{2}\right)\sin\theta + C_3 \tag{7.24a}$$

which must satisfy the boundary condition of zero velocity at the wall:

At $z = 0$, $V_x = 0$

Applying this boundary condition gives $C_3 = 0$. Thus, after simplification, the velocity is

$$V_x = \frac{\rho g}{\mu}\frac{h^2}{2}\left(\frac{2z}{h} - \frac{z^2}{h^2}\right)\sin\theta \tag{7.24b}$$

The maximum velocity occurs where $z = h$:

$$V_{x\max} = \frac{\rho g}{\mu}\frac{h^2}{2}\sin\theta \tag{7.25}$$

In terms of $V_{x\max}$, V_x is

$$\frac{V_x}{V_{x\max}} = \left(\frac{2z}{h} - \frac{z^2}{h^2}\right)\left(\begin{array}{c}\text{laminar flow}\\ \text{down an incline}\end{array}\right) \tag{7.26}$$

A dimensionless plot of $V_x/V_{x\max}$ versus z/h is provided in Figure 7.11.

The volume flow rate is obtained by integration of Equation 7.24b over the cross-sectional flow area:

$$Q = \int_0^h \int_0^b V_x \, dy \, dz$$

where b is the width of the channel

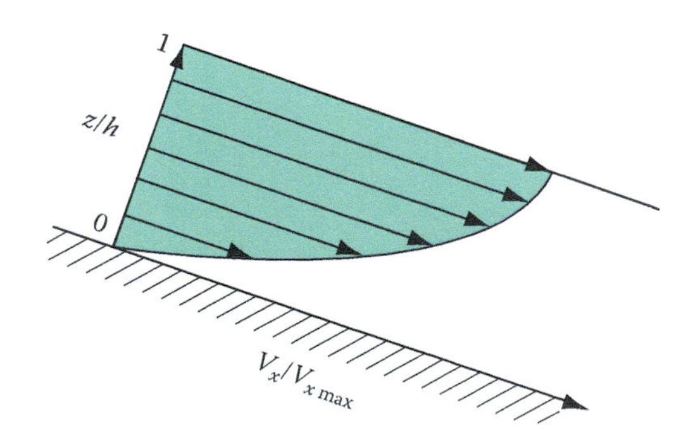

FIGURE 7.11 Dimensionless velocity distribution for laminar flow down an incline.

After substitution and integration, we obtain

$$Q = \frac{\rho g h^3 b}{3\mu}\sin\theta = \frac{g h^3 b}{3\nu}\sin\theta \quad \left(\begin{array}{c}\text{laminar flow}\\ \text{down an incline}\end{array}\right) \tag{7.27}$$

Thus, gravity directly influences the volume flow rate, whereas kinematic viscosity is inversely proportional to it. Forces due to pressure do not influence Q.

7.5.2 Reynolds Number and Transition

Just as in pipe flow, there exists a transition region in open-channel flow between the laminar and turbulent regimes. Again, the Reynolds number is the basis on which the flow can be classified. In open-channel flow, the Reynolds number is defined as

$$\mathrm{Re} = \frac{\rho V R_h}{\mu} \tag{7.28}$$

where the characteristic length R_h is the hydraulic radius defined earlier as the ratio of cross-sectional flow area to wetted perimeter. The hydraulic radius is one-fourth the hydraulic diameter, which was introduced in Chapter 5. For flow in an open channel, transition occurs over a Reynolds number in the range of 5×10^2 to 2×10^3. For our purposes, we will assume that transition occurs at a Reynolds number of 1 000. To achieve this value in a rectangular channel 3 m wide carrying water at a depth of 1 m, the velocity of flow is calculated to be

$$V = \frac{1\,000\mu}{\rho R_h} = \frac{1\,000(0.89\times 10^{-3})}{(1\,000)(3)(1)/5}$$

$$= 1.48\times 10^{-3}\text{ m/s} = 1.48\text{ mm/s}$$

where property values were obtained from Table A.5. In most open-channel flows, the velocity is usually several meters per second. Thus, most common open-channel flows are turbulent.

As with flow in pipes, open-channel flows exhibit frictional characteristics. Because of the highly turbulent nature of the flow, these frictional effects are for the most part independent of the Reynolds number. Experiments have shown that the one-dimensional approximation is suitable for modeling open-channel flows. In the following paragraphs, we will develop equations for this purpose.

7.5.3 Turbulent Open-Channel Flow

Consider a uniform open-channel flow in a prismatic channel at steady conditions. The flow is fully turbulent and is assumed to be one-dimensional; that is, the velocity is a constant at any point in a given cross section. Figure 7.12 illustrates the situation with the x-direction along the channel and the z-direction upward and normal to the channel bottom. The y-direction is into the paper. The liquid depth is h.

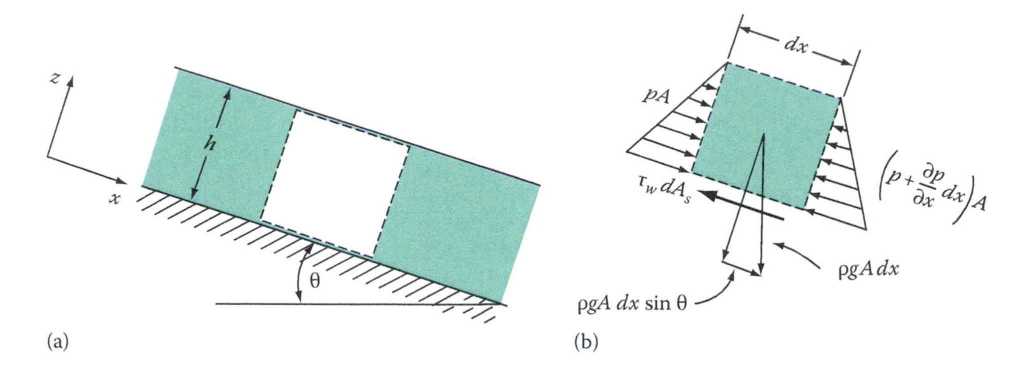

FIGURE 7.12 A control volume for one-dimensional, turbulent, uniform open-channel flow.

The control volume shown has forces due to pressure, gravity, and viscosity acting on it. The momentum equation in the x-direction is

$$\sum F_x = 0$$

$$= -\tau_w\, dA_s + \rho g A\, dx \sin\theta + pA - \left(p + \frac{\partial p}{\partial x}dx\right)A = 0$$

where

τ_w is the wall shear stress

dA_s is the wall area over which τ_w acts and is equal to the product of wetted perimeter and dx

A is the cross-sectional area of the flow

As with laminar flow, it can be shown that $\partial p/\partial x = 0$; consequently, pressure does not enter into the mathematical model for uniform flow. The x-momentum equation becomes

$$\tau_w(P\, dx) + \rho g A\, dx \sin\theta = 0$$

or

$$\tau_w = \rho g \frac{A}{P} \sin\theta = \rho g R_h \sin\theta \tag{7.29}$$

where P is the wetted perimeter. The wall shear stress is expressed in terms of a friction factor as in pipe flow:

$$\tau_w = \frac{f}{4}\frac{\rho V^2}{2} \tag{7.30}$$

Combining with Equation 7.29 and solving for velocity, we obtain

$$V = \left(\frac{8 g R_h}{f}\sin\theta\right)^{1/2} \tag{7.31}$$

The term $\sin\theta$ is the slope. For small angles, $\sin\theta = \tan\theta = \theta$, which we denote as the slope S. The previous equation may be rewritten as

$$V = C_1(R_h S)^{1/2}$$

where C_1 is a constant. The previous equation is called the Chezy formula. To better describe experimental data, it was found that the constant is

$$C_1 = \frac{C_m R_h^{1/6}}{n} \tag{7.32}$$

where n is called the Manning roughness coefficient (dimensionless). The equation for velocity becomes

$$V = C_m \frac{(R_h)^{2/3}(S)^{1/2}}{n} \tag{7.33}$$

where
$C_m = 1.0$ (SI units)
$C_m = 1.49$ (US customary units)

This constant contains the appropriate unit conversions so that the left-hand side of Equation 7.33 has dimensions of length/time. This constant is present in order to preserve the concept that the Manning roughness coefficient is dimensionless. It has the same significance in open channel flow as ε/D has in pipe flow.

Typical values of the Manning roughness coefficient for various channel shapes and beds are provided in Table 7.2. However, various researchers suggest that the roughness factor should be increased by 10%–15% for a channel with a hydraulic radius greater than 10 ft. This practice predicts a loss in capacity of large channels due to anticipated roughening of surfaces with age, marine presence, plant growth, bridge piers, and the like.

Equation 7.33 is independent of liquid properties and applies only to fully developed turbulent flow. The liquid velocity is thus a function of slope, channel geometry and an inverse function of the channel surface roughness coefficient.

Example 7.5

Water flows in a 2 m diameter culvert at a depth of 1.2 m. The culvert is uniform in cross section, made of concrete, and laid on a slope of 0.5°. Determine the volume flow rate through the culvert.

Solution

The geometry of the cross section is given in Table 7.1, where $z = 1.2$ m and $D = 2$ m. The equations for area and the other factors are given in terms of α, and α is found in terms of z by writing

$$\frac{D}{2} + \frac{D}{2}\cos\left(180 - \frac{\alpha}{2}\right) = z$$

or

$$1 + \cos\left(180 - \frac{\alpha}{2}\right) = 1.2$$

TABLE 7.2

Average Values of the Manning Roughness Coefficient for Various Channels and Bed Materials

Channel or Bed Material	Manning Coefficient, n
Metal	
Brass	0.010
Steel (smooth painted or unpainted)	0.012
Steel (welded)	0.012
Steel (riveted)	0.016
Cast iron (coated)	0.013
Cast iron (uncoated)	0.014
Wrought iron (black)	0.014
Wrought iron (galvanized)	0.016
Corrugated metal	0.024
Nonmetal	
Concrete	0.012
Wood	0.012
Wood planks	0.015
Clay	0.015
Brickwork	0.014
Rubble masonry (cemented)	0.025
Asphalt (smooth)	0.013
Excavated or dredged	
Earth (straight, uniform, weathered)	0.022
Earth (winding, sluggish, some weeds)	0.030
Channels (not maintained, weeds, brush)	0.080

Source: Adapted from Chow, V.T., *Open-Channel Hydraulics*, McGraw-Hill, Inc., New York, 1959.

Solving, we get

$$\alpha = 203° = 3.53 \text{ rad}$$

Substituting into the appropriate equations of Table 7.1 gives

$$A = \frac{D^2}{8}(\alpha - \sin\alpha) = \frac{(2)^2}{8}(3.53 - \sin 3.53)$$

$$= 1.95 \text{ m}^2$$

and

$$R_h = \frac{D}{4}\left(1 - \frac{\sin\alpha}{\alpha}\right) = \frac{2}{4}\left(1 - \frac{\sin 3.53}{3.53}\right)$$

$$= 0.55 \text{ m}$$

For a slope of 0.5°, we calculate $S = \sin\theta = \sin 0.5 = 0.008\,73$. (Note also that $\tan 0.5 = 0.008\,73$ and $0.5° = 0.008\,73$ rad.) For a concrete culvert, Table 7.2 shows $n = 0.012\ \text{m}^{1/6}$. Assuming fully turbulent flow, the Manning equation (Equation 7.33) applies

$$V = \frac{(R_h)^{2/3}(S)^{1/2}}{n} = \frac{(0.55)^{2/3}(0.008\,73)^{1/2}}{0.012}$$

$$= 5.23\,\text{m/s}$$

As a check on the turbulent-flow assumption, we find

$$\text{Re} = \frac{\rho V R_h}{\mu} = \frac{1\,000(5.23)(0.55)}{0.89\times 10^{-3}} = 3.23\times 10^{6} > 10^{3}$$

Because the Reynolds number is greater than 1 000 (the transition value), the flow is turbulent. The volume flow rate is calculated as

$$Q = AV = 1.95(5.23)$$

$$Q = 10.2\ \text{m}^3/\text{s}$$

Example 7.6

A rectangular concrete channel with a width of 18 ft conveys water at 350 ft^3/s. The channel has a slope of 0.001. Determine the liquid depth for uniform flow. Is the flow supercritical or subcritical?

Solution

From Table 7.2, we get $n = 0.012$. For a rectangular cross section, Table 7.1 shows that

$$R_h = \frac{bz}{b+2z}$$

Substitution into Equation 7.33 yields

$$V = 1.49\left(\frac{bz}{b+2z}\right)^{2/3}\frac{S^{1/2}}{n} = \frac{Q}{bz}$$

or

$$1.49\left(\frac{18z}{18+2z}\right)^{2/3}\frac{\sqrt{0.001}}{0.012} = \frac{350}{18z}$$

Simplifying, we get

$$z\left(\frac{18z}{18+2z}\right)^{2/3} = 4.952$$

Using the "goal seek" feature, we obtain the depth for uniform flow as

$$z = 2.92\ \text{ft}$$

For critical flow, we use Equation 7.17a:

$$\frac{V_{cr}^2}{g} = z_m$$

where $V_{cr} = \dfrac{Q}{bz_{cr}}$

$$z_m = z_{cr}$$

So, by substitution,

$$\frac{Q^2}{b^2 z_{cr}^2 g} = z_{cr}$$

or

$$z_{cr}^3 = \frac{(350)^2}{18^2(32.2)}$$

$$z_{cr} = 2.27 \text{ ft}$$

Because $z > z_{cr}$ we conclude that the flow is subcritical. Alternatively, we find that

$$\frac{V^2}{gz} = \frac{Q^2}{b^2 z^3 g} = \frac{350^2}{18^2(32.2)(2.92)^3} = 0.472$$

Because the Froude number is less than unity, the flow is subcritical.

As with critical flow, dimensionless curves have been prepared for uniform flow to simplify the calculation procedure. Figure 7.13 is a graph of $AR_h^{2/3}/b^{8/3}$ and $AR_h^{2/3}/D^{8/3}$ versus z/b and z/D. Thus, knowing the channel cross section and material allows us to find the depth for uniform flow.

Example 7.7

Find the uniform flow depth for the channel described in Example 7.6 (rectangular; $b = 18$ ft, $S = 0.001$, $n = 0.012$, and $Q = 350$ ft³/s). Use Figure 7.13.

Solution

The volume flow rate in the channel is given by

$$Q = AV = \frac{1.49 AR_h^{2/3} S^{1/2}}{n}$$

from which we find

$$AR_h^{2/3} = \frac{Qn}{1.49 S^{1/2}} = \frac{350(0.012)}{1.49\sqrt{0.001}} = 89.1$$

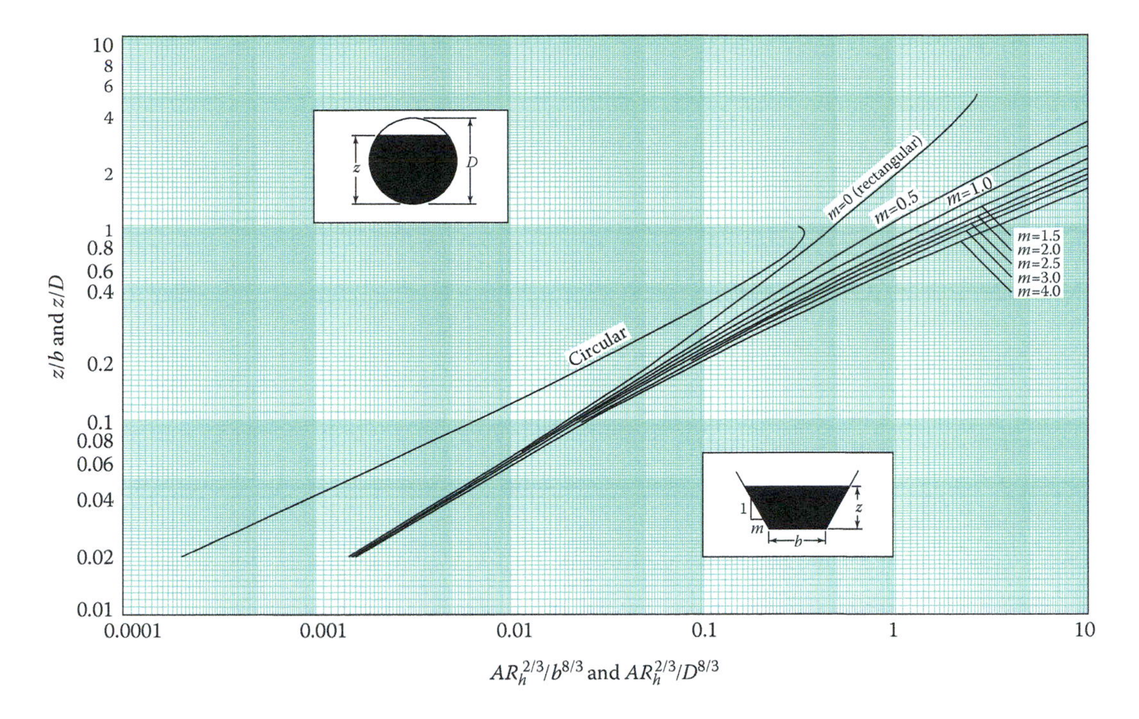

FIGURE 7.13 Curves for finding uniform depth in open-channel flow. (From Chow, V.T., *Open-Channel Hydraulics*, McGraw-Hill, Inc., New York, 1959, p. 130. With permission.)

With $b = 18$ ft,

$$\frac{AR_h^{2/3}}{b^{8/3}} = \frac{89.1}{(18)^{8/3}} = 0.040$$

Using Figure 7.13, enter the chart from the horizontal axis at 0.040; moving to the $m = 0$ line, read from the vertical axis:

$$\frac{z}{b} = 0.16$$

which yields

$$z = 0.16(18)$$

$$z = 2.88 \text{ ft}$$

This result is different by only 1.4% from the 2.92 ft found by trial and error in Example 7.6.

7.6 HYDRAULICALLY OPTIMUM CROSS SECTION

A *hydraulically optimum cross section* in open-channel flow is one that provides maximum conveyance or volume-carrying capacity for a given flow area. In the preceding discussion, we found that the discharge in an open channel can be obtained from the Manning formula for velocity:

$$Q = AV = \frac{C_m AR_h^{2/3} S^{1/2}}{n}$$

or

$$Q = \frac{C_m A^{5/3} S^{1/2}}{P^{2/3}\, n} \tag{7.34}$$

from which it is seen that for a given slope and area, the wetted perimeter P must be minimized if Q is to be maximized. Of the four cross sections in Table 7.1, the semicircle has the least perimeter with the same area. Thus, the semicircle is said to be the most hydraulically efficient of all cross sections. In practice, however, it may be impractical to construct a semicircular channel owing to conventional construction techniques and use of materials. (Digging and earth-moving methods produce planar walls more conveniently, for example, and materials such as bricks and wood are manufactured in planar configuration.) It is therefore of practical interest to examine hydraulically optimum cross sections for other shapes.

Consider a channel of rectangular cross section. The channel width is b, and the liquid depth is z. The flow area is

$$A = bz$$

or

$$b = \frac{A}{z}$$

and the wetted perimeter is

$$P = 2z + b$$

In terms of area,

$$P = 2z + \frac{A}{z}$$

For maximum discharge, the perimeter P must be a minimum; differentiating with respect to z and setting the result equal to zero yield

$$\frac{dP}{dz} = 2 - \frac{A}{z^2} = 0$$

Solving, we get

$$z^2 = \frac{A}{2} = \frac{bz}{2}$$

or

$$z = \frac{b}{2}$$

So for maximum flow through a rectangular open channel, the liquid depth should equal half the channel width.

Example 7.8

Determine the dimensions of a hydraulically optimum trapezoidal cross section using the notation of Figure 7.14.

Solution

The area for the trapezoidal section is

$$A = bz + \frac{z^2 \cos\alpha}{\sin\alpha} = bz + z^2 \cot\alpha \tag{i}$$

Rewriting, we have

$$b = \frac{A}{z} - z\cot\alpha$$

The wetted perimeter is

$$P = b + \frac{2z}{\sin\alpha} = \frac{A}{z} - z\cot\alpha + \frac{2z}{\sin\alpha}$$

Differentiating with respect to z gives

$$\frac{dP}{dz} = -\frac{A}{z^2} - \cot\alpha + \frac{2}{\sin\alpha}$$

For minimum perimeter, $dP/dz = 0$. Solving, we obtain

$$z^2 = \frac{A\sin\alpha}{2-\cos\alpha} \tag{ii}$$

Next differentiate z with respect to α and set the result equal to zero:

$$2z\frac{dz}{d\alpha} = A\left[\frac{(2-\cos\alpha)\cos\alpha - \sin\alpha(\sin\alpha)}{(2-\cos\alpha)^2}\right] = 0$$

which becomes

$$2\cos\alpha - 1 = 0$$

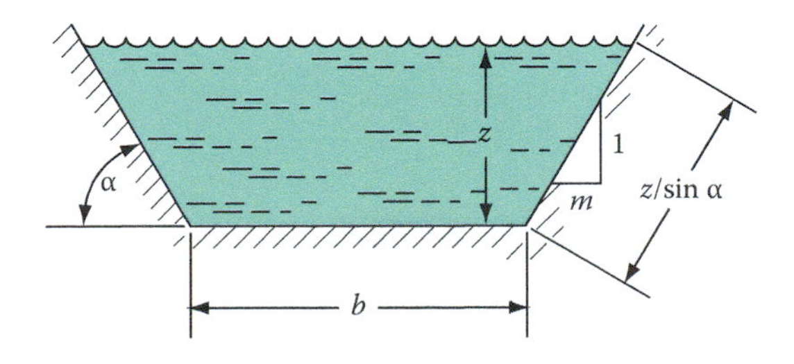

FIGURE 7.14 A trapezoidal cross section.

We find that

$$\cos\alpha = \frac{1}{2}$$

$$\alpha = 60°$$

Substituting into Equation ii for α gives

$$z^2 = \frac{A}{\sqrt{3}}$$

or

$$A = z^2\sqrt{3}$$

Combining with the equation for area (Equation i),

$$A = bz + z^2 \cot\alpha = bz + \frac{z^2}{\sqrt{3}}$$

we obtain

$$bz + \frac{z^2}{\sqrt{3}} = z^3\sqrt{3}$$

or

$$b = \frac{2\sqrt{3}}{3}z$$

The length of the sloping side is

$$\frac{z}{\sin\alpha} = \frac{z}{\sin 60} = \frac{2\sqrt{3}}{3}z = b$$

The wetted length along the sloping sides is the same as the bottom width *b*. Thus, for a given cross-sectional area, the trapezoidal section that gives a minimum perimeter is half a hexagon. This result can also be derived with the equations of Table 7.1.

7.7 NONUNIFORM OPEN-CHANNEL FLOW

As we mentioned earlier, we will confine our discussion to steady flows. Having previously examined uniform flows, in this section we focus on the two types of nonuniform flow: gradually varied flow and rapidly varied flow.

7.7.1 Gradually Varied Flow

In preceding sections, we considered only uniform flow in which the liquid surface in the open channel remained parallel to the channel bottom. We were able to develop expressions for velocity

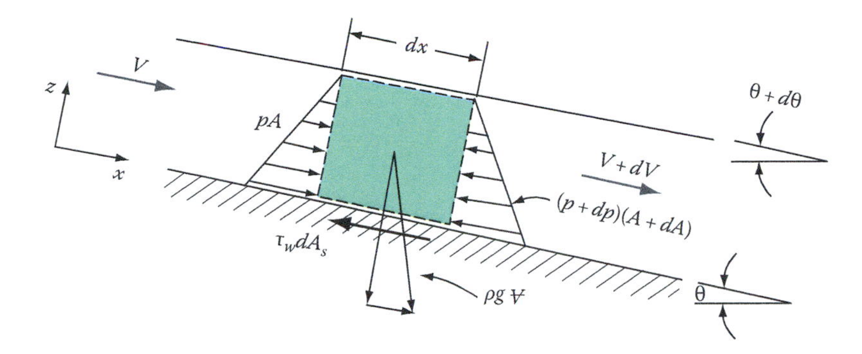

FIGURE 7.15 Gradually varied flow.

in both laminar and turbulent regimes. Although these expressions are adequate for most open-channel flows, the water depth in a river can change by as much as 100 ft over a distance of 5 m. In modeling such flows, it is important to remember that the change in depth is gradual. Here again, forces of pressure, viscosity, and gravity influence the flow. In contrast to uniform flow, we will see that pressure does influence the velocity of nonuniform flow. Because the change in flow depth is gradual, moreover, we can assume that frictional losses at any section are given by the Manning equation for uniform flow. Gravity is the primary driving force.

Figure 7.15 illustrates a control volume for analysis of gradually varied flow. The x-direction is taken to be in the direction of flow, the z-direction is normal to the channel bottom, and the y-direction is into the page. The momentum equation for the x-direction is

$$\sum F_x = \iint_{\text{CS}} V_x(\rho V_n)dA$$

Applied to Figure 7.15, we have

$$pA - (p + dp)(A + dA) - \tau_w dA_s + \rho gA\, dx \sin\theta = \rho Q(V + dV - V)$$

where

p is the pressure
A is the cross-sectional area
τ_w is the wall shear stress
dA_s is the wetted perimeter times dx
θ is the angle of inclination of the channel

Simplifying, we obtain

$$-p\, dA - A\, dp - dp\, dA - \tau_w P\, dx + \rho gA\, dx \sin\theta = \rho gA\, dV$$

Neglecting $dp\, dA$ as a second-order term leaves

$$-d(Ap) - \tau_w P\, dx + \rho gA\, dx \sin\theta = \rho gA\, dV \tag{7.35}$$

The continuity equation is

$$Q = AV = (A + dA)(V + dV)$$

or

$$AV = AV + A\,dV + V\,dA + dA\,dV$$

Again neglecting the second-order term, we rearrange this equation to solve for $V\,dV$:

$$V\,dV = -V^2 \frac{dA}{A}$$

Substituting into Equation 7.35 yields

$$-d(Ap) - \tau_w P\,dx + \rho g A\,dx \sin\theta = -\rho V^2\,dA \tag{7.36}$$

It is now necessary to evaluate the term

$$d(Ap) = A\,dp + p\,dA$$

This evaluation can be done only if a specific cross section is chosen. In this discussion, we will assume a rectangular section of width b and varying water depth z. The area is $A = bz$, and $dA = b\,dz$. The pressure at the centroid at any section is found with the hydrostatic equation

$$p = \rho g z_c$$

where z_c is the liquid depth from the surface to the centroid of the cross section. For a rectangular section of liquid depth z,

$$z_c = \frac{z}{2}\cos\theta$$

and so

$$p = \rho g \frac{z}{2}\cos\theta$$

By substitution, then,

$$d(AP) = d\left(bz\,\rho g \frac{z}{2}\cos\theta\right) = \left(\frac{b\,\rho g}{2}\cos\theta\right) d(z^2)$$

$$= \rho g b z\,dz\cos\theta$$

Substitution into Equation 7.36 yields

$$-\rho g b z\,dz\cos\theta - \tau_w P\,dx + \rho g b z\,dx\sin\theta = -\rho V^2 b\,dz$$

Dividing by bz and regrouping terms give

$$\left(\rho g \sin\theta - \tau_w \frac{P}{A}\right) dx = \left(\rho g \cos\theta - \frac{\rho V^2}{z}\right) dz$$

After further rearrangement, we have

$$\left(\cos\theta - \frac{V^2}{gz}\right)dz = \left(\sin\theta - \frac{\tau_w}{R_h \rho g}\right)dx \tag{7.37}$$

Equations 7.31 and 7.32 when combined yield the following:

$$\sqrt{\frac{8g}{f}} = \frac{1}{n}R_h^{1/6}$$

By combining this result with the following definition of τ_w,

$$\tau_w = \frac{f}{4}\frac{\rho V^2}{2} = \frac{f}{8g}\rho g V^2$$

we get

$$\tau_w = \frac{n^2}{R_h^{1/3}}\rho g V^2$$

Substitution into Equation 7.37 yields

$$\left(\cos\theta - \frac{V^2}{gz}\right)dz = \left(\sin\theta - \frac{n^2V^2}{R_h^{4/3}}\right)dx$$

or

$$\frac{dz}{dx} = \frac{\sin\theta - n^2V^2/R_h^{4/3}}{\cos\theta - V^2/gz} \tag{7.38}$$

which is the differential equation for gradually varied flow. Because this equation is nonlinear, it cannot be integrated directly to obtain a closed-form solution for the depth z. Instead, numerical means or a stepwise integration is required.

Before proceeding, however, let us investigate Equation 7.38 for uniform supercritical and subcritical flows. First, we use the approximation that $\cos\theta = 1$ for small θ. This assumption is reasonable for gradually varied flow. If the flow is uniform and $V^2/gz \neq 1$, then $dz/dx = 0$; Equation 7.38 becomes

$$0 = \sin\theta - \frac{n^2V^2}{R_h^{4/3}}$$

With $V = Q/A$, we have

$$\sin\theta = \frac{n^2Q^2}{A^2R_h^{4/3}}$$

or

$$Q = \frac{AR_h^{2/3}}{n}\sqrt{\sin\theta} \quad \text{(SI unit system)}$$

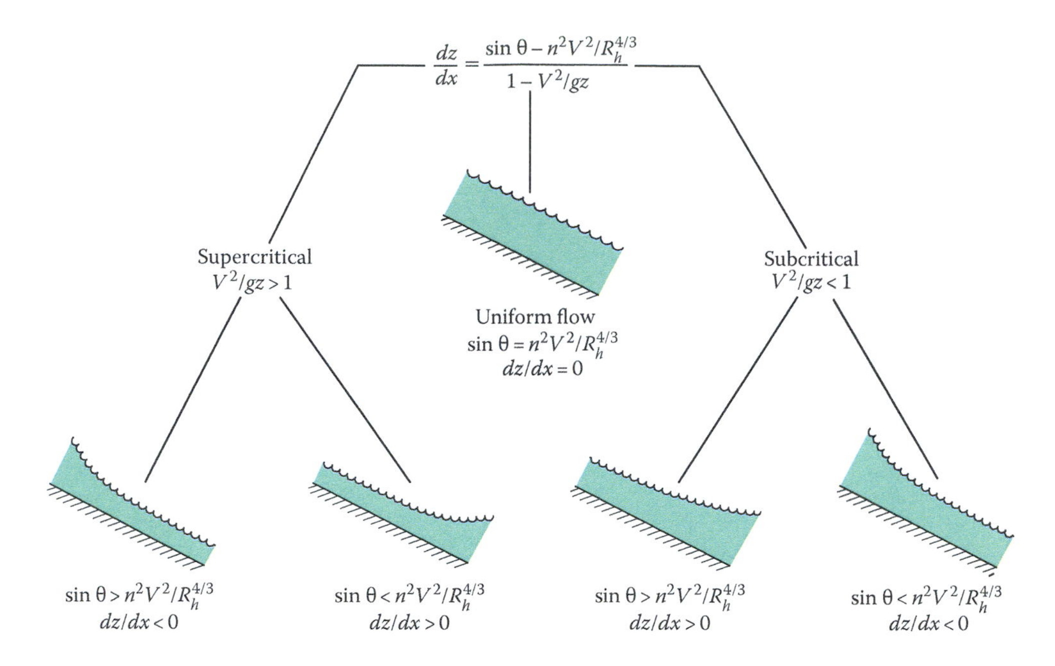

FIGURE 7.16 Analysis of Equation 7.38 with $\cos\theta \approx 1$.

which is identical to the previously derived equation (Equation 7.34) for Q for uniform flow. Only one depth corresponds to this condition of uniform flow for a given slope, flow rate, and area. If the flow is supercritical, then $V^2/gz > 1$, and the denominator of the right-hand side of Equation 7.38 is negative. If the slope $\sin\theta$ is greater than that corresponding to uniform flow, then dz/dx is negative. Thus, the depth decreases with distance downstream. Alternatively, if the slope $\sin\theta$ is less than that corresponding to uniform flow, then dz/dx is positive, and depth increases with distance downstream. For subcritical flow, $V^2/gz < 1$, and the opposite effects are predicted. These comments are summarized in Figure 7.16. The surface curves illustrated there can be calculated with Equation 7.38.

One important application of Equation 7.38 is in calculating a *backwater curve*—the liquid surface upstream of a barrier placed in the channel. The results obtained, in turn, depend on the upstream flow conditions. If the upstream flow is subcritical ($V < \sqrt{gz_{cr}}$), then Equation 7.38 predicts the situation depicted in Figure 7.17a. If the upstream flow is supercritical ($V > \sqrt{gz_{cr}}$), then Equation 7.38 predicts that height increases in the upstream direction as in Figure 7.17b. It is not physically possible for the height to increase in the upstream direction, because far upstream the backwater curve should smoothly intersect the uniform flow curve. Figure 7.17b shows the backwater curve diverging from it.

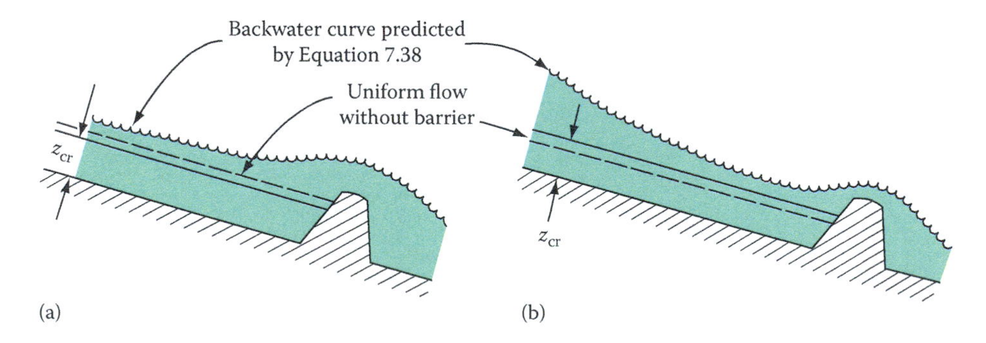

FIGURE 7.17 Flow upstream of a barrier. (a) Subcritical flow and (b) supercritical flow.

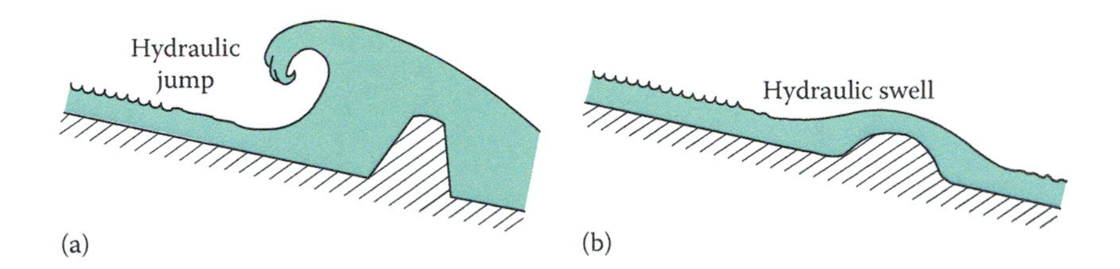

FIGURE 7.18 Supercritical flow adjustment to a barrier: (a) a hydraulic jump ahead of the barrier and (b) a hydraulic swell over the barrier.

When a supercritical flow encounters a barrier, the flow adjusts in the vicinity of the barrier, producing a rapidly varied flow such as a hydraulic jump (huge barrier with respect to the flow depth) or a rise and drop (small barrier). Figure 7.18 illustrates both of these flows.

From the previous discussion, we conclude that in backwater curve calculations, Equation 7.38 is useful only for the physical situation depicted in Figure 7.17a: subcritical flow. To use the equation for solving practical problems, we must rewrite it in a finite difference form as

$$\frac{\Delta z}{\Delta x} = \frac{\sin\theta - n^2 V_{\text{avg}}^2 / R_{h\text{avg}}^{4/3}}{\cos\theta - V_{\text{avg}}^2 / gz} \tag{7.39a}$$

where the subscript "avg" denotes an average value. It is apparent that over a channel length Δx, all the flow variables change. To overcome this difficulty, we use average values.

Example 7.9

The city of Memphis, Tennessee, is subject to heavy rainfalls during certain times of the year. Drainage ditches have been dug throughout the city. Most are concrete-lined and all drain ultimately into the Mississippi River.

Consider one such rectangular, concrete-lined channel dug out to a width of 2.5 m and inclined at a slope of 0.002. The channel is very long, and at one end there is a dam that partially restricts the flow. At the dam, the water depth is 3.5 m as sketched in Figure 7.19. Determine the variation of depth with upstream distance if the volume flow rate is 8 m³/s.

Solution

From Table 7.2 for a concrete-lined channel, $n = 0.012$. We must first determine whether the flow is subcritical. If not, a rapidly varied flow situation exists, and the backwater curve Equation 7.39a does not apply. To find critical depth, use Equation 7.17a:

$$\frac{V_{\text{cr}}^2}{g} = z_m = z_{\text{cr}}$$

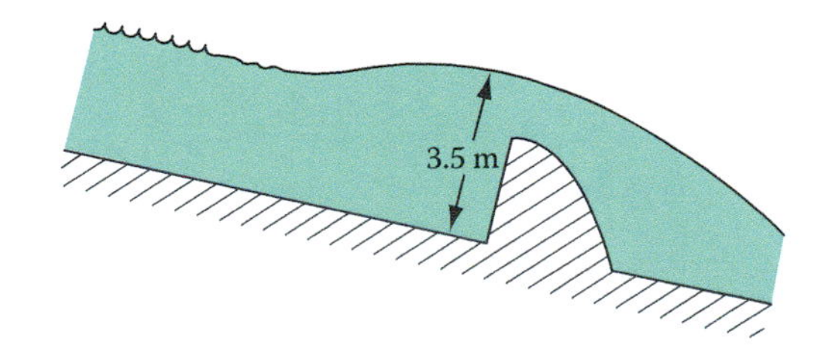

FIGURE 7.19 Backwater curve upstream of a dam.

With $V = Q/A = Q/bz_{cr}$, we have

$$\frac{Q^2}{b^2 z_{cr}^2 g} = z_{cr}$$

After rearranging and substituting,

$$z_{cr}^3 = \frac{Q^2}{b^2 g} = \frac{(8)^2}{(2.5)^2(9.81)} = 1.044$$

Solving, the critical flow depth is

$$z_{cr} = 1.014 \text{ m}$$

Next, to find the uniform flow depth, we can formulate a trial-and-error solution or we can use Manning's equation and Figure 7.13. Here we use the latter method:

$$Q = AV = A\frac{R_h^{2/3} S^{1/2}}{n} \qquad \text{(SI unit system)}$$

Rearranging and substituting,

$$AR_h^{2/3} = \frac{Qn}{S^{1/2}} = \frac{8(0.012)}{\sqrt{0.002}} = 2.147$$

So,

$$\frac{AR_h^{2/3}}{b^{8/3}} = \frac{2.147}{(2.5)^{8/3}} = 0.19$$

At this value, Figure 7.13 gives

$$\frac{z}{b} = 0.52$$

from which we find

$$z = 0.5(2.5) = 1.3 \text{ m}$$

The flow is subcritical because the uniform flow depth is greater than the critical depth; that is, $z > z_{cr}$. Equation 7.39a (which applies) is rewritten to solve for Δx:

$$\Delta x = \Delta z\left(\frac{\cos\theta - V_{avg}^2/gz_{avg}}{\sin\theta - n^2 V_{avg}^2/R_{havg}^{4/3}}\right)$$

After substituting, we get

$$\Delta x = \Delta z\left(\frac{1 - V_{avg}^2/9.81 z_{avg}}{0.002 - (0.012)^2 V_{avg}^2/R_{havg}^{4/3}}\right) \tag{i}$$

TABLE 7.3
Summary of Calculations for the Channel of Example 7.9

z	Δz	z_{avg}	$R_{havg}^{4/3} = \left(\dfrac{bz_{avg}}{b+2z_{avg}}\right)^{4/3}$	$V_{avg} = \dfrac{Q}{bz_{avg}}$	$1 - \dfrac{V_{avg}^2}{9.81 z_{avg}}$	$0.002 - \dfrac{(0.012)^2 V_{avg}^2}{R_{havg}^{4/3}}$	Δx (Equation i)
3.5							
	0.5	3.25	0.872 5	0.984 6	0.969 6	0.001 883	257
3.0							
	0.5	2.75	0.817 0	1.164	0.949 8	0.001 745	272
2.5							
	0.5	2.25	0.747 1	1.422	0.908 4	0.001 570	289
2.0							
	0.5	1.75	0.656 3	1.829	0.805 1	0.001 155	349
1.5							
	0.2	1.4	0.575 1	2.286	0.619 5	0.000 427	725
1.3							

Note: Channel data: $b = 2.5$ m, $n = 0.012$, $Q = 8$ m³/s, and $S = 0.002$.

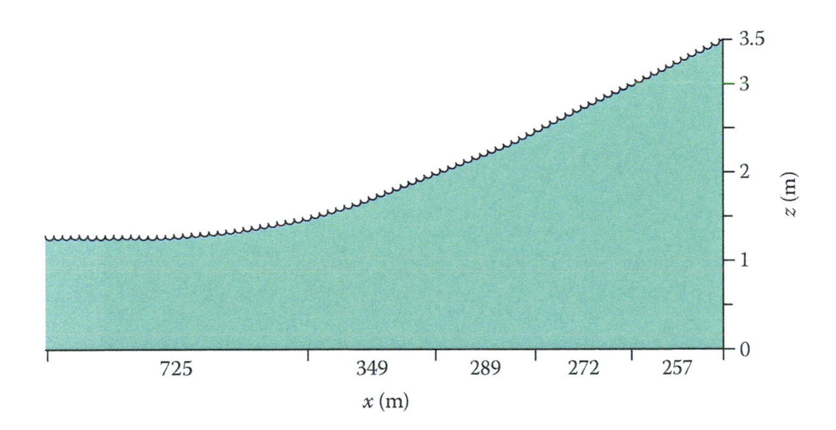

FIGURE 7.20 Backwater curve of Example 7.9.

Far upstream of the dam, the depth is 1.3 m. At the dam, the depth is given as 3.5 m. The depth z thus varies from 1.3 to 3.5 m. We arbitrarily divide this depth range into five increments, as shown in columns 1 and 2 of Table 7.3, which displays a summary of the calculations. The last column shows the values of Δx upstream of the dam. A sketch of the backwater curve (z versus x) is given in Figure 7.20.

Note that Equation 7.39a applies only to a rectangular channel. For other types, we have

$$\frac{dz}{dx} = \frac{\sin\theta - n^2V^2/R_h^{4/3}}{\cos\theta - 2V^2/gz} \quad \text{(triangular)} \tag{7.39b}$$

and

$$\frac{dz}{dx} = \frac{\sin\theta - n^2V^2/R_h^{4/3}}{\cos\theta - V^2(b+2mz)/\left[g(b+mz)z\right]} \quad \text{(trapezodial)} \tag{7.39c}$$

The derivation of each of these formulas is reserved as an exercise. By inspection of Equation 7.39 and Table 7.1, we conclude that the general form of these equations is

$$\left(\underset{\text{gravity term}}{\sin\theta} - \underset{\text{friction term}}{\frac{n^2V^2}{R_h^{4/3}}}\right)dx = \left(\underset{\text{pressure term}}{\cos\theta} - \underset{\text{momentum or inertia term}}{\frac{V^2}{gz_m}}\right)dz$$

Example 7.10

The city of New Orleans, Louisiana, is situated at a relatively low elevation. It is bounded on the north by Lake Pontchartrain and on the south by the Mississippi River. To prevent flooding, levees have been built along the shores of these water masses. Drainage canals have been dug within the city; each leads to the edge of the levee, where a pump is used at appropriate times (as during heavy rainfall) to pump water from the canal and over the levee into the lake.

Consider a single drainage canal, trapezoidal in cross section and lined with concrete. The bottom width is 11 ft, and the channel sides slope upward at 60°. During a rainfall of significant amount, the canal water rises to a point at which the pump is actuated and begins moving water over the levee. After the rainfall, water still drains into the canal from streets, and the pump continues working until the water level drops to nearly its original depth. For this example, we will examine the water surface profile upstream of the pump but neglect the runoff contribution to the flow. The channel slopes downward in the flow direction at an angle of $S = 0.0001$. The pump flow rate is 350 ft³/s. At the pump, the water depth is 5.5 ft. Determine the shape of the backwater curve (see Figure 7.21).

Solution

For a concrete-lined channel, $n = 0.012$ from Table 7.2. From Table 7.1 for a trapezoidal channel,

$$A = (b + mz)z$$

$$R_h = \frac{(b + mz)z}{b + 2z\sqrt{1 + m^2}}$$

$$z_m = \frac{(b + mz)z}{b + 2mz}$$

$$Z = \frac{[(b + mz)z]^{1.5}}{\sqrt{b + 2mz}}$$

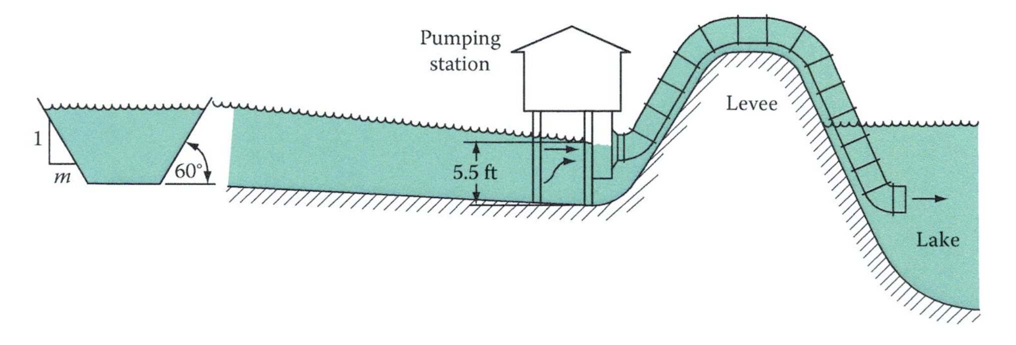

FIGURE 7.21 Schematic of the system described in Example 7.10.

From the geometry of the section,

$$\tan(90° - \alpha) = m$$

If $\alpha = 60°$, then

$$m = \tan(30°) = 0.577$$

With $b = 11$ ft, by substitution we get

$$A = (11 + 0.577z)z$$

$$R_h = \frac{(11 + 0.577z)z}{11 + 2.309z}$$

$$z_m = \frac{(11 + 0.577z)z}{11 + 1.154z}$$

$$Z = \frac{(11 + 0.577z)z}{\sqrt{11 + 1.154z}}$$

The critical depth of the flow can be calculated with Equation 7.17a:

$$\frac{V_{cr}^2}{g} = z_m$$

From continuity, $V_{cr} = Q/A$; therefore,

$$\frac{Q^2}{(11 + 0.577z_{cr})^2 z_{cr}^2 g} = \frac{(11 + 0.577z_{cr})z_{cr}}{11 + 1.154z_{cr}}$$

$$\frac{350^2}{(11 + 0.577z_{cr})^2 z_{cr}^2 (32.2)} = \frac{(11 + 0.577z_{cr})z_{cr}}{11 + 1.154z_{cr}}$$

Or, after simplifying,

$$\frac{(11 + 0.577z_{cr})^3 z_{cr}^3}{11 + 1.154z_{cr}} = 3804$$

By trial and error with a calculator or goal seek with a spreadsheet,

$$z_{cr} = 2.99 \text{ ft}$$

Alternatively, Figure 7.8 can be used to find the critical depth as follows. Equation 7.18 gives

$$Z = \frac{Q}{\sqrt{g}} = \frac{350}{\sqrt{32.2}} = 61.9$$

To use the chart, we first find

$$\frac{Z}{b^{2.5}} = \frac{61.9}{11^{2.5}} = 0.154$$

Using this value with $m = 0.577$, we find from Figure 7.8 that

$$\frac{z_{cr}}{b} = 0.26$$

So

$$z_{cr} = 0.26(11) = 2.86 \text{ ft}$$

This result is 4% different from the 2.99 ft obtained by trial and error. Although use of either value is permissible, 2.99 ft is closer to being exact. To find the uniform flow depth, we use Manning's equation and Figure 7.13:

$$Q = AV = \frac{1.49AR_h^{2/3}S^{1/2}}{n} \quad \text{(U.S. customary units)}$$

$$\frac{AR_h^{2/3}}{b^{8/3}} = \frac{Qn}{S^{1/2}b^{8/3}(1.49)}$$

$$= \frac{(350)(0.012)}{(0.0001)^{1/2}(11)^{8/3}(1.49)} = 0.471$$

Using this value with $m = 0.577$, we find from Figure 7.13 that

$$\frac{z}{b} = 0.66$$

and for uniform flow, we obtain

$$z = 0.66(11) = 7.26 \text{ ft}$$

Since the uniform flow depth is greater than the critical depth, the flow is subcritical, and Equation 7.39c applies. At the pump, the depth is 5.5 ft; far upstream, the depth is uniform at 7.26 ft. The depth range is divided into four increments arbitrarily, and Equation 7.39c is rewritten in difference form:

$$\Delta x = \Delta z\left\{\frac{1 - V_{avg}^2(11 + 0.577z_{avg})/[(32.2)(11 + 0.577z_{avg})z_{avg}]}{0.0001 - n^2V_{avg}^2/(R_{h\,avg}^{4/3}(1.49)^2)}\right\} \tag{i}$$

Table 7.4 summarizes the calculations for this equation. A plot of the results is given in Figure 7.22.

7.7.2 Rapidly Varied Flow

As previously discussed, flow in an open-channel can adjust in two ways to a barrier placed in the stream. If the flow is subcritical, the flow adjusts gradually, and Equation 7.39 can be applied. If the flow is supercritical, the flow adjusts over a relatively short channel length and is therefore called a rapidly varied flow. An example of such a flow is the hydraulic jump. Downstream of the jump, the flow becomes subcritical. Because this change occurs over a short length of channel, the energy loss is considerable. One common example of a hydraulic jump is in a kitchen sink where

TABLE 7.4

Summary of Calculations for the Channel of Example 7.10

z	Δz	z_{avg}	$A_{avg} = (11 + 0.577z_{avg})z_{avg}$	$R_{havg}^{4/3} = \left(\frac{A_{avg}}{11+2.309z_{avg}}\right)^{4/3}$	$V_{avg} = \frac{Q}{A_{avg}}$	$\frac{(1-V_{avg}^2)(11+0.577z_{avg})}{[(32.2)(11+0.577z_{avg})z_{avg}]}$	$\left(0.0001-\frac{n^2V_{avg}^2}{R_{havg}^{4/3}(1.49)^2}\right)$	Δx (Equation i)
7.26								
	0.26	7.13	107.8	6.192	3.247	0.9416	-1.171×10^{-5}	−20,906
7.0								
	0.5	6.75	100.5	5.889	3.486	0.9296	-3.516×10^{-5}	−13,219
6.5								
	0.5	6.25	91.29	5.496	3.834	0.9089	-7.548×10^{-5}	−6,020
6.0								
	0.5	5.75	82.33	5.095	4.251	0.8798	-1.327×10^{-4}	−3,311
5.5								$\Sigma\Delta x = -43{,}466$ ft

Note: Channel data: $b = 11$ ft, $m = 0.577$, $Q = 350$ ft^3/s, $n = 0.012$, and $S = 0.0001$.

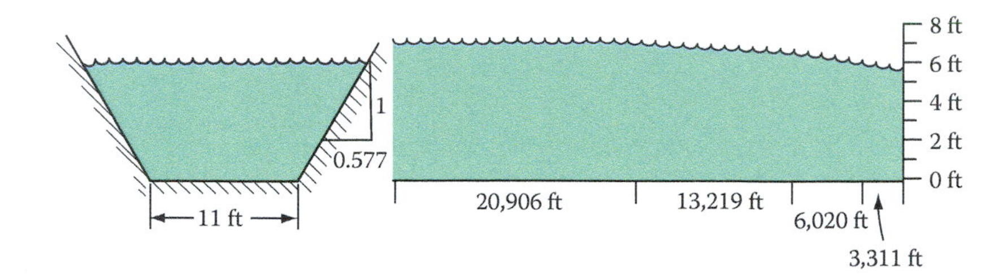

FIGURE 7.22 Backwater curve for the trapezoidal channel of Example 7.10.

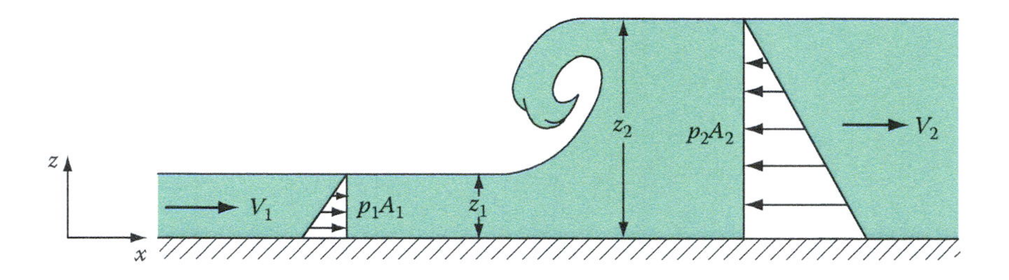

FIGURE 7.23 A hydraulic jump.

a downward-moving jet strikes the flat horizontal surface; the flow moves outward radially along the sink surface, and a jump occurs.

In this discussion, we will consider jumps occurring in rectangular cross sections; other cross sections are reserved for the problems at the end of the chapter. A hydraulic jump is illustrated in profile view in Figure 7.23. The channel bed is horizontal. The forces acting are due to gravity and a change in pressure across the jump. It is permissible to neglect the forces due to viscosity because of the relatively short channel length and the highly turbulent nature of the jump.

The momentum equation for one-dimensional flow becomes

$$\sum F_x = \rho Q(V_2 - V_1)$$

or

$$p_1A_1 - p_2A_2 = \rho Q(V_2 - V_1) \tag{7.40}$$

From hydrostatics, for a rectangular cross section,

$$p = \rho g z_c$$

where z_c is the distance from the centroid of the liquid cross section to the surface

For a rectangular section,

$$z_c = \frac{z}{2}$$

(See Table 7.1 for other cross sections.) By substitution into Equation 7.40, we obtain

$$\rho g \frac{z_1}{2}(bz_1) - \rho g \frac{z_2}{2}(bz_2) = \rho Q(V_2 - V_1) \tag{7.41}$$

From continuity,

$$Q = AV = bz_1V_1 = bz_2V_2$$

Combining with Equation 7.41 and simplifying give

$$\frac{bz_1^2}{2} + \frac{Q^2}{gbz_1} = \frac{bz_2^2}{2} + \frac{Q^2}{gbz_2} \tag{7.42}$$

Rearranging, we get

$$z_2^2 - z_1^2 = \frac{2Q^2}{gb^2}\left(\frac{1}{z_1} - \frac{1}{z_2}\right) = \frac{2Q^2}{gb^2}\left(\frac{z_2 - z_1}{z_1z_2}\right)$$

Dividing by $z_2 - z_1$ gives

$$z_2 + z_1 = \frac{2Q^2}{gb^2z_1z_2} \tag{7.43}$$

from which we get

$$z_2^2 + z_1z_2 - \frac{2Q^2}{gb^2z_1} = 0$$

This quadratic equation can be solved for the downstream height as

$$z_2 = -\frac{z_1}{2} \pm \sqrt{\frac{z_1^2}{2} + \frac{2Q^2}{gb^2z_1}} \tag{7.44a}$$

We reject the negative root because a negative z_2 has no physical meaning. The equation can now be rearranged to give dimensionless quantities:

$$\frac{z_2}{z_1} = -\frac{1}{2} + \sqrt{\frac{1}{4} + \frac{2Q^2}{gb^2z_1^3}} \tag{7.44b}$$

A graph of this equation is given in Figure 7.24, where the horizontal axis is the upstream Froude number and the vertical axis is z_2/z_1. If $z_1 = z_2$, then it is obvious that no jump has occurred, so the only portion of the graph that is physically significant is the portion corresponding to $z_2/z_1 > 1$. The upstream Froude number, as mentioned earlier, must be greater than 1 (supercritical flow) if a jump is to exist.

The energy loss across the jump, gh_j, can be calculated by using the modified Bernoulli equation applied at the free surface:

$$\frac{p_{atm}}{\rho} + \frac{V_1^2}{2} + gz_1 = \frac{p_{atm}}{\rho} + \frac{V_2^2}{2} + gz_2 + gh_j$$

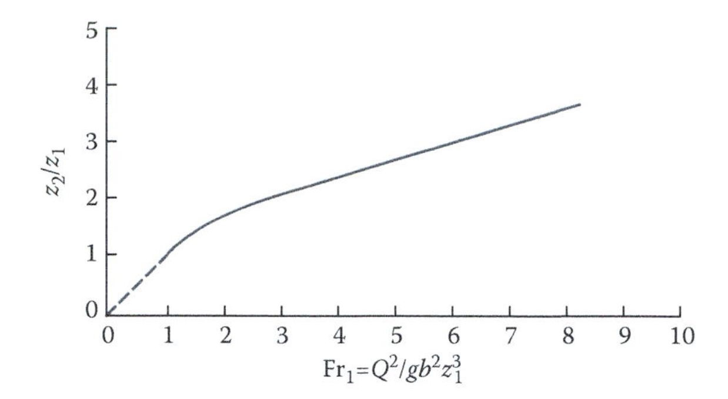

FIGURE 7.24 Depth ratio for hydraulic jump as a function of upstream Froude number.

from which we have

$$gh_j = \frac{V_1^2 - V_2^2}{2} + g(z_1 - z_2)$$

By combining with the continuity equation,

$$Q = AV = bz_1 V_1 = bz_2 V_2$$

we get

$$gh_j = \frac{Q^2}{2b^2 z_1^2} - \frac{Q^2}{2b^2 z_2^2} + g(z_1 - z_2) \tag{7.45}$$

Substitution for Q^2/b^2 from Equation 7.43 gives

$$gh_j = \frac{(z_2 + z_1)gz_2}{2(2z_1)} - \frac{(z_2 + z_1)gz_1}{2(2z_2)} + g(z_1 - z_2)$$

which finally becomes

$$h_j = \frac{(z_2 - z_1)^3}{4z_1 z_2} \tag{7.46a}$$

which is valid only for a rectangular channel. Thus, for a given volume flow, the rate of energy loss in a hydraulic jump is

$$\frac{dW}{dt} = \rho g Q h_j \tag{7.46b}$$

An alternative representation of hydraulic jump can be developed from a graph of height z as a function of the parameter appearing in Equation 7.42—namely, the momentum of the flow:

$$M = \frac{Q^2}{gbz} + \frac{bz^2}{2} \tag{7.47}$$

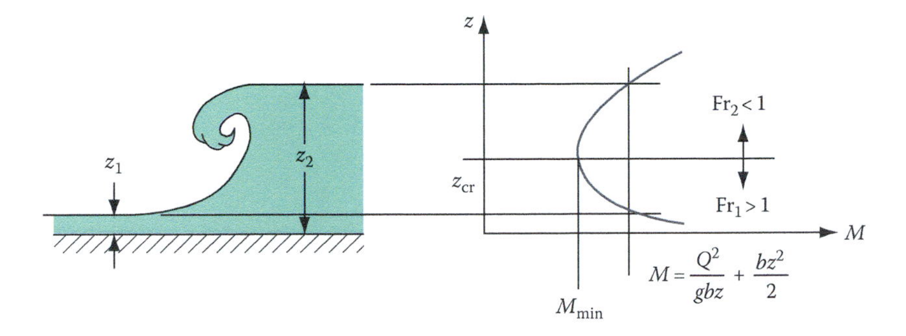

FIGURE 7.25 A momentum diagram for a hydraulic jump.

Figure 7.25 illustrates the relationship between this graph and a hydraulic jump. The curve is similar to a specific-energy diagram but should not be confused with it. The heights z_1 and z_2 are known as *conjugate depths* of the jump that lie on a vertical line through the graph. The critical depth z_{cr} occurs when M is a minimum. When $z < z_{cr}$, the flow is supercritical, and the curve corresponds to the upstream condition. When $z > z_{cr}$, the flow is subcritical, and the curve corresponds to the downstream condition. The momentum of the flow in both conditions is the same.

Example 7.11

A hydraulic jump occurs in a channel 1 m wide that conveys water at 0.6 m³/s. The upstream depth of flow is 0.2 m.

a. Determine the downstream height.
b. Calculate the energy loss in the jump.
c. Sketch a momentum diagram for the channel and show where the operating point appears.

Solution

a. For water, ρ = 1 000 kg/m³. For the hydraulic jump of the problem, Equation 7.44b applies:

$$\frac{z_2}{z_1} = -\frac{1}{2} + \sqrt{\frac{1}{4} + \frac{2Q^2}{gb^2 z_1^3}}$$

By substitution, we find

$$\frac{2Q^2}{gb^2 z_1^3} = \frac{2(0.6)^2}{9.81(1)^2(0.2)^3} = 9.17$$

and so

$$\frac{z_2}{z_1} = -\frac{1}{2} + 3.07 = 2.57$$

Therefore,

$$z_2 = 2.57(0.2)$$

$$z_2 = 0.514 \text{ m}$$

b. The rate of energy loss in the jump is found with

$$h_j = \frac{(z_2 - z_1)^3}{4z_1z_2} = \frac{(0.514 - 0.2)^2}{4(0.514)(0.2)}$$

$$h_j = 0.075\,3\,\text{m}$$

The rate of energy loss is

$$\frac{dW}{dt} = \rho g Q h_j = (1\,000\text{ kg/m}^3)(9.81\text{ m/s}^2)(0.6\text{ m/s}^3)(0.075\,3\text{ m})$$

or

$$\frac{dW}{dt} = 443\,\text{W}$$

c. The momentum for this flow situation is

$$M = \frac{Q^2}{gbz} + \frac{bz^2}{2} = \frac{(0.6)^2}{9.81(1)z} + \frac{z^2}{2}$$

or

$$M = \frac{0.036\,7}{z} + \frac{z^2}{2}$$

This equation is written for a constant flow rate. A plot of this equation is given in Figure 7.26. The operating point is shown.

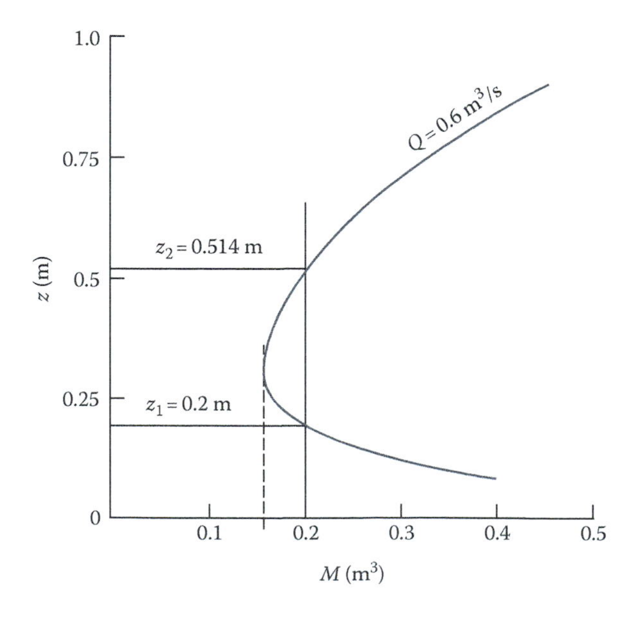

FIGURE 7.26 Graph of the momentum function for the channel of Example 7.11.

Similar developments for other than rectangular cross sections lead to the following formulas for a hydraulic jump:

Triangular channel:

$$\left(\frac{z_1^3}{z_3^2}+\frac{z_1}{z_2}+1\right)\frac{z_2^2}{z_1+z_2}=\frac{3Q^2}{m^2 g z_2^2 z_1^2} \tag{7.48}$$

Trapezoidal channel:

$$3b(z_1+z_2)+2m(z_1^2+z_1z_2+z_2^3)=\frac{6Q^2}{gz_1z_2}\left[\frac{b+m(z_1+z_2)}{(b+mz_2)(b+mz_1)}\right] \tag{7.49}$$

The derivation of these equations is left as an exercise.

7.8 SUMMARY

This chapter introduced some of the simpler topics of open-channel flow. We saw that various types of flows can exist and that geometry greatly influences the flow itself. We calculated energies associated with open-channel flows; from these, we determined that critical depth and uniform flow depth are very significant. We were able to formulate an expression for velocity in both laminar and turbulent regimes and used the turbulent expression to predict backwater surface profiles. We also found the hydraulically optimum cross section, and finally, we wrote descriptive equations for hydraulic jumps. This introduction to open-channel flow constitutes a solid foundation for further study in hydraulics.

INTERNET RESOURCES

1. Rapidly varied flow has an erosive effect on the channel downstream. How is the erosive effect minimized? Are there commercial devices for doing this? Give a brief presentation on this topic.
2. The Bonnet Carre spillway is located in Lake Charles, LA. When is it opened, and what is its purpose? Is it similar in operation as flow under a sluice gate? Give a brief presentation on spillways in general.

PROBLEMS

7.1 The right-hand side of Equation 7.7 for flow under a sluice gate in a rectangular channel is $\frac{Q^2}{b^2 z_1^3 g}$. Show that this term is really the Froude number upstream of the gate.

7.2 A rectangular channel 8 ft wide conveys water at 250 ft^3/s. A partially raised sluice gate in the channel restricts the flow. Determine the downstream liquid height if the upstream height is 6 ft.

7.3 A rectangular channel 0.5 m wide conveys water. A partially raised sluice gate in the channel restricts the flow such that the upstream height is 1 m and the downstream height is 0.25 m. Determine the volume flow rate under the gate.

7.4 A rectangular channel 3 m wide conveys water at a volume flow rate of 30 m^3/s. A partially raised sluice gate in the channel restricts the flow so that the downstream water height is 0.7 m. Determine the upstream water height.

7.5 Plot the sluice gate equation (Equation 7.7) to obtain a curve similar to that in Figure 7.4a.

a. Plot the following data on the same graph to determine how well the sluice gate equation describes actual conditions:

$Q_{ac} = 0.341$ ft/s (all case), $b = 0.5$ ft

z_1 (ft)	z_2 (ft)
0.479	0.167
0.500	0.161
0.510	0.156
0.552	0.155
0.625	0.135
0.635	0.135
0.667	0.125
0.708	0.104
0.906	0.0989

b. Use the height data with the sluice gate equation to find $\dfrac{Q_{th}^2}{b^2 z_1^3 g}$

c. Compare to $\dfrac{Q_{ac}^2}{b^2 z_1^3 g}$.

7.6 a. Derive the sluice gate equation for flow in a triangular cross-sectional channel:

$$2\left(\frac{z_2}{z_1}\right)^4\left(1-\frac{z_2}{z_1}\right)=\frac{Q^2}{m^2 z_1^5 g}$$

b. Construct a plot of z_2/z_1 versus $\dfrac{Q^2}{m^2 z_1^5 g}$

c. Differentiate the expression to find the location of z_2/z_1 that corresponds to Q_{max}.

d. Substitute this ratio into the equation and evaluate $\dfrac{Q_{max}^2}{m^2 z_1^5 g}$

7.7 Derive the sluice gate equation for a trapezoidal cross-sectional channel:

$$\frac{(bz_2+mz_2^2)^2}{(bz_1+mz_1^2)^2}\left(1-\frac{z_2}{z_1}\right)=\frac{Q^2}{2(bz_1+mz_1^2)^2 z_1 g}$$

7.8 Derive the sluice gate equation for open-channel flow in a circular channel.

7.9 A venturi flume is placed in a rectangular channel. The upstream channel width is 3.5 m, and the minimum area section is of width 2.1 m. The upstream water height is 2.2 m; the height at the throat is 1 m. Determine the volume flow rate through the channel.

7.10 A venturi flume is placed in a rectangular channel that conveys water at 250 ft^3/s. The upstream channel width is 12 ft, and the minimum width in the flume is 10 ft. The upstream water depth is 8 ft. What is the expected water depth at the minimum width?

7.11 An open-channel-flow apparatus in a hydraulics laboratory is rectangular in cross section and 16 in. wide. A venturi flume is to be placed in the channel to give the volume flow rate. The water height upstream of the flume is 18 in., and at the minimum width it is desired to make the depth about 12 in. Determine the minimum width of the venturi flume for a volume flow rate of 7 ft^3/s.

Specific-Energy Diagram

7.12 Consider a rectangular channel of width 2 ft.
- a. Plot a family of specific-energy lines for $Q = 0, 3, 6, 10$, and $20\ ft^3/s$.
- b. Draw the locus of critical depth points.
- c. Plot Q as a function of critical depth.

7.13 Consider a rectangular channel of width 12 m.
- a. Plot a family of specific-energy lines for $Q = 0, 8, 12, 20$, and $35\ m^3/s$.
- b. Draw the locus of critical depth points.
- c. Plot Q as a function of critical depth.

7.14 Consider a triangular channel of $m = 0.4$.
- a. Plot a family of specific-energy lines for $Q = 0, 4, 8, 12$, and $16\ ft^3/s$.
- b. Draw the locus of critical depth points.
- c. Plot Q versus critical depth.

7.15 Consider a triangular channel of $m = 1.2$.
- a. Plot a family of specific-energy lines for $Q = 0, 5, 10, 15$, and $20\ m^3/s$.
- b. Draw the locus of critical depth points.
- c. Plot Q versus critical depth.

7.16 Consider a trapezoidal channel with $b = 2$ ft and $m = 0.4$.
- a. Plot a family of specific-energy lines for $Q = 0, 3, 6, 12$, and $20\ ft^3/s$.
- b. Draw the locus of critical depth points.
- c. Plot Q versus critical depth.

7.17 Consider a trapezoidal channel of width $b = 12$ m and $m = 0.6$.
- a. Plot a family of specific-energy lines for $Q = 0, 5, 10, 20$, and $30\ m^3/s$.
- b. Draw the locus of critical depth points.
- c. Plot Q versus critical depth.

7.18 Consider a circular channel of diameter 2 m carrying water.
- a. Plot a family of specific-energy lines for $Q = 0, 2, 5, 10$, and $20\ m^3/s$.
- b. Draw the locus of critical depth points.
- c. Plot Q versus critical depth.
- d. Plot depth z versus angle α.

Critical Depth and Velocity

7.19 A rectangular channel conveys water at $120\ ft^3/s$ and has a width of 12 ft. Determine the critical depth and the corresponding critical velocity.

7.20 A rectangular channel conveys water at $100\ m^3/s$ and has a width of 15 m. Calculate the critical depth and corresponding critical velocity.

7.21 A triangular channel with sides having $m = 1.2$ conveys water at $50\ ft^3/s$. Determine the critical depth and the corresponding critical velocity.

7.22 A triangular channel with sides having $m = 1.5$ conveys water at $25\ m^3/s$. Determine the critical depth and the corresponding critical velocity.

7.23 A trapezoidal channel with $b = 8$ ft and $m = 1.0$ conveys water at $50\ ft^3/s$. Determine the critical depth and the corresponding critical velocity.

7.24 A trapezoidal channel with $b = 4$ m and $m = 1.2$ conveys water at $30\ m^3/s$. Determine the critical depth and the corresponding critical velocity.

7.25 A trapezoidal channel carries water at a flow of $200\ ft^3/s$. Its width is 8 ft, and $m = 1.0$. The channel is made of concrete. Calculate the critical depth of the flow.

7.26 A circular channel with $D = 3.5$ m conveys water at $12\ m^3/s$. Determine the critical depth and the corresponding critical flow velocity.

7.27 Determine the critical depth for each of the following cases assuming where applicable $b = 10$ m, $m = 1.0$, and $Q = 200$ m^3/s: (a) a triangular channel, (b) a rectangular channel, and (c) a trapezoidal channel.

7.28 Show that alternate depths in a rectangular channel are related to the critical depth by

$$z_{cr}^3 = \frac{2z_1^2 z_2^2}{z_2 + z_1}$$

7.29 A triangular channel with $m = 1.2$ is made of cement rubble masonry and conveys water at a flowrate of 50 ft^3/s. At what slope will the flow be critical?

7.30 A rectangular channel of width 6 m. Calculate the critical depth and velocity of flow for a volume flow rate of 80 m^3/s.

7.31 Figure 7.8 for determining critical depth in trapezoidal and circular sections was presented with no information about how the curves were derived. The derivation is investigated in this problem for a rectangular cross section. The graph of Figure 7.8 is of z_{cr}/b versus $Z/b^{2.5}$ and so we are seeking a relationship between these two variables for $m = 0$. In terms of the critical depth z_{cr}, the section factor Z for a rectangular channel is given in Table 7.1 as

$$Z = bz_{cr}^{1.5}$$

a. By manipulating this equation appropriately, show that

$$\frac{z_{cr}}{b} = \left(\frac{Z}{b^{2.5}}\right)^{1/1.5}$$

b. Select five different points on the graph of Figure 7.8 and show that they fit this equation.

7.32 Figure 7.8 for determining critical depth in trapezoidal and circular sections was presented with no information about how the curves were derived. The derivation is investigated in this problem for a trapezoidal cross section. The graph of Figure 7.8 is of z_{cr}/b versus $Z/b^{2.5}$, and so we are seeking a relationship between these two variables for any m. In terms of the critical depth z_{cr}, the section factor Z for a trapezoidal channel is given in Table 7.1 as

$$Z = \frac{[(b + mz_{cr})z_{cr}]^{1.5}}{(b + 2mz_{cr})^{0.5}}$$

a. Divide this equation by $b^{2.5}$ and appropriately manipulate the right-hand side to show that

$$\frac{Z}{b^{2.5}} = \frac{\left(z_{cr}/b + mz_{cr}^2/b^2\right)^{1.5}}{\left(1 + 2mz_{cr}/b\right)^{1.5}}$$

b. Select five different points on the graph of Figure 7.8 and show that they fit this equation.

Laminar Open-Channel Flow

7.33 For laminar flow down an incline, the velocity is (Equation 7.24b)

$$V_x = \frac{\rho g h^2}{2\mu}\left(\frac{2z}{h} - \frac{z^2}{h^2}\right)$$

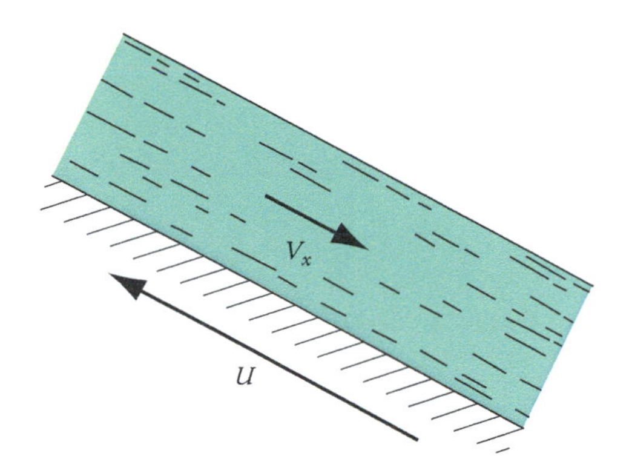

FIGURE P7.37

Integrate this equation over the area and show that the flow rate is (Equation 7.27)

$$Q = \frac{\rho g h^3 b}{3\mu} \sin\theta$$

7.34 For laminar flow down an incline, the velocity is (Equation 7.24b)

$$V_x = \frac{\rho g h^2}{2\mu}\left(\frac{2z}{h} - \frac{z^2}{h^2}\right)$$

Derive an equation for the average velocity.

7.35 A thin film of glycerine flows down an inclined surface. The film thickness is 0.2 in., and the angle of inclination is 0.18. Plot the velocity distribution in the film. Calculate the Reynolds number of the flow.

7.36 A 1 mm thick film of carbon tetrachloride flows down an incline. The free-surface velocity is 3 mm/s. Plot the velocity distribution in the film.

7.37 A thin layer of liquid is falling downward along an upward-moving inclined surface, as indicated in Figure P7.37. The liquid is Newtonian, and it is flowing under laminar conditions. Determine the wall velocity U such that the liquid surface velocity is zero.

7.38 A thin layer of liquid is falling downward along an upward-moving inclined surface, as indicated in Figure P7.37. The liquid is Newtonian and it is flowing under laminar conditions. Determine the wall velocity U such that there is no net flow of liquid down the incline.

Turbulent Open-Channel Flow

7.39 Rectangular channels made of brick and known as aqueducts were once used to convey water over land. Consider an aqueduct 6 ft wide, filled to a depth of 4 ft, and laid on a slope of 0.0008. Determine the volume carrying capacity of the channel.

7.40 Cooling water used in condensers of power plants is sometimes conveyed to cooling ponds before being returned to the body of water from which it came. Since the cooling pond gives the water opportunity to lose heat, the risk of thermal pollution is reduced. One such pond is fed by a channel that is rectangular in cross section, 3 m wide, and formed by dredging. The channel is well maintained, uniform, and weathered. If the water depth is 2 m and the channel slope is 0.09, calculate the flow rate in the channel.

7.41 A triangular channel is asphalt-lined and has $m = 1.2$. The channel is laid on a slope of 0.001 and filled to a depth of 2 m. Calculate the volume-carrying capacity of the channel.

7.42 A trapezoidal channel carries water at a flow of 200 ft^3/s. Its width is 8 ft, and $m = 1.0$. The channel is made of concrete. Calculate the slope necessary to maintain a uniform depth of 3 ft.

7.43 A circular culvert 4 ft in diameter and made of corrugated metal is filled to a depth of 3.2 ft. Determine the volume flow rate through the culvert and the critical depth of flow. Take the slope to be 0.001.

7.44 A triangular channel made of clay conveys water and has $m = 1.6$. If the channel slope is 0.002 and the water depth is 1 m, determine the volume flow rate and the critical depth of flow.

7.45 Water is flowing in a 6 m diameter culvert at a depth of 4.7 m. The culvert is uniform, made of concrete, and laid on a slope of 1°. Determine the volume flow rate through the culvert.

Uniform Open-Channel Flow

7.46 A trapezoidal channel has $b = 2$ m and conveys water at 60 m^3/s. The channel is made of wood and is laid on a slope of 0.001. Calculate m if the water depth is uniform at 4 m.

7.47 A rectangular cement channel 4 ft wide conveys water at 60 ft^3/s. The channel slope is 0.001. Determine the liquid depth for uniform flow. Is the flow supercritical or subcritical?

7.48 A rectangular brick-lined channel is 2.5 m wide and conveys water at 50 m^3/s. The channel slope is 0.000 8. Determine the liquid depth for uniform flow and for critical flow. Is the flow supercritical or subcritical?

7.49 A triangular channel with $m = 1.1$ is made of cemented rubble masonry and laid on a slope of 0.000 5. The channel conveys water at a flowrate of 10 m^3/s. Determine the depth for uniform flow and for critical flow. Is the flow supercritical or subcritical?

7.50 An excavated earth-lined trapezoidal channel that is poorly maintained conveys water at a flow of 2 m^3/s. If the channel slope is 0.001, $b = 0.4$ m, and $m = 0.6$, what is the uniform flow depth? Is the flow subcritical or supercritical?

7.51 A trapezoidal channel made of wood planks is used as a log slide so that freshly cut timber can be transported to a waterway. The channel conveys water at 2 ft^3/s, $b = 44$ ft, $m = 0.4$, and the sloping angle at which it is laid is 38°. Determine the depth of water for uniform flow. Is the flow subcritical or supercritical?

7.52 A semicircular channel of diameter 0.9 m is used as a water slide in an amusement park. The slide starts at a position 10 m high and over the course of 100 m descends to the ground. The channel is made of a plastic having $n = 0.004$. The volume flow of water is 2 m^3/s. What is the uniform flow depth? Is the flow subcritical or supercritical?

Backwater Curve

7.53 Derive the expression for gradually varied flow in triangular channels.

7.54 Derive the expression for gradually varied flow in trapezoidal channels.

7.55 Consider a rectangular brick-lined channel of width 10 ft and inclined at a slope of 0.0001. The channel is long and contains a dam at one end. The water depth just before the dam is 6.5 ft. The flowrate in the channel is 30 ft^3/s. Determine the shape of the backwater curve (see Figure 7.19).

7.56 A rectangular channel made of concrete is 3.5 m wide and conveys water at a flowrate of 12 m^3/s. A dam is placed at the channel end; the water depth there is 5 m. The channel slope is 0.000 2. Determine the shape of the backwater curve (see Figure 7.19).

7.57 A triangular channel is made of concrete ($m = 1.0$) and conveys water at a flowrate of 20 m^3/s. A dam in the channel is used to contain the flow; the water depth at the dam is 6.2 m. The channel slope is 0.000 1. Determine the shape of the backwater curve (see Figure 7.19).

7.58 A trapezoidal channel is brick-lined with $b = 4$ m, $m = 0.8$, and $S = 0.000\ 2$. The water depth over a dam placed at the downstream end is 7.2 m. For a volume flowrate over the dam of 100 m^3/s, determine the shape of the backwater curve (see Figure 7.19).

7.59 Consider a rectangular earth channel dug out to a width of 3 m and inclined at a slope of 0.001. The channel is very long, and at its end a dam partially restricts the flow. At the dam, the water depth is 5 m and the flowrate over the dam is 10 m^3/s. Determine the variation of depth upstream.

Hydraulic Jump

7.60 Equations 7.44 through 7.46 were derived for a hydraulic jump in a rectangular channel. Derive similar expressions for a triangular channel; that is, derive Equation 7.48.

7.61 Repeat Problem 7.54 for a trapezoidal channel; that is, derive Equation 7.49.

7.62 A hydraulic jump occurs in a rectangular channel 6 ft wide that conveys water at 150 ft^3/s. The upstream flow depth is 2 ft.

a. Determine the downstream height.
b. Calculate the energy loss in the jump.
c. Sketch a momentum diagram for the channel and show where the operating point appears.

7.63 A rectangular open channel is 3 m wide and contains a sluice gate. When the gate is partially raised, the flow passing underneath is supercritical. Downstream, the flow encounters a barrier and adjusts via a hydraulic jump (see Figure P7.63). The height upstream of the gate is 3 m, and the volume flow rate is 20 m^3/s. Determine the water height after the jump.

7.64 A hydraulic jump occurs in a rectangular channel 3 m wide that conveys water at 20 m^3/s. The upstream depth of flow is 1 m.

a. Determine the downstream height.
b. Calculate the energy loss in the jump.
c. Sketch a momentum diagram for the channel and show where the operating point appears.

7.65 For the prismatic section shown, perform the following calculations, assuming $b = 6$ m and $m = 2$:

a. Write geometry factor equations for each column in Table 7.1.
b. Plot a family of specific-energy lines for $Q = 0$, 10, 20, 40, and 80 m^3/s. Draw the locus of critical depth points.
c. Determine the critical depth of flow for various flow rates ranging from 0 to 80 m^3/s and plot Q versus z_{cr}.
d. Write the general backwater curve equation for the cross section.
e. Determine the depth necessary if the cross section is hydraulically optimum.

7.66 For the prismatic section shown, perform the following calculations, assuming $b = 6$ m and $m = 2$:

a. Write geometry factor equations for each column in Table 7.1.
b. Plot a family of specific-energy lines for $Q = 0$, 10, 20, 40, and 80 m^3/s. Draw the locus of critical depth points.

FIGURE P7.63

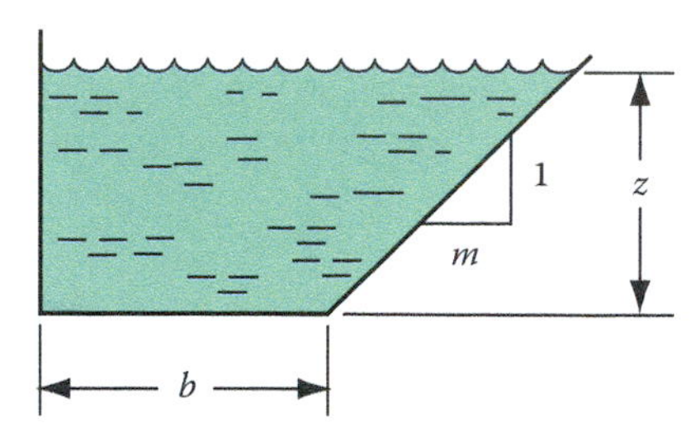

FIGURE P7.65

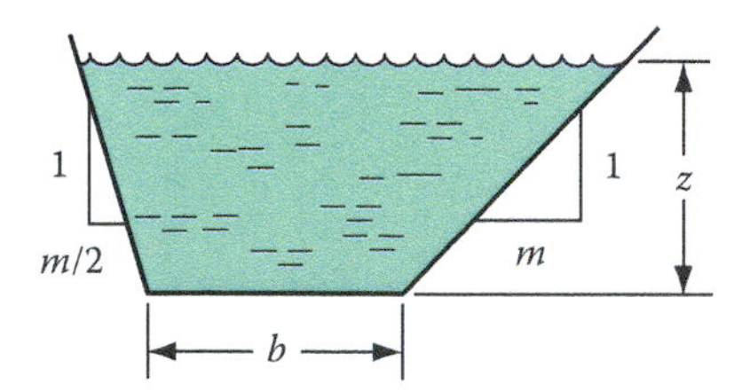

FIGURE P7.66

c. Determine the critical depth of flow for various flow rates ranging from 0 to 80 m^3/s and plot Q versus Z_{cr}.
d. Write the general backwater curve equation for the cross section.
e. Determine the depth necessary if the cross section is hydraulically optimum.

8 Compressible Flow

In previous chapters, we primarily investigated the flow of liquids. The density of a liquid remains essentially constant under the action of externally applied pressures. For example, a pressure of 200 atm exerted on water raises its density by less than 1% of that at 1 atm. It is therefore assumed with negligible error that no changes in density occur in a liquid as a result of its flow. Such flows are generally referred to as *incompressible*. On the other hand, flowing gases can experience considerable density changes as a result of externally applied pressures. Consider the ideal gas equation presented in Chapter 1:

$$\rho = \frac{p}{RT}$$

It is seen that density depends directly on pressure and inversely on temperature. Thus, density changes in the flow can in fact occur. Such flows, called *compressible* flows, are discussed in this chapter.

A study of compressible flows is important because of the wide range of examples that exist: natural gas piped from producer to consumer, high-speed flight through air, discharging of compressed gas tanks, flow of air through a compressor, flow of steam through a turbine, and many others. Mathematical modeling of compressible flows thus has many applications. The material presented here includes calculation of sonic velocity in a fluid, the concept of isentropic flow, flow through a nozzle, shock waves, and compressible flow in a pipe with friction.

However, several preliminaries should first be presented. The energy equation was presented in Chapter 3. In the study of thermodynamics, the energy equation is known as the first law, with which the idea or existence of internal energy is associated. That is, the energy equation forms a basis for the postulate that substances or systems possess a property called internal energy. There is also a second law of thermodynamics, analogous to the first law, that forms a basis for the postulate that substances or systems possess still another property called entropy (denoted as S or, on a per-unit-mass basis, s). The dimension of entropy is $(F \cdot L/t)$ or per unit mass $[F \cdot L/(t \cdot M)]$, where t is a temperature unit—for example, Btu/slug · R in engineering units or J/(kg · K) in SI units.

Entropy can be thought of as being associated with probability. As a substance changes from a less probable state to a more probable state, its entropy increases. Specifically, consider a membrane separating two chambers. One chamber contains a gas; the other is evacuated. When the membrane is ruptured, we are more likely to find gas on both sides of the membrane than on just one side with a vacuum on the other. After we rupture the membrane, the system goes from a less probable state to a more probable one with a net increase in entropy.

A concept related to entropy is that of an *irreversible* process. All naturally occurring processes are irreversible. If a cup of hot coffee is placed in a room, heat is transferred from the coffee to the surroundings. It is unlikely that the coffee will regain its heat merely by being removed from the room. There is a continuing increase in entropy in all naturally occurring processes because they tend to change from a less probable to a more probable state. The idea of efficiency is related to the concept of entropy. A higher efficiency means that we are accomplishing a given objective with a smaller total increase in entropy.

It is convenient in many problems to make calculations for a reversible process and later, if necessary, to determine the magnitude of the deviations from the irreversible process being modeled. Thus, in many of the topics presented in this chapter, we will assume a reversible process with no losses and no change in entropy. Such processes are also adiabatic. And because they involve no change in entropy, they are called *isentropic* processes. The mathematical implications associated with isentropy are not obvious but will be presented where appropriate.

8.1 SONIC VELOCITY AND MACH NUMBER

Consider a long cylinder filled with a fluid and containing a piston. The piston is suddenly moved to the right an infinitesimal amount at a velocity dV (Figure 8.1). If the fluid is incompressible, the entire liquid volume will move to the right. If the fluid is compressible, the gas molecules in layer A will first compress and in turn compress those in layer B. These will then compress the molecules in layer C, and so on. The process of compression of gaseous layers propagates to the right at a finite velocity called the *sonic velocity*; the compression process is like a wave traveling through the medium at sonic speed. After passage of the wave, the fluid velocity equals the piston velocity dV.

Figure 8.2a shows a wave traveling to the right in a duct. Upstream of the wave, the fluid is stationary. After passage of the wave, the fluid velocity is dV. To analyze the wave, we must first render it stationary by imposing a velocity equal to $-a$ on all velocities in the system (assuming one-dimensional flow, of course). The result is shown in Figure 8.2b. The wave is stationary in the new coordinate system, and the control surface is drawn about it. Fluid moves into the control volume (and into the wave) from the right at a velocity of a, with a pressure p and density ρ. After passing through the control volume, the fluid velocity becomes $a - dV$. The pressure and density downstream are $p + dp$ and $\rho + d\rho$. We write the continuity equation for the control volume of Figure 8.2b as

$$\rho A a = (\rho + d\rho)A(a - dV)$$

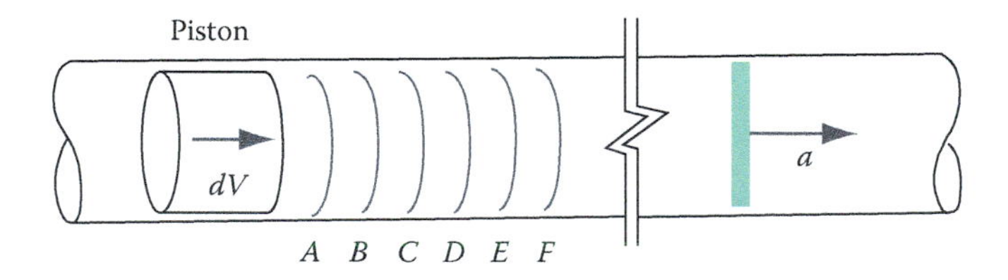

FIGURE 8.1 A piston–cylinder arrangement containing a fluid.

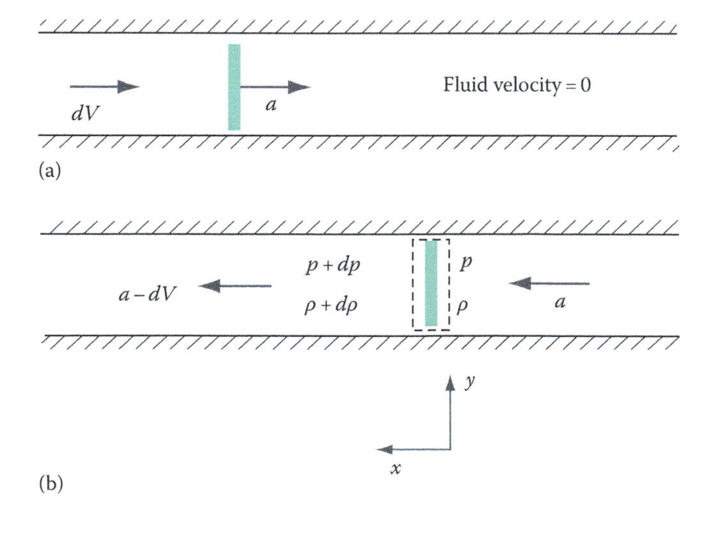

FIGURE 8.2 A moving wave rendered stationary.

Canceling area A and eliminating second-order terms leave

$$a\,d\rho = \rho\,dV$$

or

$$dV = \frac{a\,d\rho}{\rho} \tag{8.1}$$

The momentum equation in the flow direction is

$$\begin{aligned}\sum F &= \dot{m}(V_{\text{out}} - V_{\text{in}})\\ &= \rho A a(a - dV - a)\\ &= -\rho A a\,dV\end{aligned}$$

Neglecting viscous effects (shear stress at the wall), the forces acting in the flow direction on the control volume are due only to pressure differences. Thus, the momentum equation is

$$pA - (p + dp)A = -\rho A a\,dV$$

Simplifying yields

$$dp = \rho a\,dV \tag{8.2}$$

Combining with Equation 8.1 gives

$$dp = a^2 d\rho$$

or

$$a^2 = \frac{dp}{d\rho} \tag{8.3}$$

Thus, to determine sonic velocity, we must know the variation of density with pressure. We will assume that the gas with which we are concerned is ideal or perfect and that it has constant specific heats c_p and c_υ. The flow upstream and downstream is uniform. There is a temperature gradient within the control volume but not at the control surface bounding the control volume itself. Therefore, there is no heat transfer across the control surface, so the process of wave propagation is adiabatic. Changes in other properties occur across the wave, but because these are infinitesimal, the departure from reversibility is negligible. The process of wave propagation, since it is both adiabatic and reversible (with negligible error), is thus assumed to be isentropic. From thermodynamics, we get the following for a perfect gas with constant specific heats undergoing an isentropic process:

$$\frac{p}{\rho^\gamma} = C_1 \tag{8.4}$$

where
$\gamma = c_p/c_\upsilon$
C_1 is a constant

Evaluating the derivative in Equation 8.3 using Equation 8.4 gives

$$\frac{dp}{d\rho} = C_1\gamma\rho^{\gamma-1} = \frac{\gamma C_1\rho^{\gamma}}{\rho}$$

$$\frac{dp}{d\rho} = \frac{\gamma p}{\rho}$$

With the ideal gas assumption,

$$\rho = \frac{p}{RT}$$

we get

$$\frac{dp}{d\rho} = \gamma RT \tag{8.5}$$

Combining with Equation 8.3 gives the sonic velocity for an ideal gas with constant specific heats:

$$a = \sqrt{\gamma RT} \tag{8.6}$$

As an example, let us calculate the sonic velocity in air at room temperature ($T = 70°\text{F} = 530°\text{R}$):

$$a = \sqrt{1.4(1710\,\text{ft}\cdot\text{lbf/slug}\cdot°\text{R})(530°\text{R})}$$
$$= 1126\,\text{ft/s}$$

where values for γ and R can be found in Table A.6. Note that absolute temperature units must be used.

Equation 8.3 gives the sonic velocity in terms of dp and $d\rho$. For an incompressible fluid, $d\rho = 0$, and Equation 8.3 predicts an infinite sonic velocity. Generally, the more incompressible the substance, the higher the sonic velocity.

As we saw in the piston–cylinder arrangement, the presence of a disturbance in a compressible medium is signaled by the propagation of a compression or sound wave. The speed at which the wave travels is the sonic velocity. This leads us to the question of what happens if an object that is causing a disturbance travels faster than the sonic velocity. We can analyze such a situation by considering a wing traveling through air. The wing acts very much like the piston, compressing air just ahead of it and thereby sending sound waves in all directions. For a wing traveling at subsonic speed, we can sketch streamlines about the wing, as in Figure 8.3. The air layer at A senses the

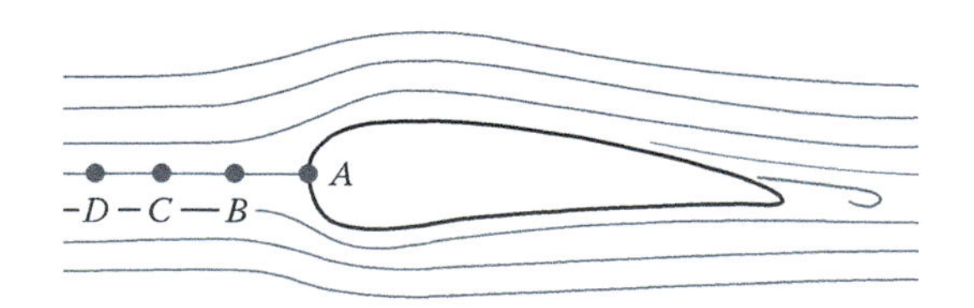

FIGURE 8.3 A wing traveling at subsonic velocity through air.

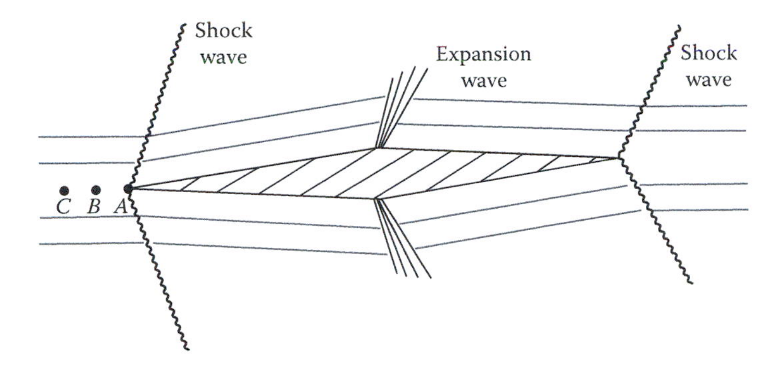

FIGURE 8.4 A diamond-shaped airfoil traveling at supersonic velocity.

presence of the wing and signals point B. The layer at B has time to adjust to the oncoming wing and in addition signals point C and so forth. The resultant streamline pattern is gradually changing and denotes smooth flow past the wing. For supersonic flow, the wing moves faster than the compression waves it is creating. The resultant flow pattern is illustrated in Figure 8.4. The air layer at point A signals point B, but the airfoil reaches point B before the signal does. Thus, point B has no time to adjust smoothly to the oncoming wing. Instead, the layer at B must adjust abruptly via formation of a shock wave. A *shock wave* is a very abrupt change in flow properties; in reality, it is a small fraction of an inch in thickness. It is a discontinuity in the flow and is by no means isentropic. After the flow passes the leading edge of the airfoil, it encounters a corner where it must turn sharply. After the turn, the air has room to expand because the wing thickness is decreasing. The abrupt turn is accomplished by what is called an expansion wave. Finally, at the trailing edge of the airfoil, the flow again adjusts via another shock wave.

From the preceding discussion, it is apparent that two different flow regimes exist in compressible flow: subsonic and supersonic. The criterion used to distinguish between the two is the *Mach number*, defined as the ratio of flow velocity to the sonic velocity in the medium:

$$M = \frac{V}{a}$$

If the Mach number is less than 1, *subsonic* flow exists. If the Mach number is greater than 1, the flow is *supersonic*. If the Mach number equals 1, the flow is *sonic*.

8.2 STAGNATION PROPERTIES AND ISENTROPIC FLOW

In this section, we discuss stagnation properties, in particular *stagnation temperature* and *stagnation pressure*. *Stagnation temperature* is the temperature attained in a flowing fluid by bringing the flow adiabatically to rest at a point. The kinetic energy of the flow is transformed entirely into enthalpy. A stagnation property is not to be confused with a static property, which is measured by an instrument moving at the local stream velocity. The two states that we are considering are the stagnation (or total) state and the static state. We can write the energy equation between these two states as

$$h_t = h + \frac{V^2}{2}$$

in which the subscript t represents the stagnation or total property, where the velocity is zero. The energy equation is written for steady flow in the absence of potential energy changes and no heat transfer or shaft work. Assuming that we have an ideal or perfect gas with constant specific heats, the change in enthalpy is

$$dh = c_p dT$$

and the energy equation becomes

$$c_p T_t = c_p T + \frac{V^2}{2}$$

This is a defining equation for stagnation temperature in terms of static temperature and kinetic energy of the flow. Dividing by c_p and rearranging give

$$T_t = T + \frac{V^2}{2c_p}$$

or

$$T_t = T\left(1 + \frac{V^2}{2c_p T}\right) \tag{8.7}$$

Combining the ideal gas law

$$\rho = \frac{p}{RT}$$

with the definition of enthalpy

$$h = u + \frac{p}{\rho}$$

yields the following:

$$h - u = RT$$

For a perfect gas with constant specific heats,

$$du = c_v \, dT$$

and, by substitution,

$$c_p T - c_v T = RT$$

or

$$c_p - c_v = R$$

Dividing by c_v and multiplying by c_p give

$$c_p\left(\frac{c_p}{c_v}-1\right)=\frac{Rc_p}{c_v}$$

or

$$c_p(\gamma-1)=\gamma R$$

Solving for c_p yields

$$c_p=\frac{\gamma R}{\gamma-1} \tag{8.8}$$

Combining Equation 8.7 and the definition $a^2 = \gamma RT$ gives

$$\begin{aligned}\frac{T_t}{T}&=1+\frac{V^2(\gamma-1)}{2\gamma RT}\\&=1+\frac{V^2}{a^2}\frac{\gamma-1}{2}\end{aligned}$$

or

$$\frac{T_t}{T}=1+\frac{\gamma-1}{2}M^2 \tag{8.9}$$

Thus, for any ideal gas with a known value of specific heat ratio, the ratio of stagnation to static temperature is known when the Mach number is given. Equation 8.9 is tabulated in Table D.1 as T/T_t versus M for $\gamma = 1.4$.

Example 8.1

Appendix Table A.6 shows that for air,

$$R = 286.8 \text{ J/(kg}\cdot\text{K)}$$

$$c_p = 1\ 005 \text{ J/(kg}\cdot\text{K)}$$

$$\gamma = 1.4$$

Calculate R for air using Equation 8.8 and compare to the tabulated value.

Solution

Equation 8.8 may be rearranged as

$$R=c_p\frac{\gamma-1}{\gamma}=1\ 005\frac{1.4-1}{1.4}$$

$$R = 287 \text{ J/kg}$$

Any discrepancies are due to round-off errors encountered when converting values in Appendix Table A.6.

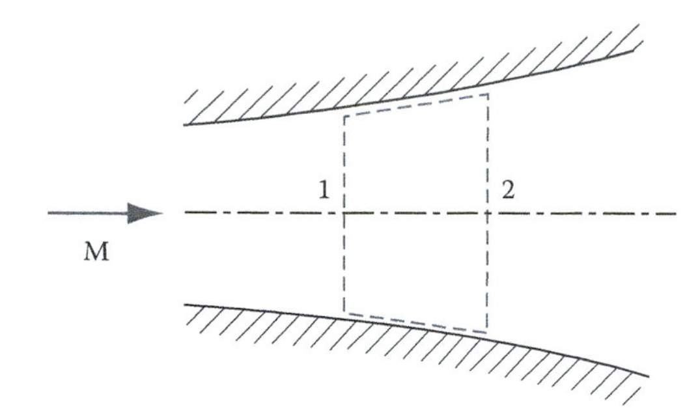

FIGURE 8.5 Adiabatic flow in a channel of varying cross section.

To illustrate the importance of stagnation temperature, consider adiabatic flow in a channel of varying cross section (Figure 8.5). The energy equation written between points 1 and 2 is

$$\frac{d\tilde{Q}}{dt} = \dot{Q} = \dot{m}_1\left(h_1 + \frac{V_1^2}{2}\right) - \dot{m}_2\left(h_2 + \frac{V_2^2}{2}\right)$$

For adiabatic flow, $\dot{Q} = 0$. Assuming a perfect gas, the equation becomes

$$c_pT_1 + \frac{V_1^2}{2} = c_pT_2 + \frac{V_2^2}{2}$$

$$c_pT_{t_1} = c_pT_{t_2}$$

or

$$T_{t_1} = T_{t_2}$$

Regardless of how the static properties and the kinetic energy vary in an adiabatic process, the stagnation temperature thus remains constant. It can therefore be used as a reference property throughout the flow field, provided that adiabatic flow exists.

Example 8.2

Four kilograms of oxygen undergo a process from conditions $p_1 = 150$ kPa (absolute), $T_1 = 25°\text{C}$ to $p_2 = 450$ kPa (absolute), $T_2 = 90°\text{C}$. Calculate the change in enthalpy.

Solution

For oxygen, $c_p = 920$ J/(kg · K) from Appendix Table A.6. The change in *specific enthalpy* is given by

$$h_2 - h_1 = c_p(T_2 - T_1) = 920(90 - 25)$$

$$h_2 - h_1 = 5.98 \times 10^4 \text{ J/kg} = 59.8 \text{ kJ/kg}$$

The change in *enthalpy* is

$$H_2 - H_1 = m(h_2 - h_1) = 4(59.8)$$

$$H_2 - H_1 = 239 \text{ kJ}$$

This result is independent of the change in pressure.

The previous example demonstrated that a fluid can undergo a process for which the enthalpy change is a function of only the temperature change. The change in pressure affects the change in entropy and provides information on how a process will occur.

From the second law of thermodynamics, we get the following relationship among the properties:

$$Tds = du + p\,dv = du + p\,d\left(\frac{1}{\rho}\right)$$

Dividing by temperature,

$$ds = \frac{du}{T} + \frac{p}{T}d\left(\frac{1}{\rho}\right)$$

For the internal energy change, we have $du = c_v dT$, and for enthalpy change, $dh = c_p dT$. Also, with $\rho = p/RT$, we have $\rho R = p/T$. Combining these equations, we get the following for the entropy change:

$$ds = c_v\frac{dT}{T} + \rho R\,d\left(\frac{1}{\rho}\right)$$

This equation can be integrated for a process between states 1 and 2 to obtain

$$s_2 - s_1 = c_\upsilon \ln\left(\frac{T_2}{T_1}\right) + R\ \ln\left(\frac{\rho_2}{\rho_1}\right)$$

Combining the ideal gas law, we can derive another form of this equation, which is

$$\frac{s_2 - s_1}{c_\upsilon} = \ln\left[\left(\frac{T_2}{T_1}\right)^\gamma\left(\frac{p_2}{p_1}\right)^{1-\gamma}\right]$$

Note that if the argument of the natural log is 1, we have ln(1) = 0, and the process is isentropic.

Example 8.3

Use the data from the previous example and determine whether the process described is isentropic. Data from that example are

$$p_1 = 150\text{ kPa}\quad T_1 = 25°\text{C} = 298\text{ K}$$

$$p_2 = 450\text{ kPa}\quad T_2 = 90°\text{C} = 363\text{ K}$$

Solution

From Appendix Table A.6, we read

$$\gamma = 1.4\quad c_p = 920\text{J}/(\text{kg}\cdot\text{K})\quad c_v = \frac{c_p}{\gamma} = \frac{920}{1.4} = 657\ \text{J}/(\text{kg}\cdot\text{K})$$

The change in entropy is given by

$$s_2 - s_1 = c_v \ln\left[\left(\frac{T_2}{T_1}\right)^{\gamma}\left(\frac{p_2}{p_1}\right)^{1-\gamma}\right]$$

$$s_2 - s_1 = 657 \ln\left[\left(\frac{363}{298}\right)^{1.4}\left(\frac{450}{150}\right)^{1-1.4}\right] = 657 \ln[1.3158(0.644)]$$

$$s_2 - s_1 = -107.3 \text{ J/(kg}\cdot\text{K)}$$

The value of the change in entropy is less significant than the negative sign, which indicates that the process is not naturally occurring.

It is interesting to calculate conditions that would yield a zero change in entropy. Given the inlet conditions, and the outlet temperature, for example, what would the final pressure have to be for the process to be isentropic? We would be searching for a value for the pressure ratio that would make the argument of the natural log equal to 1:

$$\frac{s_2 - s_1}{657} = 0 = \ln\left[\left(\frac{363}{298}\right)^{1.4}\left(\frac{p_2}{p_1}\right)^{1-1.4}\right]$$

or

$$\left(\frac{363}{298}\right)^{1.4}\left(\frac{p_2}{p_1}\right)^{1-1.4} = 1$$

$$1.318(p_2/150)^{-0.4} = 1$$

Rearranging,

$$\frac{p_2}{150} = \left(\frac{1}{1.318}\right)^{-1/0.4}$$

$$p_2 = 150(1.999)$$

Solving, the outlet pressure that corresponds to a zero change in entropy is

$$p_2 = 300 \text{ kPa}$$

Stagnation pressure or *total pressure* p_t of a flowing fluid is the pressure attained by bringing the flow isentropically to rest at a point. For an ideal gas,

$$\rho = \frac{p}{RT}$$

and at the stagnation state,

$$\rho_t = \frac{p_t}{RT_t}$$

Combining, we obtain

$$\frac{\rho_t T_t}{\rho T} = \frac{p_t}{p} \tag{8.10}$$

For an isentropic process, p/ρ^γ = a constant, or

$$\frac{p}{\rho^\gamma} = \frac{p_t}{\rho_t^\gamma}$$

which becomes

$$\frac{\rho_t}{\rho} = \left(\frac{p_t}{p}\right)^{1/\gamma}$$

Combining with Equation 8.10 gives

$$\left(\frac{T_t}{T}\right)\left(\frac{p_t}{p}\right)^{1/\gamma} = \frac{p_t}{p}$$

$$\left(\frac{p_t}{p}\right)^{(\gamma-1)/\gamma} = \frac{T_t}{T}$$

or

$$\left(\frac{T_t}{T}\right)^{\gamma/(\gamma-1)} = \frac{p_t}{p}$$

Substituting into Equation 8.9 yields

$$\frac{p_t}{p} = \left(1 + \frac{\gamma - 1}{2}M^2\right)^{\gamma/(\gamma-1)} \tag{8.11}$$

Thus, for an ideal gas with a known ratio of specific heats γ, p_t/p can be determined when the Mach number M is given. Equation 8.11 is tabulated in Table D.1 as p/p_t versus M for $\gamma = 1.4$. For isentropic flow, stagnation pressure and stagnation temperature are constant; for adiabatic flow, stagnation temperature is constant, but stagnation pressure can change.

Example 8.4

Air flows in a duct of inside area 10 in.^{2}. At a certain point in the flow, the velocity, temperature, and pressure are measured to be 1000 ft/s, 480°R, and 12 psia, respectively.

a. Determine the mass flow rate.
b. Determine the Mach number.
c. Determine the stagnation pressure.
d. Determine the stagnation temperature.

Solution

a. The mass flow rate is found with

$$\dot{m} = \rho AV$$

The density is calculated by assuming an ideal gas:

$$\rho = \frac{p}{RT} = \frac{(12\ \text{lbf/in.}^2)(144\ \text{in.}^2/\text{ft}^2)}{[1710\ \text{ft}\cdot\text{lbf/(slug}\cdot{}^\circ\text{R)}](480^\circ\text{R})}$$

where the gas constant R for air was obtained from Table A.6. The density then is

$$\rho = 0.002105\ \text{slug/ft}^3$$

The area and velocity were given in the problem statement, so

$$\dot{m} = (0.002105\ \text{slug/ft}^3(10/144\ \text{ft}^2)(1000\ \text{ft/s})$$

or

$$\dot{m} = 0.146\ \text{slug/s}$$

b. The sonic velocity is calculated with

$$a = \sqrt{\gamma RT} = \sqrt{1.4[1710\,\text{ft}\cdot\text{lbf/(slug}\cdot{}^\circ\text{R)}](480^\circ\text{R})} = 1072\,\text{ft/s}$$

The Mach number then is

$$M = \frac{V}{a} = \frac{1000}{1072}$$

$$M = 0.93$$

c. The stagnation pressure is found with Equation 8.11:

$$\frac{p_t}{p} = \left(1 + \frac{\gamma - 1}{2}M^2\right)^{\gamma/(\gamma-1)}$$

We can use this equation, or we can use instead the values found in Table D.1. Here we use the table; at a Mach number of 0.93, we read

$$\frac{p}{p_t} = 0.5721 \quad \frac{T}{T_t} = 0.8525$$

The stagnation pressure is found as

$$p_t = \frac{p}{0.5721} = \frac{14.7 \text{ psia}}{0.5721}$$

or $p_t = 25.7$ psia

d. For the stagnation temperature,

$$T_t = \frac{T}{0.8525} = \frac{480°\text{R}}{0.8525}$$

or $T_t = 563°R$

Regarding these properties, we note that sensors moving with the fluid would read 480°R and 14.7 psia (static properties). If the kinetic energy of the flow is converted without heat loss (adiabatically) to enthalpy, the properties would be measured as 563°R and 25.7 psia.

8.3 FLOW THROUGH A CHANNEL OF VARYING AREA

The type of flow considered in this section is a steady isentropic flow. Examples of channels of varying area include rocket nozzles, jet engine intake, or passage between adjacent turbine blades. Again we will assume one-dimensional flow with properties considered constant at any cross section.

Figure 8.6 illustrates flow in a channel of varying cross section. The control volume chosen shows only pressure forces acting. Viscous effects are assumed to be negligible because the flow

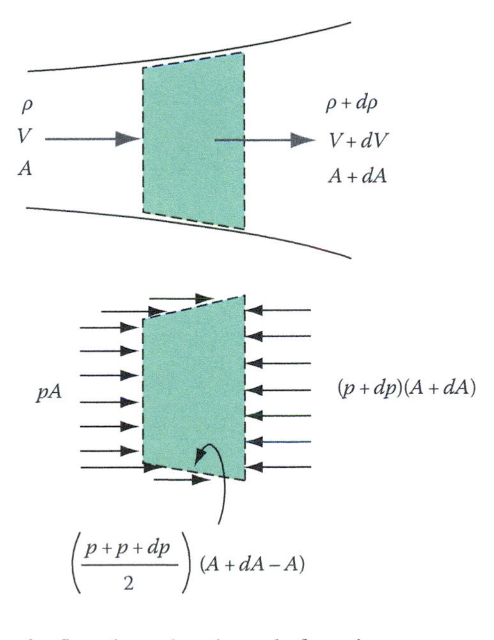

FIGURE 8.6 Control volume for flow through a channel of varying area.

is taken as being isentropic. Gravitational forces too are negligible because we are dealing with a gaseous medium. The continuity equation for the flow is

$$\iint_{CS} \rho V_n \, dA = 0$$

$$\dot{m}_{out} - \dot{m}_{in} = 0$$

$$(\rho + d\rho)(A + dA)(V + dV) - \rho AV = 0$$

Simplifying and dropping second-order terms yield

$$\frac{d\rho}{\rho} + \frac{dA}{A} + \frac{dV}{V} = 0 \tag{8.12}$$

The momentum equation in the direction of flow is

$$\sum F = \iint_{CS} V\rho V_n \, dA$$

With pressure as the only external force, we have

$$\sum F = pA + \left(p + \frac{dp}{2} \right) dA - (p + dp)(A + dA)$$

Combining, simplifying, and neglecting the $dp \, dA$ term give

$$\sum F = -A \, dp \tag{8.13a}$$

The integral term (the momentum flux) becomes

$$\iint_{CS} V\rho V_n \, dA = \rho AV(V + dV - V) = \rho AV \, dV \tag{8.13b}$$

The momentum equation is then

$$-dp = \rho V \, dV \tag{8.13c}$$

Substitution for dV from the continuity Equation 8.12 yields

$$dp + \rho V^2 \left(-\frac{d\rho}{\rho} - \frac{dA}{A} \right) = 0$$

or

$$dp - V^2 d\rho = \rho V^2 \frac{dA}{A}$$

Using the definition of sonic velocity, $a^2 = (dp/d\rho)$, the momentum equation now becomes

$$dp - \frac{V^2 d\rho}{a^2} = \rho V^2 \frac{dA}{A}$$
$$dp(1 - M^2) = \rho V^2 \frac{dA}{A} \quad (8.14)$$

This equation shows the effect of area change on the other flow variables for both subsonic and supersonic flows. For subsonic flow, $M < 1$ and $1 - M^2 > 0$. If dA increases, as in a diverging channel, pressure increases (from Equation 8.14) and velocity decreases (Equation 8.13c). If dA decreases as in a converging channel, however, pressure decreases (Equation 8.14) and velocity increases (Equation 8.13c). These are the same results as would be obtained with the equations written in Chapter 5 for incompressible flow.

For supersonic flow, $M > 1$ and $1 - M^2 < 0$. If dA increases, pressure decreases and velocity increases. If dA decreases, pressure increases and velocity decreases. This effect is the opposite of that in the subsonic case. These comments, applied to subsonic and supersonic flows in diverging and converging channels, are summarized in Figure 8.7.

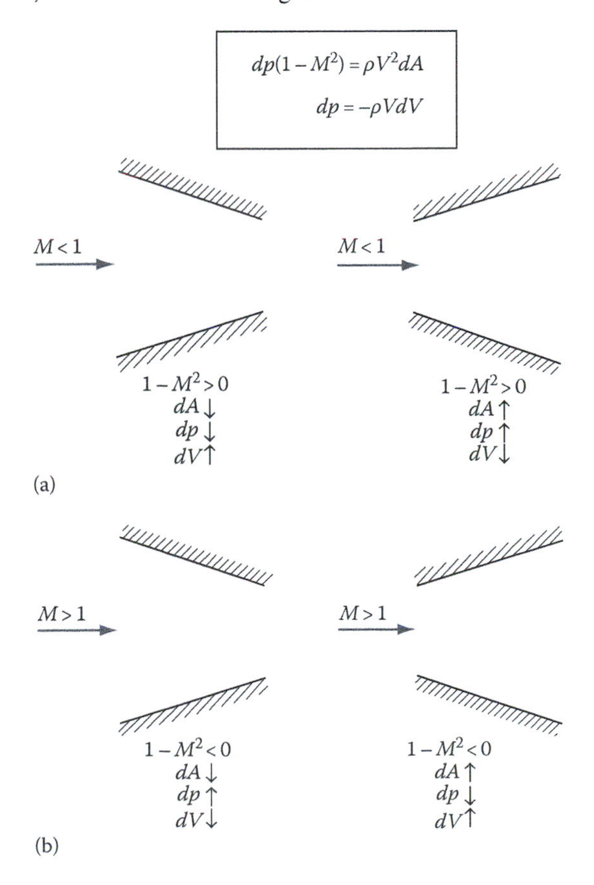

FIGURE 8.7 Effect of varying area on the flow variables for (a) subsonic and (b) supersonic flows.

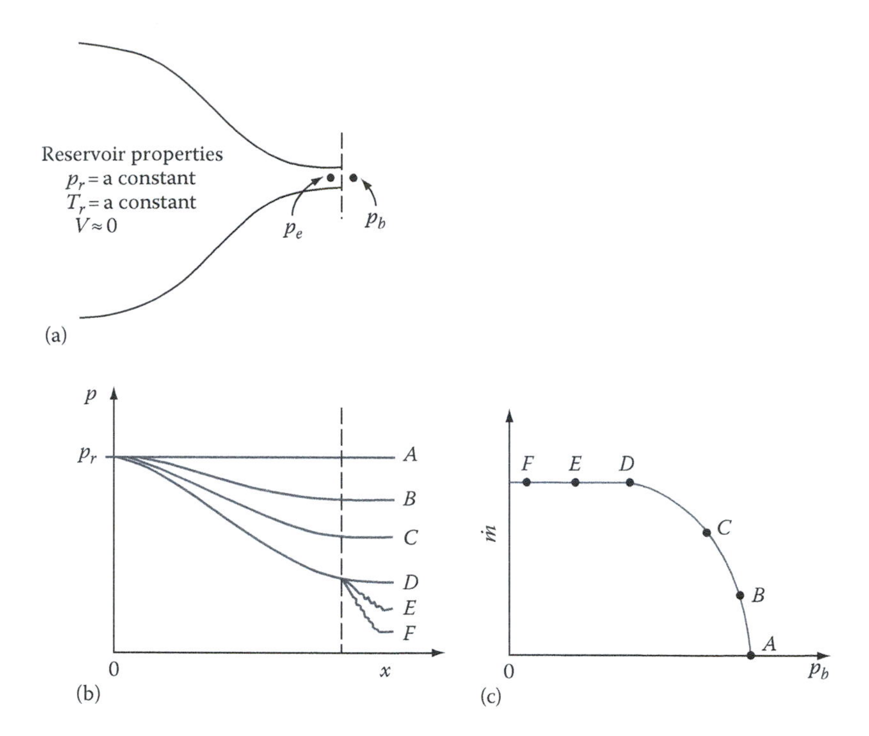

FIGURE 8.8 Flow through a converging nozzle.

It is apparent that a subsonic flow cannot be accelerated to a supersonic velocity in a converging nozzle. This procedure can be done only with a converging–diverging nozzle. To investigate this effect, let us consider a converging nozzle attached to a large reservoir in which conditions are maintained constant. Compared to flow through the nozzle, reservoir velocity is negligible; so properties in the reservoir are stagnation properties. The system is sketched in Figure 8.8, which also shows a graph of pressure versus distance from the reservoir to outside the nozzle. We will determine how reducing the back pressure p_b below the reservoir pressure p_r influences the mass flow through the nozzle. If the back pressure p_b equals the reservoir pressure, then no flow exists (curve *A*). If the back pressure is reduced slightly, some flow goes through the nozzle (curves *B* and *C*), and the pressure decreases somewhat with distance. As the back pressure is further reduced, a point is eventually reached at which sonic velocity exists at the exit plane of the nozzle (curve *D*). This point is a limiting point because it is known that, in a converging nozzle, the flow cannot be accelerated to supersonic speeds. In each case, a decrease in back pressure is sensed by the reservoir, and in turn, more flow exits through the nozzle. The decrease in back pressure is analogous to a small disturbance that transmits a signal traveling at sonic speed to the reservoir. If the velocity at the exit plane is sonic, however, any further decrease in back pressure is not sensed by the reservoir; consequently, no more flow is induced.

At point *D*, then, the nozzle is said to be *choked*. For all back pressures less than that corresponding to curve *D*, the pressure distribution remains the same as curve *D* up to the exit plane. The flow at the exit plane is at pressure p_e and expands to the back pressure p_b via expansion waves after exiting the nozzle (curves *E* and *F*).

When the nozzle is choked, the back pressure is said to be *critical*. The critical exit pressure ratio is found from Equation 8.11:

$$\frac{p_t}{p} = \left(1 + \frac{\gamma - 1}{2} M^2\right)^{\gamma/(\gamma-1)}$$

In this case, the stagnation pressure p_t, which is constant throughout the nozzle, equals the reservoir pressure p_r. Also, p equals the pressure at the exit plane p_e, where the Mach number equals 1; thus, for choked flow,

$$\frac{p_t}{p} = \left.\frac{p_r}{p_e}\right|_{cr} = \left(\frac{\gamma+1}{2}\right)^{\gamma/(\gamma-1)} \tag{8.15}$$

For $\gamma = 1.4$,

$$\left.\frac{p_e}{p_r}\right|_{cr} = 0.5283$$

Another important parameter is the mass flow rate. This can be calculated at any point in the nozzle where properties are known with

$$\dot{m} = \rho AV$$

or

$$\dot{m} = \frac{p}{RT} AM\sqrt{\gamma RT} \tag{8.16}$$

For isentropic flow, stagnation pressure and stagnation temperature are constants. By substituting Equations 8.9 and 8.11 for T_t/T and p_t/p, respectively, into Equation 8.16, we obtain an alternative expression for mass flow rate:

$$\dot{m} = \frac{p_t}{\sqrt{RT_t}} A\sqrt{\gamma}M\left(1+\frac{\gamma-1}{2}M^2\right)^{-(\gamma+1)/[2(\gamma-1)]} \tag{8.17}$$

Example 8.5

A converging nozzle of area 3 cm² is attached to the exhaust of a rocket as shown in Figure 8.9. The exhaust nozzle is supplied from a reservoir in which the pressure is 150 kPa and the temperature is 1 400 K. Calculate the mass flow rate through the nozzle for back pressures of 100 and 50 kPa. Assume that the rocket exhaust gases have a specific heat ratio of 1.4 and a mean molecular weight of 20 kg/mol. Assume isentropic flow through the nozzle.

Solution

We first find the mass flow at critical conditions—that is, $M_e = 1$. The nozzle critical pressure ratio for $\gamma = 1.4$ is

$$\frac{p_e}{p_{te}} = 0.528$$

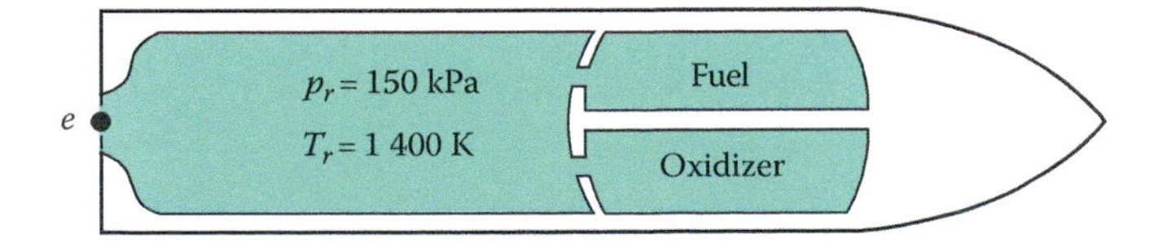

FIGURE 8.9 Rocket nozzle of Example 8.2.

from which we obtain

$$p_e = 0.528(150) = 79.2 \text{ kPa} = 79\,200 \text{ Pa}$$

From Table D.1 at $M_e = 1$, we find $T_e/T_{te} = 0.833\,3$; so

$$T_e = 0.833\,3(1\,400) = 1\,166 \text{ K}$$

The gas constant is found from

$$R = \frac{\overline{R}}{\text{MW}} = \frac{8.312 \text{ kJ/(mol}\cdot\text{K)}}{20 \text{ kJ/mol}}$$
$$= 415.7 \text{ J/(kg}\cdot\text{K)}$$

where $\overline{R}$ is the universal gas constant. The mass flow can now be calculated from

$$\dot{m} = \frac{p_e}{RT_e} AM_e\sqrt{\gamma RT_e}$$

At the critical point,

$$\dot{m} = \frac{79\,200}{415.7(1166)}(0.000\,3)(1)\sqrt{1.4(415.7)(1\,166)}$$

or

$$\dot{m} = 0.040 \text{ kg/s} \quad \text{for } p_b \le 79.2 \text{ kPa}$$

This is the mass flow rate for all back pressures that are equal to or less than 79.2 kPa. The exit pressure will not fall below this value if the reservoir conditions are maintained.

With a back pressure of 100 kPa, we find

$$\frac{p_e}{p_{te}} = \frac{100}{150} = 0.666$$

From Table D.1 at $p_e/p_{te} = 0.666$, we find $T_e/T_{te} = 0.89$ and $M_e = 0.78$. Thus,

$$T_e = 0.89(1\,400) = 1\,246 \text{ K}$$

The mass flow rate is

$$\dot{m} = \frac{p_e}{RT_e} AM_e\sqrt{\gamma RT_e}$$
$$= \frac{100\,000}{415.7(1\,246)}(0.000\,3)(0.78)\sqrt{1.4(415.7)(1\,246)}$$

or

$$\dot{m} = 0.038 \text{ kg/s} \quad \text{for } p_b = 100 \text{ kPa}$$

Example 8.6

The data provided in Appendix Table D.1 are for isentropic flow. Determine whether this is the case for data of the previous example; that is, calculate the change in entropy for the conditions in the reservoir and the conditions at the exit of the nozzle. Data from that example are

Reservoir $p_r = 150$ kPa $\quad T_r = 1\,400$ K

Exit $\quad p_e = 79.2$ kPa $\quad T_e = 1\,166$ K

$\gamma = 1.4$

Solution

The change in entropy is found with

$$\frac{s_2 - s_1}{c_v} = \ln\left[\left(\frac{T_2}{T_1}\right)^{\gamma}\left(\frac{p_2}{p_1}\right)^{1-\gamma}\right]$$

Substituting,

$$\frac{s_2 - s_1}{c_v} = \ln\left[\left(\frac{1\,166}{1\,400}\right)^{1.4}\left(\frac{79.2}{150}\right)^{1-1.4}\right] = \ln[(0.744(1.291)]$$

$$\frac{s_2 - s_1}{c_v} = 0 \quad \text{(as expected)}$$

It is now apparent that the position at which Mach number equals 1 can be used as a reference point. We have equations for critical temperature and pressure ratio. Regarding the expression for mass flow rate, we see that a third parameter—area—exists and that it is convenient to determine critical area ratio. We denote the area where Mach number equals 1 as A^*. Using Equation 8.17, we write

$$\dot{m} = \frac{p_t}{\sqrt{RT_t}} A\sqrt{\gamma} M\left(1 + \frac{\gamma - 1}{2} M^2\right)^{-(\gamma+1)/[2(\gamma-1)]}$$

and with $M = 1$,

$$\dot{m}^* = \frac{p_t}{\sqrt{RT_t}} A^* \sqrt{\gamma}(1)\left(1 + \frac{\gamma - 1}{2}\right)^{-(\gamma+1)/[2(\gamma-1)]}$$

These equations could be applied to two points having different areas in a single nozzle wherein mass flow rate and stagnation properties are constant. Taking the ratio of these equations gives

$$1 = \frac{A}{A^*} M\left(\frac{1 + \frac{\gamma - 1}{2} M^2}{\frac{\gamma + 1}{2}}\right)^{-(\gamma+1)/2(\gamma-1)}$$

or

$$\frac{A}{A^*} = \frac{1}{M}\left(\frac{\frac{\gamma+1}{2}}{1+\frac{\gamma-1}{2}M^2}\right)^{-(\gamma+1)/2(\gamma-1)} \tag{8.18}$$

This equation is tabulated in Table D.1 for $\gamma = 1.4$ as A/A^* versus M. Note that for any value of A/A^*, two values of M are possible, corresponding to subsonic and supersonic flows. This characteristic is in agreement with the statement that a converging–diverging nozzle is required to accelerate a flow to supersonic speed.

Example 8.7

Air flows into a converging duct from an inlet with cross-sectional area of 25 in.^{2} to the exit, where the area is 10 in.^{2}. If the inlet temperature is 70°F, inlet pressure is 14.7 psia, and inlet velocity is 200 ft/s, find the conditions at the exit: Mach number, pressure, and temperature.

Solution

Figure 8.10 is a sketch of the nozzle. We find

$$a_1 = \sqrt{\gamma RT} = \sqrt{1.4(1710)(70+460)}$$
$$= 1126 \text{ ft/s}$$

and

$$M_1 = \frac{V_1}{a_1} = \frac{200}{1126} = 0.17$$

If the nozzle were extended, the flow would be accelerated to the point at which $M = 1$ and the area is A^*. This hypothetical point is our reference point. From Table D.1 at $M_1 = 0.17$,

$$\frac{p_1}{p_{t_1}} = 0.9800$$

$$\frac{T_1}{T_{t_1}} = 0.9943$$

$$\frac{A_1}{A^*} = 3.4635$$

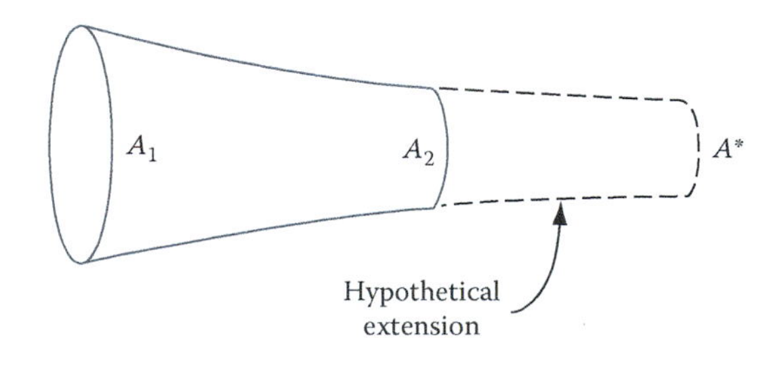

FIGURE 8.10 Converging nozzle of Example 8.3.

We now calculate A_2/A^*:

$$\frac{A_2}{A^*} = \frac{A_2}{A_1}\frac{A_1}{A^*} = \frac{10}{25}(3.4635) = 1.385$$

From Table D.1 at $A_2/A^* = 1.38$,

$$M_2 = 0.48$$

$$\frac{p_2}{p_{t_2}} = 0.8541$$

$$\frac{T_2}{T_{t_2}} = 0.9560$$

For isentropic flow, $p_{t_1} = p_{t_2}$ and $T_{t_2} = T_{t_2}$. With these ratios, we obtain

$$p_2 = \frac{p_2}{p_{t_2}}\frac{p_{t_2}}{p_{t_1}}\frac{p_{t_1}}{p_1}p_1 = 0.8541(1)\left(\frac{1}{0.9800}\right)14.7 \text{ psia}$$

$$p_2 = 12.8 \text{ psia}$$

and

$$T_2 = \frac{T_2}{T_{t_2}}\frac{T_{t_2}}{T_{t_1}}\frac{T_{t_1}}{T_1}T_1 = 0.9560(1)\left(\frac{1}{0.9943}\right)530°\text{R}$$

$$T_2 = 510°\text{R}$$

Now let us add a diverging section to the nozzle and tank of Figure 8.8 and again investigate the effect of decreasing the back pressure (Figure 8.11). Conditions in the reservoir are assumed

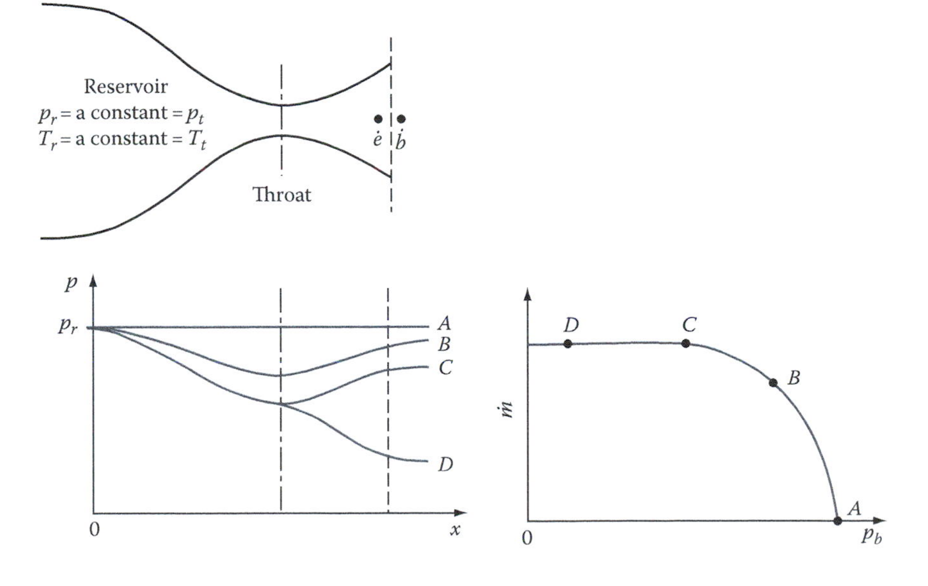

FIGURE 8.11 Isentropic flow through a converging–diverging nozzle.

to be constant. With negligible reservoir velocity, we note that reservoir temperature and pressure are stagnation properties. If the back pressure equals the reservoir pressure, then no flow goes through the nozzle, and curve *A* of pressure versus distance results. If the back pressure is slightly decreased, curve *B* results. As shown, pressure decreases to a minimum at or near the minimum area, called the throat, and increases to the back pressure at the exit. As the back pressure is further decreased, more flow is induced through the nozzle. Eventually, the point will be reached at which sonic velocity exists at the throat and the flow at the exit is subsonic (curve *C*). Any further decrease in back pressure is not sensed by the reservoir owing to sonic velocity at the throat. Therefore, the flow is maximum, and the nozzle is said to be choked.

It is possible to continue lowering the back pressure until the flow at the exit is supersonic. This condition corresponds to curve *D* and is in agreement with our previous conclusion that a stagnant fluid can be accelerated to supersonic speeds with a converging–diverging nozzle and that two solutions exist for a given area ratio with isentropic flow. For all curves shown in Figure 8.11, the exit pressure equals the back pressure. For back pressures corresponding to curves between *C* and *D*, the flow within the nozzle is not isentropic because a shock wave will exist in the diverging portion. For back pressures below those of curve *D*, the exit pressure does not equal the back pressure. Instead, the flow must adjust abruptly via expansion waves beyond the exit.

Example 8.8

Air flows through a converging–diverging nozzle as shown in Figure 8.12. At the throat, the Mach number is 1. At point 1, the Mach number is 0.35, and the area is 25 cm^2. The area at point 3 is 45 cm^2. For both subsonic and supersonic flows at the exit (corresponding to curves *C* and *D* of Figure 8.11), determine the following:

a. The throat area
b. The exit Mach number
c. The pressure ratio p_3/p_1
d. The temperature ratio T_3/T_1

Solution

From Table A.6, for air, $\gamma = 1.4$, and so Table D.1 can be used. The subsonic and supersonic calculations are presented side by side:

$$M_1 = 0.35 \rightarrow \text{Table D.1} \rightarrow p_1/p_{t_1} = 0.9188,\ T_1/T_{t_1} = 0.9761,\ A_1/A^* = 1.7780$$

$$\text{Isentropic flow} \rightarrow p_{t_1} = p_{t_2} = p_{t_3};\ T_{t_1} = T_{t_2} = T_{t_3}$$

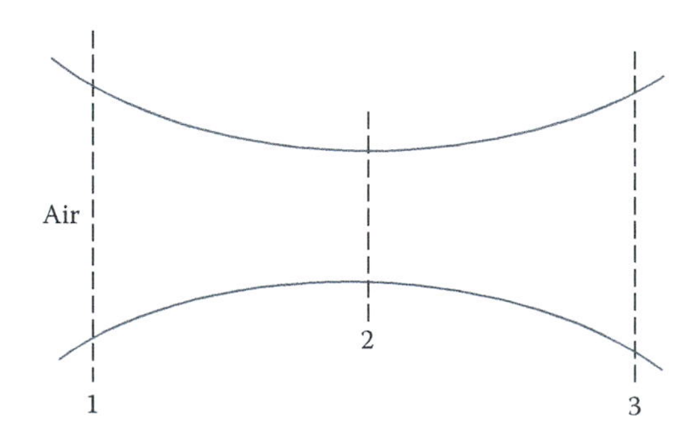

FIGURE 8.12 Converging–diverging nozzle of Example 8.4.

Subsonic

a. $A_1 = 25\text{ cm}^2$

$$A^* = \frac{A^*}{A_1}A_1 = \frac{25}{1.778\,0}$$

$A^* = 14.06\text{ cm}^2$

$A_3 = 45\text{ cm}^2$

$$\frac{A_3}{A^*} = \frac{45}{14.06} = 3.20$$

From Table D.1:

b. $M_3 = 0.18$

$$\frac{p_3}{p_{t3}} = 0.977\,6$$

$$\frac{T_3}{T_{t3}} = 0.993\,6$$

$$\frac{p_3}{p_1} = \frac{p_3}{p_{t3}}\frac{p_{t3}}{p_{t1}}\frac{p_{t1}}{p_1}$$

$$= (0.977\,6)(1)\left(\frac{1}{0.918\,8}\right)$$

c. $\dfrac{p_3}{p_1} = 1.064$

$$\frac{T_3}{T_1} = \frac{T_3}{T_{t3}}\frac{T_{t3}}{T_{t1}}\frac{T_{t1}}{T_1}$$

$$= (0.933\,6)(1)\left(\frac{1}{0.976\,1}\right)$$

d. $\dfrac{T_3}{T_1} = 1.018$

Supersonic

a. $A^* = 14.06\text{ cm}^2$

$A_3 = 45\text{ cm}^2$

$$\frac{A_3}{A^*} = 3.20$$

From Table D.1:

b. $M_3 = 2.71$

$$\frac{p_3}{p_{t3}} = 0.042\,29$$

$$\frac{T_3}{T_{t3}} = 0.405\,1$$

$$\frac{p_3}{p_1} = \frac{p_3}{p_{t3}}\frac{p_{t3}}{p_{t1}}\frac{p_{t1}}{p_1}$$

$$= (0.042\,29)(1)\left(\frac{1}{0.918\,8}\right)$$

c. $\dfrac{p_3}{p_1} = 0.046$

$$\frac{T_3}{T_1} = \frac{T_3}{T_{t3}}\frac{T_{t3}}{T_{t1}}\frac{T_{t1}}{T_1}$$

$$= (0.405\,1)(1)\left(\frac{1}{0.967\,1}\right)$$

d. $\dfrac{T_3}{T_1} = 0.415$

Example 8.9

Are the subsonic and supersonic flows in the previous example isentropic, as expected?

Solution

For the subsonic case, $p_3/p_1 = 1.064$ and $T_3/T_1 = 1.018$. The equation for entropy change is

$$\frac{s_2 - s_1}{c_v} = \ln\left[\left(\frac{T_2}{T_1}\right)^{\gamma}\left(\frac{p_2}{p_1}\right)^{1-\gamma}\right]$$

Substituting,

$$\frac{s_2 - s_1}{c_v} = \ln[(1.018)^{1.4}(1.064)^{1-1.4}] = 0 \quad \text{(isentropic)}$$

Similarly, for the supersonic case, $p_3/p_1 = 0.046$ and $T_3/T_1 = 0.415$;

$$\frac{s_2 - s_1}{c_v} = \ln[(0.415)^{1.4}(0.046)^{1-1.4}] = 0 \quad \text{(isentropic)}$$

Both cases are isentropic

Example 8.10

A huge compressed air tank is slowly being emptied through a valve into a much smaller tank. The valve itself can be considered as a converging–diverging nozzle with an exit area of 12 in.^{2} and a throat area of 6 in.^{2}. Reservoir conditions are p_r = 50 psia and T_r = 75°F. The second tank initially has a pressure of 14.7 psia; as it fills, this pressure steadily increases.

a. Find the maximum back pressure at which the nozzle or valve is choked.
b. Determine the mass flow rate at the choked condition.

Solution

Because the fluid is air (γ = 1.4), Table D.1 can be used. A choked nozzle corresponds to curve *C* or *D* of Figure 8.11. The maximum back pressure for choking occurs with curve *C*, the subsonic solution.

a. The exit area A_e = 12 in.^{2}, and the throat area for choked flow is A^* = 6 in.^{2}. From Table D.1, at A_e/A^* = 2, for subsonic flow,

$$M_e = 0.31 \quad \frac{p_e}{p_{te}} = 0.9355 \quad \frac{T_e}{T_{te}} = 0.9811$$

With $p_{te} = p_r$, for isentropic flow, the valve is choked for all back pressures less than

$$p_e = \frac{p_e}{p_{te}} p_r = 0.9355\,(50\text{ psia})$$

$$p_e = 46.8\text{ psia}$$

b. The temperature at the exit for the conditions of curve *C* is found as

$$T_e = \frac{T_e}{T_{te}} T_r = 0.9811(75 + 460)°\text{R}$$

$$T_e = 525°\text{R}$$

The mass flow rate can be calculated with the exit properties at choked conditions:

$$\dot{m} = \frac{p_e}{RT_e} A_e M_e \sqrt{\gamma R T_e}$$

$$= \frac{46.8(144)}{1710(525)} \frac{12}{144}(0.31)\sqrt{1.4(1710)(525)}$$

Solving, we get

$$\dot{m} = 0.217\text{ slug/s}$$

As a check, we can also calculate the mass flow rate with the throat properties at choked conditions:

$$p_{th} = \frac{p_{th}}{p_t} p_r = 0.528(50) = 26.4\text{ psia}$$

$$T_{th} = \frac{T_{th}}{T_t} T_r = 0.833(75 + 460) = 445.7°\text{R}$$

We thus obtain

$$\dot{m} = \frac{p_{\text{th}}}{RT_{\text{th}}} A^* M \sqrt{\gamma R T_{\text{th}}}$$

$$= \frac{26.4(144)}{1710(445.7)} \frac{6}{144} (1)\sqrt{1.4(1710)(445.7)}$$

$$\dot{m} = 0.214 \text{ slug/s}$$

The discrepancy is attributed to not interpolating values from Table D.1.

8.4 NORMAL SHOCK WAVES

A shock wave, as stated earlier, is a process wherein a sudden change in fluid properties takes place in supersonic flow. The shock itself is very thin—only a fraction of an inch in thickness. In this section, we deal with shock waves that are normal to the flow direction.

Consider steady, one-dimensional supersonic flow in which a normal shock occurs. We select a control volume that includes the shock wave and assume that all property changes occur within this control volume (Figure 8.13). In addition, the shock process is taken to be adiabatic because the temperature change occurs within the control volume. The continuity equation is

$$\rho_1 V_1 = \rho_2 V_2 \tag{8.19}$$

If we assume a perfect gas with constant specific heats, the continuity equation becomes

$$\frac{p_1}{RT_1} M_1 \sqrt{\gamma R T_1} = \frac{p_2}{RT_2} M_2 \sqrt{\gamma R T_2} \tag{8.20}$$

The momentum equation is written in the direction of flow as

$$\sum F = \iint_{\text{CS}} V \rho V_n \, dA$$

Evaluating the left-hand side, we note that the only external forces acting are pressure forces—gravitational forces are negligible in this case for the gas, and viscous forces can be omitted

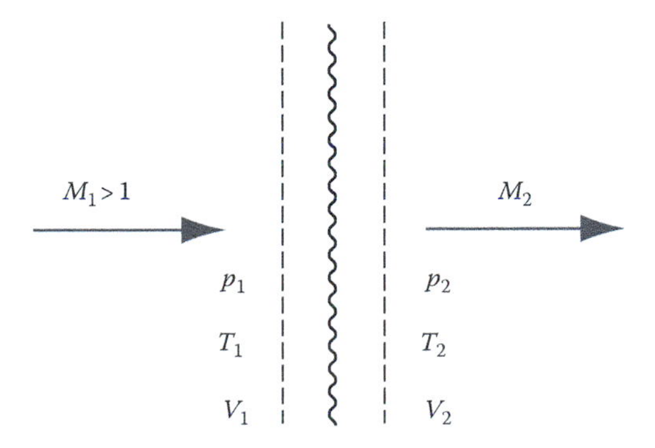

FIGURE 8.13 Supersonic flow with a normal shock wave.

with small error because most gases have a relatively low viscosity. Thus, the momentum equation becomes

$$p_1A_1 - p_2A_2 = \rho_2A_2V_2^2 - \rho_1A_1V_1^2$$

Canceling areas and rearranging yield

$$p_1 + \rho_1V_1^2 = p_2 + \rho_2V_2^2$$

or

$$p_1\left(1 + \frac{\rho_1V_1^2}{p_1}\right) = p_2\left(1 + \frac{\rho_2V_2^2}{p_2}\right)$$

For a perfect gas with constant specific heats, $\rho = p/RT$. With $a^2 = \gamma RT$, the momentum equation can be rewritten as

$$p_1\left(1 + \frac{V_1^2}{RT_1}\right) = p_2\left(1 + \frac{V_2^2}{RT_2}\right)$$
$$p_1(1 + \gamma M_1^2) = p_2\left(1 + \gamma M_2^2\right) \quad (8.21)$$

The energy equation for an adiabatic process is

$$h_1 + \frac{V_1^2}{2} = h_2 + \frac{V_2^2}{2}$$

For a perfect gas with constant specific heats, we have

$$c_pT_1 + \frac{V_1^2}{2} = c_pT_2 + \frac{V_2^2}{2}$$

$$T_1\left(1 + \frac{V_1^2}{2c_pT_1}\right) = T_2\left(1 + \frac{V_2^2}{2c_pT_2}\right)$$

Recalling that $c_p = \gamma R/(\gamma - 1)$ and that $a^2 = \gamma RT$, we rewrite the energy equation after substitution and simplification as

$$T_1\left(1 + \frac{\gamma - 1}{2}M_1^2\right) = T_2\left(1 + \frac{\gamma - 1}{2}M_2^2\right) \quad (8.22)$$

Combining Equations 8.21 and 8.22 with Equation 8.20 gives

$$\frac{M_1}{1 + \gamma M_1^2}\sqrt{1 + \frac{\gamma - 1}{2}M_1^2} = \frac{M_2}{1 + \gamma M_2^2}\sqrt{1 + \frac{\gamma - 1}{2}M_2^2} \quad (8.23)$$

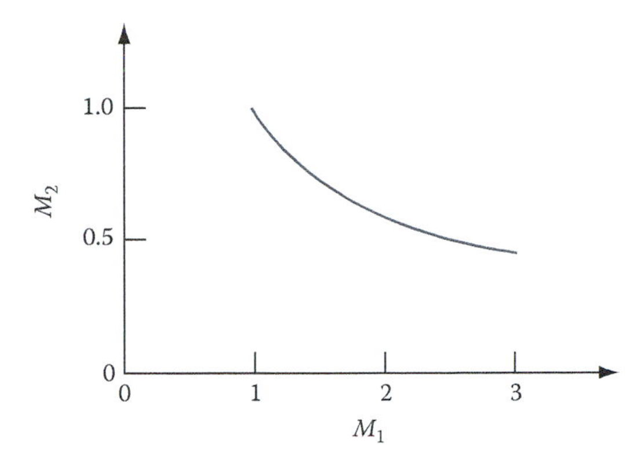

FIGURE 8.14 Mach number relationship across a shock wave (Equation 8.24).

Solving for M_2, we obtain

$$M_2^2 = \frac{M_1^2 + 2/(\gamma - 1)}{2\gamma M_1^2/(\gamma - 1) - 1} \tag{8.24}$$

This equation is plotted in Figure 8.14 for $\gamma = 1.4$ and $M_1 > 1$. A calculation of entropy change for the process (which is beyond the scope of this chapter) would show a decrease in entropy corresponding to a shock wave forming in a subsonic flow. An increase in entropy occurs for a shock wave forming in a supersonic flow. From the discussion of entropy in the introduction of this chapter, recall that entropy increases for all naturally occurring processes. We thus conclude that a shock forming in a subsonic flow is not a naturally occurring process and therefore violates the second law of thermodynamics. Equation 8.24 has no physical meaning if $M_1 < 1$. The figure shows that M_2 is always less than 1 if M_1 is supersonic. Equation 8.21 will predict that $p_2 > p_1$, which denotes what is known as a *compression shock*.

The equations thus developed describe flow of an ideal gas through a shock wave. If the ratio of specific heats γ and the inlet Mach number are both known, then M_2 can be determined from Equation 8.24, T_2/T_1 can be determined from Equation 8.22, p_2/p_1 can be determined from Equation 8.21, V_2/V_1 can be calculated by appropriately combining Equation 8.24 and the definition of sonic velocity, ρ_2/ρ_1 (=V_1/V_2) can be calculated with Equation 8.19, and p_{t2}/p_{t1} can be determined by combining Equation 8.21 with Equation 8.11. Since the flow is adiabatic through the shock, T_{t_2}/T_{t_1}. If we know γ and M_1, in other words, the equations developed so far can be used to find M_2, p_2/p_1, T_2/T_1, V_2/V_1, ρ_2/ρ_1, and p_{t_2}/p_{t_1}. These results are tabulated in Table D.2 for $\gamma = 1.4$.

Example 8.11

A stream of air travels at 500 m/s has a static pressure of 75 kPa and a static temperature of 15°C, and undergoes a normal shock. Determine air velocity and the stagnation and static conditions after the shock wave. Calculate the change in entropy for this process.

Solution

For air, $\gamma = 1.4$; from Table A.6, $R = 286.8$ J/(kg · K). By definition,

$$a_1 = \sqrt{\gamma R T_1} = \sqrt{1.4(286.8)(288)}$$

$$= 340 \text{ m/s}$$

We now find

$$M_1 = \frac{V_1}{a_1} = \frac{500}{340} = 1.47$$

From Table D.2, at $M_1 = 1.47$, we get

$$M_2 = 0.712 \quad \frac{p_2}{p_1} = 2.354 \quad \frac{T_2}{T_1} = 1.300 \quad \frac{p_{t2}}{p_{t1}} = 0.9390$$

From Table D.1 at $M_1 = 1.47$, we get

$$\frac{p_1}{p_{t_1}} = 0.284\,5 \quad \text{and} \quad \frac{T_1}{T_{t_1}} = 0.698\,2$$

With these ratios, we obtain

$$p_2 = \frac{p_2}{p_1} p_1 = 2.354(75\,\text{kPa}) = 176.55\,\text{kPa}$$

$$T_2 = \frac{T_2}{T_1} T_1 = 1.3(288) = 374.4\,\text{K} = 101.4°\text{C}$$

$$p_{t_1} = \frac{p_{t1}}{p_1} p_1 = \frac{1}{0.284\,5}(75\,\text{kPa}) = 263.6\,\text{kPa}$$

$$p_{t_2} = \frac{p_{t2}}{p_{t_1}} p_{t_1} = 0.939\,0(263.6) = 247.5\,\text{kPa}$$

$$T_{t_1} = \frac{T_{t1}}{T_1} T_1 = \frac{1}{0.698\,2}(288) = 412.5\,\text{K}$$

and finally, for adiabatic flow,

$$T_{t_2} = T_{t_1} = 412.5\,\text{K}$$

We can confirm this result with a different calculation. From Table D.1 at $M_2 = 0.712$,

$$\frac{T_2}{T_{t_2}} = 0.908$$

The stagnation temperature at Section 8.2 then is

$$T_{t_2} = \frac{T_{t2}}{T_2} T_2 = \left(\frac{1}{0.908}\right) 374.4\ \text{K}$$

$$T_{t_2} = 412.3\ \text{K}$$

The flow undergoing a shock is expected to be non-isentropic. For the shock of this example, we calculate the entropy change as

$$\frac{s_2 - s_1}{c_v} = \ln\left[\left(\frac{T_2}{T_1}\right)^{\gamma}\left(\frac{p_2}{p_1}\right)^{1-\gamma}\right]$$

$$\frac{s_2 - s_1}{c_v} = \ln[(1.3)^{1.4}(2.354)^{1-1.4}]$$

$$\frac{s_2 - s_1}{c_v} = +1.025 \quad \text{(Dimensionless)}$$

Example 8.12

A normal shock with a velocity of 3000 ft/s travels through still air at 14.7 psia (70°F), as illustrated in Figure 8.15. Determine the air velocity V_a and static properties behind the wave.

Solution

To use the equations developed in this section, we must reformulate the problem to render the wave stationary. This is illustrated in Figure 8.15, where a velocity of 3000 ft/s to the right is imposed on all velocities in the system. The inlet Mach number is

$$M_1 = \frac{V_1}{\sqrt{\gamma R T_1}} = \frac{3000}{\sqrt{1.4(1710)(530)}}$$

$$= 2.66$$

From Table D.2, at $M_1 = 2.66$,

$$M_2 = 0.4988 \quad \frac{p_2}{p_2} = 8.088 \quad \frac{\rho_2}{\rho_1} = 3.516 \quad \frac{T_2}{T_1} = 2.301$$

So the properties behind the wave are

$$p_2 = \frac{p_2}{p_1} p_1 = 8.088(14.7) = 119 \text{ psia}$$

$$T_2 = \frac{T_2}{T_1} T_1 = 2.301(530) = 1220°\text{R}$$

$$\frac{\rho_2}{\rho_1} = \frac{V_1}{V_2} = 3.516 \quad V_2 = \frac{V_1}{3.516} = \frac{3000}{3.516}$$

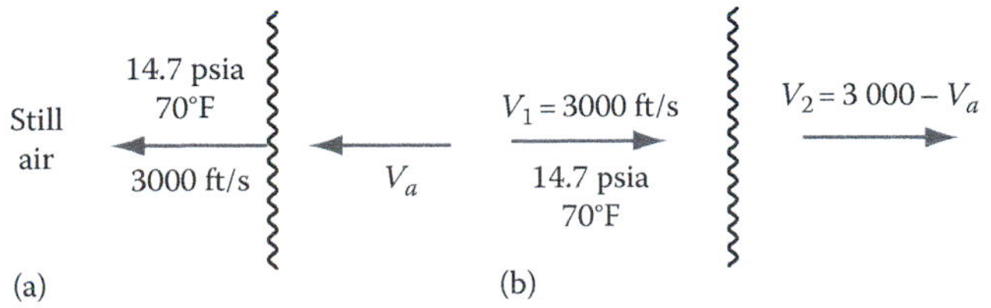

FIGURE 8.15 Moving shock of Example 8.7 rendered stationary. (a) Moving shock and (b) fixed shock.

so

$$V_2 = 853\,\text{ft/s}$$

But, as is seen in Figure 8.15,

$$V_2 = 3000 - V_a$$

or

$$V_a = 3000 - 853$$

$$V_a = 2150\ \text{ft/s}$$

We can now return to the converging–diverging nozzle of Figure 8.11 and add the results of this section. For a back pressure slightly below that of curve C, a normal shock forms in the nozzle, just downstream of the throat, as is illustrated in Figure 8.16 (curve E). The flow leaves the nozzle with an exit pressure that is equal to the back pressure and still subsonic. As the back pressure is further reduced, the normal shock moves closer to the exit until it eventually reaches the exit plane (curve F). If the back pressure is even further reduced, oblique shock waves form at the exit (curve G), and the nozzle is said to be *overexpanded*. Curve D represents supersonic exit flow; at this condition, the nozzle is said to be *perfectly expanded*. For back pressures below that of curve D, the back pressure is less than the exit pressure; expansion waves form at the exit, and the nozzle is said to be *underexpanded*.

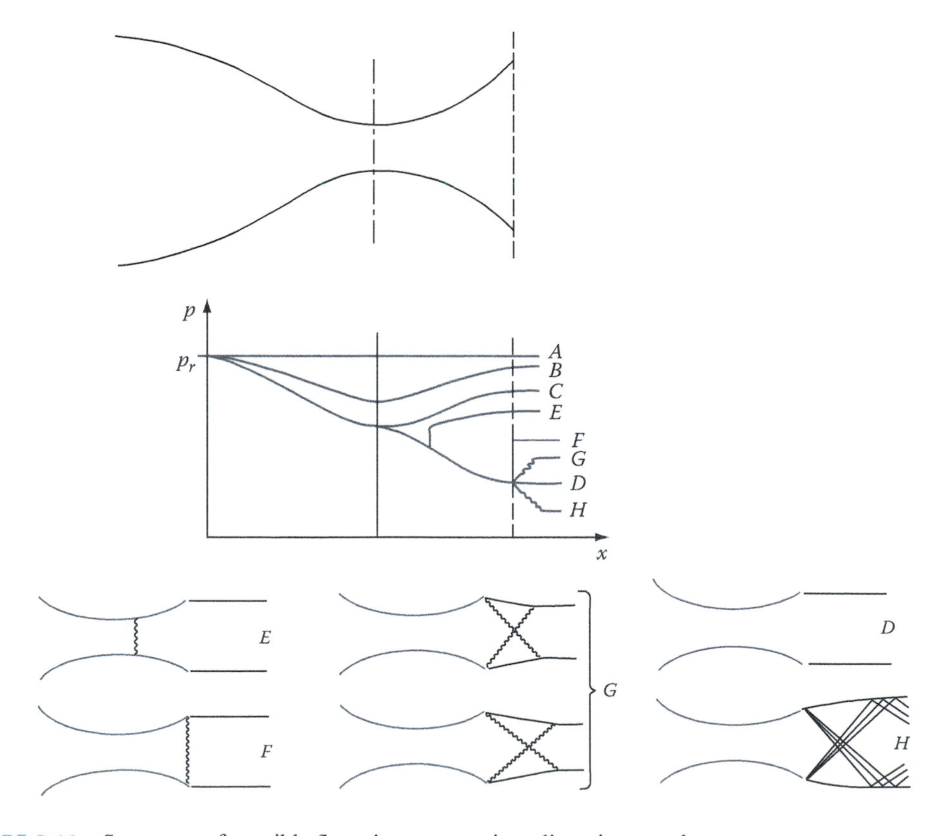

FIGURE 8.16 Summary of possible flows in a converging–diverging nozzle.

Example 8.13

A converging–diverging nozzle is supplied by a huge air reservoir. The nozzle has a throat area of 10 cm^2 and an exit area of 20 cm^2. The reservoir conditions are p_r = 300 kPa and T_r = 30°C. Determine the range of back pressures over which

a. Shocks appear in the nozzle
b. Oblique shocks form beyond the exit
c. Expansion waves form
d. Calculate the mass flow rate through the nozzle for the case in which a normal shock appears at the exit plane

Solution

A shock just downstream of the throat exists when the back pressure is slightly below that of curve C in Figure 8.16. For curve C, from Table D.1, at $A_e/A^* = 20/10 = 2$, we find

$$M_e = 0.31 \quad \frac{p_e}{p_{te}} = 0.936$$

Thus,

$$p_e = \frac{p_e}{p_{te}} p_{te} = 0.936\ (300\ \text{kPa})$$

$$= 280.8\ \text{kPa} = p_b \quad (\text{curve } C)$$

For a shock at the nozzle exit plane, curve F of Figure 8.16, we find the following from Table D.1 at $A_e/A^* = 2.0$:

$$M_e = 2.2 \quad \frac{p_e}{p_{te}} = 0.093\ 52 \quad \frac{T_e}{T_{te}} = 0.508\ 1$$

where it is assumed that isentropic flow exists up to the shock wave. From Table D.2, at $M_e = M_1 = 2.20$, we find

$$M_2 = M_b = 0.547\ 1 \quad \frac{p_2}{p_1} = \frac{p_b}{p_e} = 5.480 \quad \frac{T_2}{T_1} = \frac{T_b}{T_e} = 1.857$$

The back pressure for this case is found by

$$p_b = \frac{p_b}{p_e}\frac{p_e}{p_{te}} p_{te} = 5.480(0.093\ 52)(300\ \text{kPa})$$

$$= 153.7\,\text{kPa} \quad (\text{curve } F)$$

For curve D, we found previously that $M_e = 2.20$ and $p_e/p_{te} = 0.093\ 52$; thus,

$$p_e = p_b = 0.093\ 52(300) = 28.1\ \text{kPa (curve } D)$$

Summarizing these results, we have the following.

a. $153\ 7 < p_b < 280.8$ kPa (shock appears in nozzle)

b. $28.1 < p_b < 153.7$ kPa (oblique shocks beyond exit)

c. $p_b < 28.1$ kPa (expansion waves beyond exit)

d. For a normal shock at the exit, we found that $M_b = 0.547\,1$ and $p_b = 153.7$ kPa. Similarly,

$$T_b = \frac{T_b}{T_e}\frac{T_e}{T_{te}}T_{te} = 1.857(0.508\,1)(30+273)$$

$$T_b = 285.9 \text{ K} \quad (\text{curve } F)$$

Mass flow is

$$\dot{m} = \frac{p_b}{RT_b}AM_b\sqrt{\gamma RT_b}$$

$$= \frac{153\,700}{286.8(285.9)}\frac{20}{10^4}(0.5471)\sqrt{1.4(286.8)(285.9)}$$

$$\dot{m} = 0.695 \text{ kg/s}$$

As a check, let us calculate mass flow for conditions before the shock. First we find

$$p_e = \frac{p_e}{p_{te}}p_{te} = 0.093\,52(300) = 28.1 \text{ kPa}$$

$$T_e = \frac{T_e}{T_{te}}T_{te} = 0.508\,1(303) = 153.9 \text{ K}$$

With $M_e = 2.20$, we obtain

$$\dot{m} = \frac{p_e}{RT_e}A_eM_e\sqrt{\gamma RT_e}$$

$$= \frac{28\,100}{286.8(153.9)}\frac{20}{10^4}(2.2)\sqrt{1.4(286.8)(153.9)}$$

$$\dot{m} = 0.696 \text{ kg/s}$$

8.5 COMPRESSIBLE FLOW WITH FRICTION

In this section, we consider compressible flow with friction in insulated ducts of constant area. In real situations, frictional forces are present and may significantly affect the flow. In addition, such a study provides insight into the general effects of friction on compressible flow.

Consider flow in a constant-area duct and select a control volume like that shown in Figure 8.17. The continuity equation is

$$\rho V = C_1 = \text{a constant} \tag{8.25}$$

Implicitly differentiating yields

$$\rho dV + V d\rho = 0$$

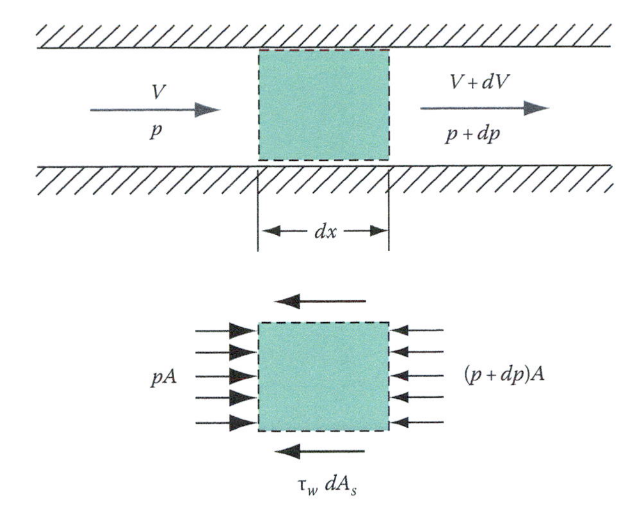

FIGURE 8.17 Control volume for flow in a constant-area duct.

Dividing by ρV, we get

$$\frac{d\rho}{\rho} + \frac{dV}{V} = 0 \tag{8.26}$$

We can obtain Equation 8.26 alternatively by logarithmic differentiation of Equation 8.25. Taking the natural logarithm of Equation 8.25 gives

$$\ln \rho + \ln V = \ln C_1 \tag{8.27}$$

Implicit differentiation gives Equation 8.26.

Returning to Figure 8.17, the momentum equation in the direction of flow is

$$\sum F_x = \iint_{CS} V_x \rho V_n dA$$

The external forces acting on the control volume are due to pressure and wall friction. By substitution, the momentum equation becomes

$$pA - (p + dp)A - \tau_w \, dA_s = \rho AV(V + dV - V) \tag{8.28}$$

where

A is a cross-sectional area

dA_s is a surface area over which the friction shear stress τ_w acts

From Chapter 5, recall the definition of hydraulic diameter:

$$D_h = \frac{4A}{P}$$

where P is the perimeter. For a circular duct,

$$D_h = \frac{4(\pi D^2/4)}{\pi D} = D$$

and the surface area is

$$dA_s = P\,dx = \pi D\,dx = \frac{4A\,dx}{D}$$

By substitution into Equation 8.28 and after simplification, we obtain

$$-dp - \tau_w\left(\frac{4dx}{D}\right) = \rho V\,dV \tag{8.29}$$

As was done for incompressible flow, we will introduce a friction factor f, which is dependent upon Reynolds number and relative roughness:

$$f = \frac{4\tau_w}{\frac{1}{2}\rho V^2}$$

Combining with Equation 8.29 gives

$$-dp - \frac{\rho V^2}{2}\frac{f\,dx}{D} = \rho V\,dV \tag{8.30}$$

We are working toward integration of this expression to obtain an equation relating duct length to velocity or Mach number. Dividing by p and substituting $\rho = p/RT$, we obtain

$$-\frac{dp}{p} - \frac{1}{2}\frac{1}{RT}V^2\frac{f\,dx}{D} = \frac{V\,dV}{RT}$$

With the definitions $a^2 = \gamma RT$ and $M = V/a$, we have

$$-\frac{dp}{p} - \frac{\gamma M^2}{2}\frac{f\,dx}{D} = \gamma M^2\frac{dV}{V} \tag{8.31}$$

Logarithmic differentiation of the ideal gas law, $p = \rho RT$, gives

$$\frac{dp}{p} = \frac{d\rho}{\rho} + \frac{dT}{T} \tag{8.32}$$

Similarly, for the definition of Mach number,

$$M = \frac{V}{\sqrt{\gamma RT}}$$

$$\frac{dM}{M} = \frac{dV}{V} - \frac{1}{2}\frac{dT}{T} \tag{8.33}$$

Combining Equations 8.26, 8.32, and 8.33 with Equation 8.31 yields

$$+\frac{1}{2}\frac{dT}{T}-\frac{dM}{M}+\frac{\gamma}{2}M^2\frac{f\,dx}{D}+\gamma M^2\frac{dM}{M}+\frac{\gamma}{2}M^2\frac{dT}{T}=0$$

For an adiabatic flow, stagnation temperature is constant, so

$$T\left(1+\frac{\gamma-1}{2}M^2\right)=\text{a constant}$$

Logarithmically differentiating, we obtain

$$\frac{dT}{T}+\frac{(\gamma-1)M^2(dM/M)}{1+((\gamma-1)/2)M^2}=0 \tag{8.34}$$

Substituting, we get

$$\frac{(\gamma-1)M^2(dM/M)}{1+((\gamma-1)/2)M^2}\left(\frac{1}{2}+\frac{\gamma M^2}{2}\right)-(\gamma M^2-1)\frac{dM}{M}=\frac{\gamma}{2}M^2\frac{f\,dx}{D}$$

Combining terms, we get

$$\frac{f\,dx}{D}=\frac{2dM}{M}\left[\frac{1-M^2}{\left(1+((\gamma-1)/2)M^2\right)\gamma M^2}\right] \tag{8.35}$$

The left-hand side of Equation 8.35 is always positive. However, the bracketed term on the right-hand side can be positive or negative, depending on whether the flow is subsonic or supersonic. It is interesting to analyze this equation for both cases. If the flow is subsonic, then the term in brackets in Equation 8.35 is positive. Consequently, dM/dx must also be positive, which means that the Mach number of the flow is predicted to increase with distance along the duct. If the flow is supersonic, the bracketed term is negative, so that dM/dx must be negative. The Mach number of the flow is thus predicted to decrease with distance. For either condition, wall friction causes the Mach number of the flow to approach unity. Therefore, a subsonic flow cannot be accelerated to supersonic speed in a duct. Sonic speed can be achieved, however, and this will occur at the end of the duct. Conversely, a supersonic flow can undergo a shock and change to subsonic flow.

From Chapter 5, we know that the friction factor depends on Reynolds number and wall roughness. If the fluid is a liquid, moreover, the Reynolds number is essentially constant along the length of the constant-area pipe, and velocity is a constant. If the fluid is a gas, the product ρV is a constant. Temperature, however, can vary by as much as 20% for subsonic airflow in a pipe, causing a viscosity variation of roughly 10%. In turn, the Reynolds number would change by about 10%, affecting the friction factor. For turbulent flow, which is usually the case in compressible flow, the change in friction factor f is quite small. It is therefore reasonable to assume that friction factor f is a constant when integrating Equation 8.35 and that f is equal to some average value. Often, the initial value in the pipe is used.

We are now ready to integrate Equation 8.35. It is convenient to choose as our limits the following:

$$0\le dx\le L_{\max}$$
$$M\le M\le 1$$

Thus, the flow at the pipe entrance (or the control volume entrance) has a Mach number of M and is accelerated to $M = 1$, where $x = L_{max}$. The location, $x = L_{max}$, can exist at the end of the pipe, or it may be at the end of a hypothetical extension. It serves merely as a reference point in our equations where the Mach number equals unity. Integrating Equation 8.35 gives

$$\frac{fL_{max}}{D} = \frac{1-M^2}{\gamma M^2} + \frac{\gamma+1}{2\gamma}\ln\left[\frac{\gamma+1}{2}M^2\left(1+\frac{\gamma-1}{2}M^2\right)^{-1}\right] \tag{8.36}$$

Calculations of fL_{max}/D as a function of M made with this equation are tabulated in Table D.3 for $\gamma = 1.4$.

A similar analysis can be developed to obtain an expression between pressure and Mach number. Using Equation 8.31 and substituting from continuity for dV/V in terms of M yields the following, after simplification:

$$\frac{dp}{p} = -\frac{dM}{M}\left[\frac{1+(\gamma-1)M^2}{1+((\gamma-1)/2)M^2}\right]$$

Integrating between p and p^* corresponding to limits M and 1 gives the following relationship for p/p^* versus M:

$$\frac{p}{p^*} = \left[\frac{2M^2}{\gamma+1}\left(1+\frac{\gamma-1}{2}M^2\right)\right]^{-1/2} \tag{8.37}$$

Calculations made with this equation are also tabulated in Table D.3 for $\gamma = 1.4$. Again $M = 1$ is a reference point, and the corresponding pressure is denoted with an asterisk.

Similarly, we can derive the following expression for T/T^* versus M from Equation 8.22 for adiabatic flow:

$$\frac{T}{T^*} = \frac{\gamma+1}{2}\left(1+\frac{\gamma-1}{2}M^2\right)^{-1} \tag{8.38}$$

Combining Equations 8.33 and 8.34 yields an integrable expression for dV versus dM. After integration, the result is

$$\frac{V}{V^*} = M\left[\frac{\gamma+1}{2}\left(1+\frac{\gamma-1}{2}M^2\right)^{-1}\right]^{1/2} \tag{8.39}$$

Finally, the stagnation pressure ratio p_t/p_t^* is found to be

$$\frac{p_t}{p_t^*} = \frac{1}{M}\left[\frac{2}{\gamma+1}\left(1+\frac{\gamma-1}{2}M^2\right)\right]^{\{(\gamma+1)/[2(\gamma-1)]\}} \tag{8.40}$$

Results of calculations made with these equations are also tabulated in Table D.3 for $\gamma = 1.4$. With the equations thus described and tabulated, one-dimensional, adiabatic, compressible flow of an ideal gas with friction, referred to as *Fanno flow*, can be adequately modeled.

Example 8.14

An airflow enters a constant-area pipe at 100 ft/s, 15 psia, and 530°R. The pipe is 500 ft long and made of 2-nominal schedule 40 PVC. Determine conditions at the pipe exit.

Solution

From the appendix tables,

$$\mu = 0.3801\times10^{-6}\ \text{lbf}\cdot\text{s/ft}^2\ \text{(Table A.3 for air)}$$

$$D = 0.1723\ \text{ft} \quad A = 0.02330\ \text{ft}^2\ \text{(Table C.1 for pipe)}$$

Also for air,

$$\rho = \frac{p}{RT} = \frac{15(144)}{1710(530)} = 0.00238\ \text{slug/ft}^3$$

The sonic velocity at inlet is

$$a_1 = \sqrt{\gamma R T_1} = \sqrt{1.4(1710)(530)} = 1126\ \text{ft/s}$$

Therefore,

$$M_1 = \frac{V_1}{a_1} = \frac{100}{1126} = 0.09$$

From Table D.3 at $M_1 = 0.09$, we get

$$\frac{T_1}{T^*} = 1.1981 \quad \frac{p_1}{p^*} = 12.162 \quad \left.\frac{fL_{max}}{D}\right|_1 = 83.496 \quad \frac{V_1}{V^*} = 0.09851$$

where L_{max} is the duct length required for the flow to achieve $M = 1$, at which the pressure is p^* and the temperature is T^*.

The Reynolds number at inlet is

$$\text{Re}_1 = \frac{\rho V_1 D}{\mu} = \frac{0.00238(100)(0.1723)}{0.3801\times10^{-6}} = 1.08\times10^5$$

For PVC pipe, assume that the surface is very smooth. At $\text{Re}_1 = 1.08\times10^5$, the Moody diagram of Chapter 5 gives

$$f = 0.0175$$

and for the actual pipe in this example, we have

$$\left.\frac{fL}{D}\right|_{ac} = \frac{0.0175(500)}{0.1723} = 50.78$$

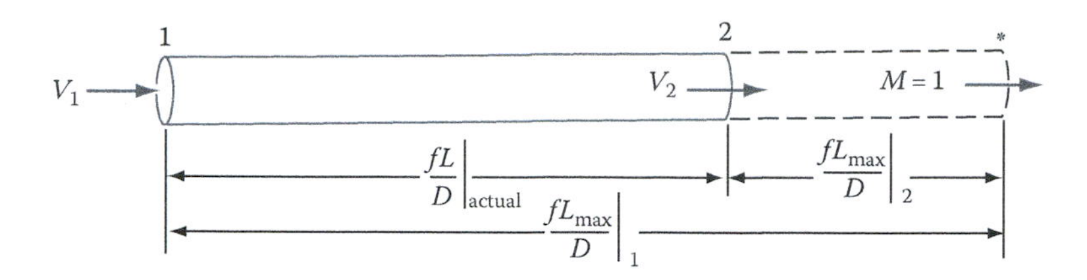

FIGURE 8.18 Sketch for Example 8.9.

To find the properties at the pipe exit, it is necessary to combine fL/D terms using $M = 1$ as a reference point. From the definition of length, we recall that $L_{1\ max}$ is the distance required to accelerate the flow from the conditions at point 1 to a Mach number equal to 1 at pipe end. Thus,

$$L_{1\ max} = L_{actual} + L_{2\ max}$$

Multiplying all terms by f/D, which we assume to be a constant, gives

$$\left.\frac{fL_{max}}{D}\right|_1 = \left.\frac{fL}{D}\right|_{actual} + \left.\frac{fL_{max}}{D}\right|_2$$

This relationship is illustrated in Figure 8.18.

Thus,

$$\left.\frac{fL_{max}}{D}\right|_2 = \left.\frac{fL_{max}}{D}\right|_1 - \left.\frac{fL_{max}}{D}\right|_{actual} = 83.496 - 50.78 = 32.71$$

From Table D.3 at $fL_{max}/D = 32.71$, we read

$$M_2 = 0.14 \qquad \frac{T_2}{T^*} = 1.1953 \qquad \frac{p_2}{p^*} = 7.8093 \qquad \frac{V_2}{V^*} = 0.15306$$

With these ratios, we find

$$p_2 = \frac{p_2}{p^*}\frac{p^*}{p_1}p_1 = (7.8093)\left(\frac{1}{12.162}\right)(15\ \text{psia}) = 9.63\ \text{psia}$$

$$T_2 = \frac{T_2}{T^*}\frac{T^*}{T_1}T_1 = (1.1953)\left(\frac{1}{1.1981}\right)(530°\text{R}) = 528.8°\text{R}$$

same format as p_2 and T_2

$$V_2 = V_1\frac{V_2}{V^*}\frac{V^*}{V_1} = 100\frac{(0.15306)}{0.09851}$$

$$V_2 = 155.4\ \text{ft/s}$$

As we see from these results, temperature did not change significantly. We were justified in assuming a constant value of friction factor, evaluated in this case at the pipe inlet.

8.6 COMPRESSIBLE FLOW WITH HEAT TRANSFER

In this section, we consider compressible flow in a constant-area duct and how heat addition or heat loss affects this flow. Figure 8.19 is a sketch of the control volume used in the analysis. Flow entering the control volume has properties that change by an infinitesimal amount upon leaving. The amount of heat added is denoted by dq, with dimensions of $F \cdot L/M$ (Btu/slug or J/kg). The continuity equation for the control volume is

$$\rho AV = (\rho + d\rho)A(V + dV)$$

Canceling area and higher-order terms gives, after rearrangement,

$$\frac{d\rho}{\rho} + \frac{dV}{V} = 0 \quad (8.41)$$

or $\rho V =$ a constant. The momentum equation written for the control volume is

$$\sum F_x = \iint V_x \rho V_n dA$$

The only forces we consider are pressure forces, so that the preceding equation becomes

$$pA - (p + dp)A = \rho AV(V + dV - V)$$

Canceling area A and simplifying give

$$dp = -\rho V\, dV$$

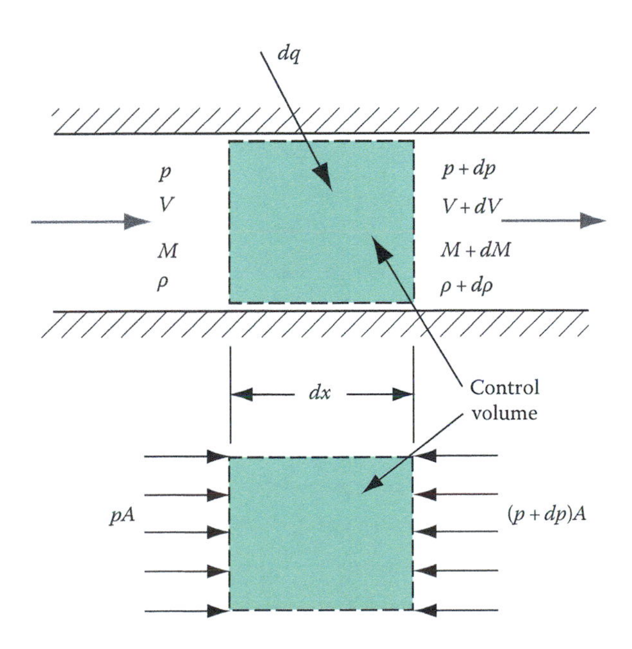

FIGURE 8.19 Control volume for compressible flow in a constant-area duct with heat transfer.

With ρV equal to a constant, the preceding equation can be integrated from point 1 to 2 to give

$$dp = (\text{constant}) \times dV$$

$$p_1 - p_2 = (\text{constant})(V_2 - V_1) = -\rho V(V_2 - V_1)$$

Noting that $\rho V = (\rho V)_1 = (\rho V)_2$, the last equation can be rewritten to yield

$$(p + \rho V^2)\big|_1 = (p + \rho V^2)\big|_2 \tag{8.42}$$

Factoring out the pressure term,

$$p\left(1 + \frac{\rho}{p}V^2\right)\Bigg|_1 = p\left(1 + \frac{\rho}{p}V^2\right)\Bigg|_2$$

With the ideal gas law ($\rho = p/RT$), the preceding becomes

$$p\left(1 + \frac{V^2}{RT}\right)\Bigg|_1 = p\left(1 + \frac{V^2}{RT}\right)\Bigg|_2$$

Using the definition of sonic velocity ($a^2 = \gamma RT$) and Mach number ($M = V/a$), the last equation is rewritten as

$$P(1 + \gamma M^2)\Big|_1 = p(1 + \gamma M^2)\Big|_2 \tag{8.43}$$

It is convenient to select as a reference point the location where the Mach number is 1. Pressure (and other properties) at this location would then be denoted with an asterisk superscript. Thus, with regard to Equation 8.43, we let point 1 be any location where the Mach number is M and point 2 be where $M = 1$. Equation 8.43 then becomes

$$p(1 + \gamma M^2) = p^*(1 + \gamma)$$

or

$$\frac{p}{p^*} = \frac{1 + \gamma}{1 + \gamma M^2} \tag{8.44}$$

We now work toward deriving similar expressions for the remaining properties.

The ideal gas law gives, for temperature,

$$T = \frac{p}{\rho R} \tag{8.45}$$

From the definition of mass flow rate, we write

$$\rho = \frac{\dot{m}}{AV}$$

and from the definition of Mach number ($M = V/a$), we have

$$V = M\sqrt{\gamma RT}$$

Combining these equations with Equation 8.45 gives

$$T = \frac{p}{\rho R} = \frac{pAV}{\dot{m}R} = \frac{pAM\sqrt{\gamma RT}}{\dot{m}R}$$

or

$$\sqrt{T} = pM \times (\text{constant})$$

Squaring both sides and writing this equation between any two points in the duct give

$$\left.\frac{T}{(pM)^2}\right|_1 = \left.\frac{T}{(pM)^2}\right|_2$$

Denoting point 1 as anywhere in the duct and point 2 (with an asterisk) as the location where $M = 1$, this equation becomes

$$\frac{T}{(pM)^2} = \frac{T}{(p^*)^2}$$

or

$$\frac{T}{T^*}\left(\frac{p}{p^*}\right)^2 M^2$$

Substituting from Equation 8.44 yields

$$\frac{T}{T^*} = \frac{(1+\gamma)^2 M^2}{(1+\gamma M^2)^2} \tag{8.46}$$

We now have relationships for static pressure and static temperature. For velocity, we use the continuity equation and the ideal gas law to obtain

$$\frac{V}{V^*} = \frac{\rho^*}{\rho} = \frac{p^*}{RT^*}\frac{RT}{p} = \frac{p^*}{p}\frac{T}{T^*}$$

Substituting from Equations 8.44 and 8.46 and simplifying, we get

$$\frac{V}{V^*} = \frac{(1+\gamma)M^2}{1+\gamma M^2} = \frac{\rho^*}{\rho} \tag{8.47}$$

For adiabatic flow, Equation 8.9 was derived to relate stagnation and static temperatures:

$$T_t = T\left(1+\frac{\gamma-1}{2}M^2\right) \tag{8.9}$$

We rewrite this equation for the reference point, where $M = 1$:

$$T_t^* = T^*\left(1+\frac{\gamma-1}{2}\right)$$

Dividing Equation 8.9 by the last equation yields

$$\frac{T_t}{T_t^*} = \frac{T}{T^*}\left(\frac{1+((\gamma-1)/2)M^2}{1+((\gamma-1)/2)}\right)$$

Combining with Equation 8.46 for T/T^*, we get

$$\frac{T_t}{T_t^*} = \frac{(1+\gamma)^2 M^2}{(1+\gamma M^2)^2}\left(\frac{1+((\gamma-1)/2)M^2}{1+((\gamma-1)/2)}\right) \tag{8.48}$$

Also from the section on isentropic flow, we derived a relationship between static and stagnation pressures:

$$p_t = p\left(1+\frac{\gamma-1}{2}M^2\right)^{\gamma/(\gamma-1)} \tag{8.11}$$

At the location where Mach number is 1, we write

$$p_t^* = p^*\left(1+\frac{\gamma-1}{2}\right)^{\gamma/(\gamma-1)}$$

Dividing the preceding equation into Equation 8.11 and substituting Equation 8.44 for p/p^*, we get the following, after some simplification:

$$\frac{p_t}{p_t^*} = \left(\frac{1+\gamma}{1+\gamma M^2}\right)\left(\frac{1+((\gamma-1)/2)M^2}{1+((\gamma-1)/2)}\right)^{\gamma/(\gamma-1)} \tag{8.49}$$

Thus far, we have relationships for p/p^*, T/T^*, V/V^*, ρ/ρ^*, T_t/T_t^*, and p_t/p_t^* for one-dimensional compressible flow of an ideal gas with heat transfer for flow in a duct. Results of calculations made with these equations are tabulated in Table D.4 for $\gamma = 1.4$.

Another equation we can derive relates the heat transferred to the stagnation temperature change. In Section 8.2, we wrote the energy equation between the static and stagnation states as

$$h_t = h + \frac{V^2}{2}$$

For an ideal gas with constant specific heats,

$$dh = c_p dT$$

In terms of specific heat, the energy equation becomes

$$c_p T_t = c_p T + \frac{V^2}{2}$$

Differentiating,

$$c_p\, dT_t = c_p\, dT + V\, dV \tag{8.50}$$

We write the energy equation again, but now for the control volume of Figure 8.19, to obtain

$$\frac{d\tilde{Q}}{dt} = \dot{Q} = \dot{m}\left(h_{\text{out}} - h_{\text{in}} + \frac{V_{\text{out}}^2}{2} - \frac{V_{\text{in}}^2}{2}\right)$$

Thus, the heat transferred per unit time reflects changes in enthalpy and in kinetic energy of the flow. Dividing by mass flow rate and differentiating yield

$$d\left(\frac{\dot{Q}}{\dot{m}}\right) = dq = (dh + V\, dV)$$

The heat transfer dq is the quantity with which we have been working in this section; as mentioned earlier, it has dimensions of $F \cdot L/M$ (Btu/slug or J/kg). The enthalpy term is rewritten in terms of temperature to give

$$dq = c_p dT + V\, dV$$

Substituting from Equation 8.45, we get

$$dq = c_p dT_t$$

So the heat transfer directly affects the stagnation temperature of the fluid. This equation can be integrated between two points in the duct to obtain the following working equation:

$$q = c_p(T_{t_2} - T_{t_1}) \tag{8.51}$$

The concept of entropy can be used to draw conclusions about the behavior of a flowing compressible fluid under various conditions. Although we do not provide proof here, we will discuss the effect of heat addition on the properties of the fluid. Consider a compressible fluid flowing at subsonic velocity into a duct that is being heated. The fluid will gain energy from the duct walls and eventually reach the point where $M = 1$. Any heat addition beyond this point serves only to reduce the mass flow rate. This means that a maximum amount of heat can be added before the flow will choke. Choked flow, remember, can also occur for flow through a converging–diverging nozzle and for flow with friction.

For supersonic flow entering a duct, the addition of sufficient heat ultimately causes the fluid to attain a Mach number of 1. Cooling (or removal of heat from the fluid) has the effect in both cases of driving the Mach number away from unity. Note that the heat addition can be in any of a number of forms and is not dependent on duct shape or length. Heat can be added over a very short distance, as in a chemical reaction, or over a long distance, as in a heat exchanger.

With the equations thus derived and tabulated, we are now in a position to solve problems dealing with heat transfer to a frictionless one-dimensional flow of an ideal gas (referred to as *Rayleigh flow*).

Example 8.15

Air flows in a constant-area duct. At the inlet, the Mach number is 0.2, the static pressure is 90 kPa, and the static temperature is 27°C. Heat is added at a rate of 120 kJ/(kg of air). Assuming a perfect gas with constant specific heats, determine the properties of the air at the end of the duct. Assume also that the flow is frictionless and that $c_p = 1\,000$ J/(kg · K).

Solution

For subsonic flow being heated in duct, the fluid will eventually reach the point where $M = 1$. This is illustrated in Figure 8.20. The properties we seek are those at section 2. From Table D.1 at $M_1 = 0.2$, we read

$$\frac{p_1}{p_{t_1}} = 0.972\,5 \quad \frac{T_1}{T_{t_1}} = 0.992\,1$$

From Table D.4 at $M_1 = 0.2$, we also read

$$\frac{T_{t1}}{T_t^*} = 0.173\,55 \quad \frac{T_1}{T^*} = 0.206\,61 \quad \frac{p_1}{p^*} = 2.272\,7$$

$$\frac{p_{t1}}{p_t^*} = 1.234\,6 \quad \frac{V_1}{V^*} = 0.090\,91$$

We can relate the stagnation temperature at 2 to that at section 1 through the heat transfer Equation 8.51:

$$q = c_p(T_{t_2} - T_{t_1})$$

or

$$T_{t_2} = \frac{q}{c_p} + T = \frac{120\,000}{1\,000} + (27 + 273)$$

$$= 420\,\text{K}$$

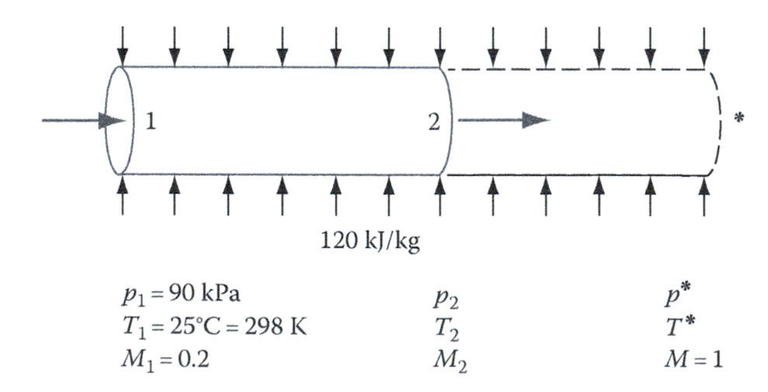

FIGURE 8.20 Flow in a duct with heat addition.

The temperature ratio we seek can be found by using the ratios already found:

$$\frac{T_{t2}}{T_t^*} = \frac{T_{t2}}{T_{t1}}\frac{T_{t1}}{T_t^*} = \frac{420}{300}0.173\,55 \approx 0.24$$

Entering Table D.4 at this value yields

$$M_2 = 0.24 \quad \frac{T_2}{T^*} = 0.284\,11 \quad \frac{p_2}{p^*} = 2.220\,9$$

$$\frac{p_{t2}}{p_t^*} = 1.221\,3 \quad \frac{V_2}{V^*} = 0.127\,92$$

Other properties can be found in a similar way:

$$T_2 = \frac{T_2}{T^*}\frac{T^*}{T_1}T_1 = (0.284\,11)\left(\frac{1}{0.206\,61}\right)(300\text{ K})$$

$$T_2 = 412\text{ K} = 139°\text{C}$$

and

$$p_2 = \frac{p_2}{p^*}\frac{p^*}{p_1}p_1 = (2.2209)\left(\frac{1}{2.272\,7}\right)(90\text{ kPa})$$

$$p_2 = 87.95\text{ kPa}$$

The static temperature changed a great deal, but static pressure changed only slightly. We can investigate the change in sonic velocity also. At inlet,

$$a_1 = \sqrt{\gamma RT} = \sqrt{1.4(286.8)(300)}$$
$$= 347.1\text{ m/s}$$

At exit,

$$a_2 = \sqrt{1.4(286.8)(412)} = 406.7\text{ m/s}$$

The velocity at each of these points is

$$V_1 = M_1 a_1 = 0.2(347.1) = 69.4\text{ m/s}$$
$$V_2 = M_2 a_2 = 0.24(406.7) = 97.7\text{ m/s}$$

Thus, heat addition has increased all properties except pressure.

8.7 SUMMARY

In this chapter, we examined some concepts associated with compressible flow. We determined an expression for sonic velocity in a compressible medium and also developed equations for isentropic flow. We examined in detail the behavior of a compressible fluid as it goes through a nozzle. We derived equations for normal shock waves and saw the effect of friction and heat transfer on compressible flow through a constant-area duct.

INTERNET RESOURCES

1. There is an analogy between normal shock wave behavior and open channel flow. Give a brief presentation on this analogy. What is the open channel counterpart to the Mach number?
2. What type of wind tunnel is used to observe supersonic flow past an object? To what does Schlieren refer?

PROBLEMS

Sonic Velocity and Mach Number

8.1 Determine the sonic velocity in air at a temperature of 0°F.

8.2 Calculate the sonic velocity in argon at a temperature of 25°C.

8.3 Calculate the sonic velocity in carbon dioxide at a temperature of 20°C.

8.4 Determine the sonic velocity in helium at a temperature of 50°C.

8.5 Calculate the sonic velocity in hydrogen at a temperature of 75°F.

8.6 Determine the sonic velocity in oxygen at a temperature of 75°F.

8.7 Sonic velocity in a solid can be calculated with what is known as the bulk modulus, defined as

$$\beta_s = \rho\left(\frac{dp}{d\rho}\right)$$

with which sonic velocity is determined from

$$a = \sqrt{\frac{\beta_s}{\rho}}$$

For copper, $\beta_s = 17.9 \times 10^6$ psi, and the density can be found in Table A.8. Calculate the velocity of sound in copper.

8.8 A wing travels at 800 ft/s in air at a temperature of 40°F and 10 psia. Determine the Mach number.

8.9 A bullet fired from a high-powered rifle travels at 3000 ft/s through air at 75°F and 14.7 psia. Is the speed of the bullet supersonic?

8.10 A wing travels at 400 m/s through air at 8°C and 50 kPa. Is the wing moving at subsonic velocity? Calculate the Mach number.

8.11 At what temperature will an object traveling at 900 ft/s have a Mach number of 1?

Ideal Gas Law

8.12 Following the development leading to the derivation of Equation 8.8, similarly derive the equation for specific heat at constant volume:

$$c_v = \frac{R}{\gamma - 1}$$

8.13 Using corresponding data for five different temperatures in Table A.3, determine if the properties in the table satisfy the ideal gas law.

8.14 Using corresponding data for any two of the gases in Table A.6, verify whether the following equation is accurate:

$$c_p = \frac{\gamma R}{\gamma - 1}$$

8.15 Using corresponding data for any two of the gases in Table A.6, verify whether the following relationship is accurate:

$$c_v = \frac{R}{\gamma - 1}$$

8.16 Using corresponding data for any two of the gases in Table A.6, (a) calculate $c_v = c_p/\gamma$ and (b) verify whether the following equation is accurate:

$$c_p - c_v = R$$

Isentropic Flow through a Channel of Varying Area

8.17 A converging nozzle of area 3 in.^{2} is attached to a reservoir containing air. The pressure and temperature in the reservoir are 50 psia and 1000°R. Calculate the mass flow rate through the nozzle for choked conditions.

8.18 A reservoir discharges air through a converging nozzle with an exit area of 8 cm^2. The reservoir pressure is 150 kPa, and the temperature is 1 000 K. Calculate the mass flow rate through the nozzle for back pressures of 100 and 50 kPa.

8.19 A converging nozzle of area 5 cm^2 is attached to a reservoir containing oxygen maintained at 150 kPa and 300 K. Determine the flow rate through the nozzle for back pressures of 100, 50, and 0 kPa.

8.20 Air flows into a converging duct from an inlet area of 15 in.^{2} to the exit, where the area is 10 in.^{2}. For inlet conditions of 85°F, 18 psia, and 400 ft/s, determine conditions at the exit.

8.21 Air flows through a diverging nozzle from an area of 20 cm^2 to the exit, where the area is 30 cm^2. For inlet conditions of 27°C, 200 kPa, and 300 m/s, determine conditions at the exit.

8.22 Hydrogen flows into a converging duct from an inlet area of 20 cm^2 to the exit, where the area is 10 cm^2. For exit conditions of 40°C, 100 kPa, and 950 m/s, determine conditions at the inlet.

8.23 Determine the mass flow rate through the nozzle of Example 8.8 for p_1 = 150 kPa and T_1 = 300 K.

8.24 A converging–diverging nozzle is attached to an air reservoir in which the conditions are 60 psia and 550°R. The throat area is 10 in.^{2}, and the exit area is 20 in.^{2}. Determine the back pressure required to choke the nozzle.

8.25 In a compressed air bottling plant, huge air reservoirs are slowly emptied through valves into smaller tanks intended for consumer use. In one such system, consider the valve as a converging–diverging nozzle of exit area 12 cm^2 and throat area 8 cm^2. The reservoir conditions are 400 kPa and 25°C. The smaller tank is initially evacuated; as it fills, its pressure steadily increases.

a. Determine the back pressure at which the valve is no longer choked.
b. Calculate the mass flow rate at the choked condition.

Normal Shock

8.26 Rederive Equation 8.24 in detail and verify the plot in Figure 8.14.

8.27 An airstream travels at 2100 ft/s and has static properties of 15 psia and 75°F. The stream undergoes a normal shock. Determine air velocity and the stagnation and static conditions after the shock wave.

8.28 A stream of oxygen travels at 600 m/s and undergoes a normal shock. Upstream properties are 50 kPa and 12°C. Determine velocity and stagnation and static conditions after the wave.

8.29 A stream of helium travels at 4000 ft/s and undergoes a normal shock. Determine helium velocity and static conditions after the shock if the upstream conditions are 15 psia and 30°F.

8.30 A normal shock with a velocity of 1 200 m/s travels through still air with properties 101.3 kPa and 25°C. Determine the air velocity and static properties behind the wave.

8.31 A normal shock of velocity 800 m/s travels through still air. Downstream properties are 600 kPa and 400 K. Determine downstream air velocity and upstream pressure and temperature.

Normal Shock in a Channel of Varying Area

8.32 A converging–diverging nozzle is supplied by a huge air reservoir in which conditions are 50 psia and 100°F. The nozzle has throat and exit areas of 12 and 18 in.^{2}, respectively. Determine the range of back pressures over which a normal shock appears in the nozzle or oblique shocks form at the exit.

8.33 A huge reservoir of air supplies a converging–diverging nozzle with throat and exit areas of 2 and 20 cm^2, respectively. The reservoir pressure and temperature are 150 kPa and 20°C, respectively. Determine the range of back pressures over which (a) shocks appear in the nozzle, (b) oblique shocks form at the exit, and (c) expansion waves form at the exit.

8.34 A converging–diverging nozzle is supplied by a huge oxygen reservoir. The nozzle has throat and exit areas of 5 and 12 in.^{2}, respectively. The reservoir conditions are 100 psia and 0°F. Calculate the mass flow rate through the nozzle for the case in which a normal shock appears at the exit plane. Make calculations both at the throat and just downstream of the shock.

8.35 A converging–diverging nozzle is supplied by a huge oxygen reservoir. The nozzle has a throat area of 10 in.^{2} and an exit area of 18 in.^{2}. Reservoir pressure and temperature are 100 psia and 500°R. Determine the range of back pressures over which (a) shocks appear in the nozzle, (b) oblique shocks form at the exit, and (c) expansion waves form at the exit.

Flow with Friction

8.36 Oxygen flows into a constant-area pipe at 60 ft/s, 40 psia, and 32°F. The pipe is 2000 ft long and made of 3-nominal galvanized iron (schedule 40). Determine the Mach number at the pipe exit.

8.37 Air flows into a 12-nominal schedule 80 asphalted cast iron pipe at 4 m/s, 200 kPa, and 30°C. What length of pipe is required for the exit Mach number to be unity? What is the exit pressure?

8.38 Oxygen flows into a 300 m long pipe at inlet conditions of 20 m/s, 400 kPa, and 6°C. The pipe is made of drawn metal (smooth walled), and the system must be capable of delivering the air at a Mach number of about 0.8. Select a suitable pipe diameter. Calculate exit pressure and temperature.

8.39 Air flows into a 6-nominal schedule 80 PVC pipe. At the exit, the velocity is 1500 ft/s, the pressure is 8 psia, and the temperature is 530°R. If the pipe length is 20 ft, calculate the inlet conditions.

8.40 Oxygen enters a smooth-walled 1-nominal schedule 40 pipe at a velocity of 100 m/s, a pressure of 210 kPa, and a temperature of 300 K. The pipe length is 4.8 m. Determine conditions at the exit.

8.41 Air enters a 12-nominal schedule 40 pipe at a velocity of 1400 ft/s, a pressure of 12 psia, and a temperature of 480°R. Determine the length required for flow to be sonic at the exit and calculate the pressure there. Take ε/D to be 0.00001.

Flow with Heat Transfer

8.42 Air at 10°F and 2 psia enters a duct at an inlet velocity of 300 ft/s. Heat is added to the fluid at a rate of 1000 Btu/(slug of air). Determine the pressure and temperature of the air at the exit. Assume specific heat at constant pressure to be 7.72 Btu/(slug·°R) and no friction.

8.43 Oxygen at 25°C and 101.3 kPa enters a duct at a velocity of 1 400 m/s. Heat is removed from the oxygen at a rate of 15 000 J/kg. Assuming frictionless flow, determine the properties (pressure, temperature, Mach number, and velocity) at the exit.

8.44 Air enters a duct at a velocity of 100 ft/s, and it is desired to determine how much heat is required such that the Mach number of the air will reach a value of 1. For inlet static conditions of 75°F and 14.7 psia, determine the conditions (pressure and temperature) at the exit and the heat transfer rate.

8.45 Nitrogen [$\gamma = 1.4$, $c_p = 1.04$ kJ/(kg · K)] enters a duct and is heated so that at the exit, the Mach number is 0.4. The heat added amounts to 40 000 J/kg. Determine pressure, temperature, and Mach number at the inlet if the exit properties are 90 kPa and 290 K.

8.46 Air at 32°F and 10 psia enters a duct and is heated by an amount equal to 2000 Btu/(slug of air). Calculate the mass flow rate of air using properties at the inlet and again using conditions at the outlet. Assume frictionless flow, a flow area of 4 in.^{2}, and an inlet velocity of 400 ft/s.

Miscellaneous Problems

8.47 Table D.1 is an isentropic flow table for a fluid having a ratio of specific heats $\gamma = 1.4$. A number of other fluids exist that have a different value for γ, as indicated in Table A.6. Using the appropriate equations, develop a generic isentropic flow table for any value of γ. That is, the user can change γ to any value, and the table automatically changes.

8.48 Table D.2 applies to a supersonic flow of a fluid having a ratio of specific heats $\gamma = 1.4$ that undergoes a normal shock. A number of other fluids exist that have a different value for γ, as indicated in Table A.6. Using the appropriate equations, generate a normal shock table for any value of γ. The user can change γ to any value, and the table automatically changes.

8.49 Table D.3 applies to a flow with a friction of a compressible fluid having a ratio of specific heats $\gamma = 1.4$ in a constant-area duct (also known as Fanno flow). A number of other fluids exist that have a different value for γ, as indicated in Table A.6. Using the appropriate equations, generate a Fanno flow table for any fluid. The user can change γ to any value, and the table automatically changes.

8.50 Table D.4 applies to a flow with heat transfer of a compressible fluid having a ratio of specific heats $\gamma = 1.4$ in a constant-area duct (also known as Rayleigh flow). A number of other fluids exist that have a different value for γ, as indicated in Table A.6. Using the appropriate equations, generate a Rayleigh flow table for any fluids. The user can change γ to any value, and the table automatically changes.

9 Turbomachinery

A turbomachine is a device that transfers energy between its own rotating parts and a moving fluid. Examples include steam turbines (as found in conventional power plants), pumps, compressors, ordinary window fans, and windmills. Our objective in this study is to learn to predict general performance characteristics of turbomachines.

A turbomachine can be either of two types: those that use external power to impart energy to a fluid (such as fans, blowers, pumps, compressors, and propellers) and those that produce external power by absorbing energy from a fluid (such as propeller turbines, impulse turbines, and windmills). In this chapter, we study various turbines and pumps and examine their performance characteristics.

The two principal components of a turbomachine are the **output shaft**, which transmits mechanical energy to or from the machine, and the **rotor**, which has vanes or blades attached. In many cases, the rotor is connected directly to the output shaft. In some designs, such as steam turbines, the rotor is housed in a casing to allow the working fluid to change pressure within the machine; the casing completely separates the fluid and machine from the surroundings. In other designs, such as windmills, the rotor is completely exposed.

The majority of turbomachines can be classified into axial-flow, radial-flow, or mixed-flow types. In an axial-flow machine, the main path of the fluid through the device is in the same direction as the rotor axis. In the radial-flow type, the main flow direction is perpendicular to the axis of rotation. In the mixed-flow type, the main flow direction is neither purely radial nor purely axial but a combination of the two. Radial- and axial-flow types are discussed in detail in this chapter.

The chapter concludes the study of turbomachines by giving an overview of general equations that are used to characterize them. Dimensional analysis is employed to obtain expressions that are useful for modeling purposes. An analysis and performance data for a centrifugal pump are also presented. This machine is the only one considered in detail here, but the information given is representative of the type of data to be expected in describing performance characteristics of turbomachines. The chapter ends with a discussion of hydraulic turbines.

In mathematically modeling flows through turbomachines, it is important to remember the common thread that runs throughout the analyses: the objective is to relate the shaft work or torque to the properties of the fluid while it flows through the machine.

9.1 EQUATIONS OF TURBOMACHINERY

In Chapter 3, we derived the linear momentum equation from Newton's law of motion. In applying this law to rotating bodies, such as parts of a turbomachine, it is necessary to consider moments relative to the axis of rotation and to evaluate moments exerted on the fluid volume. Because the axis of rotation is fixed, we can use Newton's law to develop what we will call the **angular momentum equation**.

Consider the general two-dimensional case depicted in Figure 9.1. A differential element located at a distance r from the origin is acted upon by a force dF. The force may be due to fluid moving through a turbine, for example, whereas the differential element might be part of the rotor. The component of the force acting perpendicular to r is dF_t, where the subscript t denotes a tangential force. The moment exerted on the differential element is dT_s, found from

$$dT_s = r\,dF_t \tag{9.1}$$

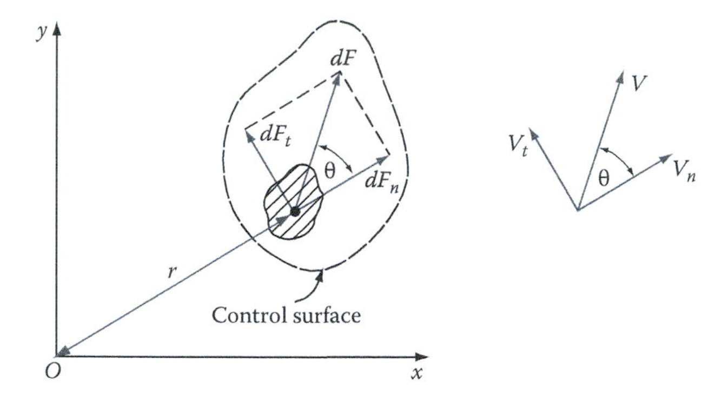

FIGURE 9.1 Force acting on a differential element at a fixed distance from the origin.

The differential force in the tangential direction is determined by applying the linear momentum equation in the tangential direction

$$\sum F_t = \frac{\partial}{\partial t}(mV)_t\bigg|_{CV} + \iint_{CS} V_t \rho V_n \, dA$$

or, in differential form,

$$dF_t = \frac{\partial}{\partial t}(V_t \, dm)\bigg|_{CV} + V_t \rho V_n \, dA \tag{9.2}$$

where V_t is the tangential velocity at radius r. Combining Equations 9.1 and 9.2 gives the following for the differential moment:

$$dT_s = r\frac{\partial}{\partial t}(V_t \, dm)\,|_{CV} + rV_t(\rho V_n \, dA)$$

The total torque is determined by integration:

$$T_s = \frac{\partial}{\partial t}\iiint_{CV} rV_t \, dm + \iint_{CS} rV_t(\rho V_n \, dA) \tag{9.3}$$

The term T_s represents the sum of all externally applied moments or torques due to pressure forces, gravity, viscous forces, and so on. The first term on the right-hand side is the rate at which angular momentum is stored in the control volume. The second term represents the net rate out (out minus in) of angular momentum from the control volume.

From Figure 9.1, we see that $V_t = V \sin\theta$. Equation 9.3 can therefore be rewritten as

$$T_s = \frac{\partial}{\partial t}\iiint_{CV} rV\sin\theta \, dm + \iint_{CS} rV\sin\theta(\rho V_n \, dA) \tag{9.4}$$

We can generalize this result into a vector form that is applicable to the three-dimensional case:

$$\mathbf{T} = \frac{\partial}{\partial t}\iiint_{CV} (\mathbf{r}\times\mathbf{V})dm + \iint_{CS} (\mathbf{r}\times\mathbf{V})(\rho\mathbf{V}\cdot\mathbf{dA}) \tag{9.5}$$

For steady flow, Equation 9.4 becomes

$$T_s = \iint_{\text{CS}} rV \sin\theta(\rho V_n \, dA) \tag{9.6}$$

The angular momentum equation is applied to a simple problem in the following example.

Example 9.1

A lawn sprinkler is shown schematically in Figure 9.2. The sprinkler consists of two nozzles at opposite ends of a rotor that rotates about its center. Water leaving the nozzles causes the rotor to spin at a constant angular velocity depending on the volume flow of water and the magnitude of the opposing torque due to bearing friction. Determine an expression for the torque in terms of the operating variables.

Solution

Select a control volume to include the entire sprinkler rotor, as shown in Figure 9.2. Next choose a constant moment arm R from the pivot to the center of the exit. All fluid particles leaving the exit do not exert the same torque, but we can make this assumption with negligible error because the nozzle exit dimension is small in comparison to R. While the rotor is rotating, the only external applied torque is due to bearing friction. No angular momentum is stored in the control volume. Further, all fluid enters in the axial direction and therefore has no angular momentum.

It is helpful to sketch a velocity diagram for the water leaving the nozzles (Figure 9.3). We have the following:

V is the absolute velocity of water leaving nozzle (as seen by stationary observer)
V_{rel} is the relative velocity of water leaving nozzle (as seen by observer moving with nozzle)
V_n is the velocity normal to control surface with $V_n = V\cos\theta = V_{rel}\cos\alpha$
V_t is the tangential velocity component of V, $V_t = V\sin\theta$

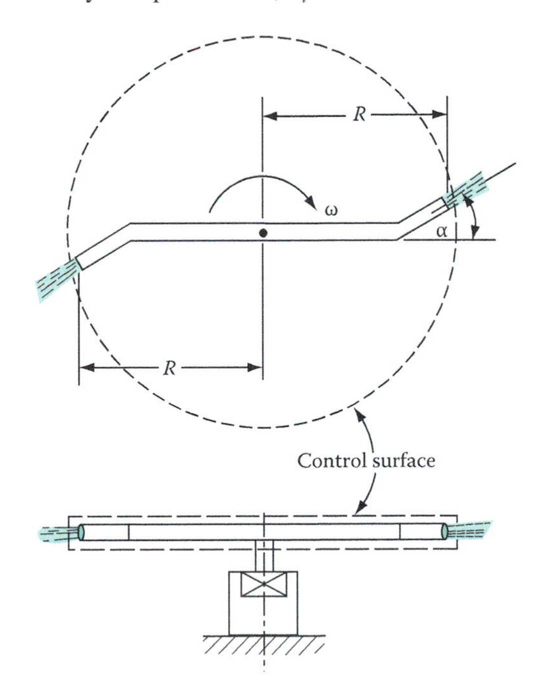

FIGURE 9.2 Lawn sprinkler of Example 9.1.

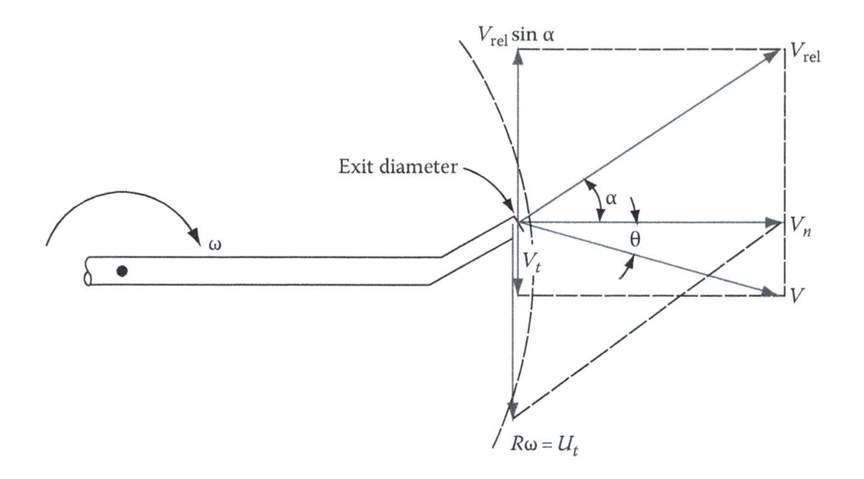

FIGURE 9.3 Velocity diagram for the water leaving the nozzle.

$V_{\text{rel}}\sin\alpha$ is the tangential velocity component of $V_{\text{rel}} = V_{\text{rel}t}$
$U_t = R\omega$ is the tip velocity of rotor in tangential direction

From the velocity diagram, we see that the tangential velocity components must balance; that is,

$$V_{\text{rel}}\sin\alpha = V_t + U_t$$

or

$$V_t = V\sin\theta = V_{\text{rel}}\sin\alpha - R\omega$$

As indicated in Figure 9.2, the relative velocity is perpendicular to the nozzle exit. Therefore, for one nozzle, we have

$$Q\Big|_{\text{1nozzle}} = Q_1 = AV_{\text{rel}} = \frac{\pi D^2}{4}V_{\text{rel}}$$

or

$$V_{\text{rel}} = \frac{4Q_1}{\pi D^2}$$

where
D is the nozzle exit diameter
Q_1 is the volume flow rate through only one nozzle (equal to half that delivered by the sprinkler)

The tangential velocity now becomes

$$V_t = V\sin\theta = \frac{4Q_1}{\pi D^2}\sin\alpha - R\omega$$

Equation 9.4 applied to one nozzle discharging liquid out of the control volume becomes

$$T_s = \iint_{\text{CS}} rV_t(\rho V_n\, dA)$$

With the efflux at $r = R$, for constant density ρ, and with no inflow in the tangential direction, the preceding equation becomes, for two identical nozzles,

$$T_s = 2\rho R \iint_{CS} V_t (V_n \, dA)$$

Substituting for tangential velocity, we get

$$T_s = 2\rho R \iint_{CS} \left(\frac{4Q_1}{\pi D^2} \sin\alpha - R\omega \right) V_n \, dA$$

Integrating, and recalling that the volume flow rate Q_1 is for one nozzle, we obtain, for the torque,

$$T_s = 2\rho Q_1 R \left(\frac{4Q_1}{\pi D^2} \sin\alpha - R\omega \right)$$

If the nozzles are discharging equal amounts of flow, then the total flow rate Q equals $2Q_1$. The torque then becomes

$$T_s = \rho Q R \left(\frac{2Q}{\pi D^2} \sin\alpha - R\omega \right)$$

We are now ready to apply the angular momentum equation to the case of a generalized turbomachine. A typical control volume is given in Figure 9.4. It is customary to take V_t as positive when it is in the direction of rotation. The normal velocities V_n are assumed to be average velocities. The continuity equation for the control volume is

$$\iint_{CS} \rho V_n \, dA = 0$$

or

$$\rho_2 A_2 V_{n_2} - \rho_1 A_1 V_{n_1} = 0 \quad (9.7a)$$

For an incompressible fluid, $\rho_1 = \rho_2$, and the continuity equation becomes

$$A_1 V_{n_1} = A_2 V_{n_2} = Q \quad (9.7b)$$

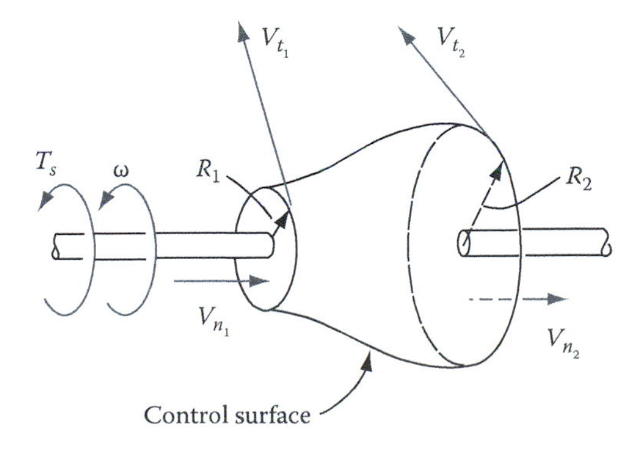

FIGURE 9.4 Control volume for a generalized turbomachine.

Applying the steady-flow angular momentum equation, we get

$$T_s = \iint_{\text{CS}} rV_t(\rho V_n\, dA) = (R_2 V_{t_2} \rho_2 Q_2 - R_1 V_{t_1} \rho_1 Q_1)$$

Combining with Equation 9.7, we get

$$T_s = \rho Q(R_2 V_{t_2} - R_1 V_{t_1}) \tag{9.8}$$

in which R_1 and R_2 are average radial distances at inlet and outlet, whereas V_{t_1} and V_{t_2} are corresponding average tangential velocities. Equation 9.8—known as the **Euler turbine equation**—relates shaft torque to fluid velocities through a turbomachine. The shaft torque is taken as positive if the turbomachine is a pump; it is taken as negative for a turbine.

By definition, the power or rate of doing work is the product of shaft torque and rotational speed:

$$-\frac{dW_s}{dt} = T_s\omega = \rho Q\omega(R_2 V_{t_2} - R_1 V_{t_1}) \tag{9.9}$$

The tangential velocity of the rotor is U_t (=$R\omega$), and by substitution, Equation 9.9 becomes

$$-\frac{dW_s}{dt} = T_s\omega = \rho Q(U_{t_2} V_{t_2} - U_{t_1} V_{t_1}) \tag{9.10a}$$

On a per-unit-mass basis, we have

$$-\frac{1}{\rho Q}\frac{dW_s}{dt} = \frac{T_s\omega}{\rho Q} = (U_{t_2} V_{t_2} - U_{t_1} V_{t_1}) \tag{9.10b}$$

Thus, the difference of the product of two tangential velocities ($U_t V_t$) is proportional to the power per unit mass of fluid passing through the turbomachine. The power in Equation 9.10 is the actual power measured at the shaft.

Applying the steady-flow energy equation (Equation 3.21) to the control volume of Figure 9.4 gives

$$\frac{d(\tilde{Q} - W)}{dt} = \left[\left(h + \frac{V^2}{2} + gz\right)\Bigg|_{\text{out}} - \left(h + \frac{V^2}{2} + gz\right)\Bigg|_{\text{in}}\right]\rho Q$$

In the absence of heat transfer (as with an adiabatic machine) and using subscript notation, the energy equation becomes

$$-\frac{dW}{dt} = \left[\left(h_2 + \frac{V_2^2}{2} + gz_2\right) - \left(h_1 + \frac{V_1^2}{2} + gz_1\right)\right]\rho Q$$

The work term W is made up of two parts—shaft work and friction work. Rewriting the energy equation, we get

$$-\frac{dW_s}{dt} - \frac{dW_f}{dt} = \rho Q\left[\left(h_2 + \frac{V_2^2}{2} + gz_2\right) - \left(h_1 + \frac{V_1^2}{2} + gz_1\right)\right] \tag{9.11a}$$

If work is done by the fluid, the shaft work W_s is a positive quantity (as in turbines). If work is done on the fluid (pumps, fans, and compressors), however, W_s is negative. The friction work W_f is always a positive quantity, because it represents shear work done by the fluid.

Equation 9.11a gives the power based on the change in fluid properties across the machine and is useful primarily for turbomachines that have a gas or vapor as the fluid medium (as in an air compressor or steam turbine). With these fluids in general, enthalpy differences are far more significant than differences in kinetic and potential energies, which are therefore usually ignored in the equation. Equation 9.11a thus becomes for a gas or vapor medium

$$-\frac{dW_s}{dt}-\frac{dW_f}{dt}=\rho Q\Delta h \tag{9.11b}$$

where $\Delta h = h_2 - h_1$, the enthalpy difference across the machine.

If the fluid medium is a liquid (as in a water turbine or pump), a modified version of Equation 9.11a is more appropriate. Recall the definition of enthalpy:

$$h=u+\frac{p}{\rho}$$

For liquids, the change in internal energy of the fluid from inlet to outlet is negligible in comparison to the changes in pressure, kinetic energy, and potential energy. Thus, Equation 9.11a becomes

$$-\frac{dW_s}{dt}-\frac{dW_f}{dt}=\rho Q\left[\left(\frac{p_2}{\rho}+\frac{V_2^2}{2}+gz_2\right)-\left(\frac{p_1}{\rho}+\frac{V_1^2}{2}+gz_1\right)\right] \tag{9.12a}$$

The quantity

$$\frac{p}{\rho g}+\frac{V^2}{2g}+z=Z$$

is called the total head Z. In terms of total head, Equation 9.12a is, for a liquid,

$$-\frac{dW_s}{dt}-\frac{dW_f}{dt}=\rho Q\Delta Z g \tag{9.12b}$$

where $\Delta Z = Z_2 - Z_1$.

Equation 9.10 is an expression for power measured at the shaft of a turbomachine, whereas Equations 9.11 and 9.12 tell how the fluid itself changes properties across the machine. The difference in these expressions is due to friction and other irreversibilities. This condition leads us to the concept of efficiency. For a fan or a pump, we put work into the system by rotating the shaft (Equation 9.10), while the output goes into overcoming frictional effects and increasing the energy of the fluid (Equations 9.11 or 9.12). By defining efficiency as

$$\eta=\frac{\text{Energy out}}{\text{Energy in}}$$

we obtain the following for a pump, fan, or compressor:

$$\eta=\frac{-(dW_s/dt)-(dW_f/dt)}{-(dW_s/dt)}$$

By substitution from Equations 9.10 through 9.12, we have

$$\text{Compressor or fan efficiency: } \eta = \frac{\Delta h}{U_{t2}V_{t2} - U_{t1}V_{t1}} \tag{9.13}$$

$$\text{Pump efficiency: } \eta = \frac{g\Delta Z}{U_{t2}V_{t2} - U_{t1}V_{t1}} \tag{9.14}$$

For a gas turbine or a water turbine, the fluid loses energy on its way through the machine. This energy goes into overcoming frictional effects and into useful power as measured at the output shaft. So for a turbine, efficiency is

$$\eta = \frac{\text{Energy out}}{\text{Energy in}}$$

$$= \frac{-(dW_s/dt)}{-(dW_s/dt) - (dW_f/dt)}$$

By substitution from Equations 9.10 through 9.12, we obtain

$$\text{Gas or vapor turbine efficiency: } \eta = \frac{U_{t2}V_{t2} - U_{t1}V_{t1}}{\Delta h} \tag{9.15}$$

$$\text{Water turbine efficiency: } \eta = \frac{U_{t2}V_{t2} - U_{t1}V_{t1}}{g\Delta Z} \tag{9.16}$$

These equations were developed for a generalized turbomachine and give an idea of what is to be expected in analyzing fluid machinery. The general equations will be modified where appropriate to make them applicable specifically to the various devices discussed in this chapter.

9.2 AXIAL-FLOW TURBINES

In an axial-flow turbine, the flow through the machine is mainly in the axial direction. A schematic of a conventional axial-flow turbine is given in Figure 9.5. It consists of a **stator**, which is a stationary housing onto which fixed blades are attached, and a rotating **rotor**, also with blades attached. The flow of fluid is through the passage between adjacent blades in the same row. The passage is sometimes referred to as a **nozzle**. A **turbine stage** is defined as one stator and one rotor set of blades. The dimensions of the rotor and stator can change with distance along the machine to make better use of the energy in the fluid from stage to stage—that is, to extract as much energy from the fluid as efficiently as possible. The multistage machine shown in Figure 9.5 is used primarily for gas and steam flows. Steam turbines are used extensively in conventional power plants for the production of electricity.

A simple approach to the study of such a machine is to assume that conditions at the mean flow radius (the pitch line) fully represent the flow at all other radii in one set of blades. This two-dimensional analysis is reasonable if the ratio of blade height to mean radius is small. If this ratio is large, however, a three-dimensional analysis is required. Because a three-dimensional analysis is beyond the range of what we wish to study, we will formulate only a two-dimensional approach.

A velocity diagram for one stage is given in Figure 9.6. It is assumed that radial velocities are zero and that the flow does not vary in the circumferential direction. The velocities of importance

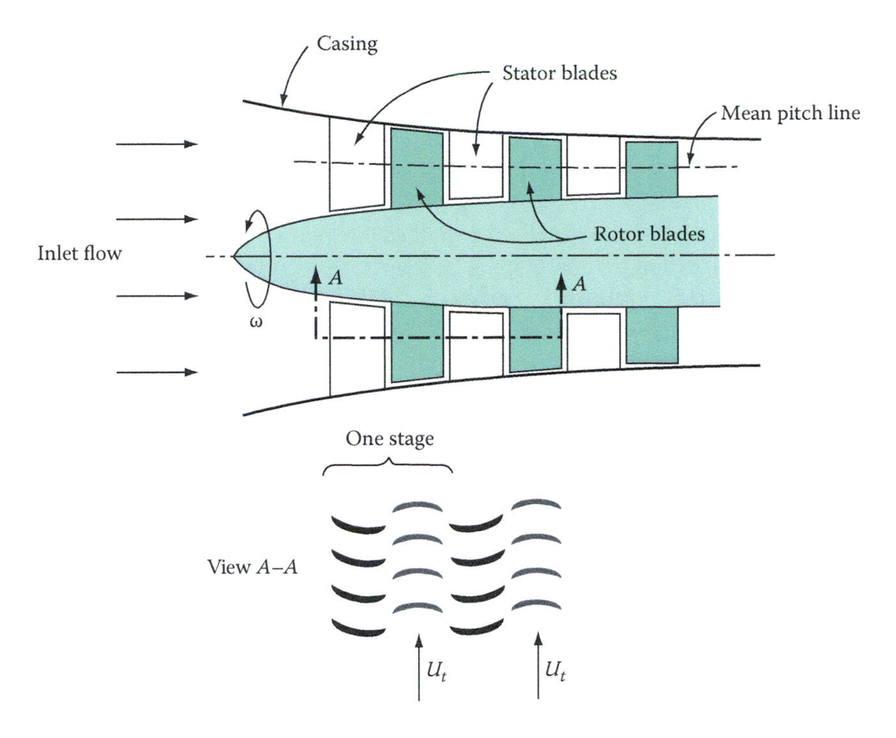

FIGURE 9.5 Schematic of an axial-flow turbine.

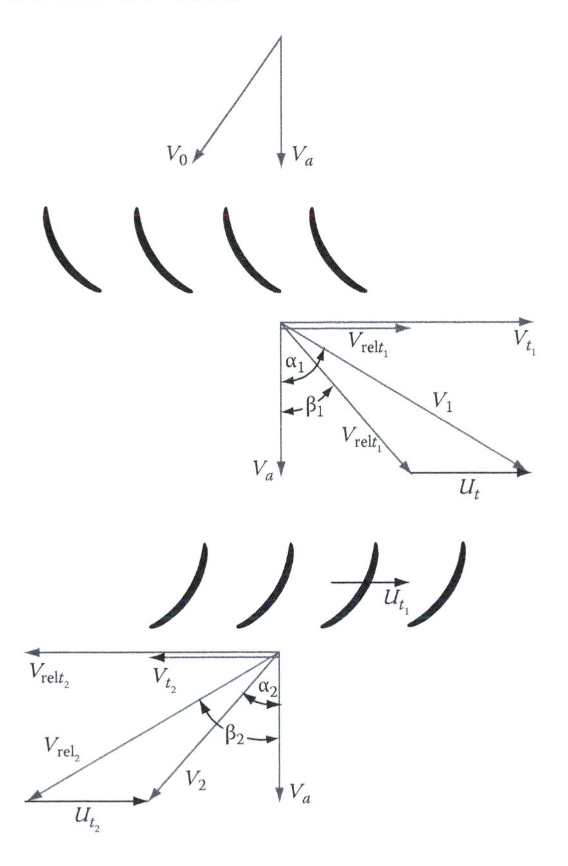

FIGURE 9.6 Velocity diagram for the axial-flow turbine stage.

are the absolute velocities V_0, V_1, and V_2, the relative velocity V_{rel}, the tangential velocity V_t, the rotor tangential velocity U_t, and the axial velocity V_a. The stage is referred to as **normal** if $V_0 = V_2$. It is customary to take the axial velocity as constant through the machine. Because the change of fluid properties within each stage is small, the fluid can be assumed to be incompressible in each stage. A different density is then selected for each successive stage to account for the compressibility effect. The controllable features of importance are the blade angles α_1 and β_2. The angle α_1 is determined by the setting of the stator blades; β_2 is determined by the setting of the moving rotor blades. Because these blade angles are independent of the flow through the machine, it is desirable to express our equations in terms of these design angles.

Equations 9.10a and 9.11a can be applied to any stage in the turbine. For the velocity diagram of Figure 9.6, Equation 9.10a becomes

$$\frac{dW_s}{dt} = T_s\omega = \rho Q U_t(V_{t_2} + V_{t_1}) \tag{9.17}$$

The sign on the shaft work term is positive, because we are dealing with a turbine. Moreover, the peripheral velocities are taken at the same radius, so $U_{t_2} = U_{t_1} = U_t$. Finally, the sign convention that the tangential velocity is positive if in the same direction as U_t requires the sign change on V_{t_1}.

Next, we apply Equation 9.11a, assuming frictionless flow through the stage and no changes in potential energy:

$$-\frac{dW_s}{dt} = \rho Q\left[\left(h_2 + \frac{V_2^2}{2}\right) - \left(h_0 + \frac{V_0^2}{2}\right)\right]$$

Because we are treating each stage as though the fluid medium passing through is incompressible, the enthalpy terms become pressure terms. We therefore obtain

$$-\frac{dW_s}{dt} = \rho Q\left[\left(\frac{p_2}{\rho} + \frac{V_2^2}{2}\right) - \left(\frac{p_0}{\rho} + \frac{V_0^2}{2}\right)\right] \tag{9.18a}$$

For a normal stage ($V_0 = V_2$), Equation 9.18a becomes

$$-\frac{dW_s}{dt} = Q(p_2 - p_0) \tag{9.18b}$$

Combining Equations 9.17 and 9.18 gives the following for the pressure drop across the stage:

$$p_0 - p_2 = \rho U_t(V_{t_2} + V_{t_1}) \tag{9.19}$$

Another important pressure-drop term that we will need is across the rotor row. Again, we apply Equations 9.10a and 9.11a. The following equations result:

$$\frac{dW_s}{dt} = \rho Q U_t(V_{t_2} + V_{t_1}) \tag{9.20}$$

and

$$-\frac{dW_s}{dt} = \rho Q\left[\left(\frac{p_2}{\rho} + \frac{V_2^2}{2}\right) - \left(\frac{p_1}{\rho} + \frac{V_1^2}{2}\right)\right] \tag{9.21}$$

Equating and regrouping give

$$\frac{p_1 - p_2}{\rho} + \frac{V_1^2 - V_2^2}{2} - U_t(V_{t_2} + V_{t_1}) = 0$$

From the velocity diagrams, we have the following for both V_1 and V_2:

$$V^2 = V_a^2 + V_t^2$$

By substitution, we get

$$\frac{p_1 - p_2}{\rho} + \frac{V_{t_1}^2 - V_{t_2}^2}{2} - U_t(V_{t_1} + V_{t_2}) = 0$$

because V_a is a constant. Further simplification gives

$$\frac{p_1 - p_2}{\rho} + \frac{1}{2}[(V_{t_1} - V_{t_2})\ (V_{t_1} + V_{t_2}) - 2U_t(V_{t_1} + V_{t_2})] = 0$$

or

$$\frac{p_1 - p_2}{\rho} + \frac{1}{2}(V_{t_1} + V_{t_2})[(V_{t_1} - U_t) - (V_{t_2} + U_t)] = 0 \tag{9.22}$$

Again from the velocity diagrams, the tangential velocities must balance

$$\begin{aligned} V_{t_1} - U_t &= V_{\text{rel}t_1} \\ V_{t_2} + U_t &= V_{\text{rel}t_2} \end{aligned} \tag{9.23}$$

Adding these gives yet another relationship:

$$V_{t_1} + V_{t_2} = V_{\text{rel}t_1} + V_{\text{rel}t_2} \tag{9.24}$$

Substituting Equations 9.23 and 9.24 into Equation 9.22 gives

$$\frac{p_1 - p_2}{\rho} + \frac{1}{2}\ (V_{\text{rel}t_1} + V_{\text{rel}t_2})\ (V_{\text{rel}t_1} - V_{\text{rel}t_2}) = 0$$

$$\frac{p_1 - p_2}{\rho} + \frac{1}{2}\left(V_{\text{rel}t_1}^2 - V_{\text{rel}t_2}^2\right) = 0$$

Adding and subtracting V_a^2 from the kinetic-energy term yield

$$\frac{p_1 - p_2}{\rho} + \frac{V_{\text{rel}t_1}^2 + V_a^2 - V_{\text{rel}t_2}^2 - V_a^2}{2} = 0$$

Simplifying and dividing by g, we obtain

$$\frac{p_1 - p_2}{\rho g} + \frac{\left(V_{\text{rel}_1}^2 - V_{\text{rel}_2}^2\right)}{2g} = 0 \tag{9.25a}$$

Thus, we have proved that, through the rotor, Bernoulli's equation using relative velocities is valid. The pressure drop across the rotor row of blades is thus

$$p_1 - p_2 = \frac{\rho}{2}\left(V_{\text{rel}_2}^2 - V_{\text{rel}_1}^2\right) \tag{9.25b}$$

Performance of an axial-flow turbine stage is conveniently described by means of what is called **degree of reaction** or **reaction ratio**, RR. The classic definition is the ratio of static pressure drop in the rotor to the static pressure drop in the stage:

$$\text{RR} = \frac{p_1 - p_2}{p_0 - p_2} \tag{9.26}$$

In terms of velocities, the reaction ratio is

$$\text{RR} = \frac{V_{\text{rel}_2}^2 - V_{\text{rel}_1}^2}{2U_t(V_{t_2} + V_{t_1})} \tag{9.27}$$

where Equations 9.19 and 9.25b for pressure drops have been substituted. With

$$V_{\text{rel}}^2 = V_a^2 + V_{\text{rel}t}^2$$

and Equation 9.24, Equation 9.27 becomes

$$\text{RR} = \frac{V_{\text{rel}t_2}^2 - V_{\text{rel}t_1}^2}{2U_t(V_{t_2} + V_{t_1})} = \frac{(V_{\text{rel}t_2} - V_{\text{rel}t_1})\,(V_{\text{rel}t_2} + V_{\text{rel}t_1})}{2U_t(V_{t_1} + V_{t_2})}$$

or

$$\text{RR} = \frac{V_{\text{rel}t_2} - V_{\text{rel}t_1}}{2U_t} \tag{9.28}$$

In terms of axial velocity, the tangential velocities can be written as

$$\begin{aligned} V_{\text{rel}t_1} &= V_a \tan\beta_1 \\ V_{\text{rel}t_2} &= V_a \tan\beta_2 \end{aligned} \tag{9.29}$$

Combining these and Equation 9.23 with Equation 9.28 gives

$$\begin{aligned} \text{RR} &= \frac{V_a}{2U_t}(\tan\beta_2 - \tan\beta_1) \\ \text{RR} &= \frac{1}{2} + \frac{V_a}{2U_t}(\tan\beta_2 - \tan\alpha_1) \end{aligned} \tag{9.30}$$

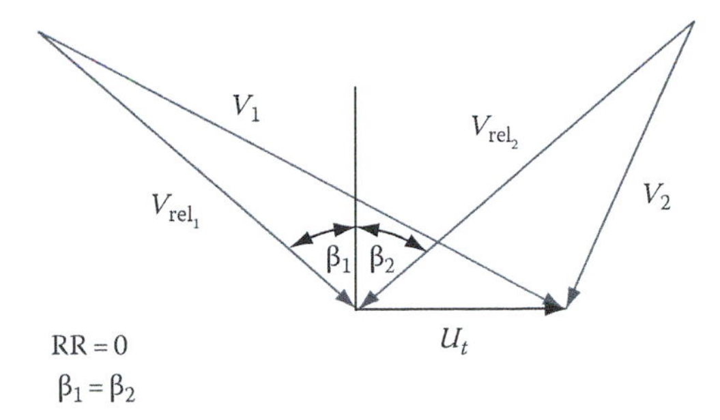

FIGURE 9.7 Combined velocity diagram for a zero reaction turbine stage.

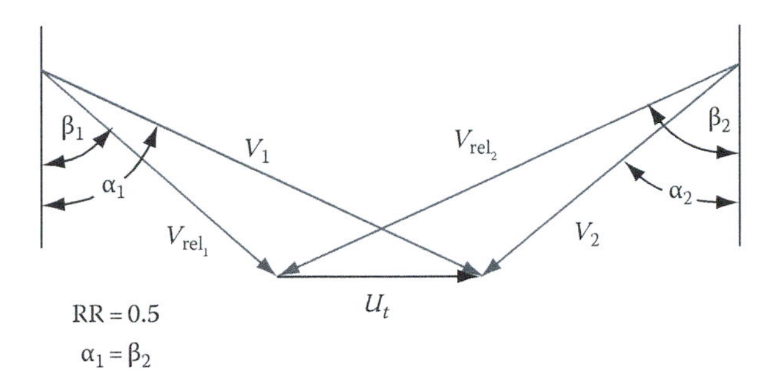

FIGURE 9.8 Combined velocity diagram for a 50% reaction turbine stage.

Two cases of note are the zero reaction stage and the 50% reaction stage. From Equation 9.30, if RR = 0, then $\beta_1 = \beta_2$. We can now resketch the velocity diagram as in Figure 9.7 to illustrate a zero reaction stage. According to Equation 9.26, this means that there is no pressure drop across the rotor row. A 50% reaction stage can be represented by the velocity diagram of Figure 9.8. From Equation 9.30, if RR = 0.5, then $\beta_2 = \alpha_1$. Half the pressure drop in the stage occurs across the rotor row.

In conventional machines, the reaction ratio falls between zero and unity. However, an undesirable phenomenon occurs if the reaction ratio exceeds unity. For this case, the absolute velocity leaving the stator row V_1 is less than the inlet velocity V_0. Accordingly, an adverse pressure gradient exists that causes separation of the flow from the blades in the stator. Since this separation results in large-scale losses, reaction ratios greater than unity are avoided in good design.

Example 9.2

Measurements on a single-stage gas turbine indicate that at stage entry, the pressure is 110 psia and temperature is 1000°R. Pressure drop across the stage is 10 psf. The rotational speed of the rotor is 180 rpm; at this speed, the shaft torque is measured to be 2.0 ft·lbf. The rotor diameter is 1.5 ft, and blade height is 1 in. The mean pitch line is at half the blade height. The mass flow of gas is 0.062 slug/s, and the specific heats are the same as those for air. Assume ideal gas behavior and that the stage is normal. The inlet flow angle between the absolute velocity and the axial velocity is 38°. Determine

a. Velocity diagrams
b. Efficiency of the turbine
c. The reaction ratio

Solution

For the following calculations, it is helpful to refer to Figure 9.6.

a. The density is found with the ideal gas law as

$$\rho = \frac{p}{RT} = \frac{110(144)}{1710(1000)} = 0.00926 \text{ slug/ft}^3$$

where R for air has been used. Density is assumed to be constant through the stage. The gas flows through the annulus formed by the rotor and the casing. The clearance is 1 in. (the blade height). The flow area is

$$A = \frac{\pi}{4}(D_0^2 - D_i^2) = \frac{\pi}{4}\left[\left(1.5 + \frac{2}{12}\right)^2 - (1.5)^2\right]$$

$$= 0.415 \text{ ft}^2$$

The axial velocity can now be determined with

$$\dot{m} = \rho A V_a$$

$$V_a = \frac{0.062}{0.00926(0.415)} = 16.2 \text{ ft/s}$$

The inlet velocity is calculated by using

$$V_a = V_0 \cos \alpha_0$$

With α_0 given as 38°, we find

$$V_0 = \frac{16.2}{\cos 38°} = 20.6 \text{ ft/s}$$

Because the stage is normal,

$$V_2 = V_0 = 20.6 \text{ ft/s}$$

For a constant axial velocity, then

$$\alpha_2 = \alpha_0 = 38°$$

Also,

$$V_{t_2} = V_2 \sin \alpha_2 = 12.7 \text{ ft/s}$$

The mean pitch line is found at a radius located at half the blade height:

$$R = \frac{1.5 + \frac{1}{12}}{2} = 0.79 \text{ ft}$$

Given a rotational speed of 180 rev/min, the tangential velocity of the rotor is

$$U_t = R\omega = 0.79 \text{ ft}(180 \text{ rev/min})(2\pi \text{ rad/rev})(1 \text{ min/60 s})$$

$$= 14.9 \text{ ft/s}$$

Summing tangential velocities, we have

$$V_{\text{rel}t_2} = V_{t_2} + U_t = 12.7 + 14.9 = 27.6 \text{ ft/s}$$

We can now find β_2 with

$$\tan\beta_2 = \frac{V_{\text{rel}t_2}}{V_a} = \frac{27.6}{16.2} = 1.71$$

$$\beta_2 = 59.6°$$

In addition, we also have

$$V_{\text{rel}_2} \sin\beta_2 = V_{\text{rel}t_2}$$

$$V_{\text{rel}_2} = 32.0 \text{ ft/s}$$

Equation 9.17 gives the power as

$$\frac{dW_s}{dt} \rho Q U_t (V_{t_2} + V_{t_1}) = T_s \omega$$

For a torque of 2.0 ft · lbf at 180 rev/min, we have

$$\frac{dW_s}{dt} = (2.0 \text{ ft} \cdot \text{lbf})(180 \text{ rev/min})(2\pi \text{ rad/rev})(1 \text{ min/60 s})$$

$$= 37.7 \text{ ft} \cdot \text{lbf/s}$$

Rearranging the preceding power equation, we find

$$V_{t_1} = \frac{dW_s}{dt} \frac{1}{\dot{m} U_t} - V_{t_2}$$

$$= \frac{37.7}{0.062(14.9)} - 12.7 = 40.7 - 12.7 = 28.0 \text{ ft/s}$$

Using the tangential velocity relationship, we obtain

$$V_{\text{rel}t_1} = V_{t_1} - U_t = 28.0 - 14.9 = 13.1 \text{ ft/s}$$

Also,

$$\tan\alpha_1 = \frac{V_{t_1}}{V_a} = \frac{28.0}{16.2} = 1.73$$

$$\alpha_1 = 59.9°$$

Next,

$$\tan\beta_1 = \frac{V_{\text{rel}t_1}}{V_a} = \frac{13.1}{16.2} = 0.808$$

$$\beta_1 = 39.0°$$

From the velocity triangles, we get

$$V_1 = \frac{V_a}{\cos \alpha_1} = \frac{16.2}{\cos 59.9°} = 32.3 \text{ ft/s}$$

and finally

$$V_{\text{rel}_t} = \frac{V_a}{\cos \beta_1} = \frac{16.2}{\cos 39.0°} = 20.9 \text{ ft/s}$$

The velocity diagrams are sketched to scale in Figure 9.9.

b. The efficiency of the turbine is the ratio of the power measured at the output shaft to the energy loss in the fluid:

$$\eta = \frac{dW_s/dt}{Q(p_0 - p_2)}$$

with $Q = \dot{m}/\rho = 0.062/0.00926 = 6.73 \text{ ft}^3/\text{s}$, we get

$$\eta = \frac{37.7}{6.73(10)}$$

$$\eta = 0.56 = 56\%$$

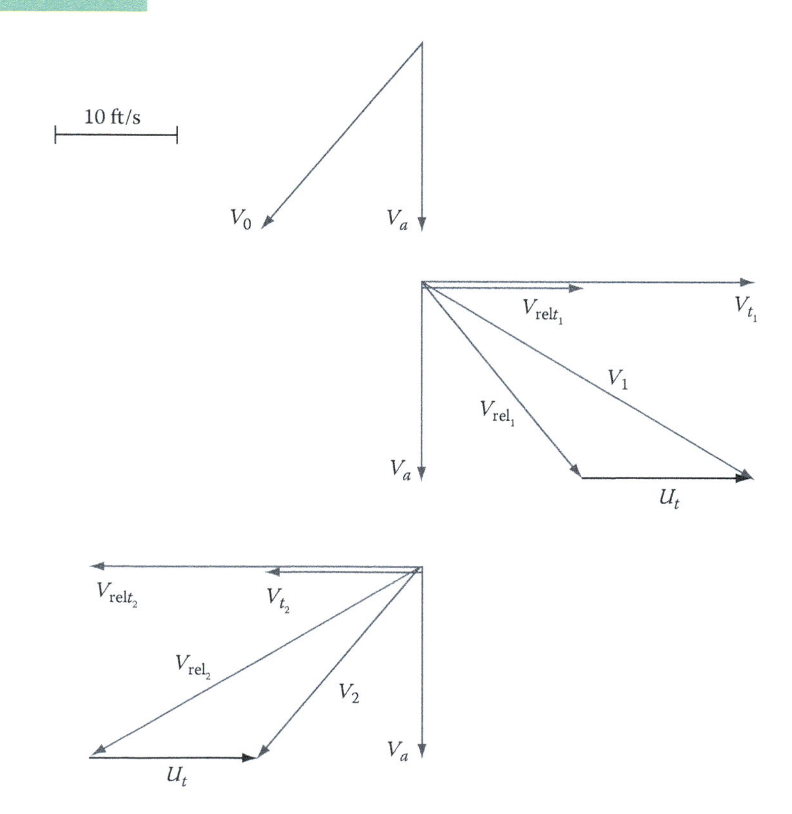

FIGURE 9.9 Velocity diagrams for the turbine of Example 9.2.

c. The reaction is found with Equation 9.28:

$$\mathrm{RR} = \frac{V_{\mathrm{rel}t_2} - V_{\mathrm{rel}t_1}}{2U_t} = \frac{27.6 - 13.1}{2(14.9)}$$

$$\mathrm{RR} = 0.49$$

As was mentioned earlier, a mean pitch line can be selected as representative of the entire stage when the blade height is small in comparison to the rotor diameter. When the blades are not short, however, we must account for the radial distance variation. This difficulty can be circumvented by dividing the blade into elements, as shown in Figure 9.10, and adding the contribution to the torque of each element. The total shaft torque thus becomes the sum of the individual contributions, and the angular momentum Equation 9.11 can be applied to each element.

Another type of axial-flow turbine is the hydraulic propeller turbine. This machine can be located in or near a dam, for example, so that water passing through provides power for the turbine to rotate a generator to produce electricity. As shown in Figure 9.11, the flow enters the system through the guide vanes, which give the liquid a tangential velocity before it reaches the turbine blades. The guide vanes cause a vortex to form. The propeller blades are in the middle of the vortex, and the transfer of energy is thus enhanced. In a fixed turbine blade runner, the guide vanes can be set to any desired angle (within limits) to accommodate changes in head water elevation and output power demand. The objective in an electrical generation installation is to achieve the constant rotational speed required by the generator to produce a constant frequency. This control could be achieved by adjusting the guide vanes. In a **Kaplan turbine** system, the turbine blades themselves and the inlet guide vanes can be made to change their orientation. The two adjustments are made simultaneously to accommodate changes in conditions. The Kaplan turbine system is suited for a wider range of operating variables than the fixed-blade type.

The turbine blades are few in number, and each is relatively flat. The blades are placed so that the relative velocity at the entrance (or exit) to the rotor is tangential to the leading edge of the blade (or trailing edge). The liquid relative velocity is high and changes slightly while passing through the turbine. Thus, a change in head requires a change in guide vane setting and, if possible, an accompanying change of the blade setting.

Analysis of the flow through the turbine is based on the same axial-flow equations that have been derived previously. The difference here, however, is that we must in addition describe the vortex flow field that is caused by the presence of the guide vanes. For this purpose, we select as our control

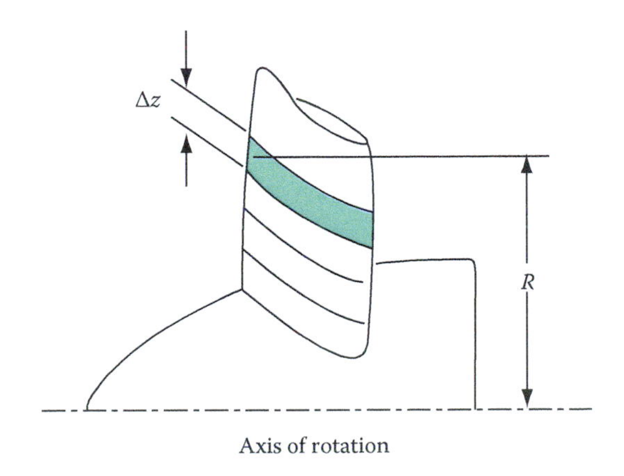

FIGURE 9.10 A rotor blade divided into a number of elements.

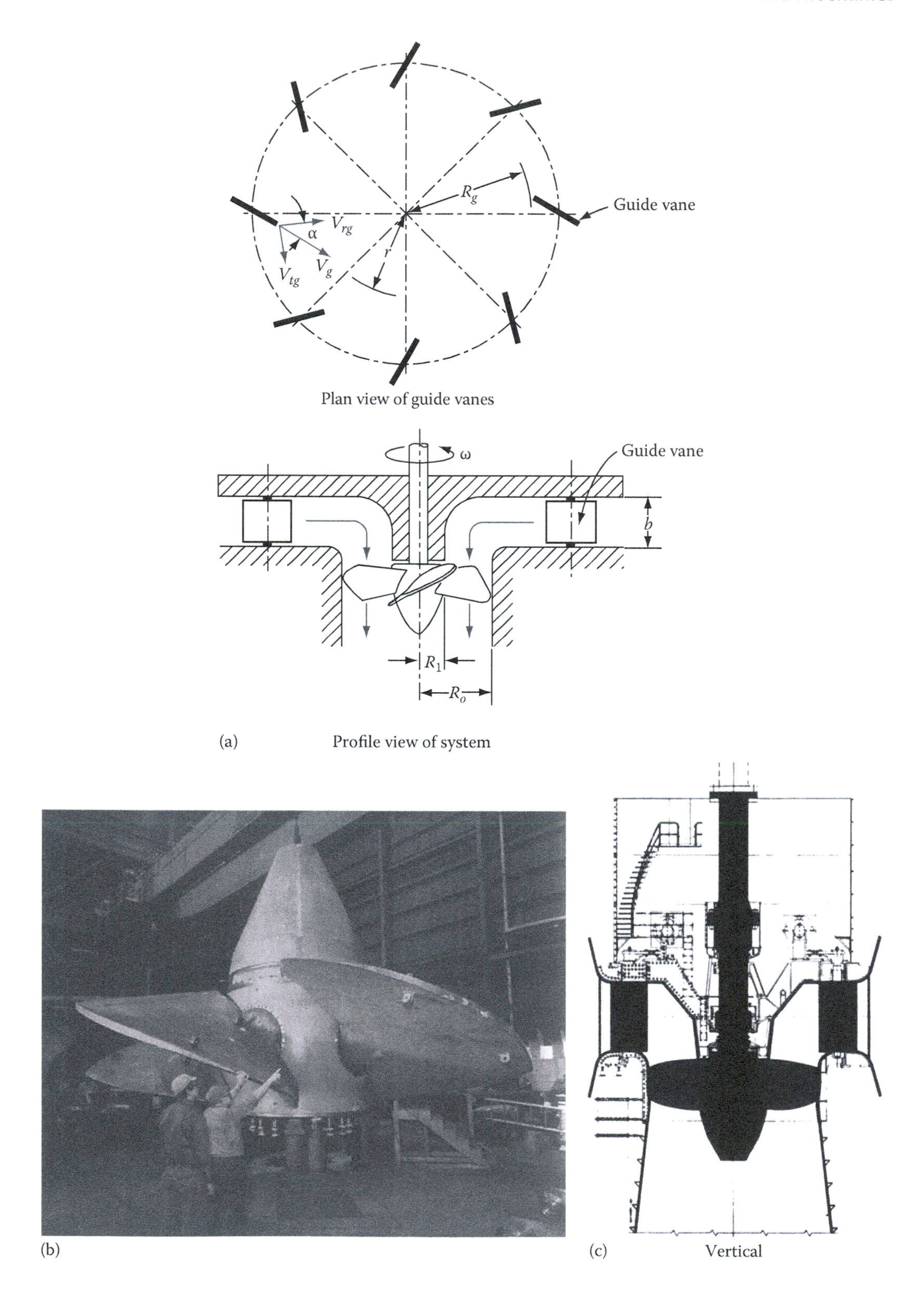

FIGURE 9.11 (a) Axial-flow propeller turbine and (b) Kaplan runner of an axial-flow turbine. (Courtesy of Allis-Chalmers Hydro Turbine Division, York, PA.) (c) Typical installation of an axial-flow turbine system. (Courtesy of Allis-Chalmers Hydro Turbine Division, York, PA.)

volume the region downstream of the guide vanes and upstream of the turbine blades. Assuming steady flow without any external torques on this fluid, the angular momentum equation becomes

$$0 = \iint_{CS} rV_t \rho V_n \, dA$$

Applying this between the entrance point at radius R_g and at any radial position r, we have

$$rV_t \rho Q - R_g V_{tg} \rho Q = 0$$

or

$$rV_t = R_g V_{tg}$$

where V_{tg} is the tangential velocity component at the radius R_g. In terms of the absolute entrance velocity V_g and the guide vane angle α, the preceding equation becomes

$$V_t = \frac{R_g V_g \sin \alpha}{r} \tag{9.31}$$

This equation gives the tangential fluid velocity input to the turbine blades. Furthermore, the flow rate through the turbine is

$$Q = 2\pi R_g b V_{rg} = 2\pi R_g b V_g \cos \alpha \tag{9.32}$$

where

b is the guide vane height
V_{rg} is the radial component of V_g
the product $2\pi R_g b$ is the inflow area (customarily discounting guide vane thickness)

Finally, the axial velocity at the turbine blades is found with

$$V_a = \frac{Q}{\pi(R_o^2 - R_i^2)} \tag{9.33}$$

where R_o and R_i are casing and hub radii, respectively. The casing and tip radii are assumed to be equal, although there must be a slight clearance.

Example 9.3

An axial-flow propeller turbine installation is sketched in Figure 9.11a. The guide vanes are set at an angle of 30° from the radial direction. The inner radius of the guide vanes is 1.5 m, and their height is 0.5 m. The guide vanes are in a location where the water head is equivalent to 0.45 m. The turbine blades have a hub radius of 0.15 m and a tip radius of 0.8 m. The rotor speed is 360 rev/min. Select appropriate blade angles for the leading edge of the turbine blades.

Solution

We can apply Bernoulli's equation at the guide vanes to calculate V_g:

$$\frac{V_g^2}{2g} = \Delta H$$

or

$$V_g = \sqrt{2g\Delta H} = \sqrt{2(9.81)(0.45)} = 2.97 \text{ m/s}$$

First make calculations for the fluid at the hub; then repeat for the tip. At the hub, $R_i = 0.15$ m. Then

$$V_{ti} = \frac{R_g V_g \sin\alpha}{R_i} = \frac{1.5(2.97)\sin 30°}{0.15}$$
$$= 14.9 \text{ m/s}$$

The radial velocity at the guide vanes is

$$V_{rg} = V_g \cos\alpha = 2.97 \cos 30° = 2.57 \text{ m/s}$$

From Equation 9.32, the volume flow rate is

$$Q = 2\pi R_b b V_{rg} = 2\pi(1.5)(0.5)(2.57)$$
$$= 12.1 \text{ m}^3/\text{s}$$

Using Equation 9.33, we determine the axial velocity to be

$$V_a = \frac{Q}{\pi(R_o^2 - R_i^2)} = \frac{12.1}{\pi(0.8^2 - 0.15^2)}$$
$$= 6.23 \text{ m/s}$$

The tangential velocity at the hub is

$$U_{ti} = R_i\omega$$
$$= 0.15(360)(2\pi/60)$$
$$= 5.65 \text{ m/s}$$

The velocity diagram at the hub can now be drawn as shown in Figure 9.12a. The blade angle can be found from

$$\tan(\beta_i - 90°) = \frac{V_\alpha}{V_{ti} - U_{ti}} = \frac{6.23}{14.9 - 5.65} = 0.674$$

Solving, we find the blade angle at the hub:

$$\beta_i = 123.9°$$

Performing similar calculations at the tip yields the following:

$$\left.\begin{array}{l} 1.\ V_{to} = \dfrac{R_g V_g \sin\alpha}{R_o} = \dfrac{1.5(2.97)\sin 30°}{0.8} = 2.78 \text{ m/s} \\ 2.\ V_{rg} = V_g \cos\alpha = 2.97\cos 30° = 2.57 \text{ m/s} \\ 3.\ Q = 2\pi R_g b V_{rg} = 2\pi(1.5)(0.5)(2.57) = 12.1 \text{ m}^3/\text{s} \\ 4.\ V_a = \dfrac{Q}{\pi(R_0^2 - R_i^2)} = \dfrac{12.1}{0.8^2 - 0.15^2} = 6.23 \text{ m/s} \\ 5.\ U_{to} = R_o\omega = 0.8(360)(2\pi/60) = 30.16 \text{ m/s} \end{array}\right\} \text{(constants)}$$

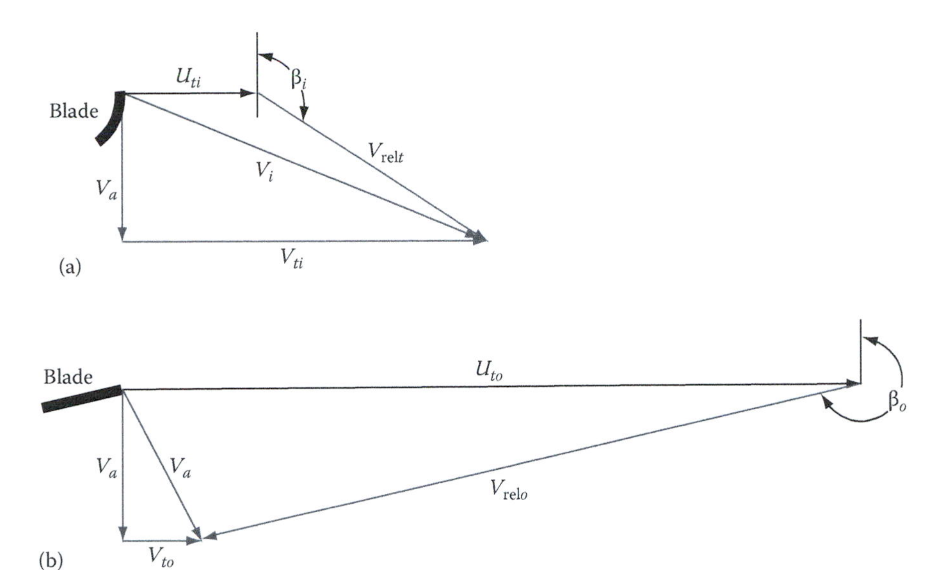

FIGURE 9.12 Velocity diagrams at the hub (a) and tip (b) of the propeller turbine of Example 9.3.

The velocity diagram is sketched in Figure 9.12b. The blade angle is found with

$$\tan(270° - \beta_o) = \frac{V_\alpha}{U_{to} - V_{to}} = \frac{6.23}{30.16 - 2.78} = 0.228$$

Solving, we find the blade angle at the tip:

$$\beta_o = 257.2°$$

A similar procedure can be followed at selected points along the blade between 0.15 and 0.8 m. With the one-dimensional assumption, the product U_tV_t at any blade location is a constant.

The third type of axial-flow turbine that we will consider is the windmill. Windmills are primarily open turbines that are not surrounded by a casing (except in some experimental or nontraditional designs). The familiar Dutch windmill with sail-like blades has been used to pump water. These huge, four-bladed types have efficiencies of ~5%. Recently, there has been a renewed interest in windmills as electrical generation devices. For our purposes, we will examine the factors that directly influence the machine's performance—specifically, the factors that control the efficiency of the conversion of wind motion to propeller rotation.

An interesting result becomes evident when the momentum equation is applied to the airflow in the vicinity of a windmill. Consider the sketch of Figure 9.13. The windspeed upstream is uniform at A_1, the area of moving air to be intercepted by the disk formed by blade rotation. The pressure sufficiently far upstream is p_1 and equals the pressure sufficiently far downstream p_4. The pressure steadily increases from p_1 to p_2, then drops across the blades to p_3, and then increases back to p_4. This pressure variation causes the moving air to take on the characteristic shape, called a slipstream, of Figure 9.13a. Selecting the **slipsteam** between points 1 and 4 as our control volume, we write the momentum equation:

$$\sum F_z = \iint_{CS} V_z \rho V_n \, dA$$

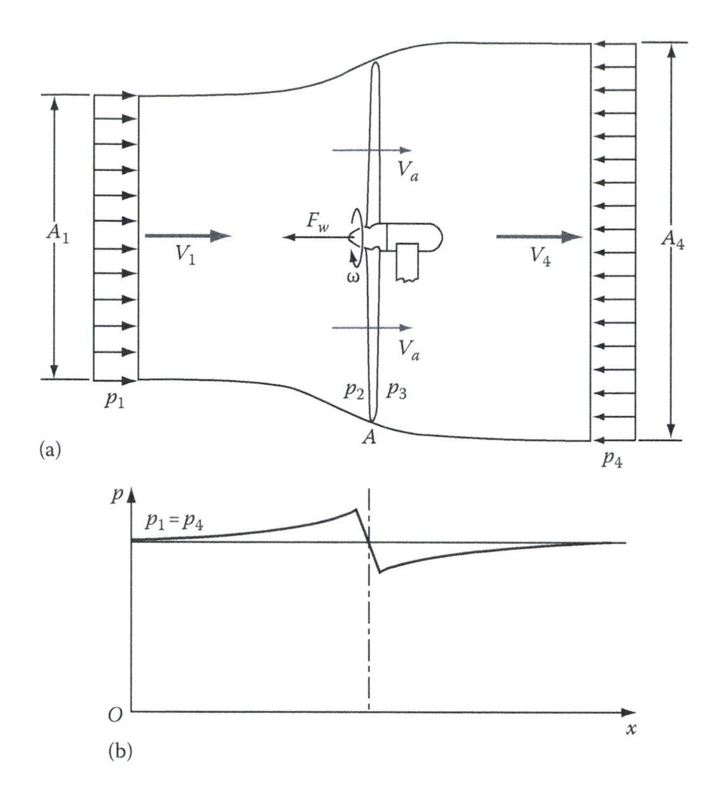

FIGURE 9.13 (a) Flow of air through the blades of a windmill (a) and corresponding pressure profile (b).

The only force acting is exerted by the windmill on the air at the propeller to keep the machine stationary. This must balance the change in momentum experienced by the fluid:

$$\sum F_z = -F_w = (p_3 - p_2)A = \rho A V_a (V_4 - V_1) \tag{9.34a}$$

or

$$p_3 - p_2 = \rho V_a (V_4 - V_1) \tag{9.34b}$$

Applying Bernoulli's equation between points 1 and 2, we get

$$\frac{p_1}{\rho g} + \frac{V_1^2}{2g} = \frac{p_2}{\rho g} + \frac{V_2^2}{2g}$$

Between points 3 and 4, we have

$$\frac{p_3}{\rho g} + \frac{V_3^2}{2g} = \frac{p_4}{\rho g} + \frac{V_4^2}{2g}$$

Adding these gives

$$\frac{p_3}{\rho g} + \frac{V_1^2}{2g} = \frac{p_2}{\rho g} + \frac{V_4^2}{2g} \tag{9.35}$$

where the relationships $p_1 = p_4$ and $V_2 = V_3 = V_a$ have been used. Equating Equations 9.34b and 9.35 gives

$$V_a = \frac{V_4 + V_1}{2} \tag{9.36}$$

This states that the velocity through the disk at the propeller is the average of the uniform velocities far ahead of and far behind the unit.

For frictionless flow through the machine, the power delivered to the windmill equals exactly the power extracted from the air. Applying the energy equation about the blade shows that this power equals the change in kinetic energy of the fluid:

$$\frac{dW_s}{dt} = \frac{\rho A V_a}{2} = (V_1^2 - V_4^2) \tag{9.37}$$

The power originally available in the wind of cross-sectional area A and velocity V_1 is

$$\frac{dW_a}{dt} = \frac{\rho A V_1^3}{2} \tag{9.38}$$

This can be found also by the application of the energy equation. Windmill efficiency is defined as

$$\eta = \frac{dW_s/dt}{dW_s/dt} = \frac{\rho A V_a (V_1^2 - V_4^2)}{\rho A V_1^3} = \frac{V_a (V_1^2 - V_4^2)}{V_1^3}$$

or, after combining with Equation 9.36,

$$\eta = \frac{(V_1 + V_4)\,(V_1^2 - V_4^2)}{2V_1^3} = \frac{1}{2}\left(1 + \frac{V_4}{V_1}\right)\left[1 - \left(\frac{V_4}{V_1}\right)^2\right] \tag{9.39}$$

The maximum efficiency is found by differentiating this expression with respect to V_4/V_1 and setting the result equal to zero. This gives

$$\left.\frac{V_4}{V_1}\right|_{\eta\,\max} = \frac{1}{3} \tag{9.40}$$

which can then be substituted into Equation 9.39 to find

$$\eta_{\max} = \frac{16}{27} = 59.3\%$$

This efficiency is the maximum that exists for frictionless flow. Because of friction and other losses, this value can only be approached in practice. The maximum efficiency for a real windmill appears to be about 50%.

Example 9.4

A two-bladed windmill with a diameter (tip-to-tip) of 35 ft is in a stream of air of velocity 70 ft/s. The air temperature is 65°F, and atmospheric pressure is 14.69 psia. Downstream the uniform flow velocity is 60 ft/s. Under these conditions, find the electrical power output. Calculate the thrust on the windmill and the overall efficiency of the installation.

Solution

The velocity at the blades is found with Equation 9.36:

$$V_a = \frac{V_1 + V_4}{2} = \frac{70 \text{ ft/s} + 60 \text{ ft/s}}{2} = 65 \text{ ft/s}$$

Assuming a constant air density and ideal gas behavior,

$$\rho = \frac{p}{RT} = \frac{14.69(144)}{1710(460+65)} = 0.00236 \text{ slug/ft}^3$$

The disk area is

$$A = \frac{\pi D^2}{4} = \frac{\pi(35)^2}{4} = 962 \text{ ft}^2$$

The thrust is calculated with Equation 9.34a as

$$-F_w = \rho A V_a (V_4 - V_1)$$

$$= 0.00236(962)(65)(60-70)$$

$$F_w = 1470 \text{ lbf}$$

The power originally available in the airstream is

$$\frac{dW_a}{dt} = \frac{\rho A V_1^3}{2} = \frac{0.00236(962)(70)^3}{2} = 3.89 \times 10^5 \text{ ft} \cdot \text{lbf/s}$$

The actual power output is, from Equation 9.37,

$$\frac{dW_s}{dt} = \frac{\rho A V_a}{2}(V_1^2 - V_4^2)$$

$$= \frac{0.00236(962)(65)}{2}(70^2 - 60^2)$$

$$= 9.59 \times 10^4 \text{ ft} \cdot \text{lbf/s}$$

Converting to kilowatts (conversion factor from Table A.2), we find the actual power output for frictionless flow to be

$$\frac{dW_s}{dt} = 9.59 \times 10^4 (1.355 \times 10^{-3}) = 130 \text{ kW}$$

The overall efficiency is

$$\eta = \frac{9.59 \times 10^4}{3.89 \times 10^5} = 0.246$$

$$\eta = 24.6\%$$

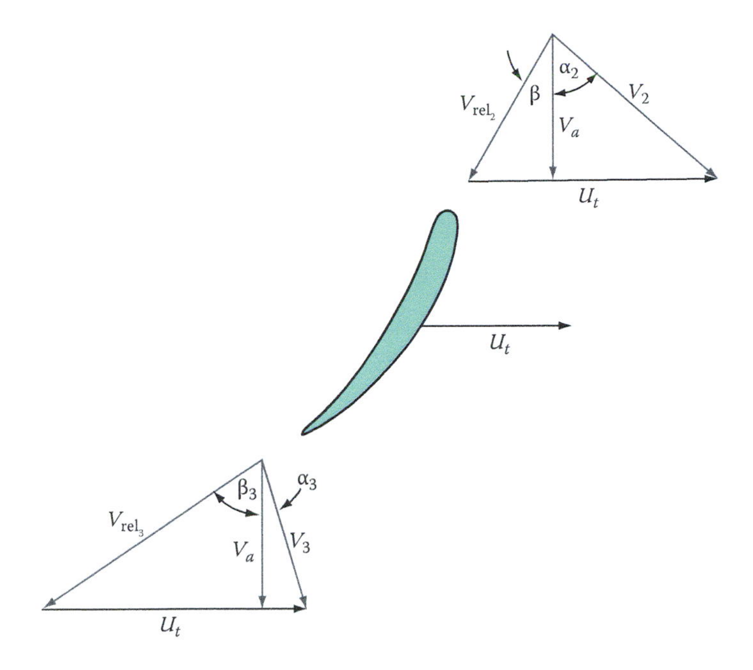

FIGURE 9.14 Flow past a moving airfoil.

In preceding discussions, we used one-dimensional flow theory to predict performance of axial-flow turbines. The one-dimensional assumption gives an accurate description of the flow when the number of blades in the machine is large. When the number of blades is few and the distance between them is large, however, each acts like an airfoil. Consequently, we must resort to airfoil theory to analyze a machine such as a two- or three-bladed windmill. To begin, consider the situation sketched in Figure 9.14, which incorporates the notation of Figure 9.13. An airfoil (represented in cross section) is moving at a tangential velocity U_t. The absolute velocity of the approach flow is V_2, whereas the flow leaves at an absolute velocity V_3. For the analysis centered about Figure 9.13, the one-dimensional assumption is made. The velocity at the blade is an axial velocity that is constant. With airfoil theory, Figure 9.14 is appropriate.

In this figure, V_2 and V_3 are no longer axial or equal as in Figure 9.13. For the moving airfoil, we therefore construct a velocity diagram before and after the blade. Note that the blade's leading and trailing edges are tangential to the local relative velocity. To simplify the analysis, we reconstruct a single velocity diagram with only one relative velocity found by

$$\bar{V}_{\text{rel}} = \frac{V_{\text{rel}\,2} + V_{\text{rel}\,3}}{2} \tag{9.41}$$

as sketched in Figure 9.15. The average relative velocity $\bar{V}_{\text{rel}}$ has an angle of incidence of δ. From Chapter 6, we know that an airfoil experiences lift and drag forces perpendicular and parallel, respectively, to the flow direction. These are shown in Figure 9.16. To determine the forces acting on the rotor, it is necessary to resolve the lift and drag forces into axial and tangential components. Taking the direction of V_a as positive, the axial force is

$$F_a = L_f \sin\beta + D_f \cos\beta \tag{9.42a}$$

The tangential force is

$$F_t = L_f \cos\beta - D_f \sin\beta \tag{9.42b}$$

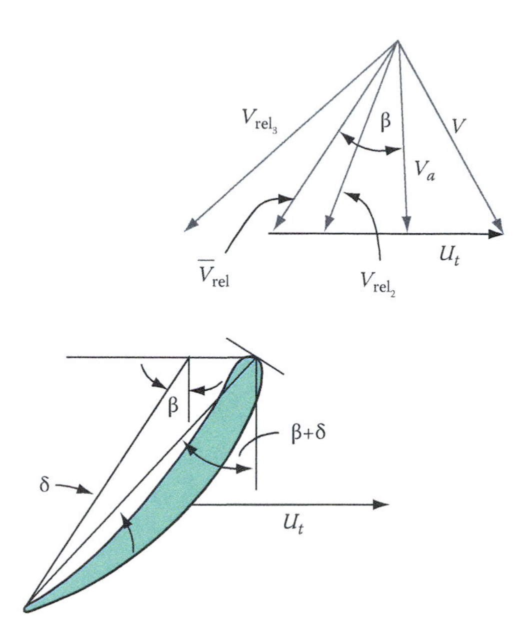

FIGURE 9.15 Average relative velocity for flow past an airfoil.

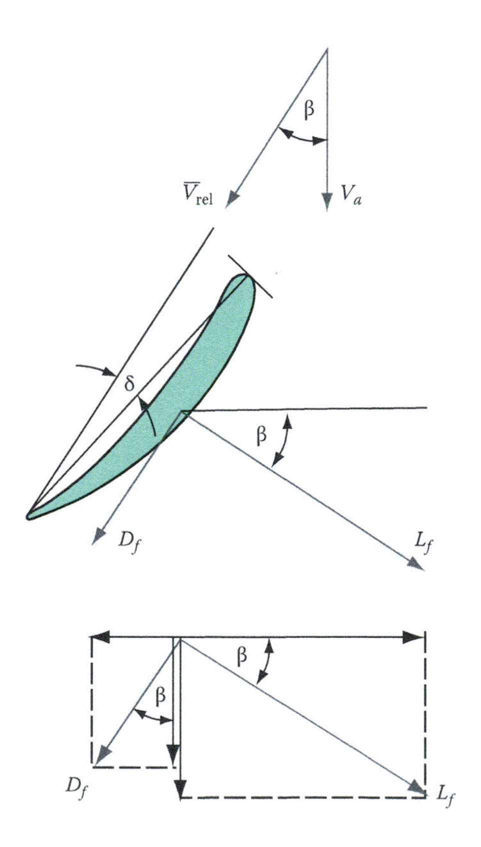

FIGURE 9.16 Forces of lift and drag acting on the airfoil resolved into axial and tangential components.

where the direction of U_t is considered positive. In terms of lift and drag coefficients, the lift and drag forces are

$$L_f = C_L \frac{\rho \overline{V}_{\rm rel}^2 A_p}{2}$$

$$D_f = C_D \frac{\rho \overline{V}_{\rm rel}^2 A_p}{2}$$

where A_p is the planform area (chord length times span) of the airfoil. Before combining these equations, note that it is necessary to account for changes in U_t with distance for the rotor and, moreover, that it may be desired to change the dimension of the airfoil with blade length. We therefore divide the blade into incremental elements Δz wide, as shown in Figure 9.17. The area becomes $A = c\ \Delta z$. Combining the lift and drag expressions with Equation 9.42 yields the following after simplification:

$$F_a = (C_L \sin\beta + C_D \cos\beta) \frac{\rho \overline{V}_{\rm rel}^2 c \Delta z}{2}$$
$$F_t = (C_L \cos\beta + C_D \sin\beta) \frac{\rho \overline{V}_{\rm rel}^2 c \Delta z}{2} \tag{9.43}$$

These forces exerted on the airfoil can now be related to the properties of the flow (static pressure and velocity). The momentum in the tangential direction is

$$\sum F_t = \iint_{\rm CS} V_t \rho V_n\, dA$$

which becomes, for the left-hand side,

$$\sum F_t = NF_t = N(C_L \cos\beta - C_D \sin\beta) \frac{\rho \overline{V}_{\rm rel}^2 c \Delta z}{2}$$

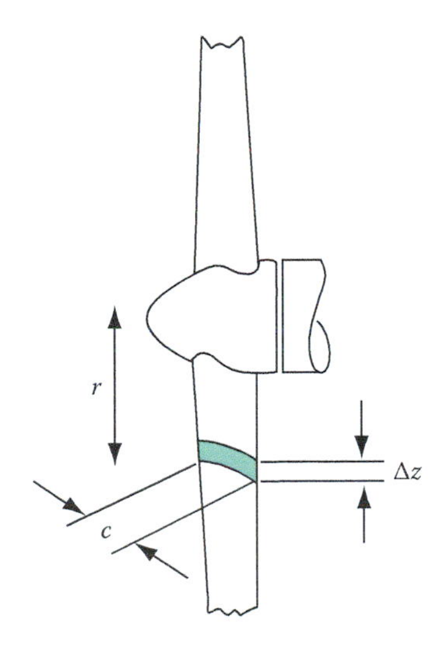

FIGURE 9.17 An element of a blade.

where N is the number of blades that contribute to the total force exerted. The change in momentum term is

$$\iint_{CS} V_t \rho V_n \, dA = \rho V_a (V_{t_3} - V_{t_2}) 2\pi r \Delta z$$

where the notation of Figure 9.13 has been used on the tangential velocity terms. Here the flow rate is the product of axial velocity, which is a constant, and area of an annulus Δz thick and located at r away from the axis of rotation. The momentum equation is

$$N(C_L \cos\beta - C_D \sin\beta)\frac{\rho \overline{V}_{rel}^2 c \Delta z}{2} = \rho V_a (V_{t_3} - V_{t_2}) 2\pi r \Delta z$$

Substituting $V_a = \overline{V}_{rel} \cos\beta$ and simplifying yield

$$V_{t_3} - V_{t_2} = \frac{\overline{V}_{rel} c N}{4\pi r}(C_L - C_D \tan\beta) \tag{9.44}$$

We now have a relationship between the tangential fluid velocities and the characteristics of the airfoil.

The momentum equation can also be applied in the axial direction:

$$\sum F_a = \iint_{CS} V_a \rho V_n \, dA$$

Using the notation of Figure 9.13, we obtain

$$NF_a = (p_2 - p_3) 2\pi r \Delta z \tag{9.45}$$

and

$$\iint_{CS} V_a \rho V_n \, dA = (\rho 2\pi r \Delta z V_a)(V_{a_3} - V_{a_2})$$

but because $V_{a_3} = V_{a_2} = V_a$, the momentum equation reduces to Equation 9.45. By substitution from Equation 9.43, then, Equation 9.45 becomes

$$N(C_L \sin\beta - C_D \cos\beta)\frac{\rho \overline{V}_{rel}^2 c \Delta z}{2} = (p_2 - p_3) 2\pi r \Delta z$$

Simplifying and rearranging give

$$p_2 - p_3 = -\frac{\rho \overline{V}_{rel}^2 c N}{4\pi r}(C_L \sin\beta + C_D \cos\beta) \tag{9.46}$$

This equation relates the pressure drop across the blade to the characteristics of the foil at a location r from the axis of rotation.

Example 9.5

Consider the windmill of the previous example. A summary of the data given there is as follows:

- Number of blades (N) = 2
- Tip-to-tip diameter (D) = 35 ft
- Disk area (A) = 962 ft^2
- Upstream air velocity (V_1) = 70 ft/s
- Downstream air velocity (V_4) = 60 ft/s
- Air velocity at blades (V) = 65 ft/s
- Air density (ρ) = 0.00236 slug/ft^3
- Shaft work (dW_s/dt) = 130 kW

It is assumed that each blade can be treated as being invariant in cross section—there are no taper and no blade twist variation with blade length. The blade makes an angle of 60° with the axial direction; the chord length is 24 in. Determine the aerodynamic properties of the airfoil, and calculate the pressure drop at the midpoint of the blade. Assume a rotational speed of 400 rev/min.

Solution

The axial velocity is the same as the air velocity at the blades:

$$V_a = 65 \text{ ft/s}$$

The relative velocity is found with

$$\bar{V}_{\text{rel}} = \frac{V_a}{\cos\beta} = \frac{65}{\cos 60°} = 130 \text{ ft/s}$$

At blade midpoint, $r = 17.5/2 = 8.75$ ft. Equation 9.46 gives the pressure in terms of lift and drag coefficients:

$$\begin{aligned} p_3 - p_2 &= -\frac{\rho \bar{V}_{\text{rel}}^2 cN}{4\pi r}(C_L \sin\beta + C_D \cos\beta) \\ &= -\frac{0.00236(130)^2(24/12)(2)}{4\pi(8.75)}(C_L \sin 60° + C_D \cos 60°) \\ &= -1.45(0.866C_L + 0.5C_D) \end{aligned}$$

In addition, the pressure drop can be calculated with Equation 9.35:

$$\begin{aligned} p_3 - p_2 &= \frac{\rho(V_4^2 - V_1^2)}{2} \\ &= \frac{0.00236(60^2 - 70^2)}{2} \\ &= -1.53 \text{ psf} \end{aligned}$$

Equating the pressure-drop equations gives

$$C_L + \frac{0.5}{0.866}C_D = -\frac{-1.53}{1.45(0.866)}$$

$$C_L + 0.577C_D = 1.22 \qquad \text{(i)}$$

The tangential velocity difference must be evaluated next. Using Equation 9.10a for the shaft work, written with the notation of Figure 9.13, we have

$$-\frac{dW_s}{dt} = \rho A V_a U_t (V_{t_3} - V_{t_2})$$

The shaft work is

$$-\frac{dW_s}{dt} = 130 \text{ kW}\left(\frac{1 \text{ ft} \cdot \text{lbf/s}}{1.355 \times 10^{-3} \text{ kW}}\right) = 9.59 \times 10^4 \text{ ft} \cdot \text{lbf/s}$$

The rotational speed of the blade at midpoint is

$$U_t = r\omega = 8.75\left(400\frac{2\pi}{60}\right) = 367 \text{ ft/s}$$

By substitution, we find

$$V_{t_3} - V_{t_2} = \frac{9.59 \times 10^4}{0.00236(962)(65)(367)} = 1.77 \text{ ft/s}$$

Combining with Equation 9.44 gives

$$C_L - C_D \tan\beta = \frac{(V_{t_3} - V_{t_2})4\pi r}{\bar{V}_{\text{rel}} c N}$$

$$C_L - C_D \tan 60° = \frac{1.77(4\pi)(8.75)}{130(24/12)(2)}$$

or

$$C_L - 1.732 C_D = 0.374 \qquad \text{(ii)}$$

Subtracting this equation from Equation i gives

$$0.577 C_D + 1.732 C_D = 1.22 - 0.374$$

or

$$C_D = 0.366$$

Using either Equation i or Equation ii, we find

$$C_L = 1.008$$

The blade has these characteristics at its midpoint, where $\bar{V}_{\text{rel}} = 130$ ft/s and $U_t = 367$ ft/s.

The preceding analysis was based on the assumption that the number of blades is large. More specifically, the analysis applies if

$$\frac{2\pi r}{cN} > 1$$

If this ratio is less than 1, the effects of the blades on the airflow are found to interfere with each other. The proximity of another airfoil reduces the lift coefficient. Thus, we would expect a multibladed windmill to have a slower rotational speed than a two-bladed type in the same airstream.

9.3 AXIAL-FLOW COMPRESSORS, PUMPS, AND FANS

The axial-flow compressor was first regarded as being the reverse of the axial-flow turbine. In early designs, efficiencies of these reversed turbines ran as high as 55%. This low value is attributed to blade stall and flow separation, or diffusion. In a turbine, the fluid pressure decreases through the machine. With a compressor, however, the pressure increases through the machine, setting up an adverse pressure gradient. If the pressure increases too abruptly, there is flow separation from the individual blades and a corresponding loss in performance. Because of this limitation, conventional designs have many stages, each contributing only a small portion of the total pressure rise. Multistage axial compressors can have as many as 20 stages. As a result, efficiencies of 90% are reported for axial-flow compressors that operate on a 6:1 or 7:1 pressure ratio (inlet to outlet).

Figure 9.18 is a schematic of an axial-flow compressor. A stage is defined as a rotor row followed by a stator row of blades. The rotor blades are attached to the rotating portion of the machine, whereas the stator blades are part of the casing. The inlet guide vanes are not considered to be part of the first stage; their function is to accelerate the flow away from the purely axial direction.

The study of axial-flow compressors requires the use of compressible flow equations. Since in this chapter we wish to examine only incompressible flow machines, we will not discuss axial-flow compressors in great detail. The preceding paragraphs were included to show that such machines do exist and to illustrate the use of axial-flow compressors in conventional gas turbines.

The next turbomachine that we will discuss is the axial-flow pump. In this case, the fluid medium is an incompressible fluid, and temperature changes are not as significant as pressure changes. Consequently, the descriptive equations can be written using static pressures. Figure 9.19 shows one design of an axial-flow pump. At the inlet or suction side are placed straightener vanes to ensure that the flow into the pump is purely axial. The impeller is made up of a hub and blades that act much like a propeller, adding energy to the liquid. The outlet guide vanes have the function of removing any tangential velocity component the liquid might have. The outflow is thus purely axial. The axial-flow pump has a high capacity and corresponding low head output.

A summary of the descriptive equations is helpful. Applying Equation 9.10a to the velocity diagram of Figure 9.19, with $V_{t_1} = 0$, gives the power as

$$\frac{dW_s}{dt} = T_s\omega = \rho Q U_t V_{t_2} \tag{9.47}$$

Equation 9.18b applied to this machine gives a second expression for the power in terms of the pressure rise:

$$\frac{dW_s}{dt} = Q(p_3 - p_1) \tag{9.18b}$$

Equating Equations 9.47 and 9.18b yields

$$p_3 - p_1 = \rho U_t V_{t_2} \tag{9.48}$$

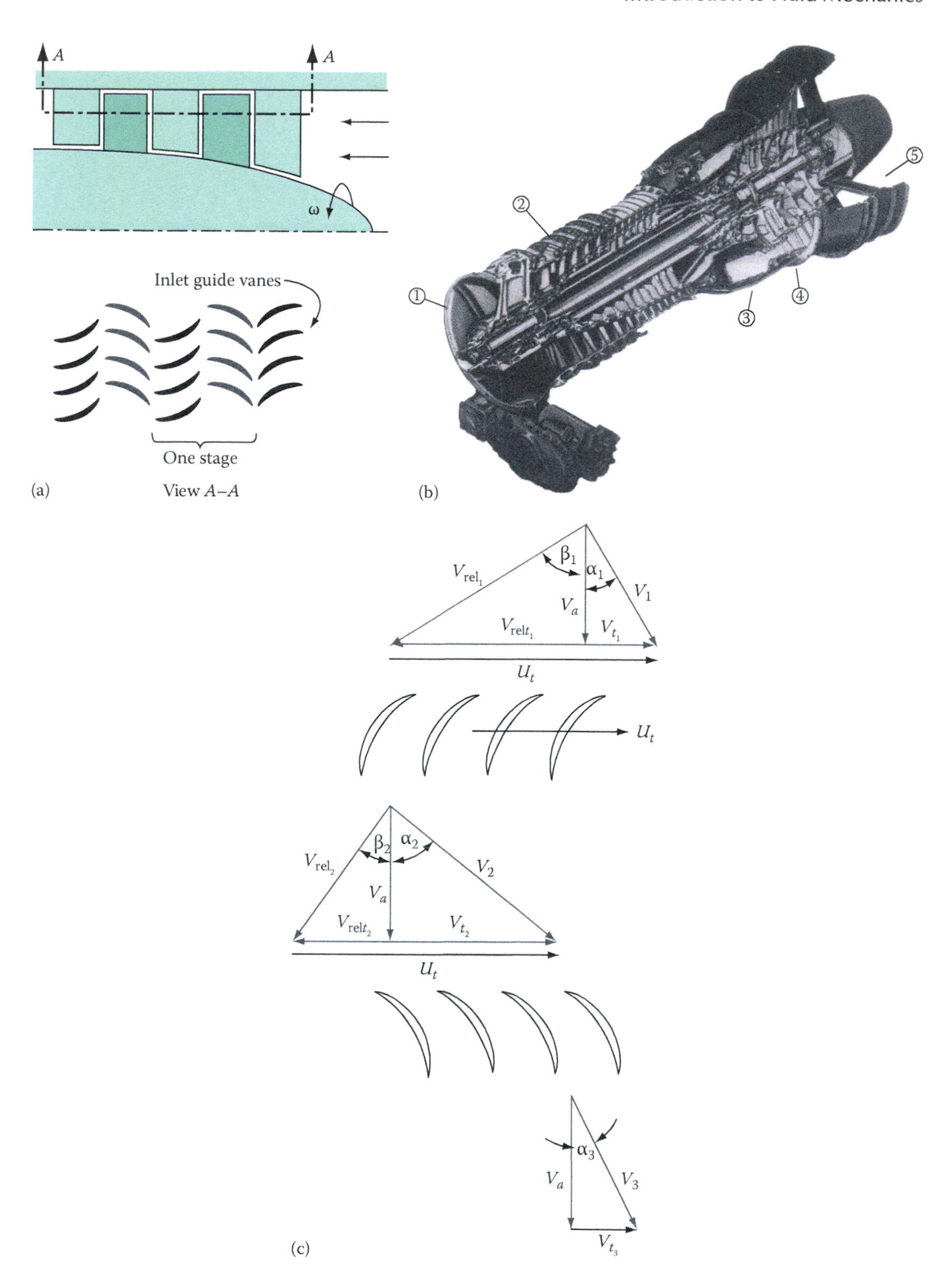

FIGURE 9.18 (a) An axial-flow compressor and (b) section view of a gas turbine. Air enters the gas turbine at 1. The axial-flow compressor (2) delivers high-pressure air to the combustion chamber (3). Hot exhaust gases travel through the axial turbine (4) to the outlet at 5. Turbine and compressor are on the same shaft. Compressor input power is from the turbine. Thrust is delivered by exhaust gases. (Courtesy Williams & Lane, Inc., Binghamton, NY.) (c) Velocity diagrams for an axial-flow compressor stage.

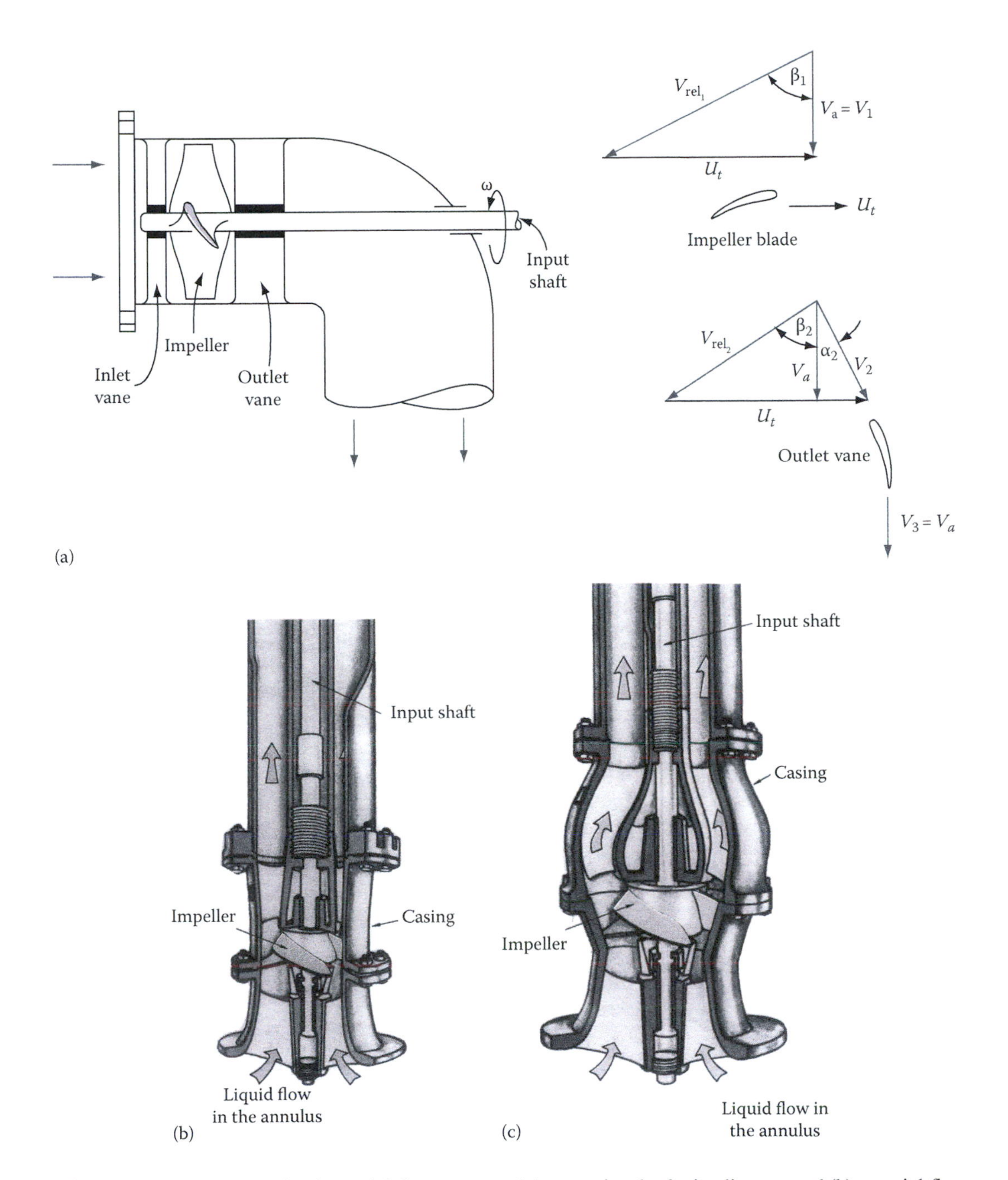

FIGURE 9.19 (a) Schematic of an axial-flow pump and the associated velocity diagram and (b) an axial-flow pump. (Courtesy of Fairbanks Morse Pump Division, Kansas City, MI.) (c) A mixed-flow pump. (Courtesy of Fairbanks Morse Pump Division, Kansas City, MI.)

For the pressure rise across the impeller blade only, we apply Bernoulli's equation with relative velocities (Equation 9.25b):

$$p_2 - p_1 = \frac{\rho}{2}\left(V_{\text{rel}_1}^2 - V_{\text{rel}_2}^2\right)$$

The reaction ratio is given by rewriting Equation 9.28 for a pump as

$$\text{RR} = \frac{p_2 - p_1}{p_3 - p_1} = \frac{V_{\text{rel}t_1} + V_{\text{rel}t_2}}{2U_t} \tag{9.28}$$

In terms of blade angles, the reaction ratio can be written as Equation 9.30 with $\alpha_1 = 0$:

$$\text{RR} = \frac{1}{2} + \frac{V_a}{2U_t}\tan\beta_2 \tag{9.49}$$

Example 9.6

An axial-flow pump rotates at 600 rev/min and conveys water at 0.12 m³/s. The hub radius is 5 cm, and the casing radius is 15 cm. The absolute velocity of the flow leaving the rotor blades makes an angle of 23° with the axial direction. Determine the angles at the leading edge of the blades (β_1 and β_2). Use the blade element procedure of dividing the blade into a number of finite widths and averaging the contributions to find the total torque and power.

Solution

As we mentioned earlier, when the blade height is small in comparison to the mean pitch radius, the properties at the radius can be taken to represent the flow characteristics in the stage. For long blades, however, the radial distance variation must be accounted for. This is done by dividing the blade into small elements and adding the contribution to the torque of each element. Total shaft power becomes the sum of the individual contributions divided by the number of elements selected. The angular momentum equation must be applied to each element of the blade. In so doing, blade angles at each location can be ascertained, and the blade can then be constructed.

At the hub, $R_i = 0.05$ m; at the tip, $R_o = 0.15$ m. The flow area is then

$$A = \frac{\pi}{4}(D_o^2 - D_i^2) = \frac{\pi}{4}(0.3^2 - 0.1^2) = 0.062\,8\ \text{m}^2$$

With $Q = 0.12$ m³/s, we have

$$V_a = \frac{Q}{A} = \frac{0.12}{0.062\,8} = 1.91\ \text{m/s}$$

which is a constant across the cross section and throughout the machine.

We will now find the appropriate parameters in terms of the blade radius. The rotational speed is 600 rev/min, so the tangential blade speed in terms of the radius is

$$U_t - R\omega = R\left(600\frac{2\pi}{60}\right) = 62.8R$$

With the angle α_2 given as 23°, we find (referring to Figure 9.19)

$$V_{t_2} = V_a \tan\alpha_2 = 1.91\tan 23° = 0.811\ \text{m/s}$$

which is also a constant and not a function of the radius. The pressure gain across the pump at any location r is determined with Equation 9.48:

$$\begin{aligned}\Delta(p_3 - p_1) &= \rho U_t V_{t_2} \\ &= 1\,000(62.8R)(0.811) \\ &= 50\,957R\end{aligned}$$

where Δ has been added to denote the blade element value. The shaft torque at any location is calculated with

$$\Delta T_s = \frac{Q}{\omega}\Delta(p_3 - p_1) = \frac{0.12}{62.8}(50\,957R)$$
$$= 97.4R$$

and the power input is

$$\Delta\frac{dW_s}{dt} = \Delta T_s \omega$$

Results of calculations made with these equations are displayed in Table 9.1. As shown in column 1, the blade is divided into five elements, each 2 cm long. The average radial value for each increment, R_{avg}, is determined and used in the expressions. The last two columns show the incremental contributions to torque and power. By summing values in these columns and then dividing by the number of elements, the total torque and total power are found to be

$$T_s = 9.73\ \text{N}\cdot\text{m}$$

and

$$\frac{dW_s}{dt} = 0.611\ \text{kW}$$

Note that these values are the same as those calculated at $R_{avg} = 0.10$ m. Thus, $R_{avg} = 0.10$ m would be the mean flow radius if it were necessary to use it in our model.

TABLE 9.1
Solution for the Axial-Flow Pump of Example 9.6

ΔR (m)	R_{avg} (m)	$U_t = R_{avg}\omega$ (m/s)	$\beta_1 = \tan^{-1} \times \left(\frac{U_t}{V_a}\right)$ (°)	$\beta_2 = \tan^{-1} \times \left(\frac{U_t - V_{t_3}}{V_a}\right)$ (°)	$\Delta(p_3 - p_1) = \rho U_t V_{t_2}$ (Pa)	$\Delta T_s = \frac{Q}{\omega} \times \Delta(p_3 - p_3)$ (N·m)	$\Delta\left(\frac{dW_s}{dt}\right) = \Delta T_s\omega$ (W)
0.05–0.07	0.06	3.76	63	57.1	3 049	5.82	366
0.07–0.09	0.08	5.03	69.2	65.7	4 079	7.79	489
0.09–0.11	0.10	6.28	73.1	70.7	5 093	9.73	611
0.11–0.13	0.12	7.54	75.8	74.2	6 114	11.68	734
0.13–0.15	0.14	8.80	77.8	76.5	7 137	13.63	856
						$T_s = \frac{48.65}{5} = 9.73$	$\frac{dW_s}{dt} = \frac{3056}{5} = 611$

Conditions: $D_0 = 0.3$ m, $D_i = 0.1$ m, $A = 0.062\,8$ m², $V_{t_2} = 0.811$ m/s, $V_a = 1.91$ m/s, $Q = 0.12$ m²/s, and $\omega = 600$ rev/min.

Just as our discussion of axial-flow turbines involved airfoil theory, there is a corresponding development for axial-flow compressors, fans, and pumps. It is necessary to use airfoil theory when the one-dimensional flow assumption is no longer adequate—that is, when the number of blades in the machine is small and the distance between them is large. To examine this case, consider the airfoil depicted in Figure 9.20. The approach flow has an absolute velocity V_1 at an angle α_1 with the axial direction. The flow leaving has an absolute velocity V_2. The velocity diagrams of Figure 9.20 can be combined into a single graph, as shown in Figure 9.21. Flow past the moving airfoil is represented as the average relative velocity $\bar{V}_{\text{rel}}$:

$$\bar{V}_{\text{rel}} = \frac{V_{\text{rel}_1} + V_{\text{rel}_2}}{2}$$

This velocity approaches the blade at an angle of incidence of δ.

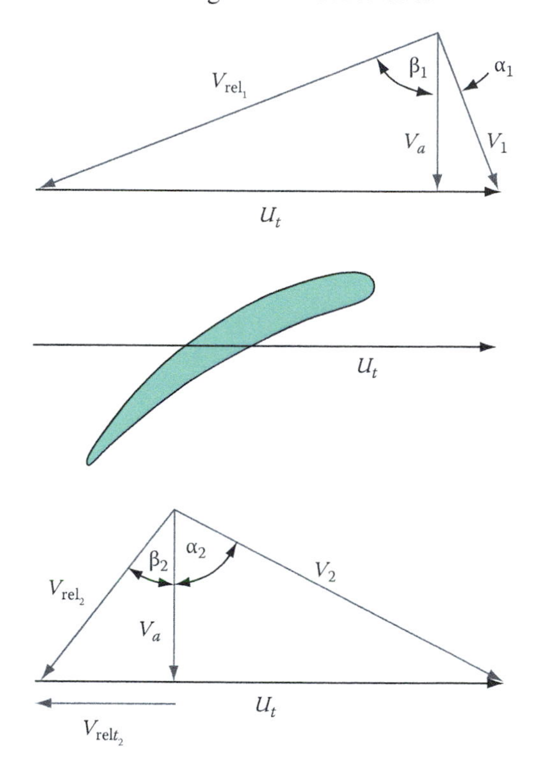

FIGURE 9.20 Flow past an isolated airfoil with associated velocity diagrams.

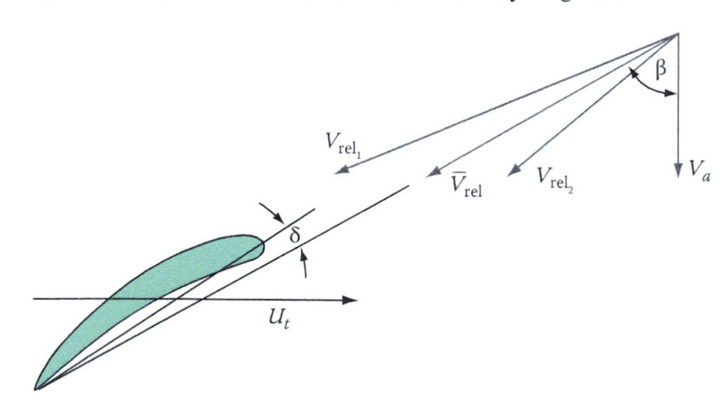

FIGURE 9.21 Combined velocity diagram.

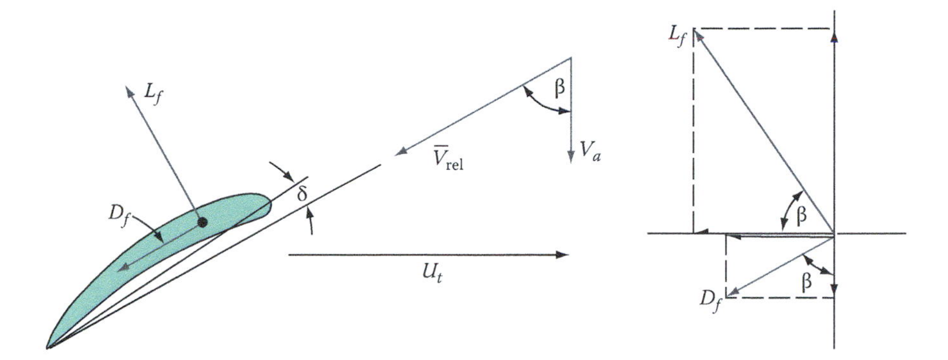

FIGURE 9.22 Forces of lift and drag resolved into axial and tangential components.

From Chapter 6, we know that flow past an airfoil causes lift and drag forces to be exerted perpendicular and parallel to the flow direction. These forces must be resolved into components (as illustrated in Figure 9.22) in the axial and tangential directions—to calculate bearing forces, for example, and so that blade aerodynamic properties can be related to fluid properties. The tangential force acting in the direction opposite of U_t (considered negative) is

$$F_t = -(L_f \cos\beta + D_f \sin\beta) \tag{9.50}$$

Taking the direction of the axial velocity as being positive, we find the axial force to be

$$F_a = -L_f \sin\beta + D_f \cos\beta \tag{9.51}$$

Lift and drag forces are usually expressed in terms of lift and drag coefficients for an airfoil:

$$L_f = C_L \frac{\rho \overline{V}_{rel}^2 A_p}{2}$$

$$D_f = C_D \frac{\rho \overline{V}_{rel}^2 A_p}{2}$$

where A_p is the planform area of the airfoil (chord length times span). Before combining these equations, it is necessary to make provision for the variation of U_t with radial distance from the rotor. The blade is therefore divided into incremental elements Δz wide, as shown in Figure 9.23. The area

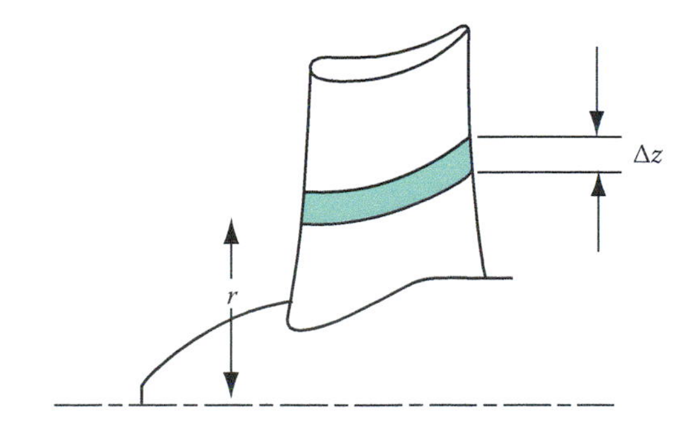

FIGURE 9.23 An incremental blade element.

of the element is $c\Delta z$, where c is the chord length. After combining the lift and drag expressions with Equations 9.50 and 9.51, we obtain the tangential and axial forces:

$$F_t = -(C_L \cos\beta + C_D \sin\beta)\frac{\rho \overline{V}_{\text{rel}}^2 c\Delta z}{2} \tag{9.52}$$

and

$$F_a = -(C_L \sin\beta - C_D \cos\beta)\frac{\rho \overline{V}_{\text{rel}}^2 c\Delta z}{2} \tag{9.53}$$

These forces can now be related to the properties of the flow before and after the blades. The momentum equation in the tangential direction is

$$\sum F_t = \iint_{\text{CS}} V_t \rho V_n \, dA$$

Evaluating each side of this equation, we obtain

$$NF_t = -N(C_L \cos\beta + C_D \sin\beta)\frac{\rho \overline{V}_{\text{rel}}^2 c\Delta z}{2}$$

where N is the number of blades and

$$\iint_{\text{CS}} V_t \rho V_n \, dA = \rho V_a (V_{t_2} - V_{t_1}) 2\pi r \Delta z$$

Equating yields

$$N(C_L \cos\beta + C_D \sin\beta)\frac{\rho \overline{V}_{\text{rel}}^2 c\Delta z}{2} = \rho V_a (V_{t_1} - V_{t_2}) 2\pi r \Delta z$$

Substituting $V_a = \overline{V}_{\text{rel}} \cos\beta$ and simplifying yield

$$V_{t_1} - V_{t_2} = (C_L + C_D \tan\beta)\frac{\overline{V}_{\text{rel}} cN}{4\pi r} \tag{9.54}$$

The momentum equation in the axial direction is

$$\sum F_a = \iint_{\text{CS}} V_a \rho V_n \, dA$$

Evaluating each term, we get

$$NF_a = (p_1 - p_2) 2\pi r \Delta z$$

$$\iint_{\text{CS}} V_a \rho V_n \, dA = (\rho 2\pi r \Delta z V_a)(V_{a_2} - V_{a_1}) \tag{9.55}$$

Because V_a is a constant, the momentum equation reduces to Equation 9.55. By substitution from Equation 9.53, we get

$$-N(C_L \sin\beta - C_D \cos\beta)\frac{\rho \overline{V}_{\text{rel}}^2 c\Delta z}{2} = (p_1 - p_2)2\pi r\Delta z$$

Simplifying and rearranging give the pressure rise as

$$p_1 - p_2 = -\frac{\rho \overline{V}_{\text{rel}}^2 cN}{4\pi r}(C_L \sin\beta - C_D \sin\beta) \tag{9.56}$$

Equations 9.54 and 9.56 relate the change in properties across the fan to aerodynamic properties of the blades.

Example 9.7

A typical three-speed window fan is placed in a square housing 56 × 56 × 15 cm wide. The fan itself has five identical evenly spaced blades, a hub diameter of 14.5 cm, and a blade tip-to-tip diameter of 48 cm. The chord length at the hub is 7.6 cm; the chord length at the tip is 17.8 cm. The inlet blade angle at the hub is 70° with respect to the axial direction. At high speed (ω = 3600 rev/min), the fan motor runs on 115 V and uses 2 A. Only 50% of this power is transferred to the air. Assuming purely axial inflow to the fan and an air density of 1.2 kg/m³, determine the pressure rise across the blades. Also calculate the aerodynamic properties of the blades.

Solution

At the hub, R_i = 14.5/2 = 7.25 cm = 0.072 5 m. So

$$U_{ti} = R_i\omega = 0.072\,5\left(3\,600\frac{2\pi}{60}\right) = 27.3 \text{ m/s}$$

Now $\beta_{i_1} = 70°$; with $\alpha_1 = 0$ (purely axial inflow), the velocity diagram shows

$$V_a = \frac{U_{ti}}{\tan\beta_i}$$

As a first approximation, assume that $\beta_{i_1} \approx \beta_i$; later, when β_{i_2} is determined, we will return to this point if a second approximation is necessary. Thus,

$$V_a \approx \frac{U_{ti}}{\tan\beta_i} = \frac{27.3}{\tan 70°} = 9.94 \text{ m/s}$$

which is a constant over the flow area. The area itself is

$$A = \frac{\pi}{4}\left(D_o^2 - D_i^2\right)$$

With D_o = 0.48 m and D_i = 0.145 m,

$$A = \frac{\pi}{4}(0.48^2 - 0.145^2) = 0.164 \text{ m}^2$$

The volume flow of air is then

$$Q = AV = 0.164(9.94) = 1.63 \text{ m}^3/\text{s}$$

The fluid is air, but there is no significant increase in stagnation temperature across the blades. Therefore, the incompressible form of the energy equation (Equation 9.12b) can be applied:

$$-\frac{dW_s}{dt} - \frac{dWf}{dt} = \rho Q\left[\left(\frac{p_1}{\rho} + \frac{V_1^2}{2}\right) - \left(\frac{p_2}{\rho} + \frac{V_2^2}{2}\right)\right] = \rho g Q \Delta Z$$

where ΔZ is the head difference across the machine. The electrical power input is

$$\frac{dW_s}{dt} = (115 \text{ V})(2 \text{ A}) = 230 \text{ W}$$

Only 50% of this power is transferred to the fluid as the head difference. Therefore,

$$-\frac{dW_s}{dt} - \frac{dW_f}{dt} = -0.5(230) = \rho g Q \Delta Z$$

or

$$\Delta Z = -\frac{115}{\rho Q g}$$

By substitution,

$$\Delta Z = \frac{-115}{1.2(1.63)(9.81)} = -5.98 \text{ m of air}$$

which is a constant over the flow area. With $V_1 = V_a$, then $V_{t_1} = 0$. Equation 9.9 for the power becomes

$$-\frac{dW_s}{dt} = T_s\omega = \rho Q\omega(-RV_{t_2}) = -\rho Q U_t V_{t_2}$$

At the hub, $U_{ti} = 27.3$ m/s; therefore,

$$V_{t_2}\Big|_i = -\frac{(-dW_s/dt)}{\rho Q U_{ti}} = \frac{115}{1.2(1.63)(27.3)} = 2.15 \text{ m/s}$$

The increase in kinetic energy across the blades is

$$\frac{V_2^2 - V_1^2}{2} = \frac{\left(V_a^2 + V_{t_2}^2\right) - \left(V_a^2 + V_{t_1}^2\right)}{2} = \frac{V_{t_2}^2}{2}$$

By definition,

$$g\Delta Z = \frac{p_1 - p_2}{\rho} + \frac{V_1^2 - V_2^2}{2}$$

After rearranging, the pressure rise across the blades is found to be

$$p_2 - p_1 = \rho\left(-\frac{V_{t2}^2}{2} - g\Delta Z\right)$$

At the hub,

$$(p_2 - p_1)\big|_i = 1.2\left[-\frac{2.15^2}{2} - 9.81(-5.98)\right]$$

$$(p_2 - p_1)\big|_t = 67.6 \text{ Pa}$$

Calculations can be formulated for the tip, where $R_o = 0.48/2 = 0.24$ m. Thus,

$$U_{to} = R_o\omega = 0.24\left(3\,600\frac{2\pi}{60}\right) = 90.47 \text{ m/s}$$

Also,

$$V_{t2}\big|_o = \frac{(-dW_s/dt)}{\rho Q U_{to}} = \frac{115}{1.2(1.63)(90.47)}$$
$$= 0.649 \text{ m/s}$$

By substitution into the expression for pressure rise,

$$p_2 - p_1 = \rho\left(-\frac{V_{t2}^2}{2} - g\Delta Z\right)$$

we find at the tip

$$(p_2 - p_1)\big|_o = 1.2\left[-\frac{0.649^2}{2} - 9.81(-5.98)\right]$$

$$(p_2 - p_1)\big|_o = 70.1 \text{ Pa}$$

The velocity diagrams at the hub and the tip are given in Figure 9.24.

To find the aerodynamic properties of the blades, we first determine the average blade angles. At the hub,

$$\beta_{i_1} = \tan^{-1}\frac{U_{ti}}{V_a} = \tan^{-1}\frac{27.3}{9.94} = 70°$$

$$\beta_{i_1} = \tan^{-1}\frac{U_{ti} - V_{t2}\big|_i}{V_a} = \tan^{-1}\frac{27.3 - 2.15}{9.94} = 68.4°$$

Because β_{i_1} and β_{i_2} are nearly equal, a second approximation is not necessary. Now

$$V_{\text{rel}i_1} = \frac{V_a}{\cos\beta_{i_1}} = \frac{9.94}{\cos 70°} = 29.1 \text{ m/s}$$

and

$$V_{\text{rel}i_2} = \frac{V_a}{\cos\beta_{i_2}} = \frac{9.94}{\cos 68.4°} = 27.0 \text{ m/s}$$

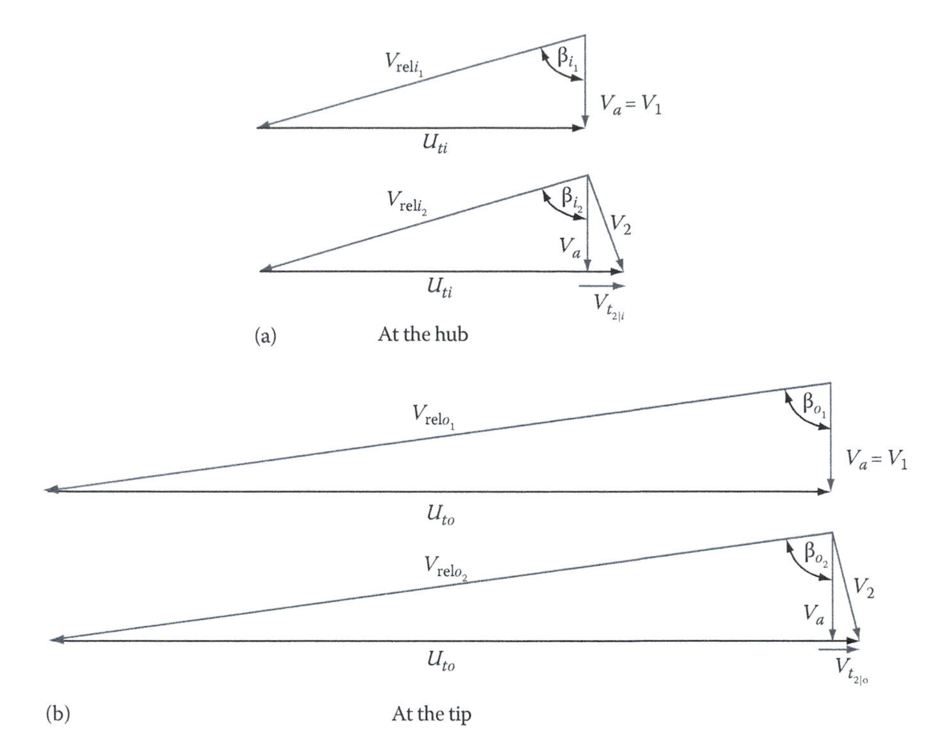

FIGURE 9.24 Velocity diagrams for the fan of Example 9.7.

Therefore,

$$\bar{V}_{\text{rel}_i} = \frac{29.1 + 27.0}{2} = 28.05 \text{ m/s}$$

which gives

$$\cos \beta_i = \frac{V_a}{V_{\text{rel}_i}} = \frac{9.94}{28.05} = 0.354$$

The average blade angle at the hub is

$$\beta_i = 69.2°$$

Similarly, at the tip,

$$\beta_{o_1} = \tan^{-1} \frac{U_{to}}{V_a} = \tan^{-1} \frac{90.47}{9.94} = 83.7°$$

$$\beta_{o_2} = \tan^{-1} \frac{U_{to} - V_{t2}\big|_o}{V_a} = \tan^{-1} \frac{90.47 - 0.649}{9.94} = 83.68°$$

These values are close, a result that implies that the blade is almost perfectly flat at the tip. It is permissible to use an average angle of

$$\beta_o = 83.69°$$

with which we obtain at the tip

$$\bar{V}_{relo} = \frac{V_a}{\cos\beta_o} = \frac{9.94}{\cos 83.69°} = 90.43 \text{ m/s}$$

Equation 9.54 can now be used. At the hub,

$$(V_{t_2} - V_{t_1})\Big|_i = \left[(C_L + C_D \tan\beta_i)\frac{\bar{V}_{rel}cN}{4\pi R}\right]\Bigg|_i$$

$$2.15 = (C_L + C_D \tan 69.2°)\frac{28.05(0.076)(5)}{4\pi(0.072\,5)}$$

or

$$C_L + 2.63C_D = 0.1838$$

Also at the hub,

$$(p_1 - p_2)\Big|_i = \left(-\frac{\rho\bar{V}_{rel}^2 cN}{4\pi R}\right)\Bigg|_i (C_L \sin\beta_i - C_D \cos\beta_i)$$

and, by substitution,

$$-67.6 = -\frac{1.2(28.05)^2(0.076)(5)}{4\pi(0.072\,5)}(C_L \sin 69.2° - C_D \cos 69.2°)$$

or

$$C_L - 0.379C_D = 0.183\,6$$

Subtracting from Equation i yields

$$263C_D + 0.379C_D = 0.183\,8 - 0.183\,6$$

or

$$C_D \approx 0$$

We also find

$$C_L \approx 0.18$$

Similarly, at the tip,

$$(p_1 - p_2)\Big|_o = \left(-\frac{\rho\bar{V}_{rel}^2 cN}{4\pi R}\right)\Bigg|_o (C_L \sin\beta_o - C_D \cos\beta_o)$$

$$-70.1 = -\frac{1.2(90.43)^2(0.178)(5)}{4\pi(0.24)}(C_L \sin 83.69° - C_D \cos 83.69°)$$

or

$$C_L - 0.111C_D = 0.024\,35$$

Also

$$(V_{t_2} - V_{t_1})\Big|_o = \left[(C_L + C_D \tan\beta_o)\frac{\bar{V}_{\text{rel}}cN}{4\pi R}\right]\Bigg|_o$$

$$0.649 = (C_L + C_D \tan 83.69°)\frac{(90.43)(0.178)(5)}{4\pi(0.24)}$$

from which we obtain

$$C_L + 9.04C_D = 0.024$$

Subtracting from Equation ii, we find the following at the tip:

$$-0.111C_D - 9.04C_D = 0.024\,35 - 0.024$$

$$C_D \approx 0$$

$$C_L \approx 0.024$$

Because the drag coefficient is quite small, C_D can usually be neglected in the equations. This is true especially if the average blade angle is greater than 45°. The implication is not that there is no drag exerted but that the drag contribution to axial and tangential forces is negligible in comparison to the lift (see Figure 9.22).

9.4 RADIAL-FLOW TURBINES

Figure 9.25a depicts a radial-flow or Francis turbine. It consists of stationary guide vanes, rotor blades that take in and discharge water in the radial direction, and a draft tube, which is a diverging circular passage on the discharge side leading to the tailwater. The guide vanes can adjust to account for variations in available water head or local power requirements. The diverging tube allows the pressure at the rotor exit to be less than atmospheric pressure, thereby increasing the total head across the turbine and thus inducing more flow to pass through. The velocity diagrams shown in Figure 9.26 correspond to flow leaving the guide vanes or gates (subscript 1) and flow leaving the turbine blades (subscript 2). In many conventional designs, the exit flow is in the axial direction. However, in this discussion, we will take inlet and outlet flows through the turbine blades to be in the radial direction.

For this turbine, the shaft torque exerted by the fluid is found with Equation 9.8:

$$T_s = \rho Q(R_2 V_{t_2} - R_1 V_{t_1})$$

From the continuity equation,

$$Q = A_1 V_{r_1} = A_2 V_{r_2}$$

or

$$Q = 2\pi R_1 b V_{r_1} = 2\pi R_2 b V_{r_2}$$

where

b is the height of the blades or gates

V_r is the radial velocity

A is the peripheral area about the rotor open to the flow and normal to V_r

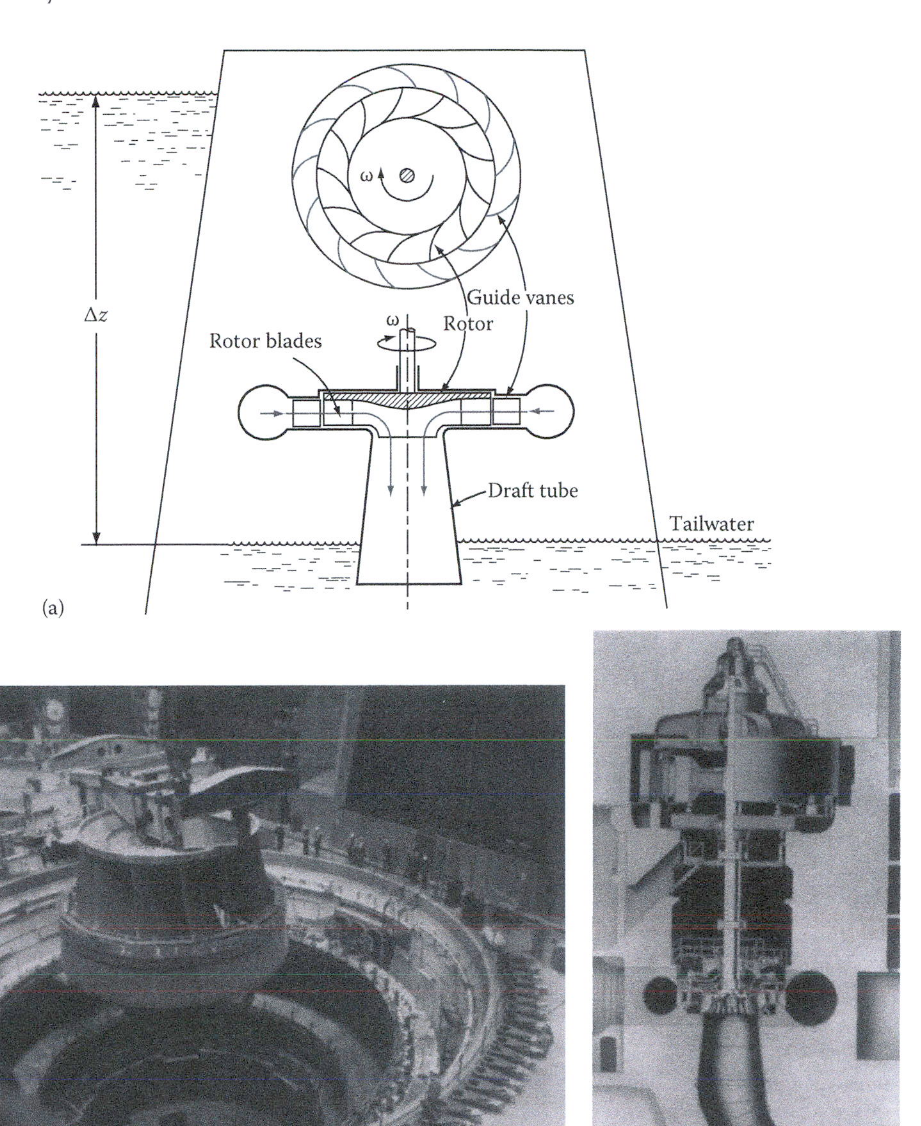

FIGURE 9.25 (a) A radial-flow turbine and (b) Francis turbine runner during installation. (Courtesy of Allis-Chalmers Hydro Turbine Division, York, PA.) (c) A Francis turbine installation. (Courtesy of Allis-Chalmers Hydro Turbine Division, York, PA.)

By simplifying, the continuity equation becomes

$$R_1 V_{r_1} = R_2 V_{r_2} \tag{9.57}$$

From the velocity diagram,

$$\frac{V_t}{V_r} = \tan\alpha$$

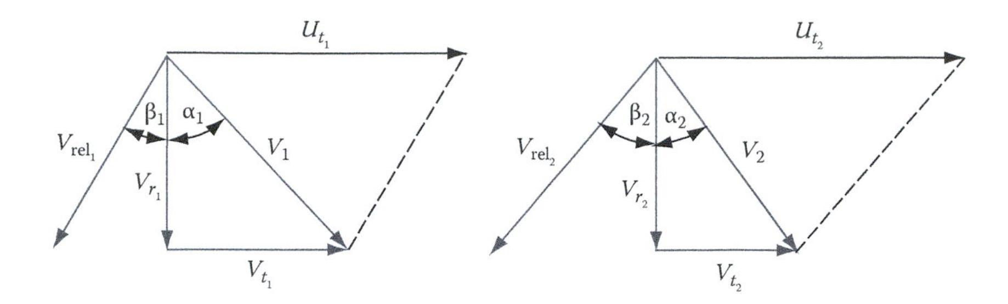

FIGURE 9.26 Velocity diagrams for a radial-flow turbine.

Combining this result with Equation 9.57 and substituting into Equation 9.8 yield

$$T_s = \rho Q(R_2 V_{r2} \tan\alpha_2 - R_1 V_{r1} \tan\alpha_1)$$
$$= \rho Q(R_2 V_{r2} \tan\alpha_2 - \tan\alpha_1)$$

which is similar to the expression for axial-flow turbines except that the radial velocity is used instead of the axial velocity.

In the radial-flow turbine, the vanes are designed such that the angle α_2 is 0°; the absolute flow velocity out of the turbine is in the radial direction. Thus, $V_{t2} = 0$, and the expression for torque becomes

$$T_s = \rho Q R_1 V_{r1} \tan\alpha_1 \tag{9.58}$$

From the velocity diagram

$$V_1 \cos\alpha_1 = V_{r1}$$

and in terms of V_1, the shaft torque becomes

$$T_s = -\rho Q R_1 V_1 \sin\alpha_1 \tag{9.59}$$

Alternatively, from the velocity diagram, we get

$$V_{t1} = U_{t1} - V_{r1} \tan\beta_1$$

and the expression for shaft torque in terms of exit blade angle is

$$T_s = \rho Q R_1 U_{t1}\left(1 - \frac{V_{r1}}{U_{r1}}\tan\beta_1\right) \tag{9.60}$$

The power or rate of doing work is found with

$$-\frac{dW_s}{dt} = T_s\omega = \rho Q\omega R_1 U_{t1}\left(1 - \frac{V_{r1}}{U_{t1}}\tan\beta_1\right) \tag{9.61}$$

Equation 9.12b gives the total power available as

$$-\frac{dW_s}{dt}-\frac{dW_f}{dt}=\rho g Q\Delta z$$

where Δz is the total head available. From Equation 9.16, we have the efficiency of the Francis turbine:

$$\eta=\frac{-dW_s/dt}{-dW_s/dt-dW_f/dt}=\frac{\omega R_1 U_{t1}}{\Delta z g}\left(1-\frac{V_{r1}}{U_{t1}}\tan\beta_1\right) \tag{9.62}$$

Example 9.8

A Francis turbine is used in an installation to generate electricity. The generator requires an input rotational speed of 1260 rev/min. For a total head of 300 ft, the volume flow rate of water through the turbine is 15 ft³/s. The absolute water velocity leaving the gates and entering the rotor makes an angle of 70° with the radial direction. The radius of the turbine rotor is 2 ft, and the inlet gate height is 0.10 ft. The absolute velocity leaving the rotor is in the radial direction. Determine the torque and power exerted by the liquid. Calculate the efficiency of the turbine.

Solution

For purely radial outflow, Equation 9.58 applies:

$$T_s=-\rho Q R_1 V_{r1}\tan\alpha_1$$

For water, ρ = 1.94 slug/ft³. The inlet area is

$$A=2\pi R_1 b=2\pi(2)(0.10)=1.26\ \text{ft}^2$$

Because this area is perpendicular to the inlet radial velocity,

$$V_{r1}=\frac{Q}{A}=\frac{15}{1.26}=11.9\ \text{ft/s}$$

By substitution into the expression for torque, we obtain

$$T_s=-1.94(15)(2)(11.9)\tan 70°$$

$$T_s=-1900\ \text{ft}\cdot\text{lbf}$$

The power is found with

$$\frac{dW_s}{dt}=-T_s\omega=-(-1900)\left(1260\frac{2\pi}{60}\right)$$

$$=250{,}700\ \text{ft}\cdot\text{lbf/s}$$

or

$$\frac{dW_s}{dt}=455\ \text{hp}=340\ \text{kW}$$

where conversion factors were obtained from Table A.2. Efficiency for a hydraulic turbine is

$$\eta = \frac{T_s\omega}{\rho g Q \Delta z} = \frac{250{,}700}{[1.94(32.2)](15)(300)}$$

or

$$\eta = 0.892 = 89.2\%$$

Note that the total head for the radial turbine is from inlet water surface to tailwater surface.

9.5 RADIAL-FLOW COMPRESSORS AND PUMPS

Radial-flow machines are those that impart centrifugal energy to the fluid. These machines consist essentially of a rotating impeller within a housing. Fluid enters the housing at the inlet to the center or eye of the impeller. As the impeller itself rotates, it adds to the energy of the fluid while it is being spun outward. The angular momentum, the static pressure, and the fluid velocity are all increased within the impeller. As the fluid leaves the impeller, it enters a volute or scroll, where it is delivered to an outlet channel. The term *compressor* refers to a device that gives a substantial pressure increase to a flowing gas. The term *fan* refers to a machine that imparts a small pressure increase to a flowing gas. (The term *blower* is often used in place of fan.) The term *pump* refers to a machine that increases the pressure of a flowing liquid.

Figure 9.27a is a schematic of a centrifugal compressor showing an impeller followed by diffuser vanes and a volute. The diffuser vanes convert some of the kinetic energy of the gas into pressure energy. Depending on the desired result, the diffusion process can be accomplished with or without diffuser vanes. At the flow entrance (section 1), the relative velocity is V_{rel_1} at an angle β_1 with respect to the axial direction. The flow velocity is turned into the purely axial direction by the inducer section, sometimes referred to as rotating guide vanes. At the exit (section 2), the absolute velocity is V_2 inclined at an angle of α_2 with the radial direction. The flow is now directed toward the diffuser vanes. An accurate description of flow through a centrifugal compressor is a three-dimensional problem beyond the range of what we wish to cover. The preceding discussion is included here merely to introduce the machine and its principle components.

The next radial-flow machine that we will consider is the centrifugal pump. Equations for a centrifugal fan are identical to those for the pump. In a fan or blower, the pressure and temperature changes are not as significant as those in a compressor. Thus, for analysis of a fan, the incompressible equations are adequate.

Figure 9.28a is a schematic of a centrifugal pump and associated velocity diagrams. A pump consists of an impeller that rotates within a housing. Fluid enters the housing in the axial direction at the eye of the impeller. The fluid then turns so that its principal direction is radial both before and after the impeller vanes. At the vane inlet, the flow is purely radial; the relative velocity is at an angle β_1 with the radial direction. At the outlet, the absolute velocity is V_2, and the relative velocity is V_{rel_2}. The continuity equation for the impeller is

$$Q = A_1 V_{r_1} = A_2 V_{r_2}$$

where A is the peripheral area of flow. For thin vanes,

$$Q = 2\pi R_1 b V_{r_1} = 2\pi R_2 b V r_2$$

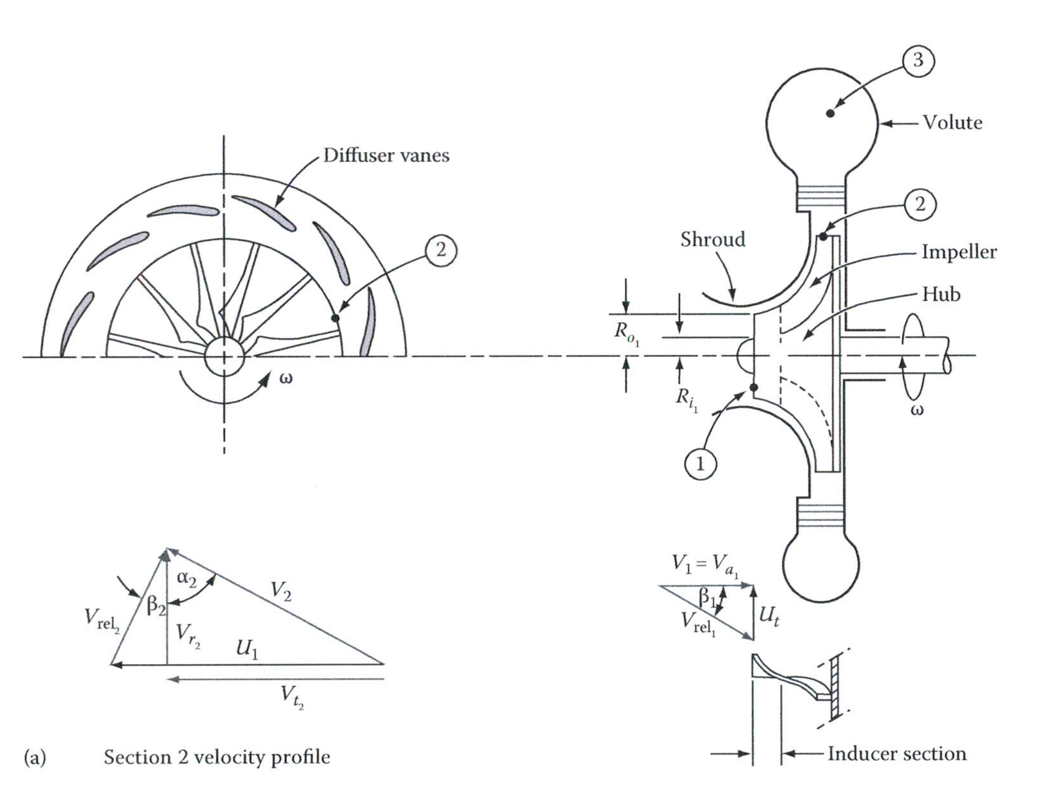

FIGURE 9.27 (a) Schematic of a centrifugal compressor and associated velocity diagrams.

(Continued)

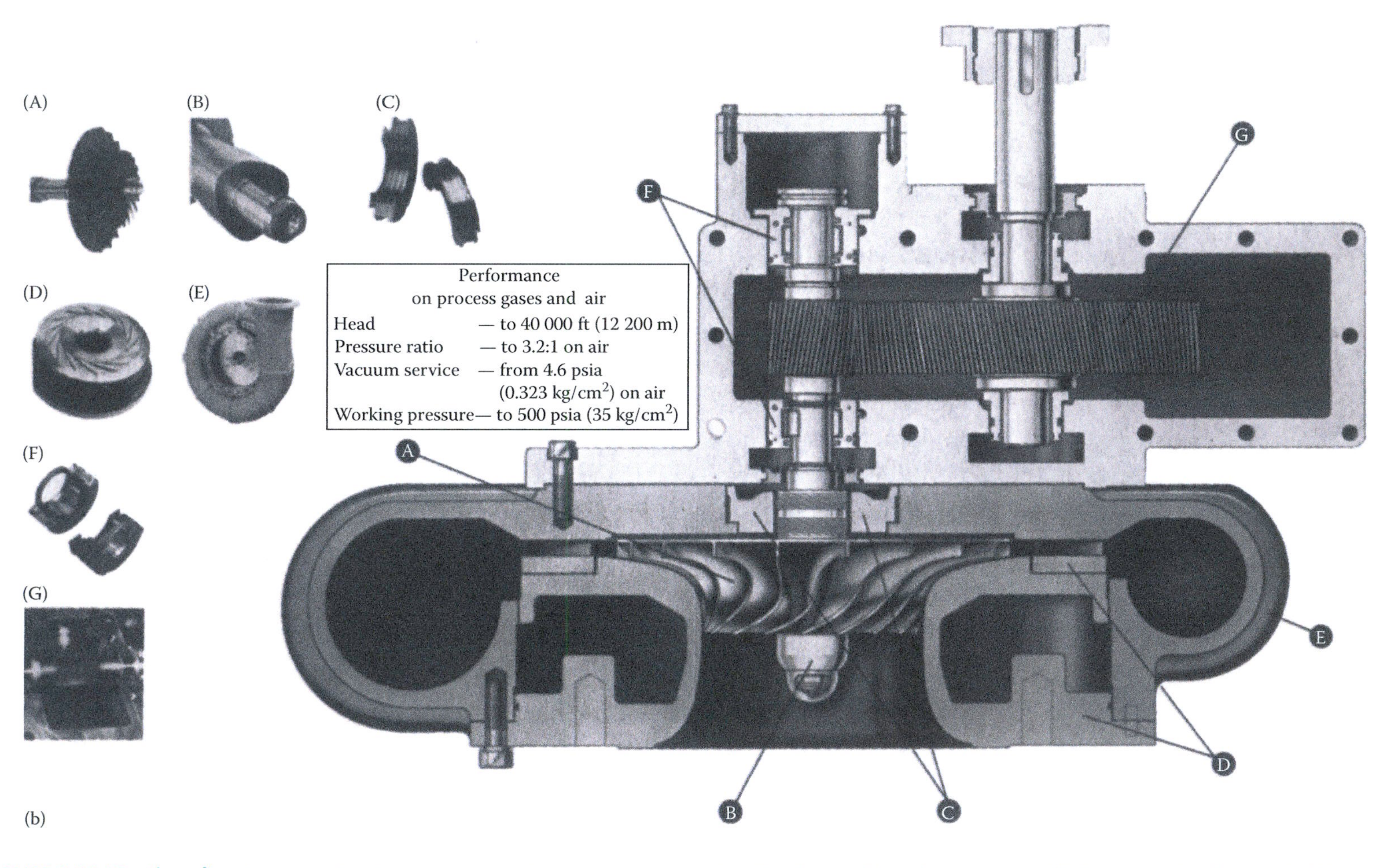

FIGURE 9.27 (*Continued*) (b) A centrifugal compressor: (A) Investment-cast impeller, (B) "Polygon" impeller mounting, (C) Process seals, (D) Vaned diffuser/inlet shroud, (E) Compressor casing, (F) Pivoting shoe bearings, and (G) Integral speed increaser. (Courtesy of Atlas Copco Turbonetics Inc., Latham, NY.)

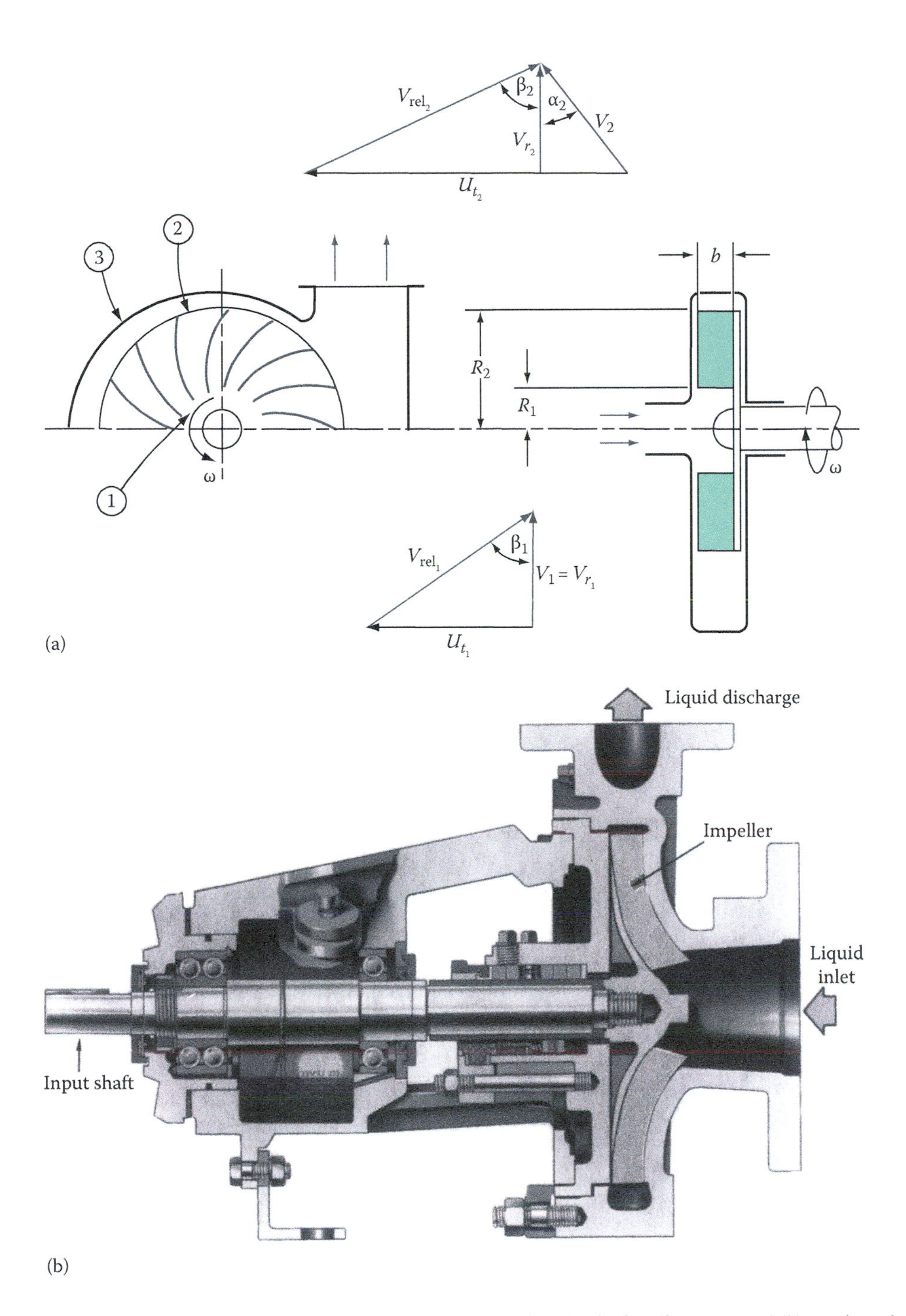

FIGURE 9.28 (a) Schematic of a centrifugal pump and associated velocity diagrams and (b) section view of a centrifugal pump and support housing. (Courtesy of Allis-Chalmers Corp., York, PA.)

or

$$R_1 V_{r_1} = R_2 V_{r_2} \tag{9.63}$$

In the following development, our objective is to express the pressure rise in terms of conditions at the impeller exit, specifically including the blade angle β_2. Equation 9.10a with $V_{t_1} = 0$ applied to the impeller gives the power delivered to the liquid:

$$\frac{dW_s}{dt} = T_s \omega = \rho Q U_{t_2} V_{t_2}$$

From the velocity diagram,

$$V_{t_2} + V_{\text{rel}t_2} = U_{t_2}$$

$$V_{t_2} = U_{t_2} - V_{r_2} \tan \beta_2 = U_{t_2} \left(1 - \frac{V_{r_2}}{U_{t_2}} \tan \beta_2 \right)$$

The power thus becomes

$$\frac{dW_s}{dt} = T_s \omega = \rho Q U_{t_2}^2 \left(1 - \frac{V_{r_2}}{U_{t_2}} \tan \beta_2 \right) \tag{9.64}$$

Equation 9.12b gives the power in terms of liquid properties for frictionless flow as

$$-\frac{dW_a}{dt} = \rho Q \left[\left(\frac{p_1}{\rho} + \frac{V_1^2}{2} \right) - \left(\frac{p_2}{\rho} + \frac{V_2^2}{2} \right) \right] \tag{9.65a}$$

or in terms of stagnation pressures,

$$\frac{dW_a}{dt} = Q(p_{t_2} - p_{t_1}) \tag{9.65b}$$

Introducing the efficiency and equating Equation 9.64 through Equation 9.65*b*, we obtain the stagnation pressure rise across the impeller:

$$p_{t_2} - p_{t_1} = \Delta p_t = \rho U_{t_2}^2 \eta \left(1 - \frac{V_{r_2}}{U_{r_2}} \tan \beta_2 \right) \tag{9.66}$$

For pumps and compressors, impeller blade designs require that $\beta_2 \geq \pi/2$. If $\beta_2 < \pi/2$, the blades would be curved forward with respect to the direction of U_t, and the flow of fluid would be inward toward the axis of rotation. This is the case for the Francis turbine of the preceding section. Equation 9.66 gives the pressure rise across the impeller in terms of conditions at the impeller exit. Note that for frictionless flow, $\eta = 1.0$.

Example 9.9

The impeller of a centrifugal water pump rotates at 900 rev/min. The impeller has an eye radius (R_1) of 2 in. and an outside diameter of 16 in. The impeller vane height is 2½ in., and measurements indicate that the vane angles are $\beta_1 = 75°$ and $\beta_2 = 83°$. Assuming radial inflow and an efficiency of 89%, determine

a. Volume flow rate through the impeller
b. Rise in stagnation pressure and increase in static pressure across the impeller
c. Pumping power transferred to the fluid
d. Input shaft power

Solution

a. Using the notation of Figure 9.28a, we have $R_1 = 2$ in., $R_2 = 8$ in., and $b = 2½$ in. The tangential impeller velocity at inlet is

$$U_{t_1} = \omega R_1 = \left(900 \times \frac{2\pi}{60}\right)\left(\frac{2}{12}\right) = 15.7 \text{ ft/s}$$

From the velocity diagram (Figure 9.28a),

$$\frac{U_{t_1}}{V_1} = \frac{U_{t_1}}{V_{r_1}} = \tan\beta_1$$

$$V_1 = V_{r_1} = \frac{15.7}{\tan 75°} = 4.2 \text{ ft/s}$$

The volume flow through the pump is

$$Q = A_1V_1 = 2\pi R_1 b V_{r_1} = 2\pi\left(\frac{2}{12}\right)\left(\frac{2.5}{12}\right)(4.2)$$

$$Q = 0.916 \text{ ft}^3\text{/s}$$

b. The continuity equation (Equation 9.63) is used to calculate V_{r_2}:

$$R_1V_{r_1} = R_2V_{r_2}$$

$$V_{r_2} = \frac{(2/12)(4.2)}{(8/12)} = 1.1 \text{ ft/s}$$

The tangential velocity at impeller outlet is

$$U_{t_2} = \omega R_2 = \left(900 \times \frac{2\pi}{60}\right)\left(\frac{8}{12}\right) = 62.8 \text{ ft/s}$$

The stagnation pressure rise is given by Equation 9.66:

$$p_{t_2} - p_{t_1} = \rho U_{t_2}^2 \eta\left(1 - \frac{V_{r_2}}{U_{t_2}}\tan\beta_2\right)$$

$$= 1.94(62.8)^2(0.89)\left(1 - \frac{1.1}{62.8}\tan 83°\right)$$

$$p_{t_2} - p_{t_1} = 5831.7 \text{ lbf/ft}^2 = 40.5 \text{ psi}$$

From the velocity diagram of Figure 9.28a at impeller exit,

$$V_{t_2} = U_{t_2} - V_{r_2} \tan\beta_2$$
$$= 62.8 - 1.1\tan 83°$$
$$= 53.8 \text{ ft/s}$$

The absolute velocity at exit is

$$V_2 = \sqrt{V_{r_2}^2 + V_{t_2}^2}$$
$$= \sqrt{1.1^2 + 53.8^2}$$
$$= 53.8 \text{ ft/s}$$

Also,

$$\tan\alpha_2 = \frac{V_{t_2}}{V_{r_2}} = \frac{53.8}{1.1} = 48.9$$
$$\alpha_2 = 88.8°$$

By definition,

$$\frac{p_{t_2} - p_{t_1}}{\rho} = \frac{p_2}{\rho} + \frac{V_2^2}{2} - \frac{p_1}{\rho} - \frac{V_1^2}{2}$$

Solving for the static pressure rise, we obtain

$$p_2 - p_1 = p_{t_2} - p_{t_1} + \frac{\rho(V_1^2 - V_2^2)}{2}$$
$$= 5831.7 + \frac{1.94(4.2^2 - 53.8^2)}{2}$$

$$p_2 - p_t = 3044 \text{ lbf/ft}^2 = 21.1 \text{ psi}$$

c. The power received by the fluid is calculated with Equation 9.65b:

$$\frac{dW_a}{dt} = Q(p_{t_2} - p_{t_1}) = 0.916(5831.7)$$

$$\frac{dW_a}{dt} = 5342 \text{ ft} \cdot \text{lbf/s}$$

Converting with 1 hp = 550 ft · lbf/s, we get

$$\frac{dW_a}{dt} = 9.71 \text{ hp}$$

d. This power is transferred directly to the water. For an efficiency of 89%, the input shaft power must be

$$\frac{dW_a}{dt} = \frac{dW_a/dt}{\eta} = \frac{9.71}{0.89}$$

$$\frac{dW_a}{dt} = 10.9 \text{ hp}$$

9.6 POWER-ABSORBING VERSUS POWER-PRODUCING MACHINES

In this section, we will compare fluid behavior in two machines that can be considered opposites: the axial-flow turbine and the axial-flow compressor. Figure 9.6 gives the velocity diagrams for a fluid flowing through a normal stage in a turbine. Flow enters the stage with an absolute velocity V_0 and constant axial velocity V_a. The stator blades turn the flow into the V_1 direction at an angle α_1 with respect to V_a. After passing through the rotor row, the flow is turned into the V_2 $(=V_0)$ direction at an angle α_2.

Figure 9.18c shows the velocity diagrams for flow through a normal axial-flow compressor stage. The flow enters the stage at an absolute velocity V_1 at an angle α_1 with respect to the axial velocity V_a. The moving rows of blades turn the flow into the V_2 direction at an angle α_2. The stator blades then redirect the flow into the V_3 direction at an angle α_3.

Comparison of these figures shows several features of note. For the turbine, $V_1 > V_2$, whereas for the compressor, $V_2 > V_1$. In addition, the turbine has $\alpha_1 > \alpha_2$ in contrast to the compressor, where $\alpha_2 > \alpha_1$. The comparison can be continued for other corresponding velocities and angles and is summarized in Table 9.2.

It is also of interest to examine the expression for torque for axial-flow machines. Equation 9.9 is a general expression applicable to either turbines or pumps. This equation relates shaft torque to flow velocities:

$$T_s = \rho QR(V_{t_2} - V_{t_1}) \tag{9.67}$$

From the velocity diagrams for the pump (or the turbine), we can write

$$V_{t_2} = U_t - V_{\text{rel}t_2} = U_t - V_a \tan\beta_2$$

$$V_{t_1} = V_a \tan\alpha_1$$

The tangential velocity difference in terms of blade angles thus becomes

$$V_{t_2} - V_{t_1} = U_t - V_a \tan\beta_2 - V_a \tan\alpha_1 \tag{9.68}$$

TABLE 9.2
Comparison of Corresponding Turbine and Compressor Velocity Diagrams

Turbine	Compressor
1. $V_1 > V_2$	$V_1 < V_2$
$\alpha_1 > \alpha_2$	$\alpha_1 < \alpha_2$
V_a = a constant	V_a = a constant
$V_{t_1} > V_{t_2}$	$V_{t_1} > V_{t_2}$
2. $V_{\text{rel}_1} < V_{\text{rel}_2}$	$V_{\text{rel}_1} > V_{\text{rel}_2}$
$\beta_1 < \beta_2$	$\beta_1 > \beta_2$
$V_{\text{rel}t_1} < V_{\text{rel}t_2}$	$V_{\text{rel}t_1} > V_{\text{rel}t_2}$
3. $V_{\text{rel}t_1} = V_{t_1} - U_t$	$V_{\text{rel}t_1} = U_t - V_{t_1}$
4. $V_{\text{rel}t_2} = U_t + V_{t_2}$	$V_{\text{rel}t_2} = U_t - V_{t_2}$

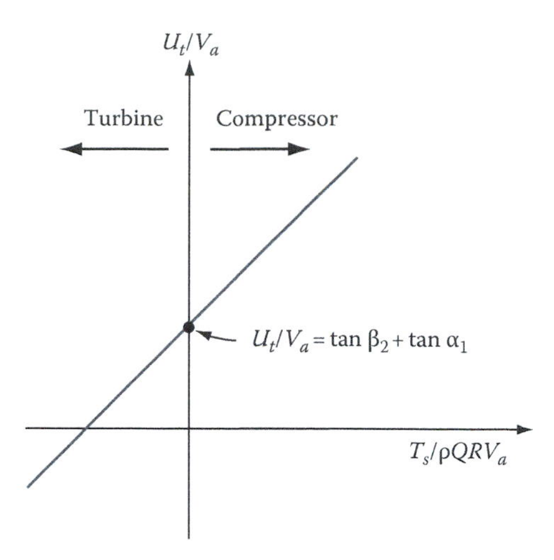

FIGURE 9.29 Dimensionless plot of shaft torque versus flow velocity.

Combining with Equation 9.67 yields

$$T_s = \rho QR(U_t - V_a \tan\beta_2 - V_a \tan\alpha_1)$$

$$= \rho QRV_a\left[\frac{U_t}{V_a} - (\tan\beta_2 + \tan\alpha_1)\right]$$

or

$$\frac{T_s}{\rho QRV_a} = \frac{U_t}{V_a} - (\tan\beta_2 + \tan\alpha_1) \quad (9.69)$$

The angles α_1 and β_2 are determined by blade settings and are not a function of flow through the machine. Thus, the parenthetical term in the preceding equation is a constant.

Equation 9.69 can be plotted as shown in Figure 9.29. The line is slanted upward and to the right at a slope of unity. The intercept is where the shaft torque is zero. To the left of the zero shaft torque point is the region describing flows through turbines; to the right is the region describing compressors.

In this section, we have qualitatively compared flow through turbines and flow through compressors. Similar comparisons for other pairs of opposite machines can be formulated.

9.7 DIMENSIONAL ANALYSIS OF TURBOMACHINERY

In the preceding sections, we derived expressions to relate power and torque to velocities through the machine, to blade angles, and to blade properties. Such equations are useful for designing the machines, but an alternative representation is required for determining the applicability of a particular machine to a specific task. The alternative is provided by methods outlined in Chapter 4 on dimensional analysis. However, it is necessary to distinguish between an incompressible flow and a compressible flow. Because our interest here is only with incompressible flows, we restrict our attention to them.

Consider an incompressible flow through a turbomachine such as a propeller turbine, Francis turbine, axial-flow pump, or centrifugal pump. For any of these machines, a dimensional analysis could be performed to express the performance parameters in terms of the geometry and the fluid properties. It is convenient to select three different dependent variables and perform a dimensional analysis three times to relate each variable to the flow quantities. The energy transfer $g\Delta Z$, the efficiency of the machine η, and the power dW/dt can all be written as functions of the flow parameters:

$$g\Delta Z = f_1(\rho, \mu, Q, \omega, D) \tag{9.70a}$$

$$\eta = f_2(\rho, \mu, Q, \omega, D) \tag{9.70b}$$

$$\frac{dW}{dt} = f_3(\rho, \mu, Q, \omega, D) \tag{9.70c}$$

We select the density ρ and viscosity μ, because they are liquid properties. The volume flow rate Q affects each independent variable, as was shown earlier. Each hydraulic machine described contains a rotating component whose rotational speed ω must be considered as an independent variable. Finally, each machine has a characteristic dimension D that differentiates it from another.

It should be mentioned that flow rate Q and rotational speed ω have special significance. Let us consider them specifically in regard to pump operation. The input rotational speed to a pump can be controlled by varying the power to the motor. The output flow of the pump can be controlled by a valve in the line. Thus, the quantities Q and ω are independently operated and, as such, are referred to as control variables.

Performing a dimensional analysis yields the following:

$$\frac{g\Delta Z}{\omega^2 D^2} = f_1\left(\frac{Q}{\omega D^3}, \frac{\rho\omega D^2}{\mu}\right) \tag{9.71a}$$

$$\eta = f_2\left(\frac{Q}{\omega D^3}, \frac{\rho\omega D^2}{\mu}\right) \tag{9.71b}$$

$$\frac{(dW/dt)}{\rho\omega^3 D^5} = \tilde{P} = f_3\left(\frac{Q}{\omega D^3}, \frac{\rho\omega D^2}{\mu}\right) \tag{9.71c}$$

where

$\dfrac{g\Delta Z}{\omega^2 D^2}$ is the energy transfer or head coefficient

$\dfrac{Q}{\omega D^3}$ is the volumetric flow coefficient or flow coefficient

$\dfrac{\rho\omega D^2}{\mu}$ is the rotational Reynolds number

$\dfrac{(dW/dt)}{\rho\omega^2 D^2} = \tilde{P}$ is the power coefficient

Experiments with turbomachines have shown that the rotational Reynolds number does not affect the dependent variables as significantly as does the flow coefficient. Thus, for incompressible flow through a turbomachine, Equations 9.71 become

$$\frac{g\Delta Z}{\omega^2 D^2} = f_1\left(\frac{Q}{\omega D^3}\right) \tag{9.72a}$$

$$\eta = f_2\left(\frac{Q}{\omega D^3}\right) \tag{9.72b}$$

$$\tilde{P} = f_3\left(\frac{Q}{\omega D^3}\right) \tag{9.72c}$$

9.8 PERFORMANCE CHARACTERISTICS OF CENTRIFUGAL PUMPS

Performance of pumps requires a lengthy discussion if each type is covered in great detail. Consequently, for the purpose of illustration, we will examine the specific characteristics of only centrifugal pumps and draw general conclusions regarding the performance of other machines.

The user of a centrifugal pump is concerned about how the device will operate in the intended application. Thus, standardized tests of performance have been devised that can be translated into a form suitable for making design calculations. Consider the test setup of Figure 9.30. A centrifugal pump circulates water from a sump tank, through a discharge line containing a flow meter and valve, and then back to the sump. The pump impeller is rotated by an electric motor that is free to pivot (within limits) about its axis of rotation. The motor rotates the impeller within the stationary pump housing. Owing to the reaction experienced by the motor, it tends to rotate in the opposite direction but can be brought to its original position by means of hanging weights. The product of weight and arm length to axis of rotation yields the input torque to the impeller. The product of torque and rotational speed is the power input from the motor:

$$\frac{dW_s}{dt} = T_s\omega \tag{9.73}$$

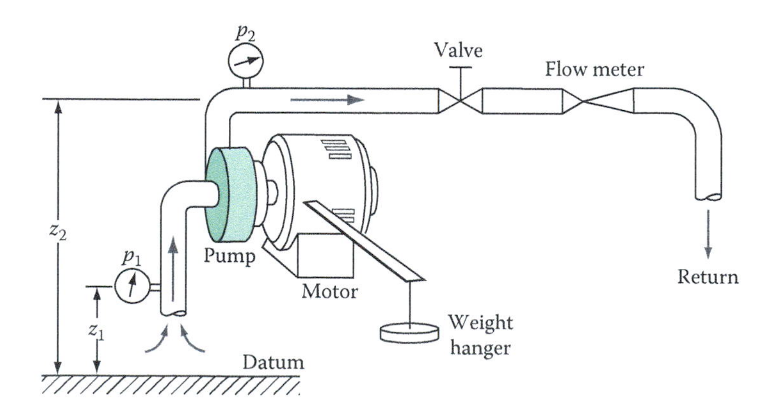

FIGURE 9.30 Schematic of a test setup for a centrifugal pump.

The flow meter provides data for calculating volume rate flow through the pump. The pressure rise across the impeller is another quantity to be measured—by pressure gauges, for example. The output power absorbed by the liquid is found with the energy equation

$$\frac{p_1}{\rho g}+\frac{V_1^2}{2g}+z_1=\frac{p_2}{\rho g}+\frac{V_2^2}{2g}+z_2+\frac{1}{g\dot{m}}\frac{dW_a}{dt} \tag{9.74a}$$

If elevation differences are neglected ($z_1 = z_2$) and the inlet and discharge pipes are the same diameter ($V_1 = V_2$), the energy equation reduces to

$$\frac{p_1}{\rho g}+\frac{p_2}{\rho g}=\frac{1}{g\rho Q}\frac{dW_a}{dt}$$

For this special case, the power thus becomes

$$\frac{dW_a}{dt}=Q(p_1-p_2) \tag{9.74b}$$

Data can be taken by varying the input electrical power (and hence rev/min) to the motor. Data are then collected and reduced to the form illustrated in Figure 9.31.

Figure 9.31 is referred to as a **performance map** of a centrifugal pump. Pressure head (as defined in Equation 9.70a) is on the vertical axis plotted as a function of flow rate on the horizontal axis.

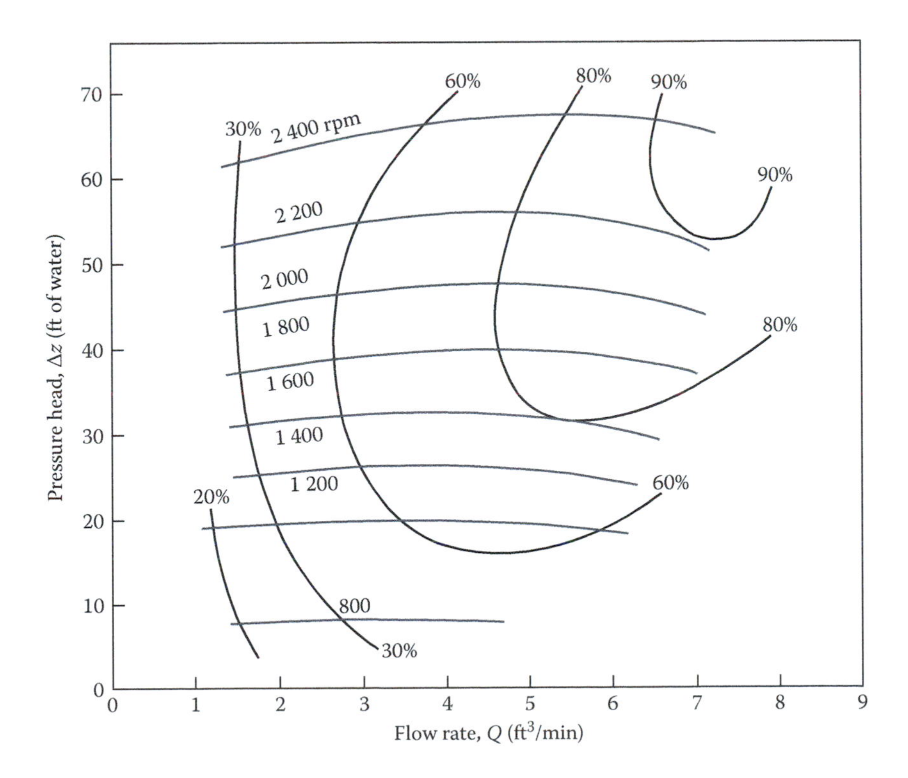

FIGURE 9.31 Centrifugal pump performance characteristics. (Data from Fluid Mechanics Laboratory, University of New Orleans, New Orleans, LA.)

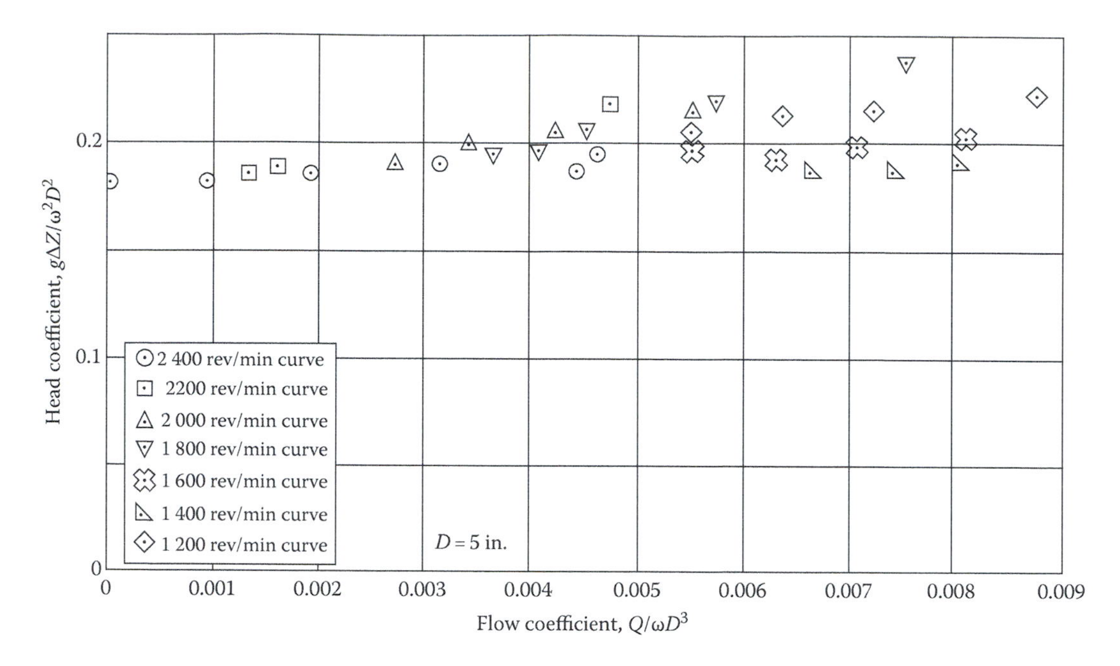

FIGURE 9.32 Dimensionless plot of the centrifugal pump data.

Next, lines of constant rotational speed are graphed. For selected points on each of these lines, efficiency is then calculated with

$$\eta = \frac{dW_a/dt \quad \text{(from Equation 9.74a)}}{dW_s/dt \quad \text{(from Equation 9.73)}} \tag{9.75}$$

Points of equal efficiency are connected to obtain the isoefficiency lines that appear in the figure. The optimum operating region for the pump is in the 80% or higher efficiency portion of the graph, a cutoff point selected arbitrarily.

According to the dimensional analysis performed for incompressible flow, the data can also be reduced by plotting g $\Delta Z/\omega^2 D^2$ versus $Q/\omega D^3$. This plot is given in Figure 9.32. For low values of the flow coefficient, the data converge onto a single line. As the flow coefficient increases, the data points diverge from a linear profile (presumably owing to the Reynolds number dependence, which was assumed to be negligible).

Because the data can be represented as shown in Figure 9.32, it is apparent that the dimensionless ratios can be used to simulate the performance of geometrically and dynamically similar pumps. Hence, data obtained on one pump provide sufficient information for predicting the performance of a similar pump.

Example 9.10

The pump of Figure 9.31 delivers 6 ft³/min of water and rotates at 2000 rev/min. The impeller on the pump is 5 in. in diameter. Determine the volume flow rate delivered if a 4 in. diameter impeller is used and the rotational speed is increased to 2200 rev/min. Calculate the new pressure head.

Solution

Assuming that similarity exists between the first and the modified pump, we can write

$$\left.\frac{Q}{\omega D^3}\right|_1 = \left.\frac{Q}{\omega D^3}\right|_2$$

For case 1, $Q = 6$ ft³/min, $D = 5$ in., and $\omega = 2000$ rev/min. Thus,

$$\left.\frac{Q}{\omega D^3}\right|_1 = \frac{6/60}{2000(2\pi/60)(5/12)^3} = 0.0066$$

For case 2, $D = 4$ in. and $\omega = 2200$ rev/min. So

$$\left.\frac{Q}{\omega D^3}\right|_2 \frac{Q_2}{2000(2\pi/60)(4/12)^3} = 0.0066$$

Solving, we get

$$Q_2 = 0.056 \text{ ft}^3/\text{s} = 3.38 \text{ ft}^3/\text{min}$$

From Figure 9.31 at $Q = 6$ ft³/min and $\omega = 2000$ rev/min, we find $\Delta Z = 46$ ft of water. The head coefficient written for both cases is

$$\left.\frac{g\Delta Z}{\omega^3 D^2}\right|_1 = \left.\frac{g\Delta Z}{\omega^3 D^2}\right|_2$$

By substitution,

$$\frac{32.2(46)}{(2000)^2(5)^2} = \frac{32.2(\Delta Z_2)}{(2200)^2(4)^2}$$

where conversion factors cancel. Solving, we get

$$\Delta Z_2 = 35.6 \text{ ft of water}$$

For centrifugal fans, axial-flow fans, and compressors, performance maps similar to that of Figure 9.31 for a pump can be developed from experimental data.

An important consideration in the design of a pump installation is the elevation of the pump itself over the level in the reservoir from which liquid is taken. The suction line contains liquid at less than atmospheric pressure. At the eye of the impeller, where the flow area is smaller, the liquid pressure is even lower. If this pressure is sufficiently low, the liquid being pumped begins to boil at the local temperature. Water, for example, boils at 92°F (33.3°C) if its pressure is 0.735 psia (5.065 kPa). This phenomenon of forming vapor bubbles through the pump is called **cavitation**. Bubbles may form near the eye of the impeller and then move outward with the surrounding liquid through the impeller vanes to a higher pressure region. The high pressure causes the bubbles to collapse, thus sending waves outward that have an erosive effect on the impeller called **cavitation erosion**. The pump's efficiency falls drastically once it starts cavitating, and subsequent failure occurs because of metal erosion and fatigue failure of seals on the shaft itself. Thus, it is important to have data on boiling or vapor pressure of various liquids as a function of temperature. These data are provided in Figure 9.33.

It is relatively simple to predict the inception of cavitation. Let the sum of pump elevation above the reservoir surface, kinetic energy head, and friction losses in the suction pipe be the

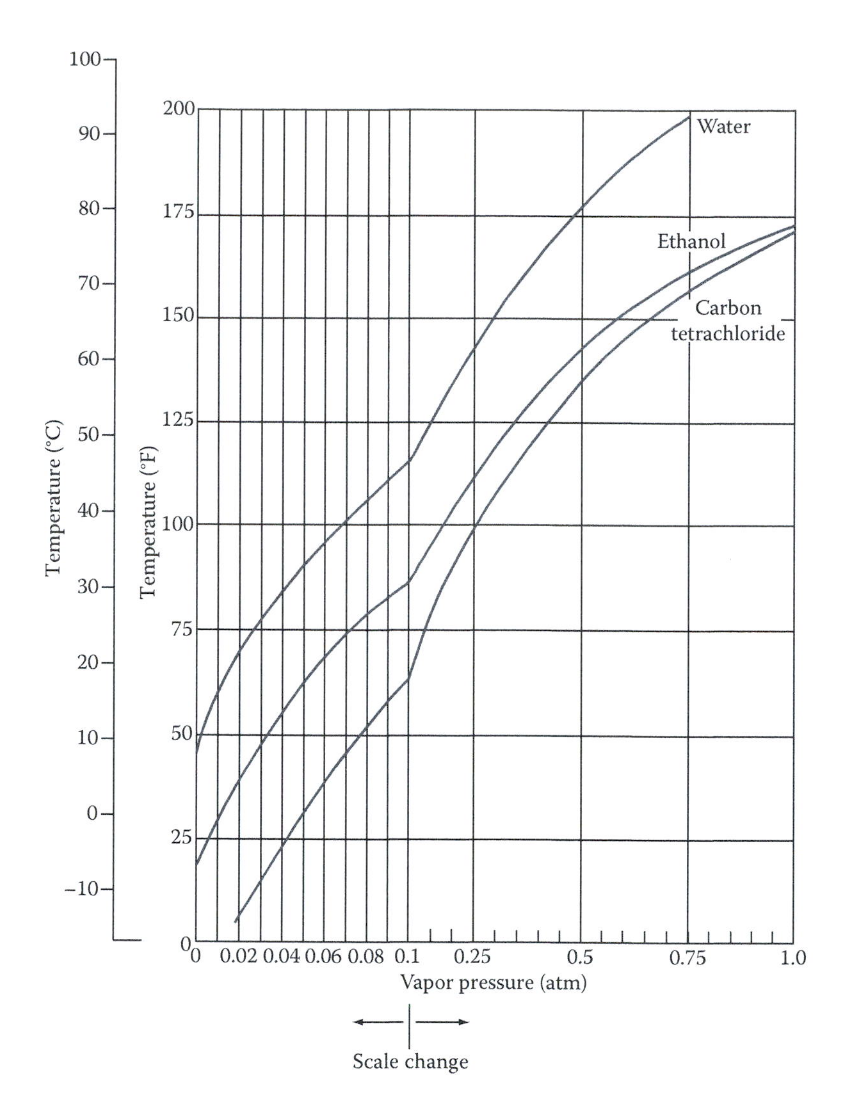

FIGURE 9.33 Vapor pressure of various liquids. (Data from Bolz, R.E. and Tuve, G.L., *CRC Handbook of Tables/or Applied Engineering Science*, 2nd edn., CRC Press, Cleveland, OH, 1973, pp. 54–55. With permission.)

total suction head Z_s. If Z_s is less than the difference between atmospheric pressure and the vapor pressure of the liquid, the pump will cavitate. The condition is commonly described by the **cavitation parameter:**

$$CP = \frac{(p_a/\rho g - p_v/\rho g) - Z_s}{\Delta Z} \tag{9.76}$$

where

p_a is the local atmospheric pressure
p_v is the vapor pressure of the liquid
ΔZ is the total dynamic head of the pump

The numerator of this expression is called the **net positive suction head** (NPSH). The cavitation parameter varies from 0.05 to 1.0 and is usually supplied by the pump manufacturer on the basis of tests performed on the pump itself. Alternatively, some manufacturers supply the net positive suction head required rather than a cavitation parameter. For pumps that produce high-volume flow rates at low heads, it is possible that net positive suction head is less than zero. This head value requires that the pump be installed below the reservoir level to eliminate cavitation.

Example 9.11

The pump represented in Figure 9.31 delivers a volume flow rate of water of 7 ft^3/min at 1800 rev/min. If the cavitation parameter at that point is 0.75, determine the maximum allowable suction head of the pump. Take the water temperature to be 150°F.

Solution

Equation 9.76 applies:

$$\text{CP} = \frac{[(p_a - p_v)/\rho g] - Z_s}{\Delta Z}$$

From Figure 9.31 at Q = 7 ft^3/min and ω = 1800 rev/min, we find ΔZ = 34 ft of water. From Figure 9.33 at T = 150°F,

$$p_v = 0.31 \text{ atm} \times \frac{14.7 \text{ psia}}{\text{atm}} = 4.56 \text{ psia}$$

By substitution into the cavitation equation, we get

$$0.75 = \frac{(14.7 - 4.56)[144/1.94(32.2)] - Z_s}{34}$$

Solving, we get

$$-Z_s = 34(0.75) - 23.4$$

$$Z_s = -2.1 \text{ ft}$$

The inlet at the eye of the impeller must therefore be 2.1 ft below the water surface.

The engineer must usually decide what type of pump is appropriate for a given application. Preliminary data such as head required and volume flow rate are usually all that is known. A dimensionless number referred to as the **specific speed** can be used as an aid in the decision-making process. This ratio is developed from the ratios in Equation 9.72 such that the characteristic diameter D is eliminated. Specific speed (also called shape number) ensures that a machine operating in its high-efficiency range is selected. The specific speed ω_{SS} is found by combining head coefficient and flow coefficient to eliminate D:

$$\omega_{SS} = \left(\frac{\omega^2 D^2}{g\Delta Z}\right)^{3/4} \left(\frac{Q}{\omega D^3}\right)^{1/2} = \frac{\omega Q^{1/2}}{(g\Delta Z)^{3/4}} \tag{9.77a}$$

It is possible to select exponents other than ¾ and ½, but these are customarily chosen for pumps. An alternative expression that is used more widely in industry is of the same form but is not dimensionally consistent. Specifically,

$$\omega_S = \frac{\omega Q^{1/2}}{\Delta Z^{3/4}} \tag{9.77b}$$

where
- ω is the rotational speed in rev/min
- Q is the volume flow rate in gal/min
- ΔZ is in feet of liquid

The specific speed found with Equation 9.77b is different by orders of magnitude from that calculated with 9.77a; ω_S is assigned the unit of rev/min.

Pumps with identical proportions but of different sizes have the same specific speed. The specific speed for pumps (ω_S) can vary from 500 to 2 000 rev/min for low-flow–high-pressure machines. Specific speed varies from 5 000 to 15 000 rev/min for high-flow–low-pressure pumps. To attach further significance to this ratio, it is assigned a value only at the point of maximum efficiency for the pump. For the performance map of Figure 9.31, the point of maximum efficiency is at approximately

$$Q = 7.25 \text{ ft}^3/\text{min} = 0.121 \text{ ft}^3/\text{s} = 54.2 \text{ gal/min}$$

$$\Delta Z = 65 \text{ ft}$$

$$\omega = 2400 \text{ rev/min} = 251.3 \text{ rad/s}$$

Equation 9.77a gives

$$\omega_{SS} = \frac{\omega Q^{1/2}}{(g\Delta Z)^{3/4}} = \frac{(251.3)(0.121)^{1/2}}{[(32.2)(65)]^{3/4}}$$

$$= 0.282$$

Equation 9.77b gives

$$\omega_S = \frac{\omega Q^{1/2}}{\Delta Z^{3/4}} = \frac{2400(54.2)^{1/2}}{65^{3/4}}$$

$$= 772 \text{ rev/min}$$

On the basis of tests performed with all types of pumps, the results of Figure 9.34 have been developed. This figure is a plot of efficiency versus specific speed over a range $500 > \omega_S > 15\,000$ rev/min. The graph relates specific speed to impeller shape and discharge, simultaneously predicting pump efficiency.

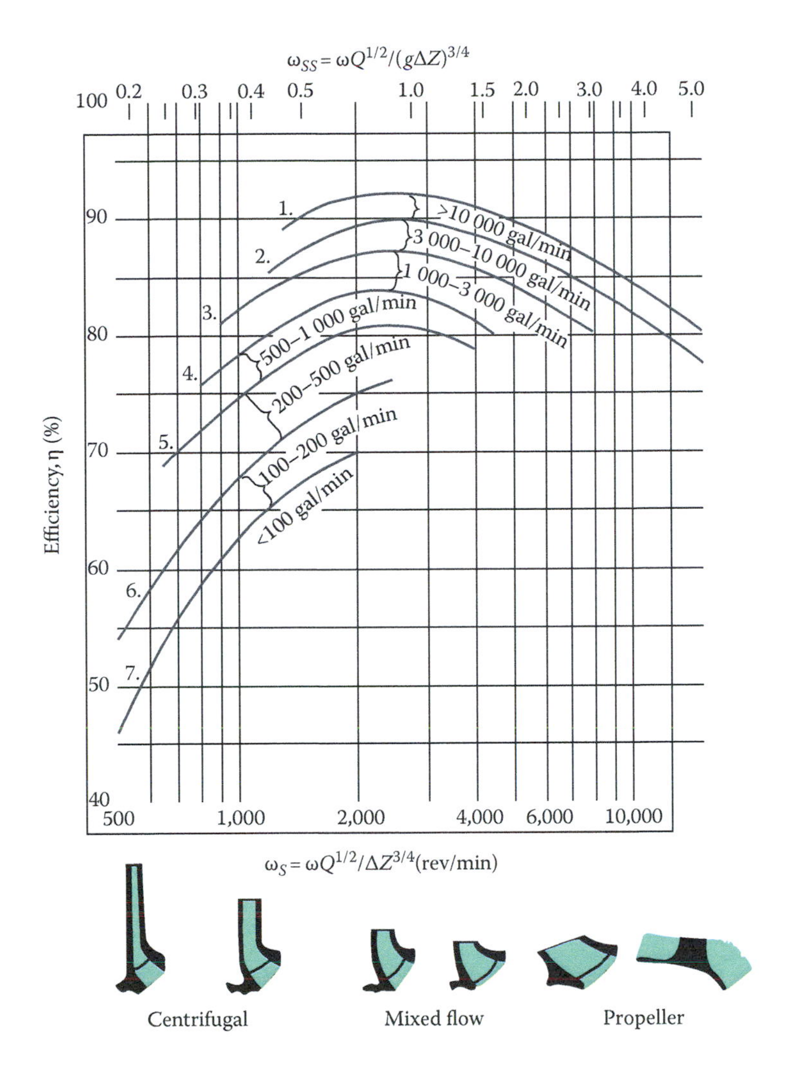

FIGURE 9.34 Relation of specific speed to impeller shape and pump efficiency. (Courtesy of Worthington Group, McGraw-Edison Company, Franksville, WI.)

Example 9.12

Determine the type of pump suited for moving 2000 gal/min (=4.456 ft³/s) of water while overcoming a 6 ft head. The available engine for the pump rotates at 800 rev/min. Calculate the power required.

Solution

Figure 9.34 is used with Equation 9.77b:

$$\omega_S = \frac{\omega Q^{1/2}}{\Delta Z^{3/4}}$$

By substitution, we get

$$\omega_S = \frac{800(2000)^{1/2}}{6^{3/4}} = 9332 \text{ rev/min}$$

At this specific speed, Figure 9.34 indicates that a propeller or axial-flow pump is a suitable choice. The efficiency is about 75%. The power required is

$$\frac{dW_a}{dt} = \rho g Q \Delta Z = 1.94(32.2)(4.456)(6)$$

$$= 1668 \text{ ft} \cdot \text{lbf/s}$$

This is the power that must be delivered to the water. The power required at the shaft is

$$\frac{dW_s}{dt} = \frac{dW_a/dt}{\eta} = \frac{1668}{0.75}$$

$$\frac{dW_s}{dt} = 2224 \text{ ft} \cdot \text{lbf/s}$$

With 550 ft · lbf/s = 1 hp,

$$\frac{dW_s}{dt} = 4.04 \text{ hp}$$

9.9 PERFORMANCE CHARACTERISTICS OF HYDRAULIC TURBINES

Turbines, both hydraulic and compressible flow types, can be analyzed in much the same way as pumps (described in the Section 9.8). A test setup can be devised and a performance map generated. In addition, dimensionless ratios can be used as illustrated in Example 9.10 to relate performance characteristics of similar devices. One significant difference, however, is in how specific speed is defined for a turbine. In this case, the power specific speed ω_{PS} is more useful. It is obtained customarily by appropriately combining Equations 9.72a and 9.72c:

$$\omega_{PS} = \frac{\tilde{p}^{1/2}}{(g\Delta z/\omega^2 D^2)^{5/4}} = \left[\frac{(dW/dt)}{\rho\omega^3 D^5}\right]^{1/2}\left(\frac{\omega^2 D^2}{g\Delta Z}\right)^{5/4}$$

$$= \frac{\omega(dW/dt)^{1/2}}{\rho^{1/2}(g\Delta Z)^{5/4}} \tag{9.78a}$$

To attach further significance to this parameter, it is assigned a value only at the point of maximum efficiency of the machine. Moreover, it is common practice to modify this equation slightly to obtain an alternative expression for power specific speed:

$$\omega_P = \frac{\omega(dW/dt)^{1/2}}{\Delta Z^{5/4}} \tag{9.78b}$$

where

ω_P is the power specific speed of the hydraulic turbine in rev/min
ω is the actual rotational speed in rev/min
dW/dt is the power in terms of hp
ΔZ is the head difference in ft of liquid

Equation 9.78b is not dimensionally consistent.

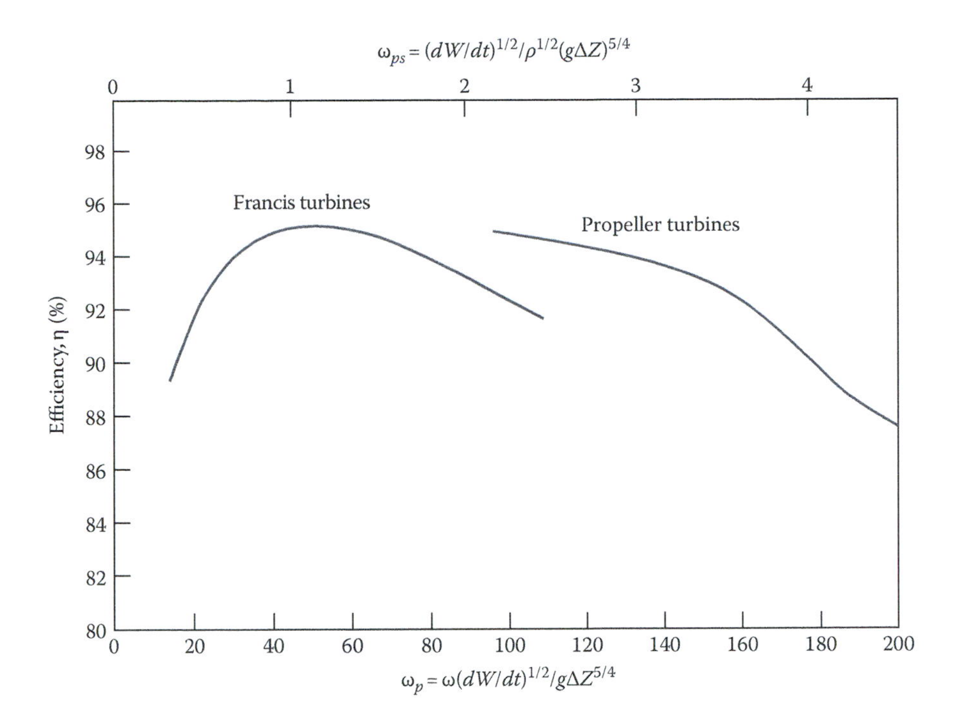

FIGURE 9.35 Hydraulic turbine efficiency ranges. (From Daugherty, R.L. and Franzini, J.B., *Fluid Mechanics with Engineering Applications*, 8th edn., McGraw-Hill Book Co., New York, 1977. With permission.)

Figure 9.35 is a graph of efficiency versus power specific speed ω_P for two hydraulic turbines. Because specific speed has significance only at the point of maximum efficiency, Figure 9.35 can be used to obtain information on which turbine is best for a given application.

Example 9.13

A turbine is to be installed in a dam where the available head varies over the course of a year; the average is 130 ft of water. The volume flow rate through the channel upstream is 1200 ft³/s. Determine the type of turbine most appropriate, assuming a rotational speed of 150 rev/min.

Solution

The power is found with Equation 9.74:

$$\frac{dW_a}{dt} = \rho g Q \Delta Z = 1.94(32.2)(1\,200)(130) = 9{,}734{,}400 \text{ ft} \cdot \text{lbf/s}$$

The power specific speed is found with Equation 9.78b:

$$\omega_P = \frac{\omega(dW_a/dt)^{1/2}}{\Delta Z^{5/4}} = \frac{150(9{,}734{,}400/550)^{1/2}}{(130)^{5/4}} = 45.46 \text{ rev/min}$$

According to Figure 9.35, a Francis turbine would be most appropriate for the application.

9.10 IMPULSE TURBINE (PELTON TURBINE)

An impulse turbine is one in which the total pressure drop of the fluid (steam or gas or water) occurs in one or more stationary nozzles. There is, by definition, no change in pressure as the fluid flows through the blades. In this section, we will consider only one type of impulse turbine: the Pelton turbine.

A Pelton turbine (or any hydraulic impulse turbine) consists of a series of vanes or buckets bolted to, or integrally cast as part of, a rotating wheel. Water under a high head is piped through a nozzle, which produces a jet of liquid aimed at the buckets. The liquid power in the jet thus produces a torque, causing the wheel to rotate.

The impulse turbine can be used to generate electricity at a location where a high head of water is available, such as in or near a dam in a river. In this type of installation, turbines are generally run at a carefully controlled constant speed, depending on the generator and local power requirements. Figure 9.36 gives a schematic of an impulse turbine installation.

The energy or head available is the vertical distance from the nozzle to the surface of the liquid supply reservoir, Δz in Figure 9.36. The water head is expended in several ways: fluid friction in the entrance pipe (frequently called the penstock), fluid friction as the liquid passes through the nozzle, fluid friction as liquid impacts the buckets, kinetic energy of the fluid as it leaves the installation, and energy delivered to the buckets. The energy equation applied from the free surface of Figure 9.36 to the region just downstream of the buckets is

$$\Delta z = \frac{f_0 L_0}{D_0}\frac{V_0^2}{2g} + C_j\left[1-\left(\frac{A_1}{A_0}\right)^2\right]\frac{V_1^2}{2g} + f_b\frac{V_{\text{rel}_2}^2}{2g} + \frac{V_2^2}{2g} + \frac{1}{\dot{m}g}\frac{dW}{dt} \tag{9.79}$$

where

Δz is the head available in feedwater
f_0 is the friction factor in the penstock
L_0 is the length of the penstock
D_0 is the penstock diameter

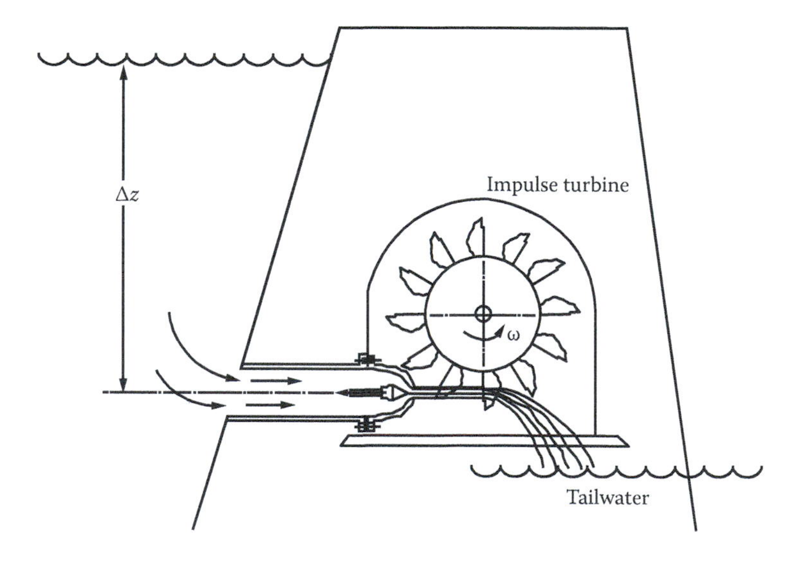

FIGURE 9.36 Impulse turbine installation, showing nozzle, turbine, head water, and tailwater.

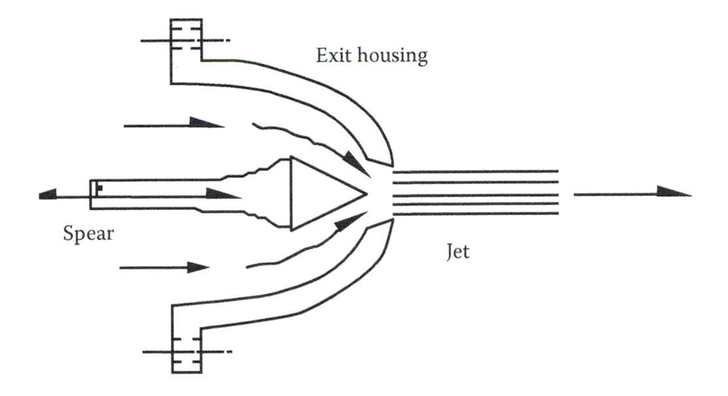

FIGURE 9.37 Schematic of the nozzle.

V_0 is the liquid velocity in pipe or penstock
C_j is the nozzle coefficient
A_1/A_0 is the penstock to jet area ratio
V_1 is the jet velocity
f_b is the liquid-bucket friction factor
V_{rel2} is the water velocity leaving the bucket measured relative to the bucket
V_2 is the liquid exit velocity (exiting from the installation and not from a bucket)
$\dot{m}$ = mass flow rate of the water
dW/dt is the power delivered to the wheel
$C_j\left[1-(A_1/A_0)^2\right]$ is the nozzle loss coefficient ≈ 0.04 for a well designed nozzle

Most of the energy delivered to the buckets is transferred to the shaft, but some is required to overcome bearing and windage losses.

The nozzle portion of the installation is used to control the rotational speed of the turbine. A schematic of the nozzle is shown in Figure 9.37. Movement of the spear in relation to the exit housing varies the flow area, and correspondingly, the velocity and area of the jet itself. The nozzle tip and spear are shaped so that the loss through the nozzle is a minimum. The nozzle must produce a jet in which all water particles move in parallel paths with no radial spreading of the jet. Thus pressure within the jet must equal that of the surroundings, that is, atmospheric pressure. The typical loss in a well-designed nozzle (represented by the coefficient of $V_1^2/2g$ in Equation 9.79) is about 0.04.

Our interest in the analysis of the impulse turbine is in its efficiency, as well as the power that can be delivered for a given Δz. To begin this analysis, we write a general equation for a given pipeline–nozzle combination. We anticipate that there is a unique jet diameter that will deliver maximum power. This can be found by rewriting the energy equation in terms of the head available at the nozzle opening:

$$\Delta z = \frac{f_0 L_0}{D_0}\frac{V_0^2}{2g} + \Delta z_a \tag{9.80}$$

where Δz_a is the energy available at nozzle just upstream of nozzle discharge. The power associated with this energy is

$$\frac{dW_a}{dt} = \dot{m}\Delta z_a = \rho A_1 V_1 g \Delta z_a$$

or, substituting from Equation 9.80,

$$\frac{dW_a}{dt} = \rho A_1 V_1 g\left(\Delta z - \frac{f_0 L_0}{D_0}\frac{V_0^2}{2g}\right) \tag{9.81}$$

The continuity equation written between the area upstream of the nozzle (subscript 0) to the jet (subscript 1) is

$$Q = A_0 V_0 = A_1 V_1$$

Solving for the upstream velocity,

$$V_0 = V_1\left(\frac{A_1}{A_0}\right)$$

Substituting into Equation 9.81,

$$\frac{dW_a}{dt} = \rho A_1 V_1 g\left(\Delta z - \frac{f_0 L_0}{D_0}\left(\frac{A_1}{A_0}\right)^2\frac{V_1^2}{2g}\right) \tag{9.82}$$

To find the maximum power, we differentiate this expression with respect to A_1 and set result equal to 0. The result is

$$\Delta z_{a|\max} = 3\frac{f_0 L_0}{D_0}\frac{V_0^2}{2g}$$

So for maximum jet power, the total head should equal 3 times the loss in the pipeline (or penstock) due to friction. Combining with Equation 9.80, we get

$$\Delta z_{a|\max} = 2\frac{f_0 L_0}{D_0}\frac{V_0^2}{2g} \tag{9.83}$$

For a well-designed nozzle, the loss is negligible, so that the energy in the jet just equals its kinetic energy:

$$\Delta z_a = \frac{V_1^2}{2g}$$

or, combining with Equation 9.83 and with continuity,

$$\frac{V_1^2}{2g} = 2\frac{f_0 L_0}{D_0}\frac{V_0^2}{2g} = 2\frac{f_0 L_0}{D_0}\left(\frac{A_1}{A_0}\right)^2\frac{V_1^2}{2g}$$

With $A = \pi D^2/4$, and canceling appropriately, we get

$$1 = 2\frac{f_0 L_0}{D_0}\left(\frac{D_1}{D_0}\right)^4$$

Solving for the jet diameter, we find that for maximum power

$$D_{1\max} = \left(\frac{D_0^5}{2 f_0 L_0}\right)^{1/4}$$

In order to proceed further, we must analyze a specific shape for the buckets. Assuming the installation is a Pelton turbine, we examine the flow in the vicinity of a single bucket. Figure 9.38 is a sketch of a Pelton bucket as seen from the nozzle. Shown are frontal, profile, and sectional views of the bucket.

Figure 9.39 is a sketch of a jet impacting a stationary bucket. The jet diameter is D_1, its area is A_1, and its velocity is V_1. The jet impacts the bucket, and is rotated through an angle of α. The external force required to keep the bucket from moving is F, which will just equal the force exerted by the jet in the absence of friction. The momentum equation for this one-dimensional flow is

$$\sum F_x = \iint V_x \rho V_n \, dA$$

Applying this equation to the flow in Figure 9.39, we get

$$-F = \rho A_1 V_1(-V_2 \cos(180 - \alpha) - V_1) \tag{9.84}$$

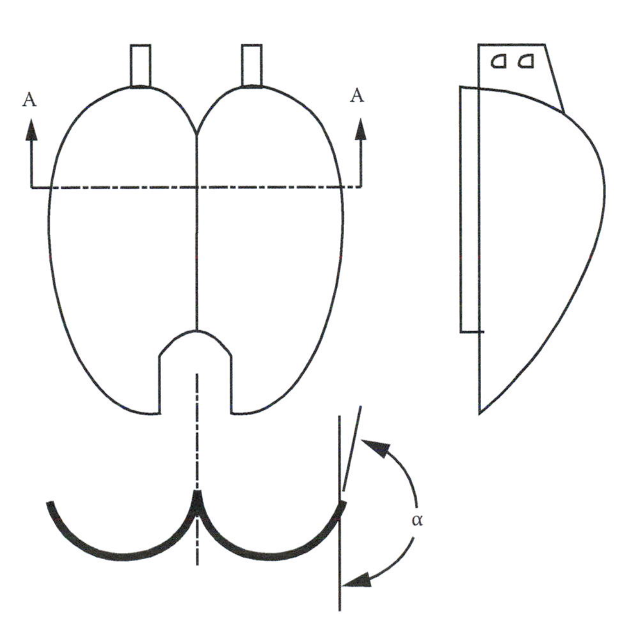

FIGURE 9.38 Schematic of a single bucket.

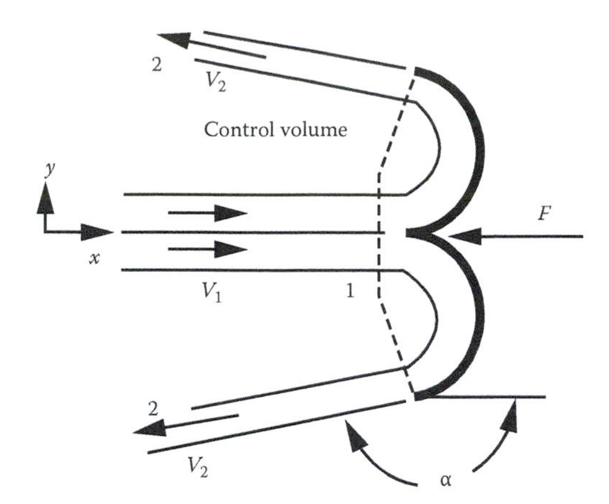

FIGURE 9.39 Schematic of the flow in the vicinity of a single stationary bucket.

For frictionless flow, the magnitude of the velocities is the same: $V_1 = V_2$. We use the trigonometric identity

$$\cos(180 - \alpha) = -\cos\alpha$$

Equation 9.84 now becomes

$$F = \rho A_1 V_1^2 (1 - \cos\alpha) \quad \text{(stationary bucket)} \tag{9.85}$$

For a moving bucket, however, we must consider the rotational speed, and this can be done with the aid of Figure 9.40. Shown is a single bucket at two different times. A liquid particle strikes the bucket first at the position labeled as J_1. The bucket moves from point 1 to 2, while deflecting the jet. During this time interval, the particle at J_1 has moved to J_2. Also shown in Figure 9.40 are the velocity diagrams at J_1 and J_2.

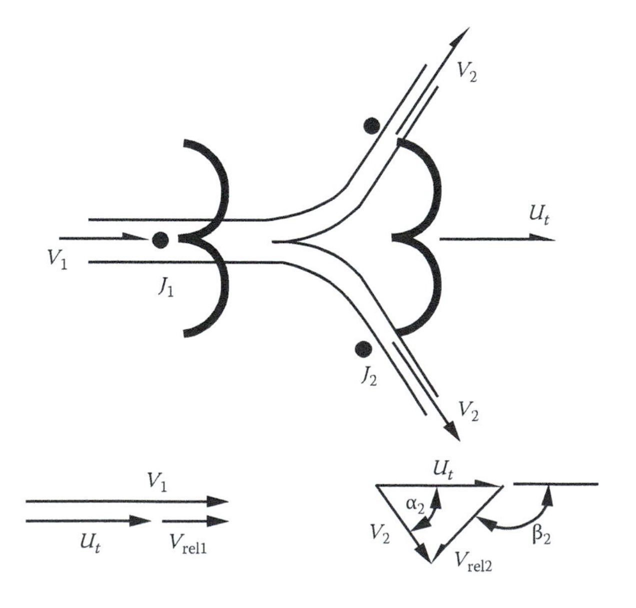

FIGURE 9.40 Flow in the vicinity of a moving bucket.

With negligible changes in potential energy, the Bernoulli equation can be written for the moving bucket of Figure 9.40. We use relative velocities, and account for bucket friction; writing the Bernoulli equation from J_1 to J_2

$$\frac{p}{\rho}+\frac{V_{\text{rel}_1}^2}{2}=\frac{p}{\rho}+\frac{V_{\text{rel}_2}^2}{2}+\frac{f_b V_{\text{rel}_2}^2}{2}$$

Assuming constant pressure, and solving for V_{rel_2}, we get

$$V_{\text{rel}_2}=\frac{V_{\text{rel}_1}}{\sqrt{1+f_b}} \tag{9.86}$$

From the velocity diagrams

$$V_{\text{rel}_1}=V_1-U_t \tag{9.87}$$

The momentum equation gives the force exerted by the jet on the moving bucket as

$$F=\rho A_1 V_1(-V_{\text{rel}_2}\cos(180-\beta_2)-V_{\text{rel}_1})$$

Substituting Equations 9.86 and 9.87, and using $\cos(180-\beta_2)=-\cos\beta_2$, we have

$$F=\rho Q\left(\frac{V_{\text{rel}_1}\cos\beta_2}{\sqrt{1+f_b}}-V_{\text{rel}_1}\right)=\rho Q V_{\text{rel}_1}\left(\frac{\cos\beta_2}{\sqrt{1+f_b}}-1\right)$$

and

$$F=\rho Q(V_1-U_t)\left(\frac{\cos\beta_2}{\sqrt{1+f_b}}-1\right)$$

The power available for doing shaft work is

$$-\frac{dW}{dt}=FU_t=\rho Q(V_1-U_t)U_t\left(\frac{\cos\beta_2}{\sqrt{1+f_b}}-1\right)$$

The power originally available in the jet is

$$+\frac{dW_a}{dt}=\frac{\rho Q V_1^2}{2}$$

The theoretical efficiency is defined as

$$\eta=\frac{dW/dt}{dW_a/dt}=\frac{\rho Q(V_1-U_t)U_t\left(1-(\cos\beta_2/\sqrt{1+f_b})\right)}{\rho Q V_1^2/2}$$

Simplifying,

$$\eta = 2\left(1-\frac{U_t}{V_1}\right)\left(\frac{U_t}{V_1}\right)\left(1-\frac{\cos\beta_2}{\sqrt{1+f_b}}\right) \tag{9.88}$$

Thus, the efficiency is independent of flow rate, but is a function of the velocity ratio U_t/V_1. To find the maximum efficiency, differentiate Equation 9.88 with respect to U_t/V_1 and set the result equal to zero. This gives

$$\eta_{max} = \frac{1}{2}\left(1-\frac{\cos\beta_2}{\sqrt{1+f_b}}\right) \tag{9.89}$$

Differentiating again, the optimum angle for maximum efficiency is 180°, but in practice, this is not good. Usually 165° or 170° is used.

Example 9.14

A 6-nominal schedule 40 cast iron pipe 300 m long delivers water from a reservoir to the nozzle of an impulse turbine installation. The difference in elevation between the water surface and the nozzle is 60 m. Graph flow rate through the nozzle versus power delivered. Assume a negligible loss of energy through the nozzle ($C_j = 0$). Find the jet diameter that corresponds to maximum power.

Solution

From the property tables,

Water	$\rho = 1\,000$ kg/m^3	$\nu = 0.914 \times 10^{-6}$ m^2/s
6 nom sch 40	$D_0 = 15.41$ cm	$A_0 = 186.5$ cm^2
Cast iron	$\varepsilon = 0.025$ cm	

From the energy equation, we have

$$\Delta z = \frac{f_0 L_0}{D_0}\frac{V_0^2}{2g} + \Delta z_a$$

With $\Delta z = 60$ m, and $L_0 = 300$ m, we substitute to get

$$60 = \frac{f_0(300)V_0^2}{0.154\,1(2)(9.81)} + \frac{V_1^2}{2g}$$

Simplifying,

$$0.605 = f_0 V_0^2 + 0.000\,514 V_1^2 \tag{i}$$

For the pipeline, $\varepsilon/D_0 = 0.025/15.41 = 0.001\,7$. From continuity

$$Q = A_1 V_1 = A_0 V_0$$

which gives

$$A_1 = 0.018\,65\frac{V_0}{V_1}$$

At this point, we select volume flow rates, and solve the equations. For example, $Q = 0.02$ m³/s, then

$$V_0 = \frac{0.02}{0.018\,65} = 1.07 \text{ m/s}$$

$$\text{Re}_0 = \frac{V_0 D_0}{\nu} = \frac{1.07(0.154\,1)}{0.914 \times 10^{-6}} = 1.81 \times 10^5$$

With this value of the Reynolds number and $\varepsilon/D_0 = 0.001\,7$, we find $f = 0.023$. Then substituting into Equation i of this example yields

$$0.605 = 0.023(1.07)^2 + 0.000\,514 V_1^2$$

Solving,

$$V_1^2 = 1\,125$$

or $V_1 = 33.5$ m/s

From continuity,

$$A_1 = 0.018\,65 \frac{V_0}{V_1} = 0.018\,65 \frac{1.07}{33.5} = 0.000\,595 \text{ m}^2$$

With $\pi D^2/4 = A$, we calculate the jet diameter to be

$$D_1 = 0.027\,5, \quad m = 2.75 \text{ cm}$$

The power in the jet is

$$+\frac{dW_a}{dt} = \frac{\rho Q V_1^2}{2} = \frac{1\,000(0.02)(33.5)^2}{2} = 11\,223 \text{ W}$$

$$+\frac{dW_a}{dt} = 11.2 \text{ kW}$$

Repeating these calculations for other flow rates gives the results displayed in Table 9.3. A graph of power as a function of flow rate is given in Figure 9.41.

TABLE 9.3

Results of Calculations Made for the Pelton Turbine

Q (m³/s)	V_0 (m/s)	Re	f_0	V_1 (m/s)	A_1 (m²)	D_1 (cm)	dW_a/dt (W)
0.02	1.07	1.81E+05	0.023	33.6	0.00 060	2.76	11 257
0.03	1.61	2.71E+05	0.022	32.7	0.00 092	3.42	15 996
0.04	2.14	3.62E+05	0.022	31.3	0.00 128	4.03	19 604
0.05	2.68	4.52E+05	0.022	29.5	0.00 170	4.65	21 734
0.057	3.06	5.15E+05	0.022	27.9	0.00 204	5.10	22 148
0.06	3.22	5.42E+05	0.022	27.1	0.00 221	5.31	22 017
0.07	3.75	6.33E+05	0.022	24.0	0.00 292	6.10	20 084
0.08	4.29	7.23E+05	0.022	19.7	0.00 406	7.19	15 565
0.09	4.83	8.14E+05	0.022	13.4	0.00 671	9.24	8 091
0.097	5.20	8.77E+05	0.022	4.3	0.02 248	16.92	903

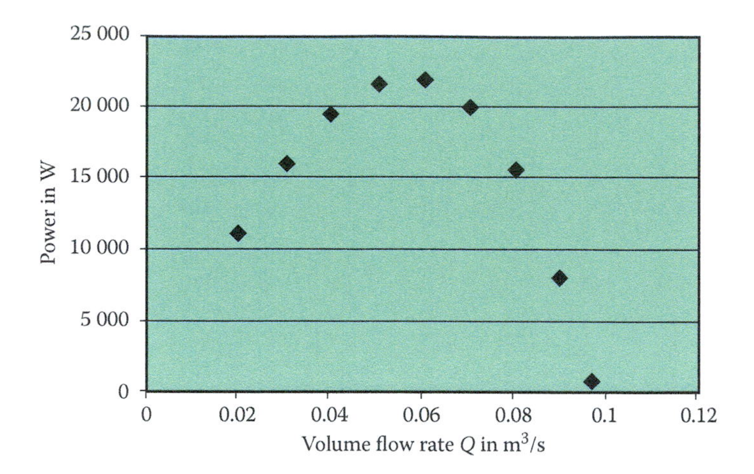

FIGURE 9.41 Graph of power as a function of flow rate for the Pelton turbine.

Shown also in Table 9.3 is the volume flow rate and corresponding parameters for maximum efficiency. These may be calculated by starting with the jet diameter for maximum power:

$$D_{1\,\text{max}} = \left(\frac{D_0^5}{2f_0L_0}\right)^{1/4} = \left(\frac{0.154\ 1^5}{2(0.022)(300)}\right)^{1/4}$$

$$D_{1\,\text{max}} = 0.050\ 7\ \text{m}$$

Next, we find

$$A_1 = \frac{\pi D^2}{4} = 0.002\ 02\ \text{m}^2$$

By combining with the continuity equation,

$$A_1 = 0.018\ 65\frac{V_0}{V_1}$$

we obtain

$$V_1 = \frac{0.018\ 65}{0.002\ 02}V_0 = 9.232V_0$$

Substitute into the energy equation (Equation i) of this example,

$$0.605 = f_0V_0^2 + 0.000\ 514V_1^2$$

$$0.605 = (0.022)V_0^2 + 0.000\ 514(9.232)^2V_0^2 \qquad \text{(i)}$$

Rearranging, and solving,

$$V_0^2 = \frac{0.605}{0.022 + 0.043} = 9.19$$

or

$$V_0 = 3.03 \text{ m/s}$$

Also,

$$V_1 = 9.323V_0 = 27.9 \text{ m/s}$$

and

$$Q = A_1V_1 = 0.002\,02(27.9) = 0.057 \text{ m}^3/\text{s}$$

$$\frac{dW_a}{dt} = \frac{\rho Q V_1^2}{2} = 22.1 \text{ kW} \quad \text{(max power)}$$

These values are shown also in Table 9.3.

Example 9.15

The nozzle-jet of the previous example is used at the maximum power setting with an impulse wheel. The jet strikes the buckets at a distance of 50 cm from the axis of rotation, and the rotational speed of the wheel is 360 rpm. The buckets are constructed such that $\beta_2 = 170°$. Determine the efficiency of the wheel and the overall efficiency of the installation. Neglect the bucket frictional effect ($f_b = 0$).

Solution

For a rotational speed of 360 rpm, at 50 cm from the axis of rotation,

$$U_t = R\omega = 0.5(360)\left(\frac{2\pi}{60}\right) = 18.8 \text{ m/s}$$

At maximum power, $V_1 = 27.9$ m/s (Table 9.3). With $\beta_2 = 170°$ and $f_b = 0$, we substitute into Equation 9.88:

$$\eta = 2\left(1 - \frac{U_t}{V_1}\right)\left(\frac{U_t}{V_1}\right)\left(1 - \frac{\cos\beta_2}{\sqrt{1+f_b}}\right) \tag{9.88}$$

$$\eta = 2\left(1 - \frac{18.8}{27.9}\right)\left(\frac{18.8}{27.9}\right)\left(1 - \frac{\cos 170}{\sqrt{1+0}}\right)$$

Solving,

$$\eta = 0.875 = 87.5\%$$

The actual power delivered to the shaft (shaft work) is

$$\frac{dW_s}{dt} = \eta\frac{dW_a}{dt}$$

where for maximum power, dW_a/dt was found in the last example to be 22.1 kW (Table 9.3). Thus

$$\frac{dW_s}{dt} = 0.875(22.1) = 19.3 \text{ kW}$$

The power existing in a 60 m head of water is found with

$$\frac{dW_s}{dt} = \rho Q g \Delta z = 1\,000(0.057)(9.81)(60) = 33.6 \text{ kW}$$

where the flow rate Q was obtained from Table 9.3. The overall efficiency of the installation (head water to shaft output) then is

$$\eta_t = \frac{19.3}{33.6}$$

$$\eta_t = 0.576 = 57.6\%$$

9.11 SUMMARY

In this chapter, we developed equations that are applicable to turbomachines. The angular momentum equation was derived and used to analyze a water sprinkler. Next we wrote general equations of turbomachinery that apply to the turbines, fans, and the like.

Equations for common types of turbines were also presented. Axial-flow turbines, radial-flow turbines, axial-flow compressors, fans, axial-flow pumps, centrifugal compressors, and centrifugal pumps were all discussed. It is unfortunate that in turbomachinery, the material presentation seems to consist of a barrage of equations. The reader is reminded that the objective is to develop equations that relate the shaft torque or work to the change in fluid properties. This theme underlies the derivations. Thus, the equations are meant to lead to an understanding of the concepts behind the design and analysis of turbomachines.

We concluded the study of turbomachinery with a brief overview of performance characteristics of selected turbomachines. Centrifugal pump characteristics were presented in some detail because these machines are very common. Characteristics of other types of machines are usually supplied by manufacturers. Finally, we formulated a dimensional analysis for the incompressible case. The results are useful for modeling purposes.

PROBLEMS

General Equations

9.1 Rework Example 9.1 for a sprinkler with three discharge nozzles.

9.2 In a typical two-nozzle lawn sprinkler (as described in Example 9.1), the radius $R = 6$ in. and the volume flow rate through the sprinkler is 1 ft^3/min. For a rotational speed of 2 rev/s, determine the shaft torque exerted on the bearings. Take the nozzle angle setting to be 30° with respect to the radial direction; take the exit diameter to be 0.3 in. Calculate also the rate of shaft work being done.

Axial-Flow Machines

9.3 A single-stage axial-flow turbine uses compressed air as the working fluid. The inlet pressure and temperature are 2 000 kPa and 25°C. The absolute velocity of the flow leaving the stator (pitch line is at a radius of 1.8 m) is 7.0 m/s. The angle between this velocity and the axial direction is 30°. The volume flow rate is 18 m^3/s, and the rotor rotates at 360 rev/min. Determine the torque exerted on the rotor and the average axial velocity. Assume that the flow leaves the rotor in the purely axial direction.

9.4 Show the equivalence for reaction ratio of the two equations labeled 9.30.

9.5 Sketch the velocity diagram for the case in which the reaction ratio for an axial-flow turbine is 1.0. Determine which blade angles are equal.

9.6 Measurements on the first stage of a multistage gas turbine indicate that pressure and temperature at stage entry are 1 000 kPa and 400 K. At stage exit, the pressure is 960 kPa. The rotational speed is 360 rev/min, and the corresponding shaft torque is 825 N · m. The rotor diameter is 1 m, and the blade height is 8 cm. The mean pitch line is at two-thirds the blade height as measured from the rotor. The mass flow of gas is 25 kg/s, and the gas constant is 300 J/(kg · K). Assume ideal gas behavior and assume that the stage is normal. The inlet flow angle between absolute and axial velocities is 40°. Determine the velocity diagrams, the stage efficiency, and the reaction ratio.

9.7 A single-stage gas turbine operates at inlet conditions of 50 psia and 600°R. The mass flow through the stage is 0.776 slug/s, while the rotor blade speed is 490 ft/s. The axial-flow velocity at rotor exit is 250 ft/s. The gas constant is 1610 ft · lbf/slug · °R. The angle between the absolute inlet flow velocity and the axial direction is 15°. The shaft power is 400 hp. If the stage is normal, determine all other blade angles and draw the velocity diagrams. Calculate the reaction ratio.

9.8 Show that the reaction ratio for a normal stage can be written as

$$RR = \frac{1}{2U_t}\left(\frac{2\dot{m}}{\rho A}\tan\alpha_0 + 2U_t - \frac{T_s\omega}{\dot{m}U_t}\right)$$

9.9 Rework the calculations of Example 9.3 for turbine blade radii of 0.2, 0.3, 0.4, 0.5, 0.6, and 0.7 m.

9.10 An axial-flow propeller turbine while in operation has guide vanes set at an angle of 35° with respect to the radial direction. The inner radius of the vanes is 15 ft; the vane height is 3 ft. The absolute inlet velocity at the vanes is 12 ft/s. The turbine blades have a tip radius of 13 ft and a hub radius of 3 ft. The rotor speed is 72 rev/min. Determine the blade angles for the leading edge of the blade at the hub and at the tip.

9.11 The inner radius of the guide vanes of a propeller turbine is 2.2 m, and the gate height is 0.8 m. The turbine itself is designed to produce power at a flow rate of 180 m^3/s and a rotational speed of 270 rev/min. The tip radius of the blades is 2 m; the inlet tangential fluid velocity at the tip is 14 m/s. Calculate the power output of the turbine. Determine the angle that the guide vanes make with the radial direction. Assume that the tangential velocity leaving the turbine blades at the tip is zero.

9.12 Examples 9.4 and 9.5 deal with a two-bladed windmill. If the number of blades is changed to three and all other conditions remain the same, determine the characteristics of the required airfoil.

9.13 Calculate the maximum power that can be developed by a windmill in a 60 km/h wind. Assume a disk radius of 25 m and ambient conditions of 110 kPa and 25°C.

9.14 A 90 ft diameter windmill operates in a wind of 60 miles/h. Assuming ideal conditions and that the windmill efficiency is 48%, determine (a) the velocity through the disk, (b) the thrust on the windmill, and (c) the pressure drop across the disk.

9.15 A single-stage axial-flow air turbine has blade angles of 50° (α_1) and 55° (β_2). The airflow through the turbine is 4 kg/s. The reaction ratio is 0.65. The inside casing diameter is 0.8 m, and the blade height is 4 cm. The mean pitch line is at three-fourths of the blade height measured from the rotor. The inlet pressure is 650 kPa, and the rotational speed is 120 rev/min. Determine the inlet air temperature, the power, the shaft torque, and the pressure drop across the stage.

9.16 Consider one stage of a three-stage gas turbine. At stage entry, the pressure is 500 kPa and the temperature is 600 K. The pressure drop across the stage is 1 kPa. Rotational speed of the rotor is 180 rev/min, at which the shaft torque is measured to be 520 N · m. Each stage contributes equally to the shaft torque. The rotor diameter is 1 m, and the blade height is 7 cm. Assume that the mean pitch line is located at the midpoint of the blades. The mass flow of gas is 10 kg/s, and its molecular mass is 40 kg/mol. The specific heats are the same as those for oxygen. Each stage is normal, and the inlet flow angle between the absolute velocity and the axial velocity is 40°. Determine the efficiency of the turbine, the reaction ratio, and the blade angles α_1 and β_2.

9.17 An axial-flow propeller turbine installation produces 15,500 hp under a water head of 37 ft, with a rotational speed of 106 rev/min. The turbine blades have a tip radius of 8 ft and a hub radius of 2 ft. The guide vanes are 9 ft high and have an inner radius of 8.2 ft. The vanes are set at an angle of 35° with respect to the radial direction. Determine angles at the leading edge of the turbine blades for the tip and for the hub. Determine the shaft torque.

9.18 An axial-flow propeller turbine installation has a runner that rotates at 94.7 rev/min. The outer radius at the runner tip is 3.048 m; the hub radius is 0.762 m. The water head available is 24.7 m, and the shaft torque is 5.376×10^6 N·m. The guide vanes are 2.6 m tall and have an inner radius of 3.1 m. Determine the angle at the leading edge of the turbine blades for the tip if the angle at the leading edge for the hub is 130°. Determine the angle setting of the guide vanes with respect to the radial direction, the volume flow rate through the system, and the power output of the installation.

9.19 A two-bladed windmill, diameter 12 ft, in an airstream of velocity 30 miles/h, delivers 12.8 kW to the shaft. The air temperature is 72°F, and pressure is 14.7 psia. Downstream of uniform flow velocity is 25 miles/h. Calculate thrust and efficiency. If the rotational speed is 500 rev/min and the blade angle (β) is 50°, sketch the velocity diagram. Take the chord length to be 6 in. Determine aerodynamic properties of the blades. Use the midpoint of each blade for the calculations.

9.20 A Clark Y airfoil has a lift coefficient of 1.2 and a drag coefficient of 0.1 at an angle of attack of 10.8°. If it is used as a windmill propeller, determine the rotational speed of the blade in a wind of 30 miles/h assumed to be at the disk. The chord length is 10 in., and the blade length is 15 ft (tip to tip). The blade is set at an angle of 61.6° ($=\beta + \delta$) from the axial direction. Calculate also the pressure drop across the blades. Perform all calculations at the blade midpoint for $N = 2$ and for a shaft power of 6918 ft·lbf/s.

9.21 An axial-flow propeller turbine installation has a turbine runner with a tip radius of 3.71 m and a hub radius of 1.62 m. The rotational speed is 85.7 rev/min under a water head of 13.11 m. The guide vanes are 2.75 m tall, and the radius to the guide vanes is 3.8 m. The vanes are set at an angle of 50° with respect to the radial direction. Determine the angles at the leading edge of the blades for radial locations of 2, 2.5, 3.0, and 3.5 m as well as at the hub and at the tip. If at each of these locations the tangential water velocity leaving the blades is zero, calculate the shaft power at each radial distance. (These should all be approximately equal.)

9.22 If the tangential velocity leaving the runner of an axial-flow turbine installation is zero, the shaft power equation becomes

$$\frac{dW_s}{dt} = \rho Q U_t V_{t_1}$$

where subscript 1 denotes an inlet condition, and the product of the tangential velocities can be taken at any radial location on a turbine blade. By substitution, show that this equation can be rewritten as

$$\frac{dW_s}{dt} = 4\pi\rho R_g^2 b\omega\Delta z g \cos\alpha \sin\alpha$$

9.23 An axial-flow propeller turbine installation has a runner with a tip radius of 3.57 m and a hub radius of 1.52 m. The water head is 24.38 m. The guide vanes are 2.6 m high, and the radius to the guide vanes is 3.7 m. The blade angle at the leading edge of the turbine runner at the hub is 130°, which is fixed. Determine and graph the variation of rotational speed and flow rate with the angle at which the guide vanes are set. Determine also the power output variation, assuming that the outlet tangential fluid velocity at the hub of the turbine blades is zero.

9.24 In an axial-flow turbine installation, the tip radius of the turbine runner is 9.82 ft and the hub radius is 3.77 ft. The water head is 45 ft. The guide vanes are located at a radius of 10 ft from the axis of rotation. The vanes themselves are 4.5 ft tall and set at an angle of 35° with respect to the radial direction. Determine the blade angle at the leading edge of the turbine blade tip if the power output is 136,000 hp. Take the outlet tangential fluid velocity at the tip to be zero. Determine also the rotational speed.

9.25 An axial-flow pump rotates at 450 rev/min and conveys 30 ft^3/s of glycerine. The hub radius is 10 in., and the casing radius is 27 in. The absolute flow velocity leaving the rotor blades is 6 ft/s. Using the blade element procedure, determine the torque and power.

9.26 The three-speed fan of Example 9.7 is turned down at 1800 rev/min. If the flow angles are the same, determine the pressure rise across the fan at the hub. Calculate the power drawn, assuming the pressure rise at the hub is constant over the cross section.

9.27 An axial-flow water pump has a rotor with blade angles that are invariant with radial distance. The rotor blade angles are $\beta_1 = 60°$ and $\beta_2 = 45°$. An outlet vane downstream removes the tangential fluid velocity component. For an inflow velocity of 12 ft/s, determine the change in pressure across the pump and the angle between the axial direction and the absolute fluid velocity leaving the rotor.

9.28 An axial-flow fan has 12 blades and is in a circular housing with an inside diameter of 2 m. The hub diameter is 1.6 m. Each blade is straight, and all are identical with a chord length of 10 cm and a mean pitch diameter of 1.8 m. All blades are set at an angle of 62° with respect to the axial direction. The inflow is purely axial with static properties of 101.3 kPa and 20°C. The pressure rise across the blades is 0.3 m of water when the rotor speed is 1800 rev/min. Determine the aerodynamic properties of the blades and the power.

Radial-Flow Machines

9.29 A Francis turbine is used in an installation for the generation of electricity. The volume flow rate through the turbine is 45 m^3/s. The absolute water velocity leaving the gates makes an angle of 60° with the radial direction. The radius of the turbine rotor is 6 m, and the inlet gate height is 0.3 m. The absolute exit velocity is in the radial direction. The turbine efficiency is 94%. Calculate the required water head, the torque, and the power exerted by the water. Take the rotational speed to be 180 rev/min.

9.30 A Francis hydraulic turbine runs at a speed of 180 rev/min and has an efficiency of 90%. The absolute velocity leaving the gates makes an angle of 80° with the radial direction. The radius of the turbine rotor is 12 ft, and the inlet gate height is 1 ft. Calculate the volume flow rate through the turbine if the head under which it acts is 120 ft.

9.31 Derive Equation 9.25b in detail.

9.32 A Francis turbine is used in an installation to generate power. The power output is 115,000 hp under a head of 480 ft. The outer radius of the rotor blades is 6.75 ft; the inner radius is 5.75 ft. The rotational speed is 180 rev/min, and the blade height is 1.5 ft. The turbine efficiency is 95%. The absolute velocity leaving the rotor is in the radial direction. Determine the angle the absolute velocity entering the rotor makes with the radial direction. Sketch the velocity diagram at the rotor inlet. The fluid medium is water.

9.33 A Francis turbine installation produces 111 MW. The blade height is 0.732 m; the outer radius at the inlet to the runner is 5.52 m. The rotational speed is 167 rev/min. The flow leaving the runner is purely radial. Determine the volume flow rate through the installation if the angle between the radial direction and the direction of the absolute velocity entering the rotor is 30°. Also find the water head to the gates.

9.34 A Francis turbine installation produces 172,000 hp under a water head of 162 ft from the reservoir surface to the tailwater. The blade height is 10 ft. The radius to the rotor blade at inlet is 18 ft. The rotational speed is 105.9 rev/min. For a volume flow through the installation of 9 851 ft^3/s, determine the angle between the absolute inlet velocity to the rotor and the axial direction. Determine also the angle between the inlet relative velocity to the rotor and the axial direction, and calculate the efficiency. Assume that the absolute exit velocity is in the radial direction.

9.35 The impeller of a centrifugal water pump rotates at 1 260 rev/min and has 2 cm high vanes. The inlet eye radius is 3 cm, and the outside radius is 7 cm. The vane angles are $\beta_1 = 60°$ and $\beta_2 = 80°$. Assuming radial inflow to the impeller vanes, determine the volume flow rate and the pumping power. Calculate also the stagnation pressure rise across the impeller.

9.36 The input power to a centrifugal water pump is 200 hp, and the shaft torque is 1150 ft · lbf. The impeller has an outside diameter of 24 in. and a diameter at the eye of 5 in. Measurements on the impeller indicate that $\beta_1 = 65°$ and $\beta_2 = 83°$. For a blade height of 2 in., determine the volume flow rate through the pump, the expected static pressure rise, and the rotational speed.

Performance Analysis and Comparisons

9.37 Compare velocity diagrams for the axial-flow propeller turbine and the axial-flow pump. Prepare a summary similar to Table 9.2 and a plot like that in Figure 9.29.

9.38 Repeat Problem 9.37 for Francis turbines versus centrifugal-flow pumps.

9.39 Derive Equation 9.71a.

9.40 Derive Equation 9.71b.

9.41 Derive Equation 9.71c.

9.42 Verify that the head coefficient, the volumetric flow coefficient, the rotational Reynolds number, and the power coefficient are all dimensionless.

9.43 The inlet of a centrifugal pump is 8-nominal schedule 40 pipe. The discharge is 6-nominal schedule 40 pipe. The pressure gauge on the outlet reads 250 kPa, and the gauge at the inlet reads 30 kPa. If the volume flow rate of water through the pump is 0.1 m^3/s and the efficiency is 85%, determine the power. Neglect elevation differences and derive an expression by applying the energy equation from inlet to outlet.

9.44 The inlet and discharge of a centrifugal pump are 1½-nominal and 1-nominal schedule 40 pipe, respectively. The pressure increase from inlet to outlet is 20 ft of water. The motor input is ½ hp, and the efficiency is 72%. Determine the volume flow rate through the pump. Neglect elevation differences.

9.45 A centrifugal water pump under test discharges 0.015 m^3/s against a head of 18 m when the rotational speed is 1800 rev/min. The impeller diameter is 32 cm, and the power input is 4.5 kW. A geometrically and dynamically similar turpentine pump has an impeller diameter of 38 cm and will run at 2200 rev/min. Assuming equal efficiencies between the two pumps, determine the head developed, the volume flow rate, and the power required for the second pump.

9.46 A large pump is to be used in a glycerine-bottling plant. The glycerine pump is to be driven by a 10 hp motor at 800 rev/min. The system is modeled with a small ¼ hp motor–pump combination that runs at 1800 rev/min and pumps water. What should be the ratio of glycerine impeller diameter to water impeller diameter?

Cavitation and Specific Speed

9.47 In a liquor-bottling plant, a centrifugal pump discharges 0.02 m^3/s against a head of 28 m. The cavitation parameter for the pump is 0.5. Where should the pump inlet location be with respect to the surface of the ethanol? Take the ethanol temperature to be 10°C.

9.48 How do the results of Example 9.11 change if the liquid is carbon tetrachloride?

9.49 The inlet of a pump is level with the surface of a water reservoir. The pump draws 70°F water at a rate of 200 gal/min and delivers it against a head of 60 ft. If the cavitation parameter is 0.5, is the pump expected to cavitate?

9.50 Verify that Equation 9.77a is dimensionless.

9.51 What are the actual units of Equation 9.77b?

9.52 What type of pump is most suitable for moving 100 gal/min of water against a 20 ft head when the rotational speed is 2400 rev/min? Determine the input power required.

9.53 What type of pump is best suited for pumping 1 m^3/s of water against a head of 12 m while rotating at 100 rev/min? Calculate the power required.

Performance of Hydraulic Turbines

9.54 What type of pump is most appropriate for pumping 6 ft^3/s of water against a head of 15 ft while rotating at 600 rev/min? What is the required power?

9.55 Verify that Equation 9.78a is dimensionless.

9.56 What are the actual dimensions of Equation 9.78b?

9.57 A water turbine is to operate under a head of 100 m of water with a flow rate of 400 m^3/s. If the rotational speed is 120 rev/min, determine the most suitable type of turbine and the power output.

9.58 A turbine operates under a head of 36 ft at a location where the volume flow rate is 110 ft^3/s. A Francis turbine is installed. For efficiency >94%, determine the allowable limits on rotational speed.

9.59 An axial-flow propeller turbine has the following characteristics: power output = 15,500 hp, water head = 37 ft, and rotational speed = 106 rev/min. Estimate the efficiency of the installation. While keeping other conditions the same, what should the rotational speed be changed to in order to obtain 95% efficiency?

9.60 An axial-flow propeller turbine has the following data: $\omega = 94.7$ rev/min, $\Delta Z = 24.7$ m, and $T_s = 5.376 \times 10^6$ N·m. Estimate the efficiency of the installation.

9.61 A Francis turbine has the following data: $\omega = 180$ rev/min, $\Delta Z = 480$ ft, and $\eta = 95\%$. Estimate the power output of the installation.

9.62 An axial-flow turbine installation produces 136,000 hp under a water head of 45 ft. For an efficiency of 90%, determine the required rotational speed.

9.63 A Francis turbine installation produces 111 MW at a runner rotational speed of 167 rev/min. The water head to the gates is calculated to be 53.3 m. Determine what the water head should be from headwater to tailwater under maximum efficiency conditions.

9.64 A Francis turbine installation produces 172,000 hp under a total head of 162 ft. The rotational speed of the runner is 105.9 rev/min. Estimate the efficiency of the installation. (*Note*: The data of this problem are from Problem 9.34; the efficiency calculated in that problem, for purposes of comparison, is 95%.)

Pelton (or Impulse) Turbine

9.65 Plot a curve of the theoretical efficiency for an impulse turbine as a function of speed ratio U_t/V_1 for a blade angle of 165° (=blade angle at exit).

9.66 A water jet 6 cm in diameter at a velocity of 60 m/s drives a 1.5 m diameter impulse wheel at 360 rpm. The jet is deflected through an angle of 165°. How much power is delivered to the turbine?

9.67 An 8 cm diameter water jet moving at 70 m/s drives a 150 cm diameter impulse turbine at 720 rpm. How much power is delivered to the turbine if the jet is deflected through an angle of 170°?

9.68 The penstock for an impulse turbine is 4000 ft long and is 2 ft in diameter. The total head upstream of the turbine is 1200 ft. The pipe is made of cast iron. Graph flow rate through the nozzle versus power delivered. What is the jet diameter corresponding to maximum power?

9.69 Start with Equation 9.82 and derive the following equation:

$$\Delta z_{max} = 3\frac{f_0 L_0}{D_0}\frac{V_0^2}{2g}$$

9.70 Beginning with Equation 9.88, derive the following equation:

$$\eta_{max} = \frac{1}{2}\left(1 - \frac{\cos\beta_2}{\sqrt{1+f_b}}\right)$$

10 Measurements in Fluid Mechanics

Fluid measurements—including the measurement of viscosity, pressure, velocity, and flow rate—are the subject of this chapter. Viscosity is a property related to how a fluid reacts under the action of an applied shear; two methods for measuring viscosity are presented. Pressure, as we will see, is measured with a pitot and pitot-static tube. Velocity of the flow can be determined by using a pitot-static tube. Flow rate in closed conduits can be measured with meters; for open-channel flow, weirs can be used.

Although it is not discussed in detail here, the accuracy of measurement is critical in fluid mechanics. Consider, for example, a soft drink–bottling company. One such operation might bottle 100,000 gal of liquid (approximately 800,000 16 oz bottles) per year. Say, the company uses a meter in the main supply line that is usually accurate to within 2%. In that case, as much as 2000 gal of liquid a year may be sold without being paid for or paid for but never sold. Now multiply this amount by the number of soft drink– and liquor–bottling companies that use conventional in-line instrumentation. If we also include oil companies, gas stations, residential and industrial water meters, and so forth, we can conclude that a significant amount of liquid may go unaccounted for each day. Accuracy of instrumentation in fluid mechanics is indeed important.

10.1 MEASUREMENT OF VISCOSITY

Several devices are commercially available for measuring the viscosity of a fluid. These devices are commonly called **viscometers** or **viscosimeters**. Such meters contain the fluid and cause it to undergo a laminar motion by the imposition of a pressure drop or by the motion of a component. The laminar motion generated can usually be described by an analytic solution with which viscosity can be calculated.

The **rotating-cup viscometer**, a device used to measure viscosity, consists of two concentric cylinders. Liquid is placed in the annulus between the cylinders, and viscosity can be calculated with descriptive equations. A rotating-cup viscometer is illustrated in Figure 10.1. The outer cylinder or cup rotates at a constant rotational speed that is carefully controlled. The inner cylinder is held stationary. Torque is transmitted from the outer cylinder through the liquid to the inner cylinder. The torque required to hold this cylinder stationary is then measured. This measurement could be done by attaching a torsion wire to the inner cylinder and determining its deflection. The viscosity of the liquid is related to the torque exerted on the inner cylinder by Newton's law of viscosity:

$$\tau = \mu \frac{dV}{dr} = \mu \frac{V_\theta}{R_2 - R_1}$$

With $V_\theta = R_2\omega$, we have

$$\tau = \mu \frac{R_2\omega}{R_2 - R_1}$$

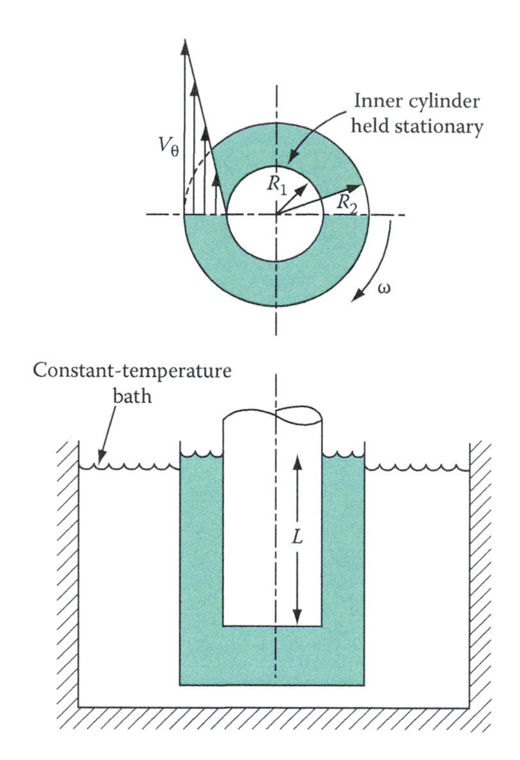

FIGURE 10.1 A rotating-cup viscometer.

The measured torque on the inner cylinder is related to the shear stress as

$$T_s = F_s \cdot R_1 = \tau(2\pi R_1 L)R_1$$

or

$$\tau = \frac{T_S}{2\pi R_1^2 L}$$

By substitution,

$$\frac{T_S}{2\pi R_1^2 L} = \mu \frac{R_2 \omega}{R_2 - R_1}$$

Solving, we get

$$\mu = \frac{T_S(R_2 - R_1)}{2\pi R_1^2 R_2 L \omega} \tag{10.1}$$

A second type of viscosity-measuring device is the **falling-sphere viscometer**. In this case, a sphere of known dimensions is dropped into a liquid medium. By determining the time required for the

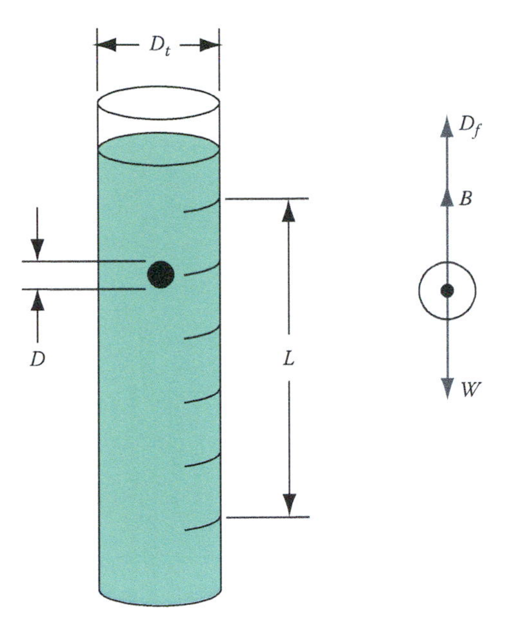

FIGURE 10.2 Falling-sphere viscometer and free-body diagram for the sphere.

sphere to fall through a certain interval, its terminal velocity can be calculated. The concept is illustrated in Figure 10.2 along with a free-body diagram for a sphere falling at terminal velocity. The forces acting are weight W, buoyancy B, and drag D_f:

$$W = \rho_S g \frac{\pi D^3}{6} \tag{10.2a}$$

$$B = \rho g \frac{\pi D^3}{6} \tag{10.2b}$$

$$D_f = C_D \frac{\rho V^2}{2} \frac{\pi D^2}{4} \tag{10.2c}$$

where

- ρ_S is the sphere density
- D is the sphere diameter
- ρ is the liquid density
- V $(=L/t)$ is the terminal velocity of the sphere falling through the liquid
- C_D is the drag coefficient determined from Figure 6.15 for flow past a sphere

Applying Newton's second law for a nonaccelerating sphere, we get

$$\sum F = \rho_S g \frac{\pi D^3}{6} - \rho g \frac{\pi D^3}{6} - C_D \frac{\rho V^2}{2} \frac{\pi D^2}{4} = 0 \tag{10.3}$$

If the sphere is sufficiently small in diameter, or if its density is not much greater than that of the liquid, it falls at a very low velocity. From Chapter 6, we know that if the Reynolds number of the sphere is less than 1, then Stokes flow exists. For this case,

$$C_D = \frac{24}{Re} = \frac{24\mu}{\rho VD} \quad \text{for Re} < 1$$

Therefore, Equation 10.2c becomes

$$D_f = \frac{24\mu}{\rho VD}\frac{\rho V^2}{2}\frac{\pi D^2}{4} = 3\pi\mu VD$$

and Equation 10.3 becomes

$$\rho_S g\frac{\pi D^3}{6} - \rho g\frac{\pi D^3}{6} - 3\pi\mu VD = 0$$

Rearranging and solving for viscosity, we have

$$\mu = \frac{g}{18}\frac{D^2}{V}(\rho_S - \rho) \tag{10.4}$$

Once viscosity is determined with this equation, the Reynolds number should be calculated to verify that it is indeed less than 1.

Equation 10.4 applies only to the case in which the sphere diameter is much smaller than the cylinder diameter. If this is not the case, then interference will exist between the wall itself and the falling motion of the sphere. Thus, the drag will be affected. To account for this wall effect, we modify the drag equation:

$$D_f = 3\pi\mu VD\zeta$$

where ζ represents a multiplication factor for the wall effect. The factor ζ has been determined analytically, and the results have been confirmed with experimental data. They are plotted in Figure 10.3 as ζ versus the ratio of the sphere diameter to the tube diameter. Equation 10.4 then becomes

$$\mu = \frac{g}{18}\frac{D^2}{V\zeta}(\rho_S - \rho) \tag{10.5}$$

Example 10.1

A commercially available shampoo is brought into the lab for viscosity measurement. A stainless steel sphere of diameter $\frac{3}{32}$ in. falls through the shampoo at a terminal velocity of 1 in. per 24.95 s. The tube holding the liquid has an inside diameter of 4 in. Determine the viscosity of the shampoo if its specific gravity is 0.998.

Solution

Equation 10.5 applies:

$$\mu = \frac{g}{18}\frac{D^2}{V\zeta}(\rho_S - \rho)$$

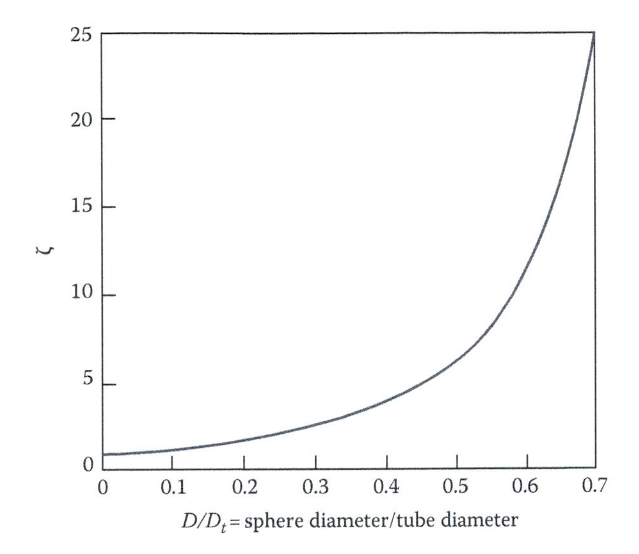

FIGURE 10.3 Wall correction factor for a falling-sphere viscometer.

The sphere diameter is $D = (3/32)(1/12) = 0.00781$ ft. The ratio of the sphere to the tube diameter is

$$\frac{D}{D_t} = \frac{3/32}{4} = 0.023$$

Figure 10.3 gives $\zeta \approx 1$. The sphere density from Table A.8 is

$$\rho_s = 8.02(1.94) = 15.6 \text{ slug/ft}^3$$

The liquid density is

$$\rho = 0.998(1.94) = 1.94 \text{ slug/ft}^3$$

The terminal velocity is

$$V = \left(\frac{1}{24.95}\right)\left(\frac{1}{12}\right) = 0.00334 \text{ ft/s}$$

By substitution, then,

$$\mu = \frac{32.2}{18}\frac{(0.00781)^2}{0.00334(1)}(15.6 - 1.94)$$

$$\mu = 0.445 \text{ lbf} \cdot \text{s/ft}^2$$

As a check, the Reynolds number is calculated as

$$\text{Re} = \frac{\rho VD}{\mu} = \frac{1.94(0.00334)(0.00781)}{0.445}$$

$$= 0.0001 < 1$$

Thus, Equation 10.5 is valid for this problem.

Another way to measure viscosity involves moving the fluid through a tube under laminar flow conditions. For a Newtonian fluid, the velocity profile (from Chapter 5) for laminar flow through a circular tube is

$$V_z = \left(-\frac{dp}{dz}\right)\frac{R^2}{4\mu}\left[1-\left(\frac{r}{R}\right)^2\right]$$

Integrating this equation over the tube cross-sectional area and dividing by the area gives the average velocity:

$$V = \left(-\frac{dp}{dz}\right)\frac{R^2}{4\mu} \simeq \frac{\Delta p}{L}\frac{R^2}{8\mu}$$

Solving for viscosity gives

$$\mu = \frac{\Delta p}{L}\frac{R^2}{8V} \tag{10.6}$$

The volume flow rate of the liquid through the tube is measured, and dividing by the area gives the average velocity V. The tube length L and inside radius R are easily measured. The pressure drop Δp over the distance L must also be measured. Once these quantities are known, the viscosity can be calculated with Equation 10.6.

10.2 MEASUREMENT OF STATIC AND STAGNATION PRESSURES

In fluid flow, there are two pressures of importance: static pressure and stagnation pressure. **Static pressure** is indicated by a measuring device that moves with the flow or by a device that causes no change in velocity within the stream. A conventional method for measuring static pressure is by means of a hole drilled through the flow conduit. The hole, which must be located normal to the wall surface, is then connected to a manometer or a pressure gauge. If it is not possible to use an existing wall, a probe that has a static pressure tap can be inserted into the flow to create an effective wall (Figure 10.4).

When measuring the static pressure in a pipe, it is often desirable or necessary to drill several static pressure taps and connect them with a tube that goes around the pipe. The tube is called a **piezometer ring**. This method will account for imperfections in the pipe wall (see Figure 10.5).

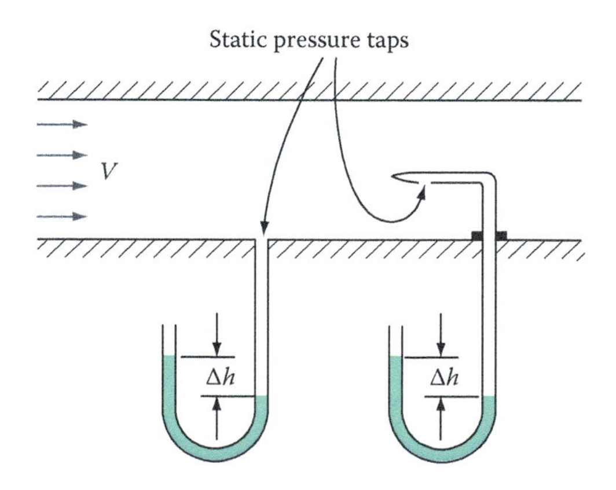

FIGURE 10.4 Methods for measuring static pressure of flow.

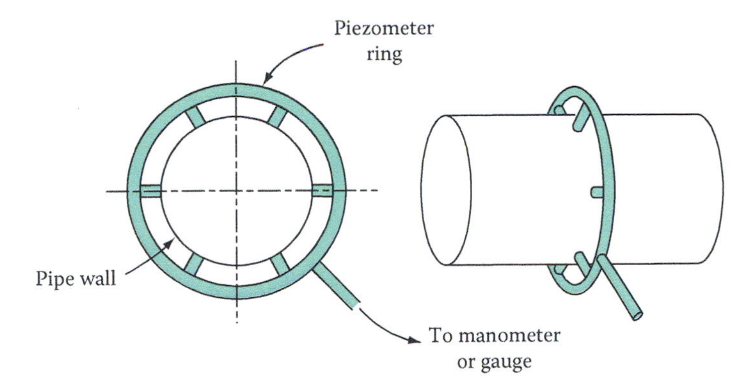

FIGURE 10.5 Piezometer ring connected to static pressure taps.

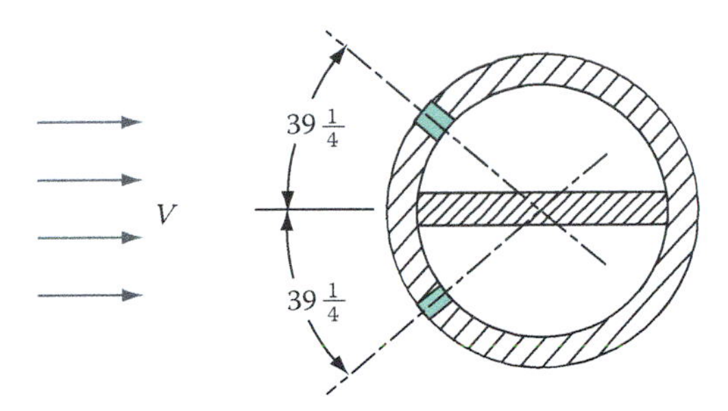

FIGURE 10.6 Direction-finding tube.

An application of the method for measuring static pressure is used in a device called a **direction-finding tube** (Figure 10.6). In two-dimensional or three-dimensional flows, the actual flow direction may not be known; a direction-finding tube will indicate it. As shown in Figure 10.6, the tube consists of two piezometer holes located at equal angles from the center. Each hole is connected to a separate manometer or gauge. The tube is then rotated in the flow until the readings on both manometers or gauges are identical. If the angle between the centerline and the holes is 39°, the indicated pressure is very close to that recorded by an instrument moving with the flow.

Stagnation pressure is the pressure indicated when bringing the flow to rest isentropically. A **pitot tube**, an open-ended tube facing the flow direction as shown in Figure 10.7, is used to measure stagnation pressure.

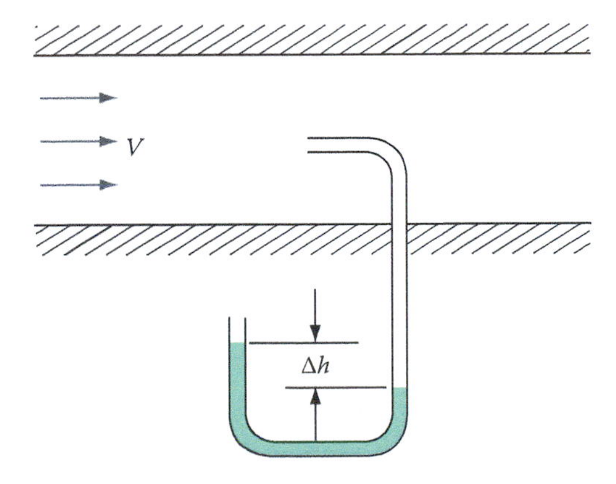

FIGURE 10.7 Measurement of stagnation pressure with a pitot tube.

10.3 MEASUREMENT OF VELOCITY

Velocity in a flow of fluid can be measured in several ways. The first method that we will discuss involves use of a **pitot-static tube**. This device consists of a tube within a tube that combines static pressure measurement with stagnation pressure measurement. A cross section of a pitot-static tube is given in Figure 10.8. Applying the Bernoulli equation to the static and stagnation holes in the tube gives

$$\frac{p_t}{\rho g} = \frac{p}{\rho g} + \frac{V^2}{2g}$$

Solving for velocity, we get

$$V = \sqrt{\frac{p_t - p}{\rho} 2}$$

The pressure difference thus gives the velocity of the flow. If the pitot-static tube is connected to opposing limbs of a differential manometer, then the reading Δh is

$$\Delta h = \frac{p_t - p}{\rho g}$$

In terms of the manometer reading (see Figure 10.8), the flow velocity is

$$V = \sqrt{2g\Delta h} \tag{10.7}$$

Example 10.2

Acetone flows through a pipe in which the velocity at the centerline is measured with a pitot-static tube. The attached manometer indicates a pressure drop of 5 cm of mercury, as shown in Figure 10.9. Calculate the velocity at the centerline.

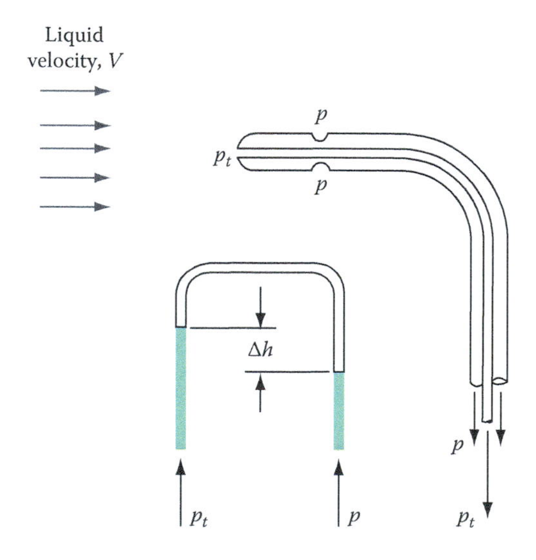

FIGURE 10.8 Cross section of a pitot-static tube and manometer connection.

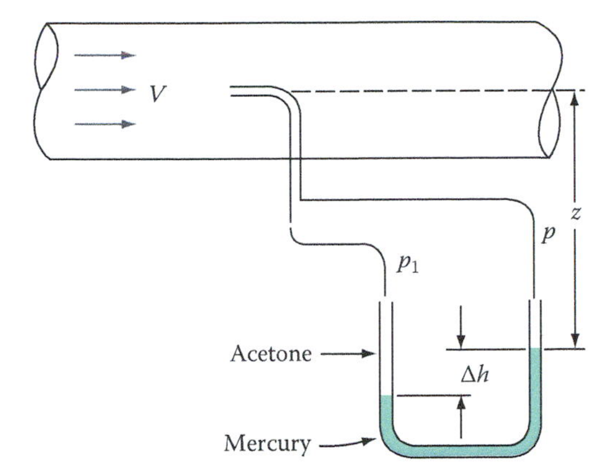

FIGURE 10.9 Measurement of velocity with a pitot-static tube (Example 10.2).

Solution

The derivation of Equation 10.7 was for an inverted, air-over-liquid, U-tube manometer. In this example, however, we have a slightly different method for measuring pressure difference, and so Equation 10.7 does not apply. Here, we are using a two-liquid manometer in a U-tube configuration. As shown in the figure, we define z as the vertical distance from the pitot-static tube to the mercury–acetone interface in the right leg of the manometer. Applying the hydrostatic equation to the manometer gives

$$p_t + \rho_a g(z + \Delta h) = p + \rho_a g_z + \rho_{\text{Hg}} g \Delta h$$

or

$$p_t - p = g\Delta h(\rho_{\text{Hg}} - \rho_a)$$

Next we write, for the pitot-static tube,

$$\frac{p_t}{\rho_a g} = \frac{p}{\rho_a g} + \frac{V^2}{2g}$$

which becomes

$$V = \sqrt{\frac{2(p_t - p)}{\rho_a}}$$

Substituting from the hydrostatic equation gives

$$V = \sqrt{2g\Delta h \frac{\rho_{\text{Hg}} - \rho_a}{\rho_a}} = \sqrt{2(9.81)(0.05)\frac{13.6(1\,000) - 0.787(1\,000)}{0.787(1\,000)}}$$

Solving,

$$V = 4.0 \text{ m/s}$$

Equation 10.7 is appropriate for velocity measurements in incompressible fluids. For a compressible fluid, a pitot tube can still be used, but consideration must be given to whether the flow is subsonic or supersonic. For subsonic flow, we apply the energy equation to a stagnation reading and to a static reading made independently. The energy equation written between the stagnation and static states for an adiabatic process with no work is

$$h_t = h + \frac{V^2}{2}$$

where h_t and h are the stagnation and static enthalpies, respectively. For an ideal gas with constant specific heats, this equation becomes

$$c_p T_t = c_p T + \frac{V^2}{2}$$

or

$$T_t = T\left(1 + \frac{V^2}{2c_p T}\right)$$

By definition,

$$c_p = \frac{\gamma R}{\gamma - 1}$$

where
- R is the gas constant
- γ is the ratio of specific heats (c_p/c_v)

Thus,

$$T_t = T\left[1 + \frac{V^2(\gamma - 1)}{2\gamma RT}\right]$$

Substituting from the definition of sonic velocity, $a^2 = \gamma RT$, we get

$$\frac{T_t}{T} = 1 + \frac{\gamma - 1}{2}\frac{V^2}{a^2} = 1 + \frac{\gamma - 1}{2}M^2 \quad (10.8)$$

where M is the Mach number of the flow, which in this case is less than 1. For an isentropic flow, we can write

$$\frac{p_t}{p} = \left(\frac{T_t}{T}\right)^{\gamma/(\gamma-1)}$$

Equation 10.8 now becomes

$$\frac{p_t}{p} = \left(1 + \frac{\gamma - 1}{2}M^2\right)^{\gamma/(\gamma-1)} \quad (10.9)$$

As was stated in Chapter 8, the values of p/p_t versus M are tabulated in Table D.1 for $\gamma = 1.4$. Use of the table will make calculations easier. By measuring the stagnation and static pressures in the flow, and by measuring the static temperature, the subsonic flow velocity can therefore be calculated with

$$V = Ma = M\sqrt{\gamma RT} \tag{10.10}$$

Example 10.3

The velocity of a commercial airliner is measured by a pitot tube attached near the front of the plane. The plane travels at an altitude of 30,000 ft, where the ambient conditions are $T = -26.2°F$ and $z = 9.38$ in. of mercury. The gauge attached to the pitot tube reads 7.0 psia. Determine the velocity.

Solution

The equivalent static pressure is calculated with the hydrostatic equation

$$p = \rho g z_{Hg} = 13.6(1.94)(32.2)\frac{(9.38)}{12} = 663.4 \text{ psfa} = 4.61 \text{ psia}$$

where ρ for mercury is obtained from Table A.5. For the pressure ratio

$$\frac{p}{p_t} = \frac{4.61}{7.0} = 0.658$$

Table D.1 gives $M = 0.80$. Assuming air as an ideal gas with constant specific heats, Equation 10.10 yields

$$V = M\sqrt{\gamma RT} = 0.8\sqrt{1.4(1710)(460-26.2)}$$

$$= 0.8(1021)$$

$$V = 816 \text{ ft/s} = 557 \text{ mi/h}$$

If the flow is supersonic, a shock wave will exist in front of the pitot probe, as in Figure 10.10. The flow velocity to be measured is upstream of the shock. Owing to the presence of the shock wave, the probe will exist in a subsonic flow, but because the shock is normal to the flow at the pitot tube, normal shock tables can be used. For the configuration of Figure 10.10, then, the pitot tube will indicate the stagnation pressure behind the shock, p_{t2}. Additionally, upstream static measurements of p_1 and T_1 are necessary to provide sufficient information to determine velocity. Table D.2 gives a tabulation of p_1/p_{t2} versus M_1 for a gas with $\gamma = 1.4$. The technique of measuring velocity thus involves using Table D.2 for convenience to determine M_1 and then substituting into Equation 10.10 to find V_1.

Example 10.4

A supersonic transport plane has a pitot tube attached to allow the pilot to determine when supersonic speed is reached. When the plane is flying at an altitude of 12 200 m (temperature is −52.1°C and pressure is 20.25 kPa), the gauge to which the pitot tube is attached reads 42 kPa. Determine the Mach number and absolute velocity of the plane.

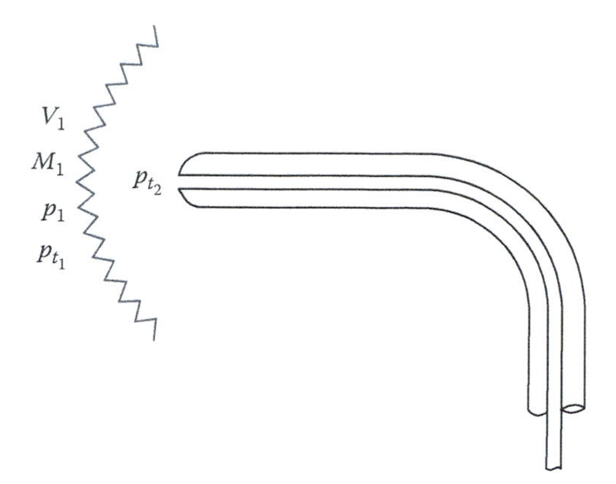

FIGURE 10.10 Pitot tube in a supersonic flow.

Solution

For the ratio

$$\frac{p_1}{p_{t_2}} = \frac{20.25}{42} = 0.482$$

Table D.2 gives $M_1 = 1.08$. After substitution into Equation 10.10, we obtain

$$V = Ma = M\sqrt{\gamma RT} = 1.08\sqrt{1.4(286.8)(273 - 52.1)}$$

$$V = 322 \text{ m/s}$$

Another device used to measure velocity in a gas is the *hot-wire anemometer*. This apparatus consists of a wire, generally made of platinum or tungsten, connected to two prongs as shown in Figure 10.11a. Electricity is passed through the wire and heats it. The heat in turn is transferred to the flowing fluid at a rate that is proportional to the flow velocity. If the current through the wire is maintained constant, the resistance of the wire can then be easily measured and the fluid velocity determined. If the wire temperature is maintained constant, the resistance will remain constant. In this case, the current is measured, and again the velocity can be determined. For either method of use, extensive calibration is required. One difference between a hot-wire

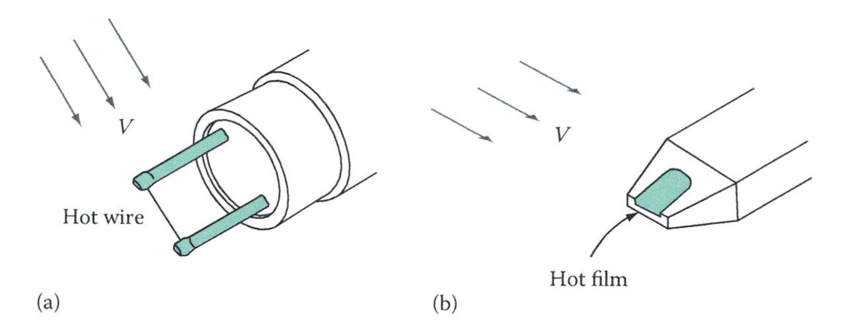

FIGURE 10.11 (a) A hot-wire anemometer used to measure flow velocity in a gas and (b) a hot-film anemometer used to measure flow velocity in a liquid.

anemometer and a pitot tube is the response time. Because the wire is thin (5×10^{-6} m characteristically), it transfers heat very quickly. Thus, sudden changes in velocity are sensed and displayed immediately by an appropriate electronic readout device; the velocity measured is the instantaneous velocity at a certain point. By comparison, a pitot-static tube is not as sensitive and indicates only a mean velocity.

A device used to measure velocity in a liquid is the **hot-film probe**. This apparatus (Figure 10.11b) works on the same principle as the hot-wire probe. (A hot wire is too delicate for use in a liquid.)

Another velocity-measuring device is the dual-beam **laser-Doppler anemometer** (LDA). This device passes two laser beams through the flow field. When particles in the fluid (liquid or gas) scatter the light at the intersection of the beams, a shift in the frequency occurs. The magnitude of this so-called Doppler shift is proportional to the flow velocity. One advantage of the LDA is that no physical disturbance of the flow occurs as with a pitot tube. One disadvantage is that the fluid must be transparent and contain impurities or particles that scatter the light. If no particles are present, the fluid must be doped. Perhaps the biggest disadvantage of the LDA is its cost. There are other devices that are useful for measuring velocity, from floats or buoyant particles to rotating anemometers, depending on the application.

10.4 MEASUREMENT OF FLOW RATES IN CLOSED CONDUITS

Readings from a pitot-static tube can be used to calculate the volume flow rate in a closed conduit. Because the pitot-static tube gives a measurement of velocity at selected points, the velocity profile can then be plotted and graphically integrated over the cross-sectional area to obtain the volume flow rate through the conduit. The integration procedure might present some difficulty, however, so an alternative method has been devised. Consider the circular duct of Figure 10.12. It is divided into five circular concentric equal areas (A_1 through A_5):

$$A = A_1 + A_2 + A_3 + A_4 + A_5 = 5A_1 \tag{10.11}$$

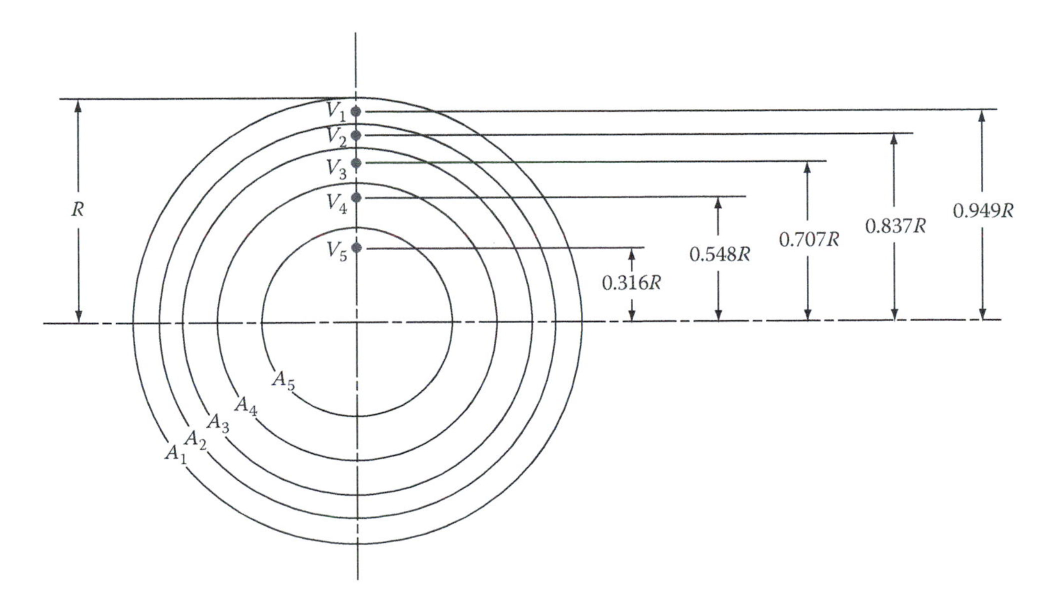

FIGURE 10.12 Circular cross section divided into five equal areas.

The velocity within each concentric area is presumed to be constant in that area. The velocity in each area must be measured with any suitable device at the prescribed point, or the velocity at these locations can be scaled from the velocity profile. The flow rate through each area is then

$$Q_1 = A_1V_1$$

$$Q_2 = A_2V_2$$

$$Q_3 = A_3V_3$$

$$Q_4 = A_4V_4$$

$$Q_5 = A_5V_5$$

The total volume flow rate through the conduit is

$$\begin{aligned} Q &= Q_1 + Q_2 + Q_3 + Q_4 + Q_5 \\ &= A_1V_1 + A_2V_2 + A_3V_3 + A_4V_4 + A_5V_5 \\ Q &= A_1(V_1 + V_2 + V_3 + V_4 + V_5) \end{aligned}$$

By combining with Equation 10.11, the volume flow rate is

$$Q = \frac{A}{5}(V_1 + V_2 + V_3 + V_4 + V_5)$$

from which we conclude that the average velocity can be found by taking the arithmetic average of the five measured values:

$$V = \frac{Q}{A} = \frac{V_1 + V_2 + V_3 + V_4 + V_5}{5} \tag{10.12}$$

A similar analysis can be developed for flows through a rectangular or square ducts where the velocities must be known at selected points in the cross section. Such points are shown in Figure 10.13. The number of equal areas chosen for the cross section varies with its dimensions. Generally, the number of readings should be between 16 and 64 at the center of equal rectangular areas. The velocities are then averaged. The product of the average velocity and the cross-sectional area gives the flow rate.

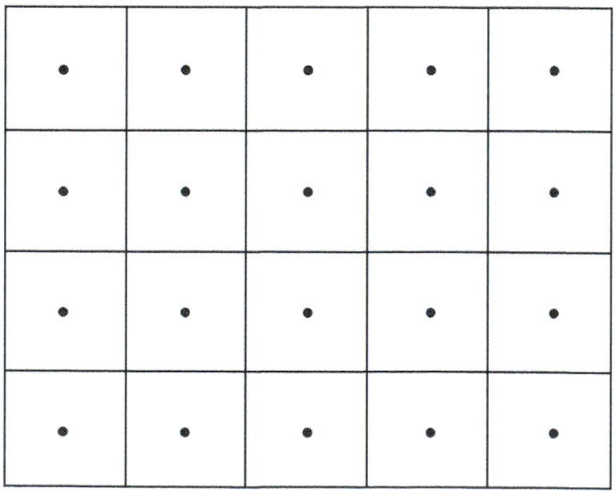

FIGURE 10.13 Division of a rectangular cross section into a number of equal areas.

Example 10.5

In a liquor-bottling plant, ethyl alcohol is piped to a mixing tank through a pipe. A pitot-static tube is used to measure a velocity profile in the pipe (24-nominal, schedule 20 stainless steel). The data are given in the following table.

Distance from Centerline (in.)	Pitot-Static Tube Reading (in. Ethyl Alcohol)
0	11.9
1	11.6
2	11.3
3	11.0
4	10.8
5	10.5
6	10.1
7	9.8
8	9.1
9	6.5
10	2.9
11	0.7

a. Sketch the velocity profile to scale.
b. Estimate the volume flow rate through the pipe.

Solution

a. Equation 10.7 applied to each reading gives the results in the following table. A sample calculation for 0 in. is

$$V = \sqrt{2g\Delta h} = \sqrt{2(32.2)(11.9/12)} = 8.0 \text{ ft/s}$$

Distance from Centerline (in.)	Velocity (ft/s)
0	8.0
1	7.9
2	7.8
3	7.7
4	7.6
5	7.5
6	7.4
7	7.25
8	7.0
9	5.9
10	3.9
11	1.9

From Table C.1, for a 24-nominal, schedule 20 pipe,

$$D = 1.938 \text{ ft} \quad \text{and} \quad A = 2.948 \text{ ft}^2$$

At the wall, R = 1.938 ft/2 = 11.63 in., at which distance the velocity is zero. A plot of the data is given in Figure 10.14.

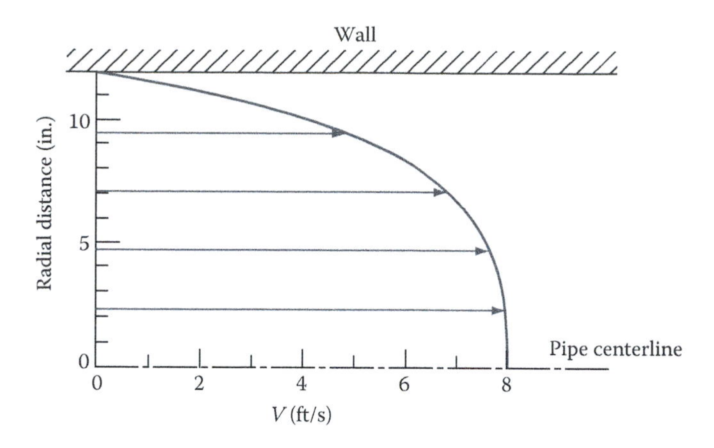

FIGURE 10.14 Velocity profile for ethyl alcohol in a pipe (Example 10.5).

b. To calculate the volume flow rate, we can use the velocity profile to estimate the average velocity. From Figure 10.12, the velocity must be known at the following locations (R = 11.63 in.):

$0.316R = 3.68 \text{ in.} \rightarrow A_5$

$0.548R = 6.37 \text{ in.} \rightarrow A_4$

$0.707R = 8.22 \text{ in.} \rightarrow A_3$

$0.837R = 9.73 \text{ in.} \rightarrow A_2$

$0.949R = 11.03 \text{ in.} \rightarrow A_1$

From the velocity diagram, the corresponding velocities are

$V_1 = 1.7 \text{ ft/s}$

$V_2 = 5 \text{ ft/s}$

$V_3 = 6.7 \text{ ft/s}$

$V_4 = 7.3 \text{ ft/s}$

$V_5 = 7.7 \text{ ft/s}$

The average velocity is

$$V = \frac{\sum V}{5} = 5.68 \text{ ft/s}$$

The volume flow rate is

$$Q = AV = 2.948(5.68)$$

$$Q = 16.7 \text{ ft}^3/\text{s}$$

Using a pitot-static tube to measure velocity can be quite cumbersome and sometimes—in the case of small tubes, for example—impossible. Fortunately, alternatives exist. Fluid meters specifically made for measuring flow rate have been proposed, tested, and developed to a high degree. Flow meters are of two types: those that measure quantity and those that measure rate. Measurements of quantity (either volume or mass) are obtained by counting successive isolated portions of flow. Rate measurements, on the other hand, are determined from the effects on a measured physical property of the flow—for example, the pressure drop within a meter related to the flow rate through it. A quantity meter reading can be modified to obtain a flow rate.

One example of a quantity-measuring device is the **nutating disk meter** illustrated in Figure 10.15. This meter is used extensively for metering cold water in service lines, both domestic and commercial. The nutating disk meter contains a metering chamber with spherical sides, a conical roof, and a radial baffle. A disk that passes through a sphere inside divides the chamber in two. A shaft attached to the sphere extends upward. As liquid flows through the meter alternately above and below the disk, the disk wobbles or nutates, thus moving the sphere and causing the shaft to generate the shape of an inverted cone. This circular motion of the shaft drives a counter that registers the total quantity of flow passing through. For gases, a **bellows meter** is used for both domestic and commercial applications. The meter contains a bellows that alternately fills and empties. The bellows is mechanically linked to a counter.

The remainder of this section is devoted to a discussion of the common types of rate meters. **Rate meters** are devices used to measure flow rate as volume per unit time or as mass per unit time. One of the most accurate commercially available types is the **turbine-type flow meter**. As shown in Figure 10.16, it consists of a pipe or tube (brass or stainless steel is commonly used) with appropriate pipe fittings. Inside the tube are flow straighteners on either side of a small propeller or turbine. Flow through the tube rotates the propeller at an angular velocity that is proportional to the flow rate.

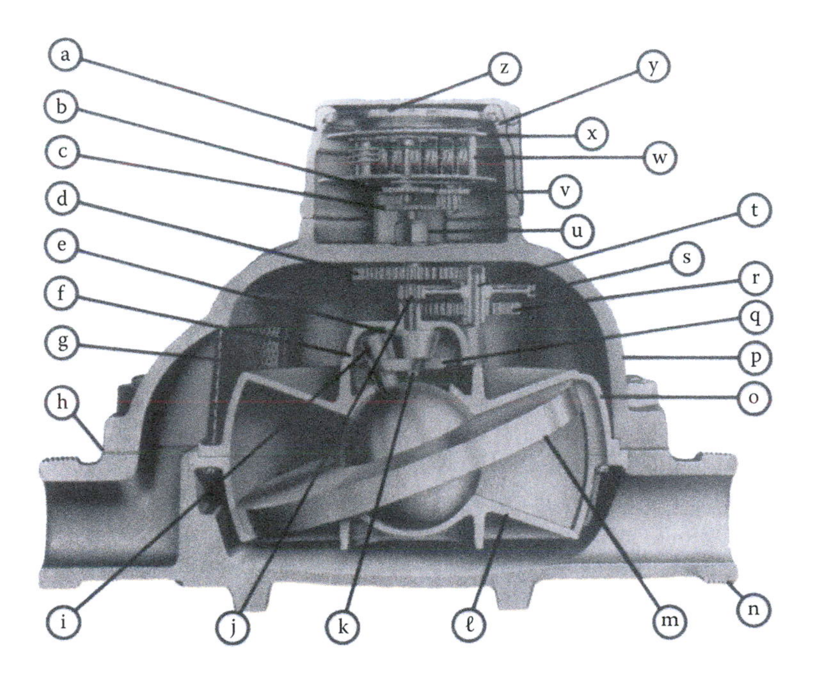

FIGURE 10.15 A nutating disk meter. (a) Register box assembly, (b) Register change gear, (c) Meter change gear, (d) Top gear and shaft, (e) Control roller, (f) Gear plate, (g) Strainer, (h) Casing gasket, (i) Disk pin, (j) Inter pinion and shaft, (k) Hex nut, driver block, (l) Bottom chamber, (m) Disk assembly, (n) Bottom casing, (o) Top chamber, (p) Top casing, (q) Driver block, (r) Inter gear 2, (s) Inter gear 1, (t) Pivot, (u) Nut, stuffing box, (v) Idler, change gear, (w) Register, (x) Register, dial face, (y) Retainer, glass, (z) Glass, register box. (Courtesy of Hersey Products, Inc., Cleveland, NC.)

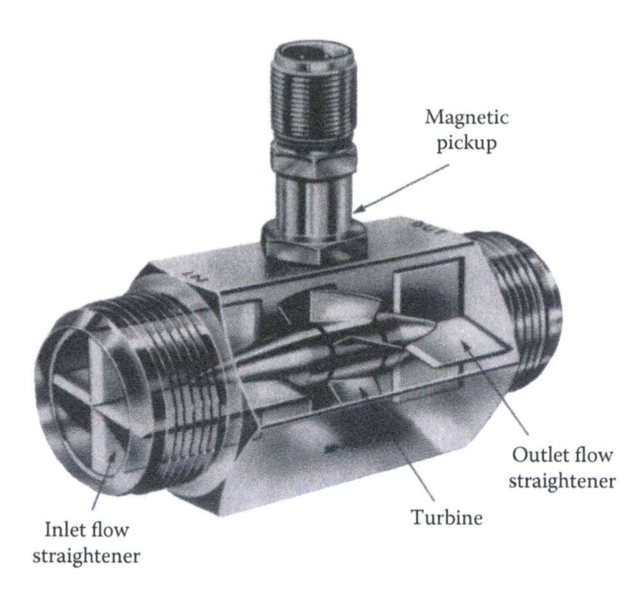

FIGURE 10.16 A turbine-type flow meter. (Courtesy of Flow Technology, Inc., Roscoe, IL.)

A magnetic pickup registers blade passages and sends a signal to a readout device that totals the pulses. Turbine-type meters are usually accurate to within 1%.

A flow can be metered by what is known as a variable-area meter, or a **rotameter**. A rotameter contains a float that is free to move within a vertical tapered glass or transparent plastic tube. The tube is etched or marked with a scale. Flow enters the meter at the bottom, raising the float within the tube (see Figure 10.17). The higher the float position, the larger the annular flow area between the float and the tube. The float reaches an equilibrium position where forces due to drag, buoyancy, and gravity are all balanced. The flow rate is determined by reading the scale at the float position. Alternatively, a sensing device can transmit a signal to a remote location. Of the direct reading types, scales are available for flow rate metering for a specific fluid and for determining the percentage of full scale; the latter requires calibration by the user. Floats are usually either spherical or cylindrical. Variable-area meters can be obtained for liquids or for gases. Accuracy is usually within 1% on expensive units and within 5% on less expensive rotameters.

Another type of rate meter introduces a flow constriction that affects a measurable property of the flow. The changed property is then related to the flow rate through the meter. The **venturi meter** is an example of this type (Figure 10.18). Basically a casting lined with a corrosion-resistant material such as bronze, this device is placed directly in the flow line. The meter consists of an upstream section that attaches to the pipeline and a convergent section that leads to a constriction or a smaller-diameter pipe. A divergent section is then connected downstream to the pipeline. Upstream and at the throat are attached static pressure taps or piezometer rings. These are connected to the two limbs of a differential manometer. It is considered good practice to provide at least 10 diameters of approach piping before the meter to ensure that a uniform, fully developed flow exists at the meter entrance. The size of a constriction meter such as a venturi is usually specified by the pipe and throat sizes. For example, an 8 × 4 venturi attaches to an 8-nominal pipe and has a throat diameter corresponding to a 4-nominal pipe.

The continuity equation can be applied to flow through the venturi meter at points 1 and 2, where the pressure change is measured. The objective here is to relate the affected property (pressure drop) to the flow through the meter. For an incompressible fluid, the continuity equation is

$$Q = A_1V_1 = A_2V_2 \tag{10.13}$$

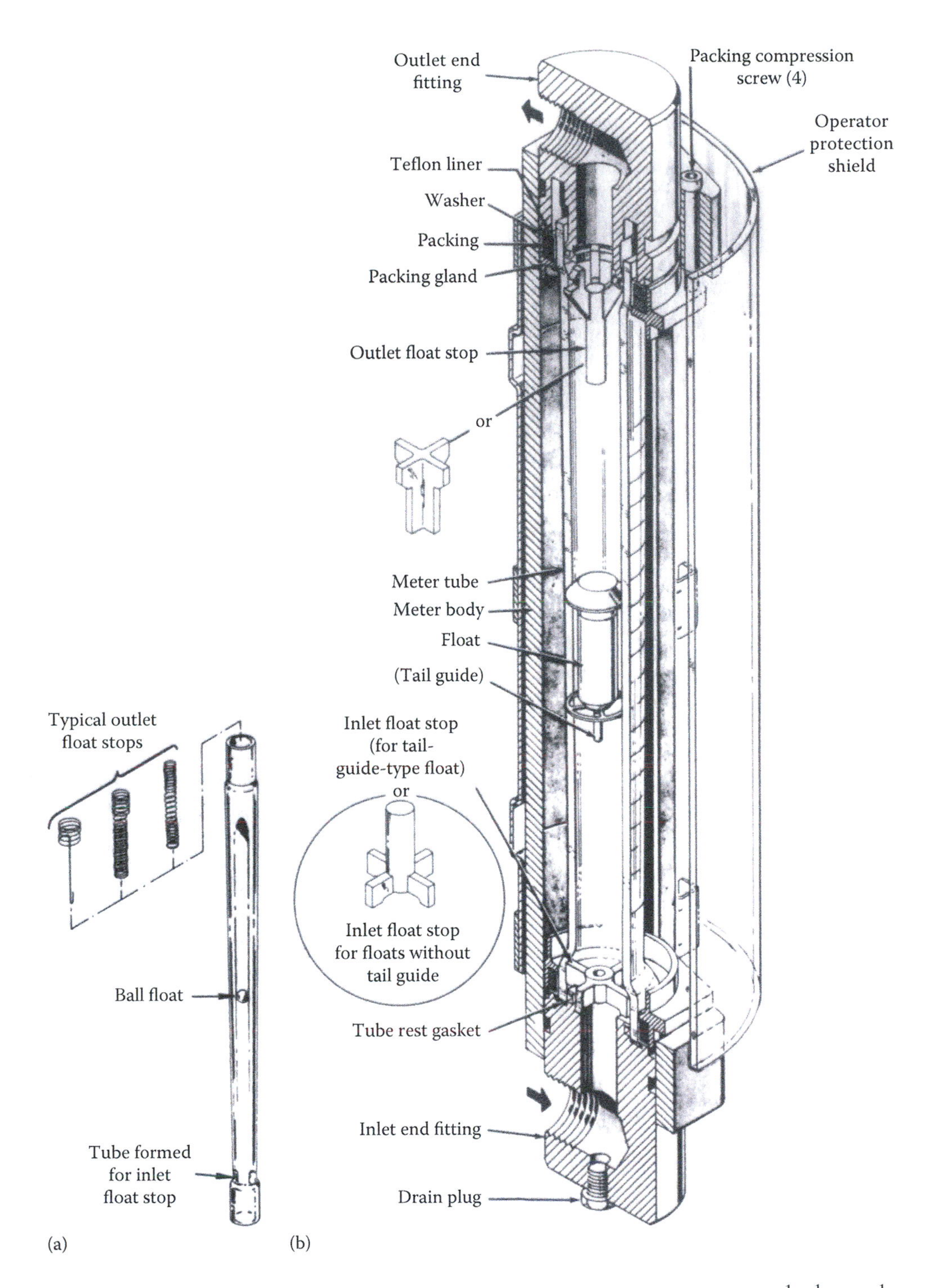

FIGURE 10.17 A rotameter. (a) Tri-flat tube: 5 in. scale length used in smaller frame $\frac{1}{16}$, $\frac{1}{8}$, and $\frac{1}{4}$ = in. meters only. (b) Beadguide tube: cutaway view of a typical meter. (Courtesy of ABB/Fischer and Porter Co., Warminster, PA.)

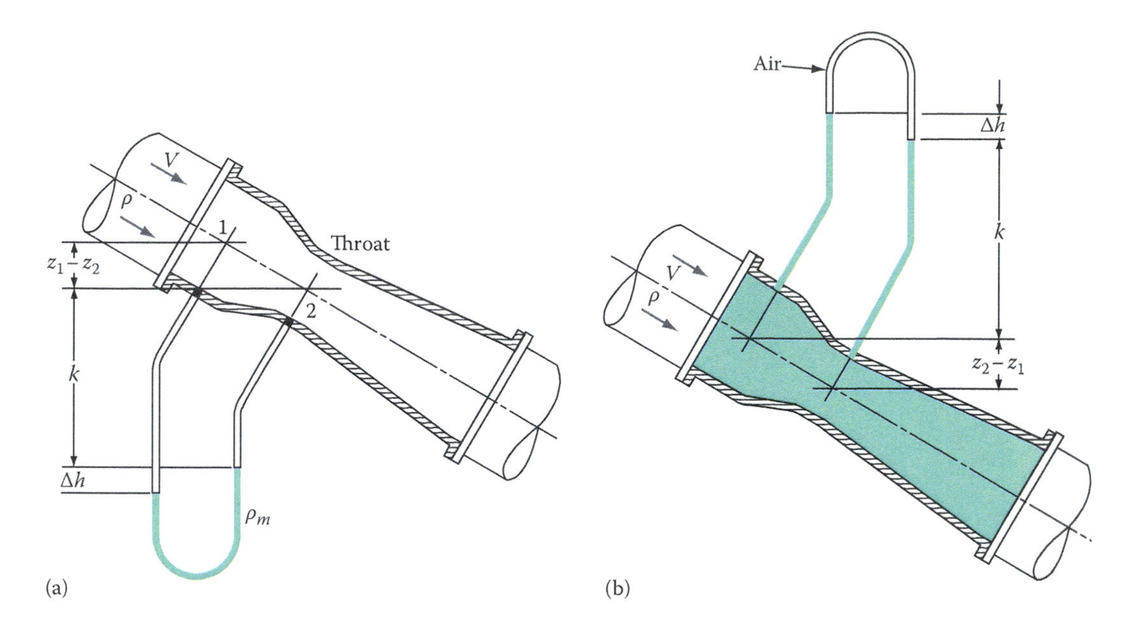

FIGURE 10.18 (a) A venturi meter with a two-liquid U-tube manometer and (b) a venturi meter with an inverted U-tube manometer.

Because the area A_2 is less than A_1, the continuity equation predicts that $V_1 < V_2$. In other words, the flow velocity must increase at the throat. The Bernoulli equation for a frictionless flow through the meter is

$$\frac{p_1}{\rho g} + \frac{V_1^2}{2g} + z_1 = \frac{p_2}{\rho g} + \frac{V_2^2}{2g} + z_2 \tag{10.14}$$

Because $V_1 < V_2$, then $p_1 > p_2$; that is, a pressure drop exists in the meter from upstream to the throat. Substituting from Equation 10.13 for velocity, Equation 10.14 becomes, after rearranging,

$$\frac{p_1 - p_2}{\rho g} + z_1 - z_2 = \frac{Q^2}{2g}\left(\frac{1}{A_2^2} - \frac{1}{A_1^2}\right)$$

$$= \frac{Q^2}{2gA_2^2}\left(1 - \frac{A_2^2}{A_1^2}\right)$$

Solving for flow rate, we get

$$Q = A_2\sqrt{\frac{2g\{[(p_1 - p_2)/\rho g] + (z_1 - z_2)\}}{1 - \left(A_2^2/A_1^2\right)}}$$

This equation is the theoretical equation for the venturi meter written for a frictionless incompressible flow. Noting that $A_2/A_1 = D_2^2/D_1^2$ and introducing the subscript "th" to denote a theoretical flow, we have

$$Q_{th} = A_2\sqrt{\frac{2g\{[(p_1 - p_2)/\rho g] + (z_1 - z_2)\}}{1 - \left(D_2^4/D_1^4\right)}} \tag{10.15}$$

The manometer reading provides the pressure drop required in the equation. From hydrostatics, we obtain the following for the manometer of Figure 10.18a:

$$p_1 + \rho g[(z_1 - z_2) + k + \Delta h] = p_2 + \rho g k + \rho_m g \Delta h$$

After rearrangement and simplification, we get

$$\frac{p_1 - p_2}{\rho g} + z_1 - z_2 = \Delta h \frac{\rho_m - \rho}{\rho} = \Delta h\left(\frac{\rho_m}{\rho} - 1\right) \quad (10.16)$$

Substitution into Equation 10.15 yields

$$Q_{\text{th}} = A_2 \sqrt{\frac{2g\Delta h(\rho_m/\rho - 1)}{1 - \left(D_2^4/D_1^4\right)}} \quad \text{[Figure 10.18a]} \quad (10.17a)$$

Thus, the theoretical flow rate through the meter is related to the manometer reading in such a manner that the meter orientation is not important; the same equation results whether the meter is horizontal, inclined, or vertical.

For the manometer of Figure 10.18b, we write

$$p_1 + \rho g(k + \Delta h) = p_2 + \rho g[(z_2 - z_1) + k] + \rho_{\text{air}} g \Delta h$$

The density of air is very small in comparison to the liquid density, so the term containing ρ_{air} can be neglected. After rearrangement and simplification, we obtain

$$\frac{p_1 - p_2}{\rho g} + z_1 - z_2 = \Delta h$$

Substitution into Equation 10.15 yields

$$Q_{\text{th}} = A_2 \sqrt{\frac{2g\Delta h}{1 - \left(D_2^4/D_1^4\right)}} \quad \text{[Figure 10.18b]} \quad (10.17b)$$

This equation differs from Equation 10.17a in that Equation 10.17a contains the term $(\rho_m/\rho - 1)$ because there are two fluids in the manometer of Figure 10.18a.

For a given combination of meter (D_1, D_2, and A_2 known), liquid, and manometer fluid (ρ and ρ_m known), a curve of pressure drop Δh versus flow rate Q_{th} can be plotted by using Equation 10.17. Consider the line labeled Q_{th} in Figure 10.19 as such a curve. Next suppose that measurements are made on the meter over a wide range of flows and that the actual data (Q_{ac} versus Δh) are plotted again in Figure 10.19, yielding the line labeled Q_{ac}. Thus, for any pressure drop Δh_i,

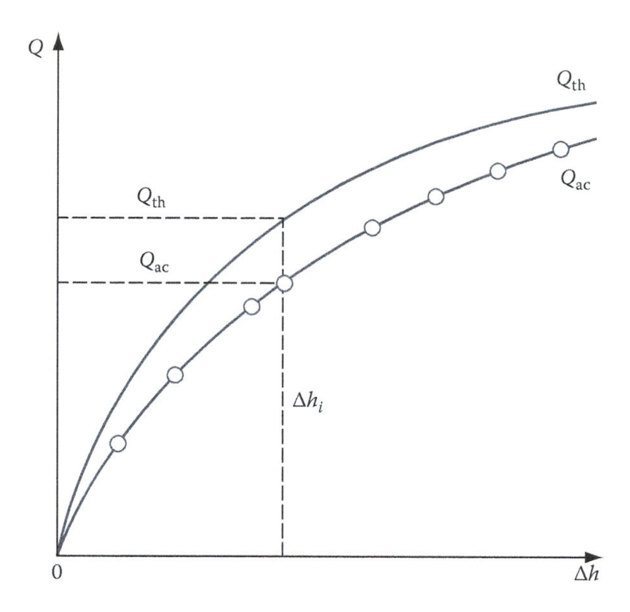

FIGURE 10.19 Flow rate versus pressure drop for a venturi meter.

there correspond two flow rates: Q_{ac} and Q_{th}. The ratio of these rates is called the **venturi discharge coefficient**, C_V:

$$C_V = \frac{Q_{ac}}{Q_{th}} \quad \text{for any } \Delta h_i \tag{10.18}$$

The coefficient C_V can be calculated for many Δhs on the plot and will vary over the range. The fact that Q_{ac} is different from Q_{th} is due to frictional effects that are not accounted for in the Bernoulli equation from which the expression for Q_{th} was obtained. For each C_V that can be determined, a corresponding upstream Reynolds number can be calculated:

$$\text{Re}_1 = \frac{V_1 D_1}{\nu} = \frac{4Q_{ac}}{\pi D_1^2}\frac{D_1}{\nu}$$

or

$$\text{Re}_1 = \frac{4Q_{ac}}{\pi D_1 \nu} \tag{10.19}$$

A graph of discharge coefficient C_V versus Reynolds number Re_1 can now be constructed. For venturi meters, the plot of Figure 10.20 applies. Such plots are available in fluid meter handbooks along with design or configuration recommendations for venturi installations. The plot of Figure 10.20 can be used with an uncalibrated meter to generate a Q_{ac} versus Δh curve from a Q_{th} versus Δh curve.

Example 10.6

An oil pipeline company is responsible for pumping kerosene overland from a refinery to a wholesale distributor. The company wants to install a meter in the line to check on the meter already installed. A 10 × 6 venturi is available but without a calibration

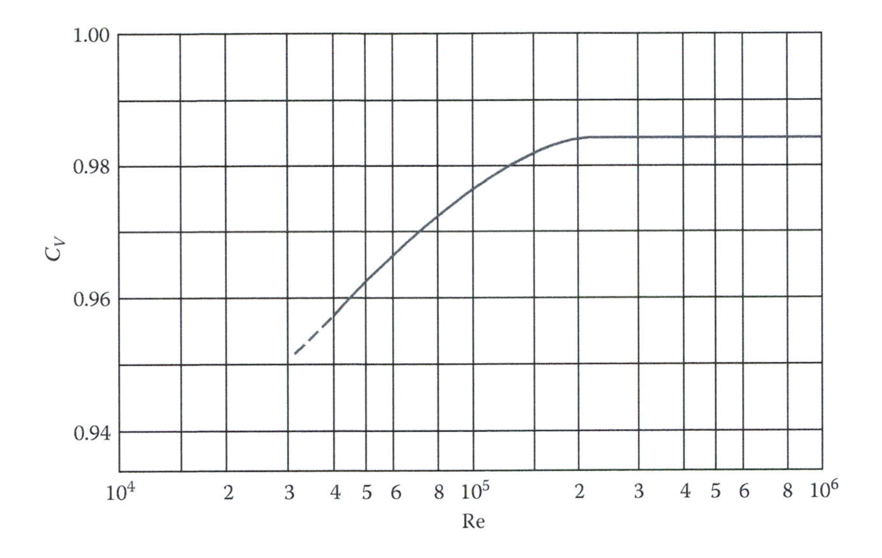

FIGURE 10.20 Discharge coefficient versus Reynolds number for venturi meters. (Reprinted from The American Society of Mechanical Engineers, *Fluid Meters—Their Theory and Application*, 5th edn., The American Society of Mechanical Engineers, New York, 1959. With permission.)

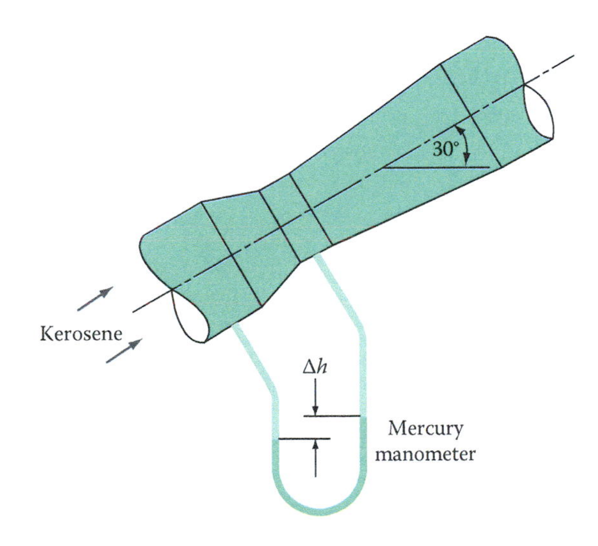

FIGURE 10.21 The venturi meter of Example 10.6.

curve—that is, without a plot of Q_{ac} versus Δh. Furthermore, facilities are not available for calibrating it. A mercury manometer is also on hand. The installation of the meter requires that it be inclined at 30°. A sketch is given in Figure 10.21. Generate a calibration curve for the meter up to a flow rate of 6 ft³/s.

Solution

From Table A.5,

Kerosene: $\rho = 0.823(1.94) = 1.60 \text{ slug/ft}^3$

$\mu = 3.42 \times 10^{-5} \text{ lbf} \cdot \text{s/ft}^2$

Mercury: $\rho_{Hg} = 13.6(1.94) = 26.4 \text{ slug/ft}^3$

For the manometer,

$$\frac{(p_1 - p_2)}{\rho g} + z_1 - z_2 = \Delta h\left(\frac{\rho_{Hg}}{\rho} - 1\right)$$

$$= \Delta h\left(\frac{13.6}{0.823} - 1\right) = 15.52\Delta h$$

From Table C.1,

$D_1 = 0.8350$ ft, $A_1 = 0.5476$ ft² (10-nominal, schedule 40; standard assumed)

$D_2 = 0.5054$ ft, $A_2 = 0.2006$ ft² (6-nominal, schedule 40)

By substitution into Equation 10.17, we get

$$Q_{th} = A_2\sqrt{\frac{2g\Delta h(\rho_{Hg}/\rho - 1)}{1-(D_2^4/D_1^4)}}$$

$$= 0.2006\sqrt{\frac{2(32.2)(15.52)\Delta h}{1-(0.5054^4/0.8350^4)}}$$

or

$$Q_{th} = 6.82\sqrt{\Delta h} \quad \text{(i)}$$

Let the manometer deflection Δh vary from 0 to 1 ft of mercury in increments of 0.2 ft. A tabulation of Q_{th} versus Δh using Equation i is given in Table 10.1.

The Reynolds number for the flowing kerosene is found with Equation 10.19:

$$\text{Re}_1 = \frac{4Q_{ac}}{\pi D_1 \nu}$$

$$= \frac{4(C_v Q_{th})(1.60)}{\pi(0.8350)(3.42\times 10^{-5})}$$

or

$$\frac{\text{Re}_1}{C_V} = 7.13\times 10^4 Q_{th}$$

Results of calculations made with this equation are given in the third column of Table 10.1.

TABLE 10.1

Summary of Calculations for Example 10.6

Δh (ft)	Q_{th} (Equation i) (ft³/s)	Re_1/C_V	C_V	$Q_{ac} = C_V Q_{th}$ (ft³/s)
0.0	0	0	—	0
0.2	3.04	2.16×10^5	0.984	2.99
0.4	4.31	3.06×10^5	0.984	4.24
0.6	5.28	3.75×10^5	0.984	5.19
0.8	6.10	4.34×10^5	0.984	6.00
1.0	6.82	4.85×10^5	0.984	6.71

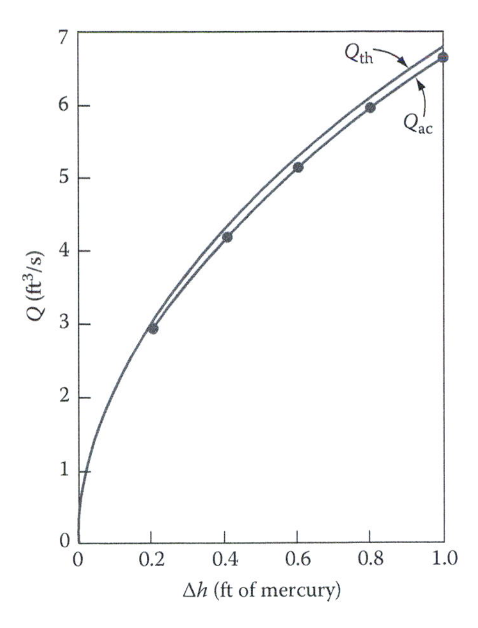

FIGURE 10.22 Calibration curve for the venturi meter of Example 10.6.

To find C_V from Figure 10.20 for each Re_1/C_V value, it is necessary to resort to a trial-and-error procedure if the range of interest lies below a Reynolds number of 2×10^5. But because the range over which our meter operates is greater than 2×10^5, Figure 10.20 shows that C_V is a constant equal to 0.984. This value is listed in Table 10.1. A plot of Q_{th} and Q_{ac} as a function of Δh is given in Figure 10.22, and the calibration curve for the meter is known.

It can be concluded that the venturi meter is a well-designed instrument because frictional effects and losses in general are small.

Another type of constriction meter is called a flow nozzle (Figure 10.23). The flow nozzle can be installed easily by cutting the pipe, attaching flanges, and inserting the nozzle. As liquid passes through, a region of flow separation and reversal exists just downstream, and this adds to the losses encountered in a flow nozzle. The Bernoulli equation can be applied to the flow nozzle as was done for the venturi. The results are identical for the theoretical flow rate. For the flow nozzle, then, we get

$$Q_{th} = A_2\sqrt{\frac{2g(p_1 - p_2)}{\rho g(1 - D_2^4/D_1^4)}} = A_2\sqrt{\frac{2g\Delta h(\rho_m/\rho - 1)}{1 - (D_2^4/D_1^4)}} \tag{10.20}$$

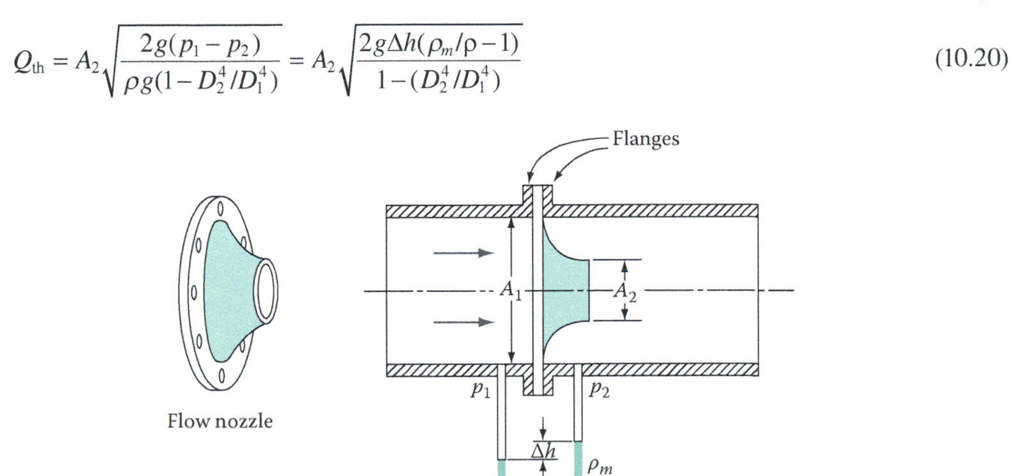

FIGURE 10.23 Flow nozzle and a conventional installation method.

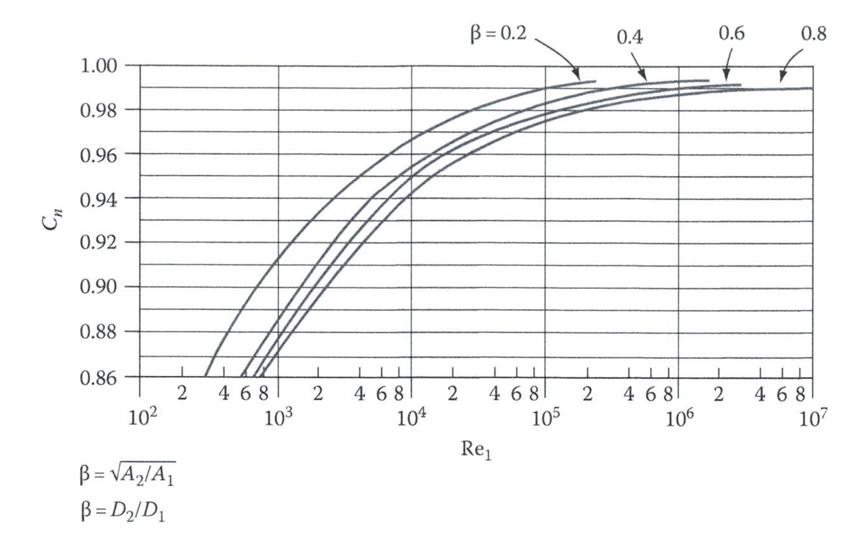

FIGURE 10.24 Discharge coefficient versus Reynolds number for flow nozzles. (Reprinted from The American Society of Mechanical Engineers, *Fluid Meters—Their Theory and Application*, 5th edn., The American Society of Mechanical Engineers, New York, 1959. With permission.)

Elevation differences are usually negligible in this case because the flow nozzle is relatively short and a small distance thus exists between the static pressure taps. It is desirable in practice to provide an upstream approach length of 10 diameters to ensure uniform flow at the meter.

Equation 10.20 gives the theoretical flow rate through the meter. Owing to losses, however, the actual flow rate is less. We therefore introduce a discharge coefficient for the nozzle defined as

$$C_n = \frac{Q_{ac}}{Q_{th}} \tag{10.21}$$

As with the venturi meter, experimental data are collected; for each value of C_n, there corresponds one pressure drop. The reduced data can be plotted as C_n versus Re_1, where $Re_1 = V_1D_1/\nu = 4Q_{ac}/\pi D_1\nu$. Tests on a series of flow nozzles have yielded the results of Figure 10.24. The variable β is the ratio of the throat diameter to the pipe diameter.

Example 10.7

Ethylene glycol is sold in retail outlets as antifreeze. A flow nozzle is to be installed in an ethylene glycol–piping system (a 12-nominal pipeline). A 12 × 6 flow nozzle is available. When the nozzle is installed with an inverted U-tube manometer, the pressure drop is measured as 42 cm of ethylene glycol, as illustrated in Figure 10.25. Determine the actual flow rate through the line.

Solution

From Table A.5, for ethylene glycol,

$$\rho = 1.1(1\ 000)\ \text{kg/m}^3$$

$$\mu = 16.2 \times 10^{-3}\ \text{N} \cdot \text{s/m}^2$$

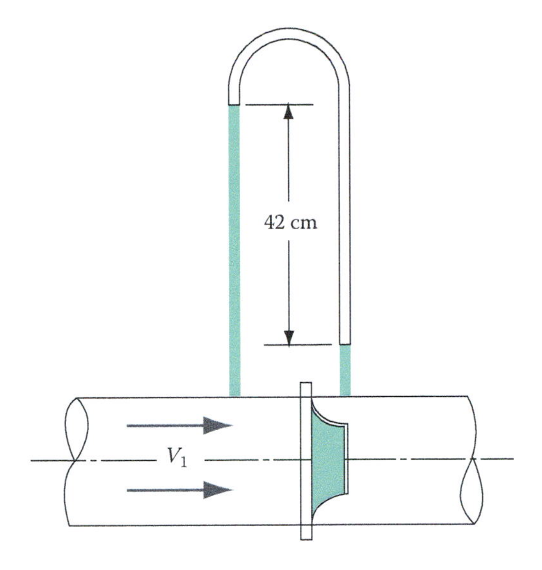

FIGURE 10.25 Flow nozzle installation of Example 10.7.

From Table C.1, for pipe,

$D_1 = 30.48$ cm, $A_1 = 729.7$ cm^2 (12-nominal standard schedule)

$D_2 = 15.41$ cm, $A_2 = 186.5$ cm^2 (6-nominal, schedule 40)

Equation 10.20 gives the theoretical flow rate through the nozzle as

$$Q_{th} = A_2\sqrt{\frac{2g(p_1 - p_2)}{\rho g(1 - D_2^4/D_1^4)}}$$

With

$$\frac{p_1 - p_2}{\rho g} = \Delta h = 0.42 \text{ m}$$

we get the following after substitution:

$$Q_{th} = \frac{186.5}{10^4}\sqrt{\frac{2(9.81)(0.42)}{1-(15.41^4/30.48^4)}}$$

$$= 0.055\,4 \text{ m}^3/\text{s}$$

The nozzle diameter ratio is

$$\beta = \frac{D_2}{D_1} = \frac{15.41}{30.48} = 0.506$$

The pipe Reynolds number is

$$\mathrm{Re}_1 = \frac{4Q_{ac}}{\pi D_1 \nu} = \frac{4Q_{ac}(1.1)(1\ 000)}{\pi(0.3048)(16.2\times10^{-3})}$$

$$= 2.83\times10^5 Q_{ac}$$

At this point, we must use Figure 10.24 to find a discharge coefficient. A trial-and-error procedure is required because Q_{ac} is unknown. If Q_{th} is substituted into the Reynolds number expression, we find

$$\mathrm{Re}_1 = 2.83\times10^5(0.055\ 4) = 1.6\times10^4.$$

This represents an upper limit. The corresponding C_n from Figure 10.24 is 0.96 for $\beta = 0.5$. To begin, assume a slightly lesser value:

$$C_n = 0.95$$

Then

$$Q_{ac} = C_n Q_{th} = 0.95(0.055\ 4) = 0.052\ 6\ \mathrm{m^3/s}$$

and

$$\mathrm{Re}_1 = 2.83\times10^5 Q_{ac} = 1.49\times10^4$$

From Figure 10.24, at this Reynolds number, the discharge coefficient is 0.955, which is our second trial value. Assume

$$C_n = 0.955$$

Then

$$Q_{ac} = C_n Q_{th} = 0.955(0.055\ 4) = 0.052\ 9\ \mathrm{m^3/s}$$

and

$$\mathrm{Re} = 2.83\times10^5 Q_{ac} = 1.49\times10^4$$

Figure 10.24 shows $C_n = 0.955$, which is the assumed value. Thus,

$Q_{ac} = 0.052\ 9\ \mathrm{m^3/s}$ of ethylene glycol

A third type of constriction meter is the **orifice meter**, shown in Figure 10.26. The **orifice plate**, merely a flat plate with a hole, is inserted into a pipeline, conventionally between flanges. The hole can be either sharp edged or square edged. As the flow goes through the plate, it follows a streamline pattern similar to that shown in Figure 10.26. Downstream of the plate, the flow reaches a point of minimum area called a vena contracta. The static pressure tap p_2 is located at

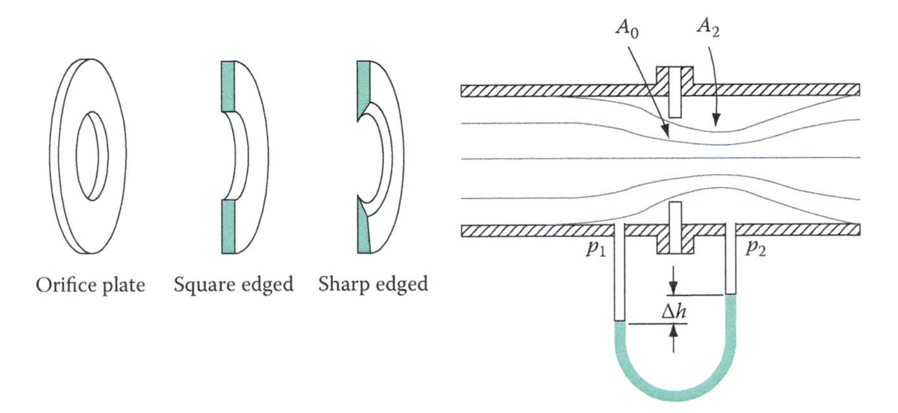

FIGURE 10.26 Orifice plates and a typical meter installation.

this point, where the streamlines are uniform and parallel. The upstream pressure is measured at point 1.

The Bernoulli equation can be applied to points 1 and 2 to obtain an expression relating flow rate to pressure drop. Although the area at point 2 (the vena contracta) is unknown, it can be expressed in terms of the orifice area A_O:

$$A_2 = C_C A_O$$

where C_C is called a **contraction coefficient**. Applying Bernoulli's equation yields the same results as before except that $C_C A_O$ is substituted for A_2 in the theoretical equation:

$$Q_{th} = C_C A_O \sqrt{\frac{2g(p_1 - p_2)}{\rho g(1 - D_2^4/D_1^4)}}$$

The actual flow rate through the meter is considerably less than the corresponding theoretical flow. By introducing an orifice discharge coefficient, defined as

$$C = \frac{Q_{ac}}{Q_{th}}$$

we obtain the following for the actual flow rate:

$$Q_{ac} = C C_C A_O \sqrt{\frac{2g(p_1 - p_2)}{\rho g(1 - D_2^4/D_1^4)}}$$

To simplify the formulation for an orifice meter, we can rewrite the equations for Q_{th} and Q_{ac} as

$$Q_{th} \approx A_O \sqrt{\frac{2g(p_1 - p_2)}{\rho g(1 - D_0^4/D_1^4)}} \tag{10.22}$$

$$Q_{ac} = C_O A_O \sqrt{\frac{2g(p_1 - p_2)}{\rho g(1 - D_0^4/D_1^4)}} \tag{10.23}$$

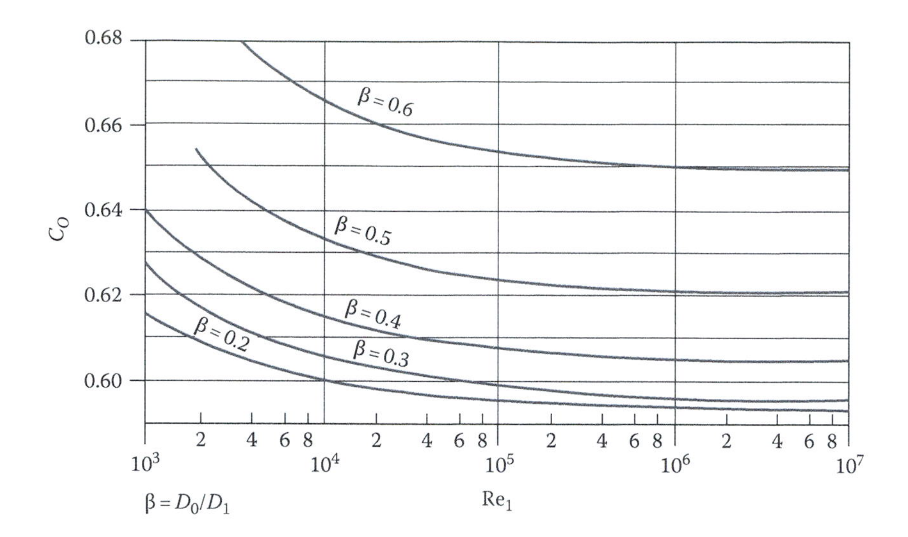

FIGURE 10.27 Orifice discharge coefficient as a function of pipe Reynolds number for circular concentric orifices. (Reprinted from The American Society of Mechanical Engineers, *Fluid Meters—Their Theory and Application*, 5th edn., The American Society of Mechanical Engineers, New York, 1959. With permission.)

The area of significance in the equation is A_O, the orifice area. Furthermore, the orifice coefficient becomes

$$C_O = CC_C = \frac{Q_{ac}}{Q_{th}}$$

It should be remembered that the static pressure p_2 is not measured at the orifice area A_O, but this discrepancy and the losses encountered due to friction and separation are accounted for in the overall coefficient C_O for the meter. Tests on a series of orifice meters yield data that can be presented in a form of C_O versus Re_1, where again

$$Re_1 = \frac{V_1 D_1}{\nu} = \frac{4Q_{ac}}{\pi D_1 \nu}$$

Such a plot is provided in Figure 10.27, where the ratio of orifice diameter to pipe diameter is an independent variable:

$$\beta = \frac{D_0}{D_1}$$

Because of the reliance on experimentally determined coefficients, it is good practice to allow at least a 10-diameter approach section before the orifice plate.

Example 10.8

An orifice meter is to be installed in a water main that supplies cooling water to a condensing unit. The water main is 24-nominal, schedule 160 pipe. The orifice hole should be drilled or turned in a lathe to make $\beta = 0.5$. The Bernoulli equation indicates that the theoretical flow rate through the line is 12,000 gpm.

a. Calculate the actual flow rate.
b. The only available manometer is 6 ft tall. Can it be used as an inverted U-tube for this installation?

Solution

a. From Table A.5, for water,

$$\rho = 1.94 \text{ slug/ft}^3$$

$$\mu = 1.9 \times 10^{-5} \text{ lbf} \cdot \text{s/ft}^2$$

From Table C.1, for a 24-nominal, schedule 160 pipe,

$$D_1 = 1.609 \text{ ft} \quad \text{and} \quad A_1 = 2.034 \text{ ft}^2$$

With $\beta = 0.5$,

$$D_0 = 0.805 \text{ ft} \quad \text{and} \quad A_O = \frac{\pi D_0^2}{4} = 0.508 \text{ ft}^2$$

Using conversion factors from Table A.2, we get

$$Q_{th} = 12{,}000 \text{ gpm} = 26.7 \text{ ft}^3/\text{s}$$

The Reynolds number upstream is

$$\text{Re}_1 = \frac{4Q_{ac}}{\pi D_1 v} = \frac{4Q_{ac}(1.94)}{\pi(1.609)(1.9 \times 10^{-5})} = 8.08 \times 10^4 Q_{ac}$$

The maximum Reynolds number we have exists at

$$\text{Re}_1 = 8.08 \times 10^4\, Q_{th} = 8.08 \times 10^4(26.7 = 2.16 \times 10^6)$$

The corresponding C_O from Figure 10.27 is 0.62, our starting value for the trial-and-error procedure. Assume

$$C_O = 0.62$$

Then

$$Q_{ac} = C_O Q_{th} = 0.62(2.67) = 16.6 \text{ ft}^3/\text{s}$$

and

$$\text{Re}_1 = 8.08 \times 10^4 Q_{ac} = 1.34 \times 10^6$$

At $\beta = 0.5$, Figure 10.24 gives $C_O \approx 0.62$, which agrees with our assumed value. Thus,

$$Q_{ac} = 16.6 \text{ ft}^3/\text{s} = 7440 \text{ gpm}$$

b. The corresponding pressure drop is found with Equation 10.22:

$$Q_{th} = A_O\sqrt{\frac{2g(p_1 - p_2)}{\rho g(1 - D_0^4/D_1^4)}} = A_O\sqrt{\frac{2g\Delta h}{1-(D_0^4/D_1^4)}}$$

By substitution, we get

$$26.7 = 0.508\sqrt{\frac{2(32.2)\Delta h}{1-(0.805/1.609)^4}}$$

$\Delta h = 40.2$ ft of water

A 6 ft tall water manometer will not suffice. A water-over-mercury manometer that is 6 ft tall will work, however.

The equation formulation for constriction meters has thus far been concerned only with incompressible flow. For compressible flows, however, gas density varies through the meter. We will, therefore, rewrite the descriptive equations to account for the change. Consider an isentropic, subsonic, steady flow of an ideal gas through a venturi meter. The continuity equation is

$$\rho_1 A_1 V_1 = \rho_2 A_2 V_2 = \dot{m}_{\text{isentropic}} = \dot{m}_s$$

If we neglect changes in potential energy, the energy equation is

$$h_1 + \frac{V_1^2}{2} = h_2 + \frac{V_2^2}{2}$$

For an ideal gas with constant specific heats,

$$h_2 - h_1 = c_p(T_2 - T_1)$$

The energy equation becomes

$$c_p T_1 + \frac{V_1^2}{2} = c_p T_2 + \frac{V_2^2}{2}$$

Combining this equation with the continuity equation, we get

$$c_p T_1 + \frac{\dot{m}_s^2}{2\rho_1^2 A_1^2} = c_p T_2 + \frac{\dot{m}_s^2}{2\rho_2^2 A_2^2}$$

or

$$\dot{m}_s^2\left(\frac{1}{\rho_2^2 A_2^2} - \frac{1}{\rho_1^2 A_1^2}\right) = 2c_p(T_1 - T_2) = 2c_p T_1\left(1 - \frac{T_2}{T_1}\right) \quad (10.24)$$

For an isentropic compression from points 1 to 2,

$$\frac{p_2}{p_1} = \left(\frac{T_2}{T_1}\right)^{\gamma/(\gamma-1)}$$

Also, $c_p = R\gamma/(\gamma - 1)$. After substitution and rearrangement, Equation 10.24 becomes

$$\frac{\dot{m}_s^2}{\rho_2^2 A_2^2}\left(1 - \frac{\rho_2^2 A_2^2}{\rho_1^2 A_1^2}\right) = 2\frac{R\gamma}{\gamma - 1}T_1\left[1 - \left(\frac{p_2}{p_1}\right)^{(\gamma-1)/\gamma}\right]$$

For an ideal gas, $\rho = p/RT$; substituting and simplifying give

$$\frac{\dot{m}_s^2}{A_2^2} = \frac{\rho_2^2 2[\gamma/(\gamma-1)](p_1/\rho_1)\left[1-(p_2/p_1)^{(\gamma-1)/\gamma}\right]}{1-\left(\rho_2^2 A_2^2/\rho_1^2 A_1^2\right)}$$

For an isentropic process,

$$\frac{p_1}{\rho_1^\gamma} = \frac{p_2}{\rho_2^\gamma}$$

Substitution yields

$$\frac{\dot{m}_s^2}{A_2^2} = \frac{\rho_1^2 (p_2/p_1)^{2/\gamma} 2[\gamma/(\gamma-1)](p_1/\rho_1)\left[1-(p_2/p_1)^{(\gamma-1)/\gamma}\right]}{1-(p_2/p_1)^{2/\gamma}(A_2/A_1)^2}$$

from which we finally obtain

$$\dot{m}_s = A_2\sqrt{\frac{2(p_1\rho_1)[\gamma/(\gamma-1)]\left[(p_2/p_1)^{2/\gamma} - (p_2/p_1)^{(\gamma-1)/\gamma}\right]}{1-(p_2/p_1)^{2/\gamma}(D_2/D_1)^4}} \quad (10.25)$$

With this isentropic equation, it is apparent that the measurements needed on a venturi are p_1, p_2 (not necessarily $p_1 - p_2$), T_1, the venturi dimensions, and the gas properties. The actual flow rate through the venturi meter is

$$\dot{m}_{ac} = C_V \dot{m}_s$$

$$= C_V A_2\sqrt{\frac{2(p_1\rho_1)[\gamma/(\gamma-1)]\left[(p_2/p_1)^{2/\gamma} - (p_2/p_1)^{(\gamma-1)/\gamma}\right]}{1-(p_2/p_1)^{2/\gamma}(D_2/D_1)^4}} \quad (10.26)$$

where C_V is the discharge coefficient given in Figure 10.20. It is convenient to rewrite Equation 10.26 in a form that is similar to the incompressible case by introducing another coefficient called the **compressibility factor Υ**:

$$\dot{m}_{ac} = C_V \Upsilon \rho_1 A_2\sqrt{\frac{2g(p_1 - p_2)}{\rho_1 g(1 - D_2^4/D_1^4)}} \quad (10.27)$$

By equating Equations 10.26 and 10.27, the compressibility factor is determined to be

$$\Upsilon = \sqrt{\frac{\gamma}{\gamma - 1}\frac{\left[(p_2/p_1)^{2/\gamma} - (p_2/p_1)^{(\gamma+1)/\gamma}\right]\left(1 - D_2^4/D_1^4\right)}{\left[1 - (D_2^4/D_1^4)(p_2/p_1)^{2/\gamma}\right]\left(1 - p_2/p_1\right)}} \tag{10.28}$$

For a given gas, the ratio of specific heats γ is known. This equation could be plotted as the compressibility factor Υ versus the pressure ratio p_2/p_1 for various values of D_2/D_1. This plot is shown in Figure 10.28 for a venturi meter with $\gamma = 1.4$. The advantage of rewriting the mass flow rate equation like the incompressible case is that a pressure drop term appears in the equation, which is more convenient to use if a manometer is available for pressure measurements. Moreover, the compressibility effect is isolated into one factor, Υ, that adds a measure of elegance and ease to the formulation. A similar analysis can be performed for a flow nozzle. The results are identical to those for the venturi meter including the compressibility factor of Equations 10.27 and 10.28 and Figure 10.28.

For the orifice meter, the area change is very abrupt, causing a flow configuration in which isentropic flow cannot be assumed. Further, the pressure p_2 is measured at A_2, which is generally not known, so the orifice area A_O is used in the equations. Therefore, a formulation like that for the venturi and flow nozzle meters does not yield an acceptable model. Consequently, Equation 10.28 cannot be used to obtain the compressibility factor for orifices. Experimental means must be resorted to, and the results are presented in Figure 10.29 for squared-edge orifices. The compressibility factor

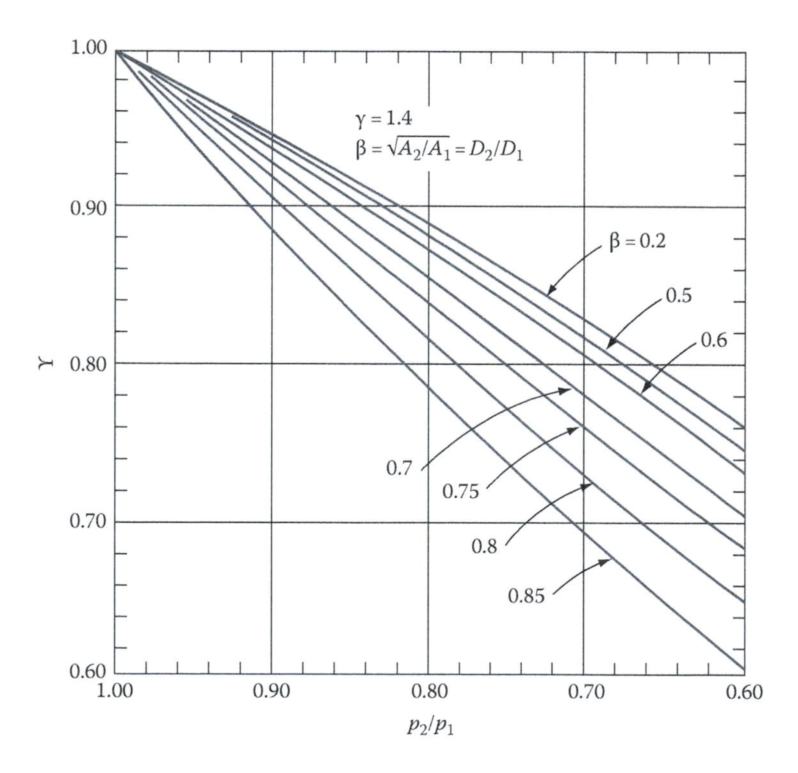

FIGURE 10.28 Compressibility factor Υ versus pressure ratio p_2/p_1 for venturi meters and flow nozzles with a compressible fluid of $\gamma = 1.4$. (Reprinted from The American Society of Mechanical Engineers, *Fluid Meters—Their Theory and Application*, 5th edn., The American Society of Mechanical Engineers, New York, 1959. With permission.)

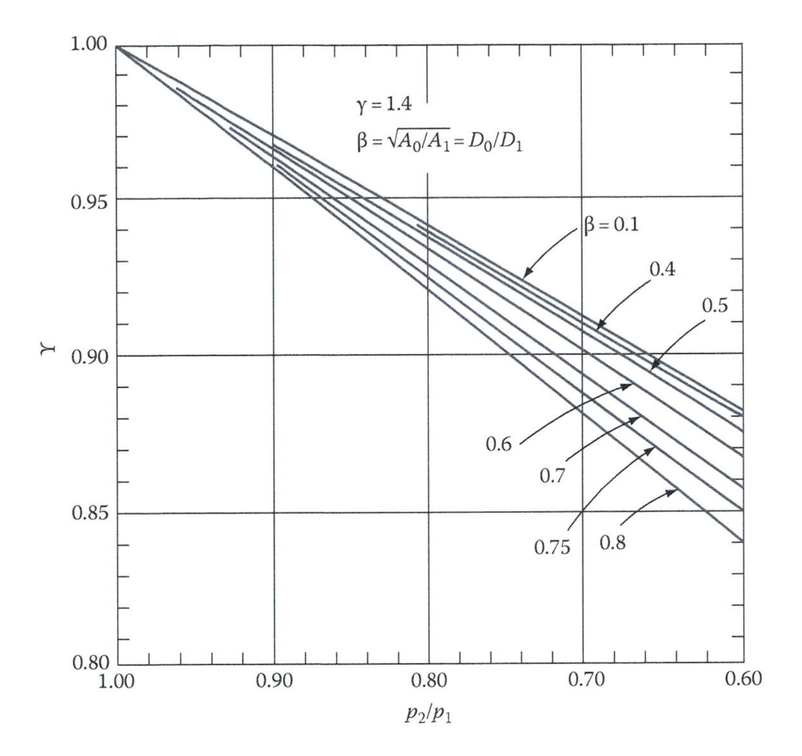

FIGURE 10.29 Compressibility factor Υ as a function of pressure ratio p_2/p_1, for square-edged orifices where the fluid has $\gamma = 1.4$. (Reprinted from The American Society of Mechanical Engineers, *Fluid Meters—Their Theory and Application*, 5th edn., The American Society of Mechanical Engineers, New York, 1959. With permission.)

Υ for a series of orifice meters is plotted as a function of pressure ratio p_2/p_1, where the fluid has $\gamma = 1.4$. For a compressible fluid with an orifice meter used to measure the flow rate,

$$\dot{m}_{ac} = C_O \Upsilon \rho_1 A_O \sqrt{\frac{2g(p_1 - p_2)}{\rho_1 g\left(1 - D_0^4/D_1^4\right)}} \tag{10.29}$$

where C_O is obtained from Figure 10.27 and Υ is found in Figure 10.29.

Example 10.9

In an oxygen-bottling plant, gaseous oxygen is fed from a tank through a 10-nominal pipe to a manifold assembly. The flow is monitored in the 10-nominal line with a 10 × 6 venturi meter. The mass flow rate of oxygen is a constant 4 kg/s. Owing to an accident in the plant, the venturi meter is rendered inoperative and must be replaced. It is desirable to continue operations, however, until another one can be obtained. An orifice plate made to the same dimensions of 10 × 6 can be installed temporarily between two flanges, static taps drilled, and gauges attached. For a line pressure of 150 kPa and an oxygen line temperature of 25°C, determine the expected reading on the pressure gauge downstream of the orifice if the flow rate of 4 kg/s can be maintained.

Solution

From Table A.6, for oxygen,

$$R = 260\ \text{J/(kg}\cdot{}^\circ\text{C)}$$

$$\mu = 20\times 10^{-6}\ \text{N}\cdot\text{s/m}^2$$

$$\gamma = 1.4$$

From Table C.1, for pipe,

$$D_1 = 25.46\ \text{cm} \quad \text{and} \quad A_1 = 509.1\ \text{cm}^2 \quad \text{(10-nominal, schedule 40)}$$

$$D_O = 15.41\ \text{cm} \quad \text{and} \quad A_O = 186.5\ \text{cm}^2 \quad \text{(6-nominal, schedule 40)}$$

Equation 10.29 gives the actual mass flow rate through an orifice meter as

$$\dot{m}_{ac} = C_O \Upsilon \rho_1 A_O \sqrt{\frac{2(p_1 - p_2)}{\rho_1\left(1 - D_0^4/D_1^4\right)}}$$

We now evaluate each term in this expression. The mass flow rate is given as

$$\dot{m}_{ac} = 4\ \text{kg/s of oxygen}$$

Assuming an ideal gas, the oxygen density in the line is

$$\rho_1 = \frac{p_1}{RT_1} = \frac{150\,000}{260(273+25)} = 1.936\ \text{kg/m}^3$$

The average velocity in the pipe is found with

$$V_1 = \frac{\dot{m}_{ac}}{\rho_1 A_1} = \frac{4}{1.936(0.050\,91)} = 40.6\ \text{m/s}$$

The Reynolds number is calculated to be

$$\text{Re}_1 = \frac{\rho_1 V_1 D_1}{\mu} = \frac{1.936(40.6)(0.254\,6)}{20\times 10^{-6}} = 1\times 10^6$$

For the orifice meter,

$$\beta = \frac{\text{orifice diameter}}{\text{pipe diameter}} = \frac{D_0}{D_1} = \frac{15.41}{25.46} = 0.605$$

At a Reynolds number of 1×10^6 and $\beta = 0.6$, Figure 10.27 shows

$$C_O = 0.65$$

We are now ready to substitute into the mass flow rate expression:

$$4 = 0.65\Upsilon(1.936)(0.018\,65)\sqrt{\frac{2(150\,000 - p_2)}{1.936[1-(0.605)^4]}}$$

This simplifies to

$$\Upsilon = \frac{156.05}{\sqrt{150\,000 - p_2}} \tag{i}$$

Because p_2 and Υ are both unknown, a trial-and-error solution is required, using Figure 10.29. Arbitrarily assume that

$$\Upsilon = 0.95$$

Then at $\beta = 0.605$, from Figure 10.29,

$$\frac{p_2}{p_1} = 0.84$$

$$p_2 = 0.84(150\,000) = 126\,000$$

The right-hand side of Equation i becomes 1.007. Next assume that

$$\Upsilon = 0.90$$

Then

$$\frac{p_2}{p_1} = 0.69$$

$$p_2 = 0.69(150\,000) = 103\,500$$

and

$$\text{RHS} = 0.723$$

Third trial:

$$\Upsilon = 0.92$$

$$\frac{p_2}{p_1} = 0.75$$

$$p_2 = 0.75(150\,000) = 112\,500$$

$$\text{RHS} = 0.805$$

Fourth trial:

$$\Upsilon = 0.93$$

$$\frac{p_2}{p_1} = 0.78$$

$$p_2 = 0.78(150\,000) = 117\,000$$

$$\text{RHS} = 0.859$$

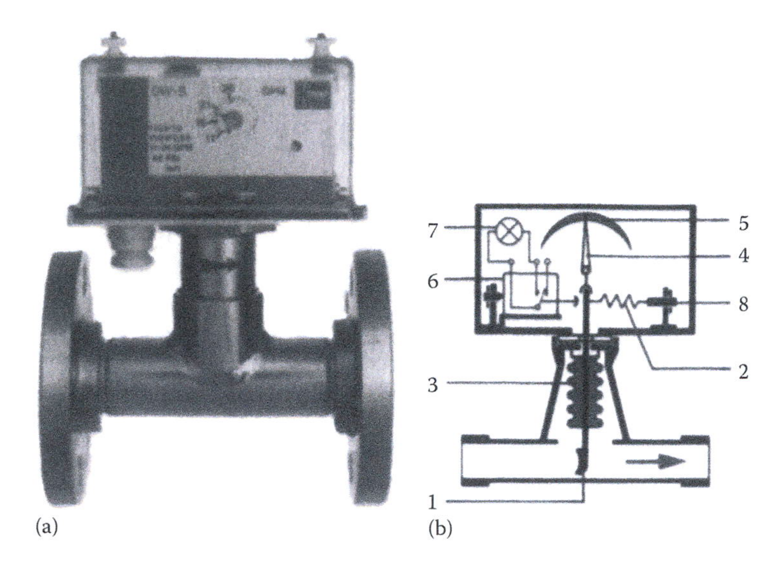

FIGURE 10.30 A paddle-bellows type of rate flow meter. (From Kobold Instruments Inc., Pittsburgh, PA. With permission.)

Fifth trial:

$$\Upsilon = 0.94$$

$$\frac{p_2}{p_1} = 0.82$$

$$p_2 = 0.82(150\,000) = 123\,000$$

$$\text{RHS} = 0.95$$

The operating point is near $\Upsilon = 0.94$; it is difficult to use Figure 10.29 and obtain more accurate results. Thus,

$$p_2 \approx 123\,000 \text{ Pa} = 123 \text{ kPa}$$

Another example of a rate meter is shown in Figure 10.30. The meter housing is made of metal and it can be attached to the piping system via flanges or threaded ends. The meter consists of a paddle and a mechanism to which the paddle is attached, as shown in Figure 10.30b. Flow through the meter causes the paddle (labeled 1) to deflect by an amount that is proportional to the flow rate. A greater flow rate causes a greater deflection, A return spring (2) acts to bring the paddle back to its equilibrium position. A bellows (3) made of metal acts as a fluid seal while allowing the paddle to deflect. The paddle (1) is attached to a lever arm, which in turn is attached to a pointer (4) such that the paddle motion is accompanied directly by the motion of the pointer. The pointer indicates the volume flow rate through the meter on a dial (5). An electrical circuit (6) can be added to operate an indicator light (7) when the flow rate exceeds a certain preset value. Such meters are typically accurate to ±3% and are used extensively for liquids.

10.5 MEASUREMENTS IN OPEN-CHANNEL FLOWS

Useful measurements in open-channel flows include velocity, flow rate, and depth. Velocity measurements are based on techniques that we have already discussed for determining flow velocity in closed conduits. Flow rate measurement in open channels usually involves introducing an obstruction in the flow and then relating a changed property to the flow rate.

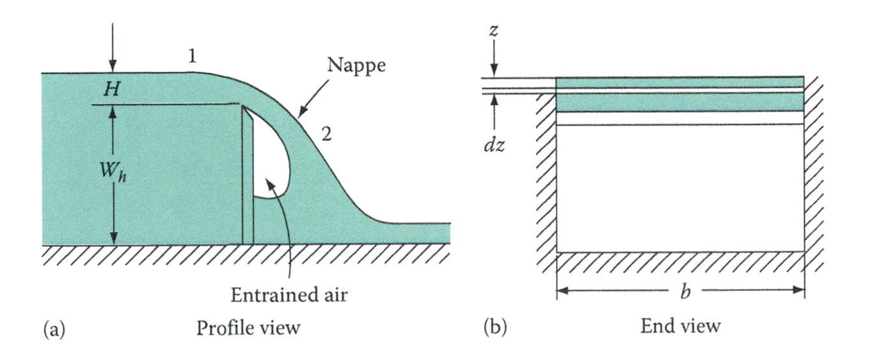

FIGURE 10.31 Flow over a suppressed sharp-crested rectangular weir placed in a rectangular channel. (a) Profile view and (b) end view.

A common method of measuring flow is to insert a vertical plate called a **weir** into the channel. The weir extends the entire width of the channel; it can be flat across its top edge or it may contain an opening. Flow over the weir is then related to the liquid height upstream. Consider, for example, the flow over a weir placed in a rectangular channel as in Figure 10.31. The weir in this figure is sharp crested and extends the entire width of the channel. Flow at the sides follows the walls and is not contracted. The end contractions are thus said to be *suppressed*. Upstream the channel sides must therefore be regular and smooth. Downstream it is necessary to extend the sides to keep the *nappe* (see Figure 10.31) laterally confined. The nappe entrains air downstream of the weir. Note that H is measured in relation to the liquid height above the crest of the weir.

To develop an equation that predicts flow rate, we begin by determining the flow through a differential area given by

$$dA = b\,dz \tag{10.30}$$

Bernoulli's equation applied to the elemental area dA is

$$\frac{p_1}{\rho g} + \frac{V_1^2}{2g} + z = \frac{p_2}{\rho g} + \frac{V_2^2}{2g}$$

Bernoulli's equation as written assumes no friction. The pressures are both equal to atmospheric pressure, and the upstream velocity is negligible in comparison to V_2. Bernoulli's equation simplifies to

$$V_2 = \sqrt{2gz}$$

The volume flow rate through the incremental area is then

$$dQ = V_2\,dA = \sqrt{2gz}b\,dz$$

The total flow is found by integration; that is,

$$Q = \int_0^H \sqrt{2g}\;bz^{1/2}dz = b\sqrt{2g}\,\frac{2z^{3/2}}{3}\bigg|_0^H$$

$$Q_{\text{th}} = \frac{2b}{3}\sqrt{2g}H^{3/2} \tag{10.31}$$

where the subscript "th" has been added to denote a theoretical flow rate. Directly over the crest of the weir, the liquid accelerates, and its height at this location is not exactly equal to H. Thus, Equation 10.31 must be modified slightly to account for this discrepancy by introducing a discharge coefficient, C_s; this coefficient also takes into account the frictional effects that are not included in the Bernoulli equation:

$$Q_{ac} = \frac{2C_s b}{3}\sqrt{2g}H^{3/2} \tag{10.32a}$$

By experiment, the coefficient of discharge has been determined to be

$$C_s = 0.605 + \frac{0.08H}{W_h} \tag{10.32b}$$

where W_h is the height of the weir.

The suppressed rectangular weir in Figure 10.31 forms a nappe that extends the entire width of the channel. A variation of this style is illustrated in Figure 10.32, where a vertical plate extends the entire width of the channel. The plate has a rectangular cutout through which liquid flows. Because the sides of the nappe are not in contact with any solid surface, they are free to contract, and the weir is thus called a **contracted rectangular weir**. The contraction effectively reduces the width of the nappe by about 1/10th of H, the height of the liquid above the crest.

To obtain an expression for flow rate, we first determine the incremental flow through an area of thickness dz and width L:

$$dA = L\,dz \tag{10.33}$$

Applying Bernoulli's equation to the incremental area gives the same expression as was obtained for the sharp-crested weir:

$$V_2 = \sqrt{2gz}$$

The incremental flow rate is

$$dQ = V_2\,dA = \sqrt{2gz}L\,dz$$

Integrating from $z = 0$ to $z = H$ gives the total flow as

$$Q_{th} = \frac{2L}{3}\sqrt{2g}H^{3/2} \tag{10.34}$$

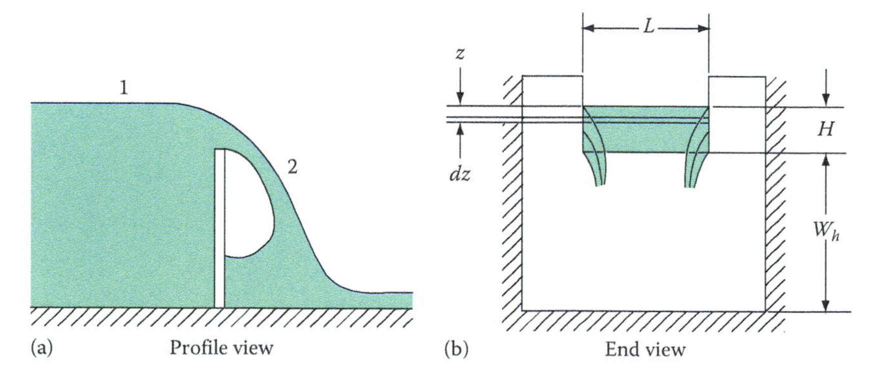

FIGURE 10.32 A contracted rectangular weir installed in a rectangular channel. (a) Profile view and (b) end view.

The contractions cause dA to be somewhat less than the ideal value given in Equation 10.33. Thus, Equation 10.34 must be modified slightly, to account for the discrepancy, by introducing a discharge coefficient, C_c:

$$Q_{ac} = \frac{2C_cL}{3}\sqrt{2g}H^{3/2} \tag{10.35}$$

By experiment, the coefficient C_c has been determined to have an average value of about 0.65.

The third type of weir that we will discuss is the *triangular*, or *V-notch*, *weir* illustrated in Figure 10.33. This weir allows liquid to flow only through the V-shaped cutout in the plate. The vertex angle θ varies in conventional designs from 10° to 90°. To obtain an expression for flow rate, we first write an equation for the incremental area dA:

$$dA = x\,dz$$

The vertex half-angle is

$$\tan\frac{\theta}{2} = \frac{x/2}{H - z}$$

or

$$x = 2(H - z)\tan\frac{\theta}{2}$$

The incremental area now becomes

$$dA = 2(H - z)\tan\frac{\theta}{2}dz \tag{10.36}$$

The Bernoulli equation applied to flow through the weir gives

$$V_2 = \sqrt{2gz}$$

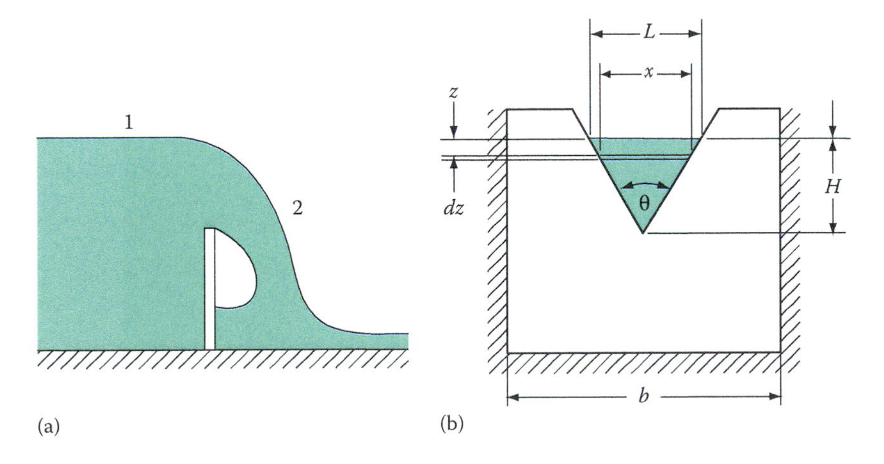

FIGURE 10.33 Flow over a triangular, or V-notch, weir placed in a rectangular channel. (a) Profile view and (b) end view.

The incremental flow rate is

$$dQ = V_2\, dA = \sqrt{2gz}\,2(H-z)\tan\frac{\theta}{2}dz$$

The total flow rate is found by integration:

$$Q = 2\sqrt{2g}\tan\frac{\theta}{2}\int_0^H z^{1/2}(H-z)dz$$

or

$$Q_{\text{th}} = \frac{8H^{5/2}}{15}\sqrt{2g}\tan\frac{\theta}{2} \quad (10.37)$$

Owing to frictional and real fluid effects, the volume flow rate is somewhat smaller. Thus, Equation 10.37 must be modified by introducing a discharge coefficient, C_{vn}:

$$Q_{\text{ac}} = \frac{8C_{\text{vn}}H^{5/2}}{15}\sqrt{2g}\tan\frac{\theta}{2} \quad (10.38)$$

By experiment, the discharge coefficient has been found to vary with head H and with θ. Results of extensive tests with V-notch weirs have produced the results given in Figure 10.34 for C_{vn} versus H with θ as an independent variable.

It should be mentioned that at low flow rates the nappe clings to the weir downstream. In other words, there is no air entrainment. When this occurs, the equations just written for various weirs no longer give an accurate descriptions of the flow rates. Other devices that are useful for measuring flow rates in open channels are the sluice gate and the venturi flume; both were discussed in Chapter 7.

As we saw in the development of the equations for weirs, the flow rate depends on the measurement of the liquid height upstream. It is essential that the liquid height be measured accurately. This can be accomplished with two conventional devices—the *point gauge* and the *hook gauge* (Figure 10.35). The sharpened end of either gauge is positioned so that it just touches the liquid surface. The vernier scale then gives the measurement of depth.

Example 10.10

A 45° V-notch weir is placed in an open channel that conveys water in an irrigation ditch. The height of the water above the vertex of the weir is 0.6 ft. Determine the volume flow rate in the channel.

Solution

Equation 10.38 applies:

$$Q_{\text{ac}} = \frac{8C_{\text{vn}}H^{5/2}}{15}\sqrt{2g}\tan\frac{\theta}{2}$$

For $H = 0.6$ ft and $\theta = 45°$, Figure 10.33 gives

$$C_{\text{vn}} = 0.592$$

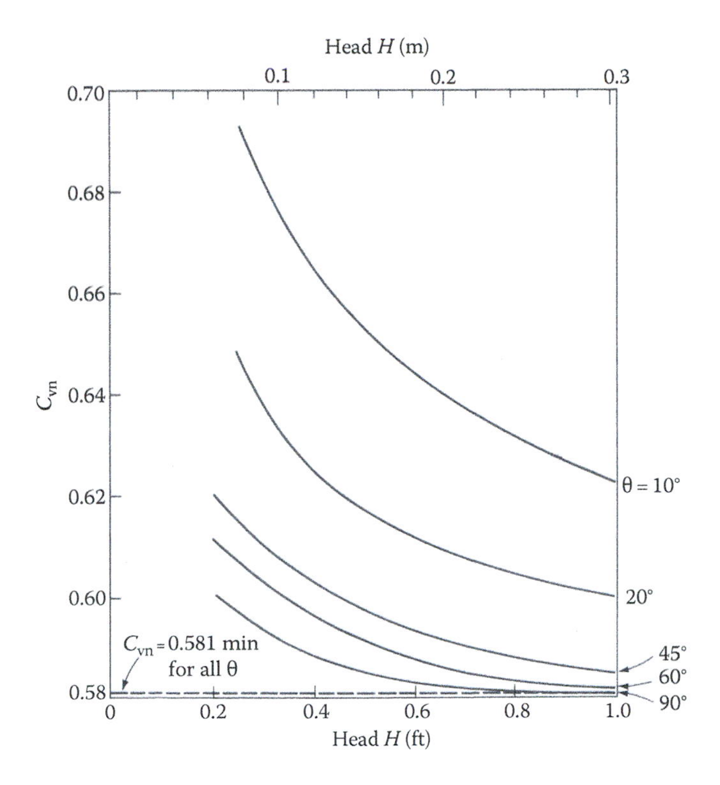

FIGURE 10.34 Discharge coefficient for V-notch weirs. (Reprinted from Lenz, A.T., *Trans. ASCE*, 108, 739, 1943. With permission.)

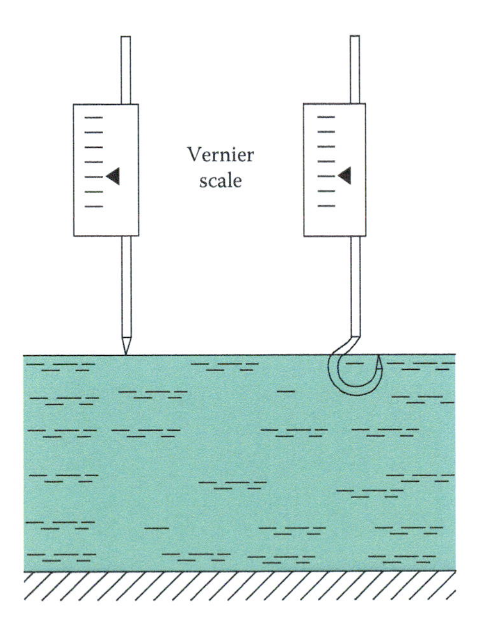

FIGURE 10.35 Point gauge and hook gauge used to measure liquid depth.

By substitution, we get

$$Q_{ac} = \frac{8(0.592)(0.6)^{5/2}}{15}\sqrt{2(32.2)}\tan\frac{45}{2}$$

$$Q_{ac} = 0.293\ \text{ft}^3/\text{s}$$

10.6 SUMMARY

In this chapter, we examined the techniques and methods of measuring various parameters in fluids. A pitot tube and a pitot-static tube were used to measure pressure and velocity. Ways of measuring fluid viscosity were presented. For closed-conduit flow, a summary of flow rate meters was given. For open-channel flow, a discussion of weirs was presented. This presentation, then, has offered some insights into theories of measurement.

PROBLEMS

Viscosity Measurement

10.1 A rotating-cup viscometer has two cylinders of radii 1.000 and 1.250 in. The cylinder depth is 4.000 in. A liquid of unknown viscosity is placed in the annulus between the two cylinders, and the outer cylinder begins rotating. Under steady-state conditions, the rotational speed is 30 rev/min. The torque exerted on the inner cylinder is 1×10^{-4} ft·lbf. Calculate the viscosity of the liquid.

10.2 In a process by which glycerine is produced, its purity is measured at various times at one location in the flow line. To measure its viscosity, for example, a rotating-cup viscometer is used. This viscometer is 15 cm high and has cups with diameters of 4.00 and 7.00 cm. If the rotational speed of the outer cylinder is 20 rev/min, determine the expected torque reading on the inner cylinder.

10.3 A rotating-cup viscometer is used to measure the viscosity of an unknown liquid. The cylinders have radii of 3 and 4 cm. A liquid of unknown viscosity is placed in the annulus between the cylinders. The liquid is tested at two different rotational speeds of the outer cup. Under steady conditions, when the rotational speed is 30 rev/min, the torque is 1×10^{-6} N·m. When the rotational speed is 40 rev/min, determine the expected torque reading if the liquid is Newtonian. Take the inner cup length to be 4 cm.

10.4 Refer to Example 10.1. If an aluminum sphere of diameter ⅛ in. is dropped into the shampoo, what is its expected terminal velocity?

10.5 A marble that is 1.5 mm in diameter falls through a jar of honey 15 cm tall. Approximately how long will it take the marble to reach the bottom? Assume the marble's properties to be the same as those of glass and the properties of honey to be equal to those of glycerine. The jar diameter is 7 cm.

10.6 A glass tube 1½ in. in diameter contains a commercial automotive oil additive. A steel sphere ⅛ in. in diameter falls through it at a rate of 1 in./4 s. Determine the viscosity of the additive if its specific gravity is 0.86.

10.7 A hard rubber sphere ⅛ in. in diameter falls through linseed oil in a cylinder. The cylinder diameter is such that the drag coefficient on the sphere is expressed by $C_D = 24/\text{Re}$. Determine the maximum cylinder diameter that is permissible for this case.

10.8 The properties of pancake syrup are being measured in a lab. The syrup's density is found to be 868 kg/m^3. A steel sphere 5 mm in diameter falls through the syrup at a rate of 3 cm/s. The tube used is 10 cm in diameter. As a check on the viscosity calculations, the syrup is placed in a rotating-cup viscometer. The cups have radii of 4 and 5 cm; the length is 12 cm. At rotational speeds of 5, 10, 20, and 30 rev/min, determine the expected torque readings.

Pressure and Velocity Measurements

10.9 If the pitot tube reading of Example 10.3 is treated as being incompressible, what is the percentage of error from the correct reading?

10.10 When making velocity measurements in a tube, it is possible to divide the tube cross section into five equal areas and calculate the volume flow rate from the data. Devise a scheme for dividing the cross section into six equal areas. Where must the velocity be known in the cross section? Rework Example 10.5 using the six-area formulation.

10.11 A huge axial-flow fan is used to force air into an underground mine shaft for ventilation. The fan is located in a circular length of ductwork 8 ft in diameter.

To measure flow rate through the fan, a traverse of the cross section is made with a pitot-static tube. The readings are given in Table P10.11.

Assuming a symmetric profile, sketch the velocity distribution and determine the mass flow rate of air if the temperature is 60°F and the static pressure is 14.8 psia.

10.12 A commercial air conditioning unit has a return air duct that is 3 m tall × 2 m wide. Air flow into the unit is to be measured with a pitot-static tube. The return air duct is thus divided into 20 equal rectangular areas as shown in Figure P10.12. In the Table following Figure P10.12.

TABLE P10.11

Distance below Centerline (ft)	Pitot-Static Tube Reading (in. Water)
0	0.7
1	0.6
2	0.42
3	0.2
4	0

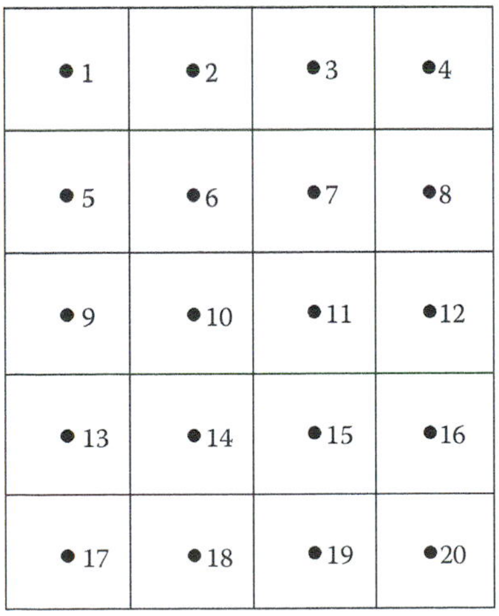

•1	•2	•3	•4
•5	•6	•7	•8
•9	•10	•11	•12
•13	•14	•15	•16
•17	•18	•19	•20

FIGURE P10.12

Determine the average flow velocity and the mass flow rate for an air temperature of 25°C and a static pressure of 100 kPa.

Position	Pressure Drop (mm Water)	Position	Pressure Drop (mm Water)
1	5	11	8.5
2	6.0	12	7.5
3	6.5	13	5.0
4	6.0	14	7.0
5	5.0	15	7.5
6	6.5	16	7.0
7	7.5	17	5.0
8	7.0	18	6.0
9	5.0	19	6.5
10	7.5	20	6.0

10.13 In Example 10.5, velocity data were used to calculate the volume flow rate through a pipe divided into five equal areas. Suppose instead that the cross section is divided into four equal areas, concentric rings with radii of 2.91 (= 11.63/4), 5.82, 8.72, and 11.63 in. Calculate the volume flow rate by taking velocities at radii of

1.45 in. (= 2.91/2)

4.36 in. (= [5.82 − 2.91]/2 + 2.91)

7.27 in.

10.18 in.

Compare your answer with the results of Example 10.5.

Volume Flow Rate Measurement

10.14 A venturi meter is calibrated in the laboratory. For a particular flow rate through the meter, the water is directed into a volume-measuring tank. Readings of pressure drops are also obtained. The data are as follows:
Upstream diameter = 1.025 in.
Throat diameter = 0.625 in.

Q_{ac} at Tank (gpm)	Δh at Meter (in. Water)
0.9	0.4
1.8	0.9
2.5	1.3
2.9	1.6
3.5	2.3
3.8	2.7
4.5	3.75
5.0	4.6
6.0	5.75

Source: Data from Fluid Mechanics Laboratory, University of New Orleans, New Orleans, LA.

a. Plot Q_{ac} versus Δh.
b. On the same graph, plot Q_{th} versus Δh.
c. Plot Re_1 versus C_V on semilog paper.

10.15 An orifice meter is calibrated in the laboratory. For a certain flow rate through the meter, water is directed into a volume-measuring tank yielding data of actual flow rate versus pressure drop in the meter. The results are as follows:

Flow Rate Measured (m^3/s)	Pressure (cm Water)
5.68×10^{-5}	0.762
1.14×10^{-4}	3.81
1.58×10^{-4}	5.72
1.83×10^{-4}	8.38
2.21×10^{-4}	11.4
2.40×10^{-4}	15.0
2.84×10^{-4}	20.3
3.15×10^{-4}	24.8
3.79×10^{-4}	29.5

Source: Data from Fluid Mechanics Laboratory, University of New Orleans, New Orleans, LA.

Upstream diameter = 2.60 in.
Throat diameter = 1.59 in.
a. Plot Q_{ac} versus Δh.
b. On the same graph, plot Q_{th} versus Δh.
c. Plot Re_1 versus C_O on semilog paper.

10.16 Water flows in an 8 × 4 venturi meter. The meter itself is in a vertical flow line with the flow direction downward. The difference in elevation between the static pressure taps is 45 cm. A mercury manometer attached to the meter reads 10 cm of mercury. Calculate the flow rate through the meter.

10.17 Repeat Problem 10.16 for a meter placed in a horizontal line. All other data and dimensions are the same.

10.18 Carbon tetrachloride flows through a 10 × 8 venturi meter inclined at an angle of 30° with the horizontal; flow is in the downward direction. A mercury manometer attached to the meter reads 5 in. Calculate the flow rate through the meter.

10.19 Linseed oil flows through a 12 × 10 venturi meter in a horizontal line. Determine the equivalent readings on a mercury manometer attached to the meter for the following flow rates: (a) 1 m^3/s, (b) 0.5 m^3/s, and (c) 2 m^3/s.

10.20 Octane is piped through a line into which a 10 × 8 venturi is installed. The line pressure is 220 kPa, and the volume flow through the meter is 0.6 m^3/s. The meter is inclined at 85° from the horizontal with flow in the downward direction. The distance between static pressure taps is 25 cm.
a. Determine the pressure at the throat.
b. If a mercury manometer is attached as in Figure 10.18a, determine Δh.
c. If an air-over-octane manometer is used in an inverted U-tube configuration, determine Δh.

10.21 Water is transported through a pipe into which a 12 × 8 flow nozzle has been installed with an inverted U-tube manometer. The pressure drop is measured as 12 cm of water. Determine the actual flow rate through the line.

10.22 Octane flows in a pipe into which a 4 × 3 flow nozzle has been installed. A water manometer attached to the static pressure taps of the nozzle reads 95 cm of water. Calculate the volume flow of octane.

10.23 Benzene flows in a pipe that has an 8 × 6 flow nozzle. A mercury manometer attached to the meter reads 4 in. Determine the volume flow rate through the meter.

10.24 Castor oil flows at 0.06 m^3/s through a pipe. A 10 × 8 flow nozzle is installed. Determine the pressure drop in terms of meters of mercury.

10.25 A 12 × 10 venturi and a 12 × 10 flow nozzle are installed in a line that conveys 12 ft^3/s of chloroform. Pressure gauges are attached to the meters. What are the expected pressure drops?

10.26 Using the charts developed for flow through constriction meters (C_V versus Re, for example) often requires a trial-and-error procedure. How can the same data be used to generate a plot that requires no trial-and-error method? Show results by equations and show logic. Assume that meter dimensions and pressure head loss are known.

10.27 In Example 10.8, the pressure drop calculated was 40.2 ft of water. Calculate the corresponding pressure drop if the following types of manometers are used in a U-tube: (a) water over mercury and (b) water over kerosene.

10.28 Glycerine flows through a 6-nominal, schedule 80 pipeline into which an orifice plate with a 4.00-cm hole has been installed. Pressure gauges are attached about the plate. What is the flow rate if the pressure drop is equivalent to 65 m of mercury?

10.29 Heptane flows through a 10-nominal, schedule 80 pipeline into which an orifice with a 6.0 in hole has been installed, A water manometer attached to the meter reads 6 ft of water. Determine the flow rate through the line.

10.30 Three different flow meters have been installed in a single pipeline: an orifice meter, a flow nozzle, and a venturi meter. All have dimensions of 16 × 8. The flow rate through the line is 0.2 m^3/s of propylene glycol. What will be the readings of a mercury manometer attached to the meters?

10.31 Derive Equation 10.28.

10.32 If the venturi meter of Example 10.9 is replaced with a 10 × 6 flow nozzle, what is the expected downstream pressure reading?

10.33 Hydrogen flows in a 1-nominal pipe. A 1 × ½ venturi meter has been installed with pressure gauges. For readings of 10 and 8 psia from the meter, determine the mass flow rate. Take the line temperature to be 60°F.

10.34 An 8 × 6 venturi is placed in a line carrying air. The line temperature is 40°F; the line pressure is 100 psia. Prepare a graph of mass flow rate as a function of throat pressure over the range of 60–90 psia.

10.35 An orifice plate with a hole diameter of 5.1 cm is placed in a 4-nominal line that conveys helium at a flow rate of 0.2 kg/s. The line pressure and temperature are 200 kPa and 25°C, respectively. Assuming that helium has a ratio of specific heats of 1.4, determine the pressure downstream of the orifice. (Note that γ for helium is not 1.4 but 1.66. Without a chart such as Figure 10.29 or an equation, the pressure found assuming $\gamma = 1.4$ is a best estimate.)

Flow over a Weir

10.36 A rectangular sharp-crested weir extends across a rectangular channel. The weir is 1 m high and 3 m wide. The upstream head measured from the channel bottom is 2.4 m. Determine the flow of water over the weir.

10.37 A rectangular sharp-crested weir is installed in an open channel that conveys water at 250 ft^3/s. The channel width is 6 ft, and the weir height is 4 ft. Determine the height of the liquid upstream.

10.38 A contracted rectangular weir is placed in a channel 12 m wide. The rectangular cutout is 6 m wide and 2 m deep. The weir height is 6.5 m. Determine the volume flow over the weir if the water head as measured from channel bottom is 5 m.

10.39 Derive Equation 10.37.

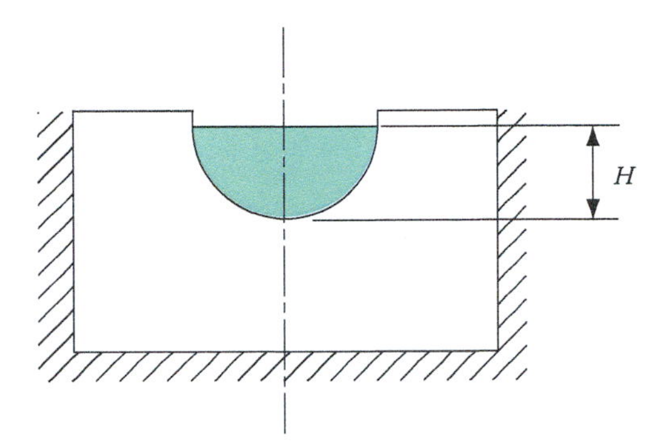

FIGURE P10.43

10.40 Repeat Example 10.10 if the weir has an included angle of 90° and all other conditions remain the same.

10.41 A 60° V-notch weir is placed in a channel that conveys water at a flow rate of 1 m³/s. Determine the water head upstream.

10.42 How do the equations for flow over weirs change with fluid properties? Rewrite all affected equations.

10.43 A weir consists of a semicircular cutout in a vertical flat plate as shown in Figure P10.43. Derive an equation for Q_{th} versus H for the weir.

10.44 A V-notch weir is placed in a channel that conveys water at a flow rate of 0.006 m³/s. Determine and graph the variation of upstream height with included angle—that is, H as a function of θ.

10.45 The flow rate in a rectangular open channel must be measured. A V-notch weir is to be placed in the channel, and it is necessary to determine which angle θ will provide the most convenient results. As the first step in the calculations, plot Q_{ac} versus H for V-notch weirs, using θ as an independent variable.

10.46 Construct a plot of the suppressed-weir equation:

$$C_s = 0.605 + \frac{0.08H}{W_h}$$

Place H on the horizontal axis, C_s on the vertical axis, and W_h as an independent variable. Let H vary from 0 to 1 ft, and let W_h take on values of 1, 2, 3, and 4 ft.

10.47 A suppressed rectangular weir is placed in a channel that conveys water at a flow rate of 0.5 ft³/s. For a channel width of 4 ft and a weir height of 2 ft, determine the height H.

10.48 A contracted rectangular weir is placed in a rectangular channel. The liquid height H is 15 cm, and the volume flow rate is 0.03 m³/s. Determine the width L of the rectangular cutout.

10.49 The flow rate of water over a suppressed rectangular weir is 100 ft³/s. The channel width is 4 ft, and the liquid height H above the weir is 2 ft. Determine the weir height W_h.

10.50 A suppressed rectangular weir is placed in a 3-m-wide channel that conveys water at 3 m³/s. The weir height is 1 m. Determine the liquid height H.

11 The Navier–Stokes Equations

Chapter 5 was concerned with flow in closed conduits. In Section 5.5, for example, we modeled laminar flow through circular, rectangular, and annular passages. For each case, a control volume was set up, and then the continuity and momentum equations were written. Solving the resulting differential equation and applying boundary conditions yielded a velocity distribution for the problem. We performed a similar analysis in Chapter 7 for laminar flow down an incline.

A control volume for a general fluid mechanics problem can be set up to obtain a *comprehensive* set of differential equations from which we can obtain differential equations for the simpler problems like those described earlier. Such differential equations are presented in this chapter.

Turbulence is an important area of study in fluid mechanics because most flows with which we are familiar (e.g., in a pipe or a channel) are turbulent flows. The equations of motion for turbulent flow will be discussed in this chapter. To show how these equations can be applied to a simple problem, the velocity profile for turbulent flow in a circular tube will be formulated.

11.1 EQUATIONS OF MOTION

A set of differential equations to describe fluid motion can be derived for the general case. Consider the control volume of Figure 11.1. It is a differential element in a fluid continuum. Forces acting on it are to include gravitational, viscous or frictional, and pressure forces to encompass the majority of fluid problems. If the continuity and momentum equations are written for all three principal directions, and the fluid is Newtonian with constant properties of density and viscosity, a set of differential equations results. The momentum equation written for each principal direction gives what are called the **Navier–Stokes equations**. Their derivation is lengthy, involved, and beyond the scope of this text. The equations will be given without derivation here, but interested readers may refer to the specific references cited at the end of the book.

For the general problem in fluid mechanics, assuming that we have a Newtonian fluid with constant properties, the governing equations in Cartesian coordinates are the following.
Continuity equation:

$$\frac{\partial \rho}{\partial t}+\frac{\partial(\rho V_x)}{\partial x}+\frac{\partial(\rho V_y)}{\partial y}+\frac{\partial(\rho V_z)}{\partial z}=0 \tag{11.1}$$

Navier–Stokes equations:
x-component:

$$\rho\left(\frac{\partial V_x}{\partial t}+V_x\frac{\partial V_x}{\partial x}+V_y\frac{\partial V_x}{\partial y}+V_z\frac{\partial V_x}{\partial z}\right)=-\frac{\partial p}{\partial x}+\mu\left(\frac{\partial^2 V_x}{\partial x^2}+\frac{\partial^2 V_x}{\partial y^2}+\frac{\partial^2 V_x}{\partial z^2}\right)+\rho g_x \tag{11.2a}$$

y-component:

$$\rho\left(\frac{\partial V_y}{\partial t}+V_x\frac{\partial V_y}{\partial x}+V_y\frac{\partial V_y}{\partial y}+V_z\frac{\partial V_y}{\partial z}\right)=-\frac{\partial p}{\partial y}+\mu\left(\frac{\partial^2 V_y}{\partial x^2}+\frac{\partial^2 V_y}{\partial y^2}+\frac{\partial^2 V_y}{\partial z^2}\right)+\rho g_y \tag{11.2b}$$

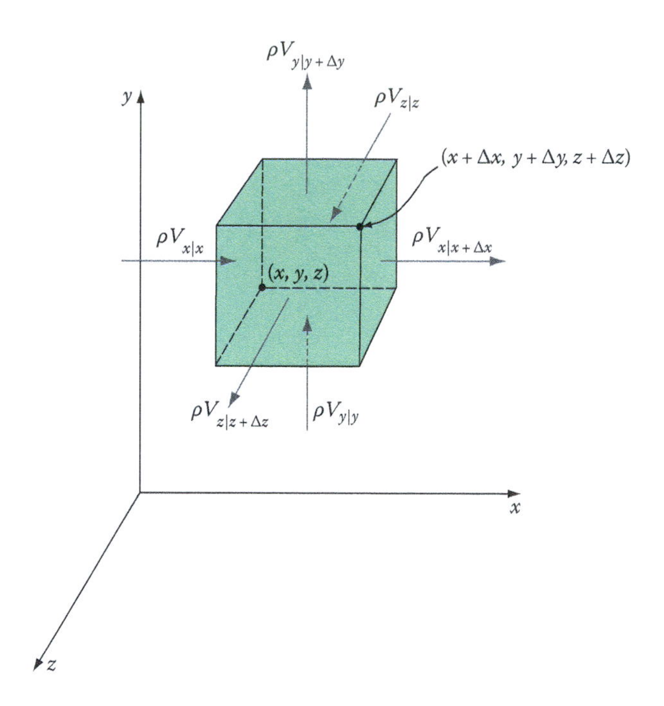

FIGURE 11.1 A differential fluid element.

z-component:

$$\rho\left(\frac{\partial V_z}{\partial t}+V_x\frac{\partial V_z}{\partial x}+V_y\frac{\partial V_z}{\partial y}+V_z\frac{\partial V_z}{\partial z}\right)=-\frac{\partial p}{\partial z}+\mu\left(\frac{\partial^2 V_z}{\partial x^2}+\frac{\partial^2 V_z}{\partial y^2}+\frac{\partial^2 V_z}{\partial z^2}\right)+\rho g_z \tag{11.2c}$$

The left-hand sides of Equations 11.2 are acceleration terms. These terms are nonlinear and present difficulties in trying to solve the equations. Even though a variety of exact solutions for specific flows have been found, the equations have not been solved in general—owing primarily to the presence of the nonlinear terms. The right-hand side of the equations includes pressure, gravitational or body, and viscous forces. In polar cylindrical coordinates, these equations are

Continuity equation:

$$\frac{\partial p}{\partial t}+\frac{1}{r}\frac{\partial(\rho r V_r)}{\partial r}+\frac{1}{r}\frac{\partial(\rho V_\theta)}{\partial \theta}+\frac{\partial(\rho V_z)}{\partial z}=0 \tag{11.3}$$

Navier–Stokes equations:

r-component:

$$\rho\left(\frac{\partial V_r}{\partial t}+V_r\frac{\partial V_r}{\partial r}+\frac{V_\theta}{r}\frac{\partial V_r}{\partial \theta}+V_z\frac{\partial V_r}{\partial z}-\frac{V_\theta^2}{r}\right)$$
$$=-\frac{\partial p}{\partial r}+\mu\left\{\frac{\partial}{\partial r}\left[\frac{1}{r}\frac{\partial(rV_r)}{\partial r}\right]+\frac{1}{r^2}\frac{\partial^2 V_r}{\partial \theta^2}+\frac{\partial^2 V_r}{\partial z^2}-\frac{2}{r^2}\frac{\partial V_\theta}{\partial \theta}\right\}+\rho g_r \tag{11.4a}$$

θ-component:

$$\rho\left(\frac{\partial V_\theta}{\partial t}+V_r\frac{\partial V_\theta}{\partial r}+\frac{V_\theta}{r}\frac{\partial V_\theta}{\partial \theta}+\frac{V_rV_\theta}{r}+V_z\frac{\partial V_\theta}{\partial z}\right)$$

$$=-\frac{1}{r}\frac{\partial p}{\partial \theta}+\mu\left\{\frac{\partial}{\partial r}\left[\frac{1}{r}\frac{\partial(rV_\theta)}{\partial r}\right]+\frac{1}{r^2}\frac{\partial^2 V_\theta}{\partial \theta^2}+\frac{2}{r^2}\frac{\partial V_r}{\partial \theta}+\frac{\partial^2 V_\theta}{\partial z^2}\right\}+\rho g_\theta \tag{11.4b}$$

z-component:

$$\rho\left(\frac{\partial V_z}{\partial t}+V_r\frac{\partial V_z}{\partial r}+\frac{V_\theta}{r}\frac{\partial V_z}{\partial \theta}+V_z\frac{\partial V_z}{\partial z}\right)$$

$$=-\frac{\partial p}{\partial z}+\mu\left[\frac{1}{r}\frac{\partial}{\partial r}\left(r\frac{\partial V_z}{\partial r}\right)+\frac{1}{r^2}\frac{\partial^2 V_z}{\partial \theta^2}+\frac{\partial^2 V_z}{\partial z^2}\right]+\rho g_z \tag{11.4c}$$

In cylindrical coordinates, the term $\rho V_\theta^2/r$ is the centrifugal force that gives the effective r-directed force resulting from fluid motion in the θ-direction. When one is transforming equations from Cartesian to cylindrical coordinates, this term arises automatically. The term $\rho V_rV_\theta/r$ is the Coriolis force, which is the effective θ-directed force resulting from flow in both the r- and θ-directions. This term also arises automatically when one is transforming from Cartesian to cylindrical coordinates.

In vector notation, the equations are

Continuity equation:

$$\frac{\partial \rho}{\partial t}+\nabla\cdot\rho\mathbf{V}=0 \tag{11.5}$$

Navier–Stokes equations:

$$\frac{\partial \mathbf{V}}{\partial t}+(\mathbf{V}\cdot\nabla)\mathbf{V}=-\frac{1}{\rho}\nabla p-\nabla g+\nu\nabla^2\mathbf{V} \tag{11.6}$$

where the vector operator del, ∇, is defined in Cartesian coordinates, for example, as

$$\nabla()=\mathbf{i}\frac{\partial()}{\partial x}+\mathbf{j}\frac{\partial()}{\partial y}+\mathbf{k}\frac{\partial()}{\partial z} \tag{11.7}$$

$$\nabla^2()=\frac{\partial^2()}{\partial x^2}+\frac{\partial^2()}{\partial y^2}+\frac{\partial^2()}{\partial z^2} \tag{11.8}$$

11.2 APPLICATIONS TO LAMINAR FLOW

The equations that we have just written are easily applied to various laminar flow problems to obtain the descriptive differential equation. The types of problems that are encountered are classified in a number of ways. For example, a **steady-flow** problem is one in which there is no time dependence. A **parallel flow** problem is one in which only one velocity exists within the fluid. Regardless of problem type, if the Navier–Stokes equations can be solved for the geometry of interest, subject to the boundary conditions (and this is not always possible), the solution will yield a velocity profile.

Once the velocity profile is known, all other parameters of the flow can be found. The volume flow rate is obtained by integrating the velocity profile over the cross-sectional area. The average velocity is obtained by dividing the volume flow rate by the cross-sectional area. The maximum velocity can be obtained either by inspection or by differentiation. The shear stress at the wall is found by differentiating the velocity according to Newton's law of viscosity. The instantaneous velocity is thus an important parameter.

In this section, we will illustrate a number of cases that involve setting up the differential equation, solving it, and deriving the pertinent flow details, which vary from problem to problem. For all problems discussed, we will follow a solution format that includes several steps:

1. Discussion of conditions and assumptions
2. Sketch of the problem with an expected velocity profile
3. Derivation of the differential equation
4. Solution for the instantaneous velocity
5. Solution for the pertinent flow details

11.2.1 Flow in a Circular Duct

Consider steady laminar flow in a tube or pipe as shown in Figure 11.2, where the flow is in the z-direction and the effect of gravity is neglected. At most, we can have three velocity components; in cylindrical coordinates, these are

$$V_r \quad V_\theta \quad V_z$$

For laminar flow, V_r and V_θ are zero. Only the axial velocity V_z is nonzero. This velocity can be a function of three space variables and of time; that is,

$$V_z = V_z(r, \theta, z, t)$$

We can eliminate the time dependence because the flow is steady. Referring to Figure 11.2, we see that the velocity is independent of z. Specifically, for two different axial locations that have the same r and θ (points 1 and 2), the velocity is the same. Similar reasoning shows that V_z is also independent of θ.

For steady laminar flow in a tube, we therefore have

$$V_z = V_z(r)$$

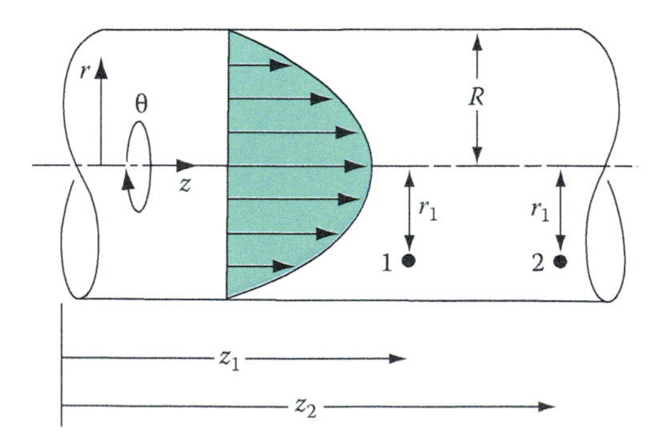

FIGURE 11.2 Laminar flow in a tube.

Equations 11.3 and 11.4 reduce to the following:

$$\text{Continuity: } \frac{\partial V_z}{\partial z} = 0 \tag{11.9a}$$

$$r\text{-component: } 0 = -\frac{\partial p}{\partial r} \tag{11.9b}$$

$$\theta\text{-component: } 0 = -\frac{1}{r}\frac{\partial p}{\partial \theta} \tag{11.9c}$$

$$z\text{-component: } 0 = -\frac{\partial p}{\partial z} + \frac{\mu}{r}\frac{\partial}{\partial r}\left(r\frac{\partial V_z}{\partial r}\right) \tag{11.9d}$$

The continuity equation states that V_z is not a function of z. The r- and θ-component equations state that pressure does not vary in the r- and θ-directions. The differential equation to be solved, then, is Equation 11.9d.

It is instructive to digress momentarily to consider steady laminar flow in an annulus as another example. Following the same lines of reasoning as for steady laminar flow in a tube, we can show that again Equation 11.9d is the equation to be solved. What differentiates these problems are the boundary conditions that must be written for each and used when solving the differential equation. For a complete statement of a problem, therefore, the differential equation and its boundary conditions must be given.

For flow in a tube, the boundary conditions are

$$r = R \qquad V_z = 0 \text{ (condition 1)}$$

$$r = 0 \qquad \frac{\partial V_z}{\partial r} = 0 \text{ (condition 2)}$$

The first of these is the nonslip condition at the wall. The second states that the slope at the centerline is zero, or that the velocity is finite at the centerline. Equation 11.9d can be solved by direct integration as in Chapter 5:

$$\frac{\mu}{r}\frac{d}{dr}\left(r\frac{dV_z}{dr}\right) = \frac{dp}{dz}$$

or

$$\frac{d}{dr}\left(r\frac{dV_z}{dr}\right) = \frac{r}{\mu}\frac{dp}{dz}$$

Integrating with respect to r gives

$$r\frac{dV_z}{dr} = \frac{r^2}{2\mu}\left(\frac{dp}{dz}\right) + C_1$$

where C_1 is a constant of integration. Applying boundary condition 2 yields

$$C_1 = 0$$

Thus, we are left with

$$\frac{dV_z}{dr} = \frac{r}{2\mu}\frac{dp}{dz}$$

Integrating, we get

$$V_z = \frac{r^2}{4\mu}\frac{dp}{dz} + C_2$$

Applying boundary condition 1 gives

$$0 = \frac{R^2}{4\mu}\frac{dp}{dz} + C_2$$

or

$$C_2 = -\frac{R^2}{4\mu}\frac{dp}{dz}$$

The velocity now becomes

$$V_z = \frac{r^2}{4\mu}\frac{dp}{dz} - \frac{R^2}{4\mu}\frac{dp}{dz}$$

$$V_z = \frac{R^2}{4\mu}\left(-\frac{dp}{dz}\right)\left[1-\left(\frac{r}{R}\right)^2\right] \tag{11.10}$$

For steady laminar flow in a tube, the velocity distribution is therefore parabolic.

The volume flow rate is found by integrating the velocity profile over the cross-sectional area:

$$Q = \iint V_z\, dA = \int_0^{2\pi}\int_0^{R} \frac{R^2}{4\mu}\left(-\frac{dp}{dz}\right)\left[1-\left(\frac{r}{R}\right)^2\right] r\, dr\, d\theta$$

Integrating and solving, we get

$$Q = \frac{\pi R^4}{8\mu}\left(-\frac{dp}{dz}\right) \tag{11.11}$$

The average velocity is

$$V = \frac{Q}{A} = \frac{R^2}{8\mu}\left(-\frac{dp}{dz}\right) \tag{11.12}$$

The shear stress in the circular duct is found with Newton's law of viscosity:

$$\tau_{rz} = -\mu \frac{dV_z}{dr} = -\frac{r}{2}\frac{dp}{dz}$$

This equation shows that for tube flow, the shear stress varies linearly with the radial coordinate r. At the centerline, $\tau_{rz}|_r = 0$; at the wall,

$$\tau_{rz}\Big|_{r=R} = \tau_w = -\frac{R}{2}\frac{dp}{dz} \tag{11.13}$$

where the notation τ_w denotes a wall shear stress.

11.2.2 Flow down an Inclined Plane

Consider steady laminar flow down an inclined plane, as shown in Figure 11.3, which also shows the coordinate system that we will use in formulating the problem. The driving force for fluid motion is gravity, and the flow is in the z-direction. Thus, only V_z is nonzero. Moreover, V_z will depend only on x. Hence,

$$V_x = 0$$

$$V_y = 0$$

$$V_z = V_z(x)$$

Equations 11.1 and 11.2 become

$$\text{Continuity equation: } \frac{\partial V_z}{\partial z} = 0$$

$$x\text{-component: } 0 = -\frac{\partial p}{\partial r} + \rho g_x$$

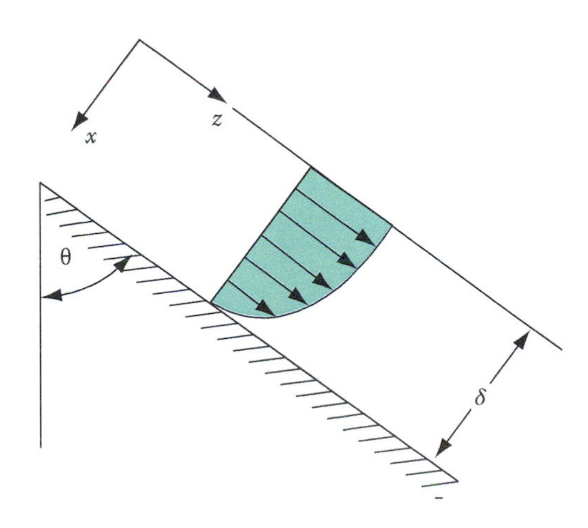

FIGURE 11.3 Laminar flow down an incline.

$$y\text{-component: } 0 = -\frac{\partial p}{\partial y}$$

$$z\text{-component: } 0 = -\frac{\partial p}{\partial z} + \mu\frac{\partial^2 V_x}{\partial x^2} + \rho g_z$$

The boundary conditions are

$$x = 0 \qquad \frac{\partial V_z}{\partial x} = 0 \text{ (condition 1)}$$

$$x = \delta \qquad V_z = 0 \text{ (condition 2)}$$

The solution to the equation, as obtained also in Chapter 7, is

$$V_z = \frac{\rho g \delta^2}{2\mu}\cos\theta\left[1 - \left(\frac{x}{\delta}\right)^2\right] \tag{11.14}$$

The volume flow rate is found to be

$$Q = \iint V\,dA = \int_0^b\int_0^\delta \frac{\rho g \delta^2}{2\mu}\cos\theta\left[1 - \left(\frac{x}{\delta}\right)^2\right] dx\,dy$$

where b is the width into the page. Integrating gives

$$Q = \frac{\rho g \delta^3 b}{3\mu}\cos\theta = \frac{g\delta^3 b}{3\nu}\cos\theta \tag{11.15}$$

Shear stress is found with

$$\tau_{xz} = \mu\frac{dV_z}{dx} = -\frac{\rho g \delta^2}{2}\cos\theta\left(-\frac{2x}{\delta^2}\right)$$

The maximum shear stress occurs when $x = \delta$ at the wall:

$$\tau_w = \rho g x \cos\theta \tag{11.16}$$

11.2.3 Flow through a Straight Channel

Consider steady laminar flow through a straight channel of width $2h$ and depth (into the page) b, as illustrated in Figure 11.4. With both walls fixed, the flow is in the z-direction, and gravity forces are neglected. We therefore write

$$V_x = V_y = 0$$

$$V_z = V_z(x)$$

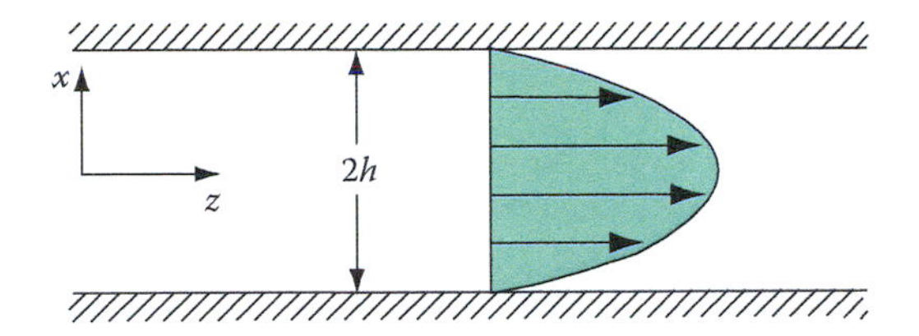

FIGURE 11.4 Steady parallel flow through a straight two-dimensional channel.

Equations 11.1 and 11.2 thus become

$$\frac{\partial V_z}{\partial z} = 0$$

$$0 = -\frac{\partial p}{\partial x}$$

$$0 = -\frac{\partial p}{\partial y}$$

$$0 = -\frac{\partial p}{\partial z} + \mu\left(\frac{\partial^2 V_z}{\partial x^2}\right)$$

The boundary conditions are

$$x = h \qquad V_z = 0$$

$$x = -h \qquad V_z = 0$$

The solution to the equation (as obtained by direct integration) is

$$V_z = \frac{h^2}{2\mu}\left(-\frac{dp}{dz}\right)\left[1-\left(\frac{x}{h}\right)^2\right] \tag{11.17}$$

The volume flow rate through the channel is

$$Q = \iint V_z\, dA = \int_0^b \int_{-h}^h \frac{x^2}{2\mu}\left(-\frac{dp}{dz}\right)\left[1-\left(\frac{x}{h}\right)^2\right] dx\, dy$$

or

$$Q = \frac{2h^3 b}{3\mu}\left(-\frac{dp}{dz}\right) \tag{11.18}$$

The average velocity is

$$V = \frac{Q}{A} = \frac{h^3}{3\mu}\left(-\frac{dp}{dz}\right) \tag{11.19}$$

The shear stress is found with

$$\tau_{xz} = \mu \frac{\partial V_z}{\partial x} = \mu \left(\frac{h^2}{2\mu} \right) \left(-\frac{dp}{dz} \right) \left(-\frac{2x}{h^2} \right)$$

$$\tau_{xz} = \left(-\frac{dp}{dz} \right) x$$

The wall shear stresses thus become

$$\tau_w \big|_{x=h} = h \left(-\frac{dp}{dz} \right)$$
$$\tau_w \big|_{x=-h} = -h \left(-\frac{dp}{dz} \right) \tag{11.20}$$

11.2.4 Plane Couette Flow

Plane Couette flow is flow through a two-dimensional channel that has one wall moving. As illustrated in Figure 11.5, the z-directed flow is steady, and gravity forces are neglected. A pressure gradient is imposed on the fluid, and the upper wall is moving at a velocity equal to U. The channel height is h, and the origin is at the lower wall (different from Figure 11.4). According to the assumptions, we have

$$V_x = V_y = 0$$

$$V_z = V_z(y)$$

Equations 11.1 and 11.2 become

$$\frac{\partial V_z}{\partial z} = 0$$

$$0 = -\frac{\partial p}{\partial x}$$

$$0 = -\frac{\partial p}{\partial y}$$

$$0 = -\frac{\partial p}{\partial z} + \mu \left(\frac{\partial^2 V_z}{\partial y^2} \right)$$

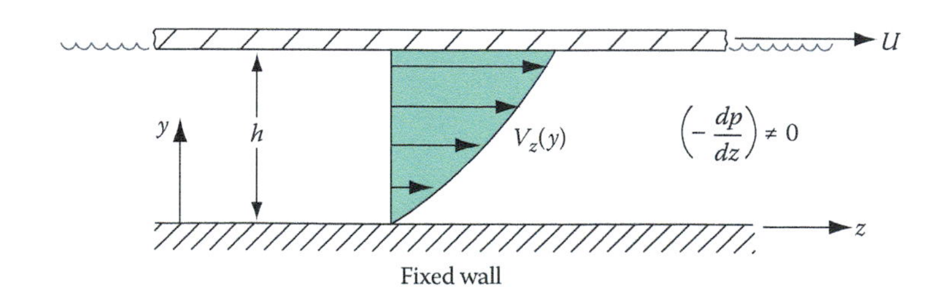

FIGURE 11.5 Plane Couette flow.

These equations are exactly what was obtained for the straight-channel problem in which both walls were stationary. The difference between the two problems will be in the boundary conditions. For plane Couette flow,

$$y = h \qquad V_z = U$$
$$y = 0 \qquad V_z = 0$$

By direct integration, the solution is easily found to be

$$\frac{V_z}{U} = \frac{y}{h} + \left(-\frac{dp}{dz}\right)\left(\frac{h^2}{2\mu U}\right)\left(\frac{y}{h}\right)\left(1 - \frac{y}{h}\right) \tag{11.21}$$

The terms on the right-hand side of Equation 11.21 represent two separate effects. The y/h term is the simple shear solution. This solution indicates a linear profile extending from $V_z = 0$ at the fixed wall to $V_z = U$ at the moving wall. The second term represents the effect of the pressure gradient. Within that term, $h^2/2\mu U$, y/h, and $(1 - y/h)$ are all positive terms. The quantity $(-dp/dz)$ tells how pressure varies with the z-direction. In other words, as z increases, p can increase or decrease. If p increases with increasing z, then $(-dp/dz)$ is actually a negative quantity; and if p decreases with increasing z, then $(-dp/dz)$ is positive. If $(-dp/dz)$ is zero, then the simple shear solution results. The effect that $(-dp/dz)$ has on the velocity profile is significant. To illustrate this effect, let us examine Equation 11.21 in detail. For purposes of simplifying the notation, define the constant C as

$$C = \left(-\frac{dp}{dz}\right)\left(\frac{h^2}{2\mu U}\right)$$

and let $\Upsilon = y/h$. The velocity profile thus becomes

$$\frac{V_z}{U} = \frac{y}{h} + C\left(\frac{y}{h} - \frac{y^2}{h^2}\right) = Y + C(\Upsilon - \Upsilon)^2 \tag{11.22}$$

or

$$\frac{V_z}{U} = (1 + C)\Upsilon - C\Upsilon^2 \tag{11.23}$$

Our objective is to graph V_z/U (the dimensionless velocity) on a horizontal axis as a function of Υ (dimensionless position) on the vertical axis. We now examine three separate cases defined by the value (or sign) of the pressure gradient and represented by C.

Case 1: $(-dp/dz) = 0$ or $C = 0$

The velocity becomes (from Equation 11.22)

$$\frac{V_z}{U} = \frac{y}{h}$$

which is graphed in Figure 11.6.

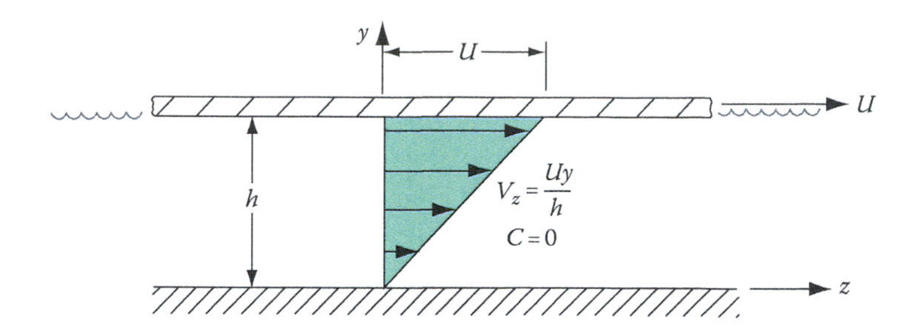

FIGURE 11.6 Velocity profile for plane Couette flow without a pressure gradient.

Case 2: $(-dp/dz) > 0$ or $C > 0$

For calculation purposes, assign C the value of +1. The velocity becomes

$$\frac{V_z}{U} = 2\Upsilon - \Upsilon^2$$

This quadratic equation is graphed in Figure 11.7.

Case 3: $(-dp/dz) < 0$ or $C < 0$

For calculation purposes, assign C the value of −1. The velocity becomes

$$\frac{V_z}{U} = -2\Upsilon + \Upsilon^2$$

which is graphed in Figure 11.8.

To represent a family of velocity profiles, we use a composite graph, as shown in Figure 11.9. Each velocity profile has an associated volume flow rate, average velocity, maximum velocity, and wall shear stress. The flow details are reserved for the exercises.

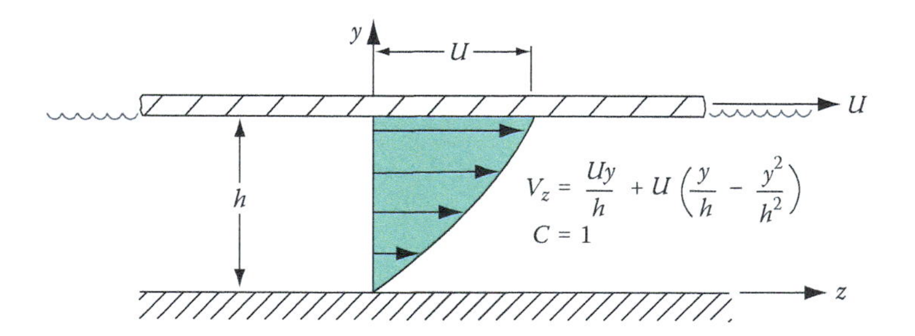

FIGURE 11.7 Velocity profile for plane Couette flow with a favorable pressure gradient.

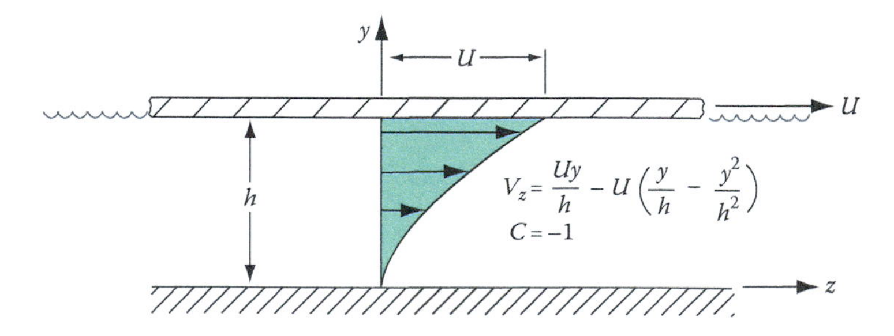

FIGURE 11.8 Velocity profile for plane Couette flow with an adverse pressure gradient.

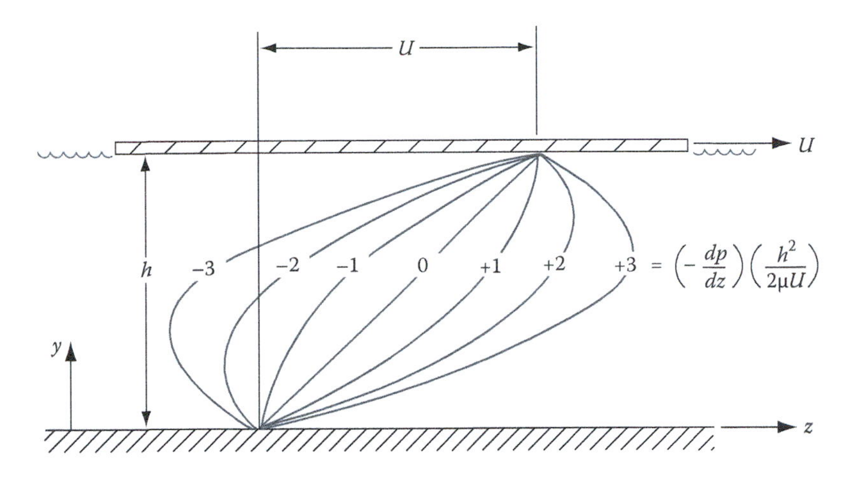

FIGURE 11.9 Family of velocity profiles for plane Couette flow.

11.2.5 Flow between Two Rotating Concentric Cylinders

Figure 11.10 shows the plan view of two concentric cylinders. The annular space between the cylinders contains a fluid. Both cylinders are rotating, and the rotation causes the fluid to flow. Assuming steady, laminar flow in the tangential direction only, with no gravity forces and no pressure gradients in the θ-direction, we conclude that

$$V_r = V_z = 0$$

$$V_\theta = V_\theta(r)$$

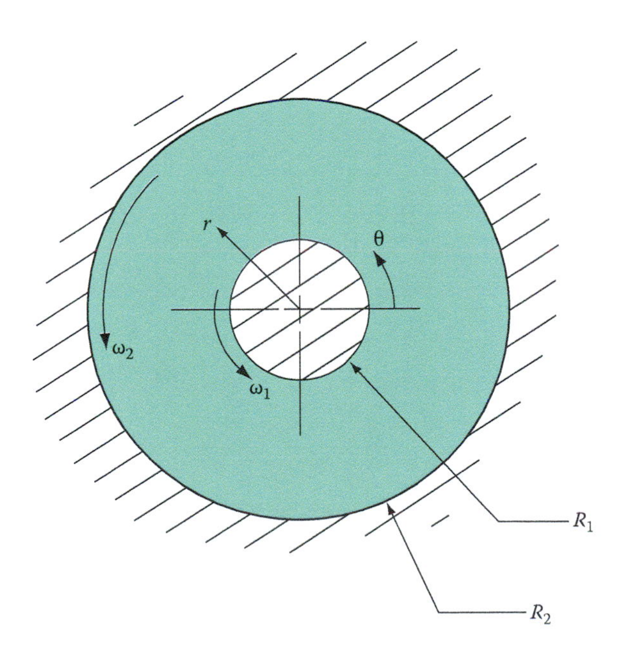

FIGURE 11.10 Flow between two rotating concentric cylinders.

Equations 11.3 and 11.4 become

$$\begin{aligned} &\frac{1}{r}\frac{\partial}{\partial\theta}(\rho V_\theta) = 0 \\ &-\frac{\rho V_\theta^2}{r} = -\frac{\partial p}{\partial r} \end{aligned} \tag{11.24}$$

$$\begin{aligned} &0 = \mu\left\{\frac{\partial}{\partial r}\left[\frac{1}{r}\frac{\partial}{\partial r}(rV_\theta)\right]\right\} \\ &0 = -\frac{\partial p}{\partial z} \end{aligned} \tag{11.25}$$

The boundary conditions are

$$\begin{aligned} r &= R_2 \qquad V_\theta = R_2\omega_2 \\ r &= R_1 \qquad V_\theta = R_1\omega_1 \end{aligned}$$

Solving Equation 11.25 subject to the boundary conditions gives

$$V_\theta = \left(\frac{\omega_2 - \omega_1 K^2}{1-K^2}\right)r + \left[\frac{(\omega_1 - \omega_2)R_1^2}{1-K^2}\right]\frac{1}{r} \tag{11.26}$$

where $K = R_1/R_2$. Several features of importance can be investigated. For example, consider that the outer cylinder is very large, such that $R_2 \to \infty$; or that the inner cylinder is very small, $R_1 \to 0$; or that one of the cylinders is fixed. These problems are easily solved by using Equation 11.26.

Equation 11.24 is the r-component of the Navier–Stokes equations applied to Figure 11.10. As is indicated, there is a pressure gradient in the radial direction that depends on the tangential velocity V_θ. If Equation 11.24 is solved subject to the boundary condition

$$r = R_1 \quad p = p_1$$

the solution becomes

$$\begin{aligned} p - p_1 = \frac{\rho}{\left(R_1^2 - R_2^2\right)^2}\Bigg[&\frac{\left(R_1^2\omega_1 - R_2^2\omega_2\right)^2\left(r^2 - R_1^2\right)}{2} + 2R_1^2R_2^2(\omega_2 - \omega_1)\left(R_1^2\omega_1 - R_2^2\omega_2\right)\ell\text{n}\frac{r}{R_1} \\ &+ R_1^4R_2^4(\omega_2 - \omega_1)^2\left(\frac{1}{r^2} - \frac{1}{R_1^2}\right)\Bigg] \end{aligned} \tag{11.27}$$

The shear stress is determined with

$$\begin{aligned} \tau_{r\theta} &= \mu\left[r\frac{\partial(V_\theta/r)}{\partial r}\right] \\ \tau_{r\theta} &= -\frac{2\mu(\omega_1 - \omega_2)VR_1^2R_2^2}{\left(R_2^2 - R_1^2\right)r^2} \end{aligned} \tag{11.28}$$

and we see that the shear stress is independent of θ.

There is a moment exerted on the cylinders by the fluid. On the inner cylinder, we have

$$M_1 = \text{tangential force} \times R_1 = [(\tau_{r\theta})|_{r=R}]A_1R_1$$

where A_1 is the surface area over which the shear stress acts. For a cylinder of length L, we have $A_1 = 2\pi R_1 L$, and the moment becomes

$$M_1 = -\frac{4\pi\mu(\omega_1 - \omega_2)R_1^2 R_2^2 L}{R_2^2 - R_1^2} \tag{11.29}$$

Under steady conditions, the magnitude of the moment exerted on the outer cylinder M_2 equals that exerted on the inner cylinder; that is, at steady state,

$$M_1 = -M_2 \tag{11.30}$$

There are many exact solutions to a number of problems described by the Navier–Stokes equations. Several more are given in the section that follows and in the exercises at the end of this chapter. We hope that this sampling has served as an inspiring introduction to the field of viscous fluid flow.

11.3 GRAPHICAL SOLUTION METHODS FOR UNSTEADY LAMINAR FLOW PROBLEMS

To this point, we have examined a number of problems, all of which were for steady laminar flow. That is, the instantaneous velocity did not vary with time. In this section, we formulate several unsteady-flow problems. The analytical solution techniques for the problems we consider are beyond the scope of this text. Alternatively, we can solve such problems using graphical solution methods.

Graphical solution methods allow a quick and easy determination of instantaneous time-dependent velocity profiles. Graphical techniques have traditionally played an important role in engineering, At one time, before numerical methods could be implemented easily with a computer, graphical solutions were the only practical means of obtaining a solution. The emphasis today is on numerical methods, and so, the need for graphical methods has been diminished. It is a mistake, however, to discard graphical solution methods as they have a place in fluid mechanics. Graphical methods can be used to solidify understanding and build confidence with only a minor expenditure of effort. They can be applied to certain more advanced (than we have heretofore considered) problems, which can be solved only by using more sophisticated mathematical techniques. Advanced problems can be discussed with graphical methods at a much earlier time during the education process.

Consider a viscous flow problem in which the velocity is a function of time and of only one space variable. Suppose that the Navier–Stokes equations for such a problem reduce to

$$\frac{\partial V_z}{\partial t} = \nu \frac{\partial^2 V_z}{\partial y^2} \tag{11.31}$$

where V_z is the instantaneous velocity in the z-direction and the kinematic viscosity is $\nu = \mu/\rho$. The graphical method (as well as the numerical method, by the way) relies on our ability to rewrite the differential equation in terms of an algebraic expression. This can be done in a number of ways, but here we rely on the **Saul'ev method**. Equation 11.31 thus becomes

$$\frac{\partial V_z}{\partial t} = \nu \frac{\partial}{\partial y}\left(\frac{\partial V_z}{\partial y}\right)$$

or

$$\frac{V_z\big|_j^{k+1} - V_z\big|_j^k}{\Delta t} = \nu \frac{\left(\left(V_z\big|_{j+1}^k - V_z\big|_j^k\right)\Big/\Delta y\right) - \left(\left(V_z\big|_j^{k+1} - V_z\big|_{j-1}^k\right)\Big/\Delta y\right)}{\Delta y}$$

where the superscript k denotes a particular time and the subscript j indicates a position in the fluid. Rearranging gives

$$V_z\big|_j^{k+1} = \left(\frac{\Delta y^2/\nu t - 1}{\Delta y^2/\nu t + 1}\right) V_z\big|_j^k + \left(\frac{1}{\Delta y^2/\nu t + 1}\right) V_z\big|_{j+1}^k + V_z\big|_{j-1}^{k+1} \tag{11.32}$$

The quantity $\Delta y^2/\nu t$ is defined as the **cell Reynolds number**, Re_c, the value of which influences what is known as the **stability** of the solution method. When calculations are performed using this equation, as time increases to infinity, the solution should approach the steady-state solution (if there is one). The value of the cell Reynolds number influences whether this occurs and how quickly. In the Saul'ev equation formulation, the preceding equation is unconditionally stable. So, any value of the cell Reynolds number can be assumed. If we assume arbitrarily that

$$\text{Re}_c = \frac{\Delta y^2}{\nu t} = 1$$

then Equation 11.32 simplifies somewhat to yield

$$V_z\big|_j^{k+1} = \frac{1}{2} V_z\big|_{j+1}^k + V_z\big|_{j+1}^{k+1}$$

or

$$V_z\big|_j^{k+1} = \frac{V_z\big|_{j+1}^k + V_z\big|_{j-1}^{k+1}}{2} \tag{11.33}$$

Equation 11.33 has a very useful graphical interpretation, which is illustrated in Figure 11.11. As indicated, the flow field is divided into segments that are Δy wide. Lines in the field (identified as **nodes**) are labeled from 0 and extend to $j + 2$ (as needed). We expect our solution to give us the value of the velocity V_z at all nodes from 0 to $j + 2$ for any time k. Shown in the diagram is the velocity in the $j - 2$ to $j + 2$ region at the time denoted by k. At the time $k + 1$, we will know $V_z\big|_{j-1}^{k+1}$ from a boundary condition or a time condition. To apply Equation 11.33, we align $V_z\big|_{j-1}^{k+1}$ with $V_z\big|_{j-1}^k$ and draw a straight line from $V_z\big|_{j-1}^{k+1}$ to the j line to obtain $V_z\big|_j^{k+1}$. Next, we align the newfound point with $V_z\big|_{j-2}^k$ and draw to the $j + 1$ line to obtain $V_z\big|_{j-1}^{k+1}$. The drawing process is continued until the boundary conditions are met or the problem requirements are satisfied. Note that the grid spacing Δy, when selected, automatically determines the time interval, because these quantities are related through the cell Reynolds number ($\text{Re}_c = \Delta y^2/\nu t$).

11.3.1 Suddenly Accelerated Flat Plate

Consider a semi-infinite region of fluid initially at rest, as shown in Figure 11.12a. As shown, the fluid extends to infinity in the y-direction and the plate lies along the x-axis. At some time that we

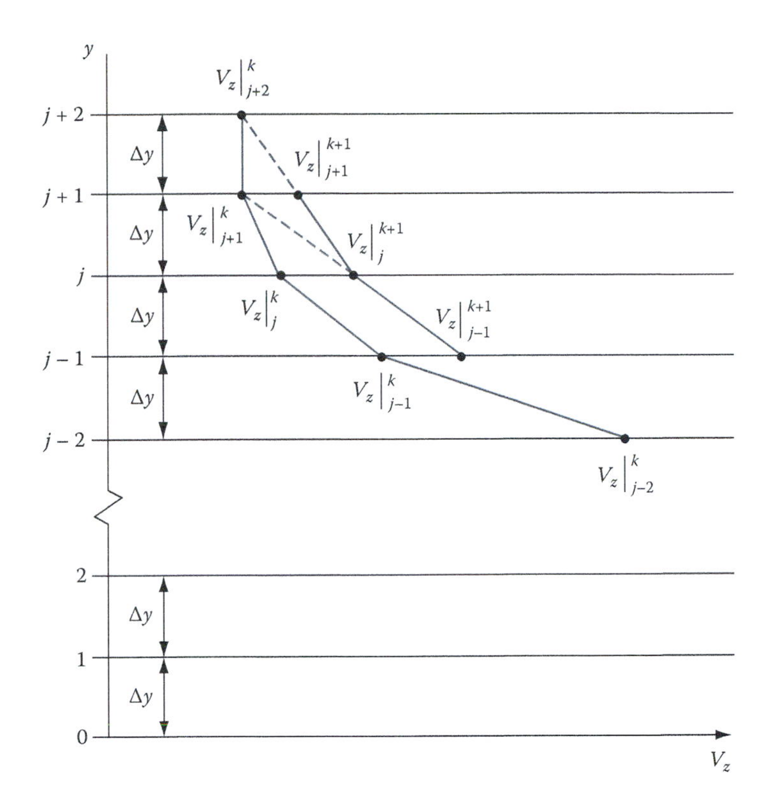

FIGURE 11.11 Graphical construction of Equation 11.33.

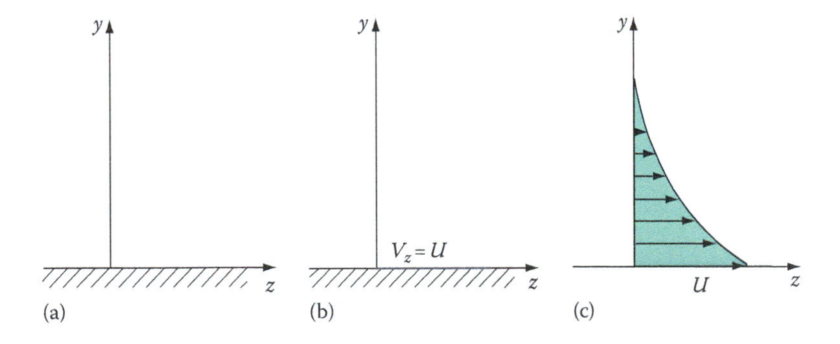

FIGURE 11.12 Semi-infinite fluid in contact with a suddenly accelerated flat plate. (a) $t < 0$, (b) $t = 0$, and (c) $t > 0$.

define as $t = 0$, the plate is suddenly accelerated to a velocity U in the x-direction. It is our objective to determine graphically the velocity within the fluid.

This problem is a classic diffusion problem. Assuming unsteady laminar flow of a Newtonian fluid with no gravity forces or pressure gradients, we conclude that

$$V_y = V_x = 0$$

$$V_z = V_z\,(y,t)$$

The continuity and momentum equations (Equations 11.1 and 11.2, respectively) become

$$\frac{\partial V_z}{\partial z} = 0$$
$$\rho \frac{\partial V_z}{\partial t} = \mu \frac{\partial^2 V_z}{\partial y^2} \tag{11.34}$$

with the following boundary conditions:

1. $t \geq 0 \quad y = 0 \quad V_z = U$ (nonslip condition)
2. $t \geq 0 \quad y = \theta \quad V_z = 0$

The solution to the differential equation subject to the boundary conditions is

$$\frac{V_z}{U} = 1 - \mathrm{erf}(\eta) \tag{11.35}$$

where U is the velocity of the boundary, $\eta = y/(2\sqrt{\nu t})$, and the **error function** is

$$\mathrm{erf}(\eta) = \frac{2}{\sqrt{\pi}} \int_0^{\eta} e^{-\xi^2} d\xi \tag{11.36}$$

The graphical solution method begins as shown in Figure 11.13. The region is divided up by a number of lines separated by a distance Δy. The drawing process is as follows:

1. Align point *a* and point *c*; draw line *ba*.
2. Align point *b* and point *d*; draw line *eb*.
3. Align point *e* and point *f*; draw line *ge*.

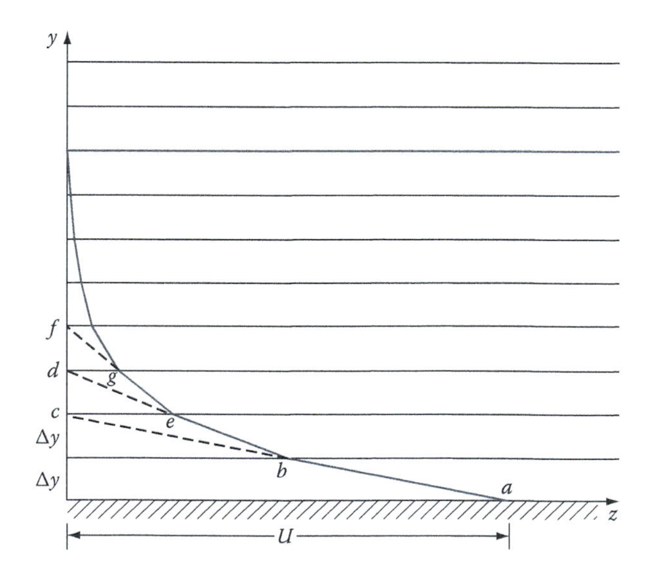

FIGURE 11.13 Graphical solution for the velocity after one time interval for the suddenly accelerated flat plate.

The process is continued until the desired height is reached. The result shown in Figure 11.13 is a velocity profile that exists after one time interval has passed. The time interval is easily calculated. Suppose that in Figure 11.13, the distance Δy is selected as 0.5 cm, and the fluid we are analyzing is glycerine (ρ = 1 263 kg/m^3 and $\mu = 950 \times 10^{-3}$ N · s/m^2 from Table A.5). The cell Reynolds number is 1. So

$$\mathrm{Re}_c = \frac{\Delta y^2}{\nu t} = \frac{\rho \Delta y^2}{\mu t}$$

Substituting, or

$$1 = \frac{1\,263(0.005)^2}{950 \times 10^{-3} t}$$

or

$$t = 30.1 \text{ s}$$

The velocity profile of Figure 11.13 will exist in the fluid at t = 30.1 s.

To obtain the velocity profile in the fluid after 60.2 s, we proceed in the same manner as before, beginning with the preceding profile. Referring to Figure 11.14, the procedure is as follows:

1. Align *a* and *c*; draw line *ga*.
2. Align *g* and *d*; draw line *hg*.
3. Align *h* and *e*; draw line *ih*.

The procedure is continued until the desired height is reached. Figure 11.15 shows the solution for the first five time intervals.

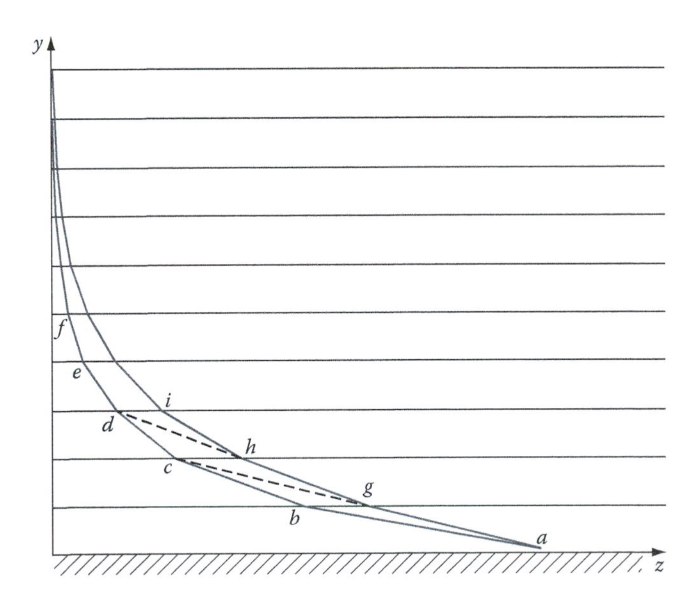

FIGURE 11.14 Graphical solution for the velocity after the first and second time intervals for the suddenly accelerated flat plate.

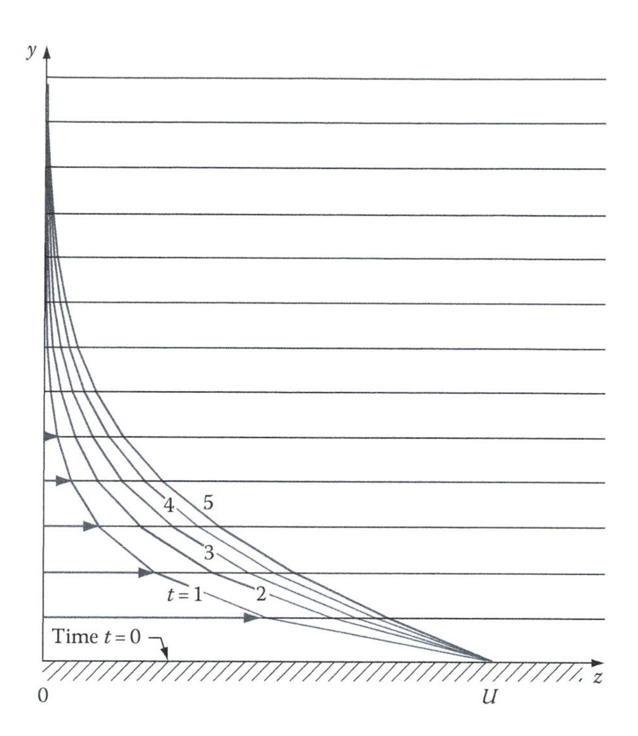

FIGURE 11.15 Solution for the velocity for the first five time intervals for the suddenly accelerated flat plate.

Note that if this problem were solved with a numerical calculation technique, the results would be in the form of numbers for the velocities at the nodes; that is, the results would give a numerical value of the velocity at the points in Figure 11.14 labeled *a, b, c, d, e, f,* and so on.

11.3.2 Unsteady Plane Couette Flow

As stated in the previous section, plane Couette flow is flow through a two-dimensional channel that has one wall moving. In this instance, however, we examine the case where there is no pressure gradient. The channel height is h, and one of the walls is given a sudden acceleration from rest to a uniform velocity U. The fluid is Newtonian with constant properties. Figure 11.16 is a sketch of the problem statement.

The formulation here is the same as in the last problem:

$$V_y = V_x = 0$$
$$V_z = V_z(y,t)$$

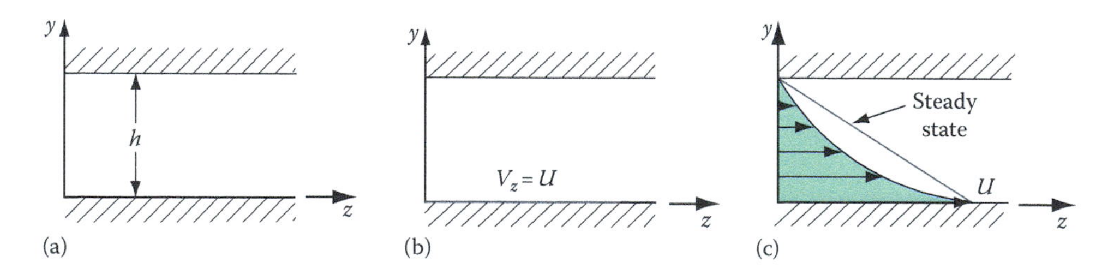

FIGURE 11.16 Unsteady plane Couette flow (unsteady flow in a two-dimensional channel). (a) $t < 0$, (b) $t = 0$, and (c) $t > 0$.

The continuity and momentum equations (Equations 11.1 and 11.2, respectively) become

$$\frac{\partial V_z}{\partial z} = 0$$

$$\rho \frac{\partial V_z}{\partial t} = \mu \frac{\partial^2 V_z}{\partial y^2} \tag{11.37}$$

with boundary conditions (both nonslip conditions):

1. $t \geq 0 \quad y = 0 \quad V_z = U$
2. $t \geq 0 \quad y = h \quad V_z = 0$

The analytical solution to this problem is usually expressed as an infinite series of error functions. Alternatively, Figure 11.17 is the graphical solution using the Saul'ev method. The channel width is divided arbitrarily into 10 Δy's. With this value, we use the cell Reynolds number and find the time interval to be

$$\mathrm{Re}_c = \frac{\rho \Delta y^2}{\mu t} = \frac{\Delta y^2}{\nu t} = 1$$

and so

$$t = \frac{\Delta y^2}{\nu}$$

With $h = 10\ \Delta y$, we find

$$t = \frac{h^2}{100\nu}$$

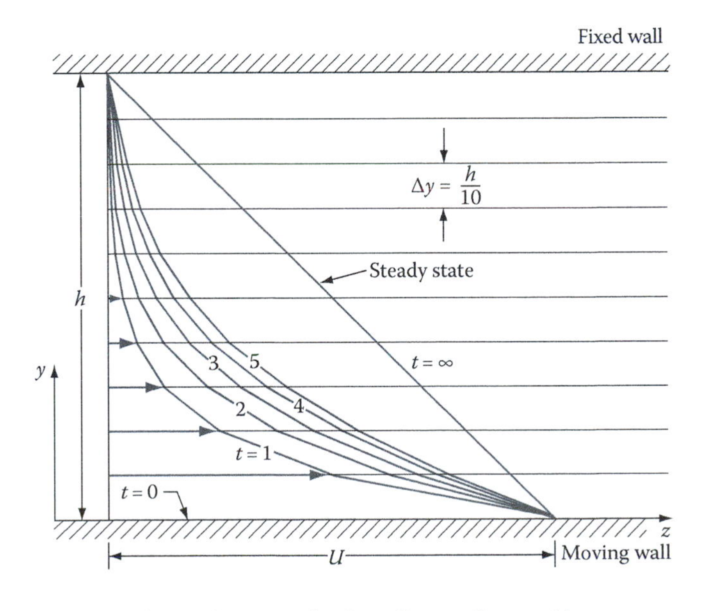

FIGURE 11.17 Graphical solution to the unsteady plane Couette flow problem.

11.3.3 Unsteady Flow between Concentric Circular Cylinders

Figure 11.18 shows the plan view of two concentric cylinders. At time = 0, the inner cylinder rotates at a fixed angular velocity. The outer cylinder remains stationary. Assuming unsteady laminar flow of a Newtonian fluid with no gravity forces or pressure gradients, we conclude that

$$V_r = V_z = 0$$

$$V_\theta = V_\theta(y,t)$$

The continuity and momentum equations become

$$\frac{1}{r}\frac{\partial}{\partial\theta}(\rho V_\theta) = 0$$

$$\rho\frac{\partial V_\theta}{\partial t} = \mu\left(\frac{\partial}{\partial r}\left[\frac{1}{r}\frac{\partial}{\partial r}(rV_\theta)\right]\right) \tag{11.38}$$

with the following conditions:

1. $r = R_2 \quad V_\theta = 0$
2. $r = R_1 \quad V_\theta = R_1\omega = U$

When the right-hand side of Equation 11.38 is differentiated, according to the chain rule, we get

$$\mu\left(\frac{\partial}{\partial r}\left[\frac{1}{r}\frac{\partial}{\partial r}(rV_\theta)\right]\right) = \mu\left(\frac{\partial^2 V_\theta}{\partial r^2} + \frac{1}{r}\frac{\partial V_\theta}{\partial r}\right)$$

The second term in parentheses on the right-hand side is known as a **curvature term**. It is this term that makes the differential equation in its present form unsuitable to solve graphically. So, further analysis is required. We introduce a new independent variable ζ, defined as

$$\zeta = \ell n(r) \tag{11.39a}$$

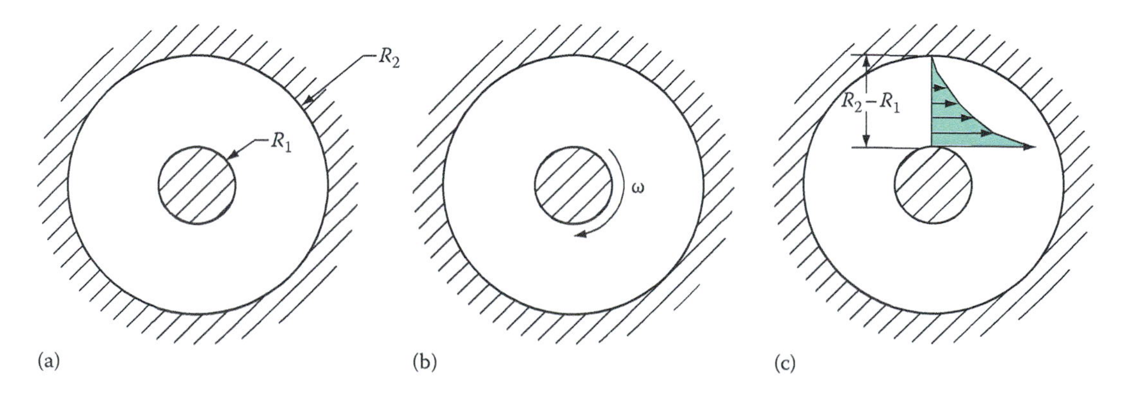

FIGURE 11.18 Unsteady flow between concentric cylinders with only the inner cylinder rotating. (a) $t < 0$, (b) $t = 0$, and (c) $t \geq 0$.

or

$$r = e^{\zeta} \tag{11.39b}$$

The derivative with respect to r, as found in the differential equation (Equation 11.38), is evaluated as

$$\frac{\partial}{\partial r} = \frac{\partial}{\partial \zeta}\frac{\partial \zeta}{\partial r} = \frac{\partial}{\partial \zeta}\frac{1}{r} = \frac{\partial}{\partial \zeta}\frac{1}{e^{\zeta}}$$

Substituting into the differential equation gives, after simplification,

$$\rho\frac{\partial V_{\theta}}{\partial t} = \frac{\mu}{r^2}\frac{\partial^2 V_{\theta}}{\partial \zeta^2} \tag{11.40}$$

This form is suitable for graphical solution.

The cell Reynolds number for the preceding equation becomes

$$\text{Re}_c = \frac{\rho(r\,\Delta\zeta)^2}{\mu\,\Delta t} = 1$$

In order to solve this problem, it is necessary to perform graphically the transformation given in Equation 11.39. This transformation is illustrated in Figure 11.19. Shown are two grids that are

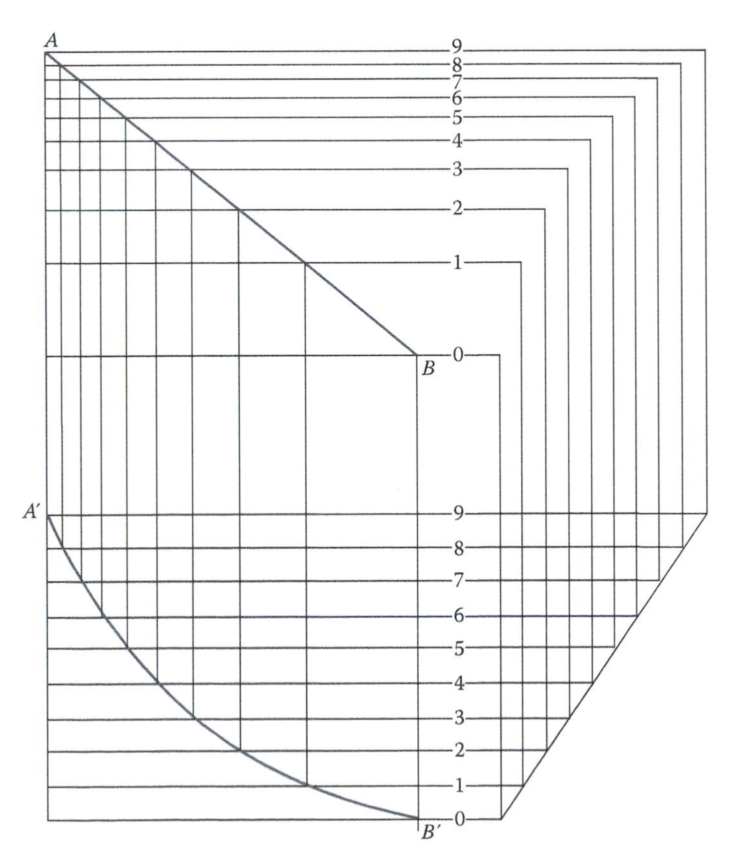

FIGURE 11.19 Effects of the logarithmic transformation function.

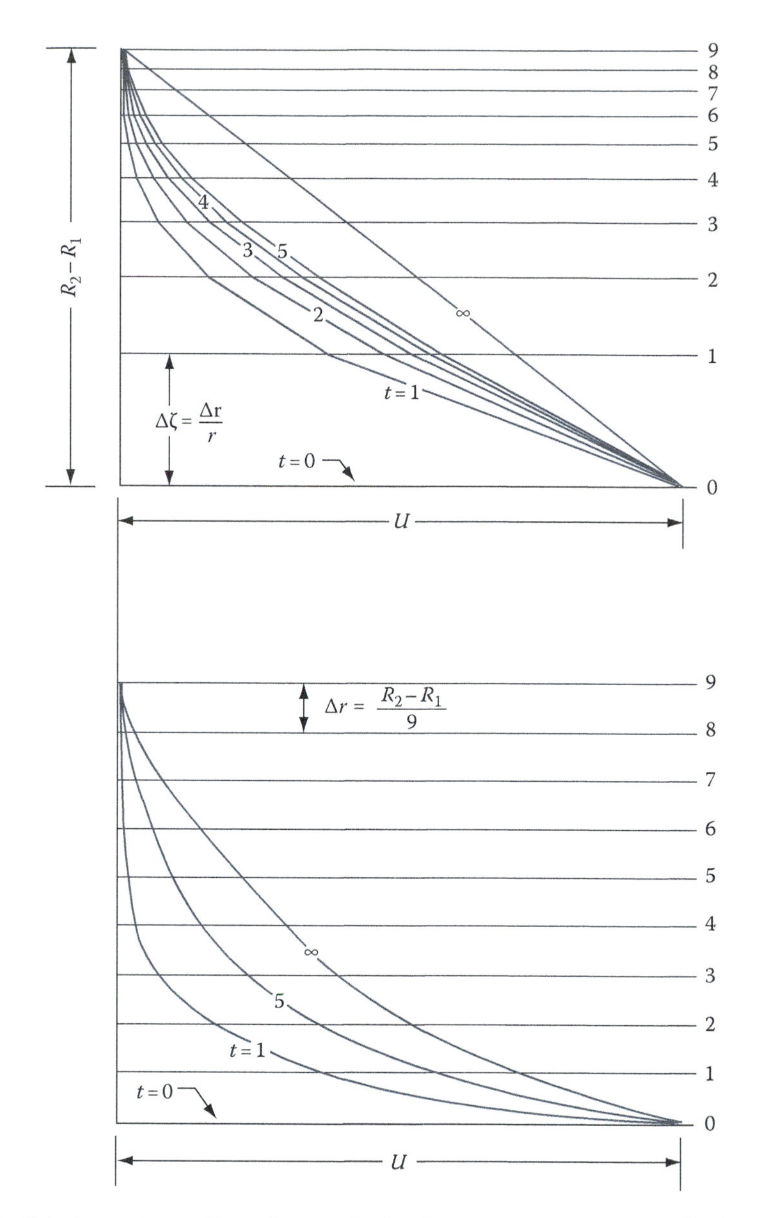

FIGURE 11.20 Solution to the problem of unsteady flow between two concentric cylinders.

aligned vertically. The top grid is logarithmic, with the widest mesh located near the inner cylinder. The bottom mesh is linearly graded. Both meshes have 10 lines, which divide the flow field into nine spaces. The number of Δy's is fixed by the logarithmic grid. Given the line AB in the log mesh, we project each nodal point downward to the linear mesh to obtain the corresponding line $A'B'$.

With regard to the viscous flow problem at hand, unsteady flow between concentric cylinders, the transformation allows us to perform the solution by drawing straight lines on the log mesh. Projecting the resulting points down the linear mesh gives us the solution in the radial plane. The solution for several times is given in Figure 11.20. With the cell Reynolds number equal to 1, the time increment is found as

$$t = \frac{\rho(R_2 - R_1)^2}{81\mu}$$

11.3.4 Unsteady Flow in a Plane Channel (Start-Up Flow)

Consider a plane channel of width $2h$ that contains a fluid. At time = 0, a pressure drop Δp is suddenly imposed on the system, and the fluid begins to flow in the direction of decreasing pressure. Figure 11.21 is a sketch of the problem statement.

The pressure gradient is given by

$$-\frac{dp}{dz} = \left(-\frac{\Delta p}{L}\right)$$

As before, we assume unsteady laminar flow of a Newtonian fluid with no gravity forces; we conclude that

$$V_y = V_x = 0$$

$$V_z = V_z(y,t)$$

The continuity and momentum equations (Equations 11.1 and 11.2, respectively) become

$$\frac{\partial V_z}{\partial_z} = 0$$

$$\rho\frac{\partial V_z}{\partial_t} = \mu\frac{\partial^2 V_z}{\partial y^2} - \frac{\Delta p}{L} \tag{11.41}$$

with the following boundary conditions:

1. $t \geq 0 \quad y = h \quad V_z = 0$ (nonslip condition)
2. $t \geq 0 \quad y = 0 \quad \dfrac{\partial v_z}{\partial y} = 0$ (finite velocity at centerline)

The pressure gradient added to the equation in this fashion means, graphically, that we add $\Delta pt/\rho L$ to each velocity profile in order to determine the succeeding one. To illustrate, consider the grid of Figure 11.22. The field (which is half the channel width) is divided into five regions Δy wide. The line labeled OO' is the initial velocity. To this initial velocity, we add $\Delta pt/\rho L$ at every node to obtain points labeled a through f. We draw the $t = 1$ line in the usual way, beginning at point O and using the points c through f. The profile is drawn perpendicular to the topmost line (the centerline of the channel), which represents the second boundary condition $dV_z/dy = 0$. Next we add $\Delta pt/\rho L$ to every node of the $t = 1$ profile to obtain the points labeled g through k. This profile, as before, should meet the centerline at a right angle.

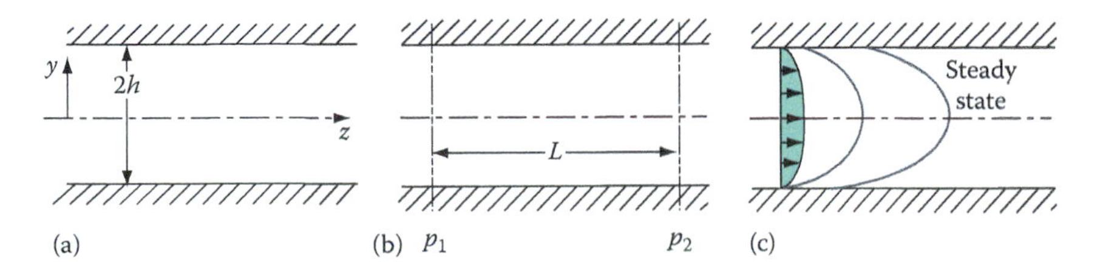

FIGURE 11.21 Unsteady start-up flow in a plane channel. (a) $t < 0$, (b) $t = 0$, and (c) $t > 0$.

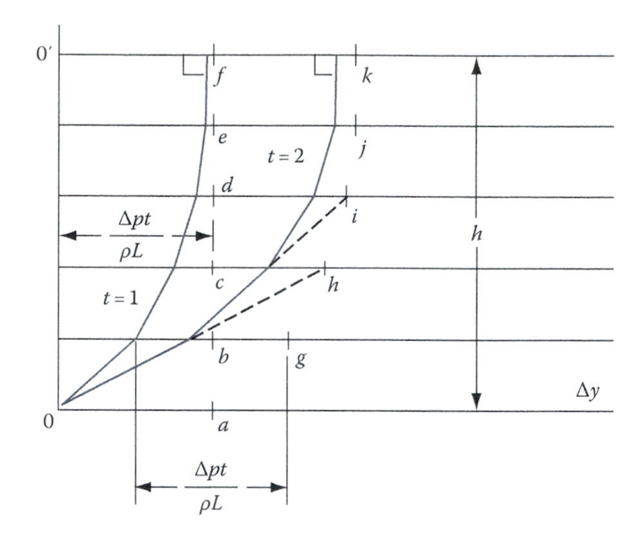

FIGURE 11.22 Illustration of the imposition of a pressure drop on a flow field.

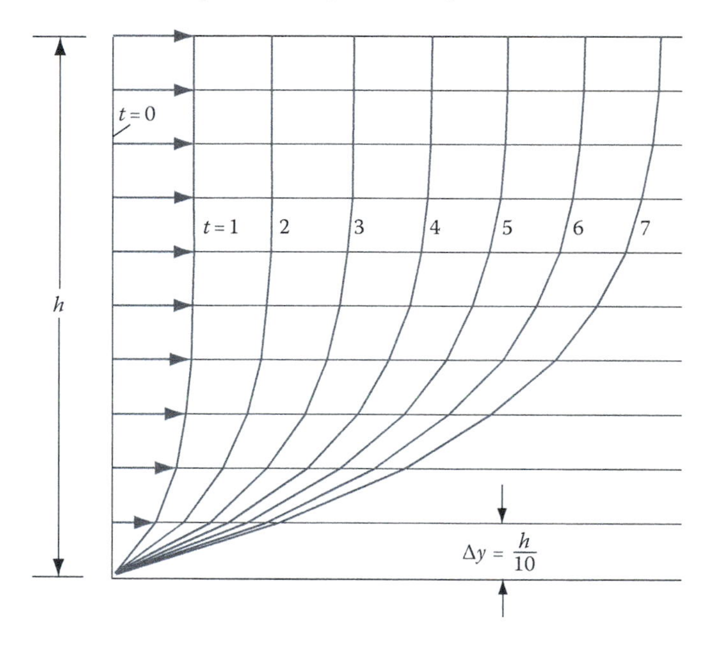

FIGURE 11.23 Solution for the first seven time intervals for unsteady start-up in a plane channel.

Figure 11.23 shows the solution for the first seven time intervals for the problem at hand—unsteady (start-up) flow in a plane channel of width $2h$. The channel half-width is divided into 10 Δy sections. At the centerline, each profile intersects the line at a 90° angle.

11.4 INTRODUCTION TO TURBULENT FLOW

In the preceding sections, we discussed only laminar flow problems and derived the differential equations for laminar flow situations from general equations. In many cases, the nonlinear terms vanish, and the remaining expression, if not too complex, can be solved. Although there is no correspondingly simple methodology for turbulent flow, the Navier–Stokes equations apply. None of the velocities in the Navier–Stokes equations vanish. Solving them, if possible, yields instantaneous

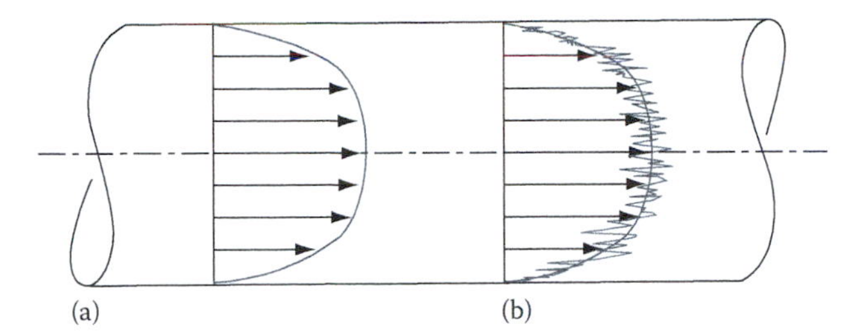

FIGURE 11.24 Turbulent flow in a tube.

values of velocity and pressure. In turbulent flow, as was indicated briefly in Chapter 5, velocity and pressure fluctuate wildly about mean values. To illustrate, consider turbulent flow in a circular tube, as shown in Figure 11.24. If the velocity at each point in the cross section were measured with a pitot-static tube, the resultant distribution would be that of Figure 11.24a. If a more sensitive instrument, such as a hot-wire anemometer, were used, a distribution such as that in Figure 11.24b would result. The instantaneous velocity at any point oscillates randomly about a mean value.

Laminar flow exists in a pipe when the Reynolds number ($\rho VD/\mu$) is less than about 2 100. Above 2 100, laminar flow can be maintained if the pipe wall is smooth and no vibrations are present. Any slight disturbance would probably induce the random motion recognized as turbulent flow.

It should be mentioned that for turbulent flow, a region exists near the wall where the flow is not random. A change in the flow exists near the wall where velocity is zero. It is customary to define three arbitrary zones within a tube as shown in Figure 11.25. In the central region of the tube, a fully developed turbulent flow exists. Near the wall, a laminar sublayer forms within which Newton's law of viscosity describes the flow. Between the two is the buffer zone, within which both laminar and turbulent effects are considered to be important.

Now we examine in detail the velocity fluctuations introduced in Figure 11.24b. Suppose we are using a sensitive instrument to measure velocity at a point in a tube in which turbulent flow exists. The results that we would obtain are shown schematically in Figure 11.26. Two velocities can be identified: the mean velocity $\bar{V}_z$, which is an average or mean value, and the instantaneous velocity V_z, which fluctuates randomly about the mean. The instantaneous velocity can be time-smoothed to obtain the mean; that is, if the instantaneous velocity is averaged over a finite time interval, the mean value results. By definition, then,

$$\bar{V}_z = \frac{1}{t}\int_0^t V_z \, dt \tag{11.42}$$

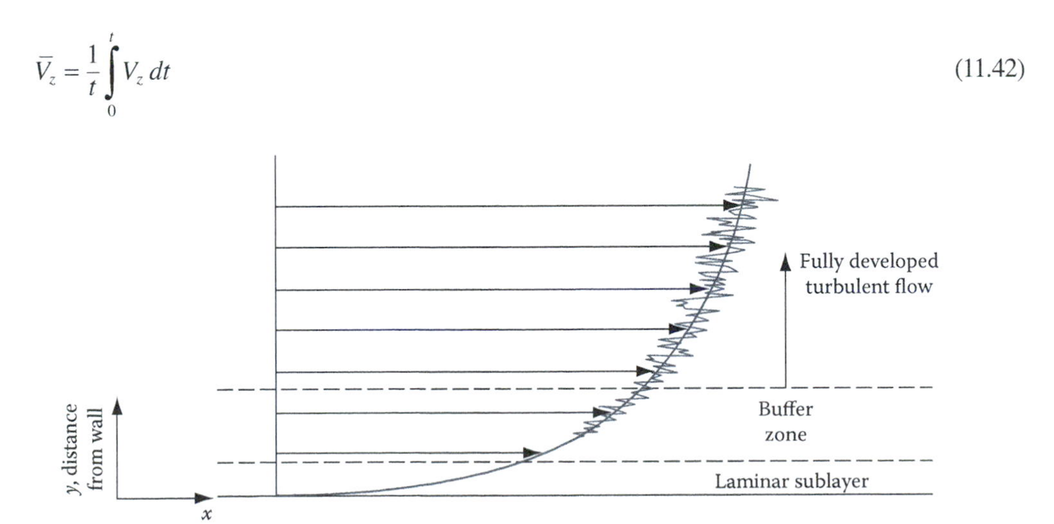

FIGURE 11.25 Flow near a pipe wall.

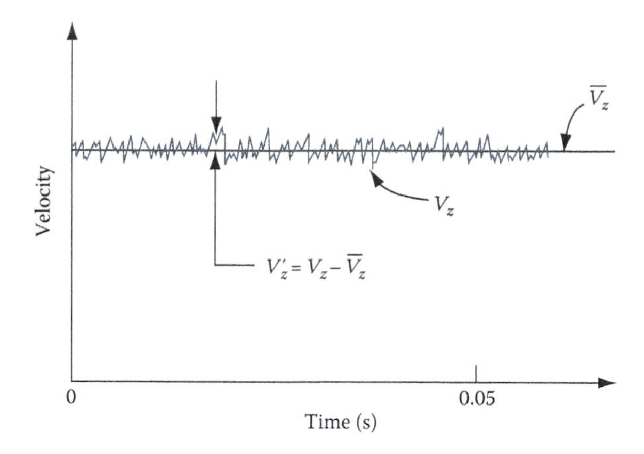

FIGURE 11.26 Velocity at a point in a turbulent flow.

The instantaneous velocity is a function of all three space variables and time. By using Equation 11.42, we have integrated out the dependence on time. The mean velocity and the instantaneous velocity differ by a velocity fluctuation V_z'. In equation form,

$$V_z = \overline{V_z} + V_z' \tag{11.43}$$

By substitution into Equation 11.42, we find

$$\overline{V_z} = \frac{1}{t}\int_0^t \overline{V_z}\,dt + \frac{1}{t}\int_0^t V_z'\,dt$$

$$= \frac{\overline{V_z}}{t}\int_0^t dt + \frac{1}{t}\int_0^t V_z'\,dt$$

or

$$\overline{V_z} = \overline{V_z} + \frac{1}{t}\int_0^t V_z'\,dt$$

We therefore conclude that the time average of the fluctuations is zero; that is,

$$\frac{1}{t}\int_0^t V_z'\,dt = \overline{V_z}' = 0 \tag{11.44}$$

Further manipulation, however, will show that $\overline{V_z'^2} \neq 0$. In fact, a measure of the magnitude of turbulence is given as

$$\text{Intensity of turbulence} = \frac{\sqrt{\overline{V_z'^2}}}{\overline{V}} \tag{11.45}$$

where $\overline{V}$ is the time-smoothed average velocity. The intensity of turbulence varies in tube flow from about 1% to 10%. A similar time-smoothing operation can be performed for pressure at a point that also fluctuates in turbulent flow.

We are now in a position to time-smooth the equations of motion—namely, the continuity and Navier–Stokes equations. To perform this task, we substitute $\overline{V_i} + V'$ for V_i and $\overline{p} + p'$ for p

everywhere in Equations 11.1 and 11.2 in Cartesian coordinates. By integrating as indicated in Equation 11.42, the results for steady incompressible turbulent flow become the following.

Continuity equation:

$$\frac{\partial(\bar{V}_x)}{\partial x}+\frac{\partial(\bar{V}_y)}{\partial y}+\frac{\partial(\bar{V}_z)}{\partial z}=0 \tag{11.46}$$

Equations of motion:

x-component:

$$\begin{aligned}\rho\left(\bar{V}_x\frac{\partial\bar{V}_x}{\partial x}+\bar{V}_y\frac{\partial\bar{V}_x}{\partial y}+\bar{V}_z\frac{\partial\bar{V}_x}{\partial z}\right)&=-\frac{\partial(\bar{p})}{\partial x}+\mu\left(\frac{\partial^2\bar{V}_x}{\partial x^2}+\frac{\partial^2\bar{V}_x}{\partial y^2}+\frac{\partial^2\bar{V}_x}{\partial z^2}\right)\\&\quad\underline{-\frac{\partial}{\partial x}\left(\rho\overline{V_x'V_x'}\right)-\frac{\partial}{\partial y}\left(\rho\overline{V_x'V_y'}\right)-\frac{\partial}{\partial z}\left(\rho\overline{V_x'V_z'}\right)+\rho g_x}\end{aligned} \tag{11.47}$$

y-component:

$$\begin{aligned}\rho\left(\bar{V}_x\frac{\partial\bar{V}_y}{\partial x}+\bar{V}_y\frac{\partial\bar{V}_y}{\partial y}+\bar{V}_z\frac{\partial\bar{V}_y}{\partial z}\right)&=-\frac{\partial(\bar{p})}{\partial x}+\mu\left(\frac{\partial^2\bar{V}_y}{\partial x^2}+\frac{\partial^2\bar{V}_y}{\partial y^2}+\frac{\partial^2\bar{V}_y}{\partial z^2}\right)\\&\quad\underline{-\frac{\partial}{\partial x}\left(\rho\overline{V_x'V_y'}\right)-\frac{\partial}{\partial y}\left(\rho\overline{V_y'V_y'}\right)-\frac{\partial}{\partial z}\left(\rho\overline{V_y'V_z'}\right)+\rho g_y}\end{aligned} \tag{11.48}$$

z-component:

$$\begin{aligned}\rho\left(\bar{V}_x\frac{\partial\bar{V}_z}{\partial x}+\bar{V}_y\frac{\partial\bar{V}_z}{\partial y}+\bar{V}_z\frac{\partial\bar{V}_z}{\partial z}\right)&=-\frac{\partial(\bar{p})}{\partial x}+\mu\left(\frac{\partial^2\bar{V}_z}{\partial x^2}+\frac{\partial^2\bar{V}_z}{\partial y^2}+\frac{\partial^2\bar{V}_z}{\partial z^2}\right)\\&\quad\underline{-\frac{\partial}{\partial x}\left(\rho\overline{V_x'V_z'}\right)-\frac{\partial}{\partial y}\left(\rho\overline{V_y'V_z'}\right)-\frac{\partial}{\partial z}\left(\rho\overline{V_z'V_z'}\right)+\rho g_z}\end{aligned} \tag{11.49}$$

These equations are similar to the Navier–Stokes equations except that the time-smoothed velocities replace the instantaneous velocities. Moreover, these equations of motion have the time-smoothed pressure. Finally, new terms have appeared. These new terms (underscored) are related to the turbulent velocity fluctuations. The dimension of the product ρVV is force per unit area. It is therefore convenient to introduce the notation

$$\begin{aligned}\bar{\tau}_{xx}^{(t)}&=\rho\overline{V_x'V_x'}\\\bar{\tau}_{xy}^{(t)}&=\rho\overline{V_x'V_y'}\\\bar{\tau}_{xz}^{(t)}&=\rho\overline{V_x'V_z'}\end{aligned} \tag{11.50}$$

and so on. These terms make up what is referred to as the **turbulent momentum flux** and are usually called **Reynolds stresses**.

In vector notation, the continuity equation and the equations of motion become the following.

Continuity: is $\nabla \cdot \mathbf{V} = 0$ (11.51)

Equation of motion:

$$\frac{\partial \mathbf{V}}{\partial t} + (\mathbf{V} \cdot \nabla)V = -\frac{\nabla p}{\rho} - \nabla g + \nu \nabla^2 \mathbf{V} + \underline{\nabla \cdot \rho \mathbf{V}'\mathbf{V}'} \tag{11.52}$$

where it is understood that the terms $\mathbf{V}$ are mean velocity vectors and $\mathbf{V}'\mathbf{V}'$ are the turbulent velocity fluctuations.

To apply Equations 11.46 through 11.49 to various flows and obtain velocity distributions, it is necessary to relate the Reynolds stresses to a strain rate in the fluid. For laminar flow, shear stresses are related to strain rate by Newton's law of viscosity: $\tau_{yx} = -\mu(dV_x/dy)$ For turbulent flow, we use an analogous formulation to write

$$\overline{\tau}_{yx}^{(t)} = -\mu^{(t)} \frac{d\overline{V_x}}{dy} \tag{11.53}$$

where $\mu^{(t)}$ is the turbulent coefficient of viscosity, also called the **eddy viscosity**. The total mean shear stress is then

$$\overline{\tau}_{yx} = \overline{\tau}_{yx}^{(l)} + \overline{\tau}_{yx}^{(t)}$$

$$\overline{\tau}_{yx} = -\mu \frac{d\overline{V_x}}{dy} - \mu^{(t)} \frac{d\overline{V_x}}{dy}$$

where

μ is a viscosity that results from molecular motions

$\mu^{(t)}$ is the eddy viscosity that results from eddying motions in turbulent flow

Physically, this motion implies that the molecular action that is responsible for viscous shear is analogous to the eddies that cause turbulent stress. This implication is a semiempirical result that was proposed early in the study of turbulent flow.

A second empirical relationship can be written by assuming that eddies move about in a fluid in the same way that molecules move about in a gas. This **mixing-length hypothesis** is illustrated in Figure 11.27. Two points, y_1 and y_2, are selected from the boundary such that they extend well into the turbulent portion of the flow. These points are separated by a distance l that equals the average size of the eddies:

$$y_1 - y_2 = l$$

The time-smoothed velocities corresponding to y_1 and y_2 are $\overline{V}_{x_1}$ and $\overline{V}_{x_2}$ It is theorized that a positive turbulent fluctuation in the vertical direction at $y_2(V'_{y_2})$ causes a corresponding reduction in the x-directed component of V_{x_1}, the instantaneous velocity at y_1. The reduction V'_{x_1} is approximately equal to the differences in mean velocities. So at y_1,

$$V'_{x_1} \approx \overline{V}_1 - \overline{V}_2 \approx \ell \frac{d\overline{V}_x}{dy}$$

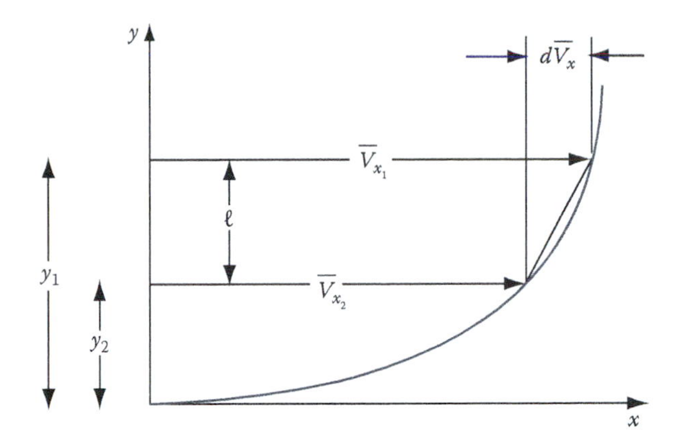

FIGURE 11.27 Illustration of mixing length ℓ.

The shear stress $\overline{\tau}_{yx}^{(t)}$ becomes (from Equation 11.50)

$$\overline{\tau}_{yx}^{(t)} = \rho\overline{V_x' V_y'} \approx \rho V_y' \ell \frac{d\overline{V}_x}{dy} \tag{11.54}$$

If it is further assumed that the turbulent fluctuation V_y' is approximately equal to the horizontal fluctuation V_x', then Equation 11.54 becomes

$$\overline{\tau}_{yx}^{(t)} \approx \rho \ell^2 \frac{d\overline{V}_x}{dy}\left|\frac{d\overline{V}_x}{dy}\right| \tag{11.55}$$

where absolute value signs are placed to ensure that the sign on the shear stress properly reflects the sign of the gradient.

A third expression for the Reynolds stresses has resulted from the performance of many experiments. This empirical expression is intended for use in the vicinity of solid surfaces:

$$\overline{\tau}_{yx}^{(t)} = -\rho n^2 \overline{V}_x y \left(1 - \exp\left\{-\frac{n^2 \overline{V}_x y}{\nu}\right\}\right)\frac{d\overline{V}_x}{dy} \tag{11.56}$$

The parameter n is a constant that has been determined to be 0.124 from measurements in turbulent flow in tubes.

The mixing-length expression can be used to derive an equation for velocity distribution for turbulent flow in a circular duct. The tube radius is R, and we will take the variable y to be

$$y = R - r$$

Thus, the independent variable y is measured from the wall as opposed to r, which is measured from the centerline.

For flow in tubes, a formulation similar to that in laminar flow can be performed by using the time-smoothed equations. The only velocity that we consider is $\overline{V}_z$, the mean axial flow velocity. Moreover, this velocity is a function of only the radial coordinate r. In terms of shear stress, the equation of motion in cylindrical coordinates is

$$0 = -\frac{dp}{dz} - \frac{1}{r}\frac{d(r\overline{\tau}_{rz})}{dr}$$

where $\bar{\tau}_{rz} = \bar{\tau}_{rz}^{(\ell)} + \bar{\tau}_{rz}^{(t)}$. Integrating the preceding equation gives

$$\frac{d(r\bar{\tau}_{rz})}{dr} = -r\frac{dp}{dz}$$

$$r\bar{\tau}_{rz} = -\frac{r^2}{2}\frac{dp}{dz} + C_1$$

At the centerline, the velocity must be finite. Therefore, one boundary condition is

$$r = 0 \qquad \bar{\tau}_{rz} = 0$$

This gives $C_1 = 0$. The equation now becomes

$$\bar{\tau}_{rz} = -\frac{r}{2}\frac{dp}{dz} = -\frac{r}{R}\frac{R}{2}\frac{dp}{dz} \tag{11.57}$$

Equation 11.13 gives the wall shear stress for flow in a tube as

$$\tau_w = -\frac{R}{2}\frac{dp}{dz}$$

Combining with Equation 11.57 gives

$$\bar{\tau}_{rz} = \frac{r}{R}\tau_w$$

With $y = R - r$, this becomes

$$\bar{\tau}_{rz} = \tau_w\left(1 - \frac{y}{R}\right)$$

Therefore,

$$\bar{\tau}_{rz}^{(\ell)} + \bar{\tau}_{rz}^{(t)} = \tau_w\left(1 - \frac{y}{R}\right) \tag{11.58}$$

Two simplifying assumptions can now be made. The first is that molecular motions in the turbulent core are negligible in comparison to eddying motion. Thus,

$$\bar{\tau}_{rz}^{(\ell)} \ll \bar{\tau}_{rz}^{(t)}$$

The second is that

$$\tau_w\left(1 - \frac{y}{R}\right) \approx \tau_w$$

which is introduced to make the mathematics easier. Equation 11.58 now reduces to

$$\bar{\tau}_{rz}^{(t)} = \tau_w$$

At this point, it is necessary to introduce Equation 11.55 written in cylindrical coordinates to relate $\overline{\tau}_{rz}$ to a strain rate. For axial tube flow,

$$\overline{\tau}_{rz}^{(t)} = \rho \ell^2 \left(\frac{dV_z}{dr} \right)^2 \tag{11.59}$$

We must have a relationship between the mixing length and the independent variable. A suitable relationship results if it is assumed that the mixing length is proportional to ℓ:

$$l = ky$$

where k is a proportionality constant. In terms of y, Equation 11.59 becomes

$$\overline{\tau}_{rz}^{(t)} = \rho k^2 y^2 \left(\frac{dV_2}{dy} \right)^2 = \tau_w$$

Rearranging to solve for the velocity gradient, we get

$$\frac{d\overline{V}_z}{dy} = \pm \sqrt{\frac{\tau_w}{\rho}} \frac{1}{ky}$$

where the negative sign will have to be rejected. The square root term has the dimensions of a velocity and may for convenience be rewritten as V^*. The differential equation now becomes

$$\frac{d\overline{V}_z}{dy} = \frac{V^*}{ky}$$

Integrating this expression gives

$$\overline{V}_z = \frac{V^*}{k} \ln(y) + C_1 \tag{11.60}$$

At the edge of the buffer layer, at y_b, the velocity is $\overline{V}_{zb}$. Applying this boundary condition, we obtain

$$\overline{V}_{zb} = \frac{V^*}{k} \ln(y_b) + C_1$$

$$C_1 = \overline{V}_{zb} - \frac{V^*}{k} \ln(y_b)$$

Substitution into Equation 11.60 gives

$$\overline{V}_z = \frac{V^*}{k} \ln(y) + \overline{V}_{zb} - \frac{V^*}{k} \ln(y)$$

or

$$\frac{\overline{V}_z}{V^*} = \frac{1}{k} \ln\left(\frac{y}{y_b} \right) + \frac{\overline{V}_{zb}}{V^*}$$

Now let $s = \rho V^* y/\mu$ and $s_b = \rho V^* y_b/\mu$, both dimensionless Reynolds numbers. The velocity distribution now becomes

$$\frac{\bar{V}_z}{V^*} = \frac{1}{k}\ln\left(\frac{s}{s_b}\right) + \frac{\bar{V}_{zb}}{V^*}$$

Experiment has shown that the best value of k is 0.36. Moreover, the outer edge of the buffer zone can be selected at $s_b = 26$, where correspondingly $\bar{V}_{zb}/V^* = 12.85$. After substitution and simplification, we finally obtain

$$\frac{\bar{V}_z}{V^*} = \frac{1}{0.36}\ln(s) + 3.8 \qquad s \geq 26 \tag{11.61}$$

This is the well-known logarithmic distribution. It has been found that for Reynolds numbers greater than 20 000, Equation 11.61 gives a good description in the turbulent core. Near the wall, Equation 11.56 can be used with success. We will not derive an expression for velocity, however, because iterative integration procedures are required, and these are beyond the scope of our study. Nevertheless, a comparison between the results given here and experiment has shown good agreement. A plot of the turbulent profile is given in Figure 11.28.

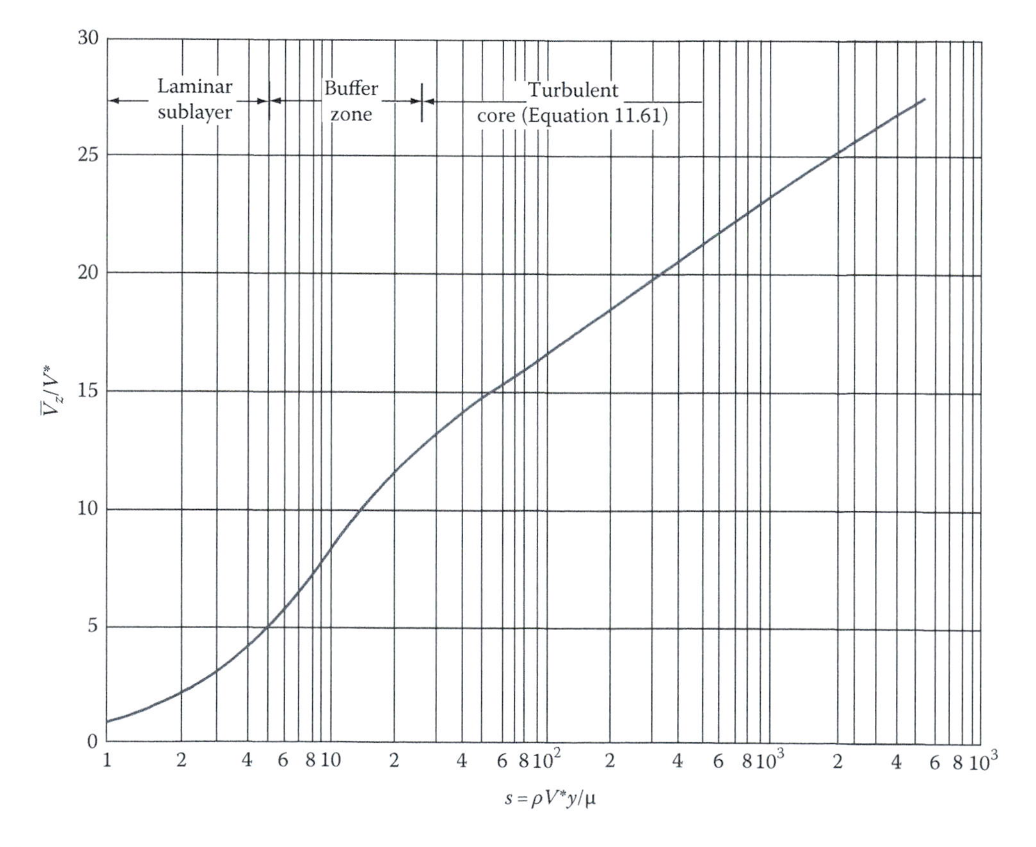

FIGURE 11.28 Velocity distribution for turbulent, incompressible flow in a tube. (Adapted from Deissler, R.G., NACA Report, 1210, 1955.)

11.5 SUMMARY

The Navier–Stokes equations for a general fluid mechanics problem were stated in this chapter. Descriptive equations were derived for various laminar flow problems. Graphical solution techniques for unsteady laminar flow problems were also presented. Next we obtained the equations of motion for turbulent flow. These equations were then simplified and applied to the problem of turbulent flow in a tube.

PROBLEMS

Steady Laminar Flow Problems

11.1 Write the boundary conditions for steady laminar flow of a Newtonian fluid in an annulus.

11.2 Show that for steady laminar flow of a Newtonian fluid in an annulus, the velocity is

$$V_z = \left(-\frac{dp}{dz}\right)\frac{R^2}{4\mu}\left[1-\left(\frac{r}{R}\right)^2+\frac{1-k^2}{\ln(1/k)}\ln\left(\frac{r}{R}\right)\right]$$

11.3 Verify the derivation of Equation 11.11.

11.4 a. Determine the maximum velocity $V_{z\,max}$ for flow in a circular duct.

b. Express the instantaneous velocity V_z in terms of $V_{z\,max}$.

c. Write the equation for volume flow rate (Equation 11.11) in terms of $V_{z\,max}$.

11.5 Verify the derivation of Equation 11.15.

11.6 a. Determine the maximum velocity $V_{z\,max}$ for flow down an incline.

b. Express the instantaneous velocity V_z in terms of $V_{z\,max}$.

c. Write the equation for volume flow rate (Equation 11.15) in terms of $V_{z\,max}$.

11.7 Solve the differential equation for flow through a straight channel and derive Equation 11.17.

11.8 Integrate the equation for instantaneous velocity V_z through a straight channel over the cross-sectional area and derive Equation 11.18.

11.9 Derive the velocity profile of Equation 11.21.

11.10 For plane Couette flow, the velocity profile is

$$\frac{V_z}{U} = \frac{y}{h}+\left(-\frac{dp}{dz}\right)\left(\frac{h^2}{2\mu U}\right)\left(\frac{y}{h}\right)\left(1-\frac{y}{h}\right)$$

Integrate this equation over the cross-sectional area and show that the volume flow rate is given by

$$Q = \frac{Ubh}{2}+\frac{bh^3}{12\mu}\left(-\frac{dp}{dz}\right)$$

where b is the width of the flow area.

11.11 For plane Couette flow, the volume flow rate is given by

$$Q = \frac{Ubh}{2}+\frac{bh^3}{12\mu}\left(-\frac{dp}{dz}\right)$$

a. Show that the required relationship between the velocity of the upper plate and the pressure drop such that the volume flow rate is zero is given by

$$U = -\frac{h^2}{6\mu}\left(-\frac{dp}{dz}\right)$$

b. Substitute this relationship into the instantaneous velocity profile and show that for $Q = 0$,

$$\frac{V_z}{U} = \frac{y}{h} - 3\left(\frac{y}{h}\right)\left(1 - \frac{y}{h}\right)$$

11.12 Verify that Figure 11.9 is a correct representation.

11.13 Derive Equation 11.26 for flow between two rotating concentric cylinders.

11.14 Consider a cylinder moving axially within a tube, as illustrated in Figure P11.14. Write the differential equation and boundary conditions for steady laminar axial flow through the annulus.

11.15 Show that for the system of Problem 11.14, the fluid velocity is given as

$$V_z = V\frac{\ln(r/R)}{\ln(k)}$$

11.16 Consider the flow of two immiscible fluids between two flat plates as shown in Figure P11.16. A pressure gradient imposed from inlet to outlet causes flow.

Write the differential equation and boundary conditions for the system if the flow is steady and laminar.

11.17 Consider two coaxial cylinders with fluid in the space between them, as shown in Figure P11.17. The outer cylinder is rotating at an angular velocity of ω_0. Write the differential equation and boundary conditions for the system if the flow is steady and laminar.

11.18 Repeat Problem 11.17 for the case in which only the inner cylinder is rotating at ω_i.

11.19 A Newtonian liquid is being rotated in a cylindrical container of radius R as shown in Figure P11.19. Write the differential equation and boundary conditions if the angular velocity is ω.

11.20 Starting with the continuity equation in Cartesian conditions, use the definitions

$$x = r\cos\theta$$

$$y = r\sin\theta\ s$$

$$z = z$$

to obtain the continuity equation in cylindrical coordinates.

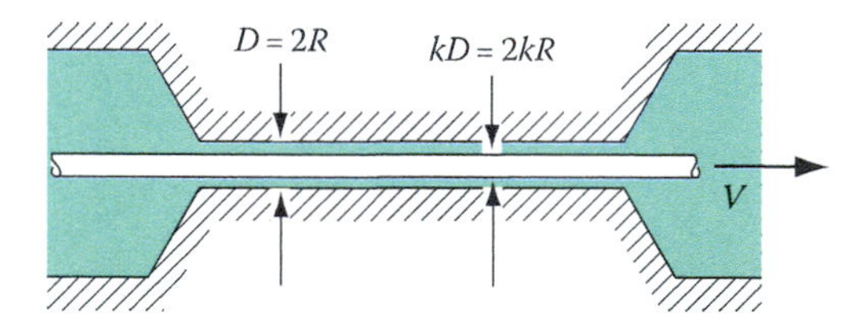

FIGURE P11.14

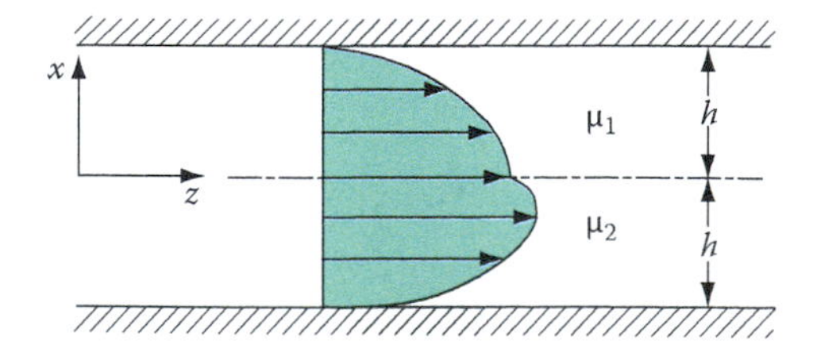

FIGURE P11.16

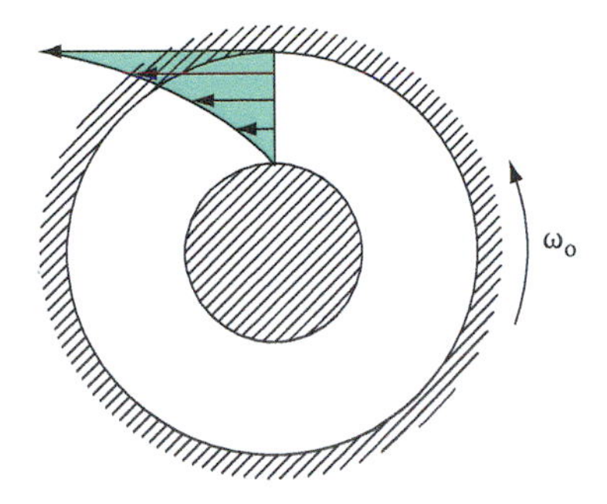

FIGURE P11.17

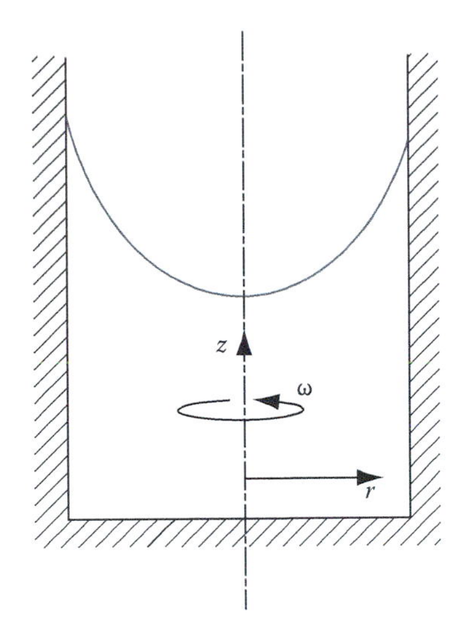

FIGURE P11.19

11.21 A wall is in contact with a fluid. Suddenly, the wall is set in motion as illustrated in Figure P11.21. Verify that the differential equation is

$$\frac{\partial V_z}{\partial t} = \frac{\mu}{\rho}\frac{\partial^2 V_z}{\partial y^2}$$

and that the initial and boundary conditions are

$$\begin{aligned}
&\text{IC}: && t \le 0, && V_z = 0 \\
&\text{BC 1}: && y = 0, && V_z = U, && t > 0 \\
&\text{BC 2}: && y = \infty, && V_z = 0, && t > 0
\end{aligned}$$

11.22 A Newtonian fluid in a pipe is initially at rest. Suddenly, a pressure drop dp/dz is imposed on the fluid (in the axial direction). Show that the differential equation, the initial condition, and the boundary conditions are

$$\rho\frac{\partial V_z}{\partial t} = -\frac{dp}{dz} + \mu\left[\frac{1}{r}\frac{\partial}{\partial r}\left(r\frac{\partial V_z}{\partial r}\right)\right]$$

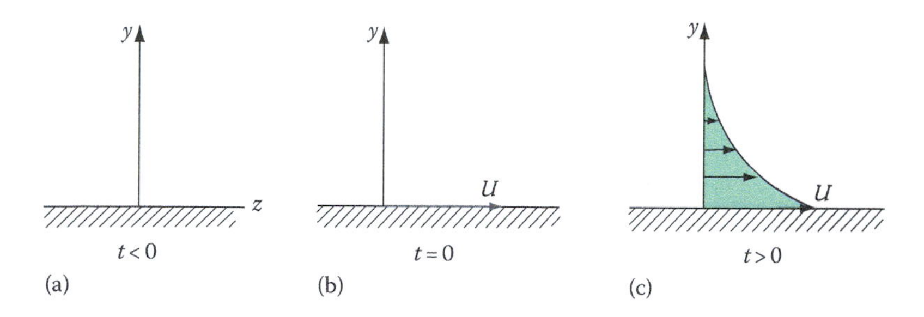

FIGURE P11.21 (a) Wall and fluid at rest, (b) wall moves at velocity V, and (c) resultant velocity profile.

$$\begin{aligned} &\text{IC}: && t = 0, && V_z = 0 \\ &\text{BC 1}: && r = 0, && \partial V_z/\partial r = 0\ (V \text{ is finite}) \\ &\text{BC 2}: && r = R, && V_z = 0 \end{aligned}$$

11.23 A Newtonian fluid in a narrow slit is initially at rest. Suddenly, a pressure drop in the longitudinal direction is imposed on the fluid. For unsteady laminar flow, determine the differential equation and boundary conditions for the system (see Problem 11.22).

Graphical Solution Methods for Unsteady Problems

11.24 Obtain a table of values for the error function (usually found in heat transfer textbooks or math handbooks) as given in Equation 11.36. Generate a graphical solution to the suddenly accelerated flat problem for the first four time intervals. Use the error function table to obtain the solution for the velocity at the fourth time interval. Plot the error function results on your solution graph, and determine how well the graphical method works for this problem. (Some error is expected due to the finite Δy distance selected.) If it is necessary to specify a fluid, use glycerine. Also, select a Δy that is different from the one used in the text. Note that

$$\text{Re}_c = \frac{\Delta y^2}{\nu t} = \frac{\Delta y^2}{\nu(4\Delta t)} = 1$$

where the 4 is inserted to represent the fourth Δt or the fourth time interval. Continuing,

$$\nu\,\Delta t = \frac{\Delta y^2}{4}$$

From the exact solution, we have defined the independent variable η as

$$\eta = \frac{y}{2\sqrt{\nu t}}$$

Combining with the previous equation gives

$$\eta = \frac{y}{4\Delta y}$$

where
- y is position
- Δy is the increment you are to select

Thus, there is a relationship between the graphical solution variables and the error function variable η.

11.25 Generate two graphical solutions to the unsteady plane Couette flow problem using two different Δy values. Compare the results obtained in the solutions by evaluating profiles at equal times.

11.26 In the problem of unsteady flow between concentric circular cylinders, the solution given was for the inner cylinder rotating. Generate a solution for when the outer cylinder rotates at a fixed angular velocity and the inner cylinder is stationary.

11.27 Verify that the time increment for unsteady flow between concentric cylinders is given by

$$t = \frac{\rho(R_2 - R_1)^2}{81\mu}$$

11.28 Using the graphical method, generate velocity profiles for the first five time intervals for unsteady start-up in a plane channel. Solve the problem using a Δy that is different from that in the text. Plot also the steady-state solution (Equation 11.17) on the same axes. Note that when this is done, the relationship between $\Delta pt/pL$ and $V_{z\,\max}$ is automatically fixed. To develop such a relationship, rewrite Equation 11.17 as

$$\frac{V_z}{V_{z\,\max}} = \left[1 - \left(\frac{y}{h}\right)^2\right]$$

Let the graph vary from 0 to 1 for velocity and from 0 (at the centerline) to 1 (at the wall) for position. Using the definition of $V_{z\,\max}$ and $\text{Re}_c = 1$, solve for $\Delta pt/pL$ in terms of $V_{z\,\max}$ and construct the graphical solution.

Turbulent Flow Problems

11.29 Ethyl alcohol flows in a ½-nominal, schedule 40 pipe. Determine the pressure drop (in Pa/m) at transition.

11.30 Repeat Problem 11.29 for glycerine.

11.31 Repeat Problem 11.29 for air.

11.32 Repeat Problem 11.29 for hydrogen.

11.33 Benzene flows in a long 6-nominal, schedule 40 tube. The pressure drop in the pipe is 0.03 lbf/ft^2 per foot of length. Calculate the wall shear stress, and plot the velocity profile in the pipe. Use Figure 11.28 if appropriate.

11.34 Water flows in a long tube made of 10-nominal, schedule 40 wrought iron. The pressure drop in the pipe is 3.0 kPa per meter of length. Plot the velocity profile. Use Figure 11.28 if appropriate.

11.35 Examine Equations 11.5 through 11.8, and determine the dimension of the del operator.

12 Inviscid Flow

In real-life situations, most flow is turbulent. In pipe flow, in open-channel flow, and in flow over immersed bodies, laminar conditions exist only rarely. To analyze problems in these areas of study, it is most beneficial to be able to obtain a velocity or a velocity distribution for the problem at hand. As we saw in the last chapter, however, solution of the Navier–Stokes equations for turbulent flow problems is no easy task. Exact solutions were obtainable only for simplified laminar flow cases. Consequently, it is often necessary to make approximations so that the engineer can formulate a working solution to a number of important problems.

Let us for the moment consider flow past an object, as illustrated in Figure 12.1. Streamlines of flow about the object also appear in the diagram. Upstream the flow is uniform. In the vicinity of the object, the flow pattern is altered from the uniform incoming flow—the object displaces the flow. Far from the surface, however, the fluid is not affected by the presence of an object. In the regions labeled A in Figure 12.1, the streamlines are therefore uniform and parallel. At the surface, the fluid adheres because of friction. Regardless of how small the viscosity of the fluid is, velocity at the wall is zero. We therefore conclude that viscous effects can be neglected in a flow field except in the vicinity of a surface. It is thus possible to divide the flow field into two regions. The first of these is a nonviscous region where viscosity need not be included in the equations of motion. The second is a viscous region for the fluid in the immediate vicinity of the object.

A study of flow of a nonviscous or inviscid fluid is the topic of interest in this chapter. Flow in the vicinity of an immersed object or near a surface is often referred to as *boundary-layer flow*, the topic of the following chapter. In this chapter, we will develop equations for steady, incompressible, inviscid flow. The continuity and Euler equations will be derived from the Navier–Stokes equations of Chapter 11. Stream and potential functions will be introduced and used to describe a multitude of flow fields. We will then use superposition to combine simpler flow fields and form more complex ones. Finally, we will discuss irrotationality of an inviscid flow and write the Laplace equation for stream and potential functions.

12.1 EQUATIONS OF TWO-DIMENSIONAL INVISCID FLOWS

The continuity and momentum (Navier–Stokes) equations were presented in the last chapter for a general three-dimensional flow of a Newtonian fluid. Now, we will reduce those equations to describe the flow of a nonviscous fluid, restricting our study to a two-dimensional, inviscid, incompressible flow. Solutions to the inviscid equations for several problems will be presented.

12.1.1 Continuity Equation

For two-dimensional flow in Cartesian coordinates, we eliminate the velocity in the z-direction, V_z, from the continuity equation of Chapter 11 (Equation 11.1). For steady, incompressible, inviscid, two-dimensional flow, Equation 11.1 becomes

$$\frac{\partial V_x}{\partial x} + \frac{\partial V_y}{\partial y} = 0 \tag{12.1}$$

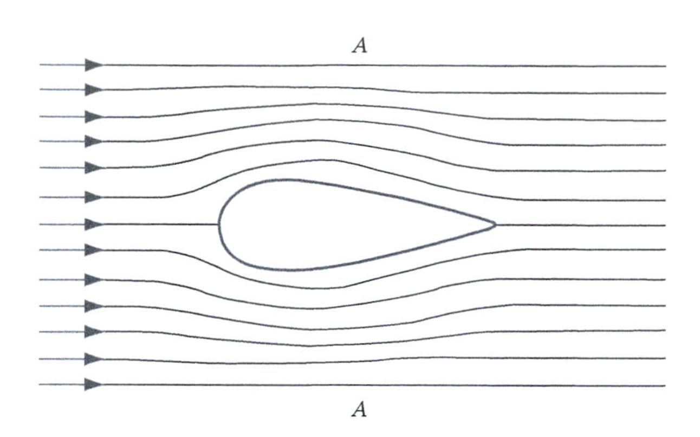

FIGURE 12.1 Streamlines of flow past an object.

Correspondingly, in polar cylindrical coordinates, the continuity equation (Equation 11.3) becomes

$$\frac{\partial(rV_r)}{\partial r}+\frac{\partial(V_\theta)}{\partial\theta}=0 \tag{12.2}$$

12.1.2 Momentum Equation

Again we refer to Chapter 11. Equations 11.2a through 11.2c are the Navier–Stokes equations. For an inviscid flow, the viscosity μ, in those equations is set equal to zero. For two-dimensional flow, $V_z = 0$; and for the steady case, all differentiation with respect to time is zero. The equations of motion become in Cartesian coordinates

$$x\text{-component: } \rho\left(V_x\frac{\partial V_x}{\partial x}+V_y\frac{\partial V_x}{\partial y}\right)=-\frac{\partial p}{\partial x}+\rho g_x \tag{12.3}$$

$$y\text{-component: } \rho\left(V_x\frac{\partial V_x}{\partial x}+V_y\frac{\partial V_y}{\partial y}\right)=-\frac{\partial p}{\partial y}+\rho g_y \tag{12.4}$$

The z-component equation vanishes. Performing similar operations for the Navier–Stokes equations in polar cylindrical coordinates (Equations 11.4a through 11.4c) yields

$$r\text{-component: } \rho\left(V_r\frac{\partial V_r}{\partial r}+\frac{V_\theta}{r}\frac{\partial V_r}{\partial\theta}-\frac{V_\theta^2}{r}\right)=-\frac{\partial p}{\partial r}+\rho g_r \tag{12.5}$$

$$\theta\text{-component: } \rho\left(V_r\frac{\partial V_\theta}{\partial r}+\frac{V_\theta}{r}\frac{\partial V_\theta}{\partial\theta}+\frac{V_rV_\theta}{r}\right)=-\frac{1}{r}\frac{\partial p}{\partial\theta}+\rho g_\theta \tag{12.6}$$

Equations 12.3 through 12.6 in differential form for frictionless flow are called the Euler equations. The continuity and Euler equations are the equations of two-dimensional, steady, inviscid flow.

12.2 STREAM FUNCTION AND VELOCITY POTENTIAL

For two-dimensional inviscid flow problems, the continuity and Euler equations must be solved subject to boundary conditions. For viscous flow, the tangential and normal velocity at a boundary is zero owing to friction; for inviscid flow, the tangential fluid velocity at a boundary is nonzero because the fluid has no viscosity. The normal fluid velocity is zero, however. These, in general, are the boundary conditions in inviscid flow.

The continuity and Euler equations contain three unknowns—V_x, V_y, and p. The equations must be solved simultaneously; and because they are nonlinear, it is difficult to apply direct solution methods successfully. This complexity can be reduced, however, by the introduction of two functions, both capable of describing the velocity components of the flow field. These are the *stream function* ψ and the *potential function* φ.

The stream function ψ is a function of x and y. It is defined in terms of the flow velocities as

$$V_x = \frac{\partial \psi}{\partial y}$$
$$V_y = -\frac{\partial \psi}{\partial x} \tag{12.7}$$

The stream function defined here satisfies the two-dimensional continuity equation. Thus, if $\psi(x, y)$ is known and is a continuously differentiable function, the velocity components V_x and V_y can be determined. The reason for defining the stream function in this manner is that the velocity vector $V\left(= \sqrt{V_x^2 + V_y^2}\right)$ is in the direction of the tangent of the streamline at any point (x, y) in the flow field. From calculus, the total differential of the stream function can be written as

$$d\psi = \frac{\partial \psi}{\partial x}dx + \frac{\partial \psi}{\partial y}dy$$

Combining this equation with Equation 12.7, we get

$$d\psi = -V_y dx + V_x dy \tag{12.8}$$

On a line of constant ψ, $d\psi = 0$, and Equation 12.8 can be rearranged to solve for the slope as

$$\left.\frac{dy}{dx}\right|_{\psi} = \frac{V_y}{V_x} \tag{12.9}$$

which is the differential equation of the streamlines. This concept is illustrated in Figure 12.2. The streamline is denoted as a "$d\psi = 0$" or a "ψ = a constant" line. The velocity vector and components are drawn at the point of interest (x, y). The slope of the tangent to the streamline at (x, y) is $dy/dx = V_y/V_x$.

The potential function φ is a function of x and y. It is defined in terms of the flow velocities as

$$V_x = \frac{\partial \phi}{\partial x}$$
$$V_y = \frac{\partial \phi}{\partial y} \tag{12.10}$$

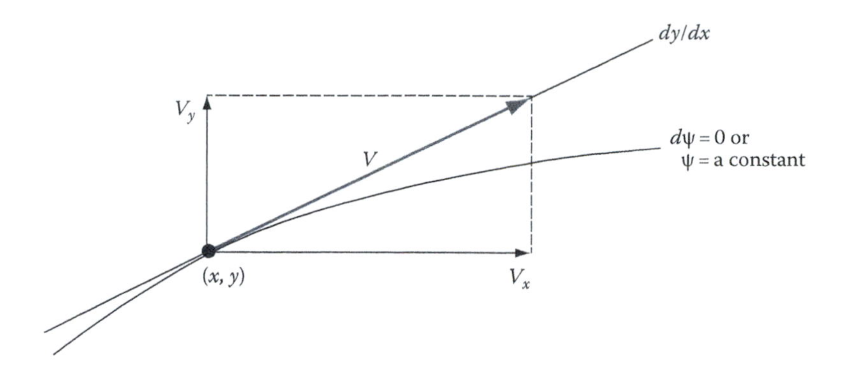

FIGURE 12.2 Streamline and velocity through a point.

Thus, if the velocity potential ϕ is known and is continuously differentiable, the velocity components V_x and V_y can be found. The differential of ϕ is written as

$$d\phi = \frac{\partial \phi}{\partial x}dx + \frac{\partial \phi}{\partial y}dy$$

Combining this equation with Equations 12.10, we get

$$d\phi = V_x dx + V_y dy \tag{12.11}$$

On a line of constant ϕ, $d\phi = 0$, and Equation 12.11 can be rearranged to solve for the slope as

$$\left.\frac{dy}{dx}\right|_{\phi} = -\frac{V_x}{V_y} \tag{12.12}$$

This is the differential equation of the potential lines. Equation 12.9 multiplied by Equation 12.12 yields

$$\left(\left.\frac{dy}{dx}\right|_{\psi}\right)\left(\left.\frac{dy}{dx}\right|_{\phi}\right) = \frac{V_y}{V_x} \cdot \frac{-V_x}{V_y} = -1$$

Therefore, at any point (x, y) in a flow field, the streamline is normal to the potential line (see Figure 12.3).

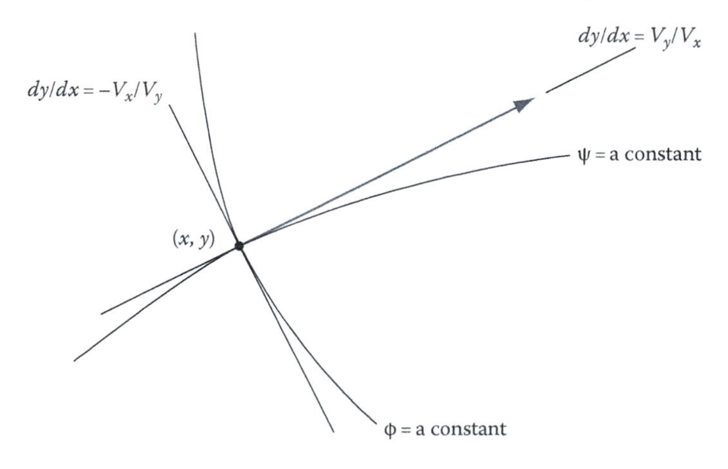

FIGURE 12.3 Orthogonality of streamlines and potential lines.

Similar definitions for ψ and ϕ can be stated for velocity components in polar cylindrical coordinates:

$$
\begin{aligned}
V_r &= \frac{1}{r}\frac{\partial \psi}{\partial \theta} = \frac{\partial \phi}{\partial r} \\
V_\theta &= -\frac{\partial \psi}{\partial r} = \frac{1}{r}\frac{\partial \phi}{\partial \theta}
\end{aligned}
\tag{12.13}
$$

It can be shown in this coordinate system that streamlines and potential lines are normal to each other.

Example 12.1

A flow field is described by the streamline equation

$$\psi = xy$$

a. Determine the velocity field.
b. Determine the velocity potential function.
c. Determine whether the flow satisfies the continuity equation.
d. Plot the streamlines and potential lines on the same set of axes.

Solution

a. The potential function is found by applying Equations 12.7 and 12.10. By definition,

$$V_x = \frac{\partial \psi}{\partial y} = x$$

$$V_y = -\frac{\partial \psi}{\partial x} = -y$$

Also,

$$\frac{\partial \phi}{\partial x} = V_x = x$$

from which we obtain by integration

$$\phi = \frac{x^2}{2} + f_1(y) + C_1$$

where
$f_1(y)$ is an unknown function of y
C_1 is a constant

Moreover,

$$\frac{\partial \phi}{\partial y} = V_y = -y$$

which gives

$$\phi = -\frac{y^2}{2} + f_2(x) + C_2$$

where
$f_2(x)$ is an unknown function of x
C_2 is a constant

Equating both expressions for ϕ yields

$$\frac{x^2}{2} + f_1(y) + C_1 = -\frac{y^2}{2} + f_2(x) + C_2$$

We thus conclude that

$$f_1(y) = -\frac{y^2}{2}$$

$$f_2(x) = \frac{x^2}{2}$$

$$C_1 = C_2$$

b. Because determining velocity involves only derivatives of ϕ or ψ, constants are generally not important and arbitrarily can be set equal to zero. The potential function, then, is

$$\phi = \frac{1}{2}(x^2 - y^2)$$

c. The continuity equation is

$$\frac{\partial V_x}{\partial x} + \frac{\partial V_y}{\partial y} = 0$$

By substitution, we get

$$+1 - 1 = 0$$

Thus, continuity is satisfied.

d. For simplicity, we will restrict the plot to the first quadrant, where $x \geq 0$ and $y \geq 0$. The stream function is

$$\psi = xy$$

Selecting constants for ψ gives various equations, each of which must be graphed:

$$\psi = 0 \quad xy = 0 \quad x = 0, \quad y = 0$$

$\psi = 8 \quad xy = 8$

y	x
8	1
4	2
2	4
1	8

$\psi = 16 \quad xy = 16$

y	x
10	1.6
8	2
4	4
2	8
1.6	10

$\psi = 24 \quad xy = 24$

y	x
10	2.4
8	3
6	4
4	6
3	8
2.4	10

$\psi = 32 \quad xy = 32$

y	x
10	3.2
8	4
4	8
3.2	10

Similarly, for the velocity potential

$$\phi = \frac{1}{2}(x^2 - y^2)$$

we have

$\phi = 0 \quad y = x \quad y = -x$

$\phi = 8 \quad y^2 = x^2 - 16$

y	x
0	4
4.47	6
6.9	8

$\phi = 8 \quad x^2 = y^2 - 16$

y	x
4	0
6	4.47
8	6.90
10	9.17

$\phi = 16 \quad y^2 = x^2 - 32$

y	x
0	5.65
2	6
5.65	8

$\phi = -16 \quad x^2 = y^2 - 32$

y	x
5.65	0
6	2
8	5.65

$\phi = 24 \quad y^2 = x^2 - 48$

y	x
0	6.92
4	8
5.74	9

$\phi = -24 \quad x^2 = y^2 - 48$

y	x
6.92	0
8	4
9	5.74

A plot of these equations is given in Figure 12.4. This type of graph is called a **flow net**. The field represents flow into a corner. If the corresponding plot for the second quadrant were also drawn (it is symmetric about the y- and x-axis), we would have a model of inviscid flow striking a flat surface.

12.3 IRROTATIONAL FLOW

Rotation is a characteristic of flow. Rotational flow occurs if a fluid element rotates while the fluid is flowing. Consider, for example, Figure 12.5a for a viscous fluid and Figure 12.5b for an inviscid fluid; in both cases the flow is between parallel boundaries. In Figure 12.5a, a fluid element at position 1 will deform as it moves to position 2. The lower horizontal line is carried downstream faster

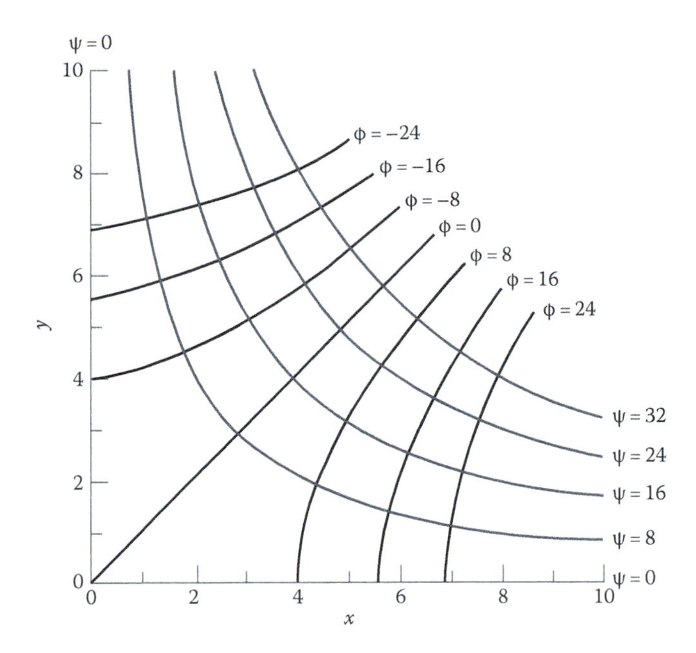

FIGURE 12.4 Graph of stream and potential functions for Example 12.1.

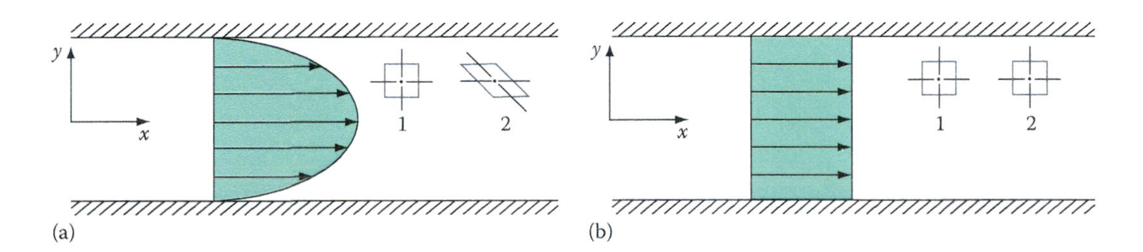

FIGURE 12.5 (a) Rotational and (b) irrotational flow in a duct.

than the upper line. The upper line is nearer the wall, where the velocity is zero. Because the sides of the element have rotated, we use the name *rotational flow*. Figure 12.5b shows a fluid element at positions 1 and 2, but the sides of the element have not rotated because the fluid is inviscid. The velocity at the wall in the main flow direction is not zero; the shape of the element has been preserved. This is called **irrotational flow**.

As a second example, consider briefly circular or vortex flow, as illustrated in Figure 12.6. In Figure 12.6a (for a viscous fluid), the flow is rotational as compared to Figure 12.6b for irrotational flow. A final example of irrotational flow is illustrated in Figure 12.7 for flow through a converging duct. As the flow accelerates, the fluid element elongates. The diagonals of the element rotate in equal amounts but in opposite directions, thus balancing each other and preserving the direction of each side of the element.

Now let us develop an expression for the rotation in a flow field. Figure 12.8 shows a fluid element in the xy plane. The element dimensions are dx by dy, and two of the sides are drawn separately to indicate the velocity at each corner. The rotation of each of the sides about point C is determined by dividing the difference in endpoint velocities by the length of the appropriate side. Taking the counterclockwise direction as positive, we obtain

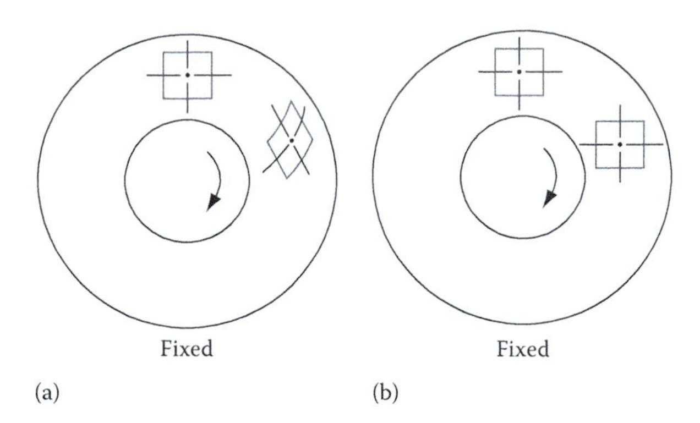

FIGURE 12.6 (a) Rotational and (b) irrotational flow.

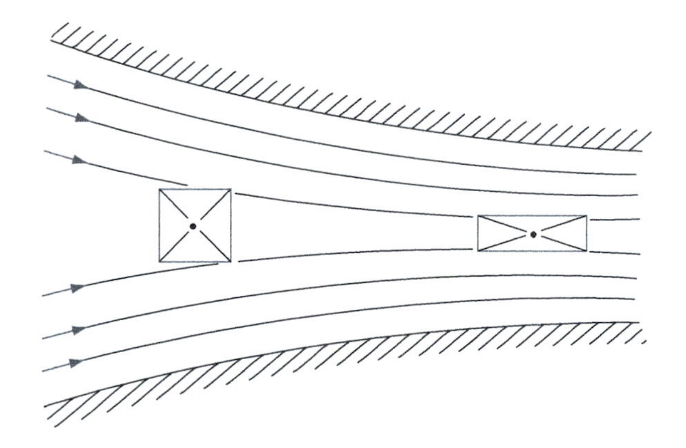

FIGURE 12.7 Irrotational flow through a converging duct.

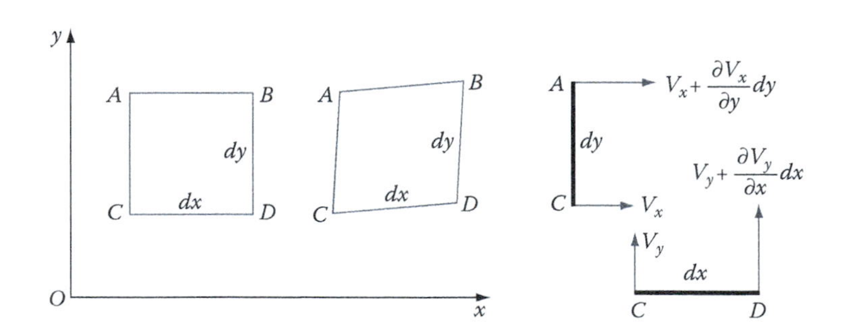

FIGURE 12.8 Rotation of a fluid element.

Rotation of *CA* about point *C*:

$$\frac{V_x - [V_x + (\partial V_x/\partial y)dy]}{dy} = -\frac{\partial V_x}{\partial y}$$

Rotation of *CD* about point *C*:

$$\frac{[Vy + (\partial V_y/\partial x)dx] - V_y}{dx} = \frac{\partial V_y}{\partial x}$$

These equations are written with the assumption that the element is not translating. The average of these two angular velocities in the plane normal to the z-axis is called the rotation. Thus,

$$\omega_z = \frac{1}{2}\left(\frac{\partial V_y}{\partial x} - \frac{\partial V_x}{\partial y}\right) \tag{12.14}$$

Similarly, for the general three-dimensional case, we also have

$$\omega_x = \frac{1}{2}\left(\frac{\partial V_z}{\partial y} - \frac{\partial V_y}{\partial z}\right)$$

and

$$\omega_y = \frac{1}{2}\left(\frac{\partial V_x}{\partial z} - \frac{\partial V_z}{\partial x}\right)$$

For our study of potential flow, we will be concerned only with ω_z. In polar cylindrical coordinates, there results

$$\omega_z = \frac{1}{2}\left[\frac{1}{r}\frac{\partial(rV_\theta)}{\partial r} - \frac{1}{r}\frac{\partial(V_r)}{\partial\theta}\right] \tag{12.15}$$

For irrotational flow, Equation 12.14 gives another useful relationship between the velocities:

$$\frac{\partial V_y}{\partial x} - \frac{\partial V_x}{\partial y} = 0 \tag{12.16}$$

Example 12.2

Determine whether the flow of Example 12.1 is rotational. That flow was described by

$$\psi = xy \quad \text{and} \quad \phi = \frac{1}{2}(x^2 - y^2)$$

Solution

For irrotational flow, Equation 12.16 applies

$$\frac{\partial V_y}{\partial x} - \frac{\partial V_x}{\partial y} = 0$$

By definition,

$$V_x = \frac{\partial \psi}{\partial y} = x$$

and

$$V_y = \frac{\partial \psi}{\partial x} = -y$$

From these definitions, we obtain

$$\frac{\partial V_y}{\partial x} = 0 \quad \text{and} \quad \frac{\partial V_x}{\partial y} = 0$$

The flow velocities thus satisfy the condition of irrotationality.

The stream function is applicable in both rotational and irrotational flows because the stream function satisfies the continuity equation. The potential function applies only to irrotational flows.

For irrotational flow, it is possible to integrate the Euler equations to obtain another relationship between the velocity and pressure in the flow. If we work in Cartesian coordinates and let gravity be nonzero only in the vertical direction, the **Euler equations** (12.3 and 12.4) are

$$V_x \frac{\partial V_x}{\partial x} + V_y \frac{\partial V_x}{\partial y} = -\frac{1}{\rho}\frac{\partial p}{\partial x} \tag{12.3}$$

$$V_x \frac{\partial V_y}{\partial x} + V_y \frac{\partial V_y}{\partial y} = -\frac{1}{\rho}\frac{\partial p}{\partial x} - g \tag{12.4}$$

Gravity is thus positive downward. By using the irrotational flow condition, namely,

$$\frac{\partial V_x}{\partial y} = \frac{\partial V_y}{\partial x}$$

the preceding equations become

$$V_x \frac{\partial V_x}{\partial x} + V_y \frac{\partial V_y}{\partial x} = -\frac{1}{\rho}\frac{\partial p}{\partial x}$$

and

$$V_x \frac{\partial V_x}{\partial y} + V_y \frac{\partial V_y}{\partial y} = -\frac{1}{\rho}\frac{\partial p}{\partial y} - g$$

These two equations can be simplified to obtain

$$\frac{1}{2}\frac{\partial\left(V_x^2\right)}{\partial x}+\frac{1}{2}\frac{\partial\left(V_y^2\right)}{\partial x}=-\frac{1}{\rho}\frac{\partial p}{\partial x}$$

and

$$\frac{1}{2}\frac{\partial\left(V_x^2\right)}{\partial y}+\frac{1}{2}\frac{\partial\left(V_y^2\right)}{\partial y}=-\frac{1}{\rho}\frac{\partial p}{\partial y}-g$$

For constant density, both equations can now be integrated to give

$$\begin{aligned}\frac{1}{2}\left(V_x^2+V_y^2\right)&=-\frac{p}{\rho}+f_1(y)\\ \frac{1}{2}\left(V_x^2+V_y^2\right)&=-\frac{p}{\rho}-gy+f_2(x)\end{aligned}\tag{12.17}$$

where
f_1 is an unknown function of y
f_2 is an unknown function of x

Equating gives

$$f_1(y)=-gy+f_2(x)$$

or

$$f_2(x)=f_1(y)+gy$$

The left-hand side is a function of only x, whereas the right-hand side is a function of only y. For this to occur, both sides must equal a constant. Combining the last equation with Equation 12.17 gives

$$\frac{1}{2}\left(V_x^2+V_y^2\right)=-\frac{p}{\rho}-gy+\text{a constant}$$

$$\frac{p}{\rho}+\frac{V^2}{2}+gy=\text{a constant}=C\ \left(\text{Bernoulli's equation}\right)\tag{12.18}$$

This equation is recognized as Bernoulli's equation derived by assuming inviscid, irrotational flow in two-dimensions. By knowing the equation of the streamlines or the velocity potential, we can determine the total velocity V. Bernoulli's equation then allows for finding the pressure in the flow.

12.4 LAPLACE'S EQUATION AND VARIOUS FLOW FIELDS

The velocity potential and the stream function can be used to solve simple problems in two-dimensional, inviscid, irrotational flow. The differential equation that ϕ and ψ must satisfy is the Laplace equation. It can be obtained by combining the continuity equation and the irrotationality relation with the definition of velocity in terms of these functions:

$$\text{Continuity: }\frac{\partial V_x}{\partial x}=\frac{\partial V_y}{\partial y}=0\tag{12.1}$$

$$\text{Potential function: } V_x = \frac{\partial \phi}{\partial x} \qquad V_y = \frac{\partial \phi}{\partial y} \tag{12.10}$$

Combining equations, we obtain

$$\frac{\partial}{\partial x}\left(\frac{\partial \phi}{\partial x}\right) + \frac{\partial}{\partial y}\left(\frac{\partial \phi}{\partial y}\right) = 0$$

or

$$\frac{\partial^2 \phi}{\partial x^2} + \frac{\partial^2 \phi}{\partial y^2} = 0 \tag{12.19}$$

This is Laplace's equation for the velocity potential; it can be rewritten as

$$\nabla^2 \phi = 0$$

A similar development can be performed for the stream function:

$$\text{Irrotationality relation: } \frac{\partial V_y}{\partial x} - \frac{\partial V_x}{\partial y} = 0 \tag{12.16}$$

$$\text{Stream function: } V_x = \frac{\partial \psi}{\partial y} \qquad V_y = -\frac{\partial \psi}{\partial x} \tag{12.17}$$

Combining equations, we obtain

$$\frac{\partial}{\partial x}\left(-\frac{\partial \psi}{\partial x}\right) - \frac{\partial}{\partial y}\left(\frac{\partial \psi}{\partial y}\right) = 0$$

or

$$\frac{\partial^2 \psi}{\partial x^2} + \frac{\partial^2 \psi}{\partial y^2} = 0 \tag{12.20}$$

which is Laplace's equation for the streamlines. In del operator notation,

$$\nabla^2 \psi = 0$$

In polar cylindrical coordinates, Laplace's equation for the velocity potential is

$$\frac{\partial^2 \phi}{\partial r^2} + \frac{1}{r}\frac{\partial \phi}{\partial r} + \frac{1}{r^2}\frac{\partial^2 \phi}{\partial \theta^2} = 0 \tag{12.21}$$

which is easily derivable from Equation 12.19. We will now examine several solutions to the equations for some simple flows.

12.4.1 Uniform Flow

Uniform flow can be described as flow parallel to, say, the x-axis moving in the positive direction at a constant velocity U. Thus,

$$V_x = U$$

and

$$V_y = 0$$

By definition,

$$V_x = \frac{\partial \phi}{\partial x} = U$$

from which we obtain

$$\phi = Ux + f_1(y)$$

Also,

$$V_y = \frac{\partial \phi}{\partial y} = 0$$

$$\phi = f_2(x)$$

Equating both expressions for ϕ gives

$$Ux + f_1(y) = f_2(x)$$

or

$$f_1(y) = Ux + f_2(x)$$

We therefore conclude that

$$\phi = Ux \text{ (uniform flow)} \tag{12.22a}$$

Following a similar procedure for the stream function, we obtain

$$\phi = Uy \text{ (uniform flow)} \tag{12.22b}$$

A plot of uniform flow is provided in Figure 12.9 for values of 0, U, $2U$, $3U$, and $4U$.

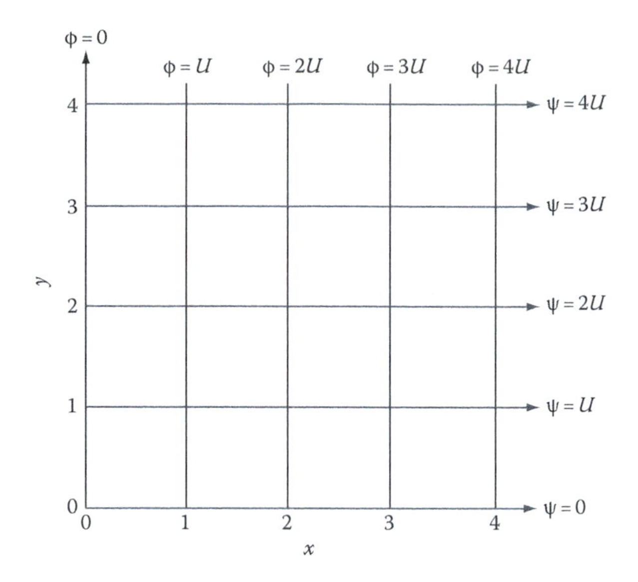

FIGURE 12.9 Streamlines and potential lines for uniform flow.

12.4.2 Source Flow

Source flow can be described as a radially outward flow emanating from a point. The velocity in the xy plane is radial at any location, as is illustrated in Figure 12.10. At any location along a circle of radius r, the radial velocity V_r is a constant. The flow rate per unit length into the page is a constant that is equal to the product of radial velocity and circumferential distance:

$$\frac{Q}{L} = q = 2\pi r V_r$$

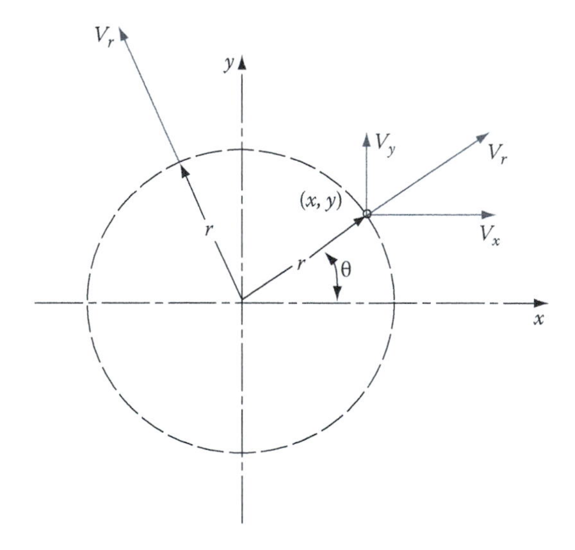

FIGURE 12.10 Flow emanating from a point or source flow.

Because q is a constant, the flow is steady. At the origin, where $r = 0$, the radial velocity V_r must be infinite. Mathematically, a point source is known as a singular point and does not exist in nature. From the preceding equation and discussion,

$$V_r = \frac{q}{2\pi r}$$

$$V_\theta = 0$$

By definition,

$$V_r = \frac{\partial \phi}{\partial r} = \frac{q}{2\pi r}$$

Integrating gives

$$\phi = \frac{q}{2\pi} \ell n(r) + f_1(\theta)$$

Moreover,

$$V_\theta = \frac{1}{r}\frac{\partial \phi}{\partial \theta} = 0$$

Integrating yields

$$\phi = f_2(r)$$

Equating both expressions for ϕ, we obtain

$$\frac{q}{2\pi} \ell n(r) + f_1(\theta) = f_2(r)$$

It is concluded, then, that for source flow

$$\phi = \frac{q}{2\pi} \ell n(r) \quad \text{(source flow)} \tag{12.23a}$$

Following a similar procedure for the stream function, we find

$$\psi = \frac{q\theta}{2\pi} \quad \text{(source flow)} \tag{12.23b}$$

For source flow, it is seen that potential lines are concentric circles and streamlines are radial lines from the origin. Potential lines and streamlines are illustrated in Figure 12.11 for various values of ϕ and ψ.

12.4.3 Sink Flow

Sink flow is radially inward flow directed at a point; it is the opposite of source flow. At any location along a circle of radius r, the radial velocity V_r is a constant. The flow rate per unit length into the page is a constant that is equal to the product of radial velocity and circumferential distance:

$$\frac{Q}{L} = q = -2\pi r V_r$$

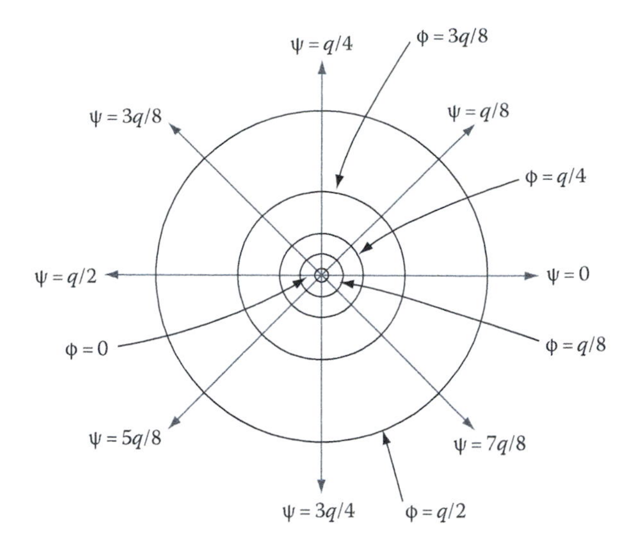

FIGURE 12.11 Potential lines and streamlines for source flow.

Again we take q as a constant and assume the flow to be steady. At the origin, where $r = 0$, V_r must therefore be infinite. This is a singular point that has no counterpart in nature. From the preceding, we have

$$V_r = -\frac{q}{2\pi r}$$

$$V_\theta = 0$$

By definition,

$$V_r = \frac{\partial \phi}{\partial r} = -\frac{q}{2\pi r}$$

Integrating gives

$$\phi = -\frac{q}{2\pi}\ell\text{n}(r) + f_1(\theta)$$

Further, with

$$V_\theta = \frac{1}{r}\frac{\partial \phi}{\partial \theta} = 0$$

we obtain

$$\phi = f_2(r)$$

Equating both expressions for ϕ, we find

$$-\frac{q}{2\pi}\ell\text{n}(r) + f_1(\theta) = f_2(r)$$

Thus, for radial flow into a sink, we get

$$\phi = -\frac{q}{2\pi}\ell n(r) \quad \text{(sink flow)} \tag{12.24a}$$

Similarly, it can be shown that for a sink,

$$\psi = -\frac{q\theta}{2\pi} \quad \text{(sink flow)} \tag{12.24b}$$

12.4.4 Irrotational Vortex Flow

A counterclockwise **irrotational vortex** (Figure 12.12) is characterized by the following velocity components:

$$V_r = 0$$

$$V_\theta = \frac{C}{r}$$

where C is a constant. By definition,

$$V_r = \frac{\partial \phi}{\partial r} = 0$$

which leads to

$$\phi = f_1(\theta)$$

Also,

$$V_\theta = \frac{1}{r}\frac{\partial \phi}{\partial \theta} = \frac{C}{r}$$

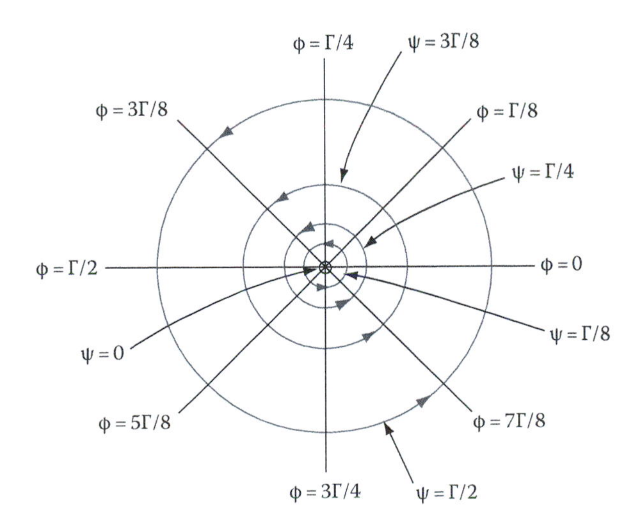

FIGURE 12.12 Potential lines and streamlines for vortex flow.

Integrating yields

$$\phi = C\theta + f_2(r)$$

Equating both expressions for ϕ, we find

$$\phi = C\theta$$

Similarly, it can be shown that for free vortex flow,

$$\psi = -C\,\ell n(r)$$

It is convenient when dealing with vortex flow to introduce the term **circulation**, denoted as Γ. The circulation is defined as the result obtained by integrating the tangential velocity around any closed contour. A simple contour that we can select is that of a circle of radius r. Thus,

$$\Gamma = \int V_\theta ds$$

where the arc length for a circle is $ds = r\,d\theta$, to be integrated from 0 to 2π. Thus,

$$\Gamma = \int_0^{2\pi} V_\theta r\,d\theta = \int_0^{2\pi} \left(\frac{C}{r}\right) r\,d\theta$$

or

$$\Gamma = 2\pi C$$

In terms of the circulation, then, we have for vortex flow

$$\phi = \frac{\Gamma\theta}{2\pi} \quad \text{(vortex flow)} \tag{12.25a}$$

$$\psi = \frac{\Gamma}{2\pi}\ell n(r) \quad \text{(vortex flow)} \tag{12.25b}$$

Streamlines and potential lines for a vortex are illustrated in Figure 12.12. Flow nets for a free vortex and for a source (or sink) have the same form; the difference is that streamlines and potential lines are exchanged. (See Equations 12.24 and 12.25.) Note that at $r = 0$, the tangential velocity V_θ becomes infinite, so the origin is a singular point in vortex flow.

Example 12.3

The eye of a tornado has a diameter of 40 m. In the eye, the tornado flow field is approximated as solid body rotation. Outside the eye, the flow is a free vortex. Determine the tangential velocity distribution in the tornado if the maximum wind velocity is 50 m/s. Neglect translational motion. Determine also the pressure variation in the tornado and calculate the minimum pressure.

Solution

The streamlines of the tornado are circular as shown in Figure 12.13, and we have for the eye

$$V_{\theta_1} = r\omega \quad (r \le R)$$

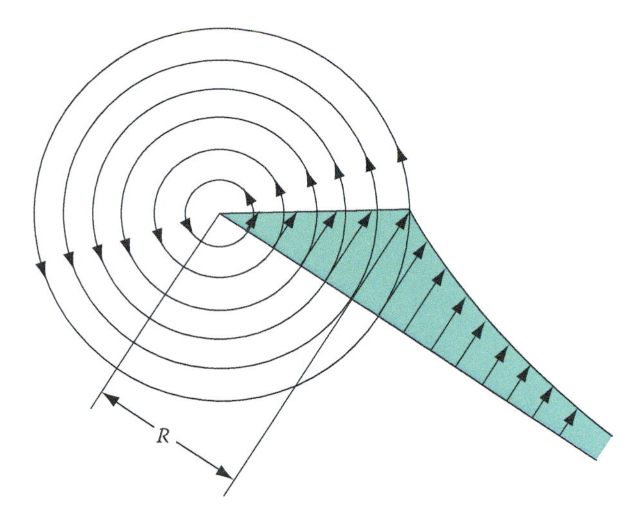

FIGURE 12.13 Streamlines and velocity distribution for flow in a tornado.

Outside the eye,

$$V_{\theta_2} = \frac{C}{r} \quad (r \geq R)$$

The maximum wind velocity occurs where $r = R$, at the edge of the eye. Given R as 20 m and $V_{\theta max}$ as 50 m/s, we calculate

$$\omega = \frac{V_{\theta max}}{R} = \frac{50}{20} = 2.5 \text{ rad/s}$$

Moreover, at the edge of the eye the two velocities must be equal. Hence,

$$R\omega = \frac{C}{R}$$

Solving for the constant C, we obtain

$$C = \omega R^2 = 2.5\,(400) = 1\,000 \text{ m}^2/\text{s}$$

The expressions for velocity thus become

$$V_{\theta_1} = 2.5r \text{ (m/s)} \quad (r < 20 \text{ m})$$

$$V_{\theta_2} = \frac{1\,000}{r} \text{ (m/s)} \quad (r < 20 \text{ m})$$

The pressure distribution within the eye is found by applying Euler's equations because flow is rotational in the eye. Outside the eye, the flow is irrotational; consequently, Bernoulli's equation can be used. Within the eye, $V_r = 0$, and V_θ is a function of only r; therefore, partial derivatives of V_θ with respect to θ vanish. The Euler equations in polar cylindrical coordinates (Equations 12.5 and 12.6) reduce to

$$\rho \frac{V_\theta^2}{r} = \frac{\partial p}{\partial r}$$

and

$$\frac{1}{r}\frac{\partial p}{\partial \theta} = 0$$

The second of these equations indicates that pressure is not a function of θ. The first equation, then, can be rewritten as an ordinary differential equation with V_{θ_1} inserted for V_θ because we are concerned with the center of the eye:

$$\frac{dp}{dr} = \frac{\rho V_{\theta_1}^2}{r} = \frac{\rho r^2\omega^2}{r} = \rho r\omega^2$$

We now obtain

$$p = \frac{\rho r^2\omega^2}{r} + C_1 \quad (r \leq R) \tag{i}$$

where C_1 is a constant of integration.

Outside the eye, Bernoulli's equation is applied to the region near R and to a location far from the tornado, where the pressure is atmospheric and the velocity is zero. We get

$$\frac{p}{\rho} + \frac{V_{\theta_2}^2}{2} = \frac{p_a}{\rho}$$

from which we obtain

$$p = p_a - \frac{\rho V_{\theta_2}^2}{2} = p_a - \frac{\rho C^2}{2r^2}$$

or

$$p = p_a - \frac{\rho\omega^2 R^4}{2r^2} \quad (r \leq R) \tag{ii}$$

At the edge of the eye, $r = R$; Equations (i) and (ii) must be equal at this point. Thus,

$$\frac{\rho R^2\omega^2}{2} + C_1 = p_a - \rho\omega^2 R^2$$

The constant of integration becomes

$$C_1 = p_a - \rho\omega^2 R^2$$

Substitution into Equation (i) gives

$$p = \frac{\rho r^2\omega^2}{2} + p_a - \frac{\rho\omega^2 R^2}{2}$$

or

$$\frac{(p_a - p)}{\rho} = R^2\omega^2 - \frac{r^2\omega^2}{2}$$

Recalling that $V_{\theta\max} = R\omega$, we find the following after substitution and simplification:

$$\frac{(p_a - p)}{\rho V_{\theta\max}^2} = 1 - \frac{r^2}{2R^2} \quad (r \le R)$$

Rewriting Equation (ii), we get

$$\frac{(p_a - p)}{\rho V_{\theta\max}^2} = \frac{R^2}{2r^2} \quad (r \ge R)$$

The pressure distribution is plotted in Figure 12.14. As we see in the figure, the minimum pressure occurs at the center of the eye. Moreover, the entire pressure profile is less than atmospheric in the region about the tornado. For this example, the minimum pressure is calculated, with $r = 0$, as

$$\frac{(p_a - p)}{\rho V_{\theta\max}^2} = 1$$

After substitution, we get

$$\frac{101\,300 - p}{1.19(50)^2} = 1$$

where ρ for air was obtained from Table A.6. Solving, we get the minimum pressure in the eye:

$$p = 98.3 \text{ kPa}$$

One of the damaging effects of a passing tornado on a dwelling is on windows. Pressure inside may still be near atmospheric while the pressure drops outside; in many cases, this pressure difference is sufficient to blow windows out.

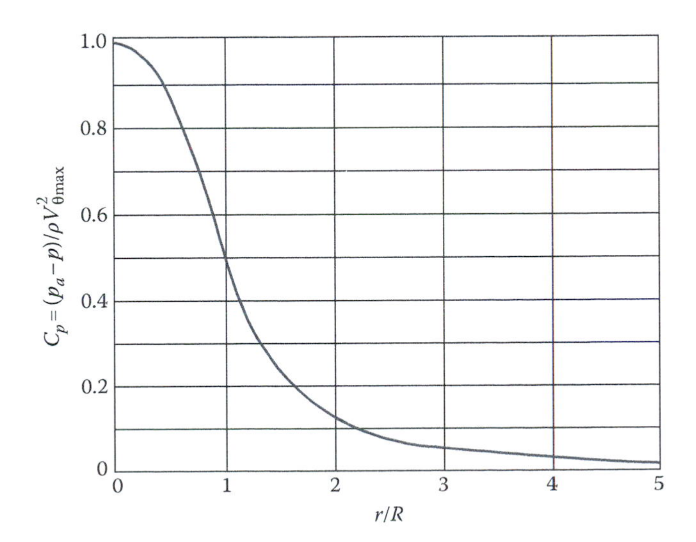

FIGURE 12.14 Pressure profile in the tornado of Example 12.3.

12.5 COMBINED FLOWS AND SUPERPOSITIONS

In each of the inviscid flows of the preceding section, the potential function satisfied Laplace's equation. It is possible to add the potential functions of various flows to obtain more complex potential flows. The prime requirement of a newly created flow is that its potential function must also satisfy Laplace's equation. It is possible to do this because Laplace's equation is linear and adding potential flows is a linear operation. To illustrate, consider two potential functions ϕ_1 and ϕ_2; both satisfy Laplace's equation:

$$\frac{\partial^2\phi_1}{\partial x^2}+\frac{\partial^2\phi_1}{\partial y^2}=0 \tag{12.26a}$$

and

$$\frac{\partial^2\phi_2}{\partial x^2}+\frac{\partial^2\phi_2}{\partial y^2}=0 \tag{12.26b}$$

The test for linearity is this: ϕ_1 and ϕ_2 are solutions, then $\phi_1 + \phi_1$ must also be a solution:

$$\frac{\partial^2(\phi_1+\phi_2)}{\partial x^2}+\frac{\partial^2(\phi_1+\phi_2)}{\partial y^2}=0$$

Because this equation equals Equations 12.26a and 12.26b added together, the Laplace equation is said to be linear. Examples of nonlinear equations are the Euler equations and the Navier–Stokes equations.

12.5.1 Flow about a Half-Body

The flow field that results from combining a source and uniform flow is known as *flow about a half-body*. The potential function is found by adding Equations 12.22a and 12.23a:

$$\phi = Ux+\frac{q}{2\pi}\ell\text{n}(r)$$

$$\phi = Ur\cos\theta+\frac{q}{2\pi}\ell\text{n}(r) \quad \text{(flow about a half-body)} \tag{12.27a}$$

$$\psi = Uy+\frac{q\theta}{2\pi}$$

$$\psi = Ur\sin\theta+\frac{q\theta}{2\pi} \quad \text{(flow about a half-body)} \tag{12.27b}$$

The resulting flow pattern is illustrated in Figure 12.15. The source is placed at the origin, and uniform flow is from left to right. At the front of the body is a stagnation point. The velocity at any point is determined by differentiation of the stream or potential function:

$$V_r = \frac{1}{r}\frac{\partial\psi}{\partial\theta} = U\cos\theta+\frac{q}{2\pi r}$$

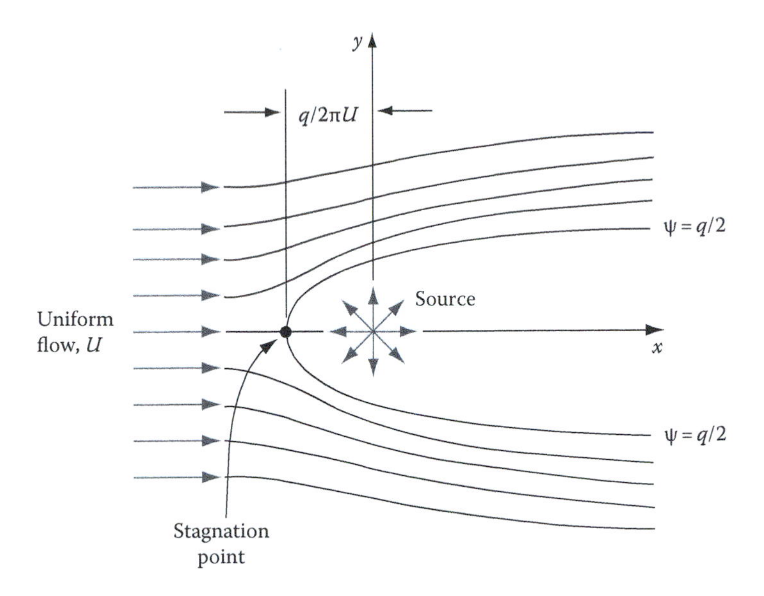

FIGURE 12.15 Flow about a half-body.

and

$$V_\theta = -\frac{\partial \psi}{\partial r} = -U \sin \theta$$

The stagnation point exists where the velocity is zero. Thus,

$$V_\theta = 0 = -U \sin \theta$$

$$\theta = 0 \text{ or } \pi$$

Moreover,

$$V_r = 0 = U \cos \theta + \frac{q}{2\pi r}$$

If $\theta = 0$, then this equation indicates a negative r, which has no meaning. Therefore, $\theta = \pi$ and

$$0 = -U + \frac{q}{2\pi r}$$

The location of the stagnation point, then, is

$$r = \frac{q}{2\pi U} \qquad \theta = \pi$$

The streamline that passes through the stagnation point is determined by substituting these values for r and θ into Equation 12.27b:

$$\psi = Ur \sin \theta + \frac{q\theta}{2\pi} = \frac{q}{2}$$

The constant $q/2$ when substituted for ψ in Equation 12.27b is the equation of the stagnation streamline that approaches from the left and divides in two to form the outline of the half-body:

$$\frac{q}{2} = Ur\sin\theta + \frac{q\theta}{2\pi}$$

The body extends to infinity on the right.

12.5.2 Source and Sink of Equal Strengths

Consider a source and sink both of strength $q/2\pi$ and located at (x, y) coordinates of $(-a, 0)$ and $(a, 0)$, respectively, as sketched in Figure 12.16a. The potential for this combined flow is obtained by adding Equations 12.23a and 12.24a:

$$\phi = \frac{q}{2\pi}\ell n(r_1) - \frac{q}{2\pi}\ell n(r_2)$$

The equations for these flows were originally written assuming the origin to be the location of the sink and the source. To accomplish an appropriate phase shift and to transform from polar to Cartesian coordinates, we use

$$r_1^2 = (x+a)^2 + y^2$$

$$r_2^2 = (x-a)^2 + y^2$$

The potential function becomes

$$\phi = \frac{q}{2\pi}\ell n\left[\frac{(x+a)^2 + y^2}{(x-a)^2 + y^2}\right]^{1/2} \quad \text{(source and sink)} \tag{12.28a}$$

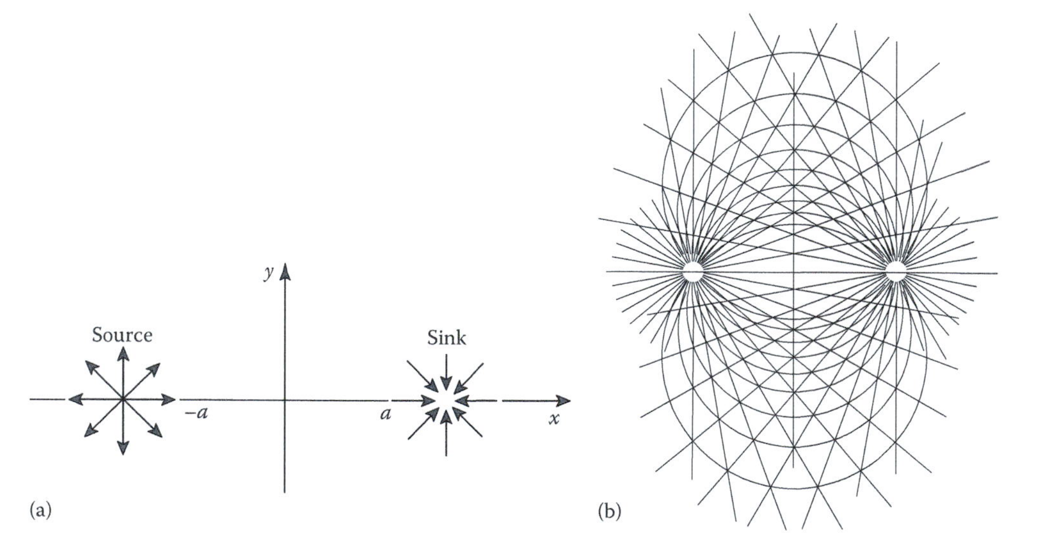

FIGURE 12.16 (a) Streamlines of flow for a source and sink of equal strength and (b) centers of circles are located at the intersection of corresponding radial lines from source and sink. Each circle passes through both singularities.

The stream function for the combined source and sink is obtained by adding Equations 12.23b and 12.24b, giving

$$\psi = \frac{q\theta_1}{2\pi} - \frac{q\theta_2}{2\pi}$$

For a source or a sink at the origin, tan $\theta = y/x$. Similar expressions can be written for the angles of the preceding equation; after substitution, the stream function would become

$$\psi = \frac{q}{2\pi}\left(\tan^{-1}\frac{y}{x+a} - \tan^{-1}\frac{y}{x-a}\right)$$

We now use the trigonometric identity

$$\tan^{-1}\alpha_1 - \tan^{-1}\alpha_2 = \tan^{-1}\frac{\alpha_1 - \alpha_2}{1+\alpha_1\alpha_2}$$

and rewrite the stream function as

$$\psi = \frac{q}{2\pi}\tan^{-1}\frac{[y/(x+a)]-[y/(x-a)]}{1-[y^2/(x^2-a^2)]}$$

Simplifying leads to

$$\psi = -\frac{q}{2\pi}\tan^{-1}\frac{2ay}{x^2+y^2-a^2} \quad \text{(source and sink)} \tag{12.28b}$$

The streamline pattern, diagrammed in Figure 12.16b, consists of circles with each center falling on the y-axis. All circles pass through the source and the sink at $(\pm a, 0)$. The potential lines are also all circles; their centers fall on the x-axis.

12.5.3 Flow about a Doublet

The flow pattern for a source and sink of equal strength was given in the preceding paragraphs. The source was located at $(-a, 0)$ and the sink at $(a, 0)$. If the distance between them is allowed to shrink to zero (that is, as $a \to 0$) and the product of qa is held constant, the resulting flow pattern is known as a *doublet*. To obtain a potential and stream function for a doublet, consider Figure 12.17. The source is denoted as point B, the sink is point C, point P is any arbitrary location in the flow field, and A is located such that CA is perpendicular to PB. The potential function is found by adding Equations 12.23a and 12.24a:

$$\phi = \frac{q}{2\pi}[\ell n(r_1) - \ell n(r_2)] = \frac{q}{2\pi}\ell n\frac{r_1}{r_2}$$

Referring to Figure 12.17, we see that the length r_1 is

$$r_1 = BA + AP = 2a\cos\theta_1 + r_2\cos(\theta_2 - \theta_1)$$

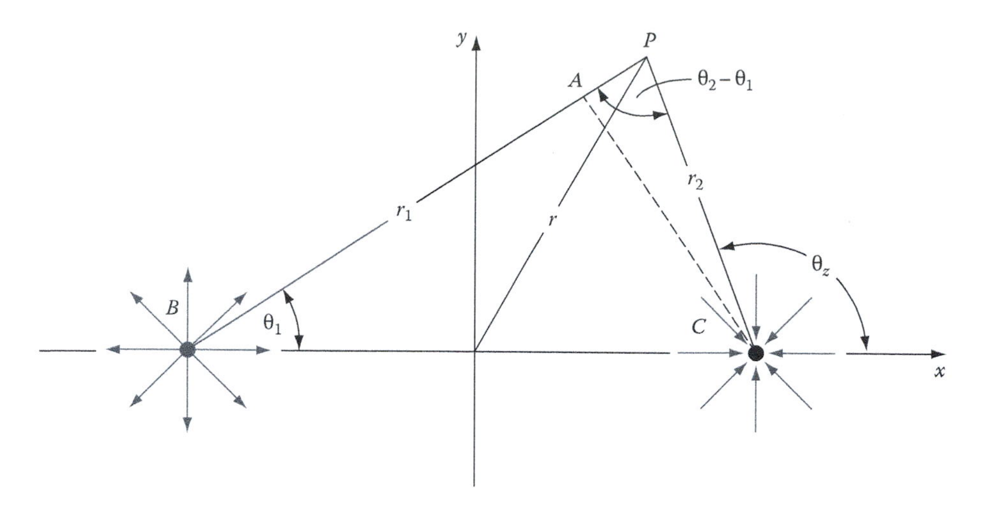

FIGURE 12.17 Source and sink flow.

By substitution, the potential function becomes

$$\phi = \frac{q}{2\pi}\ell n\left[\frac{2a\cos\theta_1}{r_2} + \cos(\theta_2 - \theta_1)\right]$$

As $a \to 0$, then,

$$r_2 \to r$$

$$\theta_2 \to \theta$$

and

$$\cos(\theta_2 - \theta_1) \to 1$$

The potential function becomes in the limit

$$\phi = \frac{q}{2\pi}\ell n\left(\frac{2a\cos\theta}{r} + 1\right)$$

For any quantity δ,

$$\ell n(\delta + 1) \approx \delta \quad \text{if } \delta \ll 1$$

The velocity potential then reduces to

$$\phi = \frac{qa\cos\theta}{\pi r} \quad \text{(doublet flow)} \tag{12.29a}$$

The stream function for a doublet is obtained by adding Equations 12.23b and 12.24b:

$$\psi = \frac{q}{2\pi}(\theta_1 - \theta_2)$$

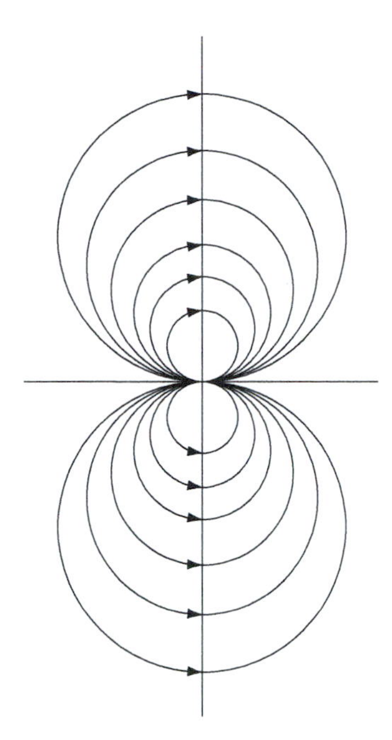

FIGURE 12.18 Streamlines of flow for a doublet.

Referring to Figure 12.17, we see the length AC is

$$AC = r_2 \sin(\theta_2 - \theta_1) = 2a \sin\theta$$

Now as $a \to 0$, $r_2 \to r$, and $\sin(\theta_2 - \theta_1) \to \theta_2 - \theta_1$. The preceding equation then becomes

$$\theta_2 - \theta_1 = \frac{2a}{r}\sin\theta$$

and the stream function is

$$\psi = -\frac{qa}{\pi r}\sin\theta \quad \text{(doublet flow)} \tag{12.29b}$$

The streamline pattern is provided in Figure 12.18.

12.5.4 Flow about a Rankine Body

The flow field that results from combining a source and a sink of equal strength with a uniform flow gives what is known as *flow about a Rankine body*. The potential function is found by adding Equations 12.22a and 12.28a:

$$\phi = Ux + \frac{q}{2\pi}\ell\text{n}\left[\frac{(x+a)^2 + y^2}{(x-a)^2 + y^2}\right]^{1/2} \quad \text{(Rankine body)} \tag{12.30a}$$

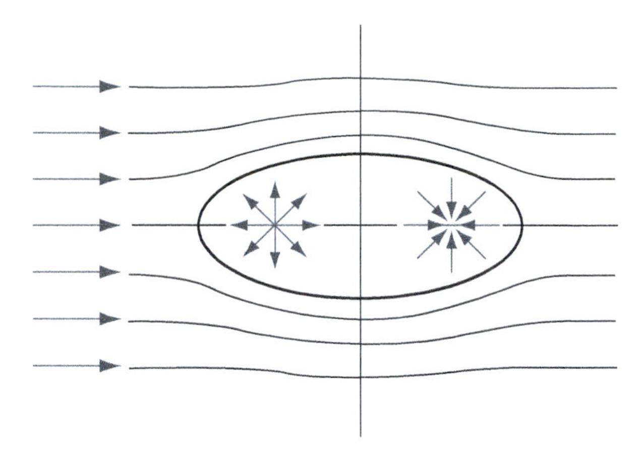

FIGURE 12.19 Streamlines of flow about a Rankine body.

The stream function plotted in Figure 12.19 is obtained by adding Equations 12.22b and 12.28b:

$$\psi = Uy - \frac{q}{2\pi}\tan^{-1}\frac{2\alpha y}{x^2 + y^2 - a^2} \quad \text{(Rankine body)} \tag{12.30b}$$

The velocity components are determined by differentiating the potential function

$$V_x = \frac{\partial \phi}{\partial x} = U + \frac{q}{2\pi}\left[\frac{x+a}{(x+a)^2 + y^2} - \frac{x-a}{(x-a)^2 + y^2}\right] \tag{12.31a}$$

This simplifies to

$$V_x = U - \frac{qa}{2\pi}\frac{x^2 - y^2 - a^2}{[(x+a)^2 + y^2][(x-a)^2 + y^2]} \tag{12.31b}$$

Also,

$$V_y = \frac{\partial \phi}{\partial y} = \frac{q}{2\pi}\left[\frac{y}{(x+a)^2 + y^2} - \frac{y}{(x-a)^2 + y^2}\right]$$

or

$$V_y = -\frac{2qaxy}{\pi}\left\{\frac{1}{[(x+a)^2 + y^2][(x-a)^2 + y^2]}\right\} \tag{12.32}$$

The stagnation points along the body are located where the velocity components are zero:

$$V_x = 0$$

and

$$V_y = 0$$

The second of these conditions, when combined with Equation 12.32, leads to the conclusion that $y = 0$. The stagnation points are expected to fall on the x-axis; the flow is symmetric about the x-axis. Setting $V_x = 0$ and $y = 0$ in Equation 12.31a gives

$$U = -\frac{q}{2\pi}\left[\frac{x+a}{(x+a)^2} - \frac{x-a}{(x-a)^2}\right]$$

$$= -\frac{q}{2\pi}\left(\frac{1}{x+a} - \frac{1}{x-a}\right)$$

$$= \frac{qa}{\pi(x^2 - a^2)}$$

Solving for x, we obtain

$$x = \pm\sqrt{a^2 + \frac{qa}{\pi U}} \tag{12.33a}$$

Thus, there are two stagnation points located at the x-axis. The stream function can now be evaluated to determine the value of the stagnation streamline. Substituting $y = 0$ into Equation 12.30b gives

$$\psi = 0 - \frac{q}{2\pi}\tan^{-1}\frac{0}{x^2 - a^2} = 0$$

The stagnation and body streamline result when ψ is set equal to zero in Equation 12.30b:

$$0 = Uy - \frac{q}{2\pi}\tan^{-1}\frac{2ay}{x^2 + y^2 - a^2}$$

or

$$\tan\frac{2U\pi y}{q} = \frac{2ay}{x^2 + y^2 - a^2} \tag{12.33b}$$

12.5.5 Flow about a Circular Cylinder

Flow about a Rankine body results when we combine uniform flow with a source and a sink. As the distance between the source and sink decreases, approaching a doublet, the length of the Rankine body approaches its width. When we combine uniform flow with a doublet, therefore, *flow about a circular cylinder* results. The potential function is obtained by adding Equations 12.22a and 12.29a:

$$\phi = Ux + \frac{qa\cos\theta}{\pi r}$$

$$\phi = Ur\cos\theta + \frac{qa\cos\theta}{\pi r} \quad \text{(circular cylinder)} \tag{12.34a}$$

Similarly, by adding Equations 12.23a and 12.29b, the stream function becomes

$$\psi = Uy - \frac{qa}{\pi r}\sin\theta$$
$$\psi = Ur\sin\theta - \frac{qa}{\pi r}\sin\theta \quad \text{(circular cylinder)} \tag{12.34b}$$

The streamline pattern is shown in Figure 12.20. As with the Rankine body, the stagnation streamline occurs at $\psi = 0$:

$$0 = Ur\sin\theta - \frac{qa}{\pi r}\sin\theta$$

Solving, we get

$$r\Big|_{\psi=0} = \pm\sqrt{\frac{qa}{U\pi}} = R$$

which is a constant and equal to the radius of the circle. Rewriting the potential and stream functions in terms of the radius gives

$$\phi = U\left(r + \frac{R^2}{r}\right)\cos\theta \tag{12.35a}$$

and

$$\psi = U\left(r - \frac{R^2}{r}\right)\sin\theta \tag{12.35b}$$

The velocity components can be determined by differentiation:

$$V_r = \frac{\partial\phi}{\partial r} = U\left(1 - \frac{R^2}{r^2}\right)\cos\theta \tag{12.36a}$$

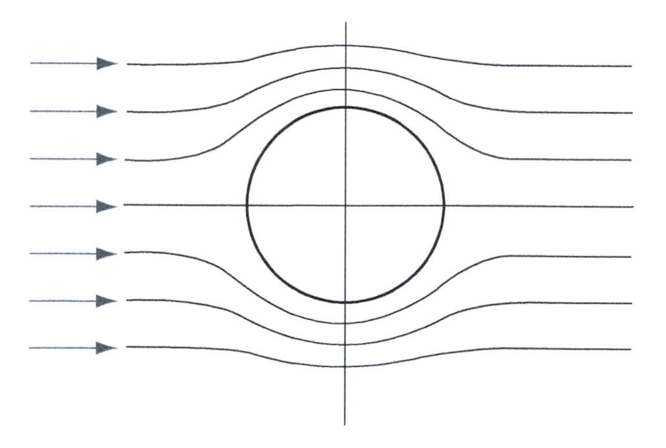

FIGURE 12.20 Flow about a circular cylinder.

$$V_\theta = \frac{1}{r}\frac{\partial \phi}{\partial \theta} = -U\left(1 + \frac{R^2}{r^2}\right)\sin\theta \tag{12.36b}$$

The stagnation points exist where $V_r = 0$ and $V_\theta = 0$ and are given by Equation 12.33. On the surface of the cylinder, where $r = R$, the velocity components are

$$\begin{aligned} V_r(R) &= 0 \\ V_\theta(R) &= -2U\sin\theta \end{aligned} \tag{12.37}$$

Thus, there is no radial flow through the cylinder surface (or through any other streamline). Moreover, V_θ varies from zero at $\theta = 0$ to a maximum of $-2U$ at $\theta = \pi/2$ and back to zero at $\theta = \pi$.

In Chapter 6, results of experiments performed with flow past a circular cylinder were presented. A drag coefficient versus Reynolds number curve was also given. It was stated that part of the drag exerted on a cylinder in a uniform flow is due to a pressure difference between the front of its surface and the rear. For purposes of comparison, let us examine the forces exerted on a cylinder in an inviscid flow. Using Bernoulli's equation, we can first determine the pressure distribution along the surface of the cylinder and then integrate the distribution to obtain the forces of interest—lift and drag. Bernoulli's equation (Equation 12.18) is written for any point on the surface of the cylinder and any other point far from the cylinder in the uniform-flow stream:

$$\frac{p}{\rho} + \frac{V^2}{2} + gy = \frac{p_\infty}{\rho} + \frac{U^2}{2} + gy$$

where

- p is the pressure on the cylinder surface
- V is the velocity along the surface made up of $V_r(R)$ and $V_\theta(R)$
- p_∞ is the free-stream pressure
- U (corresponding to U_∞ of Chapter 6) is the free-stream velocity

Rearranging and substituting, we get

$$\frac{p}{\rho} = \frac{p_\infty}{\rho} + \frac{U^2}{2} - \frac{[V_r(R)]^2 + [V_\theta(R)]^2}{2}$$

Using Equations 12.36, we obtain

$$p = p_\infty + \frac{\rho}{2}(U^2 - 4U^2\sin^2\theta) = p_\infty + \frac{\rho U^2}{2}(1 - 4\sin^2\theta) \tag{12.38a}$$

The pressure difference in dimensionless terms is

$$\frac{2(p - p_\infty)}{\rho U^2} = 1 - 4\sin^2\theta \tag{12.38b}$$

where the left-hand side was first introduced in Chapter 4 as the pressure coefficient. This equation is plotted on polar and Cartesian coordinates in Figure 12.21. Note that the pressure on the cylinder surface equals the free-stream pressure at angles of 30° with respect to the flow direction.

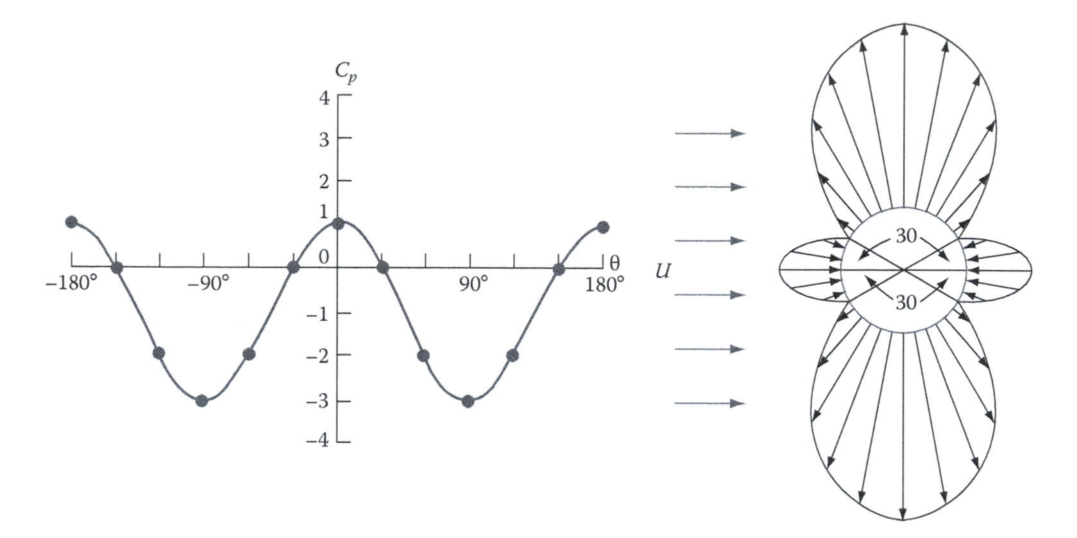

FIGURE 12.21 Pressure coefficient versus angle for flow about a circular cylinder.

To determine the lift and drag forces exerted on the cylinder, it is necessary to integrate the y- and x-components, respectively, of Equation 12.38a, the pressure distribution, about the surface. Figure 12.22 illustrates the forces of interest. The drag force is found as

$$D_f = \iint p\cos\theta\, dA_s$$

where the surface area $dA_s = L(R\,d\theta)$ and p is given by Equation 12.38b.

We obtain

$$D_f = RL\int_0^{2\pi} (p_\infty + \rho U^2 - 2\rho U^2\sin^2\theta)\cos\theta\, d\theta$$

where L is the cylinder length. Simplifying, we get

$$\frac{D_f}{RL} = \left(p_\infty + \frac{\rho U^2}{2}\right)\int_0^{2\pi} \cos\theta\, d\theta - 2\rho U^2\int_0^{2\pi} \sin^2\theta\cos\theta\, d\theta$$

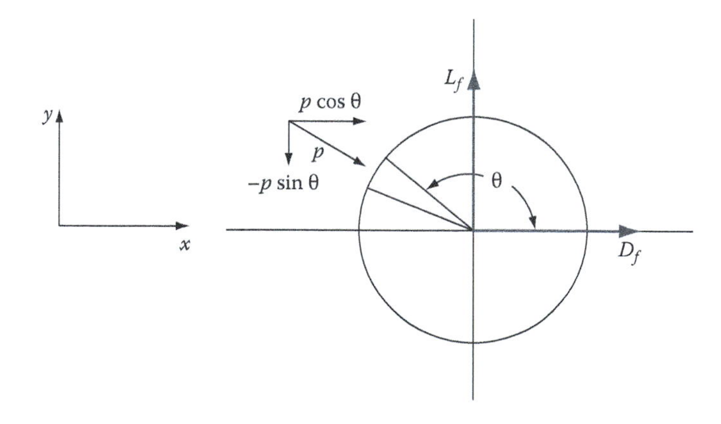

FIGURE 12.22 Pressure on the surface of a cylinder.

The values of both integrals are zero. Thus,

$$D_f = 0 \tag{12.39}$$

Similarly, for the lift force per unit length,

$$L_f = \iint -p \sin\theta \, dA_s$$

$$\frac{L_f}{RL} = \left(p_\infty + \frac{\rho U^2}{2}\right)\int_0^{2\pi} \sin\theta \, d\theta - 2\rho U^2 \int_0^{2\pi} \sin^3\theta \, d\theta$$

Again the values of both integrals are zero. Thus,

$$L_f = 0$$

Thus, the total force exerted on a cylinder immersed in a uniform flow of an inviscid fluid is zero. We might expect this result because the pressure distribution is symmetric about both the x and y axes. This result is typical for potential flow about a body. One exception, however, is for the flow field considered next.

12.5.6 Flow about a Circular Cylinder with Circulation

Uniform flow with a doublet, as we have seen, results in flow past a cylinder. We can further combine a circulatory flow with flow past a cylinder to obtain a new flow field. Streamlines for the combined flow can take on one of three configurations, as illustrated in Figure 12.23. The pattern depends on the strength of the circulation Γ in comparison with the free-stream velocity. The potential function is obtained by adding Equations 12.35a and 12.25a (modified to reflect a clockwise rotation):

$$\phi = U\left(r + \frac{R^2}{r}\right)\cos\theta - \frac{\Gamma\theta}{2\pi} \quad \text{(circular cylinder with circulation)} \tag{12.40a}$$

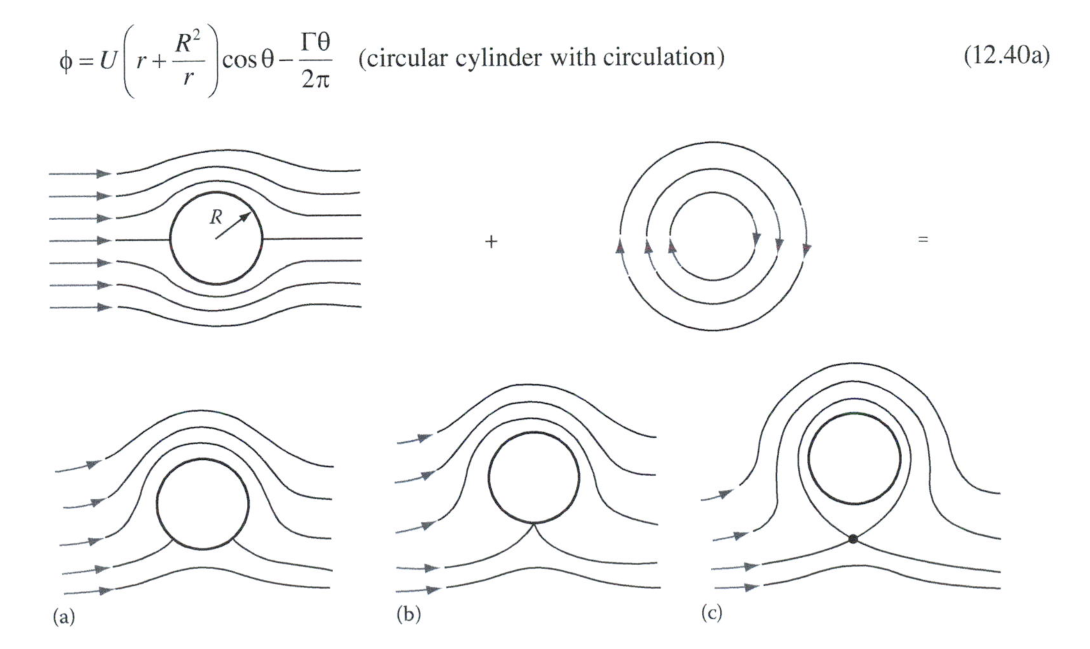

FIGURE 12.23 Flow about a circular cylinder with circulation.

The stream function can be determined by adding Equations 12.35a and 12.25b:

$$\psi = U\left(r - \frac{R^2}{r}\right)\sin\theta + \frac{\Gamma}{2\pi}\ell n(r) \quad \text{(circular cylinder with circulation)} \tag{12.40b}$$

The velocity components are found by differentiation:

$$V_r = \frac{\partial \phi}{\partial r} = U\left(1 - \frac{R^2}{r^2}\right)\cos\theta$$

and

$$V_\theta = \frac{1}{r}\frac{\partial \phi}{\partial \theta} = -U\left(1 + \frac{R^2}{r^2}\right)\sin\theta - \frac{\Gamma}{2\pi r}$$

Along the cylinder surface,

$$V_r(R) = 0$$

and

$$V_\theta(R) = -2U\,\sin\theta - \frac{\Gamma}{2\pi R}$$

For purposes of comparison, let us calculate lift and drag for this case to determine how the circulation affects these forces. Applying the Bernoulli equation to the cylinder surface and to a point far from the cylinder, we get

$$\frac{p}{\rho} + \frac{V^2}{2} + gy = \frac{p_\infty}{\rho} + \frac{U^2}{2} + gy$$

Simplifying and substituting for V, we obtain

$$p = p_\infty + \frac{\rho U^2}{2} - \frac{\rho V_\theta^2}{2} \tag{12.41}$$

or

$$\frac{2(p - p_\infty)}{\rho} = U^2 - \left(2U\sin\theta + \frac{\Gamma}{2\pi R}\right)^2$$

In dimensionless terms,

$$\frac{2(p - p_\infty)}{\rho U^2} = 1 - \left(2\sin\theta + \frac{\Gamma}{2\pi R U}\right)^2$$

The drag force per unit length is found by integration:

$$D_f = \iint p\cos\theta\, dA_s$$

$$\frac{D_f}{RL} = \int_0^{2\pi}\left[p_\infty + \frac{\rho U^2}{2} - \frac{\rho}{2}\left(2U\sin\theta + \frac{\Gamma}{2\pi R}\right)^2\right]\cos\theta\, d\theta$$

Expanding, we get

$$\frac{D_f}{RL} = p_\infty\int_0^{2\pi}\cos\theta\, d\theta + \frac{\rho U^2}{2}\int_0^{2\pi}\cos\theta\, d\theta$$

$$-2\rho U^2\int_0^{2\pi}\sin^2\theta\cos\theta\, d\theta$$

$$-\frac{\rho U\Gamma}{\pi R}\int_0^{2\pi}\sin\theta\cos\theta\, d\theta - \frac{\rho\Gamma^2}{4\pi^2R^2}\int_0^{2\pi}\cos\theta\, d\theta$$

Each integral is zero. Therefore,

$$D_f = 0 \tag{12.42}$$

The lift force per unit length is found also by integration:

$$\frac{L_f}{RL} = -\int_0^{2\pi} p\sin\theta\, d\theta$$

$$= -\int_0^{2\pi}\left[p_\infty + \frac{\rho U^2}{2} - \frac{\rho}{2}\left(2U\sin\theta + \frac{\Gamma}{2\pi R}\right)^2\right]\sin\theta\, d\theta$$

Expanding, we get

$$\frac{L_f}{RL} = -p_\infty\int_0^{2\pi}\sin\theta\, d\theta - \frac{\rho U^2}{2}\int_0^{2\pi}\sin\theta\, d\theta + 2\rho U^2\int_0^{2\pi}\sin^3\theta\, d\theta$$

$$+\frac{\rho U\Gamma}{\pi R}\int_0^{2\pi}\sin^2\theta\, d\theta + \frac{\rho\Gamma^2}{4\pi^2R^2}\int_0^{2\pi}\sin\theta\, d\theta$$

This equation reduces to

$$\frac{L_f}{RL} = \frac{\rho U\Gamma}{\pi R}\int_0^{2\pi}\sin^2\theta\, d\theta = \frac{\rho U\Gamma}{\pi R}\left(\frac{\theta}{2} - \frac{\sin 2\theta}{4}\right)\Bigg|_0^{2\pi}$$

$$= \frac{\rho U\Gamma}{\pi R}(\pi)$$

The lift force, then, is

$$\frac{L_f}{L} = \rho U \Gamma \tag{12.43}$$

Circulation is related to rotational speed. The tangential velocity in terms of circulation is

$$V_\theta = \frac{\Gamma}{2\pi R}$$

Also, $V_\theta = R\omega$, and by combining with the preceding equation, we get

$$\omega = \frac{\Gamma}{2\pi R^2}$$

The dimension of Γ is L^2/T.

The circulation can be varied independently from the free-stream velocity. The relative magnitudes of these parameters will determine whether the resulting flow pattern is represented by Figure 12.23a through c. To evaluate this feature quantitatively, we will next determine the location of the stagnation points indicated in the figure. The velocity is zero at a stagnation point. Because V_r is already zero on the surface of the cylinder, we need work only with V_θ.

$$V_\theta(R) = -2U\sin\theta - \frac{\Gamma}{2\pi R} = 0$$

$$\sin\theta = -\frac{\Gamma}{4\pi UR}$$

Three cases can be described. If $\Gamma = 4\pi UR$, then $\sin\theta = -1$, and θ becomes $3\pi/2$. The resulting flow pattern is illustrated in Figure 12.23b. If $\Gamma < 4\pi UR$, then $\sin\theta < -1$, and two stagnation points occur on the cylinder surface, as shown in Figure 12.23a. If $\Gamma > 4\pi UR$, then $\sin\theta > -1$, an imaginary solution that means there are no stagnation points on the cylinder surface; this is seen in Figure 12.23c, where the cylinder is entirely within the circulatory flow.

For a three-dimensional example of the effect of circulation on uniform flow, imagine a pitcher throwing a baseball. As the sphere rotates and translates, it tends to curve. The amount of curve depends on the magnitude of the spin given by the pitcher and the speed of the ball with respect to the air. A wind blowing from the outfield, for instance, would reduce the relative velocity and the tendency for the ball to curve.

Another example of the effect of circulation on uniform flow is the design and construction of the Flettner rotor ship. This vessel used rotating cylinders that developed a lift or thrust if wind blew past them. The original ship displaced about 300 tons and was 100 ft long. It had two cylinders or rotors that extended 50 ft above the ship's deck. Each rotor was 9 ft in diameter and rotated at 750 rev/min. The ship was built in Germany in 1924 but never gained popularity (see Problem 12.41.)

12.6 INVISCID FLOW PAST AN AIRFOIL

As we saw in the last part of the preceding section, a lift force is generated when a circulatory flow is added to flow past a cylinder. Two practical examples of this phenomenon were cited: the flight of a baseball and the design of the Flettner rotor ship. In this section, we will use those results to examine potential flow past an airfoil.

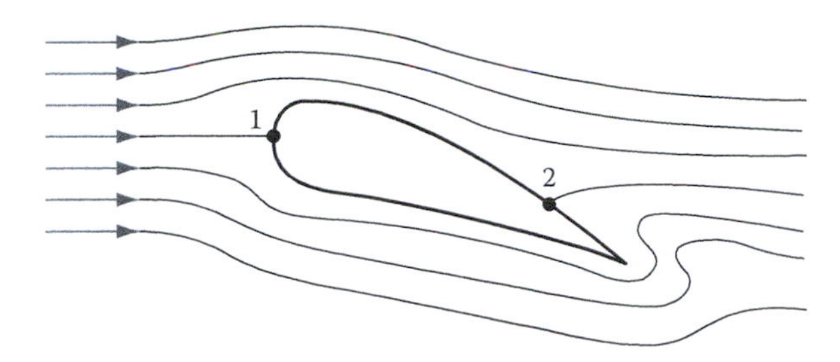

FIGURE 12.24 Streamlines of flow about an airfoil.

Figure 12.24 illustrates the streamlines of two-dimensional flow about an airfoil. The mathematics involved in determining a functional form for the streamlines will not be discussed here. We simply accept this configuration as being correct. The airfoil has a rounded front and a sharp trailing edge. Two stagnation points exist—one at the front and one on the upper surface upstream of the trailing edge (points 1 and 2 in the figure). Because the airfoil is completely immersed in the flow of an inviscid fluid, forces of lift and drag are zero.

Note that the flow must make a sharp turn at the rear edge on the upper surface and travel slightly in the upstream direction. This figure is not a good model of a real flow, however, for experiments on various airfoils indicate that point 2 should be located at the rear edge. Such a flow pattern can be generated by superimposing a clockwise rotation about the wing. The result is illustrated in Figure 12.25. The pressure distribution, if integrated about the surface of the wing, would indicate that drag is still zero but lift is not. In fact, the lift force per unit length is given by Equation 12.43:

$$\frac{L_f}{L} = \rho U \Gamma$$

where

U is the free-stream velocity

Γ is the strength of the circulatory flow

But unlike flow past a circular cylinder, here the circulation must have a certain value and cannot be controlled independently of the free-stream velocity because Γ must be just strong enough to move point 2 to the trailing edge.

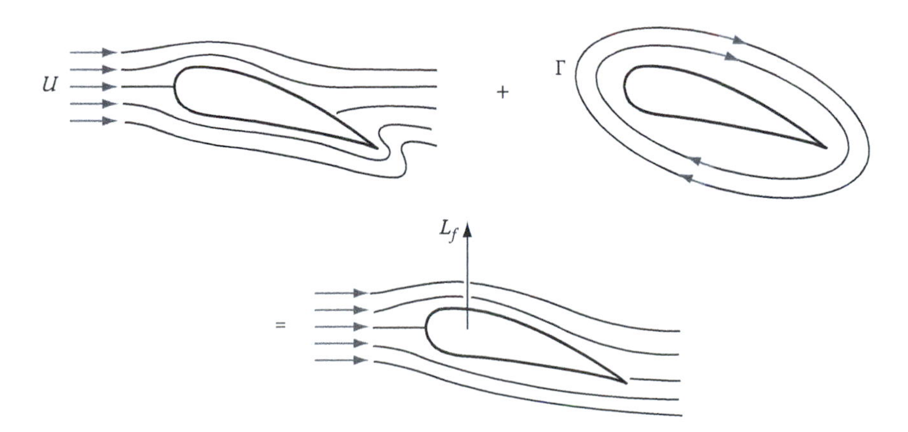

FIGURE 12.25 Superposition of uniform flow past an airfoil and circulation about an airfoil.

As was reported in Chapter 6, lift on an airfoil varies with the properties of the airfoil section and with the angle of incidence. Lift varies linearly with angle of incidence up to a certain point and then drops off somewhat.

12.7 SUMMARY

This chapter gave a brief introduction to the field of inviscid flow (flow of a nonviscous fluid). We began with the Navier–Stokes equations for steady incompressible flow and reduced them following various assumptions—no viscosity and that the flow is two-dimensional. We then obtained the continuity equation and the Euler equations. Next the stream and potential functions were introduced, and it was shown that these functions are normal to each other. The concept of irrotationality was also discussed. Laplace's equation was presented, and various inviscid flows that satisfy Laplace's equation were given. Because of the linearity of Laplace's equation, we formed various complex flows by adding simpler flows—a process called superposition. Finally, we examined inviscid flow past an airfoil.

PROBLEMS

Equations of Motion

12.1 Beginning with Equation 12.1, use the coordinate transformation

$$x = r\cos\theta$$

$$y = r\sin\theta$$

to derive Equation 12.2.

12.2 Determine the vector form of the Euler equations.

Stream Function and Velocity Potential

12.3 In Section 12.2, it was shown that in Cartesian coordinates, the stream and potential functions are normal to each other. Show this for polar cylindrical coordinates.

12.4 The stream function for a certain flow is

$$\psi = x + 2y$$

Determine the velocity potential function and the velocity components. Does the flow satisfy the continuity equation?

12.5 Construct the flow net for Problem 12.4.

12.6 Determine whether the flow of Problem 12.4 is irrotational.

12.7 A flow field can be described by following stream function:

$$\psi = xy + 2x + 3y$$

Determine the velocity components and the potential function. Does the flow satisfy continuity?

12.8 Construct the flow net for Problem 12.7

12.9 Show that the flow of Problem 12.7 is irrotational.

12.10 A certain flow field has the following potential function:

$$\phi = r\cos\theta$$

Determine the stream function and the velocity components. Does the flow satisfy continuity?

12.11 Construct the flow net for Problem 12.10.

12.12 Is the flow of Problem 12.10 irrotational?

12.13 Is the flow represented by $\psi = x^3y^3$ irrotational?

12.14 A flow is described by the following potential function:

$$\phi = \ell n(x^2 + y^2)^{1/2}$$

Show that the stream function is

$$\psi = \tan^{-1}\frac{y}{x}$$

12.15 Is the flow described in Problem 12.14 irrotational?

12.16 The stream function for a certain flow is

$$\psi = 4r\cos\theta - r\sin\theta$$

a. Determine the velocity components.
b. Find the potential function.
c. Is the flow irrotational?

12.17 Sketch the flow net for Problem 12.16.

12.18 The velocity components of a certain flow field are

$$V_x = -\frac{3}{2}y$$

$$V_y = x$$

Does the flow satisfy continuity? Is the flow rotational? Can a stream function be written for this flow? If so, find it.

12.19 The velocity components for a certain flow field are

$$V_x = x^2y$$

$$V_y = -xy^2$$

Does the flow satisfy continuity? Is the flow irrotational? Determine, if possible, the stream function for this flow.

12.20 A certain flow is given by

$$V_r = \cos\theta - \sin\theta$$

$$V_\theta = -r\cos\theta - r\sin\theta$$

Does the flow satisfy continuity? Is the flow irrotational? Determine, if possible, the stream and potential functions.

12.21 A stream function is given in polar coordinates by

$$\psi = 2r^{\pi/\alpha}\sin(\pi\theta/\alpha)$$

Show that it satisfies Laplace's equation.

12.22 A potential function is given in polar coordinates by

$$\phi = 2r^{\pi/\alpha}\cos(\pi\theta/\alpha)$$

Show that the corresponding stream function is

$$\psi = 2r^{\pi/\alpha}\sin(\pi\theta/\alpha)$$

12.23 Sketch the flow net given by the stream and potential functions of Problem 12.22. Take α to be 210° in all cases.

12.24 A flow is described by the following stream and potential functions:

$$\psi = Ur^{\pi/\alpha}\sin(\pi\theta/\alpha)$$

$$\phi = Ur^{\pi/\alpha}\cos(\pi\theta/\alpha)$$

Show that this flow reduces to simple uniform flow when $\alpha = \pi$.

12.25 Does the flow in Problem 12.24 (for any α) satisfy continuity?

Laplace's Equation and Flow Field Equations

12.26 Starting with Equation 12.19, derive Equation 12.21.

12.27 Derive Equation 12.22b.

12.28 Derive Equation 12.23b.

12.29 Derive Equation 12.24b.

12.30 Derive Equation 12.25b.

12.31 The eye of a tornado has a maximum wind velocity of 100 ft/s. Determine the minimum pressure. Determine also the pressure at the location of the maximum wind velocity.

Combined Flows and Superposition

12.32 Consider the flow field represented by adding a source and a circulation both located at the origin. Sketch the streamlines. Determine the location of any stagnation points. Let $q = \Gamma = 4\pi$.

12.33 Find the potential function of the flow described in Problem 12.32.

12.34 Show that the Euler equations are nonlinear.

12.35 Following the steps outlined below, it is possible to compose the flow about a half-body graphically. The discharge q is assigned a value of 16 units. The uniform flow is given by $q = 2\pi U$.

a. Using a plain sheet of paper, select a point for the origin and draw radial lines spaced at angular intervals of $\pi/16$ rad. (Show where $\pi/16$ comes from.)

b. The stream function becomes $\psi = 8y/\pi$. (Show its derivation.) Sketch uniform flow lines according to this equation on the same sheet where the source was drawn.

c. The result is many four-sided adjacent figures. Connect opposite corners of each appropriately to obtain the streamline pattern. Refer to Figure 12.15.

12.36 Verify Equation 12.31b.

12.37 Verify Equation 12.32.

12.38 A source and a sink having an equal strength of 5 m^3/s are immersed in a uniform flow of velocity 3 m/s. Determine the velocity at a location given by $x = 6$ m, $y = 6$ m (see Figure P12.38).

12.39 A doublet whose qa product is 100 ft^4/s is located in a uniform flow of 15 ft/s, as shown in Figure P12.39. Determine the velocity at a point located at $x = 1.0$ and $y = 1.5$. Give a sketch of the results.

12.40 A circular cylinder having a diameter of 10 ft rotates at a speed of 360 rev/min and moves through air at 25 miles/h. What is the lift per unit length of the cylinder for maximum possible circulation?

12.41 Figure P12.41 illustrates a moored Flettner rotor ship in a uniform crosswind of velocity 1500 ft/min. The rotors (length 50 ft, diameter 9 ft) rotate at 750 rev/min. Determine the magnitude of the force that tends to propel the ship forward. Calculate also the lift coefficient for each cylinder.

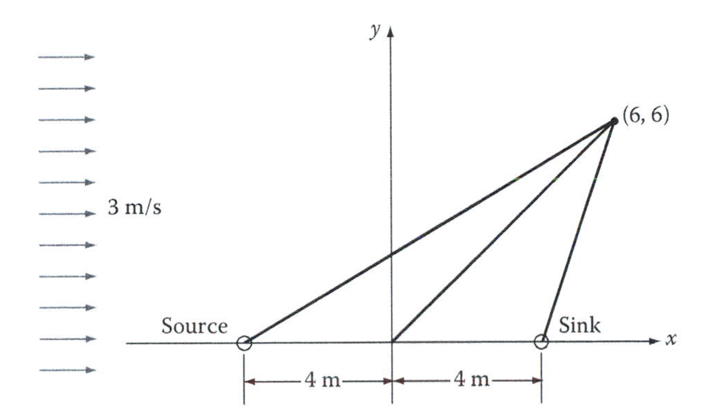

FIGURE P12.38

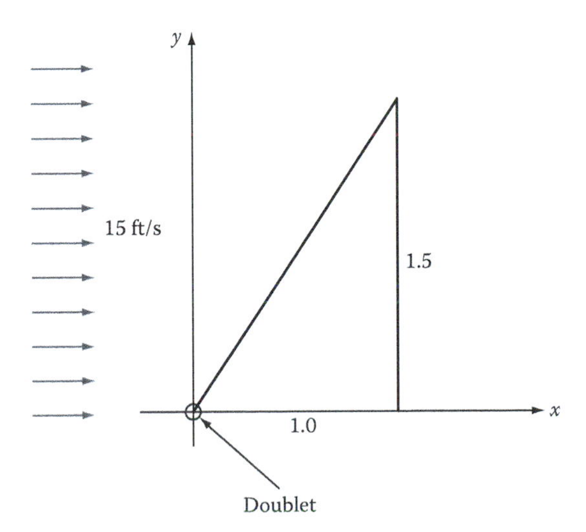

FIGURE P12.39

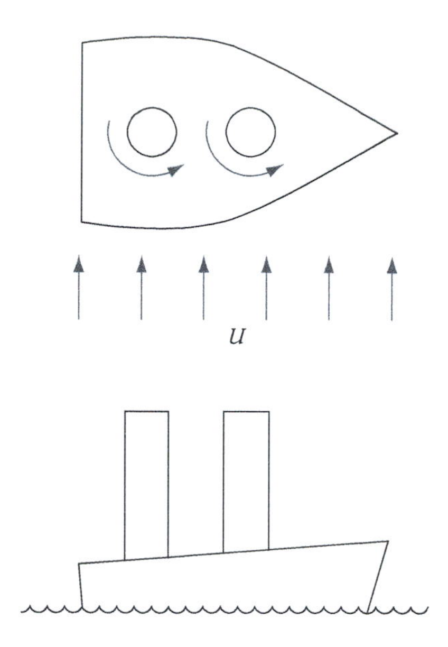

FIGURE P12.41

12.42 A pitcher throws a curve ball with a velocity of 27 m/s. For simplicity, the ball is taken to have a spin about the vertical axis. The lift coefficient is determined to be about 0.3. The catcher is 18.3 m away. If the baseball has a diameter of 7.4 cm, determine the rotational speed imparted by the pitcher and the horizontal deviation of the ball from its straight-line path at the position of the catcher. Assume that Equation 12.43 applies; for length in that expression, let L = sphere volume/surface area.

13 Boundary-Layer Flow

In the last chapter, the remark was made that most fluids problems involve turbulent flow. An exact analysis to determine velocities in the principal flow directions requires a simultaneous solution to the continuity and momentum equations. Because of the nonlinearity of the equations of motion, some simplifications have to be made. In that regard, the flow field may be divided into two portions. One of these is a nonviscous region away from any solid boundaries. Near a boundary, however, fluid adheres to the surface, and the velocity relative to the boundary at the surface is zero. Because this result is a viscous effect, the second flow region is one where viscosity is important. The flow in this region, known as boundary-layer flow, is the subject of this chapter.

Beginning with the continuity, Bernoulli, and Navier–Stokes equations, we will derive the boundary-layer equations for laminar flow. The derived expressions will then be applied to flow over a flat plate, and the Blasius solution for laminar flow will be presented. We will discuss laminar and turbulent boundary-layer flows and select a point for transition. A displacement thickness expression will also be derived. Local and total skin friction drag equations will be written for the problem as well.

Next, we will derive the momentum integral equations from the boundary-layer equations. Although approximate, the momentum integral equations are used to solve the problem of laminar flow over a flat plate, and there is good agreement between the results obtained and the exact solution. The momentum integral equation will then be applied to the problem of turbulent flow over a flat plate. The results of the laminar and turbulent flow problems will be combined to obtain a graph of skin friction drag as a function of Reynolds number at the end of the plate.

13.1 LAMINAR AND TURBULENT BOUNDARY-LAYER FLOW

Let us begin our discussion of *boundary-layer flow* by considering uniform flow over a flat plate, as illustrated in Figure 13.1a. If the fluid is inviscid, the velocity profile appears as in Figure 13.1b. The velocity at the wall is not zero, as it would be for a real fluid. A better description of a real flow is given in Figure 13.1c. Velocity at the wall is zero owing to the nonslip condition (a viscous effect). Far from the surface, the fluid is not influenced by the presence of the plate; consequently, the effect of viscosity in this region can be neglected. Everywhere outside the boundary layer, the velocity is equal to the free-stream value U, and the pressure equals the free-stream value p_∞. These are related by the Bernoulli equation if the flow is inviscid, incompressible, and Newtonian. Inside the boundary layer, the velocity varies from zero at the wall to the free-stream value U at the edge of the boundary layer. The thickness of the boundary layer increases with length downstream and is only a few thousandths of a millimeter thick. In spite of its size, its effect is extremely important in such problems as calculating resistance to motion of a body through a fluid (a ship through water) or determining heat transfer characteristics at a body surface (cold wind blowing past the window of a heated dwelling). Even the Moody diagram for pipe flow is a correlation of a surface effect (wall shear stress) for ducts with constant cross sections.

As was mentioned before, to obtain a solution for some problems, it is convenient to divide the flow into portions—the inviscid region and the boundary-layer region. At the edge of the boundary layer, Bernoulli's equation is applied to determine the pressure variation or distribution. The pressure distribution is then used in the boundary-layer equations to obtain the velocity distribution

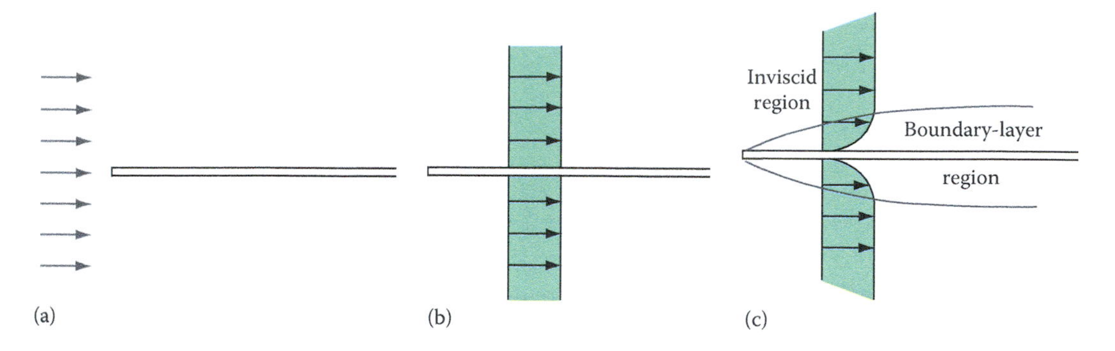

FIGURE 13.1 (a) Uniform flow over a flat plate, (b) profile for inviscid fluid, and (c) profile for real fluid.

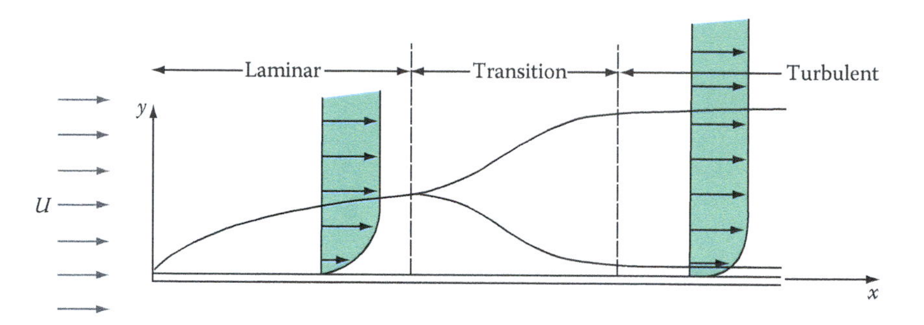

FIGURE 13.2 Laminar, transition, and turbulent flow regimes for flow over a flat plate.

close to the wall. The inviscid and boundary-layer regions are thus patched together at the edge of the boundary layer.

Now let us resort to the flat plate example. Near the leading edge, the flow in the boundary layer is laminar, and viscous forces are great. As the flow moves downstream, it becomes unstable and goes through a transition region to a region of turbulence. This process is analogous to vertical flow of smoke from a cigarette. At a certain height, the smoke changes from a smooth-flowing jet to a turbulent jet. The regimes for flow over a flat plate are illustrated in Figure 13.2. Even in the turbulent boundary layer, however, velocity at the wall is zero. Thus, near the wall, there exists a laminar sublayer, but, as will be shown, it is quite thin.

In laminar flow, shear stress is due to the sliding of one fluid layer over another. The shear stress is related to the strain rate by Newton's law of viscosity:

$$\tau_{yx}^{(\ell)} = \mu \frac{dV_x}{dy} \tag{13.1}$$

In turbulent flow, random motions of particles and velocity fluctuations are responsible for a mixing action that transports faster-moving particles into slower-moving layers and the reverse. The turbulent shear stress is related to the time-averaged velocity by

$$\tau_{yx}^{(t)} = \mu^{(t)} \frac{d\overline{V}_x}{dy} \tag{13.2}$$

where $\mu^{(t)}$ is the eddy viscosity. The eddy viscosity is many times greater than the absolute viscosity in turbulent flow. As was mentioned in Chapter 11, the absolute viscosity μ is a property of the fluid, but the eddy viscosity is not; it is a property of the flow. The velocity profiles in the laminar

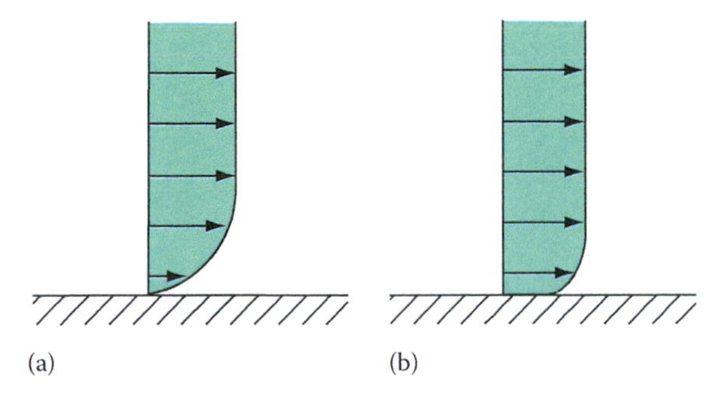

FIGURE 13.3 Velocity profiles in the boundary layer. (a) Laminar and (b) turbulent.

and turbulent boundary layers are somewhat different, as is shown in Figure 13.3. Owing primarily to the turbulent mixing effect, the turbulent profile is flatter over a greater portion of the boundary layer.

Because laminar and turbulent flows are so different, it is extremely important that a criterion be established for transition, In pipe flow, the Reynolds number based on hydraulic diameter is used to determine the type of flow. In flow over a flat plate, again the Reynolds number is used, but here it will be based on distance downstream along the plate:

$$\mathrm{Re} = \frac{Ux}{\nu} \tag{13.3}$$

Many factors affect the change: pressure gradient along the surface, surface roughness, and heat transfer to or from the fluid. Thus, no exact point can be specified for transition. Rather, a transition region exists where the flow is said to be critical. As we mentioned in Chapter 6, transition can vary over the range

$$2 \times 10^5 \leq \mathrm{Re}_{cr} \leq 3 \times 10^6$$

where Re_{cr} is the critical Reynolds number. For purposes of calculation, however, we will take the critical Reynolds number for flow over a flat plate to be 5×10^5.

13.2 EQUATIONS OF MOTION FOR THE BOUNDARY LAYER

In this section, we will derive the continuity and momentum equations for the boundary layer, beginning with the equations of Chapter 11. We will assume that the flow is two-dimensional and steady. Further, the fluid is taken to be Newtonian and incompressible with constant properties. Gravity is neglected, and the radius of curvature of the body is large. This last stipulation is included so that centrifugal forces can be neglected and also to ensure that flow separation does not occur. The boundary-layer equations that we will derive are not applicable to a region of backflow. Because the boundary layer is very thin, pressure does not vary in the direction that is normal to the surface. A schematic of the general boundary-layer problem is given in Figure 13.4. The x-coordinate is in the main flow direction along the body surface; the y-coordinate extends upward normal to the surface; the z-coordinate points out of the page.

Now we will perform what is known as an *order of magnitude analysis*. The boundary layer is thin; thus, velocities and thicknesses in the y-direction have orders of magnitude of δ. In equation form, we would write, for example,

$$V_y \sim O(\delta)$$

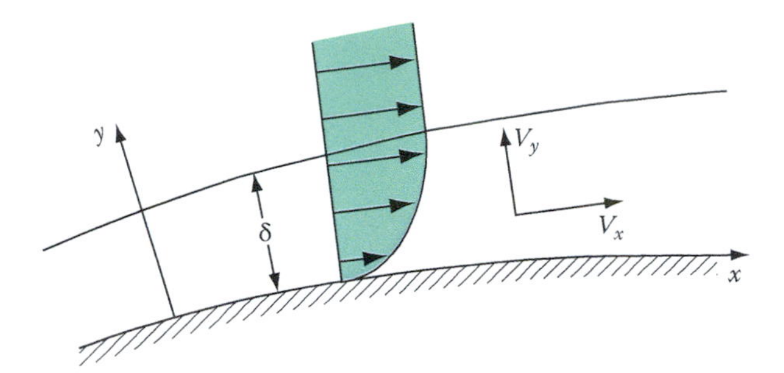

FIGURE 13.4 Flow in the boundary layer.

This states that the velocity in the y-direction has an order of magnitude of δ. Relative to the y-directed factors, velocities and distances in the x-direction have orders of magnitude of unity. For example,

$$V_x \sim O(1)$$

The values of δ and unity are important in a comparative rather than an absolute sense. In this regard, we will determine the order of magnitude of each term in the continuity, Bernoulli, and Navier–Stokes equations and discard the terms of order δ as being negligible. The continuity equation in differential form (from Chapter 11), with each term's order of magnitude term written underneath, is

$$\underset{\frac{1}{1}}{\frac{\partial V_x}{\partial x}} + \underset{\frac{\delta}{\delta}}{\frac{\partial V_y}{\partial y}} = 0$$

Thus, both terms are of order 1, and neither can be neglected.

The Bernoulli equation (from Chapter 12) applies to the region outside the boundary layer. In differential form, we have the following:

$$\underset{\frac{1}{1}1}{\frac{1}{\rho}\frac{\partial p}{\partial x}} + \underset{1\frac{1}{1}}{U\frac{\partial U}{\partial x}} = 0$$

If we assume that density is of order 1, then the pressure gradient must also be of order 1. Hence, none of the terms in this expression can be neglected.

The x-component of the Navier–Stokes equation is, neglecting gravity,

$$\underset{1\frac{1}{1}}{V_x\frac{\partial V_x}{\partial_x}} + \underset{\delta\frac{1}{\delta}}{V_y\frac{\partial V_x}{\partial y}} = \underset{1}{-\frac{1}{\rho}\frac{\partial p}{\partial x}} + \nu\left(\underset{\frac{1}{1^2}}{\frac{\partial^2 V_x}{\partial x^2}} + \underset{\frac{1}{\delta^2}}{\frac{\partial^2 V_x}{\partial y^2}}\right)$$

To ensure that the term $\partial^2 V_x/\partial y^2$ does not make the rest of the terms negligible, we assign the kinematic viscosity an order of magnitude δ^2. This value implies that the viscosity of the fluid can be extremely small and a boundary-layer flow will still exist. From the preceding equation, we find that all terms except $\partial^2 V_x/\partial x^2$ are of order 1 and thus remain.

The y-component of the Navier–Stokes equations is

$$V_x \frac{\partial V_y}{\partial x} + V_y \frac{\partial V_y}{\partial y} = \frac{1}{\rho}\frac{\partial p}{\partial y} + \nu\left(\frac{\partial^2 V_y}{\partial x^2} + \frac{\partial^2 V_y}{\partial y^2}\right)$$

$$1\frac{\delta}{1} \qquad \delta\frac{\delta}{\delta} \qquad \frac{1}{1} \qquad \delta^2\left(\frac{\delta}{1} \quad \frac{\delta}{\delta^2}\right)$$

We conclude that the pressure gradient in the y-direction must be of order δ. In fact, the entire equation must be of order of magnitude δ. Comparing to the x-component equation, which is of order 1, we can neglect the y-component equation. With $\partial p/\partial y$ negligible, we infer that the pressure gradient across the boundary layer is invariant and that the free-stream pressure equals that in the boundary layer. Summarizing the foregoing analysis, we now write the boundary-layer equations as follows:

$$\text{Continuity: } \frac{\partial V_x}{\partial x} + \frac{\partial V_y}{\partial y} = 0 \tag{13.4}$$

$$x\text{-component: } V_x \frac{\partial V_x}{\partial x} + V_y \frac{\partial V_x}{\partial y} = -\frac{1}{\rho}\frac{\partial p}{\partial x} + \nu \frac{\partial^2 V_x}{\partial y^2} \tag{13.5}$$

$$y\text{-component: } \frac{\partial p}{\partial y} = 0 \tag{13.6}$$

$$\text{Bernoulli equation: } \frac{1}{\rho}\frac{\partial p}{\partial x} = -U\frac{\partial U}{\partial x} \tag{13.7}$$

These equations are for two-dimensional, steady, incompressible flow of a Newtonian fluid with constant properties.

13.3 LAMINAR BOUNDARY-LAYER FLOW OVER A FLAT PLATE

For laminar boundary-layer flow over a flat plate, Equations 13.4 through 13.7 must be solved simultaneously for V_x, V_y, and p. Assuming uniform flow ($\partial U/\partial x = 0$), Equation 13.7 shows that there is no pressure variation with x. Equation 13.6 states that there is no pressure variation with y. The pressure everywhere, then, is a constant. The equations to be solved are

$$\frac{\partial V_x}{\partial x} + \frac{\partial V_y}{\partial y} = 0 \tag{13.8}$$

$$V_x \frac{\partial V_x}{\partial x} + V_y \frac{\partial V_x}{\partial y} = \nu \frac{\partial^2 V_x}{\partial y^2} \tag{13.9}$$

The boundary conditions are

$$y = 0, \quad V_x = 0 \text{ (condition 1)} \tag{13.10}$$

$$y = 0, \quad V_y = 0 \text{ (condition 2)} \tag{13.11}$$

$$y = \infty, \quad V_x = U \text{ (condition 3)} \tag{13.12}$$

Equation 13.12 is the patching condition between the inviscid and boundary-layer regions. The problem was solved in 1908 by Blasius by using a coordinate transformation, the mathematical details of which are beyond the scope of this discussion. The solution form is

$$\frac{V_x}{U} = f(\eta) \quad \eta = y\left(\frac{U}{\nu x}\right)^{1/2} \tag{13.13}$$

The function f has no known analytic form; it is a power series, the values of which are provided in Table 13.1 as a function of η. A plot of f versus η is given in Figure 13.5. The solution was obtained by formulating a power series expansion about $\eta = 0$ and matching it with an asymptotic solution about $\eta = \infty$. It is customary to select the boundary-layer thickness at the location where $V_x/U = 0.99$. This occurs at $\eta \approx 5$. We therefore write

$$\eta = \delta\left(\frac{U}{\nu x}\right)^{1/2} = 5.0$$

$$\delta = \frac{5.0x}{(Ux^2/\nu x)^{1/2}}$$

or, the boundary-layer thickness is

$$\delta = \frac{5.0x}{(\mathrm{Re})^{1/2}} \tag{13.14}$$

where $\mathrm{Re} = Ux/\nu$. The velocity profile in the laminar boundary layer at any location is represented by Figure 13.5. This implies that all profiles, regardless of downstream location, are similar.

TABLE 13.1

Laminar Boundary-Layer Velocity Profile for Flow Past a Flat Plate: the Blasius Solution

$\eta = y\left(\frac{U}{\nu x}\right)^{1/2}$	$f = \frac{V_x}{U}$	$\eta = y\left(\frac{U}{\nu x}\right)^{1/2}$	$f = \frac{V_x}{U}$
0.0	0.0	2.8	0.811 52
0.2	0.066 41	3.0	0.846 05
0.4	0.132 77	3.2	0.876 09
0.6	0.198 94	3.4	0.901 77
0.8	0.264 71	3.6	0.923 33
1.0	0.329 79	3.8	0.941 12
1.2	0.393 78	4.0	0.955 52
1.4	0.456 27	4.2	0.966 96
1.6	0.516 76	4.4	0.975 87
1.8	0.574 77	4.6	0.982 69
2.0	0.629 77	4.8	0.987 79
2.2	0.681 32	5.0	0.991 55
2.4	0.728 99	∞	1.000 00
2.6	0.772 46		

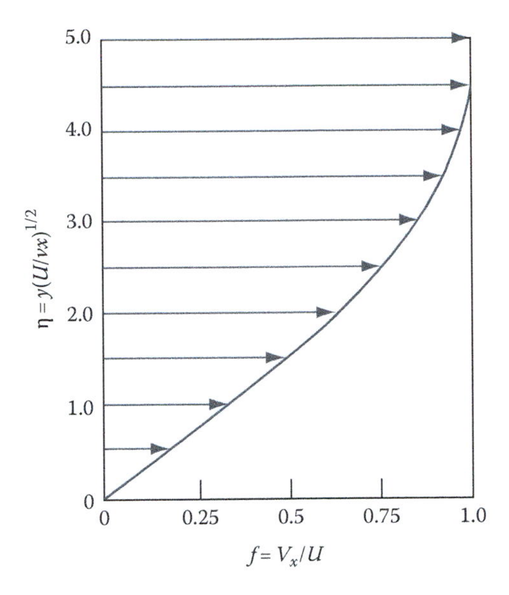

FIGURE 13.5 Blasius velocity profile for a laminar boundary layer.

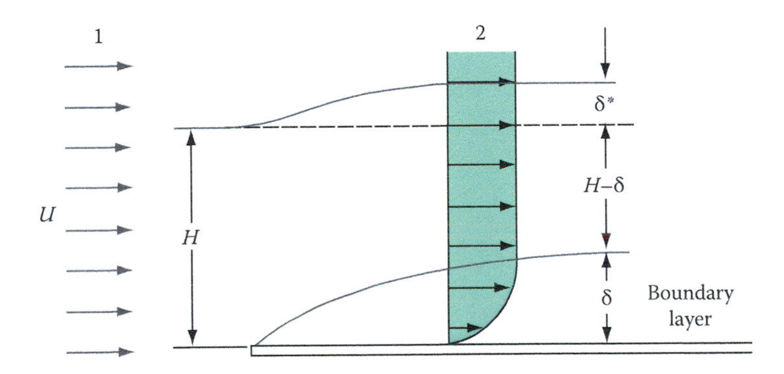

FIGURE 13.6 Displacement thickness formation due to presence of the plate.

In addition to the boundary-layer thickness, a displacement thickness can be defined and determined. Consider Figure 13.6. In the flow upstream of the plate, the velocity is uniform and equal to U. Along the plate, however, the velocity is zero, and a boundary layer has formed. Owing to the decrease in velocity in the boundary layer from the free-stream value, the flow contained in a height H upstream must be displaced upward by an amount δ^* to satisfy conservation of mass.

Mass flow at 1 = mass flow at 2:

$$UH = U\delta^* + U(H - \delta) + \int_0^{\delta} V_x\, dy$$

or

$$U\delta^* = U\delta - \int_0^{\delta} V_x\, dy$$

Dividing by U, we get

$$\delta^* = \delta - \int_0^{\delta} \frac{V_x}{U} dy$$

The right-hand side may be further manipulated:

$$\delta^* = \int_0^{\delta} dy - \int_0^{\delta} \frac{V_x}{U} dy$$

$$\delta^* = \int_0^{\delta} \left(1 - \frac{V_x}{U}\right) dy$$

Moreover, because $V_x/U = 1$ over the range $y = \delta$ to infinity, the preceding equation can be generalized as

$$\delta^* = \int_0^{\infty} \left(1 - \frac{V_x}{U}\right) dy \tag{13.15}$$

With the Blasius profile for laminar incompressible flow over a flat plate, it can be shown that

$$\delta^* = \frac{1.73x}{\sqrt{\text{Re}}} \tag{13.16}$$

To determine the force exerted on the plate, we use the Newtonian expression for shear stress applied at the wall:

$$\tau_w = \mu \left(\frac{\partial V_x}{\partial y}\right)\bigg|_{y=0}$$

From the Blasius solution, the slope of the velocity distribution at the wall is

$$\frac{\partial V_x}{\partial y}\bigg|_{y=0} = \frac{0.33U}{x}\sqrt{\text{Re}}$$

The wall shear stress thus becomes

$$\tau_w = \frac{0.33U\mu}{x}\sqrt{\text{Re}} \tag{13.17}$$

A drag coefficient can now be calculated by using this expression for wall shear stress. (Remember that the drag determined here is a surface or skin friction drag that does not include form drag.) But because τ_w varies with x, the drag coefficient too varies with x. Hence, we define a "local" drag or skin friction coefficient as

$$C_d = \frac{\tau_w}{\rho U^2/2} = \frac{2\tau_w}{\rho U^2}$$

Substituting gives

$$C_d = \frac{0.664U\mu\sqrt{\text{Re}}}{x\rho U^2}$$

or

$$C_d = \frac{0.664}{\sqrt{\text{Re}}} \tag{13.18}$$

The total drag (skin friction drag) can be obtained by integration of the wall shear stress over the length of the plate. Then, by dividing by the kinetic energy of the stream, we get the total skin friction drag coefficient. With b defined as the plate width, we thus obtain

$$D_f = \int_0^L \tau_w b\, dx = \int_0^L \frac{0.332U\mu b}{x}\left(\frac{Ux}{\nu}\right)^{1/2} dx \tag{13.19a}$$

or

$$D_f = 0.664U\mu b\sqrt{\text{Re}_L} \tag{13.19b}$$

where

$$\text{Re}_L = \frac{\rho UL}{\mu} \tag{13.20}$$

Thus the drag force varies with the square root of the length. The total skin friction drag coefficient is defined as

$$C_D = \frac{Df}{\rho U^2 bL/2} = \frac{2D_f}{\rho U^2 bL}$$

Substituting Equation 13.19 brings

$$C_D = \frac{2(0.664)U\mu b(\rho UL/\mu)^{1/2}}{\rho U^2 bL}$$

or

$$C_D = \frac{1.328}{\sqrt{\text{Re}_L}} \tag{13.21}$$

Note that C_d is a local coefficient, whereas C_D is the total obtained by integration over the length of the plate.

Example 13.1

The top view of a small water tunnel is given in Figure 13.7. An object can be placed in the test section (a free diver, e.g., testing flipper thrust) and observed as water is moved around the circuit by a propeller. Upstream of the test section is vertically placed flow straighteners. Each is submerged in 4 ft of liquid, each is 3 ft long, and all are braced across the top. For a flow velocity past the plates of 1.5 ft/s:

a. Determine boundary-layer growth with length.
b. Determine displacement-thickness growth with length.
c. Graph each of the above and also graph the velocity distribution in the boundary layer at $x = 1.5$ ft.
d. Determine total skin friction drag on each plate.

Solution

a. From Table A.5, for water, $\mu = 1.9 \times 10^{-5}$ lbf · s/ft². Therefore,

$$\nu = \frac{\mu}{\rho} = \frac{1.9 \times 10^{-5}}{1.94} = 9.6 \times 10^{-6} \text{ ft}^2/\text{s}$$

The free-stream velocity is given as 1.5 ft/s, and plate length is 3 ft. The Reynolds number at plate end is

$$\text{Re}_L = \frac{UL}{\nu} = \frac{1.5(3)}{9.6 \times 10^{-6}} = 4.7 \times 10^5$$

The boundary layer is laminar over the entire length. The boundary-layer growth is given by Equation 13.14:

$$\delta = \frac{5.0x}{(\text{Re})^{1/2}} = \frac{5.0x}{\left(\dfrac{Ux}{\nu}\right)^{1/2}}$$

or, after substitution,

$$\delta = 1.26 \times 10^{-2} x^{1/2} \quad \text{(i)}$$

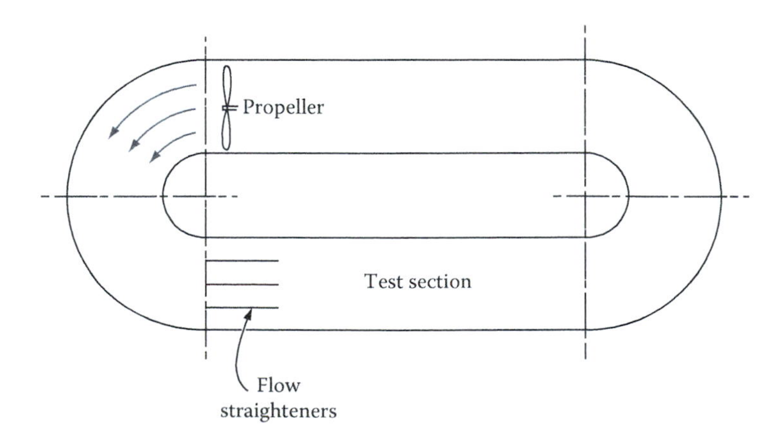

FIGURE 13.7 Water tunnel of Example 13.1.

b. The displacement thickness is given by Equation 13.16 as

$$\delta^* = \frac{1.73x}{\sqrt{\text{Re}}} = \frac{1.73x}{\left(\dfrac{Ux}{\nu}\right)^{1/2}}$$

After substitution and simplification, we get

$$\delta^* = 4.38\times10^{-3}x^{1/2} \quad \text{(ii)}$$

c. A graph of Equations i and ii is provided in Figure 13.8.
The velocity distribution is given by the data of Table 13.1. The independent variable for this problem becomes

$$\eta = y\left(\frac{U}{\nu x}\right)^{1/2} = y\left[\frac{1.5}{(9.6\times10^{-6})(1.5)}\right]^{1/2}$$

$$\eta = 3.23\times10^{2}y$$

Also,

$$V_x = Uf = 1.5f$$

Using the data, we obtain Table 13.2 and the velocity profile of Figure 13.8.

d. The total drag on one side of any of the plates is given by Equation 13.19:

$$Df\big|_1 = 0.664U\mu b\sqrt{\text{Re}_L}$$

The Reynolds number at plate end was calculated earlier as

$\text{Re}_L = 4.7\times10^5$

With $U = 1.5$ ft/s, $\mu = 1.9\times10^{-5}$ lbf · s/ft^2, and $b = 4$ ft, the drag force becomes

$$Df\big|_1 = 0.664(1.5)(1.9\times10^{-5})(4)\sqrt{4.7\times10^5}$$

$$= 0.051 \text{ lbf}$$

Owing to the presence of a boundary layer on each side of one plate, the total skin friction drag per plate is

$$D_f = 0.102 \text{ lbf}$$

TABLE 13.2
Results of Calculations for Example 13.1

η	y (ft)	$f = V_x/U$	V_x (ft/s)
0	0	0	0
1	3.10×10^{-3}	0.329 79	0.49
2	6.20×10^{-3}	0.629 77	0.94
3	9.3×10^{-3}	0.846 05	1.27
4	1.24×10^{-2}	0.955 52	1.43
5	1.55×10^{-2}	0.991 55	1.49

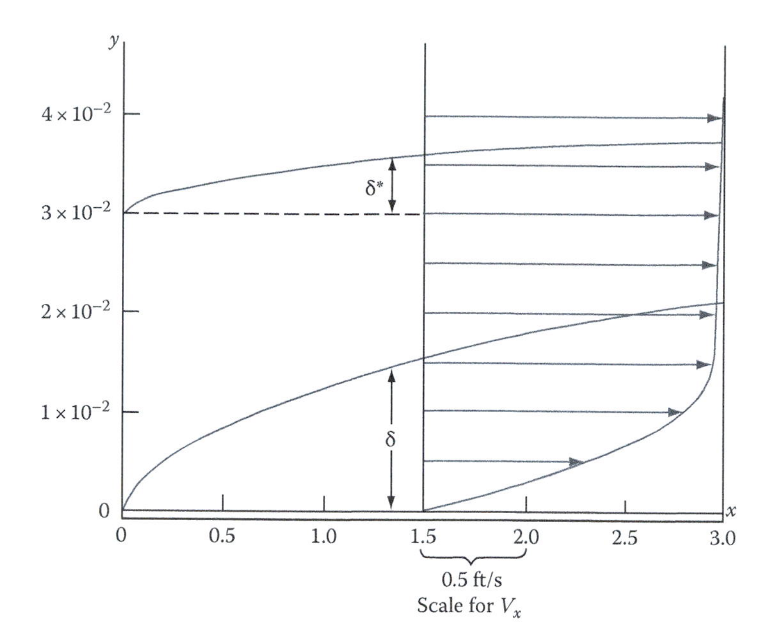

FIGURE 13.8 Boundary-layer growth and displacement-thickness growth with distance and velocity distribution at $x = 1.5$ ft.

The Blasius solution presented here is valid only for laminar flow. Further, although the results are called an exact solution, the boundary-layer equations are only approximate. The Navier–Stokes equations are the equations that apply; the boundary-layer equations are simplifications. Thus, the Blasius solution is really an exact solution to approximate equations.

The boundary-layer equations can be further modified and placed in integral form. They can then be solved for the flat-plate problem, for example, and compared to the Blasius solution. The technique, developed by von Karman, is known as the *momentum integral method.*

13.4 MOMENTUM INTEGRAL EQUATION

The momentum integral equation is derived from the boundary-layer equations; recall that for steady flow, we wrote

$$\frac{\partial V_x}{\partial x} + \frac{\partial V_y}{\partial y} = 0 \tag{13.4}$$

$$V_x \frac{\partial V_x}{\partial x} + V_y \frac{\partial V_x}{\partial y} = -\frac{1}{\rho}\frac{dp}{dx} + \nu \frac{\partial^2 V_x}{\partial y^2} \tag{13.5}$$

Integrating Equation 13.5 with respect to y from zero to δ, the edge of the boundary layer gives

$$\int_0^\delta V_x \frac{\partial V_x}{\partial x}\,dy + \int_0^\delta V_y \frac{\partial V_x}{\partial y}\,dy = -\frac{1}{\rho}\int_0^\delta \frac{dp}{dx}\,dy + \int_0^\delta \nu \frac{\partial^2 V_x}{\partial y^2}\,dy \tag{13.22}$$

The second term on the left-hand side can be integrated by parts to obtain

$$\int_0^\delta V_y \frac{\partial V_x}{\partial y}\,dy = \int_0^\delta \frac{\partial (V_x V_y)}{\partial y}\,dy - \int_0^\delta V_x \frac{\partial V_y}{\partial y}\,dy$$

Now at $y = \delta$, we have $V_x = U$; at $y = 0$, we have $V_x = 0$. The preceding equation thus becomes

$$\int_0^\delta V_y \frac{\partial V_x}{\partial y}\,dy = U\int_0^\delta \frac{\partial V_y}{\partial y}\,dy - \int_0^\delta V_x \frac{\partial V_y}{\partial y}\,dy \tag{13.23}$$

From continuity, we get

$$\frac{\partial V_y}{\partial y} = -\frac{\partial V_x}{\partial x}$$

Combining this equation with Equation 13.23, we obtain

$$\int_0^\delta V_y \frac{\partial V_x}{\partial y}\,dy = -U\int_0^\delta \frac{\partial V_x}{\partial x}\,dy + \int_0^\delta V_x \frac{\partial V_x}{\partial x}\,dy$$

Substituting into Equation 13.22 gives

$$\int_0^\delta V_x \frac{\partial V_x}{\partial x}dy - U\int_0^\delta \frac{\partial V_x}{\partial x}dy + \int_0^\delta V_x \frac{\partial V_x}{\partial x}dy = -\frac{1}{\rho}\int_0^\delta \frac{dp}{dx}dy + \nu\int_0^\delta \frac{\partial^2 V_x}{\partial y^2}dy$$

Simplifying and integrating where possible, we get

$$\int_0^\delta V_x \frac{\partial V_x^2}{\partial x}\,dy - U\int_0^\delta \frac{\partial V_x}{\partial x}\,dy = -\frac{\delta}{\rho}\frac{dp}{dx} + \frac{\mu}{\rho}\left(\frac{\partial V_x}{\partial y}\right)\Bigg|_0^\delta \tag{13.24}$$

We now evaluate each term except for pressure. The first and second terms are evaluated by using the Leibniz rule for differentiating an integral. The formula is

$$\frac{d}{dt}\int_{a(t)}^{b(t)} f(x,f) = \int_{a(t)}^{b(t)} \frac{\partial f}{\partial x}\,dx + \left[f(b,t)\frac{db}{dt} - f(a,t)\frac{da}{dt}\right]$$

The upper and lower limits of the integral on the left-hand side are functions of t. By using the Leibniz rule, the first and second terms of Equation 13.24 become

$$\int_0^\delta \frac{\partial V_x^2}{\partial x}dy = \frac{d}{dx}\int_0^\delta V_x^2\,dy - V_x^2\Big|_\delta \frac{d\delta}{dx}$$

$$= \frac{d}{dx}\int_0^\delta V_x^2\,dy - U^2\frac{d\delta}{dx}$$

and

$$U\int_0^\delta \frac{\partial V_x}{\partial x}\,dy = U\frac{d}{dx}\int_0^\delta V_x\,dy - U(V_x|_\delta)\frac{d\delta}{dx}$$

$$= U\frac{d}{dx}\int_0^\delta V_x\,dy - U^2\frac{d\delta}{dx}$$

where, at the edge of the boundary layer, $V_x = U$. Proceeding with the last term on the right-hand side, we have

$$\frac{\mu}{\rho}\left(\frac{\partial V_x}{\partial y}\right)\Bigg|_0^\delta = \frac{\mu}{\rho}\left(\frac{\partial V_x}{\partial y}\bigg|_\delta\right) - \frac{\mu}{\rho}\left(\frac{\partial V_x}{\partial y}\bigg|_0\right)$$

$$= 0 - \frac{\tau_w}{\rho}$$

The slope $\partial V_x/\partial y = 0$ at δ, and Newton's law of viscosity is applied at the wall:

$$\tau_w = \mu\left(\frac{\partial V_x}{\partial y}\right)\Bigg|_{y=0}$$

Substituting these terms back into Equation 13.24, we obtain the following after simplification:

$$\frac{d}{dx}\int_0^\delta V_x^2\,dy - U\frac{d}{dx}\int_0^\delta V_x\,dy = -\frac{\delta}{\rho}\frac{dp}{dx} - \frac{\tau_w}{\rho} \tag{13.25}$$

This equation is known as the momentum integral equation. It incorporates the continuity expression and is valid for both laminar and turbulent flows. The equation as written is independent of the y variable. The velocity V_x may contain y, but y is integrated out of each term where it might appear.

13.5 MOMENTUM INTEGRAL METHOD FOR LAMINAR FLOW OVER A FLAT PLATE

The momentum integral equation can be applied to flow over a flat plate. First, however, it is necessary to assume a velocity profile for V_x in terms of y. The more accurate the assumption of a velocity profile, the closer the results will be to the exact solution. To investigate further, let us write a general profile equation and see how it can be refined. The profiles are assumed to be similar at different locations downstream. Experimentally, it has been shown that the velocity is a function of y/δ. Assume a polynomial, then, such as

$$\frac{V_x}{U} = C_0 + C_1\frac{y}{\delta} + C_2\frac{y^2}{\delta^2} + C_3\frac{y^3}{\delta^3} + C_4\frac{y^4}{\delta^4} + C_5\frac{y^5}{\delta^5} + \cdots \tag{13.26}$$

We must now write boundary conditions to determine the constants C_i. If we write only that

$$y = 0, \quad V_X = 0 \text{ (condition 1)}$$

$$y = \delta, \quad V_X = U \text{ (condition 2)}$$

then all constants from C_2 onward are made zero. Moreover, C_0 becomes zero from condition 1, and $C_1 = 1$ from condition 2. Thus, the velocity distribution is

$$\frac{V_x}{U} = \frac{y}{\delta} \tag{13.27}$$

which is linear. Suppose next that we include a third boundary condition:

$$y = \delta, \quad \frac{\partial V_x}{\partial y} = 0 \text{ (condition 3)}$$

This ensures that the velocity profile has a vertical slope at the edge of the boundary layer and that the velocity is no longer changing with y. Combining boundary conditions 1, 2, and 3 with the first three terms of Equation 13.26, we get

$$\frac{V_x}{U} = \frac{2y}{\delta} - \frac{y^2}{\delta^2} \tag{13.28}$$

For still another boundary condition, we could refer to the boundary-layer equation for the flat plate with no pressure gradient (Equation 13.9):

$$V_x \frac{\partial V_x}{\partial x} + V_y \frac{\partial V_x}{\partial y} = \nu \frac{\partial^2 V_x}{\partial y^2}$$

Now at $y = 0$, we have $V_x = 0$ and $V_y = 0$. Thus, the preceding equation provides us with

$$y = 0, \quad \frac{\partial^2 V_x}{\partial y^2} = 0 \text{ (condition 4)}$$

Combining all four boundary conditions with the first four terms of Equation 13.26 yields

$$\frac{V_x}{U} = \frac{3y}{2\delta} - \frac{y^3}{2\delta^3} \tag{13.29}$$

The effect of including more boundary conditions is illustrated graphically in Figure 13.9, where all three distributions are plotted. Figure 13.10 shows the Blasius solution with polynomial profiles plotted for comparison. Note that we can select other forms for an assumed profile. One frequently selected profile is

$$\frac{V_x}{U} = \sin \frac{\pi y}{2\delta}$$

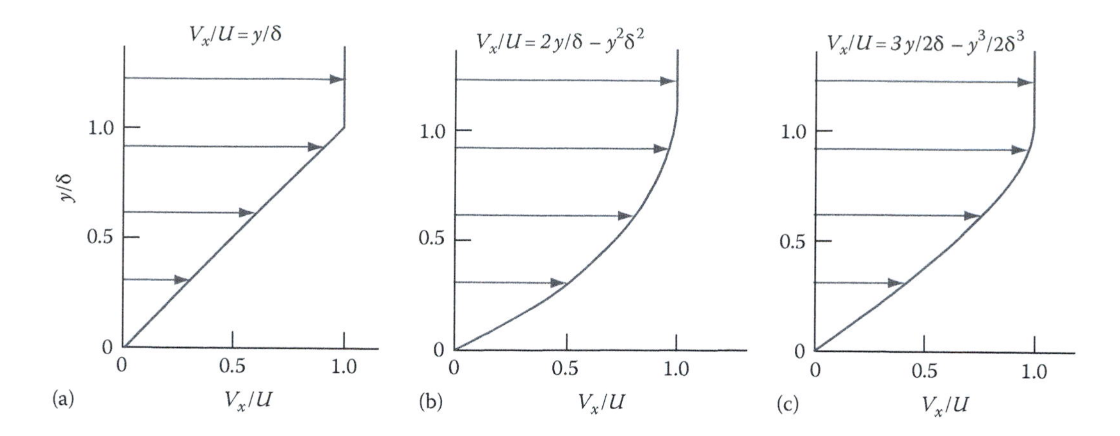

FIGURE 13.9 Three possible velocity profiles.

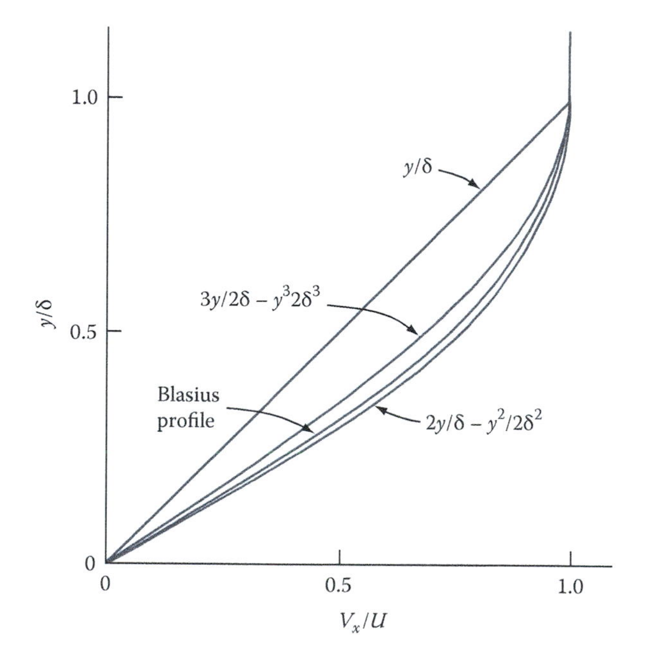

FIGURE 13.10 Blasius profile compared to the polynomial representations.

Now we will select the linear profile given by Equation 13.27 and determine expressions for boundary layer and displacement thicknesses and also the skin friction coefficient:

$$\frac{V_x}{U} = \frac{y}{\delta} \tag{13.27}$$

The momentum integral equation for flow over a flat plate is

$$\frac{d}{dx}\int_0^\delta V_x^2\,dy - U\frac{d}{dx}\int_0^\delta V_x\,dy = -\frac{\tau_w}{\rho} \tag{13.30}$$

For the linear profile,

$$\tau_w = \mu \frac{dV_x}{dy}\bigg|_{y=0} = \mu\left(\frac{U}{\delta}\right)\bigg|_{y=0} = \frac{U\mu}{\delta}$$

Substitution into Equation 13.30 gives

$$\frac{d}{dx}\int_0^\delta \frac{U^2y^2}{\delta^2}\,dy - U\frac{d}{dx}\int_0^\delta \frac{Uy^2}{\delta}\,dy = -\frac{U\mu}{\delta\rho}$$

Integrating, we get

$$\frac{d}{dx}\left(\frac{U^2y^3}{3\delta^2}\right)\bigg|_0^\delta - U\frac{d}{dx}\left(\frac{Uy^2}{2\delta}\right)\bigg|_0^\delta = -\frac{U\mu}{\delta\rho}$$

or

$$\frac{U^2}{3}\frac{d\delta}{dx} - \frac{U^2}{2}\frac{d\delta}{dx} = -\frac{U\mu}{\delta\rho}$$

Simplifying further, we finally obtain

$$\delta\frac{d\delta}{dx} = \frac{6\mu}{\rho U}$$

Integrating

$$\int_0^\delta \delta\,d\delta = \int_0^x \frac{6\mu}{\rho U}\,dx$$

we obtain

$$\frac{\delta^2}{2} = \frac{6\mu x}{\rho U}$$

The boundary-layer thickness then becomes

$$\delta = \sqrt{\frac{12\mu x^2}{\rho U x}}$$

or

$$\delta = \frac{3.46x}{\sqrt{\text{Re}}} \tag{13.31}$$

The displacement thickness is found with

$$\delta^* = \int_0^\delta \left(1 - \frac{V_x}{U}\right)dy$$

Substituting the linear profile and integrating, we get

$$\delta^* = \int_0^\delta \left(1 - \frac{y}{\delta}\right) dy = \left(y = \frac{y^2}{2\delta}\right)\Bigg|_0^\delta = \frac{\delta}{2}$$

Therefore,

$$\delta^* = \frac{1.73x}{\sqrt{\text{Re}}} \tag{13.32}$$

The skin friction or local drag coefficient is

$$C_d = \frac{2\tau_w}{\rho U^2} = \frac{2U\mu}{\delta\rho U^2} = \frac{2\mu}{\rho U}\frac{\sqrt{\text{Re}}}{3.46x}$$

or

$$C_d = \frac{0.578}{\sqrt{\text{Re}}} \tag{13.33}$$

By determining a drag force and defining a total drag coefficient, it can be shown that

$$C_D = \frac{1.156}{\sqrt{\text{Re}_L}} \tag{13.34}$$

A comparison of these results to the Blasius solution is given in Table 13.3. Also shown are the results for other profiles. It is seen that the results compare favorably to the exact solution. Remember that the von Karman momentum integral method is an approximate solution, whereas the Blasius formulation is the exact solution (to approximate equations).

TABLE 13.3

Comparison of Approximate Results to Exact Solution for Flow over a Flat Plate

Velocity Profile	δ	δ^*	C_D
Blasius (exact solution)	$\frac{5.0x}{\sqrt{\text{Re}}}$	$\frac{1.73x}{\sqrt{\text{Re}}}$	$\frac{1.328}{\sqrt{\text{Re}_L}}$
Momentum integral method			
$\frac{V_x}{U} = \frac{y}{\delta}$	$\frac{3.46x}{\sqrt{\text{Re}}}$	$\frac{1.73x}{\sqrt{\text{Re}}}$	$\frac{1.156}{\sqrt{\text{Re}_L}}$
$\frac{V_x}{U} = \frac{2y}{\delta} - \frac{y^2}{\delta^2}$	$\frac{5.48x}{\sqrt{\text{Re}}}$	$\frac{1.83x}{\sqrt{\text{Re}}}$	$\frac{1.462}{\sqrt{\text{Re}_L}}$
$\frac{V_x}{U} = \frac{3y}{2\delta} - \frac{y^3}{2\delta^3}$	$\frac{4.64x}{\sqrt{\text{Re}}}$	$\frac{1.74x}{\sqrt{\text{Re}}}$	$\frac{1.292}{\sqrt{\text{Re}_L}}$
$\frac{V_x}{U} = \sin\frac{\pi y}{2\delta}$	$\frac{4.80x}{\sqrt{\text{Re}}}$	$\frac{1.74x}{\sqrt{\text{Re}}}$	$\frac{1.310}{\sqrt{\text{Re}_L}}$

$\text{Re} = Ux/\nu; \text{Re}_L = UL/\nu.$

13.6 MOMENTUM INTEGRAL METHOD FOR TURBULENT FLOW OVER A FLAT PLATE

As we saw in the preceding section, the momentum integral method for laminar flow gives results that agree well with the exact solution. This result is true even for the crude linear profile. The success of the method thus brings us to an attempt to use it for turbulent flow. The profile that we assume must, of course, be different from that for laminar flow. As seen from Chapter 11, an exact analysis for a turbulent flow problem is not available. Consequently, we have to resort to experimental results. It has been found that a seventh-root profile correlates well with turbulent boundary-layer data over a wide range of Reynolds numbers:

$$\frac{V_x}{U} = \left(\frac{y}{\delta}\right)^{1/7} \tag{13.35}$$

A plot of this equation is provided in Figure 13.11 along with a laminar profile for purposes of comparison. Equation 13.35 does not apply at the wall surface because it predicts an infinite wall shear stress; that is,

$$\frac{dV_x}{dy} = \frac{U}{7y^{6/7}\delta^{1/7}}$$

At the wall, $y = 0$, and the gradient becomes infinite. Because this is physically impossible, the laminar sublayer along the wall is assumed to have a linear profile that becomes tangent to the seventh-root profile at the edge of the buffer layer, where $y = \delta$ (see Figure 13.11). The wall shear stress has to be obtained from experimental results. An equation that agrees well with data is the Blasius formula:

$$\tau_w = 0.0225\rho U^2 \left(\frac{\nu}{U\delta}\right)^{1/4} \tag{13.36}$$

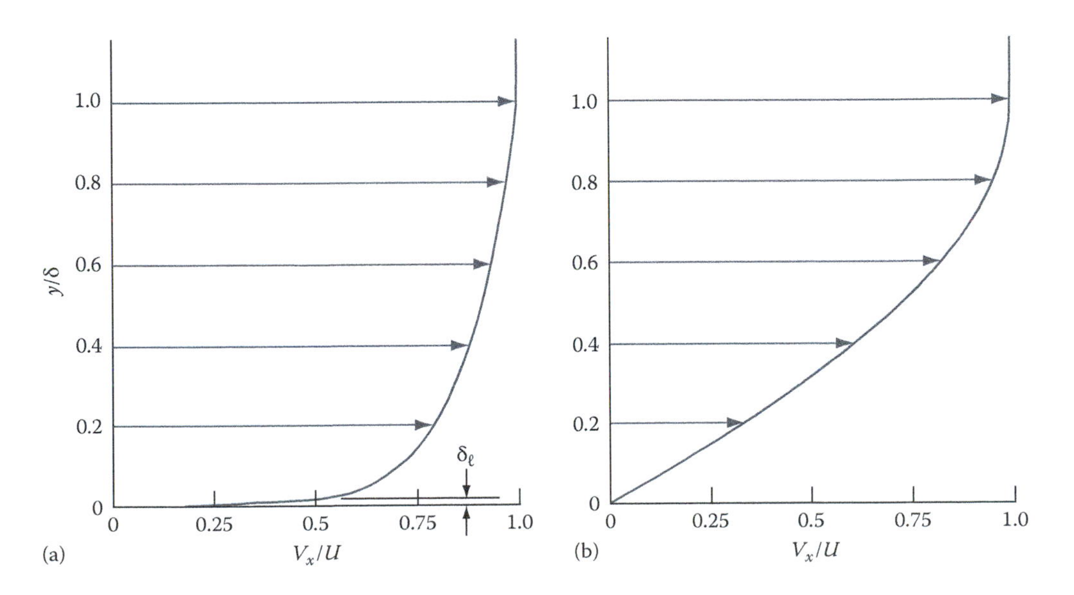

FIGURE 13.11 Turbulent boundary-layer profile compared to the laminar case. (a) $V_x/U = (y/\delta)^{1/7}$ and (b) Blasius profile.

which is valid over the range

$$5 \times 10^5 \leq \text{Re} \leq 10^7$$

This equation was developed from pipe flow tests. Equations 13.35 and 13.36 can now be substituted into Equation 13.30, the momentum integral expression for flow over a flat plate:

$$\frac{d}{dx}\int_0^\delta V_x^2\,dy - U\frac{d}{dx}\int_0^\delta V_x\,dy = -\frac{\tau_w}{\rho} \tag{13.30}$$

$$U^2\frac{d}{dx}\int_0^\delta \left(\frac{y}{\delta}\right)^{2/7} dy - U^2\frac{d}{dx}\int_0^\delta \left(\frac{y}{\delta}\right)^{1/7} dy = -0.0223U^2\left(\frac{\nu}{U\delta}\right)^{1/4}$$

Integrating and simplifying, we get

$$\frac{7}{72}\frac{d\delta}{dx} = 0.0225\left(\frac{\nu}{U\delta}\right)^{1/4}$$

$$\delta^{1/4}d\delta = 0.229\left(\frac{\nu}{U}\right)^{1/4} dx$$

Integrating, we get

$$\frac{4\delta^{5/4}}{5} = 0.229\left(\frac{\nu}{U}\right)^{1/4} x + C_1$$

where C_1 is a constant. The constant can be set equal to zero, which is equivalent to assuming that the boundary layer begins at the leading edge of the plate. Although this assumption is not strictly accurate, for long plates the laminar boundary layer exists over a very small percentage of the length, so neglecting it introduces only a small error. Moreover, a turbulent boundary layer can be induced in tests by roughening the leading edge of the plate. In cases in which the laminar boundary layer cannot be neglected, we will compensate by appropriately combining the laminar and turbulent results. With the constant equal to zero, the preceding equation becomes

$$\delta = \frac{0.368x}{(\text{Re})^{1/5}} \tag{13.37}$$

The displacement thickness is found with

$$\delta^* = \int_0^\delta \left(1 - \frac{V_x}{U}\right) dy = \int_0^\delta \left[1 - \left(\frac{y}{\delta}\right)^{1/7}\right] dy$$

Integrating, we obtain

$$\delta^* = \frac{\delta}{8}$$

$$\delta^* = \frac{0.046x}{(\text{Re})^{1/5}} \tag{13.38}$$

To find the drag force, it is first necessary to evaluate the wall shear, rewriting it in terms of x:

$$\tau_w = 0.0225\rho U^2\left(\frac{\nu}{U\delta}\right)^{1/4}$$

$$= 0.0225\rho U^2\left(\frac{\nu(\mathrm{Re})^{1/5}}{U(0.368x)}\right)^{1/4}$$

After simplification, we get

$$\tau_w = 0.0286\rho U^2\left(\frac{1}{(\mathrm{Re})}\right)^{1/5} \tag{13.39}$$

The local skin friction coefficient is then

$$C_d = \frac{2\tau_w}{\rho U^2} = \frac{0.0573}{(\mathrm{Re})^{1/5}} \tag{13.40}$$

The drag force is determined by integrating the wall shear over the surface area; with b equal to the plate width, we obtain

$$D_f = b\int_0^L \tau_w\,dx = 0.0286\rho U^2 b\int_0^L \frac{dx}{(Ux/\nu)^{1/5}}$$

Integrating and simplifying yield

$$D_f = 0.0358\rho U^2\frac{bL}{(\mathrm{Re}_L)^{1/5}} \tag{13.41}$$

The drag coefficient is

$$C_D = \frac{2D_f}{\rho U^2 bL} = \frac{0.0715}{(\mathrm{Re}_L)^{1/5}} \tag{13.42a}$$

The preceding expression is approximate because we began with the momentum integral equation and relied on empirical results. Better agreement with experimental data is obtained if the equation is altered slightly:

$$C_D = \frac{0.074}{(\mathrm{Re}_L)^{1/5}} \tag{13.42b}$$

which is valid over the range

$$5 \times 10^5 \le \mathrm{Re} \le 10^7$$

Other expressions for velocity and wall shear stress must be used if the Reynolds number is outside this range. These expressions are summarized in Table 13.4.

TABLE 13.4

Results of Momentum Integral Method Applied to Turbulent Boundary-Layer Flow over a Flat Plate

Assumed Profile, V_x/U	Wall Shear Stress, τ_w	δ	Skin Friction Drag Coefficient, C_D	Reynolds Number Range
$\left(\frac{y}{\delta}\right)^{1/7}$	$0.0225\rho U^2\left(\frac{\nu}{U\delta}\right)^{1/4}$	$\frac{0.368x}{(\text{Re})^{1/5}}$	$\frac{0.074}{(\text{Re}_L)^{1/5}}$	5×10^5 to 5×10^7
$\left(\frac{y}{\delta}\right)^{1/8}$	$0.0142\rho U^2\left(\frac{\nu}{U\delta}\right)^{1/5}$	$\frac{0.252x}{(\text{Re})^{1/6}}$	$\frac{0.045}{(\text{Re}_L)^{1/6}}$	1.8×10^5 to 4.5×10^7
$\left(\frac{y}{\delta}\right)^{1/10}$	$0.0100\rho U^2\left(\frac{\nu}{U\delta}\right)^{1/6}$	$\frac{0.201x}{(\text{Re})^{1/7}}$	$\frac{0.0305}{(\text{Re}_L)^{1/7}}$	2.9×10^7 to 5×10^8

Turbulent flow in a pipe was discussed in Chapter 11. Recall that the results of the development were expressed as a graph of $\overline{V}_z/V^*$ as a function of s—that is, dimensionless mean axial velocity in terms of a dimensionless space coordinate measured from the wall. According to Figure 11.8, the laminar sublayer exists over the interval

$$0 \le s \le 5$$

where

$$s = \frac{V^* y}{\nu}$$

At the edge of the buffer layer, we have (from Chapter 11)

$$5 = \frac{V^* \delta_\ell}{\nu}$$

$$\frac{\overline{V}_z}{V^*} = 5$$

With $\overline{V}_z = U$,

$$\delta_\ell = \frac{5\nu}{V^*} = \frac{25\nu}{U} \tag{13.43}$$

The ratio of the laminar-sublayer thickness to the boundary-layer thickness given by Equation 13.37 is

$$\frac{\delta_\ell}{\delta} = \frac{25\nu}{U}\frac{(\text{Re})^{1/5}}{0.368x}$$
$$\frac{\delta_\ell}{\delta} = \frac{67.9}{(\text{Re})^{4/5}} \tag{13.44}$$

At Re = 5×10^5, we have

$$\frac{\delta_\ell}{\delta} = 0.001\,87$$

At Re = 10^7, we have

$$\frac{\delta_\ell}{\delta} = 0.000\,17$$

The laminar sublayer is thus very thin.

13.7 LAMINAR AND TURBULENT BOUNDARY-LAYER FLOW OVER A FLAT PLATE

In the preceding section, we derived expressions for turbulent boundary-layer flow on a flat plate. It was assumed that the turbulent boundary layer started at the leading edge. For the real situation, however, the boundary layer is laminar at the leading edge and then changes to a turbulent profile. An error is introduced by assuming that the turbulent boundary layer begins at $x = 0$. An approximate method for correcting the results is to write the total drag as follows:

$$\begin{aligned} \text{Total drag} &= \text{laminar drag} && (0 \le x \le x_{cr}) \\ &+ \text{turbulent drag} && (0 \le x \le L) \\ &- \text{turbulent drag} && (0 \le x \le x_{cr}) \end{aligned}$$

The turbulent drag contribution over the laminar portion is subtracted out. The preceding expression presumes that turbulent flow begins immediately after laminar flow ends; that is, transition exists at a single point. By selecting the transition Reynolds number as $\text{Re}_{cr} = 5 \times 10^5$ and substituting from Equation 13.21 for laminar flow and Equation 13.42b for turbulent flow, the drag expression becomes

$$C_D \frac{\rho U^2}{2} bL = \frac{1.328}{(\text{Re}_{cr})^{1/2}} \frac{\rho U^2}{2} bL \frac{x_{cr}}{L} + \frac{0.074}{(\text{Re}_L)^{1/5}} \frac{\rho U^2}{2} bL - \frac{0.074}{(\text{Re}_L)^{1/5}} \frac{\rho U^2}{2} bL \frac{x_{cr}}{L}$$

Rearranging and simplifying, we get

$$C_D = \frac{1.328(\text{Re}_{cr})^{1/2}}{\text{Re}_{cr}} \frac{x_{cr}}{L} + \frac{0.074}{(\text{Re}_L)^{1/5}} \frac{0.074(\text{Re}_{cr})^{4/5}}{\text{Re}_{cr}} \frac{x_{cr}}{L}$$

It is seen that

$$\frac{\text{Re}_{cr}}{\text{Re}_L} = \frac{Ux_{cr}}{\nu} \frac{\nu}{UL} = \frac{x_{cr}}{L}$$

By substitution, then, we obtain the following for the drag coefficient:

$$C_D = \frac{1.328(5 \times 10^5)^{1/2}}{\text{Re}_L} + \frac{0.074}{(\text{Re}_L)^{1/5}} - \frac{0.074(5 \times 10^5)^{4/5}}{\text{Re}_L}$$

or

$$C_D = \frac{0.074}{(\mathrm{Re}_L)^{1/5}} - \frac{1\,743}{\mathrm{Re}_L} \tag{13.45}$$

This expression is valid up to a Reynolds number of 1 × 10^7 because the first term on the right-hand side is valid only to that point. An expression that fits experimental data over a wider range is the Prandtl–Schlichting equation:

$$C_D = \frac{0.455}{(\log \mathrm{Re}_L)^{2.58}} - \frac{1\,700}{\mathrm{Re}_L} \tag{13.46}$$

This expression is valid up to a Reynolds number of 1 × 10^9. By combining the following equations, the graph of Figure 13.12 results:

$$C_D = \frac{1.328}{(\mathrm{Re}_L)^{1/2}} \text{(laminar)} \quad x \le 5\times10^5 \tag{13.21}$$

$$C_D = \frac{0.074}{(\mathrm{Re}_L)^{1/5}} \text{(turbulent)} \quad 5\times10^5 \le x \le 1\times10^7 \tag{13.42b}$$

$$C_D = \frac{0.455}{(\log \mathrm{Re}_L)^{2.58}} - \frac{1\,700}{\mathrm{Re}_L} \text{(turbulent)} \quad 1\times10^7 \le x \le 1\times10^9 \tag{13.46}$$

The drag coefficient is graphed as a function of the Reynolds number that exists at the end of the plate. Data taken on flow over a flat plate agree well with the graph.

From Figure 13.12, it can be seen that skin friction drag is lowest at a Reynolds number of 5 × 10^5 (where the flow is laminar) or at very high Reynolds numbers. One way of effectively decreasing drag, then, is to keep the boundary layer laminar if possible. Ensuring that the plate surface is very smooth will reduce disturbances that might cause turbulence. If the transition Reynolds number can be raised to 10^6, for example, x_{cr} is effectively doubled from that corresponding to $\mathrm{Re}_{cr} = 5 \times 10^5$; the skin friction drag is cut almost in half.

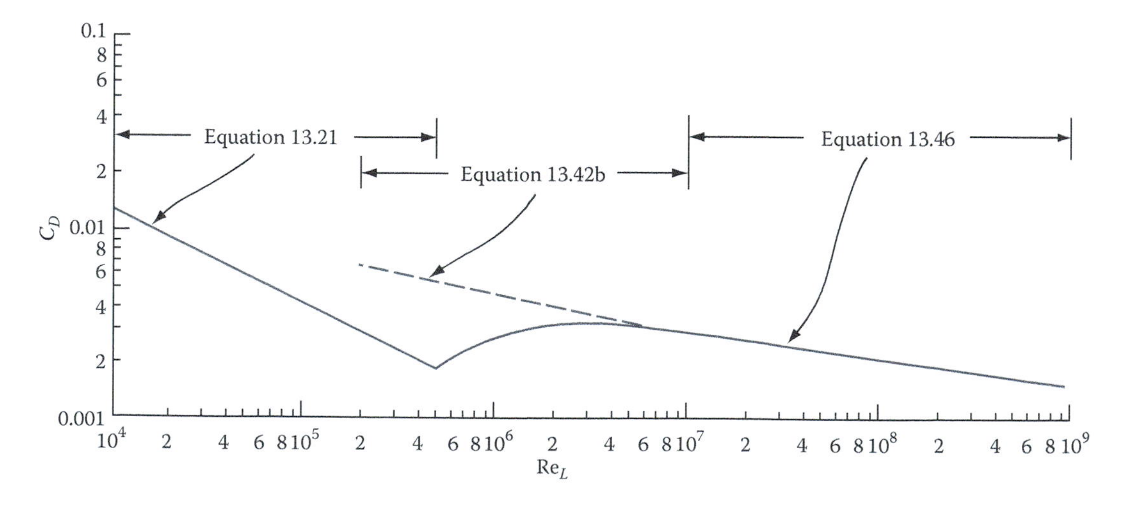

FIGURE 13.12 Drag coefficient versus Reynolds number at end of surface for flow over a flat plate.

Example 13.2

A truck trailer is pulled by a cab at 60 mph. The boxlike trailer is 40 ft long, 12 ft high, and 7.5 ft wide. Using the results of this chapter, estimate the skin friction drag on the trailer top and sides.

Solution

From Table A.6, for air, we find

$$\rho = 0.0023 \text{ slug/ft}^3 \quad \text{and} \quad \mu = 0.3758 \times 10^{-6} \text{ lbf} \cdot \text{s/ft}^2$$

The free-stream velocity is 60 mph, or 88 ft/s. The trailer length is 40 ft. The Reynolds number at the end of the trailer is

$$\text{Re}_L = \frac{\rho UL}{\mu} = \frac{0.0023(88)(40)}{0.3748 \times 10^{-6}} = 2.15 \times 10^7$$

From Figure 13.12, the skin friction drag coefficient is

$$C_D = 0.0025$$

The drag force is found with

$$D_f = C_D \frac{\rho U^2}{2} bL$$

For the trailer top, $b = 7.5$ ft and $L = 40$ ft; therefore,

$$D_f = 0.0025 \frac{0.0023(88)}{2}(7.5)(40)$$

$$D_f = 6.67 \text{ lbf (top)}$$

For either side, $b = 12$ ft. Thus,

$$D_f = 0.0025 \frac{0.0023(88)}{2}(12)(40)$$

$$D_f = 10.7 \text{ lbf (one side)}$$

13.8 SUMMARY

In this chapter, we discussed laminar and turbulent boundary-layer flows. Boundary layer and momentum integral equations were both derived. Moreover, we examined laminar and turbulent boundary-layer flows over a flat plate and presented solutions. The result of the combined problems led to a graph of skin friction drag as a function of Reynolds number at plate end.

PROBLEMS

Equations of Motion for Boundary-Layer Flow

13.1 How is Equation 13.5 affected if it is written for unsteady, two-dimensional, incompressible flow of a Newtonian fluid?

13.2 Derive Equation 13.19 in detail, beginning with Equation 13.17.

13.3 Derive Equation 13.21, beginning with Equation 13.19a.

13.4 A model airplane made of balsa wood is gliding through the air with a velocity of 1.5 m/s. The front wing is flat; it is 30 cm wide (tip to tip) and 10 cm long. Plot the variation of the boundary-layer thickness with length for the wing. Repeat for the displacement thickness. Calculate the skin friction drag on the wing.

13.5 Consider the problem of laminar flow of a thin film down a vertical wall as illustrated in Figure P13.5.* Using an order of magnitude analysis:

a. Show that the Navier–Stokes equations reduce to

$$V_x \frac{\partial V_x}{\partial x} + V_y \frac{\partial V_x}{\partial y} = -\frac{1}{\rho}\frac{dp}{dx} + g + \nu \frac{\partial^2 V_x}{\partial y^2}$$

b. Write boundary conditions for the problem.

c. Neglecting the pressure gradient, and with the aid of the continuity equation, show that the boundary-layer equations can be used to obtain

$$\frac{d}{dx}\int_0^{\delta} V_x^2\, dy - g\delta(x) = -\nu \left.\frac{\partial V_x}{\partial y}\right|_{y=0}$$

List all assumptions inherent in this equation.

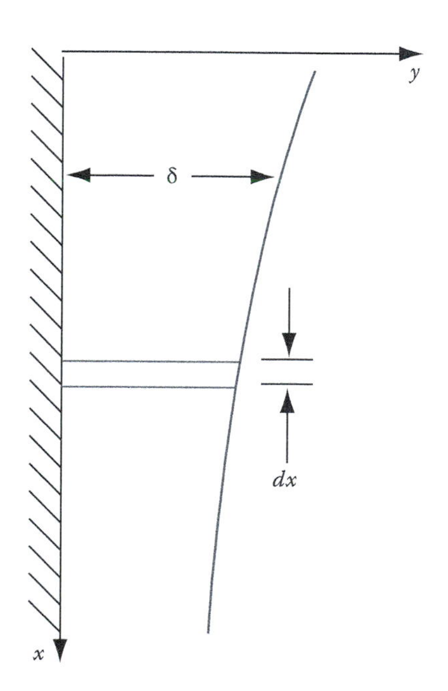

FIGURE P13.5

* Hassan, N. A. Laminar flow along a vertical wall, *Journal of Applied Mechanics*, September 1967, 135.

Momentum Integral Method for Laminar Flow

13.6 Derive Equation 13.34 in detail.

13.7 In the development that led to Equation 13.34, a linear velocity profile was used with the momentum integral boundary-layer equations. Repeat the same analysis using

$$\frac{V_x}{U} = \frac{2y}{\delta} - \frac{y^2}{\delta^2}$$

and verify the corresponding entries in Table 13.3.

13.8 Repeat Problem 13.7, using

$$\frac{V_x}{U} = \frac{3y}{2\delta} - \frac{y^3}{2\delta^3}$$

13.9 Repeat Problem 13.7, using

$$\frac{V_x}{U} = \sin\frac{\pi y}{2\delta}$$

Then plot this function on the same axes as the Blasius profile for comparison.

Momentum Integral Method for Turbulent Flow

13.10 In the development that led to Equation 13.42a, a seventh-root profile was used. Beginning with the eighth-root profile of Table 13.4 and the wall shear stress given for it, verify the corresponding entries for the skin friction drag coefficient.

13.11 Repeat Problem 13.10 for the tenth-root profile of Table 13.4.

13.12 Derive the boundary-layer thickness equation that corresponds to the eighth-root profile of Table 13.4.

13.13 Derive the boundary-layer thickness equation that corresponds to the 10th-root profile of Table 13.4.

Total Drag on a Flat Plate

13.14 A barge travels on a river with a velocity of 2 m/s, as illustrated in Figure P13.14. Estimate the skin friction drag on the bottom surface (assuming that it is flat). The barge width is 4 m.

13.15 Determine the ratio of the boundary-layer thickness just before transition to just after transition.

13.16 A smooth flat plate has air flowing over it at a rate of 20 ft/s. Estimate the airspeed at a vertical distance of 1 cm from the plate at the following locations: (a) 10 ft from the leading edge and (b) 25 ft from the leading edge.

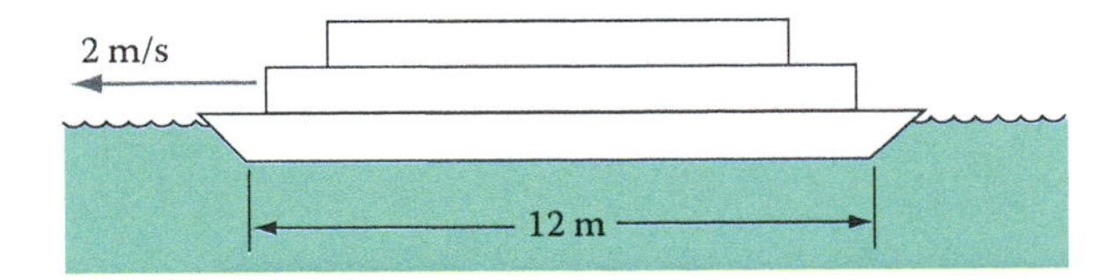

FIGURE P13.14

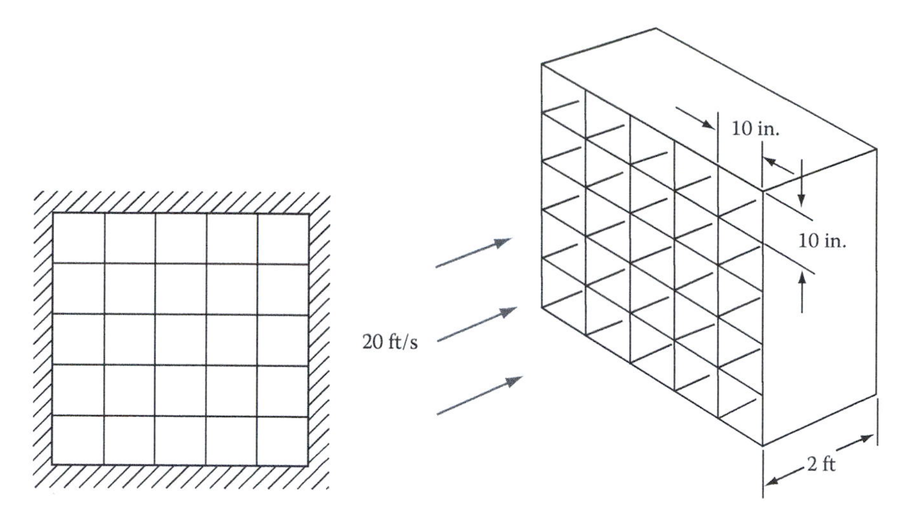

FIGURE P13.17

13.17 Flow straighteners are placed in the section upstream of the observation area in a wind tunnel. The flow straighteners consist of a series of vertical and horizontal plates as shown in Figure P13.17. The plates form 25 square ducts that are 10 in. × 10 in. × 2 ft long. The inlet flow velocity is 20 ft/s. Calculate the drag exerted on the entire bundle. Neglect edge effects.

13.18 Determine the boundary-layer thickness and the displacement thickness at the end of the straighteners of Problem 13.17.

13.19 The keel of a boat is about 42 in. deep and tapers linearly from a length of 36 in. at the hull to a length of 24 in. at the tip (see Figure P13.19). When the speed of the boat in water is 1 knot, determine the skin friction drag exerted on the keel. {*Hint*: Assume that the keel is a rectangular plate 42 in. × 30 in. [=(36 + 24)/2].} Also calculate the power required to overcome this drag.

13.20 The rudder of the boat in Figure P13.19 is 25 cm long and is 60 cm deep from hull to tip. At a velocity in water of 7 knots, determine the skin friction drag exerted on the rudder. What is the power required to overcome this drag?

13.21 Air flows at 5 m/s past a flat plate that is aligned with the flow direction. Determine the boundary-layer thickness of the flow 1 m past the leading edge of the plate, using each of the relations given in Table 13.3 if applicable.

13.22 Figure P13.22 shows a windmill located on a plateau. The structure is placed such that the plane of the blades is 90 ft past the leading edge of the plateau. For an expected wind speed of 30 miles/h, determine the height h that the blade axis of rotation should be placed so that

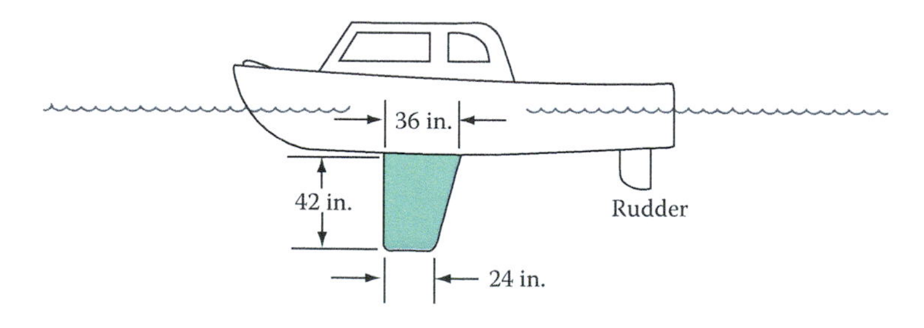

FIGURE P13.19

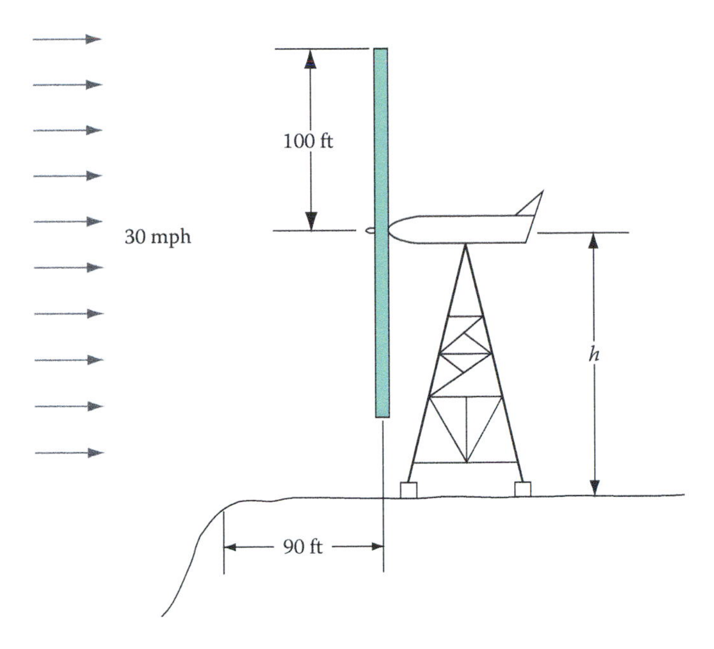

FIGURE P13.22

the tip of each blade stays at least 10 ft above the edge of the boundary layer formed along the plateau surface. Assume that the plateau behaves as a flat plate.

13.23 Air flows at 50 m/s past a flat plate aligned with the flow. Determine the boundary layer thickness 10 m past the leading edge of the plate, using all applicable equations of Table 13.4.

Appendix A: Conversion Factors and Properties of Substances

TABLE A.1
Prefixes for SI Units

Factor by Which Unit Is Multiplied	Prefix	Symbol
10^{12}	tera	T
10^{9}	giga	G
10^{6}	mega	M
10^{3}	kilo	k
10^{2}	hecto	h
10	deka	da
10^{-1}	deci	d
10^{-2}	centi	c
10^{-3}	milli	m
10^{-6}	micro	μ
10^{-9}	nano	n
10^{-12}	pico	p
10^{-15}	femto	f
10^{-18}	atto	a

Source: Mechtly, E.A., NASA SP-7012, 1969.

TABLE A.2
Conversion Factors Listed by Physical Quantity

To Convert from	To	Multiply by
Acceleration		
foot/second^2	meter/second^2	−01 3.048*
inch/second^2	meter/second^2	−02 2.54*
Area		
foot^2	meter^2	−02 9.290 304*
Inch^2	meter^2	−04 6.4516*
mile^2 (U.S. statute)	meter^2	+06 2.589 988 110 336*
yard^2	meter^2	−01 8.361 273 6*
Density		
gram/centimeter^3	kilogram/meter^3	+03 1.00*
lbm/inch^3	kilogram/meter^3	+04 2.767 990 5
lbm/foot^3	kilogram/meter^3	+01 1.601 846 3
slug/inch^3	kilogram/meter^3	+05 8.912 929 4
slug/foot^3	kilogram/meter^3	+02 5.153 79
Energy		
British thermal unit (thermochemical)	joule	+03 1.054 350 264 488
calorie (thermochemical)	joule	+00 4.184*
foot lbf	joule	+00 1.355 817 9
kilocalorie (thermochemical)	joule	+03 4.184*
kilowatt hour	joule	+06 3.60*
watt hour	joule	+03 3.60*
Energy/area time		
Btu (thermochemical)/foot^2 second	watt/meter^2	+04 1.134 893 1
Btu (thermochemical)/foot^2 minute	watt/meter^2	+02 1.891 488 5
Btu (thermochemical)/foot^2 hour	watt/meter^2	+00 3.152 480 8
Btu (thermochemical)/inch^2 second	watt/meter^2	+06 1.634 246 2
calorie (thermochemical)/centimeter^2 minute	watt/meter^2	+02 6.973 333 3
watt/centimeter^2	watt/meter^2	+04 1.00*
Force		
lbf (pound force, avoirdupois)	newton	+00 4.448 221 615 260 5*
Length		
foot	meter	−01 3.048*
inch	meter	−02 2.54*
mile (U.S. statute)	meter	+03 1.609 344*
mile (U.S. nautical)	meter	+03 1.852*
yard	meter	−01 9.144*
Mass		
gram	kilogram	−03 1.00*
lbm (pound mass, avoirdupois)	kilogram	−01 4.535 923 7*
slug	kilogram	+01 1.459 390 29

(Continued)

TABLE A.2 (*Continued*)
Conversion Factors Listed by Physical Quantity

To Convert from	To	Multiply by
Power		
Btu (thermochemical)/second	watt	+03 1.054 350 264 488
Btu (thermochemical)/minute	watt	+01 1.757 250 4
calorie (thermochemical)/second	watt	+00 4.184*
calorie (thermochemical)/minute	watt	−02 6.973 333 3
foot lbf/hour	watt	−04 3.766 161 0
foot lbf/minute	watt	−02 2.259 696 6
foot lbf/second	watt	+00 1.355 817 9
horsepower (550 foot lbf/second)	watt	+02 7.456 998 7
kilocalorie (thermochemical)/minute	watt	+01 6.973 333 3
Pressure		
atmosphere	newton/meter^2	+05 1.013 25*
bar	newton/meter^2	+05 1.00*
centimeter of mercury (0°C)	newton/meter^2	+03 1.333 22
centimeter of water (4°C)	newton/meter^2	+01 9.806 38
dyne/centimeter^2	newton/meter^2	−01 1.00*
foot of water (39.2°F)	newton/meter^2	+03 2.988 98
inch of mercury (32°F)	newton/meter^2	+03 3.386 389
inch of mercury (60°F)	newton/meter^2	+03 3.376 85
inch of water (39.2°F)	newton/meter^2	+02 2.490 82
inch of water (60°F)	newton/meter^2	+02 2.488 4
lbf/foot^2	newton/meter^2	+01 4.788 025 8
lbf/inch^2 (psi)	newton/meter^2	+03 6.894 757 2
pascal	newton/meter^2	+00 1.00*
torr (0°C)	newton/meter^2	+02 1.333 22
Speed		
foot/hour	meter/second	−05 8.466 666 6
foot/minute	meter/second	−03 5.08*
foot/second	meter/second	−01 3.048*
inch/second	meter/second	−02 2.54*
kilometer/hour	meter/second	−01 2.777 777 8
knot (international)	meter/second	−01 5.144 444 444
mile/hour (U.S. statute)	meter/second	−01 4.470 4*
mile/minute (U.S. statute)	meter/second	+01 2.682 24*
mile/second (U.S. statute)	meter/second	+03 1.609 344*
Temperature		
Celsius	Kelvin	$t_K = t_C + 273.15$
Fahrenheit	Kelvin	$t_K = (5/9)(t_F + 459.67)$
Fahrenheit	Celsius	$t_C = (5/9)(t_F - 32)$
Rankine	Kelvin	$t_K = (5/9)t_R$

(*Continued*)

TABLE A.2 (*Continued*)

Conversion Factors Listed by Physical Quantity

To Convert from	To	Multiply by
Time		
day (mean solar)	second (mean solar)	+04 8.64*
hour (mean solar)	second (mean solar)	+03 3.60*
minute (mean solar)	second (mean solar)	+01 6.00*
month (mean calendar)	second (mean solar)	+06 2.628*
second (ephemeris)	second	+00 1.000 000 000
year (calendar)	second (mean solar)	+07 3.153 6*
Viscosity		
centistokes	meters^2/second	−06 1.00*
stoke	meters/second	−04 1.00*
foot^2/second	meters/second	−02 9.290 304*
Centipoise	newton · second/meter^2	−03 1.00*
lbm/foot second	newton · second/meter^2	+00 1.488 163 9
lbf/second foot^2	newton · second/meter^2	+01 4.788 025 8
poise	newton · second/meter^2	−01 1.00*
slug foot second	newton · second/meter^2	+01 4.788 025 8
Volume		
barrel (petroleum, 42 gallons)	meter^3	−03 1.233 481 9
cup	meter^3	−04 2.365 882 365*
foot^3	meter^3	−02 2.831 684 659 2*
gallon (U.S. liquid)	meter^3	−03 3.785 411 784*
inch^3	meter^3	−051.638 706 4*
liter	meter^3	−03 1.00*
ounce (U.S. fluid)	meter^3	−05 2.957 352 956 25*
gallon (U.S. dry)	meter^3	−03 4.404 883 770 86*
pint (U.S. dry)	meter^3	−04 5.506 104 713 575*
pint (U.S. liquid)	meter^3	−04 4.731 764 73*
quart (U.S. dry)	meter^3	−03 1.101 220 942 715*
quart (U.S. liquid)	meter^3	−04 9.463 529 5
tablespoon	meter^3	−05 1.478 676 478 125*
teaspoon	meter^3	−06 4.928 921 593 75*
yard^3	meter^3	−01 7.645 548 579 84*

Source: Mechtly, E.A., NASA SP-7012, 1969.

Notes: * indicates an exact conversion, numbers not followed by asterisk (*) are only approximations of definitions or the result of physical measurements. The first two digits of each numerical entry represent a power of 10. Thus, −01 3.048 = 3.048×10^{-1} = 0.304 8.

TABLE A.3
Properties of Dry Air at Atmospheric Pressure

						μ ×10⁶	
K	°C	°R	°F	ρ(Slug/ft³)	ρ(kg/m³)	(lbf·s/ft²)	(N·s/m²)
250	−23.15	450	−10	0.00274	1.413	0.3335	15.99
260	−13.15	468	8	0.00264	1.359	0.3444	16.50
270	−3.15	486	26	0.00254	1.308	0.3550	17.00
280	6.85	504	44	0.00245	1.261	0.3652	17.50
290	16.85	522	62	0.00236	1.218	0.3752	17.98
295	21.85	531	71	0.00232	1.197	0.3801	18.22
300	26.85	540	80	0.00228	1.177	0.3854	18.46
310	36.85	558	98	0.00221	1.139	0.3950	18.93
320	46.85	576	116	0.00214	1.103	0.4047	19.39
330	56.85	594	134	0.00208	1.070	0.4143	19.85
340	66.85	612	152	0.00201	1.038	0.4236	20.30
350	76.85	630	170	0.00196	1.008	0.4329	20.75
360	86.85	648	188	0.00190	0.980 5	0.4419	21.18
370	96.85	666	206	0.00185	0.953 9	0.4509	21.60
380	106.85	684	224	0.00180	0.928 8	0.4593	22.02
390	116.85	702	242	0.00176	0.905 0	0.4683	22.44
400	126.85	720	260	0.00171	0.882 2	0.4770	22.86
410	136.85	738	278	0.00167	0.860 8	0.4854	23.27
420	146.85	756	296	0.00163	0.840	0.4938	23.66

Source: Reprinted from Bolz, R.E. and Tuve, G.L. (eds.), *CRC Handbook of Tables for Applied Engineering Science*, 2nd ed., CRC Press, Cleveland, OH, 1973. With permission.

Note on reading viscosity: For air at 250 K, $\mu \times 10^6 = 15.99$ N s/m², $\mu = 15.99 \times 10^{-6}$ N s/m².

TABLE A.4

Specific Gravity and Viscosity of Water at Atmospheric Pressure

Temperature		Specific Gravity	Absolute or Dynamic Viscosity, μ		Kinematic Viscosity, $\nu \times 10^6$	
°C	°F		(N s/m²) × 10³	(lbf s/ft²) × 10⁵	(m²/s)	(ft² s)
0	32.0	0.9999	1.787	3.730	1.787	19.22
2	35.6	1.0000	1.671	3.490	1.671	17.98
4	39.2	1.0000	1.567	3.273	1.567	16.86
6	42.8	1.0000	1.472	3.074	1.472	15.84
8	46.4	0.9999	1.386	2.895	1.386	14.91
10	50.0	0.9997	1.307	2.730	1.307	14.06
12	53.6	0.9995	1.235	2.579	1.236	13.30
14	57.2	0.9998	1.169	2.441	1.170	12.59
16	60.8	0.9990	1.109	2.316	1.110	11.94
18	64.4	0.9986	1.053	2.199	1.054	11.34
20	68.0	0.9982	1.002	2.093	1.004	10.80
22	71.6	0.9978	0.9548	1.994	0.9569	10.30
24	75.2	0.9973	0.9111	1.903	0.9135	9.829
26	78.8	0.9968	0.8705	1.818	0.8732	9.396
28	82.4	0.9963	0.8327	1.739	0.8358	8.993
30	86.0	0.9957	0.7975	1.665	0.8009	8.618
32	89.6	0.9951	0.7647	1.597	0.7685	8.269
34	93.3	0.9944	0.7340	1.533	0.7381	7.942
36	96.8	0.9937	0.7052	1.473	0.7097	7.636
38	100	0.9930	0.6783	1.417	0.6831	7.350
40	104	0.9922	0.6529	1.364	0.6580	7.080
42	108	0.9915	0.6291	1.314	0.6345	6.827
44	111	0.9907	0.6067	1.267	0.6124	6.589
46	115	0.9898	0.5856	1.223	0.5916	6.366
48	118	0.9890	0.5656	1.181	0.5719	6.154
50	122	0.9881	0.5468	1.142	0.5534	5.955
52	126	0.9871	0.5290	1.105	0.5359	5.766
54	129	0.9862	0.5121	1.069	0.5193	5.588
56	133	0.9852	0.4961	1.036	0.5036	5.419
58	136	0.9842	0.4809	1.005	0.4886	5.257
60	140	0.9832	0.4665	0.9744	0.4745	5.106
62	144	0.9822	0.4528	0.9458	0.4610	4.960
64	147	0.9811	0.4398	0.9184	0.4483	4.824
66	151	0.9800	0.4273	0.8923	0.4360	4.691
68	154	0.9789	0.4155	0.8678	0.4245	4.5677
70	158	0.9778	0.4042	0.8442	0.4134	4.448
72	162	0.9766	0.3934	0.8218	0.4028	4.334
74	165	0.9755	0.3831	0.8000	0.3927	4.225
76	169	0.9743	0.3732	0.7795	0.3830	4.121
78	172	0.9731	0.3638	0.7599	0.3738	4.022

(Continued)

TABLE A.4 (*Continued*)

Specific Gravity and Viscosity of Water at Atmospheric Pressure

Temperature		Specific Gravity	Absolute or Dynamic Viscosity, μ		Kinematic Viscosity, $\nu \times 10^6$	
°C	°F		$(N\ s/m^2) \times 10^3$	$(lbf\ s/ft^2) \times 10^5$	(m^2/s)	$(ft^2\ s)$
80	176	0.9718	0.3537	0.7388	0.3640	3.917
82	180	0.9706	0.3460	0.7226	0.3565	3.836
84	183	0.9693	0.3377	0.7052	0.3484	3.749
86	187	0.9680	0.3297	0.6884	0.3406	3.665
88	190	0.9667	0.3221	0.6726	0.3332	3.588
90	194	0.9653	0.3147	0.6574	0.3260	3.508
92	198	0.9640	0.3076	0.6424	0.3191	3.434
94	201	0.9626	0.3008	0.6281	0.3125	3.363
96	205	0.9612	0.2942	0.6145	0.3061	3.294
98	208	0.9584	0.2879	0.6011	0.3000	3.228
100	212	0.9584	0.2818	0.5887	0.2940	3.163

Source: Reprinted from Bolz, R.E. and Tuve, G.L. (eds.), *CRC Handbook of Tables for Applied Engineering Science,* 2nd ed., CRC Press, Cleveland, OH, 1973. With permission.

Notes on reading the table: Density, ρ = (sp. gr. × 1.94) slug/ft³ = (sp. gr. × 1 000) kg/m³ = (sp. gr. × 62.4) lbm/ft³ and kinematic viscosity at 0°C, $\nu \times 10^6 = 1.787$ m²/s, $\nu = 1.787 \times 10^{-6}$ m²/s.

TABLE A.5
Properties of Common Liquids at 1.0 atm Pressure, 77°F (25°C)

Name	Specific Gravity	Viscosity $(lbf\ s/ft^2) \times 10^5$	Viscosity $(N \cdot s/m^2) \times 10^3$	Sound Velocity (m/s)	Surface Tension $[(N/m) \times 10^3]$
Acetone	0.787	0.659	0.316	1 174	23.1
Alcohol, ethyl	0.787	2.29	1.095	1 144	22.33
Alcohol, methyl	0.789	1.17	0.56	1 103	22.2
Alcohol, propyl	0.802	4.01	1.92	1 205	23.5
Benzene	0.876	1.26	0.601	1 298	28.18
Carbon disulfide	1.265	0.752	0.36	1 149	32.33
Carbon tetrachloride	1.59	1.90	0.91	924	26.3
Castor oil	0.960	1 356	650	1 474	—
Chloroform	1.47	1.11	0.53	995	27.14
Decane	0.728	1.79	0.859	—	23.43
Dodecane	—	2.87	1.374	—	—
Ether	0.715	0.466	0.223	985	16.42
Ethylene glycol	1.100	33.8	16.2	1 644	48.2
Fluorine refrigerant R-11	1.480	0.876	0.42	—	18.3
Fluorine refrigerant R-12	1.315	—	—	—	—
Fluorine refrigerant R-22	1.197	—	—	—	8.35
Glycerin	1.263	1 983	950	1 909	63.0
Heptane	0.681	0.786	0.376	1 138	19.9
Hexane	0.657	0.622	0.297	1 203	18.0
Kerosene	0.823	3.42	1.64	1 320	—
Linseed oil	0.93	69.0	33.1	—	—
Mercury	13.6	3.20	1.53	1 450	484
Octane	0.701	1.07	0.51	1 171	21.14
Propane	0.495	0.23	0.11	—	6.6
Propylene	0.516	0.19	0.09	—	7.0
Propylene glycol	0.968	88	42	—	36.3
Seawater	1.03	—	—	1 535	—
Turpentine	0.87	2.87	1.375	1 240	—
Water	1.00	1.9	0.89	1 498	71.97

Source: Reprinted from Bolz, R.E. and Tuve, G.L. (eds.), *CRC Handbook of Tables for Applied Engineering Science,* 2nd ed., CRC Press, Cleveland, OH, 1973. With permission.

Notes: Density, ρ = (sp. gr. × 1.94) slug/ft^3 = (sp. gr. × 1 000) kg/m^3 = (sp. gr. × 62.4) lbm/ft^3; viscosity of acetone, m × 10^3 = 0.316 N · s/m^3, m = 0.316 × 10^{-3} N · s/m^2.

TABLE A.6

Physical Properties of Gases at Room Temperature and Pressure

Gas	Gas Constant, R		Density, ρ		Dynamic Viscosity, $\mu \times 10^6$		Specific Heat, C_p		$C_p/C_v = \gamma$
	(ft·lbf/slug·°R)	[J/(kg·K)]	(slug/ft^2)	(kg/m^3)	(lbf·s/ft^2)	(N·s/m^2)	(Btu/slug·°R)	[J/(kg·K)]	
Air	1,710	286.8	0.0023	1.19	0.3758	18.0	7.72	1 005	1.40
Argon	1,250	208	0.00318	1.61	0.4162	20.0	4.02	523	1.67
Carbon dioxide	1,130	189	0.00354	1.82	0.2919	14.0	6.60	876	1.30
Helium	12,400	2 077	0.000317	0.164	0.4161	20.0	39.9	5 188	1.66
Hydrogen	24,700	4 126	0.000160	0.082 6	0.1879	9.0	110	14 310	1.405
Oxygen	1,550	260	0.00254	1.31	0.4161	20.0	7.08	920	1.40

Notes: $R = \bar{R}/MW$, where $\bar{R} = 1545$ ft·lbf/lb mol·°R = 49,700 ft·lbf/slug mol·°R = 8 312 N·m/(mol·K) and MW = molecular weight (engineering units) = molecular mass (SI units); 1 Btu = 778 ft·lbf.

TABLE A.7
Specific Gravity of Various Solids at Ordinary Atmospheric Temperature

Substance	Specific Gravity	Substance	Specific Gravity
Amber	1.06–1.11	Porcelain	2.3–2.5
Asbestos slate	1.8	Quartz	2.65
Asphalt	1.1–1.5	Rock salt	2.18
Beeswax	0.96–0.97	Rubber (hard)	1.19
Bone	1.7–2.0	Rubber (soft)	
Brick	1.4–2.2	Commercial	1.1
Butter	0.86–0.87	Pure gum	0.91–0.93
Camphor	0.99	Slate	2.6–3.3
Cardboard	0.69	Sugar	1.59
Cement (set)	2.7–3.0	Talc	2.7–2.8
Chalk	1.9–2.8	Topaz	3.5–3.6
Clay	1.8–2.6	Wax (sealing)	1.8
Cocoa butter	0.89–0.91	Wood (seasoned)	
Cork	0.22–0.26	Ash	0.65–0.85
Diamond	3.01–3.52	Balsa	0.11–0.14
Emery	4.0	Bamboo	0.31–0.40
Gelatin	1.27	Basswood	0.32–0.59
Glass	2.4–2.8	Birch	0.51–0.77
Glue	1.27	Cedar	0.49–0.57
Granite	2.64–2.76	Cherry	0.70–0.90
Graphite	2.30–2.72	Dogwood	0.76
Gypsum	2.31–2.33	Elm	0.54–0.60
Ice	0.917	Hickory	0.60–0.93
Ivory	1.83–1.92	Maple	0.62–0.75
Linoleum	1.18	Oak	0.60–0.90
Marble	2.6–2.84	Walnut	0.64–0.70
Paraffin	0.87–0.91		

Sources: Reprinted from Bolz, R.E. and Tuve, G.L. (eds.), *CRC Handbook of Tables for Applied Engineering Science,* 2nd ed., CRC Press, Cleveland, OH, 1973. With permission.

Note: Density, ρ = (sp. gr. × 1.94) slug/ft^3 = (sp. gr. × 1 000) kg/m^3 = (sp. gr. × 62.4) lbm/ft^3.

TABLE A.8
Specify Gravity of Common Metals and Alloys

Common Name	Specific Gravity
Aluminum	2.70
Aluminum alloy 2017, annealed (ASTM B221)	2.8
Aluminum bronze	7.8
Beryllium	1.85
Beryllium copper	8.25
Cast gray iron (ASTM A48–48, class 25)	7.2
Chromium	7.2
Copper	8.96
Gold	19.32
Ingot iron	7.86
Iridium	22.42
Iron	7.87
Lead	11.35
Magnesium	1.74
Magnesium alloy (AZ31 B)	1.77
Mercury	13.546
Nickel	8.90
Plain carbon steel (AISI-SAE 1020)	7.86
Platinum	21.45
Potassium	0.86
Silicon	2.33
Silver	10.50
Sodium	0.97
Solder 50–50	8.89
Stainless steel (type 304)	8.02
Tin	7.31
Titanium	4.54
Tungsten	19.3
Uranium	18.8
Vanadium	6.1
Yellow brass (high brass)	8.47
Zinc	7

Source: Reprinted from Bolz, R.E. and Tuve, G.L. (eds.), *CRC Handbook of Tables for Applied Engineering Science,* 2nd ed., CRC Press, Cleveland, OH, 1973. With permission.

Note: Density, ρ = (sp. gr. × 1.94) slug/ft^3 = (sp. gr. × 1 000) kg/m^3 = (sp. gr. × 62.4) lbm ft^3.

TABLE A.9

Surface Tension of Liquids at Atmospheric Pressure and Room Temperature

Name	In Contact with	Surface Tension [(N/m) × 10^3]
Acetone	Air or vapor	23.1
Benzene	Air	28.2
Butyl alcohol	Air or vapor	24.6
Carbon tetrachloride	Vapor	26.3
Chloroform	Air	27.1
Ethyl alcohol	Vapor	22.4
Ethyl ether	Vapor	16.5
Glycerol	Air	63.0
n-Hexane	Air	18.4
Isobutyl alcohol	Vapor	23.0
Isopropyl alcohol	Air or vapor	21.7
Mercury	Air	484
Methyl alcohol	Air	22.2
Octane	Vapor	21.8
Propyl alcohol	Vapor	23.8
Water	Air	72

Source: Reprinted from Bolz, R.E. and Tuve, G.L. (eds.), *CRC Handbook of Tables for Applied Engineering Science*, 2nd ed., CRC Press, Cleveland, OH, 1973. With permission.

Note: Surface tension of acetone, $\sigma \times 10^3 = 23.1$ N/m, $\sigma = 23.1 \times 10^3$ N/m.

Appendix B: Geometric Elements of Plane Areas

TABLE B.1

Areas, Centroidal Distances, and Area Moments of Inertia about Centroidal Axes for Various Plane Areas

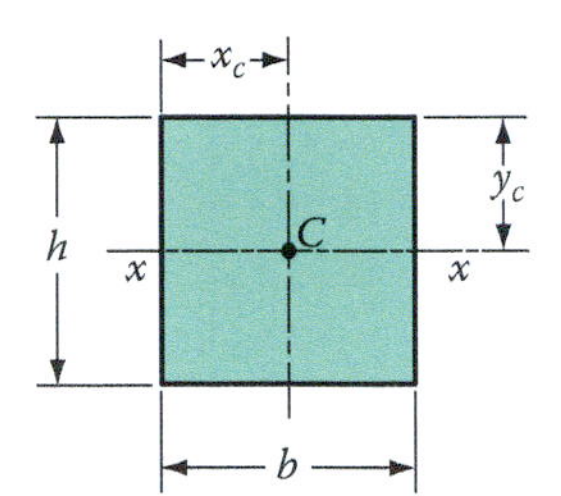

$A = bh$
$x_c = b/2$
$y_c = h/2$
$I_{xxc} = bh^3/12$
$I_{xyc} = 0$

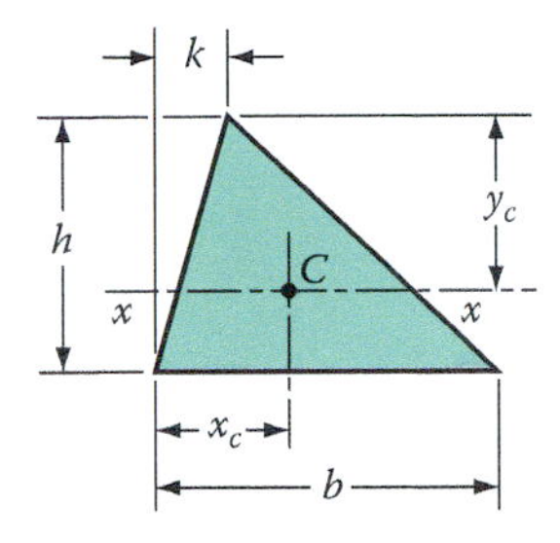

$A = bh/2$
$x_c = (b + k)/3$
$y_c = 2h/3$
$I_{xxc} = bh^3/36$
$I_{xyc} = bh^2\ (2k = b)/72$

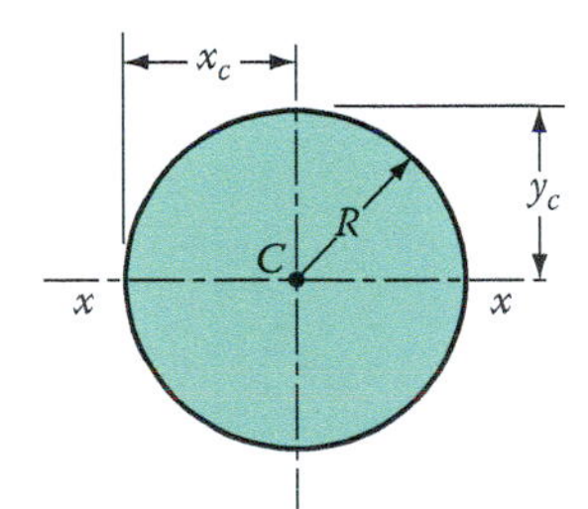

$A = \pi R^2$
$x_c = R$
$y_c = R$
$I_{xxc} = \pi R^4/4$
$I_{xyc} = 0$

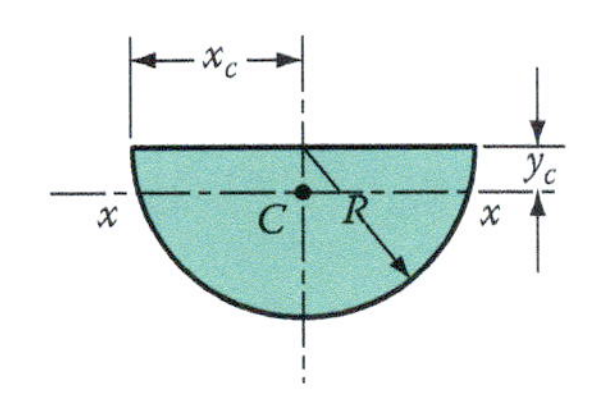

$A = \pi R^2/2$
$x_c = R$
$y_c = 4R/3\pi$
$I_{xxc} = R^4(9\pi^2 - 64)/72\pi$
$I_{xyc} = 0$

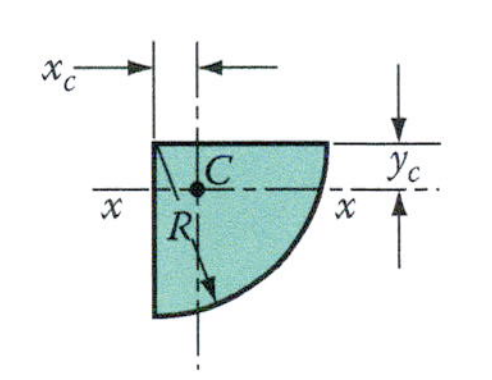

$A = \pi R^2/4$
$x_c = 4R/3\pi$
$y_c = 4R/3\pi$
$I_{xxc} = R^4(9\pi^2 - 64)/144\pi$
$I_{xyc} = R^4(31 - 9\pi)/72\pi$

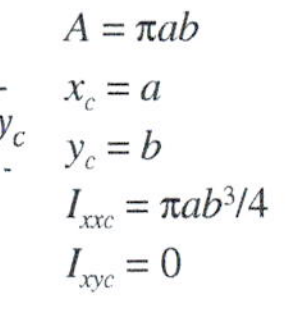

$A = \pi ab$
$x_c = a$
$y_c = b$
$I_{xxc} = \pi ab^3/4$
$I_{xyc} = 0$

$A = \pi ab/2$
$x_c = a$
$y_c = 4b/3\pi$
$I_{xxc} = ab^3(9\pi^2 - 64)72\pi$
$I_{xyc} = 0$

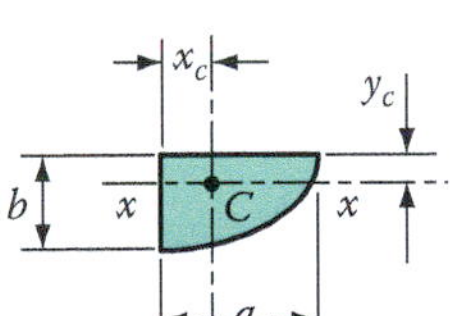
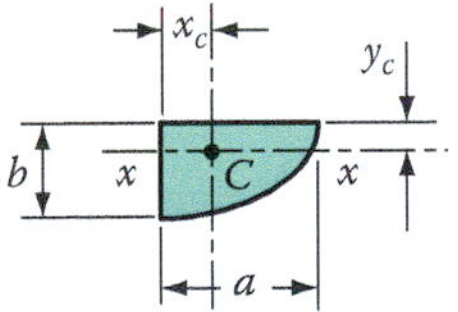

$A = \pi ab/4$
$x_c = 4a/3\pi$
$y_c = 4b/3\pi$
$I_{xxc} = ab^3(9\pi^2 - 64)/144\pi$
$I_{xyc} = a^2b^2(32 - 9\pi)/72\pi$

Appendix C: Pipe and Tube Specifications

TABLE C.1
Dimensions of Wrought Steel and Wrought Iron Pipe

Pipe Size	Outside Diameter		Schedule	Internal Diameter		Flow Area	
	in.	cm		ft	cm	ft²	cm²
⅛	0.405	1.029	40 (STD)	0.02242	0.683	0.0003947	0.366 4
			80 (XS)	0.01792	0.547	0.0002522	0.235 0
¼	0.540	1.372	40 (STD)	0.03033	0.924	0.0007227	0.670 6
			80 (XS)	0.02517	0.768	0.0004974	0.463 2
⅜	0.675	1.714	40 (STD)	0.04108	1.252	0.001326	1.233
			80 (XS)	0.03525	1.074	0.0009759	0.905 9
½	0.840	2.134	40 (STD)	0.05183	1.580	0.002110	1.961
			80 (XS)	0.04550	1.386	0.001626	1.508
			160	0.03867	1.178	0.001174	1.090
			(XXS)	0.02100	0.640	0.0003464	0.322
¾	1.050	2.667	40 (STD)	0.06867	2.093	0.003703	3.441
			80 (XS)	0.06183	1.883	0.003003	2.785
			160	0.05100	1.555	0.002043	1.898
			(XXS)	0.03617	1.103	0.001027	0.956
1	1.315	3.340	40 (STD)	0.08742	2.664	0.006002	5.574
			80 (XS)	0.07975	2.430	0.004995	5.083
			160	0.06792	2.070	0.003623	3.365
			(XXS)	0.04992	1.522	0.001957	1.815
1¼	1.660	4.216	40 (STD)	0.1150	3.504	0.01039	9.643
			80 (XS)	0.1065	3.246	0.008908	8.275
			160	0.09667	2.946	0.007339	6.816
			(XXS)	0.07467	2.276	0.004379	4.069
1½	1.900	4.826	40 (STD)	0.1342	4.090	0.01414	13.13
			80 (XS)	0.1250	3.810	0.01227	11.40
			160	0.1115	3.398	0.009764	9.068
			(XXS)	0.09167	2.794	0.006600	6.131
2	2.375	6.034	40 (STD)	0.1723	5.252	0.02330	21.66
			80 (XS)	0.1616	4.926	0.02051	19.06
			160	0.1406	4.286	0.01552	14.43
			(XXS)	0.1253	3.820	0.01232	11.46
2½	2.875	7.303	40 (STD)	0.2058	6.271	0.03325	30.89
			80 (XS)	0.1936	5.901	0.02943	27.35
			160	0.1771	5.397	0.02463	22.88
			(XXS)	0.1476	4.499	0.01711	15.90

(Continued)

TABLE C.1 (*Continued*)
Dimensions of Wrought Steel and Wrought Iron Pipe

	Outside Diameter			Internal Diameter		Flow Area	
Pipe Size	in.	cm	Schedule	ft	cm	ft²	cm²
3	3.500	8.890	40 (STD)	0.2557	7.792	0.05134	47.69
			80 (XS)	0.2417	7.366	0.04587	42.61
			160	0.2187	6.664	0.03755	34.88
			(XXS)	0.1917	5.842	0.02885	26.80
3½	4.000	10.16	40 (STD)	0.2957	9.012	0.06866	63.79
			80 (XS)	0.2803	8.544	0.06172	57.33
4	4.500	11.43	40 (STD)	0.3355	10.23	0.08841	82.19
			80 (XS)	0.3198	9.718	0.07984	74.17
			120	0.3020	9.204	0.07163	66.54
			160	0.2865	8.732	0.06447	59.88
			(XXS)	0.2626	8.006	0.05419	50.34
5	5.563	14.13	40 (STD)	0.4206	12.82	0.1389	129.10
			80 (XS)	0.4011	12.22	0.1263	117.30
			120	0.3803	11.59	0.1136	105.50
			160	0.3594	10.95	0.1015	94.17
			(XXS)	0.3386	10.32	0.09004	83.65
6	6.625	16.83	40 (STD)	0.5054	15.41	0.2006	186.50
			80 (XS)	0.4801	14.64	0.1810	168.30
			120	0.4584	13.98	0.1650	153.50
			160	0.4322	13.18	0.1467	136.40
			(XXS)	0.4081	12.44	0.1308	121.50
8	8.625	21.91	20	0.6771	20.64	0.3601	334.60
			30	0.6726	20.50	0.3553	330.10
			40 (STD)	0.6651	20.27	0.3474	322.70
			60	0.6511	19.85	0.3329	309.50
			80 (XS)	0.6354	19.37	0.3171	294.70
			100	0.6198	18.89	0.3017	280.30
			120	0.5989	18.26	0.2817	261.90
			140	0.5834	17.79	0.2673	248.60
			(XXS)	0.5729	17.46	0.2578	239.40
			160	0.5678	17.31	0.2532	235.30
10	10.750	27.31	20	0.8542	26.04	0.5730	532.60
			30	0.8447	25.75	0.5604	520.80
			40 (STD)	0.8350	25.46	0.5476	509.10
			60 (XS)	0.8125	24.77	0.5185	481.90
			80	0.7968	24.29	0.4987	463.40
			100	0.7760	23.66	0.4730	439.70
			120	0.7552	23.02	0.4470	416.20
			140 (XXS)	0.7292	22.23	0.4176	388.10
			160	0.7083	21.59	0.3941	366.10

(*Continued*)

TABLE C.1 (*Continued*)

Dimensions of Wrought Steel and Wrought Iron Pipe

	Outside Diameter			Internal Diameter		Flow Area	
Pipe Size	in.	cm	Schedule	ft	cm	ft^2	cm^2
12	12.750	32.39	20	1.021	31.12	0.8185	760.60
			30	1.008	30.71	0.7972	740.71
			(STD)	1.000	30.48	0.7854	729.70
			40	0.9948	30.33	0.773	722.50
			(XS)	0.9792	29.85	0.7530	699.80
			60	0.9688	29.53	0.7372	684.90
			80	0.9478	28.89	0.7056	655.50
			100	0.9218	28.10	0.6674	620.20
			120 (XXS)	0.8958	27.31	0.6303	585.80
			140	0.8750	26.67	0.6013	558.60
			160	0.8438	25.72	0.5592	519.60
14	14.000	35.56	30 (STD)	1.104	33.65	0.9575	889.30
			160	0.9323	28.42	0.6827	634.40
16	16.000	40.64	30 (STD)	1.271	38.73	1.268	1 178.00
			160	1.068	32.54	0.8953	831.60
18	18.000	45.72	(STD)	1.438	43.81	1.623	1 507.00
			160	1.203	36.67	1.137	1 056.00
20	20.000	50.80	20 (STD)	1.604	48.89	2.021	1 877.00
			160	1.339	40.80	1.407	1 307.00
22	22.000	55.88	20 (STD)	1.771	53.97	2.463	2 288.00
			160	1.479	45.08	1.718	1 596.00
24	24.000	60.96	20 (STD)	1.938	59.05	2.948	2 739.00
			160	1.609	49.05	2.034	1 890.00
26	26.000	66.04	(STD)	2.104	64.13	3.477	3 230.00
28	28.000	71.12	(STD)	2.271	69.21	4.050	3 762.00
30	30.000	76.20	(STD)	2.438	74.29	4.666	4 335.00
32	32.000	81.28	(STD)	2.604	79.34	5.326	4 944.00
34	34.000	86.36	(STD)	2.771	84.45	6.030	5 601.00
36	36.000	91.44	(STD)	2.938	89.53	6.777	6 295.00
38	38.000	96.52	_	3.104	94.61	7.568	7 030.00
40	40.000	101.6	_	3.271	99.69	8.403	7 805.00

Source: Dimensions in English units obtained from ANSI B36.10-1979, *American National Standard Wrought Steel and Wrought Iran Pipe*, The American Society of Mechanical Engineers, New York.

Notes: STD implies standard; XS is extra strong; and XXS is double extra strong.

TABLE C.2
Dimensions of Seamless Copper Tubing

Standard Size	Outside Diameter		Type	Internal Diameter		Flow Area	
	in.	cm		ft	cm	ft^2	cm^2
¼	0.375	0.953	K	0.02542	0.775	0.0005074	0.471 7
			L	0.02625	0.801	0.0005412	0.503 9
⅜	0.500	1.270	K	0.03350	1.022	0.0008814	0.820 3
			L	0.03583	1.092	0.001008	0.936 6
			M	0.03750	1.142	0.001104	1.024
½	0.625	1.588	K	0.04392	1.340	0.001515	1.410
			L	0.04542	1.384	0.001620	1.505
			M	0.04742	1.446	0.001766	1.642
⅝	0.750	1.905	K	0.05433	1.657	0.002319	2.156
			L	0.05550	1.691	0.002419	2.246
¾	0.875	2.222	K	0.06208	1.892	0.003027	2.811
			L	0.06542	1.994	0.003361	3.123
			M	0.06758	2.060	0.003587	3.333
1	1.125	2.858	K	0.08292	2.528	0.005400	5.019
			L	0.08542	2.604	0.005730	5.326
			M	0.08792	2.680	0.006071	5.641
1¼	1.375	3.493	K	0.1038	2.163	0.008454	7.858
			L	0.1054	3.213	0.008728	8.108
			M	0.1076	3.279	0.009090	8.444
1½	1.625	4.128	K	0.1234	3.762	0.01196	11.12
			L	0.1254	3.824	0.01235	11.48
			M	0.1273	3.880	0.01272	11.82
2	2.125	5.398	K	0.1633	4.976	0.02093	19.45
			L	0.1654	5.042	0.02149	19.97
			M	0.1674	5.102	0.02201	20.44
2½	2.625	6.668	K	0.2029	6.186	0.03234	30.05
			L	0.2054	6.262	0.03314	30.80
			M	0.2079	6.338	0.03395	40.17
3	3.125	7.938	K	0.2423	7.384	0.04609	42.82
			L	0.2454	7.480	0.04730	43.94
			M	0.2484	7.572	0.04847	45.03
3½	3.625	9.208	K	0.2821	8.598	0.06249	58.06
			L	0.2854	8.700	0.06398	59.45
			M	0.2883	8.786	0.06523	60.63
4	4.125	10.48	K	0.3214	9.800	0.08114	75.43
			L	0.3254	9.922	0.08317	77.32
			M	0.3279	9.998	0.08445	78.51
5	5.125	13.02	K	0.4004	12.21	0.1259	117.10
			L	0.4063	12.38	0.1296	120.50
			M	0.4089	12.47	0.1313	122.10

(*Continued*)

TABLE C.2 (*Continued*)

Dimensions of Seamless Copper Tubing

Standard Size	Outside Diameter		Type	Internal Diameter		Flow Area	
	in.	cm		ft	cm	ft^2	cm^2
6	6.125	15.56	K	0.4784	14.58	0.1798	167.00
			L	0.4871	14.85	0.1863	173.20
			M	0.4901	14.39	0.1886	175.30
8	8.125	20.64	K	0.6319	19.26	0.3136	291.50
			L	0.6438	19.62	0.3255	302.50
			M	0.6488	19.78	0.3306	307.20
10	10.125	25.72	K	0.7874	24.00	0.4870	452.50
			L	0.8021	24.45	0.5053	469.50
			M	0.8084	24.64	0.5133	476.80
12	12.125	30.80	K	0.9429	28.74	0.6983	648.80
			L	0.9638	29.38	0.7295	677.90
			M	0.9681	29.51	0.7361	684.00

Source: Dimensions in English units obtained from ANSI = ASTM B88-78, *Standard Specifications for Seamless Copper Water Tube*, American Society for Testing Materials, West Conshohocken, PA.

Notes: Type K is for underground service and general plumbing, type L is for interior plumbing, and type M is for use only with soldered fittings.

Appendix D: Compressible Flow Tables

TABLE D.1

Isentropic Flow Tables for a Gas Having $\gamma = 1.4$

M	p/p_t	T/T_t	A/A^*
0.00	1.0000	1.0000	∞
0.01	0.9999	1.0000	57.8738
0.02	0.9997	0.9999	28.9421
0.03	0.9994	0.9998	19.3005
0.04	0.9989	0.9997	14.4815
0.05	0.9983	0.9995	11.5914
0.06	0.9975	0.9993	9.6659
0.07	0.9966	0.9990	8.2915
0.08	0.9955	0.9987	7.2616
0.09	0.9944	0.9984	6.4613
0.10	0.9930	0.9980	5.8218
0.11	0.9916	0.9976	5.2992
0.12	0.9900	0.9971	4.8643
0.13	0.9883	0.9966	4.4969
0.14	0.9864	0.9961	4.1824
0.15	0.9844	0.9955	3.9103
0.16	0.9823	0.9949	3.6727
0.17	0.9800	0.9943	3.4635
0.18	0.9776	0.9936	3.2779
0.19	0.9751	0.9928	3.1123
0.20	0.9725	0.9921	2.9635
0.21	0.9697	0.9913	2.8293
0.22	0.9668	0.9904	2.7076
0.23	0.9638	0.9895	2.5968
0.24	0.9607	0.9886	2.4956
0.25	0.9575	0.9877	2.4027
0.26	0.9541	0.9867	2.3173
0.27	0.9506	0.9856	2.2385
0.28	0.9470	0.9846	2.1656
0.29	0.9433	0.9835	2.0979
0.30	0.9395	0.9823	2.0351
0.31	0.9355	0.9811	1.9765
0.32	0.9315	0.9799	1.9219
0.33	0.9274	0.9787	1.8707
0.34	0.9231	0.9774	1.8229
0.35	0.9188	0.9761	1.7780
0.36	0.9143	0.9747	1.7358
0.37	0.9098	0.9733	1.6961
0.38	0.9052	0.9719	1.6587
0.39	0.9004	0.9705	1.6234
0.40	0.8956	0.9690	1.5901
0.41	0.8907	0.9675	1.5587
0.42	0.8857	0.9659	1.5289
0.43	0.8807	0.9643	1.5007
0.44	0.8755	0.9627	1.4740
0.45	0.8703	0.9611	1.4887
0.46	0.8650	0.9594	1.4246
0.47	0.8596	0.9577	1.4018
0.48	0.8541	0.9560	1.3801
0.49	0.8486	0.9542	1.3595
0.50	0.8430	0.9524	1.3398
0.51	0.8374	0.9506	1.3212
0.52	0.8317	0.9487	1.3034
0.53	0.8259	0.9468	1.2865
0.54	0.8201	0.9449	1.2703
0.55	0.8142	0.9430	1.2550
0.56	0.8082	0.9410	1.2403
0.57	0.8022	0.9390	1.2263
0.58	0.7962	0.9370	1.2130
0.59	0.7901	0.9349	1.2003
0.60	0.7840	0.9328	1.1882
0.61	0.7778	0.9307	1.1767
0.62	0.7716	0.9286	1.1657
0.63	0.7654	0.9265	1.1552
0.64	0.7591	0.9243	1.1452
0.65	0.7528	0.9221	1.1356
0.66	0.7465	0.9199	1.1265
0.67	0.7401	0.9176	1.1179
0.68	0.7338	0.9153	1.1097
0.69	0.7274	0.9131	1.1018

(Continued)

TABLE D.1 (*Continued*)

Isentropic Flow Tables for a Gas Having $\gamma = 1.4$

M	p/p_t	T/T_t	A/A^*
0.70	0.7209	0.9107	1.0944
0.71	0.7145	0.9084	1.0873
0.72	0.7080	0.9061	1.0806
0.73	0.7016	0.9037	1.0742
0.74	0.6951	0.9013	1.0681
0.75	0.6886	0.8989	1.0624
0.76	0.6821	0.8964	1.0570
0.77	0.6756	0.8940	1.0519
0.78	0.6691	0.8915	1.0471
0.79	0.6625	0.8890	1.0425
0.80	0.6560	0.8865	1.0382
0.81	0.6495	0.8840	1.0342
0.82	0.6430	0.8815	1.0305
0.83	0.6365	0.8789	1.0270
0.84	0.6300	0.8763	1.0237
0.85	0.6235	0.8737	1.0207
0.86	0.6170	0.8711	1.0179
0.87	0.6106	0.8685	1.0153
0.88	0.6041	0.8659	1.0129
0.89	0.5977	0.8632	1.0108
0.90	0.5913	0.8606	1.0089
0.91	0.5849	0.8579	1.0071
0.92	0.5785	0.8552	1.0056
0.93	0.5721	0.8525	1.0043
0.94	0.5658	0.8498	1.0031
0.95	0.5595	0.8471	1.0022
0.96	0.5532	0.8444	1.0014
0.97	0.5469	0.8416	1.0008
0.98	0.5407	0.8389	1.0003
0.99	0.5345	0.8361	1.0001
1.00	0.5283	0.8333	1.000
1.01	0.5221	0.8306	1.000
1.02	0.5160	0.827	1.000
1.03	0.5099	0.8250	1.001
1.04	0.5039	0.8222	1.001
1.05	0.4979	0.8193	1.002
1.06	0.4919	0.8165	1.003
1.07	0.4860	0.8137	1.004
1.08	0.4800	0.8108	1.005
1.09	0.4742	0.8080	1.006
1.10	0.4684	0.8052	1.008
1.11	0.4626	0.8023	1.010
1.12	0.4568	0.7994	1.011
1.13	0.4511	0.7966	1.013
1.14	0.4455	0.7937	1.015
1.15	0.4398	0.7908	1.017
1.16	0.4343	0.7879	1.020
1.17	0.4287	0.7851	1.022
1.18	0.4232	0.7822	1.025
1.19	0.4178	0.7793	1.026
1.20	0.4124	0.7764	1.030
1.21	0.4070	0.7735	1.033
1.22	0.4017	0.7706	1.037
1.23	0.3964	0.7677	1.040
1.24	0.3912	0.7648	1.043
1.25	0.3861	0.7619	1.047
1.26	0.3809	0.7590	1.050
1.27	0.3759	0.7561	1.054
1.28	0.3708	0.7532	1.058
1.29	0.3658	0.7503	1.062
1.30	0.3609	0.7474	1.066
1.31	0.3560	0.7445	1.071
1.32	0.3512	0.7416	1.075
1.33	0.3464	0.7387	1.080
1.34	0.3417	0.7358	1.084
1.35	0.3370	0.7329	1.089
1.36	0.3323	0.7300	1.094
1.37	0.3277	0.7271	1.099
1.38	0.3232	0.7242	1.104
1.39	0.3187	0.7213	1.109
1.40	0.3142	0.7184	1.115
1.41	0.3098	0.7155	1.120
1.42	0.3055	0.7126	1.126
1.43	0.3012	0.7097	1.132
1.44	0.2969	0.7069	1.138
1.45	0.2927	0.7040	1.144
1.46	0.2886	0.7011	1.150
1.47	0.2845	0.6982	1.156
1.48	0.2804	0.6954	1.163
1.49	0.2764	0.6925	1.169

(*Continued*)

TABLE D.1 *(Continued)*

Isentropic Flow Tables for a Gas Having $\gamma = 1.4$

M	p/p_t	T/T_t	A/A^*	M	p/p_t	T/T_t	A/A^*
1.50	0.2724	0.6897	1.176	1.90	0.1492	0.5807	1.555
1.51	0.2685	0.6868	1.183	1.91	0.1470	0.5782	1.568
1.52	0.2646	0.6840	1.190	1.92	0.1447	0.5756	1.580
1.53	0.2608	0.6811	1.197	1.93	0.1425	0.5731	1.593
1.54	0.2570	0.6783	1.204	1.94	0.1403	0.5705	1.606
1.55	0.2533	0.6754	1.212	1.95	0.1381	0.5680	1.619
1.56	0.2496	0.6726	1.219	1.96	0.1360	0.5655	1.633
1.57	0.2459	0.6698	1.227	1.97	0.1339	0.5630	1.646
1.58	0.2423	0.6670	1.234	1.98	0.1318	0.5605	1.660
1.59	0.2388	0.6642	1.242	1.99	0.1298	0.5580	1.674
1.60	0.2353	0.6614	1.250	2.00	0.1278	0.5556	1.688
1.61	0.2318	0.6586	1.258	2.01	0.1258	0.5531	1.702
1.62	0.2284	0.6558	1.267	2.02	0.1239	0.5506	1.716
1.63	0.2250	0.6530	1.275	2.03	0.1220	0.5482	1.730
1.64	0.2217	0.6502	1.284	2.04	0.1201	0.5458	1.745
1.65	0.2184	0.6475	1.292	2.05	0.1182	0.5433	1.760
1.66	0.2151	0.6447	1.301	2.06	0.1164	0.5409	1.775
1.67	0.2119	0.6419	1.310	2.07	0.1146	0.5385	1.790
1.68	0.2088	0.6392	1.319	2.08	0.1128	0.5361	1.806
1.69	0.2057	0.6364	1.328	2.09	0.1111	0.5337	1.821
1.70	0.2026	0.6337	1.338	2.10	0.1094	0.5313	1.837
1.71	0.1996	0.6310	1.347	2.11	0.1077	0.5290	1.853
1.72	0.1966	0.6283	1.357	2.12	0.1060	0.5266	1.869
1.73	0.1936	0.6256	1.367	2.13	0.1043	0.5243	1.885
1.74	0.1907	0.6229	1.376	2.14	0.1027	0.5219	1.902
1.75	0.1878	0.6202	1.386	2.15	0.1011	0.5196	1.919
1.76	0.1850	0.6175	1.397	2.16	0.9956^{-1}	0.5173	1.935
1.77	0.1822	0.6148	1.407	2.17	0.9802^{-1}	0.5150	1.953
1.78	0.1794	0.6121	1.418	2.18	0.9649^{-1}	0.5127	1.970
1.79	0.1767	0.6095	1.428	2.19	0.9500^{-1}	0.5104	1.987
1.80	0.1740	0.6068	1.439	2.20	0.9352^{-1}	0.5081	2.005
1.81	0.1714	0.6041	1.450	2.21	0.9207^{-1}	0.5059	2.023
1.82	0.1688	0.6015	1.461	2.22	0.9064^{-1}	0.5036	2.041
1.83	0.1662	0.5989	1.472	2.23	0.8923^{-1}	0.5014	2.059
1.84	0.1637	0.5963	1.484	2.24	0.8785^{-1}	0.4991	2.078
1.85	0.1612	0.5936	1.495	2.25	0.8648^{-1}	0.4969	2.096
1.86	0.1587	0.5910	1.507	2.26	0.8154^{-1}	0.4947	2.115
1.87	0.1563	0.5884	1.519	2.27	0.8382^{-1}	0.4925	2.134
1.88	0.1539	0.5859	1.531	2.28	0.8251^{-1}	0.4903	2.154
1.89	0.1516	0.5833	1.543	2.29	0.8123^{-1}	0.4881	2.173

(Continued)

TABLE D.1 (*Continued*)

Isentropic Flow Tables for a Gas Having $\gamma = 1.4$

M	p/p_t	T/T_t	A/A^*	M	p/p_t	T/T_t	A/A^*
2.30	0.7997^{-1}	0.4859	2.193	2.70	0.4295^{-1}	0.4068	3.183
2.31	0.7873^{-1}	0.4837	2.213	2.71	0.4229^{-1}	0.4051	3.213
2.32	0.7751^{-1}	0.4816	2.233	2.72	0.4165^{-1}	0.4033	3.244
2.33	0.7631^{-1}	0.4794	2.254	2.73	0.4102^{-1}	0.4015	3.275
2.34	0.7512^{-1}	0.4773	2.274	2.74	0.4039^{-1}	0.3998	3.306
2.35	0.7396^{-1}	0.4752	2.295	2.75	0.3978^{-1}	0.3980	3.338
2.36	0.7281^{-1}	0.4731	2.316	2.76	0.3917^{-1}	0.3963	3.370
2.37	0.7168^{-1}	0.4709	2.338	2.77	0.3858^{-1}	0.3945	3.402
2.38	0.7057^{-1}	0.4688	2.359	2.78	0.3799^{-1}	0.3928	3.434
2.39	0.6948^{-1}	0.4668	2.381	2.79	0.3742^{-1}	0.3911	3.467
2.40	0.6840^{-1}	0.4647	2.403	2.80	0.3685^{-1}	0.3894	3.500
2.41	0.6734^{-1}	0.4626	2.425	2.81	0.3629^{-1}	0.3877	3.534
2.42	0.6630^{-1}	0.4606	2.448	2.82	0.3574^{-1}	0.3860	3.567
2.43	0.6527^{-1}	0.4585	2.471	2.83	0.3520^{-1}	0.3844	3.601
2.44	0.6426^{-1}	0.4565	2.494	2.84	0.3467^{-1}	0.3827	3.636
2.45	0.6327^{-1}	0.4544	2.517	2.85	0.3415^{-1}	0.3810	3.671
2.46	0.6229^{-1}	0.4524	2.540	2.86	0.3363^{-1}	0.3794	3.706
2.47	0.6133^{-1}	0.4504	2.564	2.87	0.3312^{-1}	0.3777	3.741
2.48	0.6038^{-1}	0.4484	2.588	2.88	0.3263^{-1}	0.3761	3.777
2.49	0.5945^{-1}	0.4464	2.612	2.89	0.3213^{-1}	0.3745	3.813
2.50	0.5853^{-1}	0.4444	2.637	2.90	0.3165^{-1}	0.3729	3.850
2.51	0.5762^{-1}	0.4425	2.661	2.91	0.3118^{-1}	0.3712	3.887
2.52	0.5674^{-1}	0.4405	2.686	2.92	0.3071^{-1}	0.3696	3.924
2.53	0.5586^{-1}	0.4386	2.712	2.93	0.3025^{-1}	0.3681	3.961
2.54	0.5500^{-1}	0.4366	2.737	2.94	0.2980^{-1}	0.3665	3.999
2.55	0.5415^{-1}	0.4347	2.763	2.95	0.2935^{-1}	0.3649	4.038
2.56	0.5332^{-1}	0.4328	2.789	2.96	0.2891^{-1}	0.3633	4.076
2.57	0.5250^{-1}	0.4309	2.815	2.97	0.2848^{-1}	0.3618	4.115
2.58	0.5169^{-1}	0.4289	2.842	2.98	0.2805^{-1}	0.3602	4.155
2.59	0.5090^{-1}	0.4271	2.869	2.99	0.2764^{-1}	0.3587	4.194
2.60	0.5012^{-1}	0.4252	2.896	3.00	0.2722^{-1}	0.3571	4.235
2.61	0.4935^{-1}	0.4233	2.923	3.01	0.2682^{-1}	0.3556	4.275
2.62	0.4859^{-1}	0.4214	2.951	3.02	0.2642^{-1}	0.3541	4.316
2.63	0.4784^{-1}	0.4196	2.979	3.03	0.2603^{-1}	0.3526	4.357
2.64	04711^{-1}	0.4177	3.007	3.04	0.2564^{-1}	0.3511	4.399
2.65	0.4639^{-1}	0.4159	3.036	3.05	0.2526^{-1}	0.3496	4.441
2.66	0.4568^{-1}	0.4141	3.065	3.06	0.2489^{-1}	0.3481	4.483
2.67	0.4498^{-1}	0.4122	3.094	3.07	0.2452^{-1}	0.3466	4.526
2.68	0.4429^{-1}	0.4104	3.123	3.08	0.2416^{-1}	0.3452	4.570
2.69	0.4362^{-1}	0.4086	3.153	3.09	0.2380^{-1}	0.3437	4.613

(*Continued*)

TABLE D.1 (*Continued*)
Isentropic Flow Tables for a Gas Having $\gamma = 1.4$

M	p/p_t	T/T_t	A/A^*	M	p/p_t	T/T_t	A/A^*
3.10	0.2345^{-1}	0.3422	4.657	3.50	0.1311^{-1}	0.2899	6.790
3.11	0.2310^{-1}	0.3408	4.702	3.51	0.1293^{-1}	0.2887	6.853
3.12	0.2276^{-1}	0.3393	4.747	3.52	0.1274^{-1}	0.2875	6.917
3.13	0.2243^{-1}	0.3379	4.792	3.53	0.1256^{-1}	0.2864	6.982
3.14	0.2210^{-1}	0.3365	4.838	3.54	0.1239^{-1}	0.2852	7.047
3.15	0.2177^{-1}	0.3351	4.884	3.55	0.1221^{-1}	0.2841	7.113
3.16	0.2146^{-1}	0.3337	4.930	3.56	0.1204^{-1}	0.2829	7.179
3.17	0.2114^{-1}	0.3323	4.977	3.57	0.1188^{-1}	0.2818	7.246
3.18	0.2083^{-1}	0.3309	5.025	3.58	0.1171^{-1}	0.2806	7.313
3.19	0.2053^{-1}	0.3295	5.073	3.59	0.1155^{-1}	0.2795	7.382
3.20	0.2023^{-1}	0.3281	5.121	3.60	0.1138^{-1}	0.2784	7.450
3.21	0.1993^{-1}	0.3267	5.170	3.61	0.1123^{-1}	0.2773	7.519
3.22	0.1964^{-1}	0.3253	5.219	3.62	0.1107^{-1}	0.2762	7.589
3.23	0.1936^{-1}	0.3240	5.268	3.63	0.1092^{-1}	0.2751	7.659
3.24	0.1908^{-1}	0.3226	5.319	3.64	0.1076^{-1}	0.2740	7.730
3.25	0.1880^{-1}	0.3213	5.369	3.65	0.1062^{-1}	0.2729	7.802
3.26	0.1853^{-1}	0.3199	5.420	3.66	0.1047^{-1}	0.2718	7.874
3.27	0.1826^{-1}	0.3186	5.472	3.67	0.1032^{-1}	0.2707	7.947
3.28	0.1799^{-1}	0.3173	5.523	3.68	0.1018^{-1}	0.2697	8.020
3.29	0.1773^{-1}	0.3160	5.576	3.69	0.1004^{-1}	0.2686	8.094
3.30	0.1748^{-1}	0.3147	5.629	3.70	0.9903^{-2}	0.2675	8.169
3.31	0.1722^{-1}	0.3134	5.682	3.71	0.9767^{-2}	0.2665	8.244
3.32	0.1698^{-1}	0.3121	5.736	3.72	0.9633^{-2}	0.2654	8.320
3.33	0.1673^{-1}	0.3108	5.790	3.73	0.9500^{-1}	0.2644	8.397
3.34	0.1649^{-1}	0.3095	5.845	3.74	0.9370^{-2}	0.2633	8.474
3.35	0.1625^{-1}	0.3082	5.900	3.75	0.9242^{-2}	0.2623	8.552
3.36	0.1602^{-1}	0.3069	5.956	3.76	0.9116^{-2}	0.2613	8.630
3.37	0.1579^{-1}	0.3057	6.012	3.77	0.8991^{-2}	0.2602	8.709
3.38	0.1557^{-1}	0.3044	6.069	3.78	0.8869^{-2}	0.2592	8.789
3.39	0.1534^{-1}	0.3032	6.126	3.79	0.8748^{-2}	0.2582	8.870
3.40	0.1512^{-1}	0.3019	6.184	3.80	0.8629^{-2}	0.2572	8.951
3.41	0.1491^{-1}	0.3007	6.242	3.81	0.8512^{-2}	0.2562	9.032
3.42	0.1470^{-1}	0.2995	6.301	3.82	0.8396^{-2}	0.2552	9.115
3.43	0.1449^{-1}	0.2982	6.360	3.83	0.8283^{-2}	0.2542	9.198
3.44	0.1428^{-1}	0.2970	6.420	3.84	0.8171^{-2}	0.2532	9.282
3.45	0.1408^{-1}	0.2958	6.480	3.85	0.8060^{-2}	0.2522	9.366
3.46	0.1388^{-1}	0.2946	6.541	3.86	0.7951^{-2}	0.2513	9.451
3.47	0.1368^{-1}	0.2934	6.602	3.87	0.7844^{-2}	0.2503	9.537
3.48	0.1349^{-1}	0.2922	6.664	3.88	0.7739^{-2}	0.2493	9.624
3.49	0.1330^{-1}	0.2910	6.727	3.89	0.7635^{-2}	0.2484	9.711

(*Continued*)

TABLE D.1 (*Continued*)

Isentropic Flow Tables for a Gas Having $\gamma = 1.4$

M	p/p_t	T/T_t	A/A^*	M	p/p_t	T/T_t	A/A^*
3.90	0.7532^{-2}	0.2474	9.799	4.30	0.4449^{-2}	0.2129	13.95
3.91	0.7431^{-1}	0.2464	9.888	4.31	0.4393^{-2}	0.2121	14.08
3.92	0.7332^{-2}	0.2455	9.977	4.32	0.4337^{-2}	0.2113	14.20
3.93	0.7233^{-2}	0.2446	10.07	4.33	0.4282^{-2}	0.2105	14.32
3.94	0.7137^{-2}	0.2436	10.16	4.34	0.4228^{-2}	0.2098	14.45
3.95	0.7042^{-2}	0.2427	10.25	4.35	0.4174^{-2}	0.2090	14.57
3.96	0.6948^{-2}	0.2418	10.34	4.36	0.4121^{-2}	0.2083	14.70
3.97	0.6855^{-2}	0.2408	10.44	4.37	0.4069^{-2}	0.2075	14.82
3.98	0.6764^{-2}	0.2399	10.53	4.38	0.4018^{-2}	0.2067	14.95
3.99	0.6675^{-2}	0.2390	10.62	4.39	0.3968^{-2}	0.2060	15.08
4.00	0.6586^{-2}	0.2381	10.72	4.40	0.3918^{-2}	0.2053	15.21
4.01	0.6499^{-2}	0.2372	10.81	4.41	0.3868^{-2}	0.2045	15.34
4.02	0.6413^{-2}	0.2363	10.91	4.42	0.3820^{-2}	0.2038	15.47
4.03	0.6328^{-2}	0.2354	11.01	4.43	0.3772^{-2}	0.2030	15.61
4.04	0.6245^{-2}	0.2345	11.11	4.44	0.3725^{-2}	0.2023	15.74
4.05	0.6163^{-2}	0.2336	11.21	4.45	0.3678^{-2}	0.2016	15.87
4.06	0.6082^{-2}	0.2327	11.31	4.46	0.3633^{-2}	0.2009	16.01
4.07	0.6002^{-2}	0.2319	11.41	4.47	0.3587^{-2}	0.2002	16.15
4.08	0.5923^{-2}	0.2310	11.51	4.48	0.3543^{-2}	0.1994	16.28
4.09	0.5845^{-2}	0.2301	11.61	4.49	0.3499^{-2}	0.1987	16.42
4.10	0.5769^{-2}	0.2293	11.71	4.50	0.3455^{-2}	0.1980	16.56
4.11	0.5694^{-2}	0.2284	11.82	4.51	0.3412^{-2}	0.1973	16.70
4.12	0.5619^{-2}	0.2275	11.92	4.52	0.3370^{-2}	0.1966	16.84
4.13	0.5546^{-2}	0.2267	12.03	4.53	0.3329^{-2}	0.1959	16.99
4.14	0.5474^{-2}	0.2258	12.14	4.54	0.3288^{-2}	0.1952	17.13
4.15	0.5403^{-2}	0.2250	12.24	4.55	0.3247^{-2}	0.1945	17.28
4.16	0.5333^{-2}	0.2242	12.35	4.56	0.3207^{-2}	0.1938	17.42
4.17	0.5264^{-2}	0.2233	12.46	4.57	0.3168^{-2}	0.1932	17.57
4.18	0.5195^{-2}	0.2225	12.57	4.58	0.3129^{-2}	0.1925	17.72
4.19	0.5128^{-2}	0.2217	12.68	4.59	0.3090^{-2}	0.1918	17.87
4.20	0.5062^{-2}	0.2208	12.79	4.60	0.3053^{-2}	0.1911	18.02
4.21	0.4997^{-2}	0.2200	12.90	4.61	0.3015^{-2}	0.1905	18.17
4.22	0.4932^{-2}	0.2192	13.02	4.62	0.2978^{-2}	0.1898	18.32
4.23	0.4869^{-2}	0.2184	13.13	4.63	0.2942^{-2}	0.1891	18.48
4.24	0.4806^{-2}	0.2176	13.25	4.64	0.2906^{-2}	0.1885	18.63
4.25	0.4745^{-2}	0.2168	13.36	4.65	0.2871^{-2}	0.1878	18.79
4.26	0.4684^{-2}	0.2160	13.48	4.66	0.2836^{-2}	0.1872	18.94
4.27	0.4624^{-2}	0.2152	13.60	4.67	0.2802^{-2}	0.1865	19.10
4.28	0.4565^{-2}	0.2144	13.72	4.68	0.2768^{-2}	0.1859	19.26
4.29	0.4507^{-2}	0.2136	13.83	4.69	0.2734^{-2}	0.1852	19.42

(*Continued*)

TABLE D.1 (*Continued*)

Isentropic Flow Tables for a Gas Having $\gamma = 1.4$

M	p/p_t	T/T_t	A/A^*	M	p/p_t	T/T_t	A/A^*
4.70	0.2701^{-1}	0.1846	19.58	4.85	0.2255^{-2}	0.1753	22.15
4.71	0.2669^{-2}	0.1839	19.75	4.86	0.2229^{-2}	0.1747	22.33
4.72	0.2637^{-2}	0.1833	19.91	4.87	0.2202^{-2}	0.1741	22.51
4.73	0.2605^{-2}	0.1827	20.07	4.88	0.2177^{-2}	0.1735	22.70
4.74	0.2573^{-2}	0.1820	20.24	4.89	0.2151^{-2}	0.1729	21.88
4.75	0.2543^{-2}	0.1814	20.41	4.90	0.2126^{-2}	0.1724	23.07
4.76	0.2512^{-2}	0.1808	20.58	4.91	0.2101^{-2}	0.1718	23.25
4.77	0.2482^{-2}	0.1802	20.75	4.92	0.2076^{-2}	0.1712	23.44
4.78	0.2452^{-2}	0.1795	20.92	4.93	0.2052^{-2}	0.1706	23.63
4.79	0.2423^{-2}	0.1789	21.09	4.94	0.2028^{-2}	0.1700	23.82
4.80	0.2394^{-2}	0.1783	21.26	4.95	0.2004^{-2}	0.1695	24.02
4.81	0.2366^{-2}	0.1777	21.44	4.96	0.1981^{-2}	0.1689	24.21
4.82	0.2338^{-2}	0.1771	21.61	4.97	0.1957^{-2}	0.1683	24.41
4.83	0.2310^{-2}	0.1765	21.79	4.98	0.1985^{-2}	0.1678	24.60
4.84	0.2283^{-2}	0.1759	21.97	4.99	0.1912^{-2}	0.1672	24.80
				5.00	0.1890^{-2}	0.1667	25.00

Source: NACA, Equations, tables and charts for compressible flow, NACA Report 1135, 1953.

Notes on reading the table: $0.9956^{-1} = 0.9956 \times 10^{-1} = 0.09956$. $0.7214^{-1} = 0.7214 \times 10^{-1} = 0.07214$.

TABLE D.2

Normal Shock Tables for a Gas Having $\gamma = 1.4$

M_1	M_2	p_2/p_1	ρ_2/ρ_1	T_2/T_1	p_{t_2}/p_{t_1}	p_1/p_{t_2}
1.00	1.000	1.000	1.000	1.000	1.000	0.5283
1.01	0.9901	1.023	1.017	1.007	1.000	0.5221
1.02	0.9805	1.047	1.033	1.013	1.000	0.5160
1.03	0.9712	1.071	1.050	1.020	1.000	0.5100
1.04	0.9620	1.095	1.067	1.026	0.9999	0.5039
1.05	0.9531	1.120	1.084	1.033	0.9999	0.4980
1.06	0.9444	1.144	1.101	1.039	0.9997	0.4920
1.07	0.9360	1.169	1.118	1.046	0.9996	0.4861
1.08	0.9277	1.194	1.135	1.052	0.9994	0.4803
1.09	0.9196	1.219	1.152	1.059	0.9992	0.4746
1.10	0.9116	1.245	1.169	1.065	0.9989	0.4689
1.11	0.9041	1.271	1.186	1.071	0.9986	0.4632
1.12	0.8966	1.297	1.203	1.078	0.9982	0.4576
1.13	0.8892	1.323	1.221	1.084	0.9978	0.4521
1.14	0.8820	1.350	1.238	1.090	0.9973	0.4467
1.15	0.8750	1.376	1.255	1.097	0.9967	0.4413
1.16	0.8682	1.403	1.272	1.103	0.9961	0.4360
1.17	0.8615	1.430	1.290	1.109	0.9953	0.4307
1.18	0.8549	1.458	1.307	1.115	0.9946	0.4255
1.19	0.8485	1.485	1.324	1.122	0.9937	0.4204
1.20	0.8422	1.513	1.342	1.128	0.9928	0.4154
1.21	0.8360	1.541	1.359	1.134	0.9918	0.4104
1.22	0.8300	1.570	1.376	1.141	0.9907	0.4055
1.23	0.8241	1.598	1.394	1.147	0.9896	0.4006
1.24	0.8183	1.627	1.411	1.153	0.9884	0.3958
1.25	0.8126	1.656	1.429	1.159	0.9871	0.3911
1.26	0.8071	1.686	1.446	1.166	0.9857	0.3865
1.27	0.8016	1.715	1.463	1.172	0.9842	0.3819
1.28	0.7963	1.745	1.481	1.178	0.9827	0.3774
1.29	0.7911	1.775	1.498	1.185	0.9811	0.3729
1.30	0.7860	1.805	1.516	1.191	0.9794	0.3685
1.31	0.7809	1.835	1.533	1.197	0.9776	0.3642
1.32	0.7760	1.866	1.551	1.204	0.9758	0.3599
1.33	0.7712	1.897	1.568	1.210	0.9738	0.3557
1.34	0.7664	1.928	1.585	1.216	0.9718	0.3516
1.35	0.7618	1.960	1.603	1.223	0.9697	0.3475
1.36	0.7572	1.991	1.620	1.229	0.9676	0.3435
1.37	0.7527	2.023	1.638	1.235	0.9653	0.3395
1.38	0.7483	2.055	1.655	1.242	0.9630	0.3356
1.39	0.7440	2.087	1.672	1.248	0.9607	0.3317

(Continued)

TABLE D.2 (*Continued*)
Normal Shock Tables for a Gas Having $\gamma = 1.4$

M_1	M_2	p_2/p_1	ρ_2/ρ_1	T_2/T_1	p_{t_2}/p_{t_1}	p_1/p_{t_2}
1.40	0.7397	2.120	1.690	1.255	0.9582	0.3280
1.41	0.7355	2.153	1.707	1.261	0.9557	0.3242
1.42	0.7314	2.186	1.724	1.268	0.9531	0.3205
1.43	0.7274	2.219	1.742	1.274	0.9504	0.3169
1.44	0.7235	2.253	1.759	1.281	0.9476	0.3133
1.40	0.7397	2.120	1.690	1.255	0.9582	0.3280
1.41	0.7355	2.153	1.707	1.261	0.9557	0.3242
1.42	0.7314	2.186	1.724	1.268	0.9531	0.3205
1.43	0.7274	2.219	1.742	1.274	0.9504	0.3169
1.44	0.7235	2.253	1.759	1.281	0.9476	0.3133
1.45	0.7196	2.286	1.776	1.287	0.9448	0.3098
1.46	0.7157	2.320	1.793	1.294	0.9420	0.3063
1.47	0.7120	2.354	1.811	1.300	0.9390	0.3029
1.48	0.7083	2.389	1.828	1.307	0.9360	0.2996
1.49	0.7047	2.423	1.845	1.314	0.9329	0.2962
1.50	0.7011	2.458	1.862	1.320	0.9298	0.2930
1.51	0.6976	2.493	1.879	1.327	0.9266	0.2898
1.52	0.6941	2.529	1.896	1.334	0.9233	0.2366
1.53	0.6907	2.564	1.913	1.340	0.9200	0.2835
1.54	0.6874	2.600	1.930	1.347	0.9166	0.2804
1.55	0.6841	2.636	1.947	1.354	0.9132	0.2773
1.56	0.6809	2.673	1.964	1.361	0.9097	0.2744
1.57	0.6777	2.709	1.981	1.367	0.9061	0.2714
1.58	0.6746	2.746	1.998	1.374	0.9026	0.2685
1.59	0.6715	2.783	2.015	1.381	0.8989	0.2656
1.60	0.6684	2.820	2.032	1.388	0.8952	0.2628
1.61	0.6655	2.857	2.049	1.395	0.8915	0.2600
1.62	0.6625	2.895	2.065	1.402	0.8877	0.2573
1.63	0.6596	2.933	2.082	1.409	0.8838	0.2546
1.64	0.6568	2.971	2.099	1.416	0.8799	0.2519
1.65	0.6540	3.010	2.115	1.423	0.8760	0.2493
1.66	0.6512	3.048	2.132	1.430	0.8720	0.2467
1.67	0.6485	3.087	2.148	1.437	0.8680	0.2442
1.68	0.6458	3.126	2.165	1.444	0.8640	0.2417
1.69	0.6431	3.165	2.181	1.451	0.8598	0.2392
1.70	0.6405	3.205	2.198	1.458	0.8557	0.2368
1.71	0.6380	3.245	2.214	1.466	0.8516	0.2344
1.72	0.6355	3.285	2.230	1.473	0.8474	0.2320
1.73	0.6330	3.325	2.247	1.480	0.8431	0.2296
1.74	0.6305	3.366	2.263	1.487	0.8389	0.2273

(*Continued*)

TABLE D.2 (*Continued*)
Normal Shock Tables for a Gas Having $\gamma = 1.4$

M_1	M_2	p_2/p_1	ρ_2/ρ_1	T_2/T_1	p_{t_2}/p_{t_1}	p_1/p_{t_2}
1.75	0.6281	3.406	2.279	1.495	0.8346	0.2251
1.76	0.6257	3.447	2.295	1.502	0.8302	0.2228
1.77	0.6234	3.488	2.311	1.509	0.8259	0.2206
1.78	0.6210	3.530	2.327	1.517	0.8215	0.2184
1.79	0.6188	3.571	2.343	1.524	0.8171	0.2163
1.80	0.6165	3.613	2.359	1.532	0.8127	0.2142
1.81	0.6143	3.655	2.375	1.539	0.8082	0.2121
1.82	0.6121	3.698	2.391	1.547	0.8038	0.2100
1.83	0.6099	3.740	2.407	1.554	0.7993	0.2080
1.84	0.6078	3.783	2.422	1.562	0.7948	0.2060
1.85	0.6057	3.826	2.438	1.569	0.7902	0.2040
1.86	0.6036	3.870	2.454	1.577	0.7857	0.2020
1.87	0.6016	3.913	2.469	1.585	0.7811	0.2001
1.88	0.5996	3.957	2.485	1.592	0.7765	0.1982
1.89	0.5976	4.001	2.500	1.600	0.7720	0.1963
1.90	0.5956	4.045	2.516	1.608	0.7674	0.1945
1.91	0.5937	4.089	2.531	1.616	0.7627	0.1927
1.92	0.5918	4.134	2.546	1.624	0.7581	0.1909
1.93	0.5899	4.179	2.562	1.631	0.7535	0.1891
1.94	0.5880	4.224	2.577	1.639	0.7488	0.1873
1.95	0.5862	4.270	2.592	1.647	0.7442	0.1856
1.96	0.5844	4.315	2.607	1.655	0.7395	0.1839
1.97	0.5826	4.361	2.622	1.663	0.7349	0.1822
1.98	0.5808	4.407	2.637	1.671	0.7302	0.1806
1.99	0.5791	4.453	2.652	1.679	0.7255	0.1789
2.00	0.5774	4.500	2.667	1.688	0.7209	0.1773
2.01	0.5757	4.547	2.681	1.696	0.7162	0.1757
2.02	0.5740	4.594	2.696	1.704	0.7115	0.1741
2.03	0.5723	4.641	2.711	1.712	0.7069	0.1726
2.04	0.5707	4.689	2.725	1.720	0.7022	0.1710
2.05	0.5691	4.736	2.740	1.729	0.6975	0.1695
2.06	0.5675	4.784	2.755	1.737	0.6928	0.1680
2.07	0.5659	4.832	2.769	1.745	0.6882	0.1665
2.08	0.5643	4.881	2.783	1.754	0.6835	0.1651
2.09	0.5628	4.929	2.798	1.762	0.6789	0.1636
2.10	0.5613	4.978	2.812	1.770	0.6742	0.1622
2.11	0.5598	5.027	2.826	1.779	0.6696	0.1608
2.12	0.5583	5.077	2.840	1.787	0.6649	0.1594
2.13	0.5568	5.126	2.854	1.796	0.6603	0.1580
2.14	0.5554	5.176	2.868	1.805	0.6557	0.1567

(*Continued*)

TABLE D.2 (*Continued*)

Normal Shock Tables for a Gas Having $\gamma = 1.4$

M_1	M_2	p_2/p_1	ρ_2/ρ_1	T_2/T_1	p_{t_2}/p_{t_1}	p_1/p_{t_2}
2.15	0.5540	5.226	2.882	1.813	0.6511	0.1553
2.16	0.5525	5.277	2.896	1.822	0.6464	0.1540
2.17	0.5511	5.327	2.910	1.831	0.6419	0.1527
2.18	0.5498	5.378	2.924	1.839	0.6373	0.1514
2.19	0.5484	5.429	2.938	1.848	0.6327	0.1502
2.20	0.5471	5.480	2.951	1.857	0.6281	0.1489
2.21	0.5457	5.531	2.965	1.866	0.6236	0.1476
2.22	0.5444	5.583	2.978	1.875	0.6191	0.1464
2.23	0.5431	5.636	2.992	1.883	0.6145	0.1452
2.24	0.5418	5.687	3.005	1.892	0.6100	0.1440
2.25	0.5406	5.740	3.019	1.901	0.6055	0.1428
2.26	0.5393	5.792	3.032	1.910	0.6011	0.1417
2.27	0.5381	5.845	3.045	1.919	0.5966	0.1405
2.28	0.5368	5.898	3.058	1.929	0.5921	0.1394
2.29	0.5356	5.951	3.071	1.938	0.5877	0.1382
2.30	0.5344	6.005	3.085	1.947	0.5833	0.1371
2.31	0.5332	6.059	3.098	1.956	0.5789	0.1360
2.32	0.5321	6.113	3.110	1.965	0.5745	0.1349
2.33	0.5309	6.167	3.123	1.974	0.5702	0.1338
2.34	0.5297	6.222	3.136	1.984	0.5658	0.1328
2.35	0.5286	6.276	3.149	1.993	0.5615	0.1317
2.36	0.5275	6.331	3.162	2.002	0.5572	0.1307
2.37	0.5264	6.386	3.174	2.012	0.5529	0.1297
2.38	0.5253	6.442	3.187	2.021	0.5486	0.1286
2.39	0.5242	6.497	3.199	2.031	0.5444	0.1276
2.40	0.5231	6.553	3.212	2.040	0.5401	0.1266
2.41	0.5221	6.609	3.224	2.050	0.5359	0.1257
2.42	0.5210	6.666	3.237	2.059	0.5317	0.1247
2.43	0.5200	6.722	3.249	2.069	0.5276	0.1237
2.44	0.5189	6.779	3.261	2.079	0.5234	0.1228
2.45	0.5179	6.836	3.273	2.088	0.5193	0.1218
2.46	0.5169	6.894	3.285	2.098	0.5152	0.1209
2.47	0.5159	6.951	3.298	2.108	0.5111	0.1200
2.48	0.5149	7.009	3.310	2.118	0.5071	0.1191
2.49	0.5140	7.067	3.321	2.128	0.5030	0.1182
2.50	0.5130	7.125	3.333	2.138	0.4990	0.1173
2.51	0.5120	7.183	3.345	2.147	0.4950	0.1164
2.52	0.5111	7.242	3.357	2.157	0.4911	0.1155
2.53	0.5102	7.301	3.369	2.167	0.4871	0.1147
2.54	0.5092	7.360	3.380	2.177	0.4832	0.1138

(*Continued*)

TABLE D.2 (*Continued*)
Normal Shock Tables for a Gas Having $\gamma = 1.4$

M_1	M_2	p_2/p_1	ρ_2/ρ_1	T_2/T_1	p_{t_2}/p_{t_1}	p_1/p_{t_2}
2.55	0.5083	7.420	3.392	2.187	0.4793	0.1130
2.56	0.5074	7.479	3.403	2.198	0.4754	0.1122
2.57	0.5065	7.539	3.415	2.208	0.4715	0.1113
2.58	0.5056	7.599	3.426	2.218	0.4677	0.1105
2.59	0.5047	7.659	3.438	2.228	0.4639	0.1097
2.60	0.5039	7.720	3.449	2.238	0.4601	0.1089
2.61	0.5030	7.781	3.460	2.249	0.4564	0.1081
2.62	0.5022	7.842	3.471	2.259	0.4526	0.1074
2.63	0.5013	7.908	3.483	2.269	0.4489	0.1066
2.64	0.5005	7.965	3.494	2.280	0.4452	0.1058
2.65	0.4996	8.026	3.505	2.290	0.4416	0.1051
2.66	0.4988	8.088	3.516	2.301	0.4379	0.1043
2.67	0.4980	8.150	3.527	2.311	0.4343	0.1036
2.68	0.4972	8.213	3.537	2.322	0.4307	0.1028
2.69	0.4964	8.275	3.548	2.332	0.4271	0.1021
2.70	0.4956	8.338	3.559	2.343	0.4236	0.1014
2.71	0.4949	8.401	3.570	2.354	0.4201	0.1007
2.72	0.4941	8.465	3.580	2.364	0.4166	0.9998^{-1}
2.73	0.4933	8.528	3.591	3752	0.4131	0.9929^{-1}
2.74	0.4926	8.592	3.601	2.386	0.4097	0.9860^{-1}
2.75	0.4918	8.656	3.612	2.397	0.4062	0.9792^{-1}
2.76	0.4911	8.721	3.622	2.407	0.4028	0.9724^{-1}
2.77	0.4903	8.785	3.633	2.418	0.3994	0.9658^{-1}
2.78	0.4896	8.850	3.643	2.429	0.3961	0.9591^{-1}
2.79	0.4889	8.915	3.653	2.440	0.3928	0.9526^{-1}
2.80	0.4882	8.980	3.664	2.451	0.3895	0.9461^{-1}
2.81	0.4875	9.045	3.674	2.462	0.3862	0.9397^{-1}
2.82	0.4868	9.111	3.684	2.473	0.3829	0.9334^{-1}
2.83	0.4861	9.177	3.694	2.484	0.3797	0.9271^{-1}
2.84	0.4854	9.243	3.704	2.496	0.3765	0.9209^{-1}
2.85	0.4847	9.310	3.714	2.507	0.3733	0.9147^{-1}
2.86	0.4840	9.376	3.724	2.518	0.3701	0.9086^{-1}
2.87	0.4833	9.443	3.734	2.529	0.3670	0.9026^{-1}
2.88	0.4827	9.510	3.743	2.540	0.3639	0.8966^{-1}
2.89	0.4820	9.577	3.753	2.552	0.3608	0.8906^{-1}
2.90	0.4814	9.645	3.763	2.563	0.3577	0.8848^{-1}
2.91	0.4807	9.713	3.773	2.575	0.3547	0.8790^{-1}
2.92	0.4801	9.781	3.782	2.586	0.3517	0.8732^{-1}
2.93	0.4795	9.849	3.792	2.598	0.3487	0.8675^{-1}
2.94	0.4788	9.918	3.801	2.609	0.3457	0.8619^{-1}

(*Continued*)

TABLE D.2 (*Continued*)

Normal Shock Tables for a Gas Having $\gamma = 1.4$

M_1	M_2	p_2/p_1	ρ_2/ρ_1	T_2/T_1	p_{t_2}/p_{t_1}	p_1/p_{t_2}
2.95	0.4782	9.986	3.811	2.621	0.3428	0.8563^{-1}
2.96	0.4776	10.06	3.820	2.632	0.3398	0.8507^{-1}
2.97	0.4770	10.12	3.829	2.644	0.3369	0.8453^{-1}
2.98	0.4764	10.19	3.839	2.656	0.3340	0.8398^{-1}
2.99	0.4758	10.26	3.848	2.667	0.3312	0.8345^{-1}
3.00	0.4752	10.33	3.857	2.679	0.3283	0.8291^{-1}
3.01	0.4746	10.40	3.866	2.691	0.3255	0.8238^{-1}
3.02	0.4740	10.47	3.875	2.703	0.3227	0.8186^{-1}
3.03	0.4734	10.54	3.884	2.714	0.3200	0.8134^{-1}
3.04	0.4729	10.62	3.893	2.726	0.3172	0.8083^{-1}
3.05	0.4723	10.69	3.902	2.738	0.3145	0.8032^{-1}
3.06	0.4717	10.76	3.911	2.750	0.3118	0.7982^{-1}
3.07	0.4712	10.83	3.920	2.762	0.3091	0.7932^{-1}
3.08	0.4706	10.90	3.929	2.774	0.3065	0.7882^{-1}
3.09	0.4701	10.97	3.938	2.786	0.3038	0.7833^{-1}
3.10	0.4695	11.05	3.947	2.799	0.3012	0.7785^{-1}
3.11	0.4690	11.12	3.955	2.811	0.2986	0.7737^{-1}
3.12	0.4685	11.19	3.964	2.823	0.2960	0.7689^{-1}
3.13	0.4679	11.26	3.973	2.835	0.2935	0.7642^{-1}
3.14	0.4674	11.34	3.981	2.848	0.2910	0.7595^{-1}
3.15	0.4669	11.41	3.990	2.860	0.2885	0.7549^{-1}
3.16	0.4664	11.48	3.998	2.872	0.2860	0.7503^{-1}
3.17	0.4659	11.56	4.006	2.885	0.2835	0.7457^{-1}
3.18	0.4654	11.63	4.015	2.897	0.2811	0.7412^{-1}
3.19	0.4648	11.71	4.023	2.909	0.2786	0.7367^{-1}
3.20	0.4643	11.78	4.031	2.922	0.2762	0.7323^{-1}
3.21	0.4639	11.85	4.040	2.935	0.2738	0.7279^{-1}
3.22	0.4634	11.93	4.048	2.947	0.2715	0.7235^{-1}
3.23	0.4629	12.01	4.056	2.960	0.2691	0.7192^{-1}
3.24	0.4624	12.08	4.064	2.972	0.2668	0.7149^{-1}
3.25	0.4619	12.16	4.072	2.985	0.2645	0.7107^{-1}
3.26	0.4614	12.23	4.080	2.998	0.2622	0.7065^{-1}
3.27	0.4610	12.31	4.088	3.011	0.2600	0.7023^{-1}
3.28	0.4605	12.38	4.096	3.023	0.2577	0.6982^{-1}
3.29	0.4600	12.46	4.104	3.036	0.2555	0.6941^{-1}
3.30	0.4596	12.54	4.112	3.049	0.2533	0.6900^{-1}
3.31	0.4591	12.62	4.120	3.062	0.2511	0.6860^{-1}
3.32	0.4587	12.69	4.128	3.075	0.2489	0.6820^{-1}
3.33	0.4582	12.77	4.135	3.088	0.2468	0.6781^{-1}
3.34	0.4578	12.85	4.143	3.101	0.2446	0.6741^{-1}

(*Continued*)

TABLE D.2 (*Continued*)

Normal Shock Tables for a Gas Having $\gamma = 1.4$

M_1	M_2	p_2/p_1	ρ_2/ρ_1	T_2/T_1	p_{t_2}/p_{t_1}	p_1/p_{t_2}
3.35	0.4573	2.93	4.151	3.114	0.2425	0.6702^{-1}
3.36	0.4569	3.00	4.158	3.127	0.2404	0.6664^{-1}
3.37	0.4565	3.08	4.166	3.141	0.2383	0.6626^{-1}
3.38	0.4560	3.16	4.173	3.154	0.2363	0.6588^{-1}
3.39	0.4556	3.24	4.181	3.167	0.2342	0.6550^{-1}
3.40	0.4552	13.32	4.188	3.180	0.2322	0.6513^{-1}
3.41	0.4548	13.40	4.196	3.194	0.2302	0.6476^{-1}
3.42	0.4544	13.48	4.203	3.207	0.2282	0.6439^{-1}
3.43	0.4540	13.56	4.211	3.220	0.2263	0.6403^{-1}
3.44	0.4535	13.64	4.218	3.234	0.2243	0.6367^{-1}
3.45	0.4531	13.72	4.225	3.247	0.2224	0.6331^{-1}
3.46	0.4527	13.80	4.232	3.261	0.2205	0.6296^{-1}
3.47	0.4523	13.88	4.240	3.274	0.2186	0.6261^{-1}
3.48	0.4519	13.96	4.247	3.288	0.2167	0.6226^{-1}
3.49	0.4515	14.04	4.254	3.301	0.2148	0.6191^{-1}
3.50	0.4512	14.13	4.261	3.315	0.2129	0.6157^{-1}
3.51	0.4508	14.21	4.268	3.329	0.2111	0.6123^{-1}
3.52	0.4504	14.29	4.275	3.343	0.2093	0.6089^{-1}
3.53	0.4500	14.37	4.282	3.356	0.2075	0.6056^{-1}
3.54	0.4496	14.45	4.289	3.370	0.2057	0.6023^{-1}
3.55	0.4492	14.54	4.296	3.384	0.2039	0.5990^{-1}
3.56	0.4489	14.62	4.303	3.398	0.2022	0.5957^{-1}
3.57	0.4485	14.70	4.309	3.412	0.2004	0.5925^{-1}
3.58	0.4481	14.79	4.316	3.426	0.1987	0.5892^{-1}
3.59	0.4478	14.87	4.323	3.440	0.1970	0.5861^{-1}
3.60	0.4474	14.95	4.330	3.454	0.1953	0.5829^{-1}
3.61	0.4471	15.04	4.336	3.468	0.1936	0.5798^{-1}
3.62	0.4467	15.12	4.343	3.482	0.1920	0.5767^{-1}
3.63	0.4463	15.21	4.350	3.496	0.1903	0.5736^{-1}
3.64	0.4460	15.29	4.356	3.510	0.1887	0.5705^{-1}
3.65	0.4456	15.38	4.363	3.525	0.1671	0.5675^{-1}
3.66	0.4453	15.46	4.369	3.539	0.1855	0.5645^{-1}
3.67	0.4450	15.55	4.376	3.553	0.1839	0.5615^{-1}
3.68	0.4446	15.63	4.382	3.568	0.1823	0.5585^{-1}
3.69	0.4443	15.72	4.388	3.582	0.1807	0.5556^{-1}
3.70	0.4439	15.81	4.395	3.596	0.1792	0.5526^{-1}
3.71	0.4436	15.89	4.401	3.611	0.1777	0.5497^{-1}
3.72	0.4433	15.98	4.408	3.625	0.1761	0.5469^{-1}
3.73	0.4430	16.07	4.414	3.640	0.1746	0.5440^{-1}
3.74	0.4426	16.15	4.420	3.654	0.1731	0.5412^{-1}

(Continued)

TABLE D.2 (*Continued*)

Normal Shock Tables for a Gas Having $\gamma = 1.4$

M_1	M_2	p_2/p_1	ρ_2/ρ_1	T_2/T_1	p_{t_2}/p_{t_1}	p_1/p_{t_2}
3.75	0.4423	16.24	4.426	3.669	0.1717	0.5384^{-1}
3.76	0.4420	16.33	4.432	3.684	0.1702	0.5356^{-1}
3.77	0.4417	16.42	4.439	3.698	0.1687	0.5328^{-1}
3.78	0.4414	16.50	4.445	3.713	0.1673	0.5301^{-1}
3.79	0.4410	16.59	4.451	3.728	0.1659	0.5274^{-1}
3.80	0.4407	16.68	4.457	3.743	0.1645	0.5247^{-1}
3.81	0.4404	16.77	4.463	3.758	0.1631	0.5220^{-1}
3.82	0.4401	16.86	4.469	3.772	0.1617	0.5193^{-1}
3.83	0.4398	16.95	4.475	3.787	0.1603	0.5167^{-1}
3.84	0.4395	17.04	4.481	3.802	0.1589	0.5140^{-1}
3.85	0.4392	17.13	4.487	3.817	0.1576	0.5114^{-1}
3.86	0.4389	17.22	4.492	3.832	0.1563	0.5089^{-1}
3.87	0.4386	17.31	4.498	3.847	0.1549	0.5063^{-1}
3.88	0.4383	17.40	4.504	3.863	0.1536	0.5038^{-1}
3.89	0.4380	17.49	4.510	3.878	0.1523	0.5012^{-1}
3.90	0.4377	17.58	4.516	3.893	0.1510	0.4987^{-1}
3.91	0.4375	17.67	4.521	3.908	0.1497	0.4962^{-1}
3.92	0.4372	17.76	4.527	3.923	0.1435	0.4938^{-1}
3.93	0.4369	17.85	4.533	3.939	0.1472	0.4913^{-1}
3.94	0.4366	17.94	4.538	3.954	0.1460	0.4889^{-1}
3.95	0.4363	18.04	4.544	3.969	0.1448	0.4865^{-1}
3.96	0.4360	18.13	4.549	3.985	0.1435	0.4841^{-1}
3.97	0.4358	18.22	4.555	4.000	0.1423	0.4817^{-1}
3.98	0.4355	18.31	4.560	4.016	0.1411	0.4793^{-1}
3.99	0.4352	18.41	4.566	4.031	0.1399	0.4770^{-1}
4.00	0.4350	18.50	4.571	4.047	0.1388	0.4747^{-1}
4.01	0.4347	18.59	4.577	4.062	0.1376	0.4723^{-1}
4.02	0.4344	18.69	4.582	4.078	0.1364	0.4700^{-1}
4.03	0.4342	18.78	4.588	4.094	0.1353	0.4678^{-1}
4.04	0.4339	18.88	4.593	4.110	0.1342	0.4655^{-1}
4.05	0.4336	18.97	4.598	4.125	0.1330	0.4633^{-1}
4.06	0.4334	19.06	4.604	4.141	0.1319	0.4610^{-1}
4.07	0.4331	19.16	4.609	4.157	0.1308	0.4588^{-1}
4.08	0.4329	19.25	4.614	4.173	0.1297	0.4566^{-1}
4.09	0.4326	19.35	4.619	4.189	0.1286	0.4544^{-1}
4.10	0.4324	19.45	4.624	4.205	0.1276	0.4523^{-1}
4.11	0.4321	19.54	4.630	4.221	0.1265	0.4501^{-1}
4.12	0.4319	19.64	4.635	4.237	0.1254	0.4480^{-1}
4.13	0.4316	19.73	4.640	4.253	0.1244	0.4459^{-1}
4.14	0.4314	19.83	4.645	4.269	0.1234	0.4438^{-1}

(*Continued*)

TABLE D.2 *(Continued)*

Normal Shock Tables for a Gas Having $\gamma = 1.4$

M_1	M_2	p_2/p_1	ρ_2/ρ_1	T_2/T_1	p_{t_2}/p_{t_1}	p_1/p_{t_2}
4.15	0.4311	19.93	4.650	4.285	0.1223	0.4417^{-1}
4.16	0.4309	20.02	4.655	4.301	0.1213	0.4396^{-1}
4.17	0.4306	20.12	4.660	4.318	0.1203	0.4375^{-1}
4.18	0.4304	20.22	4.665	4.334	0.1193	0.4355^{-1}
4.19	0.4302	20.32	4.670	4.350	0.1183	0.4334^{-1}
4.20	0.4299	20.41	4.675	4.367	0.1173	0.4314^{-1}
4.21	0.4297	20.51	4.680	4.383	0.1164	0.4294^{-1}
4.22	0.4295	20.61	4.685	4.399	0.1154	0.4274^{-1}
4.23	0.4292	20.71	4.690	4.416	0.1144	0.4255^{-1}
4.24	0.4290	20.81	4.694	4.432	0.1135	0.4235^{-1}
4.25	0.4288	20.91	4.699	4.449	0.1126	0.4215^{-1}
4.26	0.4286	21.01	4.704	4.466	0.1116	0.4196^{-1}
4.27	0.4283	21.11	4.709	4.482	0.1107	0.4177^{-1}
4.28	0.4281	21.20	4.713	4.499	0.1098	0.4158^{-1}
4.29	0.4279	21.30	4.718	4.516	0.1089	0.4139^{-1}
4.30	0.4277	21.41	4.723	4.532	0.1080	0.4120^{-1}
4.31	0.4275	21.51	4.728	4.549	0.1071	0.4101^{-1}
4.32	0.4272	21.61	4.732	4.566	0.1062	0.4082^{-1}
4.33	0.4270	21.71	4.737	4.583	0.1054	0.4064^{-1}
4.34	0.4268	21.81	4.741	4.600	0.1045	0.4046^{-1}
4.35	0.4266	21.91	4.746	4.617	0.1036	0.4027^{-1}
4.36	0.4264	22.01	4.751	4.633	0.1028	0.4009^{-1}
4.37	0.4262	22.11	4.755	4.651	0.1020	0.3991^{-1}
4.38	0.4260	22.22	4.760	4.668	0.1011	0.3973^{-1}
4.39	0.4258	22.32	4.764	4.685	0.1003	0.3956^{-1}
4.40	0.4255	22.42	4.768	4.702	0.9948^{-1}	0.3938^{-1}
4.41	0.4253	22.52	4.773	4.719	0.9867^{-1}	0.3921^{-1}
4.42	0.4251	22.63	4.777	4.736	0.9787^{-1}	0.3903^{-1}
4.43	0.4249	22.73	4.782	4.753	0.9707^{-1}	0.3886^{-1}
4.44	0.4247	22.83	4.786	4.771	0.9628^{-1}	0.3869^{-1}
4.45	0.4245	22.94	4.790	4.788	0.9550^{-1}	0.3852^{-1}
4.46	0.4243	23.04	4.795	4.805	0.9473^{-1}	0.3835^{-1}
4.47	0.4241	23.14	4.799	4.823	0.9396^{-1}	0.3818^{-1}
4.48	0.4239	23.25	4.803	4.840	0.9320^{-1}	0.3801^{-1}
4.49	0.4237	23.35	4.808	4.858	0.9244^{-1}	0.3785^{-1}
4.50	0.4236	23.46	4.812	4.875	0.9170^{-1}	0.3768^{-1}
4.51	0.4234	23.56	4.816	4.893	0.9096^{-1}	0.3752^{-1}
4.52	0.4232	23.67	4.820	4.910	0.9022^{-1}	0.3735^{-1}
4.53	0.4230	23.77	4.824	4.928	0.8950^{-1}	0.3719^{-1}
4.54	0.4228	23.88	4.829	4.946	0.8878^{-1}	0.3703^{-1}
4.55	0.4226	23.99	4.833	4.963	0.8806^{-1}	0.3687^{-1}
4.56	0.4224	24.09	4.837	4.981	0.8735^{-1}	0.3671^{-1}
4.57	0.4222	24.20	4.841	4.999	0.8665^{-1}	0.3656^{-1}
4.58	0.4220	24.31	4.845	5.017	0.8596^{-1}	0.3640^{-1}
4.59	0.4219	24.41	4.849	5.034	0.8527^{-1}	0.3624^{-1}

(Continued)

TABLE D.2 (*Continued*)

Normal Shock Tables for a Gas Having $\gamma = 1.4$

M_1	M_2	p_2/p_1	ρ_2/ρ_1	T_2/T_1	p_{t_2}/p_{t_1}	p_1/p_{t_2}
4.60	0.4217	24.52	4.853	5.052	0.8459^{-1}	0.3609^{-1}
4.61	0.4215	24.63	4.857	5.070	0.8391^{-1}	0.3593^{-1}
4.62	0.4213	24.74	4.861	5.088	0.8324^{-1}	0.3578^{-1}
4.63	0.4211	24.84	4.865	5.106	0.8257^{-1}	0.3563^{-1}
4.64	0.4210	24.95	4.869	5.124	0.8192^{-1}	0.3548^{-1}
4.65	0.4208	25.06	4.873	5.143	0.8126^{-1}	0.3533^{-1}
4.66	0.4206	25.17	4.877	5.160	0.8062^{-1}	0.3518^{-1}
4.67	0.4204	25.28	4.881	5.179	0.7998^{-1}	0.3503^{-1}
4.68	0.4203	25.39	4.885	5.197	0.7934^{-1}	0.3488^{-1}
4.69	0.4201	25.50	4.889	5.215	0.7871^{-1}	0.3474^{-1}
4.70	0.4199	25.61	4.893	5.233	0.7809^{-1}	0.3459^{-1}
4.71	0.4197	25.71	4.896	5.252	0.7747^{-1}	0.3445^{-1}
4.72	0.4196	25.82	4.900	5.270	0.7685^{-1}	0.3431^{-1}
4.73	0.4194	25.94	4.904	5.289	0.7625^{-1}	0.3416^{-1}
4.74	0.4192	26.05	4.908	5.307	0.7564^{-1}	0.3402^{-1}
4.75	0.4191	26.16	4.912	5.325	0.7505^{-1}	0.3388^{-1}
4.76	0.4189	26.27	4.915	5.344	0.7445^{-1}	0.3374^{-1}
4.77	0.4187	26.38	4.919	5.363	0.7387^{-1}	0.3360^{-1}
4.78	0.4186	26.49	4.923	5.381	0.7329^{-1}	0.3346^{-1}
4.79	0.4184	26.60	4.926	5.400	0.7271^{-1}	0.3333^{-1}
4.80	0.4183	26.71	4.930	5.418	0.7214^{-1}	0.3319^{-1}
4.81	0.4181	26.83	4.934	5.437	0.7157^{-1}	0.3305^{-1}
4.82	0.4179	26.94	4.937	5.456	0.7101^{-1}	0.3292^{-1}
4.83	0.4178	27.05	4.941	5.475	0.7046^{-1}	0.3278^{-1}
4.84	0.4176	27.16	4.945	5.494	0.6991^{-1}	0.3265^{-1}
4.85	0.4175	27.28	4.948	5.512	0.6936^{-1}	0.3252^{-1}
4.86	0.4173	27.39	4.952	5.531	0.6882^{-1}	0.3239^{-1}
4.87	0.4172	27.50	4.955	5.550	0.6828^{-1}	0.3226^{-1}
4.88	0.4170	27.62	4.959	5.569	0.6775^{-1}	0.3213^{-1}
4.89	0.4169	27.73	4.962	5.588	0.6722^{-1}	0.3200^{-1}
4.90	0.4167	27.85	4.966	5.607	0.6670^{-1}	0.3187^{-1}
4.91	0.4165	27.96	4.969	5.626	0.6618^{-1}	0.3174^{-1}
4.92	0.4164	28.07	4.973	5.646	0.6567^{-1}	0.3161^{-1}
4.93	0.4163	28.19	4.976	5.665	0.6516^{-1}	0.3149^{-1}
4.94	0.4161	28.30	4.980	5.684	0.6465^{-1}	0.3136^{-1}
4.95	0.4160	28.42	4.983	5.703	0.6415^{-1}	0.3124^{-1}
4.96	0.4158	28.54	4.987	5.723	0.6366^{-1}	0.3111^{-1}
4.97	0.4157	28.65	4.990	5.742	0.6317^{-1}	0.3099^{-1}
4.98	0.4155	28.77	4.993	5.761	0.6268^{-1}	0.3087^{-1}
4.99	0.4154	28.88	4.997	5.781	0.6220^{-1}	0.3075^{-1}
5.00	0.4152	29.00	5.000	5.800	0.6172^{-1}	0.3062^{-1}

Source: NACA, Equations, tables and charts for compressible flow, NACA Report 1135, 1953.

Note on reading the table: $0.7214^{-1} = 0.7214 \times 10^{-1} = 0.07214$.

TABLE D.3

Fanno Flow (Flow with Friction) Tables for a Gas Having γ = 1.4

M	T/T^*	p/p^*	p_t/p_t^*	V/V^*	fL_{max}/D
0.00	1.2000	∞	∞	0	∞
0.01	1.2000	109.544	57.874	0.01095	7134.40
0.02	1.1999	54.770	28.942	0.02191	1778.45
0.03	1.1998	36.511	19.300	0.03286	787.08
0.04	1.1996	27.382	14.482	0.04381	440.35
0.05	1.1994	21.903	11.5914	0.05476	280.02
0.06	1.1991	18.251	9.6659	0.06570	193.03
0.07	1.1988	15.642	8.2915	0.07664	140.66
0.08	1.1985	13.684	7.2616	0.08758	106.72
0.09	1.1981	12.162	6.4614	0.09851	83.496
0.10	1.1976	10.9435	5.8218	0.10943	66.922
0.11	1.1971	9.9465	5.2992	0.12035	54.688
0.12	1.1966	9.1156	4.8643	0.13126	45.408
0.13	1.1960	8.4123	4.4968	0.14216	38.207
0.14	1.1953	7.8093	4.1824	0.15306	32.511
0.15	1.1946	7.2866	3.9103	0.16395	27.932
0.16	1.1939	6.8291	3.6727	0.17482	24.198
0.17	1.1931	6.4252	3.4635	0.18568	21.115
0.18	1.1923	6.0662	3.2779	0.19654	18.543
0.19	1.1914	5.7448	3.1123	0.20739	16.375
0.20	1.1905	5.4555	2.9635	0.21822	14.533
0.21	1.1895	5.1936	2.8293	0.22904	12.956
0.22	1.1885	4.9554	2.7076	0.23984	11.596
0.23	1.1874	4.7378	2.5968	0.25063	10.416
0.24	1.1863	4.5383	2.4956	0.26141	9.3865
0.25	1.1852	4.3546	2.4027	0.27217	8.4834
0.26	1.1840	4.1850	2.3173	0.28291	7.6876
0.27	1.1828	4.0280	2.2385	0.29364	6.9832
0.28	1.1815	3.8820	2.1656	0.30435	6.3572
0.29	1.1802	3.7460	2.0979	0.31504	5.7989
0.30	1.1788	3.6190	2.0351	0.32572	5.2992
0.31	1.1774	3.5002	1.9765	0.33637	4.8507
0.32	1.1759	3.3888	1.9219	0.34700	4.4468
0.33	1.1744	3.2840	1.8708	0.35762	4.0821
0.34	1.1729	3.1853	1.8229	0.36822	3.7520
0.35	1.1713	3.0922	1.7780	0.37880	3.4525
0.36	1.1697	3.0042	1.7358	0.38935	3.1801
0.37	1.1680	2.9209	1.6961	0.39988	2.9320
0.38	1.1663	2.8420	1.6587	0.41039	2.7055
0.39	1.1646	2.7671	1.6234	0.42087	2.4983

(*Continued*)

TABLE D.3 (*Continued*)
Fanno Flow (Flow with Friction) Tables for a Gas Having γ = 1.4

M	T/T^*	p/p^*	p_t/p_t^*	V/V^*	fL_{max}/D
0.40	1.1628	2.6958	1.5901	0.43133	2.3085
0.41	1.1610	2.6280	1.5587	0.44177	2.1344
0.42	1.1591	2.5634	1.5289	0.45218	1.9744
0.43	1.1572	2.5017	1.5007	0.46257	1.8272
0.44	1.1553	2.4428	1.4739	0.47293	1.6915
0.45	1.1533	2.3865	1.4486	0.48326	1.5664
0.46	1.1513	2.3326	1.4246	0.49357	1.4509
0.47	1.1492	2.2809	1.4018	0.50385	1.3442
0.48	1.1471	2.2314	1.3801	0.51410	1.2453
0.49	1.1450	2.1838	1.3595	0.52433	1.1539
0.50	1.1429	2.1381	1.3399	0.53453	1.06908
0.51	1.1407	2.0942	1.3212	0.54469	0.99042
0.52	1.1384	2.0519	1.3034	0.55482	0.91741
0.53	1.1362	2.0112	1.2864	0.56493	0.84963
0.54	1.1339	1.9719	1.2702	0.57501	0.78662
0.55	1.1315	1.9341	1.2549	0.58506	0.72805
0.56	1.1292	1.8976	1.2403	0.59507	0.67357
0.57	1.1268	1.8623	1.2263	0.60505	0.62286
0.58	1.1244	1.8282	1.2130	0.61500	0.57568
0.59	1.1219	1.7952	1.2003	0.62492	0.53174
0.60	1.1194	1.7634	1.1882	0.63481	0.49081
0.61	1.1169	1.7325	1.1766	0.64467	0.45270
0.62	1.1144	1.7026	1.1656	0.65449	0.41720
0.63	1.1118	1.6737	1.1551	0.66427	0.38411
0.64	1.1091	1.6456	1.1451	0.67402	0.35330
0.65	1.10650	1.6183	1.1356	0.68374	0.32460
0.66	1.10383	1.5919	1.1265	0.69342	0.29785
0.67	1.10114	1.5662	1.1179	0.70306	0.27295
0.68	1.09842	1.5413	1.1097	0.71267	0.24978
0.69	1.09567	1.5170	1.1018	0.72225	0.22821
0.70	1.09290	1.4934	1.09436	0.73179	0.20814
0.71	1.09010	1.4705	1.08729	0.74129	0.18949
0.72	1.08727	1.4482	1.08057	0.75076	0.17215
0.73	1.08442	1.4265	1.07419	0.76019	0.15606
0.74	1.08155	1.4054	1.06815	0.76958	0.14113
0.75	1.07865	1.3848	1.06242	0.77893	0.12728
0.76	1.07573	1.3647	1.05700	0.78825	0.11446
0.77	1.07279	1.3451	1.05188	0.79753	0.10262
0.78	1.06982	1.3260	1.04705	0.80677	0.09167
0.79	1.06684	1.3074	1.04250	0.81598	0.08159

(*Continued*)

TABLE D.3 (*Continued*)

Fanno Flow (Flow with Friction) Tables for a Gas Having $\gamma = 1.4$

M	T/T^*	p/p^*	p_t/p_t^*	V/V^*	fL_{max}/D
0.80	1.06383	1.2892	1.03823	0.82514	0.07229
0.81	1.06080	1.2715	1.03422	0.83426	0.06375
0.82	1.05775	1.2542	1.03047	0.84334	0.05593
0.83	1.05468	1.2373	1.02696	0.85239	0.04878
0.84	1.05160	1.2208	1.02370	0.86140	0.04226
0.85	1.04849	1.2047	1.02067	0.87037	0.03632
0.86	1.04537	1.1889	1.01787	0.87929	0.03097
0.87	1.04223	1.1735	1.01529	0.88818	0.02613
0.88	1.03907	1.1584	1.01294	0.89703	0.02180
0.89	1.03589	1.1436	1.01080	0.90583	0.01793
0.90	1.03270	1.12913	1.00887	0.91459	0.014513
0.91	1.02950	1.11500	1.00714	0.92332	0.011519
0.92	1.02627	1.10114	1.00560	0.93201	0.008916
0.93	1.02304	1.08758	1.00426	0.94065	0.006694
0.94	1.01978	1.07430	1.00311	0.94925	0.004815
0.95	1.01652	1.06129	1.00215	0.95782	0.003280
0.96	1.01324	1.04854	1.00137	0.96634	0.002056
0.97	1.00995	1.03605	1.00076	0.97481	0.001135
0.98	1.00664	1.02379	1.00033	0.98324	0.000493
0.99	1.00333	1.01178	1.00008	0.99164	0.000120
1.00	1.00000	1.00000	1.00000	1.00000	0
1.01	0.99666	0.98844	1.00008	1.00831	0.000114
1.02	0.99331	0.97711	1.00033	1.01658	0.000458
1.03	0.98995	0.96598	1.00073	1.02481	0.001013
1.04	0.98658	0.95506	1.00130	1.03300	0.001771
1 05	0.98320	0.94435	1.00203	1.04115	0.002712
1.06	0.97982	0.93383	1.00291	1.04925	0.003837
1.07	0.97642	0.92350	1.00394	1.05731	0.005129
1.08	0.97302	0.91335	1.00512	1.06533	0.006582
1.09	0.96960	0.90338	1.00645	1.07331	0.008185
1.10	0.96618	0.89359	1.00793	1.08124	0.009933
1.11	0.96276	0.88397	1.00955	1.08913	0.011813
1.12	0.95933	0.87451	1.01131	1.09698	0.013824
1.13	0.95589	0.86522	1.01322	1.10479	0.015949
1.14	0.95244	0.85608	1.01527	1.11256	0.018187
1.15	0.94899	0.84710	1.01746	1.1203	0.02053
1.16	0.94554	0.83827	1.01978	1.1280	0.02298
1.17	0.94208	0.82958	1.02224	1.1356	0.02552
1.18	0.93862	0.82104	1.02484	1.1432	0.02814
1.19	0.93515	0.81263	1.02757	1.1508	0.03085

(*Continued*)

TABLE D.3 (*Continued*)
Fanno Flow (Flow with Friction) Tables for a Gas Having γ = 1.4

M	T/T^*	p/p^*	p_t/p_t^*	V/V^*	fL_{max}/D
1.20	0.93168	0.80436	1.03044	1.1583	0.03364
1.21	0.92820	0.79623	1.03344	1.1658	0.03650
1.22	0.92473	0.78822	1.03657	1.1732	0.03942
1.23	0.92125	0.78034	1.03983	1.1806	0.04241
1.24	0.91777	0.77258	1.04323	1.1879	0.04547
1.25	0.91429	0.76495	1.04676	1.1952	0.04858
1.26	0.91080	0.75743	1.05041	1.2025	0.05174
1.27	0.90732	0.75003	1.05419	1.2097	0.05494
1.28	0.90383	0.74274	1.05809	1.2169	0.05820
1.29	0.90035	0.73556	1.06213	1.2240	0.06150
1.30	0.89686	0.72848	1.06630	1.2311	0.06483
1.31	0.89338	0.72152	1.07060	1.2382	0.06820
1.32	0.88989	0.71465	1.07502	1.2452	0.07161
1.33	0.88641	0.70789	1.07957	1.2522	0.07504
1.34	0.88292	0.70123	1.08424	1.2591	0.07850
1.35	0.87944	0.69466	1.08904	1.2660	0.08199
1.36	0.87596	0.68818	1.09397	1.2729	0.08550
1.37	0.87249	0.68180	1.09902	1.2797	0.08904
1 38	0.86901	0.67551	1.10419	1.2864	0.09259
1.39	0.86554	0.66931	1.10948	1.2932	0.09616
1.40	0.86207	0.66320	1.1149	1.2999	0.09974
1.41	0.85860	0.65717	1.1205	1.3065	0.10333
1.42	0.85514	0.65122	1.1262	1.3131	0.10694
1.43	0.85168	0.64536	1.1320	1.3197	0.11056
1.44	0.84822	0.63958	1.1379	1.3262	0.11419
1.45	0.84477	0.63387	1.1440	1.3327	0.11782
1.46	0.84133	0.62824	1.1502	1.3392	0.12146
1.47	0.83788	0.62269	1.1565	1.3456	0.12510
1.48	0.83445	0.61722	1.1629	1.3520	0.12875
1.49	0.83101	0.61181	1.1695	1.3583	0.13240
1.50	0.82759	0.60648	1.1762	1.3646	0.13605
1.51	0.82416	0.60122	1.1830	1.3708	0.13970
1.52	0.82075	0.59602	1.1899	1.3770	0.14335
1.53	0.81734	0.59089	1.1970	1.3832	0.14699
1.54	0.81394	0.58583	1.2043	1.3894	0.15063
1.55	0.81054	0.58084	1.2116	1.3955	0.15427
1.56	0.80715	0.57591	1.2190	1.4015	0.15790
1.57	0.80376	0.57104	1.2266	1.4075	0.16152
1.58	0.80038	0.56623	1.2343	1.4135	0.16514
1.59	0.79701	0.56148	1.2422	1.4195	0.16876

(*Continued*)

TABLE D.3 (*Continued*)

Fanno Flow (Flow with Friction) Tables for a Gas Having $\gamma = 1.4$

M	T/T^*	p/p^*	p_t/p_t^*	V/V^*	fL_{max}/D
1.60	0.79365	0.55679	1.2502	1.4254	0.17236
1.61	0.79030	0.55216	1.2583	1.4313	0.17595
1.62	0.78695	0.54759	1.2666	1.4371	0.17953
1.63	0.78361	0.54308	1.2750	1.4429	0.18311
1.64	0.78028	0.53862	1.2835	1.4487	0.18667
1.65	0.77695	0.53421	1.2922	1.4544	0.19022
1.66	0.77363	0.52986	1.3010	1.4601	0.19376
1.67	0.77033	0.52556	1.3099	1.4657	0.19729
1.68	0.76703	0.52131	1.3190	1.4713	0.20081
1.69	0.76374	0.51711	1.3282	1.4769	0.20431
1.70	0.76046	0.51297	1.3376	1.4825	0.20780
1.71	0.75718	0.50887	1.3471	1.4880	0.21128
1.72	0.75392	0.50482	1.3567	1.4935	0.21474
1.73	0.75067	0.50082	1.3665	1.4989	0.21819
1.74	0.74742	0.49686	1.3764	1.5043	0.22162
1.75	0.74419	0.49295	1.3865	1.5097	0.22504
1.76	0.74096	0.48909	1.3967	1.5150	0.22844
1.77	0.73774	0.48527	1.4070	1.5203	0.23183
1.78	0.73453	0.48149	1.4175	1.5256	0.23520
1.79	0.73134	0.47776	1.4282	1.5308	0.23855
1.80	0.72816	0.47407	1.4390	1.5360	0.24189
1.81	0.72498	0.47042	1.4499	1.5412	0.24521
1.82	0.72181	0.46681	1.4610	1.5463	0.24851
1.83	0.71865	0.46324	1.4723	1.5514	0.25180
1.84	0.71551	0.45972	1.4837	1.5564	0.25507
1.85	0.71238	0.45623	1.4952	1.5614	0.25832
1.86	0.70925	0.45278	1.5069	1.5664	0.26156
1.87	0.70614	0.44937	1.5188	1.5714	0.26478
1.88	0.70304	0.44600	1.5308	1.5763	0.26798
1.89	0.69995	0.44266	1.5429	1.5812	0.27116
1.90	0.69686	0.43936	1.5552	1.5861	0.27433
1.91	0.69379	0.43610	1.5677	1.5909	0.27748
1.92	0.69074	0.43287	1.5804	1.5957	0.28061
1.93	0.68769	0.42967	1.5932	1.6005	0.28372
1.94	0.68465	0.42651	1.6062	1.6052	0.28681
1.95	0.68162	0.42339	1.6193	1.6099	0.28989
1.96	0.67861	0.42030	1.6326	1.6146	0.29295
1.97	0.67561	0.41724	1.6461	1.6193	0.29599
1.98	0.67262	0.41421	1.6597	1.6239	0.29901
1.99	0.66964	0.41121	1.6735	1.6284	0.30201

(*Continued*)

TABLE D.3 (*Continued*)

Fanno Flow (Flow with Friction) Tables for a Gas Having $\gamma = 1.4$

M	T/T^*	p/p^*	p_t/p_t^*	V/V^*	fL_{max}/D
2.00	0.66667	0.40825	1.6875	1.6330	0.30499
2.01	0.66371	0.40532	1.7017	1.6375	0.30796
2.02	0.66076	0.40241	1.7160	1.6420	0.31091
2.03	0.65783	0.39954	1.7305	1.6465	0.31384
2.04	0.65491	0.39670	1.7452	1.6509	0.31675
2.05	0.65200	0.39389	1.7600	1.6553	0.31965
2.06	0.64910	0.39110	1.7750	1.6597	0.32253
2.07	0.64621	0.38834	1.7902	1.6640	0.32538
2.08	0.64333	0.38562	1.8056	1.6683	0.32822
2.09	0.64047	0.38292	1.8212	1.6726	0.33104
2.10	0.63762	0.38024	1.8369	1.6769	0.33385
2.11	0.63478	0.37760	1.8528	1.6811	0.33664
2.12	0.63195	0.37498	1.8690	1.6853	0.33940
2.13	0.63914	0.37239	1.8853	1.6895	0.34215
2.14	0.62633	0.36982	1.9018	1.6936	0.34488
2.15	0.62354	0.36728	1.9185	1.6977	0.34760
2.16	0.62076	0.36476	1.9354	1.7018	0.35030
2.17	0.61799	0.36227	1.9525	1.7059	0.35298
2.18	0.61523	0.35980	1.9698	1.7099	0.35564
2.19	0.61249	0.35736	1.9873	1.7139	0.35828
2.20	0.60976	0.35494	2.0050	1.7179	0.36091
2.21	0.60704	0.35254	2.0228	1.7219	0.36352
2.22	0.60433	0.35017	2.0409	1.7258	0.36611
2.23	0.60163	0.34782	2.0592	1.7297	0.36868
2.24	0.59895	0.34550	2.0777	1.7336	0.37124
2.25	0.59627	0.34319	2.0964	1.7374	0.37378
2.26	0.59361	0.34091	2.1154	1.7412	0.37630
2.27	0.59096	0.33865	2.1345	1.7450	0.37881
2.28	0.58833	0.33641	2.1538	1.7488	0.38130
2.29	0.58570	0.33420	2.1733	1.7526	0.38377
2.30	0.58309	0.33200	2.1931	1.7563	0.38623
2.31	0.58049	0.32983	2.2131	1.7600	0.38867
2.32	0.57790	0.32767	2.2333	1.7637	0.39109
2.33	0.57532	0.32554	2.2537	1.7673	0.39350
2.34	0.57276	0.32342	2.2744	1.7709	0.39589
2.35	0.57021	0.32133	2.2953	1.7745	0.39826
2.36	0.56767	0.31925	2.3164	1.7781	0.40062
2.37	0.56514	0.31720	2.3377	1.7817	0.40296
2.38	0.56262	0.31516	2.3593	1.7852	0.40528
2.39	0.56011	0.31314	2.3811	1.7887	0.40760

(*Continued*)

TABLE D.3 (*Continued*)

Fanno Flow (Flow with Friction) Tables for a Gas Having γ = 1.4

M	T/T^*	p/p^*	p_t/p_t^*	V/V^*	fL_{max}/D
2.40	0.55762	0.31114	2.4031	1.7922	0.40989
2.41	0.55514	0.30916	2.4254	1.7956	0.41216
2.42	0.55267	0.30720	2.4479	1.7991	0.41442
2.43	0.55021	0.30525	2.4706	1.8025	0.41667
2.44	0.54776	0.30332	2.4936	1.8059	0.41691
2.45	0.54533	0.30141	2.5168	1.8092	0.42113
2.46	0.54291	0.29952	2.5403	1.8126	0.42333
2.47	0.54050	0.29765	2.5640	1.8159	0.42551
2.48	0.53810	0.29579	2.5880	1.8192	0.42768
2.49	0.53571	0.29395	2.6122	1.8225	0.42983
2.50	0.53333	0.29212	2.6367	1.8257	0.43197
2.51	0.53097	0.29031	2.6615	1.8290	0.43410
2.52	0.52862	0.28852	2.6865	1.8322	0.43621
2.53	0.52627	0.28674	2.7117	1.8354	0.43831
2.54	0.52394	0.28498	2.7372	1.8386	0.44040
2.55	0.52163	0.28323	2.7630	1.8417	0.44247
2.56	0.51932	0.28150	2.7891	1.8448	0.44452
2.57	0.51702	0.27978	2.8154	1.8479	0.44655
2.58	0.51474	0.27808	2.8420	1.8510	0.44857
2.59	0.51247	0.27640	2.8689	1.8541	0.45059
2.60	0.51020	0.27473	2.8960	1.8571	0.45259
2.61	0.50795	0.27307	2.9234	1.8602	0.45457
2.62	0.50571	0.27143	2.9511	1.8632	0.45654
2.63	0.50349	0.26980	2.9791	1.8662	0.45850
2.64	0.50127	0.26818	3.0074	1.8691	0.46044
2.70	0.48820	0.25878	3.1830	1.8865	0.47182
2.65	0.49906	0.26658	3.0359	1.8721	0.46237
2.66	0.49687	0.26499	3.0647	1.8750	0.46429
2.67	0.49469	0.26342	3.0938	1.8779	0.46619
2.68	0.49251	0.26186	3.1234	1.8808	0.46807
2.69	0.49035	0.26032	3.1530	1.8837	0.46996
2.71	0.48606	0.25726	3.2133	1.8894	0.47367
2.72	0.48393	0.25575	3.2440	1.8922	0.47551
2.73	0.48182	0.25426	3.2749	1.8950	0.47734
2.74	0.47971	0.25278	3.3061	1.8978	0.47915
2.75	0.47761	0.25131	3.3376	1.9005	0.48095
2.76	0.47553	0.24985	3.3695	1.9032	0.48274
2.77	0.47346	0.24840	3.4017	1.9060	0.48452
2.78	0.47139	0.24697	3.4342	1.9087	0.48628
2.79	0.46933	0.24555	3.4670	1.9114	0.48803

(*Continued*)

TABLE D.3 (*Continued*)
Fanno Flow (Flow with Friction) Tables for a Gas Having γ = 1.4

M	T/T^*	p/p^*	p_t/p_t^*	V/V^*	fL_{max}/D
2.80	0.46729	0.24414	3.5001	1.9140	0.48976
2.81	0.46526	0.24274	3.5336	1.9167	0.49148
2.82	0.46324	0.24135	3.5674	1.9193	0.49321
2.83	0.46122	0.23997	3.6015	1.9220	0.49491
2.84	0.45922	0.23861	3.6359	1.9246	0.49660
2.85	0.45723	0.23726	3.6707	1.9271	0.49828
2.86	0.45525	0.23592	3.7058	1.9297	0.49995
2.87	0.45328	0.23458	3.7413	1.9322	0.50161
2.88	0.45132	0.23326	3.7771	1.9348	0.50326
2.89	0.44937	0.23196	3.8133	1.9373	0.50489
2.90	0.44743	0.23066	3.8498	1.9398	0.50651
2.91	0.44550	0.22937	3.8866	1.9423	0.50812
2.92	0.44358	0.22809	3.9238	1.9448	0.50973
2.93	0.44167	0.22682	3.9614	1.9472	0.51133
2.94	0.43977	0.22556	3.9993	1.9497	0.51291
2.95	0.43788	0.22431	4.0376	1.9521	0.51447
2.96	0.43600	0.22307	4.0763	1.9545	0.51603
2.97	0.43413	0.22185	4.1153	1.9569	0.51758
2.98	0.43226	0.22063	4.1547	1.9592	0.51912
2.99	0.43041	0.21942	4.1944	1.9616	0.52064
3.0	0.42857	0.21822	4.2346	1.9640	0.52216
3.5	0.34783	0.16850	6.7896	2.0642	0.58643
4.0	0.28571	0.13363	10.719	2.1381	0.63306
4.5	0.23762	0.10833	16.562	2.1936	0.66764
5.0	0.20000	0.08944	25.000	2.2361	0.69381
6.0	0.14634	0.06376	53.180	2.2953	0.72987
7.0	0.11111	0.04762	104.14	2.3333	0.75281
8.0	0.08696	0.03686	190.11	2.3591	0.76820
9.0	0.06977	0.02935	327.19	2.3772	0.77898
10.0	0.05714	0.02390	535.94	2.3905	0.78683
∞	0	0	∞	2.4495	0.82153

$$\frac{T}{T^*} = \frac{\gamma+1}{2}\left(1+\frac{\gamma-1}{2}M^2\right)^{-1} \quad \frac{p}{p^*} = \left[\frac{2M^2}{\gamma+1}\left(1+\frac{\gamma-1}{2}M^2\right)\right]^{-1/2}$$

$$\frac{p_t}{p_t^*} = \frac{1}{M}\left[\frac{2}{\gamma+1}\left(1+\frac{\gamma-1}{2}M^2\right)\right]^{(\gamma+1)/[2(\gamma-1)]}$$

$$\frac{V}{V^*} = M\left[\frac{\gamma+1}{2}\left(1+\frac{\gamma-1}{2}M^2\right)^{-1}\right]^{1/2}$$

$$\frac{fL_{max}}{D} = \frac{1-M^2}{\gamma M^2} + \frac{\gamma+1}{2\gamma}\ell n\left[\frac{\gamma+1}{2}M^2\left(1+\frac{\gamma-1}{2}M^2\right)^{-1}\right]$$

TABLE D.4

Rayleigh Flow (Flow with Heat Addition) Tables for a Gas Having $\gamma = 1.4$

M	T/T^*	T_t/T_t^*	p/p^*	p_t/p_t^*	v/v^*
0	0	0	2.4000	1.2679	0
0.01	0.000576	0.000480	2.3997	1.2678	0.000240
0.02	0.00230	0.00192	2.3987	1.2675	0.000959
0.03	0.00517	0.00431	2.3970	1.2671	0.00216
0.04	0.00917	0.00765	2.3946	1.2665	0.00383
0.05	0.01430	0.01192	2.3916	1.2657	0.00598
0.06	0.02053	0.01712	2.3880	1.2647	0.00860
0.07	0.02784	0.02322	2.3836	1.2636	0.01168
0.08	0.03621	0.03022	2.3787	1.2623	0.01522
0.09	0.04562	0.03807	2.3731	1.2608	0.01922
0.10	0.05602	0.04678	2.3669	1.2591	0.02367
0.11	0.06739	0.05630	2.3600	1.2573	0.02856
0.12	0.07970	0.06661	2.3526	1.2554	0.03388
0.13	0.09290	0.07768	2.3445	1.2533	0.03962
0.14	0.10695	0.08947	2.3359	1.2510	0.04578
0.15	0.12181	0.10196	2.3267	1.2486	0.05235
0.16	0.13743	0.11511	2.3170	1.2461	0.05931
0.17	0.15377	0.12888	2.3067	1.2434	0.06666
0.18	0.17078	0.14324	2.2959	1.2406	0.07439
0.19	0.18841	0.15814	2.2845	1.2377	0.08247
0.20	0.20661	0.17355	2.2727	2.2346	0.09091
0.21	0.22533	0.18943	2.2604	1.2314	0.09969
0.22	0.24452	0.20574	2.2477	1.2281	0.10879
0.23	0.26413	0.22244	2.2345	1.2247	0.11821
0.24	0.28411	0.23948	2.2209	1.2213	0.12792
0.25	0.30440	0.25684	2.2069	1.2177	0.13793
0.26	0.32496	0.27446	2.1925	1.2140	0.14821
0.27	0.34573	0.29231	2.1777	1.2102	0.15876
0.28	0.36667	0.31035	2.1626	1.2064	0.16955
0.29	0.38774	0.32855	2.1472	1.2025	0.18058
0.30	0.40887	0.34686	2.1314	1.1985	0.19183
0.31	0.43004	0.36525	2.1154	1.1945	0.20329
0.32	0.45119	0.38369	2.0991	1.1904	0.21495
0.33	0.47228	0.40214	2.0825	1.1863	0.22678
0.34	0.49327	0.42056	2.0657	1.1822	0.23879
0.35	0.51413	0.43894	2.0487	1.1779	0.25096
0.36	0.53482	0.45723	2.0314	1.1737	0.26327
0.37	0.55529	0.47541	2.0140	1.1695	0.27572
0.38	0.57553	0.49346	1.9964	1.1652	0.28828
0.39	0.59549	0.51134	1.9787	1.1609	0.30095

(Continued)

TABLE D.4 (*Continued*)

Rayleigh Flow (Flow with Heat Addition) Tables for a Gas Having γ = 1.4

M	T/T^*	T_t/T_t^*	p/p^*	p_t/p_t^*	v/v^*
0.40	0.61515	0.52903	1.9608	1.1566	0.31373
0.41	0.63448	0.54651	1.9428	1.1523	0.32658
0.42	0.65346	0.56376	1.9247	1.1480	0.33951
0.43	0.67205	0.58076	1.9065	1.1437	0.35251
0.44	0.69025	0.59748	1.8882	1.1394	0.36556
0.45	0.70804	0.61393	1.8699	1.1351	0.37865
0.46	0.72538	0.63007	1.8515	1.1308	0.39178
0.47	0.74228	0.64589	1.8331	1.1266	0.40493
0.48	0.75871	0.66139	1.8147	1.1224	0.41810
0.49	0.77466	0.67655	1.7962	1.1182	0.43127
0.50	0.79012	0.69136	1.7778	1.1141	0.44444
0.51	0.80509	0.70581	1.7594	1.1099	0.45761
0.52	0.81955	0.71990	1.7409	1.1059	0.47075
0.53	0.83351	0.73361	1.7226	1.1019	0.48387
0.54	0.84695	0.74695	1.7043	1.0979	0.49696
0.55	0.85987	0.75991	1.6860	1.0940	0.51001
0.56	0.87227	0.77249	1.6678	1.0901	0.52302
0.57	0.88416	0.78468	1.6496	1.0863	0.53597
0.58	0.89552	0.79648	1.6316	1.0826	0.54887
0.59	0.90637	0.80789	1.6136	1.0789	0.56170
0.60	0.91670	0.81892	1.5957	1.0753	0.57447
0.61	0.92653	0.82957	1.5780	1.0717	0.58716
0.62	0.93584	0.83983	1.5603	1.0682	0.59978
0.63	0.94466	0.84970	1.5428	1.0648	0.61232
0.64	0.95298	0.85920	1.5253	1.0615	0.62477
0.65	0.96081	0.86833	1.5080	1.0582	0.63713
0.66	0.96816	0.87708	1.4908	1.0550	0.64941
0.67	0.97503	0.88547	1.4738	1.0519	0.66158
0.68	0.98144	0.89350	1.4569	1.0489	0.67366
0.69	0.98739	0.90118	1.4401	1.0460	0.68564
0.70	0.99290	0.90850	1.4235	1.0431	0.69751
0.71	0.99796	0.91548	1.4070	1.0403	0.70928
0.72	1.00260	0.92212	1.3907	1.0376	0.72093
0.73	1.00682	0.92843	1.3745	1.0350	0.73248
0.74	1.01062	0.93442	1.3585	1.0325	0.74392
0.75	1.01403	0.94009	1.3427	1.0301	0.75524
0.76	1.01706	0.94546	1.3270	1.0278	0.76645
0.77	1.01970	0.95052	1.3114	1.0255	0.77755
0.78	1.02198	0.95528	1.2961	1.0234	0.78853
0.79	1.02390	0.95975	1.2809	1.0213	0.79939

(*Continued*)

TABLE D.4 (*Continued*)

Rayleigh Flow (Flow with Heat Addition) Tables for a Gas Having $\gamma = 1.4$

M	T/T^*	T_t/T_t^*	p/p^*	p_t/p_t^*	v/v^*
0.80	1.02548	0.96395	1.2658	1.0193	0.81013
0.81	1.02672	0.96787	1.2510	1.0175	0.82075
0.82	1.02763	0.97152	1.2362	1.0157	0.83125
0.83	1.02823	0.97492	1.2217	1.0140	0.84164
0.84	1.02853	0.97807	1.2073	1.0124	0.85190
0.85	1.02854	0.98097	1.1931	1.0109	0.86204
0.86	1.02826	0.98363	1.1791	1.0095	0.87207
0.87	1.02771	0.98607	1.1652	1.0082	0.88197
0.88	1.02689	0.98828	1.1515	1.0070	0.89175
0.89	1.02583	0.99028	1.1380	1.0059	0.90142
0.90	1.02452	0.99207	1.1246	1.0049	0.91097
0.91	1.02297	0.99366	1.1115	1.0039	0.92039
0.92	1.02120	0.99506	1.0984	1.0031	0.92970
0.93	1.01922	0.99627	1.0856	1.0024	0.93889
0.94	1.01702	0.99729	1.0728	1.0017	0.94797
0.95	1.01463	0.99814	1.0603	1.0012	0.95693
0.96	1.01205	0.99883	1.0479	1.0008	0.96577
0.97	1.00929	0.99935	1.0357	1.0004	0.97450
0.98	1.00636	0.99971	1.0236	1.0002	0.98311
0.99	1.00326	0.99993	1.0117	1.0000	0.99161
1.00	1.00000	1.00000	1.0000	1.0000	1.00000
1.01	0.99659	0.99993	0.98841	1.0000	1.00828
1.02	0.99304	0.99973	0.97698	1.0002	1.01645
1.03	0.98936	0.99940	0.96569	1.0004	1.02450
1.04	0.98554	0.99895	0.95456	1.0006	1.03246
1.05	0.98161	0.99838	0.94358	1.0012	1.04030
1.06	0.97755	0.99769	0.93275	1.0017	1.04804
1.07	0.97339	0.99690	0.92206	1.0024	1.05567
1.08	0.96913	0.99601	0.91152	1.0031	1.06320
1.09	0.96477	0.99501	0.90112	1.0039	1.07063
1.10	0.96031	0.99392	0.89087	1.0049	1.07795
1.11	0.95577	0.99275	0.88075	1.0059	1.08518
1.12	0.95115	0.99148	0.87078	1.0070	1.09230
1.13	0.94645	0.99013	0.86094	1.0082	1.09933
1.14	0.94169	0.98871	0.85123	1.0095	1.10626
1.15	0.93685	0.98721	0.84166	1.0109	1.11310
1.16	0.93196	0.98564	0.83222	1.0124	1.11984
1.17	0.92701	0.98400	0.82292	1.0140	1.12649
1.18	0.92200	0.98230	0.81374	1.0157	1.13305
1.19	0.91695	0.98054	0.80468	1.0175	1.13951

(*Continued*)

TABLE D.4 (*Continued*)

Rayleigh Flow (Flow with Heat Addition) Tables for a Gas Having γ = 1.4

M	T/T^*	T_t/T_t^*	p/p^*	p_t/p_t^*	v/v^*
1.20	0.91185	0.97872	0.79576	1.0194	1.14589
1.21	0.90671	0.97684	0.78695	1.0214	1.15218
1.22	0.90153	0.97492	0.77827	1.0235	1.15838
1.23	0.89632	0.97294	0.76971	1.0257	1.16449
1.24	0.89108	0.97092	0.76127	1.0279	1.17052
1.25	0.88581	0.96886	0.75294	1.0303	1.17647
1.26	0.88052	0.96675	0.74473	1.0328	1.18233
1.27	0.87521	0.96461	0.73663	1.0354	1.18812
1.28	0.86988	0.96243	0.72865	1.0380	1.19382
1.29	0.86453	0.96022	0.72078	1.0408	1.19945
1.30	0.85917	0.95798	0.71301	1.0437	1.20499
1.31	0.85380	0.95571	0.70536	1.0466	1.21046
1.32	0.84843	0.95341	0.69780	1.0497	1.21585
1.33	0.84305	0.95108	0.69036	1.0528	1.22117
1.34	0.83766	0.94873	0.68301	1.0561	1.22642
1.35	0.83227	0.94637	0.67577	1.0594	1.23159
1.36	0.82689	0.94398	0.66863	1.0629	1.23669
1.37	0.82151	0.94157	0.66158	1.0664	1.24173
1.38	0.81613	0.93914	0.65464	1.0701	1.24669
1.39	0.81076	0.93671	0.64778	1.0738	1.25158
1.40	0.80539	0.93425	0.64103	1.0777	1.25641
1.41	0.80004	0.93179	0.63436	1.0816	1.26117
1.42	0.79469	0.92931	0.62779	1.0856	1.26587
1.43	0.78936	0.92683	0.62130	1.0898	1.27050
1.44	0.78405	0.92434	0.61491	1.0940	1.27507
1.45	0.77874	0.92184	0.60860	1.0983	1.27957
1.46	0.77346	0.91933	0.60237	1.1028	1.28402
1.47	0.76819	0.91682	0.59623	1.1073	1.28840
1.48	0.76294	0.91431	0.59018	1.1120	1.29273
1.49	0.75771	0.91179	0.58421	1.1167	1.29700
1.50	0.75250	0.90928	0.57831	1.1215	1.30120
1.51	0.74732	0.90676	0.57250	1.1265	1.30536
1.52	0.74215	0.90424	0.56676	1.1315	1.30945
1.53	0.73701	0.90172	0.56111	1.1367	1.31350
1.54	0.73189	0.89920	0.55552	1.1419	1.31743
1.55	0.72680	0.89669	0.55002	1.1473	1.32142
1.56	0.72173	0.89418	0.54458	1.1527	1.32530
1.57	0.71669	0.89168	0.53922	1.1583	1.32913
1.58	0.71168	0.88917	0.53393	1.1640	1.33291
1.59	0.70669	0.88668	0.52871	1.1697	1.33663

(*Continued*)

TABLE D.4 (*Continued*)

Rayleigh Flow (Flow with Heat Addition) Tables for a Gas Having γ = 1.4

M	T/T^*	T_t/T_t^*	p/p^*	p_t/p_t^*	v/v^*
1.60	0.70174	0.88419	0.52356	1.1756	1.34031
1.61	0.69680	0.88170	0.51848	1.1816	1.34394
1.62	0.69190	0.87922	0.51346	1.1877	1.34753
1.63	0.68703	0.87675	0.50851	1.1939	1.35106
1.64	0.68219	0.87429	0.50363	1.2002	1.35455
1.65	0.67738	0.87184	0.49880	1.2066	1.35800
1.66	0.67259	0.86939	0.49405	1.2131	1.36140
1.67	0.66784	0.86696	0.48935	1.2197	1.36475
1.68	0.66312	0.86453	0.48472	1.2264	1.36806
1.69	0.65843	0.86212	0.48014	1.2333	1.37133
1.70	0.65377	0.85971	0.47562	1.2402	1.37455
1.71	0.64914	0.85731	0.47117	1.2473	1.37774
1.72	0.64455	0.85493	0.46677	1.2545	1.38088
1.73	0.63999	0.85256	0.46242	1.2618	1.38398
1.74	0.63545	0.85019	0.45813	1.2692	1.38705
1.75	0.63095	0.84784	0.45390	1.2767	1.39007
1.76	0.62649	0.84551	0.44972	1.2843	1.39306
1.77	0.62205	0.84318	0.44559	1.2920	1.39600
1.78	0.61765	0.84087	0.44152	1.2999	1.39891
1.79	0.61328	0.83857	0.43750	1.3078	1.40179
1.80	0.60894	0.83628	0.43353	1.3159	1.40462
1.81	0.60464	0.83400	0.42960	1.3241	1.40743
1.82	0.60036	0.83174	0.42573	1.3324	1.41019
1.83	0.59612	0.82949	0.42191	1.3409	1.41292
1.84	0.59191	0.82726	0.41813	1.3494	1.41562
1.85	0.58774	0.82504	0.41440	1.3581	1.41829
1.86	0.58359	0.82283	0.41072	1.3669	1.42092
1.87	0.57948	0.82064	0.40708	1.3758	1.42351
1.88	0.57540	0.81845	0.40349	1.3849	1.42608
1.89	0.57136	0.81629	0.39994	1.3940	1.42862
1.90	0.56734	0.81414	0.39643	1.4033	1.43112
1.91	0.56336	0.81200	0.39297	1.4127	1.43359
1.92	0.55941	0.80987	0.38955	1.4222	1.43604
1.93	0.55549	0.80776	0.38617	1.4319	1.43845
1.94	0.55160	0.80567	0.38283	1.4417	1.44083
1.95	0.54774	0.80358	0.37954	1.4516	1.44319
1.96	0.54392	0.80152	0.37628	1.4616	1.44551
1.97	0.54012	0.79946	0.37306	1.4718	1.44781
1.98	0.53636	0.79742	0.36988	1.4821	1.45008
1.99	0.53263	0.79540	0.36674	1.4925	1.45233

(*Continued*)

TABLE D.4 (*Continued*)

Rayleigh Flow (Flow with Heat Addition) Tables for a Gas Having γ = 1.4

M	T/T^*	T_t/T_t^*	p/p^*	p_t/p_t^*	v/v^*
2.00	0.52893	0.79339	0.36364	1.5031	1.45455
2.01	0.52525	0.79139	0.36057	1.5138	1.45674
2.02	0.52161	0.78941	0.35754	1.5246	1.45890
2.03	0.51800	0.78744	0.35454	1.5356	1.46104
2.04	0.51442	0.78549	0.35158	1.5467	1.46315
2.05	0.51087	0.78355	0.34866	1.5579	1.46524
2.06	0.50735	0.78162	0.34577	1.5693	1.46731
2.07	0.50386	0.77971	0.34291	1.5808	1.46935
2.08	0.50040	0.77782	0.34009	1.5924	1.47136
2.09	0.49696	0.77593	0.33730	1.6042	1.47336
2.10	0.49356	0.77406	0.33454	1.6162	1.47533
2.11	0.49018	0.77221	0.33182	1.6282	1.47727
2.12	0.48684	0.77037	0.32912	1.6404	1.47920
2.13	0.48352	0.76854	0.32646	1.6528	1.48110
2.14	0.48023	0.76673	0.32382	1.6653	1.48298
2.15	0.47696	0.76493	0.32122	1.6780	1.48484
2.16	0.47373	0.76314	0.31865	1.6908	1.48668
2.17	0.47052	0.76137	0.31610	1.7037	1.48850
2.18	0.46734	0.75961	0.31359	1.7168	1.49029
2.19	0.46418	0.75787	0.31110	1.7300	1.49207
2.20	0.46106	0.75613	0.30864	1.7434	1.49383
2.21	0.45796	0.75442	0.30621	1.7570	1.49556
2.22	0.45488	0.75271	0.30381	1.7707	1.49728
2.23	0.45184	0.75102	0.30143	1.7846	1.49898
2.24	0.44882	0.74934	0.29908	1.7986	1.50066
2.25	0.44582	0.74768	0.29675	1.8128	1.50232
2.26	0.44285	0.74602	0.29446	1.8271	1.50396
2.27	0.43990	0.74438	0.29218	1.8416	1.50558
2.28	0.43698	0.74276	0.28993	1.8562	1.50719
2.29	0.43409	0.74114	0.28771	1.8710	1.50878
2.30	0.43122	0.73954	0.28551	1.8860	1.51035
2.31	0.42838	0.73795	0.28333	1.9012	1.51190
2.32	0.42555	0.73638	0.28118	1.9165	1.51344
2.33	0.42276	0.73482	0.27905	1.9319	1.51496
2.34	0.41998	0.73326	0.27695	1.9476	1.51646
2.35	0.41723	0.73173	0.27487	1.9634	1.51795
2.36	0.41451	0.73020	0.27281	1.9794	1.51942
2.37	0.41181	0.72868	0.27077	1.9955	1.52088
2.38	0.40913	0.72718	0.26875	2.0119	1.52232
2.39	0.40647	0.72569	0.26676	2.0284	1.52374

(*Continued*)

TABLE D.4 (*Continued*)

Rayleigh Flow (Flow with Heat Addition) Tables for a Gas Having γ = 1.4

M	T/T^*	T_t/T_t^*	p/p^*	p_t/p_t^*	v/v^*
2.40	0.40384	0.72421	0.26478	2.0451	1.52515
2.41	0.40122	0.72275	0.26283	2.0619	1.52655
2.42	0.39864	0.72129	0.26090	2.0789	1.52793
2.43	0.39607	0.71985	0.25899	2.0962	1.52929
2.44	0.39352	0.71842	0.25710	2.1136	1.53065
2.45	0.39100	0.71699	0.25522	2.1311	1.53198
2.46	0.38850	0.71558	0.25337	2.1489	1.53331
2.47	0.38602	0.71419	0.25154	2.1669	1.53461
2.48	0.38356	0.71280	0.24973	2.1850	1.53591
2.49	0.38112	0.71142	0.24793	2.2033	1.53719
2.50	0.37870	0.71006	0.24615	2.2218	1.53846
2.51	0.37630	0.70871	0.24440	2.2405	1.53972
2.52	0.37392	0.70736	0.24266	2.2594	1.54096
2.53	0.37157	0.70603	0.24093	2.2785	1.54219
2.54	0.36923	0.70471	0.23923	2.2978	1.54341
2.55	0.36691	0.70340	0.23754	2.3173	1.54461
2.56	0.36461	0.70210	0.23587	2.3370	1.54581
2.57	0.36233	0.70081	0.23422	2.3569	1.54699
2.58	0.36007	0.69952	0.23258	2.3770	1.54816
2.59	0.35783	0.69826	0.23096	2.3972	1.54931
2.60	0.35561	0.69700	0.22936	2.4177	1.55046
2.61	0.35341	0.69575	0.22777	2.4384	1.55159
2.62	0.35122	0.69451	0.22620	2.4593	1.55272
2.63	0.34906	0.69328	0.22464	2.4805	1.55383
2.64	0.34691	0.69206	0.22310	2.5018	1.55493
2.65	0.34478	0.69084	0.22158	2.5233	1.55602
2.66	0.34266	0.68964	0.22007	2.5451	1.55710
2.67	0.34057	0.68845	0.21857	2.5671	1.55816
2.68	0.33849	0.68727	0.21709	2.5892	1.55922
2.69	0.33643	0.68610	0.21562	2.6117	1.56027
2.70	0.33439	0.68494	0.21417	2.6343	1.56131
2.71	0.33236	0.68378	0.21273	2.6571	1.56233
2.72	0.33035	0.68264	0.21131	2.6802	1.56335
2.73	0.32836	0.68150	0.20990	2.7035	1.56436
2.74	0.32638	0.68037	0.20850	2.7270	1.56536
2.75	0.32442	0.67926	0.20712	2.7508	1.56634
2.76	0.32248	0.67815	0.20575	2.7748	1.56732
2.77	0.32055	0.67705	0.20439	2.7990	1.56829
2.78	0.31864	0.67595	0.20305	2.8235	1.56925
2.79	0.31674	0.67487	0.20172	2.8482	1.57020

(*Continued*)

TABLE D.4 (*Continued*)

Rayleigh Flow (Flow with Heat Addition) Tables for a Gas Having γ = 1.4

M	T/T^*	T_t/T_t^*	p/p^*	p_t/p_t^*	v/v^*
2.80	0.31486	0.67380	0.20040	2.8731	1.57114
2.81	0.31299	0.67273	0.19910	2.8982	1.57207
2.82	0.31114	0.67167	0.19780	2.9237	1.57300
2.83	0.30931	0.67062	0.19652	2.9493	1.57391
2.84	0.30749	0.66958	0.19525	2.9752	1.57482
2.85	0.30568	0.66855	0.19399	3.0014	1.57572
2.86	0.30389	0.66752	0.19275	3.0278	1.57661
2.87	0.30211	0.66651	0.19151	3.0544	1.57749
2.88	0.30035	0.66550	0.19029	3.0813	1.57836
2.89	0.29860	0.66450	0.18908	3.1084	1.57923
2.90	0.29687	0.66350	0.18788	3.1359	1.58008
2.91	0.29515	0.66252	0.18669	3.1635	1.58093
2.92	0.29344	0.66154	0.18551	3.1914	1.58178
2.93	0.29175	0.66057	0.18435	3.2196	1.58261
2.94	0.29007	0.65960	0.18319	3.2481	1.58343
2.95	0.28841	0.65865	0.18205	3.2768	1.58425
2.96	0.28675	0.65770	0.18091	3.3058	1.58506
2.97	0.28512	0.65676	0.17979	3.3350	1.58587
2.98	0.28349	0.65583	0.17867	3.3646	1.58666
2.99	0.28188	0.65490	0.17757	3.3944	1.58745
3.00	0.28028	0.65398	0.17647	3.4245	1.58824
3.50	0.21419	0.61580	0.13223	5.3280	1.61983
4.00	0.16831	0.58909	0.10256	8.2268	1.64103
4.50	0.13540	0.56982	0.08177	12.502	1.65588
5.00	0.11111	0.55556	0.06667	18.634	1.66667
6.00	0.07849	0.53633	0.04669	38.946	1.68093
7.00	0.05826	0.52438	0.03448	75.414	1.68966
8.00	0.04491	0.51647	0.02649	136.62	1.69536
9.00	0.03565	0.51098	0.02098	233.88	1.69930
10.00	0.02897	0.50702	0.01702	381.61	1.70213
∞	0.00000	0.48980	0.00000	∞	1.71430

$$\frac{p}{p^*}=\frac{1+\gamma}{1+\gamma M^2}\qquad \frac{T}{T^*}=\frac{(1+\gamma)^2M^2}{(1+\gamma M^2)^2}\qquad \frac{V}{V^*}=\frac{(1+\gamma)M^2}{1+\gamma M^2}$$

$$\frac{T_t}{T_t^*}\frac{(1+\gamma)^2M^2}{(1+\gamma M^2)^2}\left(\frac{1+((\gamma-1)/2)M^2}{1+((\gamma-1)/2)}\right)\qquad \frac{p_t}{p_t^*}=\left(\frac{1+\gamma}{1+\gamma M^2}\right)\left(\frac{1+((\gamma-1)/2)M^2}{1+((\gamma-1)/2)}\right)^{\gamma/(\gamma-1)}$$

Appendix E: Miscellaneous

TABLE E.1
Greek Alphabet

English Spelling	Capital Greek Letters	Lowercase Greek Letters
Alpha	Α	α
Beta	Β	β
Gamma	Γ	γ
Delta	Δ	δ
Epsilon	Ε	ε
Zeta	Ζ	ζ
Eta	Η	η
Theta	Θ	θ
Lota	Ι	ι
Kappa	Κ	κ
Lambda	Λ	λ
Mu	Μ	μ
Nu	Ν	υ
Xi	Ξ	ζ
Omicron	Ο	ο
Pi	Π	π
Rho	Ρ	ρ
Sigma	Σ	σ
Tau	Τ	τ
Upsilon	ϒ	ϒ
Phi	Φ	Φ
Chi	Χ	Χ
Psi	Ψ	Ψ
Omega	Ω	Ω

Bibliography

Bird, R. B., W. E. Stewart, and E. N. Lightfoot. *Transport Phenomena*. New York: John Wiley & Sons, 1960.

Bober, W. and R. A. Kenyon. *Fluid Mechanics*. New York: John Wiley & Sons, 1980.

Bolz, R. E. and G. L. Tuve, editors. *CRC Handbook of Tables for Applied Engineering Science*, 2nd edn. Cleveland, OH: CRC Press, 1973.

Chen, N. H. An explicit equation for friction factor in pipe. *Industrial & Engineering Chemistry Fundamentals*, 18(3), 296–297, 1979.

Chow, V. T. *Open-Channel Hydraulics*. New York: McGraw Hill, Inc., 1959.

Daugherty, R. L. and J. B. Franzini. *Fluid Mechanics with Engineering Applications*, 8th edn. New York: McGraw-Hill Book Co., 1977.

Deissler, R. G. Analysis of turbulent heat transfer, mass transfer, and friction in smooth tubes at high Prandtl and Schmidt numbers, NACA report 1210, 1955.

Dimensions in English units obtained from ANSI/ASTM B88-78. Standard specifications for seamless copper water tube. Washington, DC: American Society for Testing and Materials, 1978.

Dimensions in English units obtained from ANSI B36.10-1979. American National Standard wrought steel and wrought iron pipe. New York: American Society of Mechanical Engineers, 1979.

Dixon, S. L. *Fluid Mechanics: Thermodynamics of Turbomachinery*, 3rd edn. New York: Pergamon Press, 1978.

Forsythe, W. E., editor. *Smithsonian Physical Tables*, 9th edn. Washington, DC: The Smithsonian Institution, 1956.

Fox, J. A. *An Introduction to Engineering Fluid Mechanics*. New York: McGraw-Hill Book Co., 1974.

Fox, R. W. and A. T. McDonald. *Introduction to Fluid Mechanics*, 3rd edn. New York: John Wiley & Sons, 1985.

Geonkoplis, C. J. *Transport Processes and Unit Operations*. Boston, MA: Allyn & Bacon, 1978.

Gerhart, P. M. and R. J. Gross. *Fundamentals of Fluid Mechanics*. Reading, MA: Addison Wesley Publishing Co., 1985.

Granger, R. A. *Fluid Mechanics*. New York: Holt, Rinehart and Winston, Division of CBS College Publishing, 1985.

Gross, A. C., C. R. Kyle, and D. J. Malewicki. The aerodynamics of human powered land vehicles. *Scientific American*, 249(6), 142–152, December 1983.

John, J. E. A. and W. Haberman. *Introduction to Fluid Mechanics*, 2nd edn. Englewood Cliffs, NJ: Prentice Hall, 1980.

Lenz, A. T. Viscosity and surface tension effects on V-notch weir coefficients. *Transactions of ASCE*, 108, 739–802, 1943.

Lissaman, P. B. S. and C. A. Shollenberger. Formation flight of birds. *Science*, 168(3934), 1003–1005, May 1970.

Mechtly, E. A. NASA SP-701Z, 1969.

Moody, L. F. Friction factors in pipe flow. *Transactions of ASME*, 68, 672, 1944.

Murdock, J. W. *Fluid Mechanics and Its Applications*. Boston, MA: Houghton Mifflin Co., 1976.

NACA. Equations, tables and charts for compressible flow, NACA report 1135. Moffett, CA: Ames Aeronautical Laboratory, 1953.

Olson, R. M. *Essentials of Engineering Fluid Mechanics*, 4th edn. New York: Harper & Row, 1980.

Prasuhn, A. L. *Fundamentals of Fluid Mechanics*. Englewood Cliffs, NJ: Prentice Hall, 1980.

Roberson, J. A. and C. T. Crowe. *Engineering Fluid Mechanics*, 3rd edn. Boston, MA: Houghton Mifflin Co., 1985.

Schlichting, H. *Boundary Layer Theory*, 7th edn. New York: McGraw-Hill, Inc., 1979.

Selby, S. M., editor. *Standard Mathematical Tables*, 16th edn. Cleveland, OH: Chemical Rubber Co., 1968.

Shames, I. H. *Mechanics of Fluids*, 2nd edn. New York: McGraw-Hill Book Co., 1982.

Silverstein, A. NACA report 502, p. 15, 1934.

Standard for metric practice, E 380-76. Philadelphia, PA: American Society for Testing and Materials, 1976.

Streeter, V. L. and E. B. Wylie. *Fluid Mechanics*, 8th edn. New York: McGraw-Hill Book Co., 1985.

Student ME's letter. *ASME*, 2(1), I.

The American Society of Mechanical Engineers. *Fluid Meters: Their Theory and Application*, 5th edn. New York: The American Society of Mechanical Engineers, 1959.

Vennard, J. K. and R. L. Street. *Elementary Fluid Mechanics*, 6th edn. New York: John Wiley & Sons, 1982.

White, F. M. *Fluid Mechanics*, 2nd edn. New York: McGraw-Hill Book Co., 1986.

Yuan, S. W. *Foundations of Fluid Mechanics*. Englewood Cliffs, NJ: Prentice Hall, 1967.

Index

L

M

N

O

P

Q

R

S

T

U

V

W